Johnson Marimuthu • Helena Fernández •
Ashwani Kumar • Shibila Thangaiah
Editors

# Ferns

## Biotechnology, Propagation, Medicinal Uses and Environmental Regulation



Springer

*Editors*
Johnson Marimuthu
Department of Botany, Centre for Plant Biotechnology
St. Xavier's College (Autonomous)
Palayamkottai, India

Helena Fernández
Department of Organisms and Systems Biology, Area of Plant Physiology
University of Oviedo
Oviedo, Spain

Ashwani Kumar
Department of Botany and Biotechnology
University of Rajasthan
Jaipur, Rajasthan, India

Shibila Thangaiah
Sri Paramakalyani Centre for Excellence in Environmental Sciences
Manonmaniam Sundaranar University
Alwarkurichi, Tamil Nadu, India

ISBN 978-981-16-6172-3 ISBN 978-981-16-6170-9 (eBook)
https://doi.org/10.1007/978-981-16-6170-9



This Springer imprint is published by the registered company Springer Nature Singapore Pte Ltd.
The registered company address is: 152 Beach Road, #21-01/04 Gateway East, Singapore 189721, Singapore

***Dedication to Professor S.C. Verma on his 90th Birthday***

***Dr. Satish Chander Verma*** *(b. 19 December, 1931), FNASC, FLS, FBS, FSCG, FIFS, former Head and Professor of Botany at Punjab University, Chandigarh, is known (since 1956) for his contributions in cytology, cytogenetics, genetics, and reproductive and evolutionary biology of homosporous ferns and Isoetes. His contribution to the genetics of self-incompatibility in Brassicaceae, nuclear DNA estimations in species of Brassica of U's triangle, Giemsa C-banding studies in Rye, and Lathyrus are exemplary. Supported by the*

*British Council he had the distinction of working with Professors Hugh Rees, FRS, and Dan Lewis, FRS. Professor Verma has vast experience of teaching and research, has over 200 publications, and has participated and lectured in several national and international conferences/symposia. He was awarded the Commonwealth Academic staff Fellowship, Royal Society, London, Bursary (UK); Visitorship of British Council (UK) at University College (London) and Queen Mary College (London); and Visiting Professorship at Hiroshima, Tokyo, and Sendai (Japan), University College, London (UK), Colorado College (Colorado Springs, USA), and University of Arizona (Tempe, USA). He was Honorary Research Fellow of University College, London (1977–1984), UGC National Lecturer (1986–87), UGC Emeritus Fellow (1995 & 1996), and recipient of PN Mehra Memorial Medal and DD Pant Oration Medal.*

# Foreword

I am delighted and honored to be asked to write a foreword to this book on "*Biotechnology, Propagation, Medicinal Uses and Environmental Regulation in Ferns*". Although numerous books have been written over recent decades and centuries about various aspects of the biology, ecology, evolution, and horticulture of ferns, to my knowledge this is the first volume focusing on applied research in ferns spanning the broad fields of biotechnology, propagation, ethnobotany, phytochemistry, and the practical aspects of fern ecology. Despite the overall applied and practical focus of the volume, there are also broad overviews of basic ecology, phylogeny, and evolution of ferns. The editors have assembled a diverse suite of international scholars representing a broad spectrum of expertise in the research areas represented. The chapters comprise various combinations of in-depth reviews of published literature as well as presentations of original research.

The first part of the book is devoted to the wide field of biotechnology, which includes chapters ranging from genomics and molecular phylogenetics to gene evolution in the model fern species *Ceratopteris richardii*, the domestication and uses of *Azolla* spp., and ferns as sources of pesticides. Under propagation of ferns, there are chapters on propagation of various species, including micropropagation, propagation of epiphytic species, gametophyte development, morphogenesis, and the effects of phytohormones and other bioactive compounds on gametophytes. The part on the ethnomedicinal uses of ferns covers diverse areas of investigation including the use of ferns and lycophytes by various human cultures, as well as the description of detailed studies on specific uses of ferns as nutraceuticals, in pharmacology and cosmetics, the potential use for colon cancer treatment, as well as the in vitro antioxidant and antidiabetic activities of some fern extracts. The last part of the book includes chapters dealing with various aspects of environmental regulation.

Every fern expert and enthusiast will find much of interest in this highly informative and thoughtfully devised collection of articles. I recommend this book for academic researchers and students, as well as professionals in applied fields of biotechnology and agriculture.

School of Life Sciences, University of Hawai'i at Mānoa
Honolulu, HI, USA

Tom A. Ranker

# Preface

In our earlier book *Working with Ferns: Issues and Applications* by Helena Fernández, Ashwani Kumar, and Maria Angeles Revilla (eds), some years ago, the focus was on a unique group of plants that provided wider knowledge and applications. The classical pteridological studies are shifting from classical studies such as taxonomy, paleobotany, and morphology, to experimental areas such as ecology, physiology, development, genetics, and biochemistry. Our book offers a glimpse of new Pteridology. The arrival of new and innovative techniques applied to the study of ferns has permitted a more widespread use of sophisticated instrumentation opening up new horizons in the cellular and molecular biology approach to the study of ferns as a result. With the growing knowledge base, particularly the renewed interest in fern genome and platogenomic, DNA marker, evolution, propagation, traditional medicinal uses, and environmental role of ferns prompted us to bring out this second book with additional information. Distinguished contributors and authors, who are masters in the field of Pteridology and Biotechnology, have provided up-to-date information woven in beautiful illustrations, tables, and text to bring all the information under one head. Ferns were first colonizers and are oldest to develop vascular system. However their use in biotechnological investigations involving recombinant DNA technology is very limited. Besides, their use in food or medicine has also remained unexplored. The second largest group of plants after Angiosperms has less frequently been used for genetic manipulations. Ferns have sustained high $CO_2$ levels and harsh climate of million almost 400 million years ago and may contain repository of genes which might help developing plants having resistance to climate change. The present book provides detailed information about their genomic and phylogenomic studies which might be helpful in locating some important genetic material in ferns to be used to develop plants resistant to climate change. Propagation of ferns in vitro and in vivo is essential for mass multiplication, and the part on propagation outlines achievements in multiplying difficult fern species. Traditional medicinal use of ferns and the secondary metabolites are highlighted in the part on medicinal uses of ferns. Climate change is a biggest problem of mankind in 2021, and ferns are hardy plants and can help us fight problems of climate change in several ways by environmental protection. *Azolla* is one such example to remove heavy metals and provide environmental protection.

The book is dedicated to Professor Emeritus S.C. Verma, oldest Pteridologist on his 90th birthday, for his distinguished services to ferns. We also thank Professor Emeritus S. P. Khullar for his valuable contribution to ferns and acknowledge the *Indian Fern Journal* to provide resource material.

We, the editors, acknowledge with thanks the support of our senior colleagues, fellow colleagues, and students in the preparation of this manuscript and authors for contributing their best articles to the subject. We sincerely hope that book will fill the aspirations of a wide range of readers in the field of "Monilophytes," scholars, researchers, and industry people and general readers.

Palayamkottai, Tamil Nadu, India — Johnson Marimuthu
Oviedo, Spain — Helena Fernández
Jaipur, Rajasthan, India — Ashwani Kumar
Alwarkurichi, Tamil Nadu, India — Shibila Thangaiah

# Acknowledgments

We, the editors, are thankful to Professor Emeritus S.C. Verma, senior most Pteridologist from India, and most respected Professor Emeritus S.P. Khullar, Editor of *Indian Fern Journal* from Punjab University, Chandigarh, for suggesting the concept of the book. The editors also thank the distinguished authors who have contributed valuable chapters to this book.

# Contents

# Editors and Contributors

## About the Editors

**Johnson Marimuthu**, F.I.F.S., completed his Doctorate in Botany (Plant Tissue Culture) from Manonmaniam Sundaranar University, Tirunelveli District, Tamil Nadu, and is working as Assistant Professor of Botany and Director, Centre for Plant Biotechnology, St. Xavier's College (autonomous), Palayamkottai, Tamil Nadu. He has 21 years of research experience and 18 years of teaching experience. He has published 223 papers in reputed journals, 27 papers in proceedings, and 37 papers as book chapters. He is a recipient of the prestigious SERB Young Scientist—2013, Prof. S.S. Bir Gold Medal—2017, and Certificate of Merit 2014 by Indian Fern Society, India.

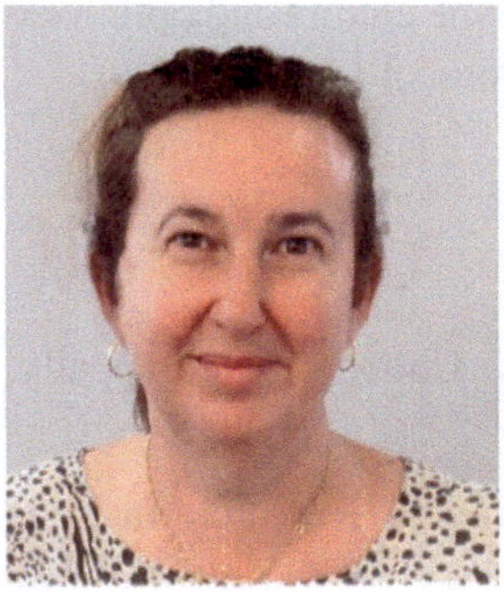

**Helena Fernández** is Associate Professor of Plant Physiology at the University of Oviedo, Spain. She received her PhD in Biology in 1993 from Oviedo University, followed by an EU-funded postgraduate research sojourn at the INRA in Orléans, France, training in HPLC-MS techniques. She got the Ramón y Cajal National Excellence Research Program, dealing with reproduction in ferns. She is guest researcher at the Institute of Plant Biology, University of Zurich, working on apomixes, through omics techniques. She is a member of the dean's team, in the Faculty of Biology at the University of Oviedo. She publishes in relevant journals, having edited two books with Springer.

**Ashwani Kumar** Alexander von Humboldt Fellow (Germany), MSc (Botany) (Gold Medalist) (1967)), PhD (1971), postdoc AvH Fellow (Germany) and JSPS fellow (Japan) Selected in the Indian Administrative Services (IAS: IPS) (1972) but he opted career in botany. 37 years of teaching & research experience. Guided 39 PhD. research students and published 320 research papers and 26 books. Elected fellow FBS, FPSI, FIMPS and FIFS. Editor in Chief IJLS. V. Puri medal (2008) CEE award (2015). President of CHEC (India), Indian Botanical Society, Society for Promotion of Plant Sciences. Professor Emeritus, Department of University of Rajasthan, Jaipur. Presently, Chair Professor at JVWU, Jaipur and Adj. Professor Modi University, Sikar.

**Shibila Thangaiah** obtained her PhD in Botany from Manonmaniam Sundaranar University, Tirunelveli District, Tamil Nadu, and is working as an Assistant Professor at Sri Paramakalyani Centre for Excellence in Environmental Sciences, Manonmaniam Sundaranar University, Alwarkurichi, Tamil Nadu. She has published more than two book chapters, 20 proceedings, and 25 research papers in international journals. She has attended 18 national and international conferences and seminars as well as received best paper presentation award in a national conference. She has 7 years of research experience in plant tissue culture and optimized the large-scale propagation protocol for four ferns.

## Contributors

**S. Abraham** Department of Botany, St. Dominic's College Kanjirappally, Kanjirappallay, Kerala, India

**R. Antony** Jawaharlal Nehru Tropical Botanic Garden and Research Institute (JNTBGRI), Thiruvananthapuram, Kerala, India

**M. Ayyanar** Department of Botany, A.V.V.M. Sri Pushpam College (Autonomous) Bharathidasan University, Thanjavur, India

**A. Balkrishna** Patanjali Herbal Research Department, Patanjali Research Foundation Trust, Haridwar, Uttarakhand, India

**M. Bhatia** Department of Botany, Daulat Ram College, University of Delhi, Delhi, India

**I. Bigot** Molecular Plant Physiology, Biology Department, Utrecht University, Utrecht, The Netherlands

**A. Bujak** The Azolla Foundation, Blackpool, UK

**J. Bujak** The Azolla Foundation, Blackpool, UK

**M. J. Cañal** Area of Plant Physiology, Department of BOS, The University of Oviedo, Oviedo, Spain

**S. Catharin Sara** PG and Research Department of Botany, Holy Cross College (Autonomous), Tiruchirappalli, Tamil Nadu, India

**J. Chandra Saleride** Department of Botany, Women's Christian College, Nagercoil, Tamil Nadu, India

**C. -L. Cheng** Department of Biology, The University of Iowa, Iowa City, IA, USA

**A. Correas Grifoll** Molecular Plant Physiology, Biology Department, Utrecht University, Utrecht, The Netherlands

**A. P. Della** Institute of Biosciences, University of São Paulo, São Paulo, Brazil

**L. W. Dijkhuizen** Molecular Plant Physiology, Biology Department, Utrecht University, Utrecht, The Netherlands

**Helena Fernández** Department of Organisms and Systems Biology, Area of Plant Physiology, University of Oviedo, Oviedo, Spain

**P. Giri** Department of Botany, University of Delhi, Delhi, India

**R. Gnana Deepa Ruby** PG and Research Department of Botany, Holy Cross College (Autonomous), Tiruchirappalli, Tamil Nadu, India

**S. Granados** Laboratorio de Fisiología Vegetal, Dpto. B.O.S., Facultad de Biología, Universidad de Oviedo, Oviedo, Asturias, Spain

**P. A. N. M. Gudde** Molecular Plant Physiology, Biology Department, Utrecht University, Utrecht, The Netherlands

**E. Güngör** Molecular Plant Physiology, Biology Department, Utrecht University, Utrecht, The Netherlands

**T. G. Harish Kumar** Department of Horticulture, University of Agricultural Sciences, Bangalore, Karnataka, India

**S. Hegde** Nitte (Deemed to be University), Nitte University Centre for Science Education and Research, Mangalore, Karnataka, India

**E. E. Irish** Department of Biology, The University of Iowa, Iowa City, IA, USA

**N. Janakiraman** Department of Botany, The Madura College (Autonomous), Madurai, Tamil Nadu, India

**Johnson Marimuthu** Department of Botany, Centre for Plant Biotechnology, St. Xavier's College (Autonomous), Palayamkottai, India

**B. S. Kholia** Botanical Survey of India, Dehradun, Uttarakhand, India

**Ashwani Kumar** Former Head of the Department of Botany and Biotechnology, University of Rajasthan, Jaipur, Rajasthan, India

**R. A. Lavate** Department of Botany, RajeRam Rao College, Jath, Maharashtra, India

**R. C. Mathias** Nitte (Deemed to be University), Nitte University Center for Science Education and Research, Mangalore, India

**K. Mehltreter** Red de Ecología Funcional, Instituto de Ecología, A.C., El Haya, Xalapa, Veracruz, Mexico

**A. Menéndez** Area of Plant Physiology, Department of BOS, Oviedo, Spain

**S. Morajkar** Goa State Wetland Authority, O/o Goa State Biodiversity Board, DSTE Complex, Bardez, Goa, India
Post Graduate Department of Biotechnology and Research, St. Aloysius College (Autonomous), Mangalore, India

**M. Narayani** Harvest University, Lusaka, Zambia
Centre for Plant Biotechnology, Department of Botany, St. Xavier's College (Autonomous), Palayamkottai, Tamil Nadu, India

**S. M. Patil** Department of Botany, Faculty of Science, The Maharaja Sayajirao University of Baroda, Vadodara, Gujarat, India
Modern College of Arts, Science and Commerce, Ganeshkhind, Pune, India

**K. Paulraj** Nesamony Memorial Christian College, Marthandam, Tamil Nadu, India

**Smruthi G. Prabhu** Post Graduate Department of Biotechnology and Research, St. Aloysius College (Autonomous), Mangalore, India

**K. Rajput** Department of Botany, Faculty of Science, The Maharaja Sayajirao University of Baroda, Vadodara, Gujarat, India

**V. M. Raole** Department of Botany, Faculty of Science, The Maharaja Sayajirao Univerity of Baroda, Vadodara, Gujarat, India

**N. Rijken** Molecular Plant Physiology, Biology Department, Utrecht University, Utrecht, The Netherlands

**A. Rivera** Area of Plant Physiology, Department of BOS, Oviedo, Spain

**P. T. Roshni** Post Graduate Department of Biotechnology and Research, St. Aloysius College (Autonomous), Mangalore, India

**K. Sahayaraj** Department of Zoology, St. Xavier's College, Crop Protection Research Centre, Palayamkottai, Tamil Nadu, India

**S. Sajeev** Post Graduate Department of Biotechnology and Research, St. Aloysius College (Autonomous), Mangalore, India

**E. Sánchez** Area of Plant Physiology, Department of BOS, Oviedo, Spain

**H. Schluepmann** Molecular Plant Physiology, Biology Department, Utrecht University, Utrecht, The Netherlands

**Shibila Thangaiah** Sri Paramakalyani Centre for Excellence in Environmental Sciences, Manonmaniam Sundaranar University, Alwarkurichi, Tamil Nadu, India

**B. Shivananthini** Department of Botany, Centre for Plant Biotechnology, St. Xavier's College (Autonomous), Palayamkottai, India

**M. Sierra** Department of Functional Biology (Genetic Area), Oncology University Institute (IUOPA) and Institute of Sanitary Research of Asturias (ISPA), University of Oviedo, Oviedo, Spain

**A. Sivaraman** Maria Research Foundations, Tirunelveli, Tamil Nadu, India

**Chowdaiah Suneetha** Department of Horticulture, University of Agricultural Sciences, Bangalore, Karnataka, India

**S. Suresh** Division of Conservation Biology, Jawaharlal Nehru Tropical Botanic garden and Research Institute (JNTBGRI), Thiruvananthapuram, Kerala, India

**J. Sureshkumar** Department of Botany, Annai Vailankanni Arts and Science College, Bharathidasan University, Thanjavur, India

**T. Thomas** Department of Botany, St. Thomas College Palai, Pala, Kerala, India

**P. L. Uniyal** Department of Botany, University of Delhi, Delhi, India

**Vallinayagam Sambantham** Department of Plant Science, Manonmaniam Sundaranar University, Tirunelveli, Tamil Nadu, India

**C. E. Youngstrom** Department of Biology, The University of Iowa, Iowa City, IA, USA

# Introduction 1

Helena Fernández and Ashwani Kumar

Pteridophytes are non-flowering, primitive vascular plants widely used by people all across the globe as they are easily available in large biomass in wild and often grow on relatively poor soils. Ferns, are one of the earliest groups of vascular plants, originated around 400 Mya, attained remarkable levels of diversity and abundance from the Carboniferous period to the Jurassic period (from c. 300 to 150 million years ago). They were especially prevalent in Paleozoic terrestrial ecosystems, before the rise of seed plants and maintained their dominance even into the Early Cretaceous (~145 Mya). Establishing the timescale of early land plant evolution is essential for testing hypotheses on the coevolution of land plants and Earth's System (Qi et al. 2018).

Our earlier books (Fernández et al. 2011; Fernández 2018) have covered basic aspects of ferns, their propagation, molecular diversity, use in medicines, and their role in climate protection. However, this book deals with the latest developments in fern biotechnology, propagation, ethnobotanical uses, and role in climate protection. The content of the present book is distributed in four parts: biotechnology, propagation, medicinal uses, and environmental applications.

Undoubtedly, ferns represent a genetic legacy of great value, being descendants of the first plants that developed vascular tissue, about 470 Mya. This change was due to the need to adapt to the new conditions imposed by the exit of water to the earth, such as the availability of water, the variation in temperatures or the increased exposure to solar radiation. We can say that ferns have been very little studied from a biological point of view. Only a handful of species have been used to delve into

H. Fernández (✉)
Department of Organisms and Systems Biology, Area of Plant Physiology, University of Oviedo, Oviedo, Spain
e-mail: fernandezelena@uniovi.es

A. Kumar
Former Head of the Department of Botany and Biotechnology, University of Rajasthan, Jaipur, Rajasthan, India





J. Marimuthu et al. (eds.), *Ferns*, https://doi.org/10.1007/978-981-16-6170-9_1

aspects of plant development, such as photomorphogenesis (Wada 2007), germination (Salmi et al. 2005), cellular polarity (Salmi and Bushart 2010), cell wall composition (Eeckhout et al. 2014), or reproduction (von Aderkas 1984; Wen et al. 1999; Fernández and Revilla 2003; Kaźmierczak 2010; López and Renzaglia 2014; Valledor et al. 2014; de Vries et al. 2016; Grossmann et al. 2017; Wyder et al. 2020), among others. However, it must be said that in recent years we are witnessing a greater approach to this plant group by researchers, since they are plants that have much to offer us. On the one hand, ferns were able to survive atmospheric conditions in which the amount of carbon dioxide was much higher than it is now, and whose levels are increasing (Perez-Martin et al. 2014; Tosens et al. 2016). Likewise, they appealed for their resistance in situations of drought, salinity or contamination of the soil with heavy metals (Rathinasabapathi 2006). Besides this they are generally not attacked by insects (Markham et al. 2006). These species also contain a wide arsenal of secondary metabolites, such as flavonoids, alkaloids, or phenols, sometimes displaying antibacterial, antidiabetic, anticancer, and antioxidant activity (Cao et al. 2017).

Traditionally, taxonomical studies of variation and discrimination of species were established based on their morphological and anatomical characteristics. The limitations in morphological identification systems and the declining pool of taxonomists signal the need for a new approach to recognize taxa.

In the part devoted to Biotechnology, there are seven chapters, including issues such as genomics and plastogenomics, evolution and development, and practical proposals on the use of ferns as fertilizers or alternative to harmful environmental practices as chemical pesticides.

Once again, we must emphasize the need for their application in monilophytes because they have limited blunt variations in morphological characters and high species diversity. Definitively, molecular marker studies have important repercussions in breeding for trait improvement, and detection of adulteration in compounded forms of herbal extracts, among others. The advancement of molecular techniques has generated information on the genetic basis for diversity and variability in organisms. The recent development of molecular markers revolutionized the entire scenario of life sciences, being DNA barcoding an emerging global standard for recognition of taxa, which has drastically simplified the identification process. Indeed, barcoding studies on pteridophytes mostly utilized data from several chloroplast markers. The improvement of molecular techniques has generated information on the genetic basis for diversity and variability in organisms. Several DNA-based fingerprinting techniques have been used such as plastid and nuclear SSRs, RAPD, AFLP, DNA sequencing, and classical taxonomy tools, to study population dynamics, species delimitation, hybridization, and phylogenetics. The gene *rbc*L has been used extensively by various researchers for analyzing the evolutionary relationship of ferns at both generic and familial levels. To study the plastogenomics, will help in understanding the process of evolution in one of the oldest living plants, the impact of reticulate evolution in the early evolution of fern and their interrelationship with angiosperms.

Genomics and plastogenomics supply useful information to revise both the taxonomy and phylogeny based mainly on morphological evidences. New classifications and evolutionary schemes have been proposed, reflecting different interpretations than previous data. The phylogenetic relationships among these four basal fern orders are the most debated topics in fern phylogeny (Kumar et al. 2022a, b; Janakiraman et al. 2022; Morajikar et al. 2022 this volume). The study of plastogenomics will help in understanding the process of evolution in one of the oldest living plants, the impact of reticulate evolution in the early evolution of fern and their interrelationship with angiosperms. Further analyses of the fern chloroplast genomes should provide new insights into the plastid genome evolution. Phylogenomics based on chloroplast genomes has shown many advantages in plant phylogenetics in the recent years. Integration of evidences from all living and fossil species in morphological as well as molecular data and evolutionary models will provide increased precision. With more nuclear data available recently, chloroplast phylogenomics can provide a framework for testing the impact of reticulate evolution in the early evolution of ferns (Kumar et al. 2022b this volume).

A very interesting chapter using the fern model *Ceratopteris richardii*, included in this part of the book, is committed to stress the high value of this plant group, in elucidating evolutionary developmental questions in land plants, in addition to other aspects of plant biology. As it is well known, plants have the competence to produce new organs throughout their life due to persistent proliferation of tissue-specific stem cell populations, being the family of WUSCHEL-related homeobox (WOX) genes recognized as determinant transcription factors for stem cell maintenance. The fact that WOX genes are found in all land plants and some extant algae was the basis to revise the current understanding of the evolution of the WOX gene family in controlling cell proliferation of various meristem domains in distantly related land plant models, *Physcomitrella patens*, *C. richardii*, and *Arabidopsis thaliana* (Youngstrom et al. 2022 this volume).

The next chapter of this part is a chapter devoted to the symbiosis between *Azolla* genus and the filamentous cyanobacteria, *Nostoc azollae,* allowing $N_2$-fixation and transferring reproductive structures for generations. Ferns from the *Azolla* genus are highly productive without nitrogen fertilizer because filamentous cyanobacteria, *Nostoc azollae,* associated with the shoot stem cells, invade leaf cavities for $N_2$-fixation, and reproductive structures for generational transfer concluded that for rapid breeding, the next vital development will be genome editing of fern host and cyanobacterial symbiont and described the first steps towards this end (Schluepmann et al. 2022 this volume). The focus is to take benefit of this co-operation for circular economy, including the sustainable production of plant protein, diving into the most recent and also the new proposals, on molecular research. The authors start describing novel investigation areas required to integrate agro system development with domestication, describing the first successes to control the life cycle of the symbiosis in relation to dissemination, storage and pre-breeding, and ending with the identification of key traits of the symbiosis needed to achieve yield stability. Next chapter describes *Azolla* as a biofertilizer and feed for animals, poultry, waterfowl, freshwater fish, and crustacea. Apart from it, *Azolla* increases its biomass so meaningfully that it is able to sequester large amounts of atmospheric $CO_2$, acquiring the capacity to mitigate anthropogenic climate change through Carbon Capture and Storage

(CCS). Moreover, *Azolla*'s biomass can also be useful as biofuel and a range of products that are urgently needed as our population increase by a billion every 12 years (Bujak and Bujak 2022 this volume).

The phytoecdysteroids have the same structural features as ecdysteroids found in insects or other arthropods. Sahayaraj (2022 this volume) described that ferns are a source of Phytoecdysones and suggested their applications in insect management. As it is well known, insects cause severe damage to numerous economically important crops and their control relies on pesticides. Although chemical pesticides may have beneficial action on agricultural production, but they can also have adverse effect on environment. In concrete, ecdysteroids are very common in Polypodiaceae, and they are reported to act as repellents from these harmful organisms. The chapter revises in depth the potential use of ferns for research on selective biodegradable substances which can be used as green insecticides (Sahayaraj 2022 this volume)

The second part of this book gathers chapters on propagation, which can be used for different purposes: ornamental, medicinal, etc. No doubts, their exquisite foliage patterns and resilient growing make them popular ornamental plants worldwide. Those demanded species for its beauty can be propagated by vegetative or sexual means either by traditional or tissue culture methods (Suneetha and Hegde 2022; Shibila and Johnson 2022; Vallinayagam 2022; Bhatia and Uniyal 2022; Johnson et al. 2022a; Granados et al. 2022; Sánchez et al. 2022).

For the last half of century, micropropagation has emerged as a powerful tool for the rapid propagation of rare and endangered plants, being welcome to employ scientific methods such as tissue culture to conserve and propagate these pteridophytes. A much superior and uniform quality of ferns can be produced independent of the season by this technology (Suneetha and Hegde 2022 this volume). An exhaustive revision is supplied in this work. In line with it, a protocol for the mass multiplication of two epiphytic species *Lepisorus nudus* (Hook.) Ching and *Elaphoglossum stelligerum* (Wall. ex Baker) T. Moore, using in vitro spore culture, is detailed next (Shibila and Johnson 2022 this volume) and also for the ornamental fern *Histiopteris incisa* (Thunb.) J. Sm., which has been identified as metallophytes recently, from spore culture until field establishment (Vallinayagam 2022 this volume).

Bhatia and Uniyal (2022 this volume) studied the reproductive biology in a few ferns and the effects of Nitric oxide on the development of gametophyte and gametangial production in *Ceratopteris thalictroides*. In the same work, the effect of sodium nitroprusside (SNP—a nitric oxide donor) on spore germination is also reported as well as the effect on vegetative and sexual organ development.

Different morphogenetic events of ferns, viz., vegetative propagation, apogamy, apospory, somatic embryogenesis, and polyembryony have been reviewed. Finally, a molecular approach on some of these processes is uncovered, to give evidence about new promising research lines, committed to deciphering the molecular clues operating behind them, and also to envisage important biotechnological repercussions (Johnson et al. 2022a this volume). Apogamy is a peculiar case of apomixis, very frequent in ferns, in which an asexual embryo derived from somatic cells of the gametophyte generation. Two reports of this event in the obligate apogamous species *Dryopteris affinis* ssp. *affinis* are included (Sánchez et al. 2022 this volume). The first one talks us about the effect of several phytohormones and inhibitors of their biosynthesis or transport on the gametophyte and the second brings to us some amazing results about the effect of yucasine, an inhibitor of the auxin biosynthesis, and the solvent DMSO, on vegetative gametophyte development and this kind of asexual reproduction. The effect on vegetative and apogamous gametophyte development of 14 phytohormones and inhibitors of their biosynthesis or transport (HBTIs) was analyzed in homogenized gametophytes cultured in liquid Murashige and Skoog (MS) medium (Granados et al. 2022 this volume).

The third part of the book is dedicated to the medicinal uses of ferns. Ethnic communities all over the world use ferns for various ailments such as dysentery, malaria, stomach-ache, urinary disorders, burns, etc. (Abraham and Thomas 2022; Kholia and Balkrishnan 2022; Antony and Suresh 2022; Johnson et al. 2022b, c; Giri et al. 2022; Sara and Ruby 2022; Patil et al. 2022; Sureshkumar and Ayyanar 2022 this volume). In recent times, many phytochemical and pharmacological studies of ferns are being carried out providing new information about the bioactive components with properties such as antimicrobial, anti-inflammatory, antidiabetic, anticancerous, etc. Certainly, India is one of the countries with a longer tradition in the use of pteridophytes, probably due to the fact to have around 1200 taxa, of which nearly 800 occur in the Himalayan region (Kholia and Balkrishna 2022 this volume). Recent development and modernization of societies completely change the social status and way of life of people and this traditional knowledge is being diminishing generation by generation. Keeping this in mind, the socioeconomically useful of ferns are documented (Raju and Suresh this volume 2022).

Consistent with it, an update on the phytochemistry of Indian pteridophytes is provided, covering the recent findings concerning the phytochemical composition of crude extracts, their histochemical, spectroscopic, and chromatographic profiles . A report of 168 species of preliminary phytochemical analysis, quantitative profiles of 115 species, histochemical profiles of 61 species, chromatographic profile (amino acids and sugar) of 43 species, TLC profile of 14 species, HPLC and HPTLC profiles of 23 species, and GC-MS profiles of 32 species are included in the review (Johnson et al. 2022b this volume). The available literature on phytochemistry confirmed that Indian Pteridophytes are a pool of therapeutic agents. In addition, another review summarizes the available biopotential of Pteridophytes from 2000 to 2021 from a total of 244 species, hoping it might be useful for pteridologist, phytochemist, and pharmacist further research (Johnson et al. 2022b this volume). In this study, the antioxidant, antibacterial, and antifungal activities of several pteridophytes, as well cytotoxic and hepatoprotective properties, their anticancer, anti-inflammatory, antidiabetic, and wound healing potential, and larvicidal activity are recorded.

The successive chapters left of this third part of the book focus either on some species or some particular compounds. Thus, it is included a work focusing on the Omega-3 and Omega-6 fatty acids (FAs) distribution in pteridophytes and its significance in *nutraceutical,* pharmaceutical, and cosmetic industry (Giri et al. 2022 this volume). These macromolecules act as a source of energy as well as play a vital role in human health and nutrition. The most commonly considered plant sources of polyunsaturated fatty acids PUFAs are micro- and macroalgae. Among the terrestrial plants, pteridophytes have some intriguing long chain PUFAs (arachidonic and eicosapentaenoic acids) (Giri et al. 2022 this volume). However, the cultivation and handling of algae have some complication. Monilophytes can be a great source of fatty acids, which are starting to be evaluated in major depth, and to be taken into account for the welfare of human beings.

The book stares at important concerns in the world population today as it is cancer. A study was carried out in *Angiopteris evecta* (G. Forst.) Hoffm., commonly called as giant or king fern, and its anti-proliferative activity of extracts against malignant Colon cancer cells (HT-29) and non-malignant colon cancer cells (L929) (Sara and Ruby 2022 this volume). The findings support the ethnomedicinal observations of these plants for management of cancer, and open new projects committed to isolation and purification of the active elements in the extracts, to investigate the synergy and additive pharmacological effect in killing cancer cells.

Another example of ferns with important bioactivity of the secondary metabolites is the genus *Adiantum*, belonging to the Adiantaceae family, and very popular of folk medicines (Patil et al. 2022 this volume). It is a rich source of many compounds with multiple pharmacological activities such as analgesic, antibacterial, anticancer, antifungal, antidiabetic, and antipyretics. *Nephrolepis auriculata* (L.) has also been checked for phytochemical analysis in vitro antioxidant activity using eight different assays, and antidiabetic activity by α-glucosidase and α-amylase inhibitory assays (Sureshkumar and Ayaanar 2022 this volume). *Nephrolepis auriculata* (L.) was reported to have strong antioxidant and metal chelating activity. In addition, the α-glucosidase and α-amylase inhibitory assays also showed potent antidiabetic properties.

The fourth and last block of the book accumulates knowledge about the useful of ferns as environmental evaluators. To start with, it is the consideration of ferns as potential ecological indicators due to the correlation between their geographic distribution and abiotic variables. They can be easily measurable substitutes for unmeasured ecological values or when extensive studies (with big scales) are nonviable due to budget and time constraints. In the next chapter, a brief introduction on the discovery of fern diversity is provided, and it pursuits to highlight important questions such as giving an overview of fern species richness at a global scale or discussing their distribution across several environmental gradients. The author worries about how the habitat destruction and land use change are the major threats to fern diversity (Della 2022; Mehltreter 2022 this volume).

The fact that pteridophytes have grown efficiently in ever-changing environmental conditions and soils rich in toxic metals. The adaptation of the pteridophytes to anthropogenic activities indicates its evolutionary resilience to various abiotic stress factors, including their high hyperaccumulation capacity, particularly in the fronds, which is called as phytoremediation. Regarding to this topic, there are two contributions. In the first one, the use of ferns as phytoremediators of organic and inorganic environmental pollutants from soil and water is reviewed (Della 2022; Mehltreter 2022; Sajeev et al. 2022; Prabhu and Hegde 2022; Fernández and Sierra

2022 this volume). The wide geographical spread, high environmental adaptation, resilience in toxicity, and bioaccumulation potential of pteridophytes facilitate broad application in the field of phytoremediation since two decades ago. Sajeev et al. (2022 this volume) reported an alternative technique to phytoremediation. In this technique the grounded, non-living, processed pteridophyte-based biosorbents, are applied to remediate the point-source of pollution, in situ. Biosorbents are efficient when the biomass is cost-effective, easy to grow, or harvest, plentiful in availability, and usage of the same quality. Pteridophytes qualify these criteria, making them good biosorbent candidates (Smruthi and Hegde 2022 this volume).

In this book we also bringout the threat which is generally not taken into account: the fern *Pteridium aquilinum,* commonly named bracken, which has been distributed worldwide for ~23.8 Mys. It belonged to open forest communities long before human impact on forests and landscapes took place. However, today the prevalence of bracken has expanded considerably partly not only due to human land use, but also due to the natural aggressiveness of the fern towards competing grasses, herbs, and trees. In addition, bracken is toxic to livestock and humans. The overall geographical distribution and the local abundance of bracken seem to be increasing in many places of the world due to several causes like cultural modifications. Climate change seems to favor bracken spread, somewhere, and given the negative consequences this fern might cause to humans and animals, care must be taken to avoid over-exposition to its dangerous chemicals (Fernández and Sierra 2022 this volume).

## References

Abraham S, Thomas T (2022) Ferns: a potential of source medicine and future prospects. In: Johnson M, Fernández H, Kumar A, Shibila A (eds) Ferns—biotechnology, propagation, medicinal uses and environmental regulation. Singapore, Singapore. (this volume)

Antony R, Suresh S (2022) Ethnobotanical uses of ferns and lycophytes of Kerala. In: Johnson M, Fernández H, Kumar A, Shibila A (eds) Ferns—biotechnology, propagation, medicinal uses and environmental regulation. Singapore, Singapore. (this volume)

Bhatia M, Uniyal PL (2022) In vitro gametophyte development, reproductive biology and nitric oxide signaling in ferns. In: Johnson M, Fernández H, Kumar A, Shibila A (eds) Ferns—biotechnology, propagation, medicinal uses and environmental regulation. Singapore, Singapore. (this volume)

Bujak A, Bujak J (2022) Azolla's use as a biofertilizer and livestock feed. In: Johnson M, Fernández H, Kumar A, Shibila A (eds) Ferns—biotechnology, propagation, medicinal uses and environmental regulation. Singapore, Singapore. (this volume)

Cao H, Chai TT, Wang X et al (2017) Phytochemicals from fern species: potential for medicine applications. Phytochem Rev 16:379–440. https://doi.org/10.1007/s11101-016-9488-7

de Vries J, Fischer AM, Roettger M, Rommel S, Schluepmann H, Bräutigam A, Carlsbecker A, Gould SB (2016) Cytokinin-induced promotion of root meristem size in the fern *Azolla* supports a shoot-like origin of euphyllophyte roots. New Phytol 209(2):705–720

Della AP (2022) Ferns as ecological indicators. In: Johnson M, Fernández H, Kumar A, Shibila A (eds) Ferns—biotechnology, propagation, medicinal uses and environmental regulation. Singapore, Singapore. (this volume)

Eeckhout S, Leroux O, Willats WGT, Popper ZA, Viane RLL (2014) Comparative glycan profiling of *Ceratopteris richardii* C-Fern gametophytes and sporophytes links cell-wall composition to functional specialization. Ann Bot 114(6):1.295–1.307

Fernández H (ed) (2018) Current advances in fern research. Springer, Cham, p 544. ISBN 978-3-319-75102-3

Fernández H, Revilla MA (2003) *In vitro* culture of ornamental ferns. Plant Cell Tissue Organ Cult 73:1–13

Fernández H, Sierra M (2022) Pteridium aquilinum: a threat for biodiversity, and human and animal health. In: Johnson M, Fernández H, Kumar A, Shibila A (eds) Ferns—biotechnology, propagation, medicinal uses and environmental regulation. Singapore, Singapore. (this volume)

Fernández H, Kumar A, Revilla MA (eds) (2011) Working with ferns: issues and applications. Springer, New York, p 386. ISBN 978-1-4419-7161-6

Giri P, Kumar A, Uniyal PL (2022) Omega-3 and Omega-6 fatty acids distribution in Pteridophytes and its significance in nutraceutical, pharmacology and cosmetic industry. In: Johnson M, Fernández H, Kumar A, Shibila A (eds) Ferns—biotechnology, propagation, medicinal uses and environmental regulation. Singapore, Singapore. (this volume)

Granados S, Alejandro Rivera M, Cañal J, Fernández H (2022) The Effect of phytohormones and inhibitors on apogamous gametophytes of Dryopteris affinis ssp. affinis. In: Johnson M, Fernández H, Kumar A, Shibila A (eds) Ferns—biotechnology, propagation, medicinal uses and environmental regulation. Singapore, Singapore. (this volume)

Grossmann J, Fernández H, Chaubey PM, Valdés AE, Gagliardini V, Cañal MJ, Russo G, Grossniklaus U (2017) Proteogenomic analysis greatly expands the identification of proteins related to reproduction in the apogamous fern *Dryopteris affinis* ssp. *affinis*. Front Plant Sci 8(336):1–16

Janakiraman N, Narayani M, Johnson M (2022) A review on molecular phylogeny of Pteridophytes using DNA barcoding. In: Johnson M, Fernández H, Kumar A, Shibila A (eds) Ferns—biotechnology, propagation, medicinal uses and environmental regulation. Singapore, Singapore. (this volume)

Johnson M, Fernández H, Shibila T (2022a) Morphogenesis in ferns. In: Johnson M, Fernández H, Kumar A, Shibila A (eds) Ferns—biotechnology, propagation, medicinal uses and environmental regulation. Singapore, Singapore. (this volume)

Johnson M, Janakiraman N, Chandra Saleride J, Sivaraman A, Shivananthini B, Paulraj K (2022b) Phytochemistry of Indian Pteridophytes—a review. In: Johnson M, Fernández H, Kumar A, Shibila A (eds) Ferns—biotechnology, propagation, medicinal uses and environmental regulation. Singapore, Singapore. (this volume)

Johnson M, Janakiraman N, Chandra Saleride J, Sivaraman A, Shivananthini B, Paulraj K (2022c) Biopotency of ferns. In: Johnson M, Fernández H, Kumar A, Shibila A (eds) Ferns—biotechnology, propagation, medicinal uses and environmental regulation. Singapore, Singapore. (this volume)

Kaźmierczak A (2010) Gibberellic acid and ethylene control male sex determination and development of Anemia phyllitidis gametophytes. In: Fernández H, Kumar A, Revilla MA (eds) Working with ferns. Issues and applications. Springer, New York, pp 49–65

Kholia RS, Balkrishnan A (2022) Pteridophytes used by peoples of Indian Himalayan region and northern India. In: Johnson M, Fernández H, Kumar A, Shibila A (eds) Ferns—biotechnology, propagation, medicinal uses and environmental regulation. Singapore, Singapore. (this volume)

Kumar A, Giri P, Uniyal PL (2022a) Genome evolution in ferns: molecular phylogenomics—a review. In: Johnson M, Fernández H, Kumar A, Shibila A (eds) Ferns—biotechnology, propagation, medicinal uses and environmental regulation. Springer, Singapore. (this volume)

Kumar A, Giri P, Uniyal PL (2022b) Plastogenomics provides a tool to study phylogenetic interrelationships of monilophytes: a review. In: Johnson M, Fernández H, Kumar A, Shibila A (eds) Ferns—biotechnology, propagation, medicinal uses and environmental regulation. Singapore, Singapore. (this volume)

López RA, Renzaglia KS (2014) Multiflagellated sperm cells of *Ceratopteris richardii* are bathed in arabinogalactan proteins throughout development. Am J Bot 101(12):2.052–2.061

Markham K, Chalk T, Neal Stewart C Jr (2006) Evaluation of fern and moss protein-based defenses against phytophagous insects. Int J Plant Sci 167(1):111–117

Mehltreter K (2022) Diversity of ferns and lycophytes at different spatial scales, along environmental gradients and in the anthropogenic landscape. In: Johnson M, Fernández H, Kumar A, Shibila A (eds) Ferns—biotechnology, propagation, medicinal uses and environmental regulation. Singapore, Singapore. (this volume)

Morajikar S, Suneetha C, Harish Kumar TG, Hedge S (2022) Molecular markers in Pteridophytes. In: Johnson M, Fernández H, Kumar A, Shibila A (eds) Ferns—biotechnology, propagation, medicinal uses and environmental regulation. Singapore, Singapore. (this volume)

Patil S, Lavate R, Raole V, Rajput K (2022) Adiantum L.: Overview on taxonomy, distribution, conservation status ethnomedicinal uses, phytochemistry and pharmacognosy from India. In: Johnson M, Fernández H, Kumar A, Shibila A (eds) Ferns-biotechnology, propagation, medicinal uses and environmental regulation. Singapore, Singapore. (this volume)

Perez-Martin A, Michelazzo C, Torres-Ruiz JM, Flexas J, Fernández JE, Sebastiani L, Diaz-Espejo A (2014) Regulation of photosynthesis and stomatal and mesophyll conductance under water stress and recovery in olive trees: correlation with gene expression of carbonic anhydrase and aquaporins. J Exp Bot 65(12):3143–3156. https://doi.org/10.1093/jxb/eru160

Prabhu S, Hegde S (2022) Pteridophytes as effective biosorption agents of heavy metals. In: Johnson M, Fernández H, Kumar A, Shibila A (eds) Ferns—biotechnology, propagation, medicinal uses and environmental regulation. Singapore, Singapore. (this volume)

Qi X, Kuo L-Y, Guo C, Li H, Li Z, Qi J et al (2018) A well-resolved fern nuclear phylogeny reveals the evolution history of numerous transcription factor families. Mol Phylogenet Evol 127:961–977. https://doi.org/10.1016/j.ympev.2018.06.043

Rathinasabapathi B (2006) Ferns represent an untapped biodiversity for improving crops for environmental stress tolerance. New Phytol 172:385–390. https://doi.org/10.1111/j.1469-8137.2006.01889.x

Sahayaraj K (2022) Ferns: a source of phytoecdysones and their applications in pestiferous insect's management. In: Johnson M, Fernández H, Kumar A, Shibila A (eds) Ferns—biotechnology, propagation, medicinal uses and environmental regulation. Singapore, Singapore. (this volume)

Sajeev S, Roshni PT, Mathias RC, Morajkar S, Prabhu S, Hegde S (2022) Pteridophytes: effective agents of phytoremediation. In: Johnson M, Fernández H, Kumar A, Shibila A (eds) Ferns—biotechnology, propagation, medicinal uses and environmental regulation. Singapore, Singapore. (this volume)

Salmi ML, Bushart TJ (2010) Cellular, molecular, and genetic changes during the development of *Ceratopteris richardii* gametophytes. In: Fernández H, Kumar A, Revilla MA (eds) Working with ferns. Issues and applications. Springer, New York, pp 11–24

Salmi ML, Bushart TJ, Stout SC, Roux SJ (2005) Profile and analysis of gene expression changes during early development in germinating spores of *Ceratopteris richardii*. Plant Physiol 138: 1.734–1.745

Sara C, Ruby GD (2022) In vitro antiproliferative effect of Angiopteris evecta (G. Forst.) Hoffm extracts against cultured Ht-29 colon cancer cells. In: Johnson M, Fernández H, Kumar A, Shibila A (eds) Ferns—biotechnology, propagation, medicinal uses and environmental regulation. Singapore, Singapore. (this volume)

Sánchez E, Alejandro Menéndez M, Cañal J, Fernández H (2022) Antagonistic action of yucasine and DMSO on apogamy in the fern Dryopteris affinis ssp. affinis. In: Johnson M, Fernández H, Kumar A, Shibila A (eds) Ferns—biotechnology, propagation, medicinal uses and environmental regulation. Singapore, Singapore. (this volume)

Schluepmann H, Bigot I, Rijken N, Grifoll AC, Gudde P, Dijkhuizen LW, Güngör E (2022) domestication of the floating fern symbiosis Azolla. In: Johnson M, Fernández H, Kumar A, Shibila A (eds) Ferns—biotechnology, propagation, medicinal uses and environmental regulation. Singapore, Singapore. (this volume)

Shibila T, Johnson M (2022) In vitro propagation of epiphytic fern species. In: Johnson M, Fernández H, Kumar A, Shibila A (eds) Ferns—biotechnology, propagation, medicinal uses and environmental regulation. Singapore, Singapore. (this volume)
Suneetha C, Hegde S (2022) Micropropagation of Pteridophytes. In: Johnson M, Fernández H, Kumar A, Shibila A (eds) Ferns—biotechnology, propagation, medicinal uses and environmental regulation. Singapore, Singapore. (this volume)
Sureshkumar J, Ayyanar M (2022) Phytochemical composition, in vitro antioxidant and antidiabetic activities of Nephrolepis auriculata (L.) Trimen, an unexplored ethnomedicinal fern. In: Johnson M, Fernández H, Kumar A, Shibila A (eds) Ferns—biotechnology, propagation, medicinal uses and environmental regulation. Singapore, Singapore. (this volume)
Tosens T, Nishida K, Gago J, Coopman RE, Cabrera HM, Carriquí M, Laanisto L, Morales L, Nadal M, Rojas R, Talts E, Tomas M, Hanba Y, Niinemets Ü, Flexas J (2016) The photosynthetic capacity in 35 ferns and fern allies: mesophyll $CO_2$ diffusion as a key trait. New Phytol 209:1576–1590. https://doi.org/10.1111/nph.13719
Valledor L, Menéndez V, Canal MJ, Revilla A, Fernández H (2014) Proteomic approaches to sexual development mediated by antheridiogen in the fern *Blechnum spicant* L. Proteomics 14(17–18):2.061–2.071
Vallinayagam S (2022) In vitro propagation of Histiopteris incisa (Thunb.) J. Sm.—an ornamental fern. In: Johnson M, Fernández H, Kumar A, Shibila A (eds) Ferns—biotechnology, propagation, medicinal uses and environmental regulation. Singapore, Singapore. (this volume)
von Aderkas P (1984) Promotion of apogamy in *Matteuccia struthiopteris*, the ostrich fern. Am Fern J 74(1):1–6
Wada M (2007) The fern as a model system to study photomorphogenesis. J Plant Res 120(1):3–16
Wen CK, Smith R, Banks JA (1999) ANI1. A sex pheromone-induced gene in Ceratopteris gametophytes and its possible role in sex determination. Plant Cell 11:1.307–1.317
Wyder S, Rivera A, Valdés AE, Cañal MJ, Gagliardini V, Fernández H, Grossniklaus U (2020) Differential gene expression profiling of one- and two-dimensional apogamous gametophytes of the fern Dryopteris affinis ssp. affinis. Plant Physiol Biochem 148:302–311
Youngstrom CE, Irish EE, Cheng C-L (2022) The crucial role of Ceratopteris richardii in understanding the evolution of the WOX gene family. In: Johnson M, Fernández H, Kumar A, Shibila A (eds) Ferns—biotechnology, propagation, medicinal uses and environmental regulation. Singapore, Singapore. (this volume)

# Part I

# Biotechnology

# Genome Evolution in Ferns: Molecular Phylogenomics – A Review

# 2

Ashwani Kumar, Priti Giri, and Prem Lal Uniyal

**Abstract**

Ferns are one of the earliest groups of vascular plants, originated around 400 Ma, that attained remarkable levels of diversity and abundance from the Carboniferous period to the Jurassic period (from c. 300 to 150 million years ago). They were especially prevalent in Paleozoic terrestrial ecosystems, before the rise of seed plants, and maintained their dominance even into the Early Cretaceous (~145 Mya). Many classifications and evolutionary schemes have been proposed reflecting different interpretations based on morphological evidences. Ferns are the only major lineage of vascular plants not represented by a sequenced nuclear genome. However, recent genomic evidences have helped to strengthen the basic understanding of the ideas of possible relationships, and classification has changed accordingly. Now more nuclear data are being available, and new systems of classification are emerging. We have examined such data in the light of available earlier data and systems. This mini review provides an overview of the available information based on nuclear data for future researchers. In another review in this volume, we have attempted to look into chloroplast genome data for the study of the interrelationships of monilophytes.

**Keywords**

Genome · Ferns · Polypods · Land plants · Phylogenomics

A. Kumar (✉)
Former Head of the Department of Botany and Biotechnology, University of Rajasthan, Jaipur, Rajasthan, India

P. Giri · P. L. Uniyal
Department of Botany, University of Delhi, Delhi, India





J. Marimuthu et al. (eds.), *Ferns*, https://doi.org/10.1007/978-981-16-6170-9_2

## 2.1 Introduction

Ferns are a diverse lineage of vascular plants that occupy a wide range of ecological niches and are an important part of the world's land flora (Smith et al. 2006). Ferns are the closest surviving relatives of the seed plants (Pryer et al. 2001) with over 12,000 species (Moran 2008). Ferns (monilophytes) have a unique role in the development and ecology of land plants due to their species richness and ecological effects (Rothfels et al. 2015).

During the mid-Ordovician to lower Silurian (480–430 million years ago), the origin and early evolution of land plants (embryophytes) initiated the establishment of the modern terrestrial ecosystems (Kenrick and Crane 1997; Rubinstein et al. 2010; Steemans et al. 2009; Wellman et al. 2003; Qiu et al. 2006). The origin and evolution of embryophytes resulted in a huge variety of morphological, physiological, reproductive features in the ecological system (Lenton et al. 2012). Between ~350 and 210 million years ago (Mya), ferns were the most dominant plant types on Earth (Soltis et al. 1999; Cai et al. 2021). From the Carboniferous period to the Jurassic period approximately 300–150 million years ago (Mya), they displayed amazing levels of diversity and richness (Tang 2020). Before the development of seed plants, ferns and lycophytes were extremely common in Paleozoic terrestrial ecosystems, and they retained their dominance well into the Early Cretaceous (~145 Mya) (Barreda et al. 2019). During the Late Cretaceous, Nagalingum et al. (2002) observed a drop in the relative diversity and abundance of free-sporing plants, which coincided with an increase in angiosperms (ca. 150 Mya), but the relative contribution of gymnosperms remained unchanged (Schneider et al. 2004).

Pryer et al. (2001) discovered that current vascular plants are divided into three monophyletic groups such as lycophytes, equisetophytes (horsetails), and psilotophytes (whisk ferns), including all eusporangiate and leptosporangiate ferns and seed plants. Leptosporangiate ferns, on the other hand, grew to become the second largest group of vascular plants (Crane 1985; Schneider et al. 2004; Schuettpelz and Pryer 2009; Rai and Graham 2010; Christenhusz et al. 2011; PPG I 2016). Psilotoids (whisk ferns) ophioglossoids, equisetoids (horsetails), marattioids, and leptosporangiates are the four primary clades of ferns (Sessa et al. 2014; Fig. 2.1).

Ferns as a whole include lineages that diverged from one another prior to the divergence of the major seed plant clades. The most recent common ancestor of all leptosporangiates arose approximately 280 Mya (Schneider et al. 2004; Smith et al. 2010). The ancestors of *Ceratopteris* and *Azolla* diverged from each other ca. 200 Mya, well before the divergence of monocots and eudicots. Dates quoted from TimeTree (*TimeTree* http://timetree.org; Hedges et al. 2006).

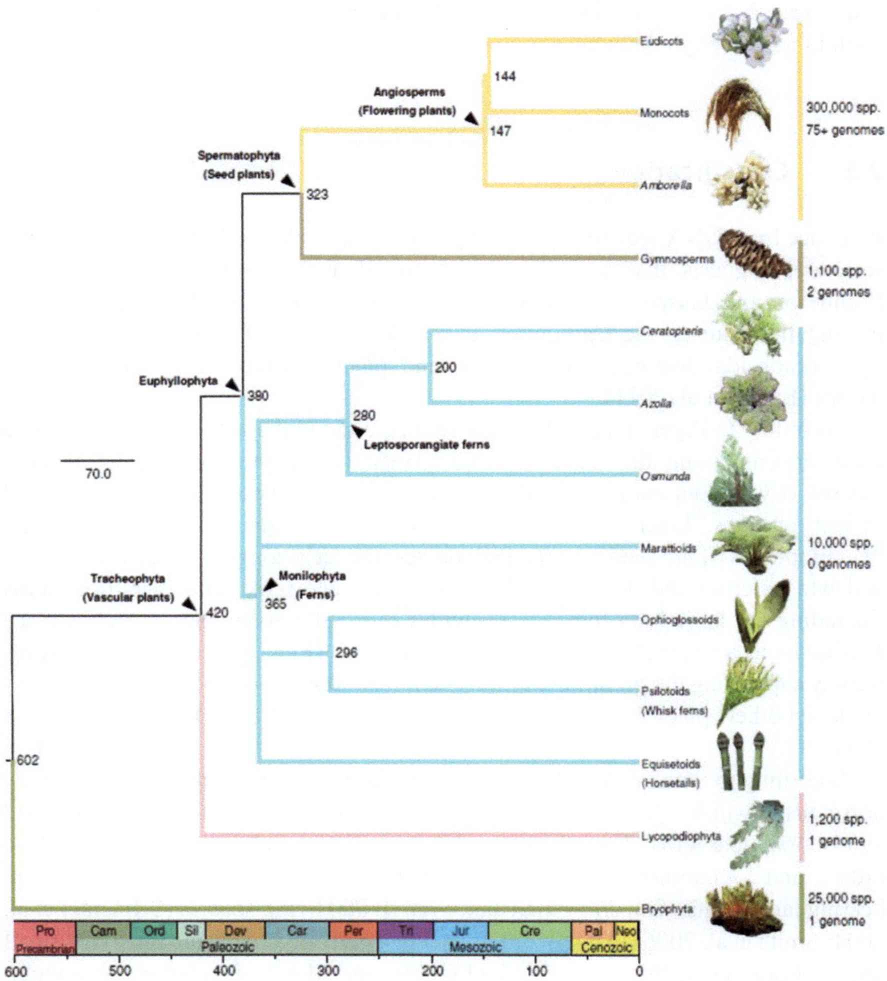


**Fig. 2.1** Sessa et al. (2014) projected the phylogeny of major groups of land plants. (Based on Pryer et al. 2001; Pryer et al. 2004; Rai and Graham 2010). Approximate numbers of species and available genome sequences are given, and approximate times of major divergences are indicated. (Source: Sessa et al. (2014) Between Two Fern Genomes, *GigaScience*, 3, (1) 2047–217X–3–15. This is an open-access article distributed under the terms of the Creative Commons CC BY license)

## 2.2 General Characters: Alternation of Generation

Lindsay (1794) provided the ferns with two distinct generations. Sori on sporophytes produce haploid spores, which germinate and produce prothallus that is independent of the sporophyte. It has sex organs that produce sporophytes after fertilization (Christenhusz and Chase 2014). Ferns have been classified based on spore and gametophyte morphology, as well as the number of sperm flagella (Christenhusz and Chase 2014). Spores in a favorable habitat from a photosynthetic free-living haploid gametophyte known as prothalli.

## 2.3 Classification

Evidence from DNA sequences of multiple genes, as well as information on genome structure, suggests that a major split occurred around 400 million years ago (Raubeson and Jansen 1992). Although lycophytes were abundant and dominant in land flora during the Carboniferous era (Kenrick and Crane 1997), the class Lycopodiopsida diverged shortly after land plants evolved to acquire vascular tissues (Banks et al. 2011).

According to Pryer et al. (2004), this split resulted in the extant lycophytes and a clade encompassing the other vascular plants. Lycophytes include living club mosses (Lycopodiaceae) and spike mosses (Selaginellaceae), as well as several extinct lineages. Later, the remaining vascular plant lineage was divided into "monilophytes" and seed plants. Psilophytes (adder's tongue ferns, moonworts, and whisk ferns) and the horsetails, leptosporangiate ferns, and marattioid ferns (including the large king fern) are examples of monilophytes. Some studies place *Equisetum* as a sister clade to leptosporangiate ferns (Renzaglia et al. 2000), others weakly supporting the grouping of *Equisetum* with Marattiopsida (Pryer et al. 2001), while yet others place *Equisetum* as the sister group to all ferns (Kenrick and Crane 1997).

According to Smith et al. (2006), ferns (monilophytes) in the broad sense include horsetails, whisk ferns and ophioglossoid ferns, marattioid ferns, and leptosporangiate ferns. They recognized 4 monophyletic classes, 11 monophyletic orders, and 37 families, 32 of which are monophyletic. They are the most spore-bearing land plants, with 250 genera and over 10,000 living species (Schneider et al. 2004; Smith et al. 2006). All fern orders except 4 early divergent and relatively small orders (Equisetales, Psilotales, Ophioglossales, and Marattiales) belong to a group termed leptosporangiate ferns (also defined as subclass Polypodidae), which includes about 10,000 (~94%) of extant fern species. The spermatophytes (seed plants) or also commonly called phanerogams, which number over 260,000 species (Thorne 2002; Scotland and Wortley 2003). However, the monilophytes (ferns, sensu Pryer et al. 2004; Schuettpelz and Pryer 2007), have around 9000 species. Despite the fact that ferns (monilophytes) are the most varied plant lineage on Earth, second only to angiosperms, almost all of these species are found in the monophyletic (leptosporangiate) polypod fern lineage, which has 267 genera and around 9000 species (Vanneste et al. 2015).

Earlier attempts of classification published in the molecular era by Smith et al. (2006) focused exclusively on ferns; these plants were also referred to as "monilophytes" in some publications (Pryer et al. 2004). Classifications by Christenhusz et al. (2011) and Rothfels et al. (2012) reflected in subsequent higher-level schemes.

Monilophyte phylogeny is poorly understood. The branching order of the earliest splits in monilophytes, leptosporangiate ferns, and polypod fern phylogeny, respectively, have been studied by Rai and Graham (2010) who firmly established filmy ferns (Hymenophyllales) as the sister group of all leptosporangiates except Osmundaceae, resolving the second deepest split in leptosporangiate-fern phylogeny. Rai and Graham (2010) also examined the utility of a large plastid-based data set in inferring backbone relationships for monilophytes and found it highly congruent with earlier multigene studies, corroborating clades in common across studies (Knie et al. 2015).

Christenhusz and Chase (2014, Fig. 2.2) published a new classification, consistent with an understanding of fern phylogeny that expands Aspleniaceae, Cyatheaceae, Polypodiaceae, and Schizaeaceae in comparison to recent classification and which is a modification of Smith et al. (2006, 2008) and Christenhusz et al. (2011). The classification of Christenhusz and Chase (2014) represented a considerable departure in terms of stability, hence it has not been widely adopted.

Lycopods are similar to ferns in this regard, but ferns are the sister group of the seed plants, whereas the lycopods are sister to all other vascular plants (ferns and the seed plants) (Christenhusz and Chase 2014). Ruggiero et al. (2015) treated lycophytes and ferns as distinct subphyla (Lycopodiophytina and Polypodiophytina), each with a single class (Lycopodiopsida and Polypodiopsida).

Further, the PPG I classification agrees with Ruggiero et al. (2015) in recognizing four subclasses within ferns (Equisetidae, Ophioglossidae, Marattiidae, and Polypodiidae). Present taxonomy of ferns includes 48 families in 11 orders (PPG I: 2016). Shmakov (2016) provided a modern, comprehensive classification for lycophytes and ferns. They used monophyly as the primary criterion for the recognition of taxa. In total, this classification treats an estimated 11,916 species in 337 genera, 51 families, 14 orders, and 2 classes. Shmakov (2016) provided modern classification based on tracheophyte phylogeny, depicting relationships among lycophyte and fern families recognized in their PPG I classification (Fig. 2.3). Composite topology is derived from the results of numerous phylogenetic studies (e.g., Pryer et al. 2001, 2004; Korall et al. 2007; Schuettpelz and Pryer 2007; Rai and Graham 2010; Lehtonen 2011; Rothfels et al. 2012, 2015; Knie et al. 2015; Zhang and Zhang 2012). Most nodes have received consistently strong support; dotted lines indicate areas of considerable uncertainty (Fig. 2.3).

Shmakov (2016) recognized two classes Lycopodiopsida (lycophytes) and Polypodiopsida (ferns) which form distinct evolutionary lineages in the tracheophyte phylogenetic tree. For each family, Shmakov (Shmakov 2016; Fig. 2.3) noted the total number of genera recognized in PPG I and the sum of species estimates for these genera (terminal clade height is roughly proportional to diversity for families with more than 100 species).

Szövényi et al. (2021) reported that during the past few years, several high-quality genomes have been published from charophyte algae, bryophytes, lycophytes, and ferns. Szövényi et al. (2021) compared patterns of various levels of genome and epigenomic organization found in seed-free plants to those of seed plants. They reported that in stark contrast to other land plants, *Selaginella* and most liverworts

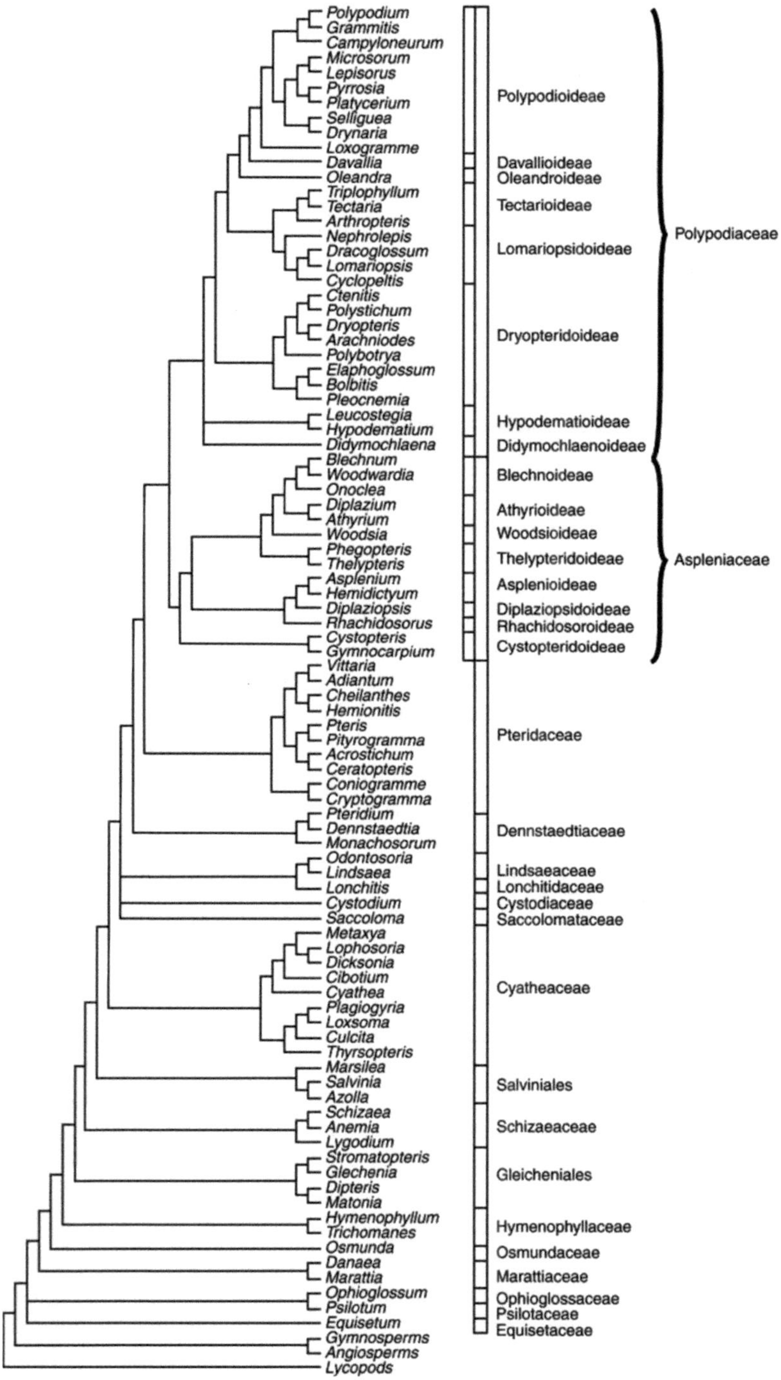


**Fig. 2.2** Phylogenetic tree showing relationships of a representative selection of fern genera based on molecular (DNA) data, modified from Schuettpelz and Pryer (2007), Lehtonen (2011), Rothfels et al. (2012), and Schneider et al. (2013). (Source: Christenhusz and Chase (2014) Trends and

are devoid of whole-genome duplication. Phylogenetic evolution of ferns poses several unanswered answers.

## 2.4 Cytological Studies

Two groups of extant plant such as lycopsids and psilopsids, according to Raubeson and Jansen (1992), are living representatives of the earliest divergent lineage in vascular plant evolution. Polyploidy contributes to *c.* 31% of speciation events in ferns compared to *c.* 15% in angiosperms, which supports this notion (Wood et al. 2009). Divergences in the genome of ferns were linked to significant karyological changes (Haufler 2014). Sessa et al. (2014) conducted cytological studies and revealed that ferns, particularly homosporous species (which include up to 99% of all extant ferns) (Smith et al. 2006), have much higher chromosome numbers than other plants (Britton 1974). There is considerable variation (80-fold) in chromosome numbers (ranging from $2n = 18$ in *Salvinia natans* to $2n = 1440$ in *Ophioglossum reticulatum*), and 94-fold variation in genome size (ranging from $1C = 72.68$ pg in *Psilotum nudum* to $1C = 0.77$ pg in *Azolla microphylla*) (Haufler 2002). Besides this, considerable variation in the average chromosome numbers in homosporous ferns and lycophytes ($n = 57.05$), as compared to angiosperms ($n = 15.99$), and for heterosporous ferns and lycophytes ($n = 13.62$) has been reported (Klekowski and Baker 1966; Henry et al. 2015). Haufler (2002, 2014) proposed that ferns went through multiple cycles of polyploidy (whole-genome duplications) followed by subsequent diploidization via gene silencing, but with no apparent chromosome loss, resulting in high chromosome numbers.

Many ferns have unusually large chromosome numbers, presumably due to whole-genome duplications (WGDs) that occurred near the end of the Cretaceous (~145–66 Mya). The homosporous fern *Ophioglossum reticulatum* has the highest chromosome ($2n = 1440$) of any multicellular organism (Ghatak 1977). The fundamental cause of the link between homospory and high chromosome number is not unknown.

Based on a large-scale transcriptome assembly performed by next-generation sequencing of *E. giganteum*, Vanneste et al. (2015) revealed unambiguous evidence for a paleopolyploidy event within the horsetails. Although transcriptome data only cover genes that are actively transcribed, they are appropriate for exploratory analysis of the gene space (Matasci and Zhixiang 2014).

Although nuclear genome size is highly variable among vascular plants, ferns are the only land plant lineage in which chromosome number and genome size have a substantial positive correlation (Nakazato et al. 2008; Bainard et al. 2011). Despite significant variance in both parameters, Clark et al. (2016) found a correlation

**Fig. 2.2** (continued) concepts in fern classification. Ann Bot. Mar;113(4):571–94. doi: https://doi.org/10.1093/aob/mct299. Reproduced under license no. 5085950855981 dated 11 June)

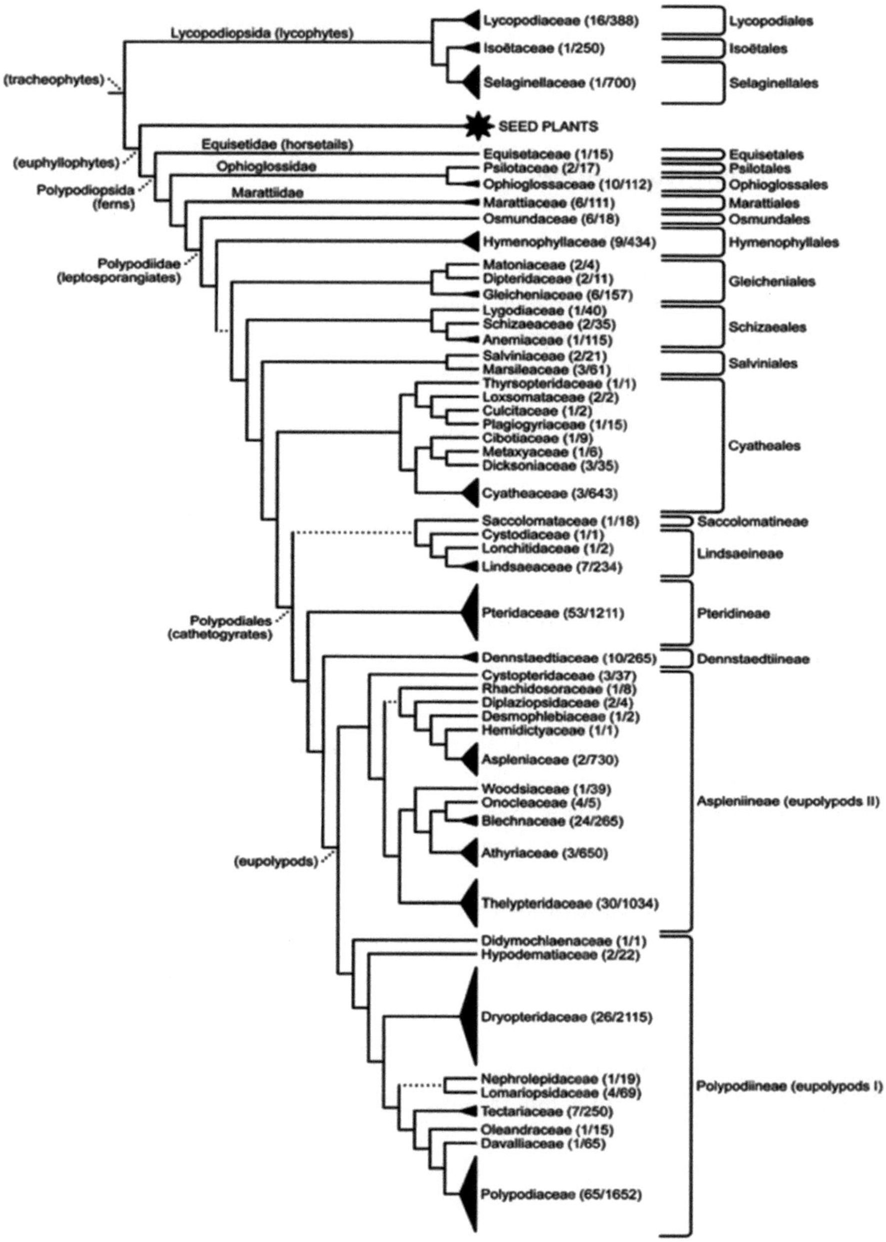


**Fig. 2.3** Shmakov (2016) provided modern classification. (Source: PPG I Shmakov (2016), A community-derived classification for extant lycophytes and ferns. Jnl of Sytematics Evolution, 54: 563–603. https://doi.org/10.1111/jse.12229). Reproduced under rights link license 5094460819755 dated 22 June 2021

between genome size and chromosome numbers across all ferns. Long terminal repeat retrotransposons (LTR-RTs) are major contributor to long-term change in the amount of nuclear DNA. LTR-RT dynamics in ferns and lycophytes are poorly understood in comparison to flowering plants (Soltis and Soltis 2013; Baniaga and Barker 2019).

## 2.5 Molecular Phylogenetic Relationship

### 2.5.1 Ancestral Plant Genome Structure

Considerable progress has been achieved in inferring phylogenetic relationships among the three major branches of vascular-plant phylogeny including the lycophytes, monilophytes, and seed plants (Doyle 1998). The phylogenetic relationships among these four basal fern orders are the most debated topics in fern phylogeny. Pryer et al. (2001) suggested that maximum likelihood analysis showed unambiguously that horsetails and ferns together are the closest relatives to seed plants. In general, divergence time estimates are consistent with the hypothesis of a major replacement of the Carboniferous leptosporangiate ferns with new (extant) lineages at the end of the Paleozoic (Stewart and Rothwell 1993; Rothwell 1999).

However, this refutes the prevailing view that horsetails and ferns are transitional evolutionary grades between bryophytes and seed plants and has important implications for our understanding of the development and evolution of plants. Pryer et al. (2004) have explored the phylogenetic structure of ferns (= monilophytes) with a special focus on the early divergences among leptosporangiate lineages, as comprehensive analysis of the early leptosporangiate divergences was lacking. In the late Devonian, the early branches of monilophyte phylogeny may have diverged from each other very quickly (Smith et al. 2006; Pryer and Schuettpelz 2009).

The fossil record indicates that leptosporangiate ferns first appeared in the earliest Carboniferous (Tournaisian) and soon after diversified to give rise to roughly six independent lineages, all of which have ambiguous relationships to extant lineages. Some of these Carboniferous leptosporangiate ferns have been put forward as putative stem groups of extant lineages (Stewart and Rothwell 1993; Rothwell 1999), for example, Anachoropteridaceae for osmundaceous ferns, Tedeleaceae for schizaeoid ferns, and Sermyaceae for gleichenioid ferns (Collinson 1996; Rothwell 1999). Further studies are needed to elucidate relationships among Carboniferous and extant leptosporangiate ferns. Phylogenetic analyses of sequenced genomes for 62 taxa and more than 5000 bp from the plastid (*rbcL*, *atpB*, *rps4*) and the nuclear (18S rDNA) used by Pryer et al. (2004) confirmed that Osmundaceae are sister to the rest of the leptosporangiates. They resolved a diverse set of ferns formerly thought to be a subsequent grade as possibly monophyletic (Dipteridaceae, Matoniaceae, Gleicheniaceae, Hymenophyllaceae). Moreover, they placed schizaeoid ferns as sister to a large clade of "core leptosporangiates" that includes heterosporous ferns, tree ferns, and polypods. Grewe et al. (2013) reported that phylogenetic affinities were revealed by mapping rare genomic structural changes in a phylogenetic context: the presence of a unique mitochondrial *atp1* intron argues strongly for a sister relationship between Polypodiopsida and Marattiopsida, and the absence of the *rps16 gene* and the *rps12i346* intron from *Equisetum*, *Psilotum*, and *Ophioglossum* indicates that Equisetopsida is sister to Psilotopsida (Grewe et al. 2013).

Sphenophyta (horsetails) dating back to the Upper Devonian shed light on how their larger ancestors lived and how this ancient lineage has managed to thrive in tropical climates (Husby 2013). Horsetails, which belong to the monilophytes, have preserved multiple ancient features dating back to the Jurassic (Channing et al. 2011) and probably even the Triassic (Hauke 1963), and *Equisetum* may be the oldest extant genus among vascular plants (Husby 2013). The Equisetopsida are thought to have diverged from other monilophytes about ~354 million years ago (Schneider et al. 2004). Sphenophyta was an abundant and diversified clade among the monilophytes (Husby 2013).

Vanneste et al. (Vanneste et al. 2015) resolved the relationships between the different horsetail with *Equisetum bogotense* is basal to both the subgenera *Hippochaete* and *Equisetum*, each of which contains seven species (Guillon 2004, 2007), whereas the Equisetopsida are most likely sister to both the whisk ferns (Psilotales) and ophioglossoid ferns (Grewe et al. 2013).

des Marais et al. (2003) reported that *Equisetum* with a rich fossil record has two subgenera (1) subg. *Equisetum* (eight species; superficial stomates; stems branched) and (2) subg. *Hippochaete* (seven species; sunken stomates; stems generally unbranched). *Equisetum*, with 15 species, is regarded to be the oldest extant genus among vascular plants, probably going back to the Triassic (Vanneste et al., Vanneste et al. 2015).

Horsetails (*Equisetum* L.) are the only survivors of the formerly more diverse Sphenopsida, free-sporing plants distinguished by articulate stems bearing whorls of leaves at each node. Sphenophyta has giant horsetails and has a history dating back to the Upper Devonian (Guillon 2004). Horsetails have retained multiple ancient features dating back to the Jurassic (Channing et al. 2011) and possibly even the Triassic. This is supported by fossil findings and characterstics of *Equisetum.*

*Equisetum* crown group appears to have diversified in the early Cenozoic, although the Equisetaceae total group is thought to have a Paleozoic origin (des Marais et al. 2003). *Equisetum* diverged at 343 Mya (Christenhusz et al. 2021) in phylogenetic molecular clock with the first major split among extant species occurring 170 Mya (Middle Jurassic). Hauke (1963) suggested that *Equisetum* may be the oldest extant genus within the vascular plants (Guillon 2004; Husby 2013). *Equisetum* architecture and anatomy enabled them to adapt a wide range of conditions, including disturbance tolerance, soil anoxia, high metals, and salinity, as well as effective nutrient uptake and nitrogen fixation (Husby 2013). Extant horsetails diverged in the Early Cenozoic, not long after the Cretaceous-Paleogene boundary ~66 million years ago, according to both molecular dating studies (des Marais et al. 2003; Pryer et al. 2004) and fossil evidence (Stewart and Rothwell 1993).

Based on a large-scale transcriptome assembly performed by next-generation sequencing of *E. giganteum*, Vanneste et al. (Vanneste et al. 2015) revealed unambiguous evidence for a paleopolyploidy event within the horsetails (Matasci and Zhixiang 2014). Their phylogenetic and biogeographic patterns, however, are still poorly understood.

Despite the high chromosome number of *E. giganteum* ($n = 108$; Leitch and Leitch 2013), it was confirmed that high chromosome number in homosporous ferns did not evolve through repeated rounds of recent polyploidy (Haufler and Soltis 1986; Haufler 1987; Leitch and Leitch 2013, Kevin et al. 2015). Vanneste et al. (Vanneste et al. 2015) confirmed that horsetails had an independent paleopolyploidy (high chromosome count; $n = 108$) during the Late Cretaceous, prior to the genus diversification, but that no recent polyploidizations could account for their high chromosome number. If the ancestor of ferns and seed plants had a low chromosome number, Vanneste et al. (Vanneste et al. 2015) stated that a single paleopolyploidy could not have resulted in a chromosome number of $n = 108$, requiring additional explanations to resolve this phenomenon.

The number of gene duplicates has been suggested by Nakazato et al. (2006) as a possible mechanism for previous polyploidization in *Ceratopteris richardii.* They created a high-resolution genetic linkage map of *C. richardii* ($n = 39$), a homosporous fern model species. Despite significant variance in both parameters, Clark et al. (2016) found a correlation between genome size and chromosome number across all ferns. Because the earliest ferns had high chromosome numbers, it is possible that the earliest fern divergences were correlated with significant karyological changes.

Based on both genomic and cytogenetic data, Marchant et al. (2019) reported a single ancient polyploidy event and the distribution of repeat elements in the evolutionary history of C-Fern (*Ceratopteris richardii*).

### 2.5.2 Molecular Phylogenetics

Molecular phylogenetics, according to Christenhusz and Chase (2014), has increased our understanding of fern relationships. Unsurprisingly, "fern allies" (e.g., Lycopodiales, Psilotaceae, Equisetaceae) were found to be polyphyletic. The lycopods (Lycopodiaceae, Selaginellaceae, and Isoetaceae) form a clade which are sister to all other vascular plants, whereas the whisk ferns (Psilotaceae) are sister to Ophioglossaceae (Fig. 2.5).

Knie et al. (2015) reported that under dense taxon sampling, phylogenetic information from a prudent choice of loci is currently superior to character-rich phylogenomic approaches at low taxon sampling. Knie et al. (2015) ultimately obtained a well-supported molecular phylogeny placing Marattiales as sister to leptosporangiate ferns and horsetails as sister to all remaining monilophytes. The "Monilophyte" clade comprising ferns, horsetails, and whisk ferns receives unequivocal support from molecular data as the sister clade to seed plants (Knie et al. 2015; Fig. 2.4).

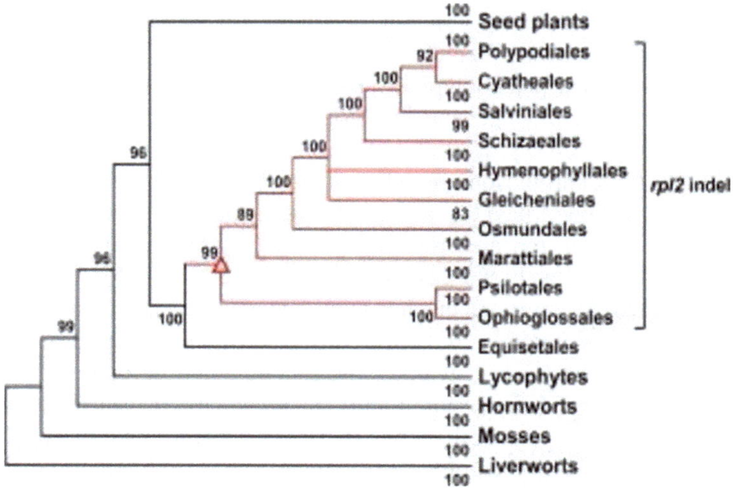


**Fig. 2.4** The "Monilophyte" clade comprising ferns, horsetails, and whisk ferns receives unequivocal support from molecular data as the sister clade to seed plants. (Source: Knie et al. (2015). Horsetails are the sister group to all other monilophytes and Marattiales are sister to leptosporangiate ferns. *Molecular Phylogenetics and Evolution*, *90*, 140–149. https://doi.org/https://doi.org/10.1016/j.ympev.2015.05.008. Reproduced under license number 5080631320156 dated 2 June 2021)

#### 2.5.2.1 Pteridaceae

Chao et al. (2014) concluded that *Pteris* diversified around 47 million years ago, and when species colonized new geographical areas, they generated new lineages, which are associated with morphological character transitions. *Pteris* (Pteridaceae) with 250 species was once thought to be monotypic, but 3 genera *Neurocallis*, *Ochropteris*, and *Platyzoma* have been included (Chao et al. 2014; Fig. 2.5).

Chao et al. (2014) studied molecular phylogeny of *Pteris* in 135 species, based on cpDNA *rbcL* and *matK*. The monophyly of *Pteris* remains uncertain, especially regarding the relationship of *Pteris* with *Actiniopteris* + *Onychium* and *Platyzoma*.

### 2.5.3 Whole-Genome Duplications in Ferns

One of the mechanisms of evolving polyploidy is Whole-genome duplication (WGD). It is a frequent event during the evolution of the plant kingdom, the majority of species of angiosperms are now polyploids or cryptic polyploids. WGD has been associated with the induction of various alterations to both genome sequence and patterns of gene expression, and some of these changes likely favored evolutionary adaption (Adams and Wendel 2005; Chen 2007). The changes in both DNA sequence and methylation can result in the differential expression of homoeologous genes, leading to the adaptive phenotypic variation (Chen 2007) though the mechanisms underlying these changes remain poorly understood.

**Fig. 2.5** Chao et al. (2014) studied selected morphologies of *Pteris*. (**a**) *P. dactylina*, pedate, pinnate, with exaggerated basiscopic pinnules, lithophytic, Taiwan. (**b**) *P. nipponica*, single axis, simple pinnae, with exaggerated basiscopic pinnules, with pale marks, lithophytic, Japan. (**c**) *P.* aff. *Heteroclita*, single axis, irregularly bipinnatifid, without exaggerated basiscopic pinnules, terrestrial, Madagascar. (**d**) *P. grevilleana* var. *ornata*, single axis, bipinnatifid, with exaggerated

Whole-genome duplication (WGD) is frequently suggested as a cause of species diversification (Landis et al. 2018). Whole-genome duplications (WGDs) have been uncovered in many angiosperm clades and have been associated with the success of angiosperms, in terms of both species richness and biomass dominance (Parisod et al. 2010). However WGD has remained understudied in non-angiosperm clades. *Equisetum* appears to be an exception, as with only 15 extant species the genus hardly evidences a link between WGD and diversification. Vanneste et al. (Vanneste et al. 2015) demonstrated that horsetails underwent an independent paleopolyploidy during the Late Cretaceous prior to the diversification of the genus but did not experience any recent polyploidization that could account for their high chromosome number. In lieu of high species diversity, Vanneste et al. (2013) suggested that the WGD event may have contributed to the longevity of the lineage, despite estimating a relatively recent *Equisetum* WGD. Christenhusz et al. (2021) concluded that extant plant groups with a long fossil history are key elements in understanding vascular plant evolution. Horsetails (*Equisetum*, Equisetaceae) have a nearly continuous fossil record dating back to the Carboniferous, but their phylogenetic and biogeographic patterns are still poorly understood.

Sessa et al. (2014) reported that ferns also lack chromosome-level evidence of substantial ancient polyploidy, such as syntenic chromosomal blocks (Nakazato et al. 2006; Barker 2009). Recent evidence has changed this opinion, and homosporous ferns with the putative base chromosome are said to be genetically like diploids (Haufler and Soltis 1986; Wolf et al. 1987; Nakazato et al. 2006). Soltis and Soltis (2000) considered that this combination of high chromosome numbers and lack of evidence for substantial polyploidy in homosporous ferns as the "polyploidy paradox" and whole-genome data are needed to resolve it (Sessa et al. 2014).

**Fig. 2.5** (continued) basiscopic pinnules, with pale marks, terrestrial, China. (**e**) *P. terminalis*, single axis, bipinnatifid, with exaggerated basiscopic pinnules, terrestrial, Taiwan. (**f**) *P. geminata*, single axis, bipinnatifid, with exaggerated basiscopic pinnules, terrestrial, Comoros. (**g**) *P. tremula*, single axis, tripinnatifid, with exaggerated basiscopic pinnules, terrestrial, Australia. (**h**) *P. tripartita*, tripartite, bipinnatifid, with exaggerated basiscopic pinnules, terrestrial, Vanuatu. (**i**) *P. vittata*, single axis, simple pinnae, without exaggerated basiscopic pinnules, lithophytic, Taiwan. (**j**) *P. dimidiata*, single axis, dimidiate, without exaggerated basiscopic pinnules, terrestrial, Taiwan. (**k**) *P. fauriei* var. *minor*, single axis, bipinnatifid, with exaggerated basiscopic pinnules, terrestrial, Taiwan. (**l, m**) *P. wallichiana*, tripartite, bipinnatifid, with exaggerated basiscopic pinnules, anastomosing vein, terrestrial, Taiwan. (**n**) *P. biaurita*, bipinnatifid, anastomosing vein, Taiwan. (**o**) *P. warburgii*, pinnate, anastomosing vein, Solomon Islands. (**p**) *P. pacifica*, bipinnatifid, free vein, Vanuatu. (**q**) False veinlets (arrows) of *P. grevilleana*, Taiwan. (**a**), (**b**), (**d**), (**e**), (**g**), (**i**), (**j**), (**k**), (**m**), (**n**), and (**q**) photographed by Yi-Shan Chao; (**c**), (**f**), and (**h**) photographed by Germinal Rouhan; (**o**) and (**p**) photographed by Cheng-Wei Chen; (**l**) photographed by Chien-Yu Lin. (Source: Chao et al. (2014) Molecular phylogeny and biogeography of the fern genus *Pteris* (Pteridaceae), *Annals of Botany*, 114 (1): 109–124, https://doi.org/10.1093/aob/mcu086. Reproduced under RightsLink license number 5097800369850 dated 28 June 2021)

## 2.6 Discussion

Establishing the timescale of early land plant evolution is essential for testing hypotheses on the coevolution of land plants and Earth's system (Morris et al. 2018). According to Banks et al. (2011), vascular plants appeared ~410 million years ago and then diverged into several lineages, of which only two survive: the euphyllophytes (ferns and seed plants) and the lycophytes. Rothwell (1999) reviewed that the homosporous eusporangiate have extremely high chromosome numbers and large sporangia that develop from several sporangial initials, e.g., Ophioglossales and Marattiales. Homosporous ferns because of their high chromosome numbers (Chiarugi 1960) were initially assumed to have experienced many rounds of ancient whole-genome duplication (polyploidy) (Klekowski and Baker 1966). Psilotales (which includes *Psilotum* and *Tmesipteris*) are closer to ferns (Hendy and Penny 1989; Pryer et al. 2001; Qiu et al. 2006, 2007; Korall et al. 2010; Zhong et al. 2011, 2014).

In addition to this, a third order of homosporous ferns, the Filicales, is the most species rich of non-angiospermous vascular plant groups (Rothwell 1996) and is widely regarded as leptosporangiate (Gifford and Foster 1989). The remaining living ferns consist of heterosporous leptosporangiate genera that are either amphibious (Marsileales) or floating aquatics (Salviniales; Gifford and Foster 1989). Recent phylogenetic studies have provided compelling evidence that confirms the once disputed hypothesis of monophyly for heterosporous leptosporangiate ferns (Marsileaceae and Salviniaceae) (Pryer 1999). Recent phylogenetic studies have revealed a basal dichotomy within vascular plants, separating the lycophytes (less than 1% of extant vascular plants) from the euphyllophytes (Pryer et al. 2001, 2004).

The extant monilophytes or "ferns" comprise of Equisetaceae (horsetails), Ophioglossaceae (ophioglossoid ferns), Psilotaceae (whisk ferns), Marattiaceae (marattioid ferns), and leptosporangiate ferns (Kenrick and Crane 1997).

Only three orders are currently recognized within Lycopodiopsida, including Lycopodiales, Isoetales, and Selaginellales (Shmakov 2016). Order Lycopodiales includes 1 family and 16 genera, whereas orders Isoetales and Selaginellales each contain a single genus *Isoetes* and *Selaginella*, respectively (Zhou and Zhang 2015; Weststrand and Koral Weststrand and Korall 2016a, b).

Establishment of the Pteridophyte Phylogeny Group (PPG) began through promotion of the concept at international conferences (including the 2015 *Next Generation Pteridology* conference) and posts to society websites and e-mail lists, including those of the American Fern Society (AFS) and the International Association of Pteridologists (IAP) (PPG I 2016 https://doi.org/10.1111/jse.12229). PPG I (2016 https://doi.org/10.1111/jse.12229) classification presented here reflects our current, collective understanding of lycophyte and fern phylogeny.

According to PPG ( 2016), within Polypodiopsida are four subclasses: Equisetidae (horsetails), Ophioglossidae, Marattiidae, and Polypodiidae (leptosporangiates). Extant Equisetidae includes a single order, a single small family, and a single genus (*Equisetum* L.). Subclass Ophioglossidae encompasses 2 orders, each with a single family, and a total of 12 genera. Marattiidae includes just one order, one family, and six genera.

In subclass Polypodiidae of Polypodiopsida, Pteridophyte Phylogeny Group (PPG) (2016; https://doi.org/10.1111/jse.12229) recognized seven orders (Osmundales, Hymenophyllales, Gleicheniales, Schizaeales, Salviniales, Cyatheales, and Polypodiales), with the Polypodiales subsequently divided into six suborders (Saccolomatineae, Lindsaeineae, Pteridineae, Dennstaedtiineae, Aspleniineae, and Polypodiineae). In Osmundales, they recognized a single small family with six genera; order Hymenophyllales also includes just one family with nine genera; Gleicheniales and Schizaeales have three relatively small families each. The order Salviniales comprises two families and five genera, and the order Cyatheales encompasses seven small families, with a total of ten genera, plus the larger Cyatheaceae with three genera.

Clark et al. (2016) suggested that genome size was correlated with chromosome number across all ferns despite some substantial variation in both traits. Marchant et al. (2019) reported a single ancient polyploidy event and spread of repeat elements in the evolutionary history of C-Fern (*Ceratopteris richardii*) based on both genomic and cytogenetic data.

The high chromosome count of horsetails could then indeed be caused by a few paleopolyploidies, of which a large fraction of genetic material has been retained (Haufler 1987). Alternatively, the ancestor of all vascular plants could have exhibited a relatively high chromosome number (Soltis and Soltis 1987) so that a single paleopolyploidy could have resulted in the very high chromosome number in horsetails. It is emphasized that both theories are not mutually exclusive. WGD is often proposed as a driver of species diversification (Landis et al. 2018). *Equisetum* seems to be an exception, as with only 15 extant species, the genus hardly evidences a link between WGD and diversification. In lieu of high species diversity, Vanneste et al. (2013) have suggested that the WGD event may have contributed to the longevity of the lineage, despite estimating a relatively recent *Equisetum* WGD. Nakazato et al. (2006) suggested that abundance of gene duplicates is a potential mechanism for the past polyploidization in *Ceratopteris richardii*. They constructed a high-resolution genetic linkage map of the homosporous fern model species, *C. richardii* ($n = 39$). WGD is often proposed as a driver of species diversification (Landis et al. 2018).

Clark et al. (2016) reported that genome size was correlated with chromosome number across all ferns despite some substantial variation in both traits. They observed a trend toward conservation of the amount of DNA per chromosome, although Osmundaceae and Psilotaceae have substantially larger chromosomes. Reconstruction of the ancestral genome traits suggested that the earliest ferns were already characterized by possessing high chromosome numbers and that the earliest divergences in ferns were correlated with substantial karyological changes. Evidence for repeated whole-genome duplications was found across the phylogeny. This paradox led to the hypothesis that ferns underwent multiple cycles of polyploidy (whole-genome duplications (WGDs)) accompanied by subsequent diploidization involving gene silencing, but without apparent chromosome loss, so high chromosome numbers were retained (Haufler 2002, 2014). Support for this hypothesis has been provided by observations that polyploidy contributes to c. 31% of speciation events in ferns compared with c. 15% in angiosperms (Wood et al. 2009).

Ferns have especially high chromosome numbers, *Ophioglossum reticulatum* possessing the highest known chromosome count among extant eukaryotes ($n > 600$; Khandelwal 1990). However, a distinction should be made between heterosporous ferns that have chromosome counts similar to angiosperms ($n = 15.99$ on average) and homosporous ferns where the chromosome count is more than three times higher ($n = 57.05$ on average; Klekowski and Baker 1966).

Christenhusz et al. (2021) concluded that extant plant groups with a long fossil history are key elements in understanding vascular plant evolution. Horsetails (*Equisetum*, Equisetaceae) have a nearly continuous fossil record dating back to the Carboniferous, but their phylogenetic and biogeographic patterns are still poorly understood (Christenhusz et al. 2019).

## 2.7 Conclusion

Almost all recognized extant fern families and nearly all monilophyte genera at the early diverging nodes have now been sampled in published molecular phylogenetic studies, with few exceptions. Further analyses of the fern chloroplast genomes should provide new insights into the plastid genome evolution. Phylogenomics based on chloroplast genomes has shown many advantages in plant phylogenetics in recent years. With more nuclear data available recently, chloroplast phylogenomics can provide a framework for testing the impact of reticulate evolution in the early evolution of ferns. The examination of plastid genomic features, such as gene content, gene order, intron gain/loss, genome size, nucleotide composition, and codon usage, may also offer independent tests of hypotheses derived from analysis of DNA sequence data. Further plastome sequencing of marattioid ferns and early diverging leptosporangiate ferns will likely be necessary to solidify the sister relationship between these two lineages, but the position of *Equisetum* is unlikely to be resolvable with more plastome data. However, Grewe et al. (2013) concluded that Equisetopsida is sister to Psilotopsida.

## 2.8 Prospects

Mosses and liverworts have been resolved as monophyletic in phylogenetic analyses of complete plastomes (Karol et al. 2010), multigene data sets (Nickrent et al. 2000), and morphological analyses (Renzaglia et al. 2000). The position of the hornworts relative to a mosses + liverwort clade and tracheophytes, however, has varied in these studies, and sparse taxon sampling may have influenced resulting topologies (Karol et al. 2010). However, integration of evidences from all living and fossil species in morphological as well as molecular data and evolutionary models will provide increased precision (Ronquist et al. 2012).

Morris et al. (2018) established that the Cambrian origin of Embryophyta is the highest probability and the estimated ages for crown tracheophytes range from Late Ordovician to late Silurian (425 million years ago). Two important events happened in evolution in this period: (1) plants descended from charophyte algae and (2) the dominant generation changed from a haploid gametophyte to a diploid sporophyte (Bower 1908). The first event provided underexplored niche of high-intensity solar radiation and abundant $CO_2$ to photosynthetic life. The second event conferred on plants two abilities to adapt to a life in a water-deficient and UV-abundant terrestrial environment (Qiu et al. 2006). This facilitated the ability to produce a large number of genetically diverse gametes to ensure fertilization on land and the ability to mask deleterious mutations through the dominant-recessive interaction of alleles (Charlesworth et al. 1993). Qiu et al. (2006) suggested that our understanding of these events hinges on our knowledge of relationships between the organisms involved in these major evolutionary transitions. Debates over whether SSRs play any functional role in organism development, adaptation, survival, and evolution are never-ending (Powell et al. 1996; Li et al. 2000, 2004). Despite numerous studies using diverse approaches analyzing morphological and/or molecular characters, plastome analysis, and transcriptomics, the relationships among early land plants remain controversial (Kenrick and Crane 1997). Fossil evidence, although increasingly improved, has not helped to resolve the issues decisively (Wellman et al. 2003).

**Acknowledgments** This chapter is dedicated to my (AK) teacher Dr. P.D. Sharma for introducing the subject of pteridophytes now "monilophytes" during several field visits during master's study at the University of Rajasthan, Jaipur. We are also thankful to Professor S.K. Verma and Professor S.P. Khullar noted pteridologists for the support and during the preparation of this MS. We acknowledge the authors and the publishers of the following figures used in this paper:

- Figure 1: Sessa E B et al. (2014), Between Two Fern Genomes, *GigaScience*, 3, (1) 2047–217X–3–15. This is an open-access article distributed under the terms of the Creative Commons CC BY license
- Figure 2: Christenhusz MJ, Chase MW. (2014) Trends and concepts in fern classification. Ann Bot. Mar;113(4):571–94. doi: https://doi.org/10.1093/aob/mct299. Reproduced under license no. 5085950855981 dated 11 June
- Figure 3: PPG I Shmakov (2016), A community-derived classification for extant lycophytes and ferns. Jnl of Sytematics Evolution, 54: 563–603. https://doi.org/10.1111/jse.12229. Reproduced under RightsLink license 5094460819755 dated 22 June 2021
- Figure 4: Knie, N., Fischer, S., Grewe, F., Polsakiewicz, M., & Knoop, V. (2015). Horsetails are the sister group to all other monilophytes and Marattiales are sister to leptosporangiate ferns. *Molecular Phylogenetics and Evolution*, *90*, 140–149. https://doi.org/10.1016/j.ympev.2015.05.008. Reproduced under license number 5080631320156 dated 2 June 2021
- Figure 5: Chao Yi-Shan, Germinal Rouhan, Victor B. Amoroso, Wen-Liang Chiou (2014) Molecular phylogeny and biogeography of the fern genus *Pteris* (Pteridaceae), *Annals of Botany*, 114(1): 109–124, https://doi.org/10.1093/aob/mcu086. Reproduced under RightsLink license number 5097800369850 dated 28 June 2021

AK is thankful to his wife Professor Dr. Vijay Rani Kumar Retd Professor of Botany for critically examining the MS and valuable suggestions.

## References


Adams KL, Wendel JF (2005) Novel patterns of gene expression in polyploid plants. Trends Genet 21:539–543

Bainard JD, Henry TA, Bainard LD, Newmaster SG (2011) DNA content variation in monilophytes and lycophytes: large genomes that are not endopolyploid. Chromosom Res 19(6):763–775. https://doi.org/10.1007/s10577-011-9228-1

Baniaga AE, Barker MS (2019) Nuclear genome size is positively correlated with median LTR-RT insertion time in fern and lycophyte genomes. Am Fern J 109(3):248–266

Banks JA, Nishiyama T, Hasebe M, Bowman JL, Gribskov M, Depamphilis CW, Albert VA, Aono N, Aoyama T, Ambrose BA (2011) The *Selaginella* genome identifies genetic changes associated with the evolution of vascular plants. Science 332:960–963

Barker MS (2009) Evolutionary genomic analyses of ferns reveal that high chromosome numbers are a product of high retention and fewer rounds of polyploidy relative to angiosperms. Am Fern J 99:136–141

Barreda VD, Palazzesi L, Olivero EB (2019) When flowering plants ruled Antarctica: Evidence from Cretaceous pollen grains. New Phytol 223:1023–1030

Bower FO (1908) The origin of land flora, a theory based upon the facts of alternation. Macmillan, London

Britton DM (1974) The significance of chromosome numbers in ferns. Ann Mo Bot Gard 61:310–317. https://doi.org/10.2307/2395059

Cai S, Huang Y, Chen F, Zhang X, Sessa E, Zhao C, Marchant DB, Xue D, Chen G, Dai F, Leebens-Mack JH, Zhang G, Shabala S, Christie JM, Blatt MR, Nevo E, Soltis PS, Soltis DE, Franks PJ, Wu F, Chen ZH (2021) Evolution of rapid blue-light response linked to explosive diversification of ferns in angiosperm forests. New Phytol 230(3):1201–1213. https://doi.org/10.1111/nph.17135

Channing A, Zamuner A, Edwards D, Guido D (2011) Equisetum thermale sp. nov. (Equisetales) from the Jurassic San Agustín hot spring deposit, Patagonia: anatomy, paleoecology, and inferred paleoecophysiology. Am J Bot 98:680–697

Chao YS, Rouhan G, Amoroso VB, Chiou WL (2014) Molecular phylogeny and biogeography of the fern genus *Pteris* (Pteridaceae). Ann Bot 114(1):109–124. https://doi.org/10.1093/aob/mcu086

Charlesworth B, Morgan MT, Charlesworth D (1993) The effect of deleterious mutations on neutral molecular variation. Genetics 134(4):1289–1303

Chen ZJ (2007) Genetic and epigenetic mechanisms for gene expression and phenotypic variation in plant polyploids. Annu Rev Plant Biol 58:377–406

Chiarugi A (1960) Tavole cromosomiche delle Pteridophyta. Caryologia 13:27–150

Christenhusz MJM, Chase MW (2014) Trends and concepts in fern classification. Ann Bot 113(4): 571–594

Christenhusz MJM, Zhang X-C, Schneider H (2011) A linear sequence of extant lycophytes and ferns. Phytotaxa 19:7–54

Christenhusz MJM, Chase M, Fay MF, Bangiolo L (2019) Phylogenetics, classification and typification of extant horsetails (Equisetum, Equisetaceae). Bot J Linn Soc 189(4):311–352

Christenhusz MJM, Chase MW, Hay F, Hidalgo O, Leitch IJ, Pellicer J, Viruel J (2021) Biogeography and genome size evolution of the oldest extant vascular plant genus, *Equisetum* (Equisetaceae). Ann Bot 127(5):681–695

Clark J, Hidalgo O, Pellicer J, Liu H, Marquardt J, Robert Y, Christenhusz M, Zhang S, Gibby M, Leitch IJ, Schneider H (2016) Genome evolution of ferns: evidence for relative stasis of genome size across the fern phylogeny. New Phytol 210(3):1072–1082

Collinson ME (1996) "What use are fossil ferns?" 20 years on: with a review of the fossil history of extant pteridophyte families and genera. In: Camus JM, Gibby M, Johns RJ (eds) Pteridology in perspective. Royal Botanic Gardens, Kew, pp 349–394

Crane PR (1985) Phylogenetic analysis of seed plants and the origin of angiosperms. Ann Missouri Bot Gard 72:716–793

des Marais DL, Smith AR, Britton DM, Pryer KM (2003) Phylogenetic relationships and evolution of extant horsetails, Equisetum, based on chloroplast DNA sequence data (rbcL and trnL-F). Int J Plant Sci 164:737–751

Doyle JA (1998) Phylogeny of vascular plants. Annu Rev Ecol Syst 29:567–599

Ghatak J (1977) Biosystematic survey of Pteridophyta from Shevaroy hills, South India. Nucleus 20:105–108

Gifford EM, Foster AS (1989) Morphology and evolution of vascular plants. WH Freeman, New York

Grewe F, Guo W, Gubbels EA, Hansen AK, Mower JP (2013) Complete plastid genomes from *Ophioglossum californicum*, *Psilotum nudum*, and *Equisetum hyemale* reveal an ancestral land plant genome structure and resolve the position of Equisetales among monilophytes. BMC Evol Biol 13:8

Guillon JM (2004) Phylogeny of horsetails (*Equisetum*) based on the chloroplast rps4 gene and adjacent noncoding sequences. Syst Bot 29:251–259

Guillon JM (2007) Molecular phylogeny of horsetails (Equisetum) including chloroplast atp B sequences. J Plant Res 120:569–574

Haufler CH (1987) Electrophoresis is modifying our concepts of evolution in homosporous pteridophytes. Am J Bot 74:953–966

Haufler CH (2002) Homospory 2002: an odyssey of progress in pteridophyte genetics and evolutionary biology. Bioscience 52:1081–1093

Haufler CH (2014) Ever since Klekowski: testing a set of radical hypotheses revives the genetics of ferns and lycophytes. Am J Bot 101:2036–2042

Haufler CH, Soltis DE (1986) Genetic evidence suggests that homosporous ferns with high chromosome numbers are diploid. Proc Natl Acad Sci U S A 83:4389–4393

Hauke RL (1963) A taxonomic monograph of the genus *Equisetum* subgenus *Hippochaete*. Nova Hedwigia 8:1–123

Hedges SB, Dudley J, Kumar S (2006) TimeTree: a public knowledge-base of divergence times among organisms. Bioinformatics 22(23):2971–2972

Hendy MD, Penny D (1989) A framework for the quantitative study of evolutionary trees. Syst Zool 38:297–309

Henry TA, Bainard JD, Newmaster SG (2015) Genome size evolution in Ontario ferns (Polypodiidae): evolutionary correlations with cell size, spore size, and habitat type and an absence of genome downsizing. Genome 57:555–566

Husby C (2013) Biology and functional ecology of Equisetum with emphasis on the giant horsetails. Bot Rev 79:147–177

Karol KG, Arumuganathan K, Boore JL, Duffy AM, Everett KD, Hall JD, Hansen SK, Kuehl JV, Mandoli DF, Mishler BD, Olmstead RG, Renzaglia KS, Wolf PG (2010) Complete plastome sequences of Equisetum arvense and Isoetes flaccida: implications for phylogeny and plastid genome evolution of early land plant lineages. BMC Evol Biol 10:321. https://doi.org/10.1186/1471-2148-10-321

Kenrick P, Crane PR (1997) The origin and early diversification of land plants: a cladistic study. Smithsonian Inst Press, Washington, DC

Khandelwal S (1990) Chromosome evolution in the genus *Ophioglossum* L. Bot J Linn Soc 102:205–217

Klekowski EJ, Baker HG (1966) Evolutionary significance of polyploidy in the Pteridophyta. Science 153:305–307. https://doi.org/10.1126/science.153.3733.305

Knie N, Fischer S, Grewe F, Polsakiewicz M, Knoop V (2015) Horsetails are the sister group to all other monilophytes and Marattiales are sister to leptosporangiate ferns. Mol Phylogenet Evol 90:140–149. https://doi.org/10.1016/j.ympev.2015.05.008

Korall P, Conant D, Metzgar J, Schneider H, Pryer K (2007) A molecular phylogeny of scaly tree ferns (Cyatheaceae). Am J Bot 94(5):873–886. Retrieved July 1, 2021, from http://www.jstor.org/stable/27733249

Korall P, Schuettpelz E, Pryer KM (2010) Abrupt deceleration of molecular evolution linked to the origin of arborescence in ferns. Evolution 64:2786–2792

Kumar A, Giri P, Uniyal PL (2022a) Genome evolution in Ferns: molecular phylogenomics – a review. In: Johnson M, Fernandez H, Kumar A, Shibila A (eds) Ferns-biotechnology, propagation, medicinal uses and environmental regulation. Springer, Switzerland (this volume)

Kumar A, Giri P, Uniyal PL (2022b) Plastogenomics provides a tool to study phylogenetic interrelationships of monilophytes: a review. In: Johnson M, Fernandez H, Kumar A, Shibila A (eds) Ferns-biotechnology, propagation, medicinal uses and environmental regulation. Springer, Switzerland (this volume)

Landis JB, Soltis DE, Li Z, Marx HE, Barker MS, Tank DC, Soltis PS (2018) Impact of whole-genome duplication events on diversification rates in angiosperms. Am J Bot 105:348–363. https://doi.org/10.1002/ajb2.1060

Lehtonen S (2011) Towards resolving the complete fern tree of life. PLoS One 6(10):e24851. https://doi.org/10.1371/journal.pone.0024851

Leitch I, Leitch A (2013) Genome size diversity and evolution in land plants. In: Leitch IJ (ed) Plant genome diversity, vol 2. Springer pp, Vienna, pp 307–322

Lenton TM, Crouch M, Johnson M, Pires N, Dolan L (2012) First plants cooled the Ordovician. Nat Geosci 5(2):86–89

Li YC, Korol AB, Fahima T, Beiles A, Nevo E (2000) Microsatellites: genomic distribution, putative functions and mutational mechanisms: a review. Mol Ecol 11:2453–2465

Li YC, Korol AB, Fahima T, Nevo E (2004) Microsatellites within genes: structure, function and evolution. Mol Biol Evol 21:991–1007

Lindsay J (1794) XIII. Account of the germination and raising of ferns from the seed. Transactions of the Linnean Society of London 2:93–100. https://doi.org/10.1111/j.1096-3642.1794.tb00247.x

Marchant DB, Sessa EB, Wolf PG, Heo K, Barbazuk WB, Soltis PS, Soltis DE (2019) The C-Fern (*Ceratopteris richardii*) genome: insights into plant genome evolution with the first partial homosporous fern genome assembly. Sci Rep 9(1):18181. https://doi.org/10.1038/s41598-019-53968-8

Matasci LHH, Zhixiang Y, et al. (2014) Data access for the 1,000 Plants (1KP) project. Giga Sci 3(1):2047–217X–3–17. https://doi.org/10.1186/2047-217X-3-17

Moran RC (2008) Diversity, biogeography, and floristics. In: Ranker TA, Haufler CH (eds) Biology and evolution of ferns and lycophytes. Cambridge University Press, Cambridge, pp 367–394

Morris JL, Puttick MN, Clark JW, Edwards D, Kenrick P, Pressel S, Donoghue PCJ (2018) The timescale of early land plant evolution. Proc Nat Acad Sci U S A 115(10):E2274–E2283. https://doi.org/10.1073/pnas.1719588115

Nagalingum NS, Drinnan AN, Lupia R, McLoughlin S (2002) Fern spore diversity and abundance in Australia during the Cretaceous. Rev Palaeobot Palynol 119(1):69–92. https://doi.org/10.1016/S0034-6667(01)00130-0

Nakazato T, Jung MK, Housworth EA, Rieseberg LH, Gastony GJ (2006) Genetic map-based analysis of genome structure in the homosporous fern Ceratopteris richardii. Genetics 173:1585–1597. https://doi.org/10.1534/genetics.106.055624

Nakazato T, Barker MS, Rieseberg LH, Gastony GJ (2008) Evolution of the nuclear genome of ferns and lycophytes. In: Ranker TA, Haufler CH (eds) Biology and Evolution of Ferns and Lycophytes. Cambridge University Press, Cambridge, pp 175–198

Nickrent DL, Parkinson CL, Palmer JD, Duff RJ (2000) Multigene phylogeny of land plants with special reference to bryophytes and the earliest land plants. Mol Biol Evol 17(12):1885–1895

Parisod C, Holderegger R, Brochmann C (2010) Evolutionary consequences of autopolyploidy. New Phytol 186:5–17
Powell W, Machray GC, Provan J (1996) Polymorphism revealed by simple sequence repeats. Trends Plant Sci 1:215–222
PPG I (2016) A community-derived classification for extant lycophytes and ferns. J Syst Evol 54(6):563–603. https://doi.org/10.1111/jse.12229
Pryer KM (1999) Phylogeny of Marsileaceous Ferns and Relationships of the Fossil *Hydropteris pinnata* Reconsidered. Int J Plant Sci 160(5):931–954. https://doi.org/10.1086/314177
Pryer KM, Schuettpelz E (2009) Ferns (Monilophyta). In: Hedges SB, Kumar S (eds) The timetree of life. Oxford University Press, Oxford, pp 153–156
Pryer KM, Schneider H, Smith AR, Cranfill R, Wolf PG, Hunt JS, Sipes SD (2001) Horsetails and ferns are a monophyletic group and the closest living relatives to seed plants. Nature 409:618–622. https://doi.org/10.1038/35054555
Pryer KM, Schuettpelz E, Wolf PG, Schneider H, Smith AR, Cranfill R (2004) Phylogeny and evolution of ferns (monilophytes) with a focus on the early leptosporangiate divergences. Am J Bot 91:1582–1598
Qiu Y, Li L, Wang B, Chen Z, Knoop V, Groth-Malonek M, Dombrovska O, Lee J, Kent L, Rest J, Estabrook GF, Hendry TA, Taylor DW, Testa CM, Ambros M, Crandall-Stotler B, Duff RJ, Stech M, Frey W, Quandt D, Davis CC (2006) The deepest divergences in land plants inferred from phylogenomic evidence. Proc Natl Acad Sci U S A 103:15511–15516
Qiu YL, Li L, Wang B, Chen Z, Dombrovska O, Lee J, Kent L, Li R, Jobson RW, Hendry TA, Taylor DW, Testa CM, Ambros M (2007) A non-flowering land plant phylogeny inferred from nucleotide sequences of seven chloroplast, mitochondrial and nuclear genes. Int J Plant Sci 168: 691–708
Rai HS, Graham SW (2010) Utility of a large, multigene plastid data set in inferring higher-order relationships in ferns and relatives (monilophytes). Am J Bot 97:1444–1456. https://doi.org/10.3732/ajb.0900305
Raubeson LA, Jansen RK (1992) Chloroplast DNA evidence on the ancient evolutionary split in vascular land plants. Science 255:1697–1699
Renzaglia KS, Nickrent DL, Garbary DJ, Garbary DJ, Duff RJT (2000) Vegetative and reproductive innovations of early land plants: Implications for a unified phylogeny. Philos Trans R Soc Lond Ser B Biol Sci 355(1398):769–793
Ronquist F, Klopfstein S, Vilhelmsen L, Schulmeister S, Murray DL, Rasnitsyn AP (2012) A total-evidence approach to dating with fossils, applied to the early radiation of the hymenoptera. Syst Biol 61:973–999
Rothfels CJ, Larsson A, Kuo L-Y, Korall P, Chiou W-L, Pryer KM (2012) Overcoming deep roots, fast rates, and short internodes to resolve the ancient rapid radiation of eupolypod II ferns. Syst Biol 61:490–509
Rothfels CJ, Li F-W, Sigel EM, Huiet L, Larsson A, Burge DO, Ruhsam M, Deyholos M, Soltis DE, Stewart CN Jr, Shaw SW, Pokorny L, Chen T, de Pamphilis C, DeGironimo L, Chen L, Wei X, Sun X, Korall P, Stevenson DW, Graham SW, Wong GK-S, Pryer KM (2015) The evolutionary history of ferns inferred from 25 low-copy nuclear genes. American J Bot 102: 1089–1107. https://doi.org/10.3732/ajb.1500089
Rothwell GW (1996) Pteridophytic evolution: an often under appreciated phytological success story. Rev Palaeobot Palynol 90:209–222
Rothwell GW (1999) Fossils and ferns in the resolution of land plant phylogeny. Bot Rev 65(3): 188–218. Retrieved June 1, 2021, from http://www.jstor.org/stable/4354346
Rubinstein CV, Gerrienne P, de la Puente GS, Astini RA, Steemans P (2010) Early Middle Ordovician evidence for land plants in Argentina (eastern Gondwana). New Phytol 188(2): 365–369
Ruggiero MA, Gordon DP, Orrell TM, Bailly N, Bourgoin T, Brusca RC, Cavalier-Smith T, Guiry MD, Kirk PM (2015) A higher level classification of all living organisms. PLoS One 10: e0119248

Schneider H, Schuettpelz E, Pryer KM, Cranfill R, Magallon S, Lupia R (2004) Ferns diversified in the shadow of angiosperms. Nature 428:553–557. https://doi.org/10.1038/nature02361

Schneider H, He L, Hennequin S, Zhang X-C (2013) Towards a natural classification of Pteridaceae: inferring the relationships of enigmatic pteridoid fern species occurring in the Sino-Himalaya and Afro-Madagascar. Phytotaxa 77:49–60

Schuettpelz E, Pryer KM (2007) Fern phylogeny inferred from 400 leptosporangiate species and three plastid genes. Taxon 56:1037–1050

Schuettpelz E, Pryer KM (2009) Evidence for a Cenozoic radiation of ferns in an angiosperm-dominated canopy. Proc Natl Acad Sci U S A 106:11200–11205

Scotland RW, Wortley AH (2003) How many species of seed plants are there? Taxon 52:101–104. https://doi.org/10.2307/3647306

Sessa EB, Jo Ann B, Michael SB, Der Joshua P, Aaron MD, Sean WG, Mitsuyasu H, Jane L, Fay-Wei L, Blaine MD, Kathleen MP, Carl JR, Stanley JR, Mari LS, Erin MS, Douglas ES, Pamela SS, Dennis WS, Paul GW (2014) Between two fern genomes, GigaSci 3(1): 2047–217X–3–15. https://doi.org/10.1186/2047-217X-3-15

Shmakov A (2016) A community-derived classification for extant lycophytes and ferns. J Syst Evol 54:563–603

Smith AR, Pryer KM, Schuettpelz E, Korall P, Schneider H, Wolf P (2006) A classification for extant ferns. Taxon 55:705–731. https://doi.org/10.2307/25065646

Smith AR, Pryer KM, Schuettpelz E, Korall P, Schneider H, Wolf PG (2008) Fern classification. In: Ranker TA, Haufler CH (eds) Biology and evolution of ferns and lycophytes. Cambridge University Press, Cambridge, pp 417–467

Smith SA, Beaulieu JM, Donoghue MJ (2010) An uncorrelated relaxed-clock analysis suggests an earlier origin for flowering plants. Proc Natl Acad Sci 107:5897–5902. https://doi.org/10.1073/pnas.1001225107

Soltis DE, Soltis PS (1987) Polyploidy and breeding systems in homosporous pteridophyta - a reevaluation. Am Nat 130:219–232

Soltis PS, Soltis DE (2000) The role of genetic and genomic attributes in the success of polyploids. Proc Natl Acad Sci 97:7051–7057. https://doi.org/10.1073/pnas.97.13.7051

Soltis PS, Soltis DE (2013) A conifer genome spruces up plant phylogenomics. Genome Biol 14(6): 122

Soltis PS, Soltis DE, Chase MW (1999) Angiosperm phylogeny inferred from multiple genes as a tool for comparative biology. Nature 402:402–404

Steemans P et al (2009) Origin and radiation of the earliest vascular land plants. Science 324(5925): 353

Stewart WN, Rothwell GW (1993) Paleobotany and the evolution of plants. Cambridge University Press, Cambridge

Szövényi P, Gunadi A, Li FW (2021) Charting the genomic landscape of seed-free plants. Nat Plants 7:554–565. https://doi.org/10.1038/s41477-021-00888-z

Tang CM (2020) Jurassic Period. Encyclopedia Britannica, 12 Nov. https://www.britannica.com/science/Jurassic-Period. Accessed 1 July 2021

Thorne RF (2002) How many species of seed plants are there? Taxon 51:511–522

Vanneste K, Van de Peer Y, Maere S (2013) Inference of genome duplications from age distributions revisited. Mol Biol Evol 30:177–190

Vanneste K, Sterck L, Myburg AA, Van de Peer Y, Mizrachi E (2015) Horsetails are ancient polyploids: evidence from *Equisetum giganteum*. Plant Cell 27(6):1567–1578. https://doi.org/10.1105/tpc.15.00157

Wellman CH, Osterloff PL, Mohiuddin U (2003) Fragments of the earliest land plants. Nature 425(6955):282–285

Weststrand S, Korall P (2016a) Phylogeny of Selaginellaceae: there is value in morphology after all! Am J Bot 103:2136–2159

Weststrand S, Korall P (2016b) A subgeneric classification of *Selaginella* (Selaginellaceae). Am J Bot 103:2160–2169

Wolf PG, Haufler CH, Sheffield E (1987) Electrophoretic evidence for genetic diploidy in the bracken fern (*Pteridium aquilinum*). Science 236:947–949. https://doi.org/10.1126/science.236.4804.947

Wood TE, Takebayashi N, Barker MS, Mayrose I, Greenspoon PB, Rieseberg LH (2009) The frequency of polyploid speciation in vascular plants. Proc Natl Acad Sci U S A 106:13875–13879

Zhang L-B, Zhang L (2012) The inclusion of Acrophorus, Diacalpe, Nothoperanema, and Peranema in Dryopteris: the molecular phylogeny, systematics, and nomenclature of Dryopteris subg. Nothoperanema (dryopteridaceae). Taxon 61:1199–1216

Zhong B et al (2011) Systematic error in seed plant phylogenomics. Genome Biol Evol 3:1340–1348

Zhong B, Fong R, Collins LJ, McLenachan PA, Penny D (2014) Two new fern chloroplasts and decelerated evolution linked to the long generation time in tree ferns. Genome Biol Evol 6: 1166–1173

Zhou XM, Zhang LB (2015) A classification of *Selaginella* (Selaginellaceae) based on molecular (chloroplast and nuclear), macromorphological, and spore features. Taxon 64:1117–1140

**Ashwani Kumar** Alexander von Humboldt Fellow (Germany), MSc (Botany) (Gold Medalist) (1967)), PhD (1971), postdoc AvH Fellow (Germany) and JSPS fellow (Japan) Selected in the Indian Administrative Services (IAS: IPS) (1972) but he opted career in botany. 37 years of teaching & research experience. Guided 39 PhD. research students and published 320 research papers and 26 books. Elected fellow FBS, FPSI, FIMPS and FIFS. Editor in Chief IJLS. V. Puri medal (2008) CEE award (2015). President of CHEC (India), Indian Botanical Society, Society for Promotion of Plant Sciences. Professor Emeritus, Department of University of Rajasthan, Jaipur. Presently, Chair Professor at JVWU, Jaipur and Adj. Professor Modi University, Sikar.

**Priti Giri** is a research scholar in the Department of Botany, University of Delhi. She has completed M.Sc. and M.Phil. from the Department of Botany, University of Delhi, Delhi. Her area of specialization is conservation and phytochemical and biochemical screening of ethnomedicinally important ferns (pteridophytes), and she has published many research paper and book chapters on her subject.

**Dr. Prem Lal Uniyal** is presently working as a professor in the Department of Botany, University of Delhi. He did his masters from HNB Garhwal University, Srinagar, Uttarakhand, and Ph.D. in Botany from the Panjab University, Chandigarh, and has been teaching since the last 27 years. Dr. Uniyal is trained in bryologist and specialized in the Biology of Archegoniates. His other areas of research are conservation biology, gametophyte development and reproductive biology of ferns, population biology of conifers, biology of Podostemaceae, and biological activity of lichens. His research group has discovered a few new taxa of mosses and *Ephedra* from India. He organized a few conservation programs with indigenous societies in Himachal Pradesh. He has published 4 books and more than 105 original research papers in peer-reviewed national and international journals; 35 review articles in proceedings, book, and scientific magazine; and 25 popular scientific articles. He has supervised 17 Ph.D. scholars and 22 M.Phil. students.

# A Review on Molecular Phylogeny of Pteridophytes Using DNA Barcoding

3

N. Janakiraman, M. Narayani, and M. Johnson

**Abstract**

The limitations in morphological identification systems and the declining pool of taxonomists signal the need for a new approach to recognize taxa. The recent development of molecular markers revolutionized the entire scenario of life sciences. This article focuses on the utility of DNA-based markers for fern identification. DNA barcoding is an emerging global standard for recognition, and it has simplified the identification process. The Consortium for the Barcode of Life (CBoL) has progressively defined the barcode markers to be used for each taxonomic group. Efforts to identify a correct DNA barcode for discriminating species have been successfully studied in all plant groups. The arrival of molecular phylogenetics has rapidly improved our understanding of fern relationships through DNA sequence data. Barcoding studies on pteridophytes mostly utilized data from several chloroplast markers. The gene *rbc*L has been used extensively by various researchers for analyzing the evolutionary relationship of ferns at both generic and familial levels. However, this method of fern identification should be used in combination with other approaches for effective identification.

N. Janakiraman
Department of Botany, The Madura College (Autonomous), Madurai, Tamil Nadu, India

M. Narayani
Harvest University, Lusaka, Zambia

Centre for Plant Biotechnology, Department of Botany, St. Xavier's College (Autonomous), Palayamkottai, Tamil Nadu, India

M. Johnson (✉)
Centre for Plant Biotechnology, Department of Botany, St. Xavier's College (Autonomous), Palayamkottai, Tamil Nadu, India





J. Marimuthu et al. (eds.), *Ferns*, https://doi.org/10.1007/978-981-16-6170-9_3

**Keywords**

Phylogeny · Barcode · Pteridophytes · *rbc*L · Chloroplast

## 3.1 Introduction

Plant genomics has been studied extensively to bring out revolution in the area of research over the last few decades. During the early period, classical strategies including comparative morphology, anatomy, physiology, and embryology were employed in genetic analysis to determine inter- and intra-species variability. In recent times, molecular markers have rapidly complemented the traditional strategies. Molecular taxonomy is an aspect of molecular systematics, a broader term that includes the use of molecular data in phylogenetics and biogeography (Schneider and Schuettpelz 2006). The impact of molecular data on the field of taxonomy can hardly be overstated. In combination with explicit methods for phylogenetic analysis, molecular data have reshaped the concepts of relationships and circumscriptions at all levels of the taxonomical hierarchy (Soltis and Soltis 1999). As molecular phylogenetic studies have accumulated, it has become apparent that different molecular tools are required for different questions because of varying rates of sequence evolution among genomes, genes, and gene regions. The choice of molecular tool is very important to ensure that an appropriate level of variation is recovered to answer the phylogenetic questions at hand. However, the plant systematic community is using only a small fraction of the available molecular tools (Small et al. 2005). In general, closely related organisms, which look morphologically similar, have a high degree of agreement in the molecular structure of these substances, while the molecules of distantly related organisms usually show a pattern of dissimilarity. Conserved sequences are expected to accumulate mutations over time, and assuming a constant rate of mutation provides a molecular regulator for dating divergence. Such molecular composition can be studied by genetic markers that represent genetic differences between individual organisms or species. These markers do not affect the trait of interest because they are located only near or linked to genes controlling the trait.

Information on genetic variation within and among populations is crucial for conservation of rare and endangered species. A species may be considered as a group of individuals organized into populations that share an amalgamation of indicative characters, which are not found outside the group. Survival chance of a species is indicated in genetic diversity within the population (Tsuda et al. 2009). Molecular markers have been used for assessing genetic diversity and generating baseline information (Pertoldi et al. 2007; Mirialili et al. 2009). In order to deduct a concrete, reliable data, a marker should meet many criteria; a marker should be inheritable and have the power to discriminate between individuals. It should be easy to generate and interpret and highly polymorphic in nature and frequently distributed throughout the genome. Moreover, a marker should be easy to detect and comparable with similar characters (Hillis and Moritz 1990). Three classes of markers based on morphological, biochemical, and molecular traits are routinely explored. Morphological markers are usually visually characterized phenotypic characters such as sorus type, shape, and arrangement, indusium attachment, rhizome, type of fertile segments, vein branching, rachis scales, and microphyllous leaves (Lasalita-Zapico et al. 2011). Biochemical markers are differences in protein and enzymes that are detected by electrophoresis and specific staining (Pillai et al. 2000). The major disadvantages of these two markers are they may be limited in number and are influenced by environmental factors or the developmental stages of the plant (Winter and Kahl 1995). Early attempts at molecular systematics make use of proteins, enzymes, carbohydrates, and other molecules that were separated and characterized using biochemical techniques such as chromatography and electrophoresis. Electrophoretic banding pattern of polypeptides can be an efficient approach for the assessment of different plant samples, while the results were not quantitative and did not initially improve on morphological classification; they simply provided tantalizing hints that are long-held notion of the classifications of plants (Dai et al. 2006). Microsatellites are used as a powerful genetic marker for pteridophytes (Jimenez 2011). Molecular approaches are used to characterize proteins mediating morphogenetic changes in plants (Kumar and Fernandez 2019).

## 3.2 DNA Barcoding

DNA barcoding is an innovative molecular technique which uses short and agreed DNA sequences from either nuclear or cytoplasmic genome for the rapid identification of biological specimen at species level. Based on their study on 200 allied lepidopteran species, Hebert et al. (2003) were the first to propose the use of short DNA sequences as "taxon barcodes" for species-level identification. Realizing the importance of this technology, an international "Consortium for the Barcode of Life" (CBOL) was established in 2004 (Lahaye et al. 2008). CBOL's mission is to promote the exploration and development of DNA barcoding as global standard for species identification, through rapid compilation of high-quality DNA barcodes in a public library of DNA sequences (CBOL 2016). International Barcode of Life (IBOL) contains DNA barcodes of more than 97,42,000 sequences from 200 different nations in July 2021.

Taxonomy is based on a detailed understanding of morphology, physiology, and behavioral attributes (Ebach and Holdrege 2005; Arvind et al. 2007). Taxonomists consider that the traditional morphology-based identification of a plant species would diminish and result in incorrect species identification because of the morphologically similar fronds (Kress et al. 2005). Molecular taxonomical studies using DNA barcoding generate genetic information and inferred phylogenetic relationship based on cpDNA (Shu et al. 2017). DNA barcoding is a fast-developing technique to identify species by using short and standard DNA sequences (Wang et al. 2016; Patel and Reddy 2018; Park et al. 2021). It helps to identify endangered species and to test the identity and purity of botanical products such as commercial herbal medicines and dietary supplements. They are also being used to address ecological, evolutionary, and conservation issues such as the ecological rules controlling the assembly of species in plant communities (Kress et al. 2009).

Although efforts to identify a DNA barcode for discriminating plant species are less successful than animals, in recent few years, researchers have reported many efficient cases for plant DNA barcoding (Kress 2017; Patel and Reddy 2018). It is an emerging global standard for the recognition and identification of eukaryotic species through comparison of sequences of a short DNA fragment called DNA barcode from an unknown specimen to a library of reference sequences from known species (Nitta et al. 2020). In addition, to assigning specimens to known species, DNA barcoding will accelerate the ace of species discovery by allowing taxonomists to rapidly sort specimens and by highlighting divergent taxa that may represent new species. By augmenting their capabilities in these ways, DNA barcoding offers taxonomists an opportunity to greatly expand and eventually complete a global inventory of life's diversity (Shu et al. 2017). DNA barcodes are being increasingly integrated into biological surveys to document biodiversity at unprecedented scale (Nitta et al. 2020). Recent studies demonstrated the successful application of DNA barcoding to identify the pteridophytes (Nitta et al. 2017; Dong et al. 2020).

## 3.3 DNA Barcoding in Pteridophytes

Pteridophytes have a longer evolutionary history than other vascular land plants. The first molecular systematic studies on ferns were published in the mid-1990s (Hasebe et al. 1994, 1995) and set the direction for modern fern systematics. The arrival of phylogenetics and molecular phylogenetics, in particular, has rapidly improved our understanding of fern relationships through phylogenetic analyses of DNA sequence data (Pryer et al. 2004; Schneider et al. 2004a), morphological data alone (Schneider et al. 2009), or combined analyses of molecular and morphological evidence (Lehtonen et al. 2010). Since then, numerous molecular phylogenetic studies have focused on certain classically defined fern groups by sampling members from the group studied or tested the backbone fern classification by sampling exemplar species of higher taxa. However, both kinds of studies have specific limitations to recover the complete fern tree of life.

DNA barcoding in ferns is potentially of great value when they generally lack the complexity for morphology-based identification and underappreciated in ecological studies. Molecular phylogenetic hypotheses for extant ferns have utilized data from several chloroplast markers (*rbc*L, *atp*A, *atp*B, *acc*D, *rps*4, 16S rDNA, ITS), one nuclear gene (18S rDNA), and three mitochondrial genes (*atp*1, *nad*2, *nad*5). A natural outgrowth of these one-gene or few-gene studies on a wide array of ferns has led to broader and increasingly robust multiple gene phylogenetic analyses. All these studies have given rise to growing confidence in relationships and correspondingly to the composition of fern taxa at familial and ordinal ranks. Nowadays, it is widely accepted that any valid plant barcode will be multi-locus, preferably existing of a conservative coding region like *rbc*L, in combination with a more rapidly evolving region, which is most likely non-coding (Kress et al. 2009). The non-coding *trn*L intron and *trn*L-F intergenic spacer have been repeatedly suggested for this purpose (Taberlet et al. 2006; Hollingsworth et al. 2009) for the identification of a mysterious aquatic gametophyte. Besides the technical issues of primer universality, sequence quality, and complexity, Schneider and Schuettpelz (2006) mentioned three potential difficulties for any tested marker to overcome: incomplete sampling of the online records to be used as a reference for identification (GenBank), the occurrence of mis-identified and erroneous sequences in these online databases, and the potential inability of the marker to discriminate among species.

Molecular studies on ferns have generally relied on a subset of the sequences used in angiosperm systematics (Pryer et al. 2004). The gene *rbc*L has been used extensively in studies for both higher-level and lower-level taxa (Ranker et al. 2004; Schneider et al. 2004a). Phylogenetic studies using nucleotide sequences of the gene encoding the large subunit of *rbc*L have successfully revealed the relationships of ferns at both generic and familial levels (Hasebe et al. 1995; Skog et al. 2004). Intergenic spacer sequences such as *trn*L-F and *rps*4-*trn*S were more effective for resolving recent divergence events in the ferns (Small et al. 2005). *trn*L-F was proved to be useful in studies of Ophioglossaceae (Hauk et al. 1996, 2003), Schizaeaceae (Skog et al. 2002), Polypodiaceae (Haufler et al. 2003; Ranker et al. 2004; Schneider et al. 2004b), *Asplenium* (Schneider et al. 2004a), and *Cyrtomium* (Lu et al. 2005). Another intergenic spacer, *rps*4-*trn*S, has been applied to infrageneric phylogenetic work in the ferns *Hymenophyllum* (Hennequin et al. 2003), *Elaphoglossum* (Rouhan et al. 2004; Skog et al. 2004), and *Polystichum* (Perrie et al. 2003).

Green (1971) isolated DNA from three genera of ferns (*Angiopteris*, *Ophioglossum*, *Cibotium*) and three genera of fern allies (*Psilotum*, *Equisetum*, *Selaginella*). Hasebe et al. (1994) sequenced 1206 nucleotides of the large subunit of the *rbc*L gene from 58 species representing almost all families of leptosporangiate ferns. Phylogenetic trees proved monophyletic relationship of the tree ferns with 98% of bootstrap probability in the neighbor-joining method and 73% in the parsimony method. Two morphologically distinct heterosporous water ferns, viz., *Marsilea* and *Salvinia*, are sister genera, the tree ferns (Cyatheaceae, Dicksoniaceae, and Metaxyaceae) are monophyletic, and polypodioids are distantly related to the gleichenioids in spite of the similarity of their exindusiate soral morphology. Wolf et al. (1994) clarified the phylogenetic relationships among dennstaedtioid ferns by sequencing 1320 bp of *rbcL* from 45 species representing 13 families. Sequence divergence for *rbcL* averaged 0.9% among species within genera, 10.3% among genera within families, and 14.8% among families. Tree ferns form a single clade and Hymenophyllaceae appear as sister groups to the dennstaedtioid ferns on all shortest trees. Polypodiaceae and adiantoid fern groups are considered separate from the dennstaedtioid families emerged within the dennstaedtioid clade. Pryer et al. (1995) recorded a broad sampling of 50 existing pteridophyte taxa, with representatives of all major fern groups. Kolukisaoglu et al. (1995) indicated that *Selaginella* and *Equisetum* emerge earlier than *Psilotum* based on phytochrome gene. Dubuisson (1997) used *rbc*L sequences as a promising tool for 18 species of the fern genus *Trichomanes* to test the ability of this gene for resolving relationships within this taxon and to reveal the major phylogenetic tendencies. Gastony and Ungerer (1997) determined nucleotide sequences of the chloroplast-encoded *rbc*L gene for all five species of the onocleoid ferns including both varieties of *Onoclea sensibilis* and for outgroup member *Blechnum glandulosum.*

Wolf et al. (1999) suggested based on *rbc*L sequence from two *Hymenophyllopsis* species, the family was monophyletic and sister to the single Cyatheaceae species. Yatabe et al. (2001) crossed three sympatric *rbc*L sequence types of *Asplenium nidus* and observed that the molecularly distinct types were reproductively isolated because hybrids failed to form between at least two pairs of *rbc*L types. Korall and Kenrick (2002) developed a phylogenetic framework for the clubmoss family Selaginellaceae based on *rbcL* sequences. The analysis supports to distinguish the monophyly of subgenera *Selaginella* and *Tetragonostachys*. Heede et al. (2003) investigated the phylogenetic relationships among 20 taxa of the fern genus *Asplenium* subgenus *Ceterach* using DNA sequence data from the nuclear ribosomal internal transcribed spacers and plastid *trn*L-F intergenic spacer. In addition, a single sample per taxon was used in analysis of the plastid *rbc*L gene. The trees produced from the separate plastid and nuclear matrices agree in the recognition of identical groups of accessions corresponding to *A. dalhousiae*, *A. ceterach*, *A. aureum*, *A. cordatum*, *A. phillipsianum*, and *A. haughtonii*. Ting et al. (2003) sequenced chloroplast *trn*L intron and *trn*L-*trn*F intergenic spacers of *Alsophila spinulosa* and *Cyathea tsangii*. The clade showed *Sphaeropteris brunoniana*, *Sphaeropteris hainanensis*, and *Cyathea contaminans* diverged from the rest of the members and the latter was further separated into two subclades corresponding to the subgenera *Alsophila* and *Gymnosphaera*. The genus *Sphaeropteris* was placed in the basal position of Cyatheaceae, whereas the genus *Alsophila* was placed as the derived sister group, which supported Tryon's hypothesis accounting for the evolutionary relationships within Cyatheaceae and the derivation of their indusium.

Ranker et al. (2003) conducted a molecular phylogenetic analysis of the fern genus *Adenophorus* (Grammitididaceae) by employing DNA sequence variation from three cpDNA fragments, viz., *rbcL*, *atpB*, and the *trnL-trnF* intergenic spacer (IGS). Phylogenetic analyses of individual DNA fragments resulted in single strong supported phylogenetic hypothesis. The primary features of hypothesis are *Adenophorus* is monophyletic, subgenus *Oligadenus* is paraphyletic, and the endemic Hawaiian species *Grammitis tenella* is strongly supported as the sister taxon to *Adenophorus*. Hauk et al. (2003) represented the phylogenetic studies of 36 species of Ophioglossaceae using *rbcL*, *trnL*, and *trnF* plastid DNA sequences. Individual and combined analyses of the three datasets revealed two main clades of ophioglossoid and botrychioid within the family. In the botrychioid clade, *Helminthostachys* was sister to a broadly defined *Botrychium*, and in some cases, *Sceptridium* was sister to *Botrychium*. In the ophioglossoid clade, *Ophioglossum* was sister to *Cheiroglossa* and *Ophioderma*, but relationships within *Ophioglossum* were not well supported. Schneider et al. (2004a) explored the phylogeny of the polygrammoid ferns using nucleotide sequences derived from 3 plastid loci for each of 98 selected species. The analyses revealed four major monophyletic lineages: the loxogrammoids, two clades, and a largely Neotropical clade which includes the pantropical Grammitidaceae. Pryer et al. (2004) studied the phylogeny and evolution of ferns (monilophytes) with a focus on the early leptosporangiate divergences. More than 5000 bp from the plastid (*rbcL, atpB, rps4*) and the nuclear (18S rDNA) genomes were sequenced for 62 taxa. Phylogenetic analyses revealed that Osmundaceae are sister to the leptosporangiates and Dipteridaceae, Matoniaceae, Gleicheniaceae, and Hymenophyllaceae which are monophyletic and schizaeoid ferns are sister to a large clade of "core leptosporangiates" including heterosporous ferns, tree ferns, and polypods.

Korall and Kenrick (2004) evaluated the phylogenetic history of Selaginellaceae based on DNA sequences from the plastid (*rbcL*) and nucleus (26S rDNA gene) sequences. 26S rDNA and *rbcL* regions represent the main elements of species diversity in Selaginellaceae family and were analyzed by means of maximum parsimony and Bayesian inference. Substitution rates in the 26S rDNA were found to be high (26% informative) but lower than *rbcL* (37% informative). Wang et al. (2004) used RAPD analysis and sequences of cpDNA *atp*B-*rbc*L intergenic spacers to characterize the pattern of genetic variation and phylogenetic relationships of *Alsophila spinulosa*. 28 random primers generated 118 bands, out of which 26 (22.03%) were polymorphic loci, distinguishing 17 different RAPD phenotypes. AMOVA showed that 47.44% of the variance was partitioned among regions, and 34.01% attributed among populations within regions, whereas only 18.55% occurred within populations. Low level of intra-specific diversity was maintained in *A. spinulosa* with Shannon diversity and gene diversity merely 0.0560 and 0.0590, respectively. Baracaldo (2004) generated robust phylogeny of the Neotropical fern genera *Jamesonia* and *Eriosorus* based on sequence data of the nuclear ETS of 18S-26S rDNA, the plastid gene *rps*4, and the intergenic spacer *rps*4-*trn*S. Su et al. (2004) sequenced cpDNA *atp*B-*rbc*L intergenic spacers of individuals of a relict tree fern *Alsophila spinulosa* collected from ten populations in southern China. Sequence length varied from 724 to 731 bp showing length polymorphism, and base composition was with high A+T content between 63.17% and 63.95%. A total of 19 haplotypes were identified based on nucleotide variation. High levels of haplotype diversity and nucleotide diversity were detected in *A. spinulosa*, which allowed the accumulation of genetic variation within lineages.

Su et al. (2005) inferred genetic differentiation and phylogeographical pattern of *A. spinulosa* in southern China using sequence variations of *trn*L-F non-coding regions of cpDNA. AMOVA analysis indicated that most of the genetic variation was partitioned among regions. Zhang et al. (2005) carried out phylogenetic analysis of cryptogrammoid ferns and related taxa based on *rbc*L sequences. The resulting cladogram places *Coniogramme*, *Cryptogramma*, and *Llavea* into a moderately supported clade, constituting a cryptogrammoid group distantly related to the cheilanthoid ferns. Korall and Taylor (2006) indicated that the genus *Selaginella* is monophyletic based on *rbc*L map; however, at subgenus level, it is monophyletic or paraphyletic. Smith et al. (2006) provided the most recent arrangement of Pteridaceae into five monophyletic groups based on morphological and molecular data. Many *rbc*L sequences of Pteridaceae are now available in GenBank which showed that they are variable enough to provide good resolution across the taxonomic diversity of fern taxa. Schneider and Schuettpelz (2006) tested the principle of DNA barcoding of fern gametophytes using plastid *rbc*L sequence and successfully identified a cultivated gametophyte as *Osmunda regalis*. However, whether *rbc*L shows sufficient variation to allow general identification below genus level remains uncertain. Korall et al. (2006) investigated phylogenetic relationships within tree ferns Cyatheaceae, Plagiogyriaceae, and Hymenophyllopsidaceae based on analyses of four protein-coding, plastid loci (*atpA*, *atpB*, *rbcL*, and *rps4*). Four well-supported clades with genera of Dicksoniaceae interspersed among Loxomataceae, *Culcita*, Plagiogyriaceae, *Calochlaena*, *Dicksonia*, and Lophosoriaceae where *Cibotium* and Cyatheaceae nested within Hymenophyllopsidaceae.

Schuettpelz and Pryer (2007) analyzed the most inclusive molecular dataset for leptosporangiate ferns using three plastid genomes (*rbcL*, *atpB*, and *atpA*). More than 4000 bp were sequenced for each of 400 leptosporangiate fern species and 5 outgroups. Maximum likelihood analysis yielded strong phylogeny (80%), and the nodes were supported by a maximum likelihood bootstrap percentage $\geq$70. Liu et al. (2007) carried out phylogenetic analysis of Dryopteridaceae family using chloroplast *rbc*L and *atp*B genes. The results indicate that Dryopteridaceae form a monophyletic group with the exception of *Didymochlaena*, *Hypodematium*, and *Leucostegia*. They are sister to a large clade comprising Lomariopsidaceae, Tectariaceae, Polypodiaceae, Davalliaceae, and Oleandraceae. Lu et al. (2007) studied the molecular phylogeny of the polystichoid ferns in Asia based on *rbc*L sequences. Guillon (2007) used original (*atp*B) and published (*rbc*L, *trn*L-*trn*F, *rps*4) sequence data to investigate the phylogeny of the genus *Equisetum*. Analyses of *atp*B sequences give an unusual topology with *Equisetum bogotense* branching within *Hippochaete*. Korall et al. (2007) investigated the phylogenetic relationships of scaly tree ferns based on DNA sequence data from five plastid regions (*rbc*L, *rbc*L-*acc*D IGS, *rbc*L-*atp*B IGS, *trn*G-*trn*R, and *trn*L-*trn*F). A basal dichotomy resolves *Sphaeropteris* with conform scales is sister to all other taxa having marginate scales. The marginate-scaled clade consists of a basal trichotomy with the three groups, viz., *Cyathea*, *Alsophila*, and *Gymnosphaera*. In recent phylogenetic analyses, tree ferns were shown to be the sister group of polypods, the most diverse group of living ferns. Nagalingam et al. (2008) conducted a broad species-level relationship for *Pilularia* and *Salvinia* using coding (*atpB*, *rbcL*, and *rpsA*) and non-coding plastid regions (*trnL*, *trnGR*, and *rps4-trnS*). Their analyses resolved Marsileaceae, Salviniaceae, and all of the component genera as monophyletic. Salviniaceae incorporate *Salvinia* and *Azolla*; in Marsileaceae, *Marsilea* is sister to the clade of *Regnellidium* and *Pilularia*. Individual species-level investigations for *Pilularia* and *Salvinia* provide phylogenies within all genera of heterosporous ferns.

Madeira et al. (2008) conducted the molecular phylogeny of the genus *Lygodium* using *trn*L intron and *trn*L-F intergenic spacer of cpDNA to determine the relationship of *Lygodium microphyllum* and other *Lygodium* species. Three major clades appeared, one with *Lygodium palmatum* and *Lygodium articulatum*, second with *Lygodium reticulatum* and *L. microphyllum*, and a third comprising the other species examined. Nitta (2008) explored the utility of three plastid loci (*rbc*L, *trn*SGG, and *trn*H-*psb*A) for biocoding the filmy ferns (Hymenophyllaceae) of Moorea. *trn*H-*psb*A has the greatest utility as a potential marker for DNA-based identification because of its high inter-specific variability and high degree of amplification success. *rbc*L and *trn*H-*psb*A were successfully used in combination with morphological characters to identify *Polyphlebium borbonicum*. Gao et al. (2009) sequenced the complete chloroplast genome of a scaly tree fern *Alsophila spinulosa.* It shares some unusual characteristics with the previously sequenced genome of the polypod fern *Adiantum capillus-veneris*, including the absence of five tRNA genes that exist in most other chloroplast genomes. The genome shows a high degree of synteny with that of *Adiantum*, but differs considerably from two basal ferns (*Angiopteris evecta* and *Psilotum nudum*). Song et al. (2009) authenticated the members of Polygonaceae in Chinese pharmacopoeia using DNA barcoding. The amplification efficiency of six DNA barcodes (*rbc*L, *trn*H- *psb*A, *ndh*J, *rpo*B, *rpo*C1, *acc*D) was 100%. Inter-specific divergence was highest for the *trn*H-*psb*A (20.05%) followed by the *nr*ITS (14.01%) across all species pairs. Schneider et al. (2009) used morphological dataset of 136 vegetative and reproductive characters to infer the tracheophyte phylogeny with an emphasis on early divergences of ferns (monilophytes). Independent phylogenetic analyses of morphological evidence recover the same phylogenetic relationships among tracheophytes utilizing DNA sequence data but differ within seed plants and ferns.

Grusz et al. (2009) studied the origin of apomictic polyploids in the *Cheilanthes yavapensis* complex (Pteridaceae) by plastid and nuclear DNA sequencing. Plastid and nuclear DNA sequencing revealed *Cheilanthes lindheimeri* is an autopolyploid derived from a rare and undetected sexual diploid. The apomictic triploid *Cheilanthes wootonii* is an inter-specific hybrid between *Cheilanthes fendleri* and *Cheilanthes lindheimeri*, whereas the apomictic tetraploid *Cheilanthes yavapansis* comprised two cryptic and geographically distinct lineages. Ebihara et al. (2010) identified the pteridophytic flora of Japan (733 taxa including subspecies and varieties) in molecular level to test the utility of two plastid DNA barcode regions (*rbcL* and *trnH-psbA*) with the intention of developing an identification system for native gametophytes. DNA sequences were obtained from each of 689 taxa for *rbcL* and 617 taxa for *trnH-psbA*. Mean inter-specific divergence values across all taxon pairs did not reveal a significant difference in rate between *trnH-psbA* and *rbcL*, but mean genetic distances of each genus showed significant heterogeneity according to systematic position. Pryer et al. (2010) exposed DNA barcoding approaches to identity a mistaken fern in the horticultural trade. Plastid *rbcL, atpA*, and *trnG-R* sequence data demonstrated that a fern marketed as *Cheilanthes wrightii* in the horticultural trade is in fact *Cheilanthes distans*. Labiak et al. (2010) resolved molecular phylogeny, character evolution, and biogeography of the grammitid fern genus *Lellingeria* (Polypodiaceae) among 61 species and sequences were obtained for 2 genes (*atpB* and *rbcL*) and 4 intergenic spacers (*atpB-rbcL*, *rps4-trnS*, *trnG-trnR*, and *trnL-trnF*). *Lellingeria* is composed of two main clades; the *L. myosuroides* and the *Lellingeria* clades together are sister to *Melpomene*. Sister to all three of these is a clade with two species of the polyphyletic genus *Terpsichore*. In the *L. myosuroides* clade, several dispersal events occurred between the neotropics, Africa, and the Pacific Islands, whereas *Lellingeria* is restricted to the neotropics, with about 60% of its diversity in the Andes.

Ma et al. (2010) analyzed five DNA sequence markers (*psb*A-*trn*H intergenic region, *rbc*L, *rpo*B, *rpo*C1, and *mat*K) using six chloroplast genomic sequences from GenBank and found *psb*A-*trn*H intergenic region was a suitable DNA marker for species identification in medicinally important pteridophytes. Li et al. (2010) identified diminutive and featureless stages of ferns using tissue-direct PCR combined with amplifying plant barcodes. It was very helpful for large-scale ecological studies surveying distribution and population structure. Gao et al. (2009) sequenced the complete chloroplast genome of a scaly tree fern *Alsophila spinulosa*. The *Alsophila* cp genome is 156,661 base pairs (bp) in size and has a typical quadripartite structure with the large (LSC, 86,308 bp) and small single copy (SSC, 21,623 bp) regions separated by two copies of an inverted repeat (IRs, 24,365 bp each). This genome contains 117 different genes encoding 85 proteins, 4 rRNAs, and 28 tRNAs. The *Alsophila* genome shares some unusual characteristics with the previously sequenced cp genome of the polypod fern *Adiantum capillus-veneris*, including the absence of five tRNA genes that exist in most other cp genomes. The genome shows a high degree of synteny with that of *Adiantum*, but differs considerably from two basal ferns (*Angiopteris evecta* and *Psilotum nudum*). Beck et al. (2011) estimated the ages of sexual and asexual diploids in the fern genus *Astrolepis* using a well-supported plastid phylogeny *trnGR*. The 50 asexual polyploidy samples were estimated to comprise 19 distinct lineages, including a variety of auto- and allopolyploid genomic combinations. The confounding association between asexuality and polyploidy precludes regarding the effect of asexuality suggested that asexuality limits evolutionary potential in *Astrolepis*.

Li et al. (2011) assessed the discriminatory power of the core plant DNA barcode (*rbcL* and *matK*) as well as an alternative proposed fern barcodes (*trnH-psbA* and *trnL-F*) among two fern groups, viz., *Deparia* (Woodsiaceae) and the *Cheilanthes marginata* group (Pteridaceae). With its high sequence variation, *matK* complements *rbcL* provided a strong resolving power of the loci. With sequence variation, *matK* and *trnL-F* appear to be suitable alternative barcode regions in ferns if universal primer development of *matK* fails. *trnH-psbA* showed reduced sequence variation for the majority of ferns (Fig. 3.1).

Wolf et al. (2011) described evolutionary patterns and processes in fern plastid genomes (plastomes) and compared plastid organization and patterns of evolution in ferns to those in seed plants. A large clade of ferns is characterized by a plastome reorganized with respect to the ancestral gene order similar to seed plants and also observed high levels of RNA in fern plastomes. Groot et al. (2011) evaluated the combination of *rbc*L with a non-coding plastid marker *trn*L-F to obtain DNA identifications for 86 fern species. All species with non-equal chloroplast genomes formed their own well-supported monophyletic clade, indicating a high discriminatory power. Inter-specific distances were larger than intra-specific distances for all the tested taxa. Banks et al. (2011) reported the genome sequence of the lycophyte *Selaginella moellendorffii*. By comparing gene content in evolutionary diverse taxa, secondary metabolic genes expanded extensively and in parallel to the lycophyte and

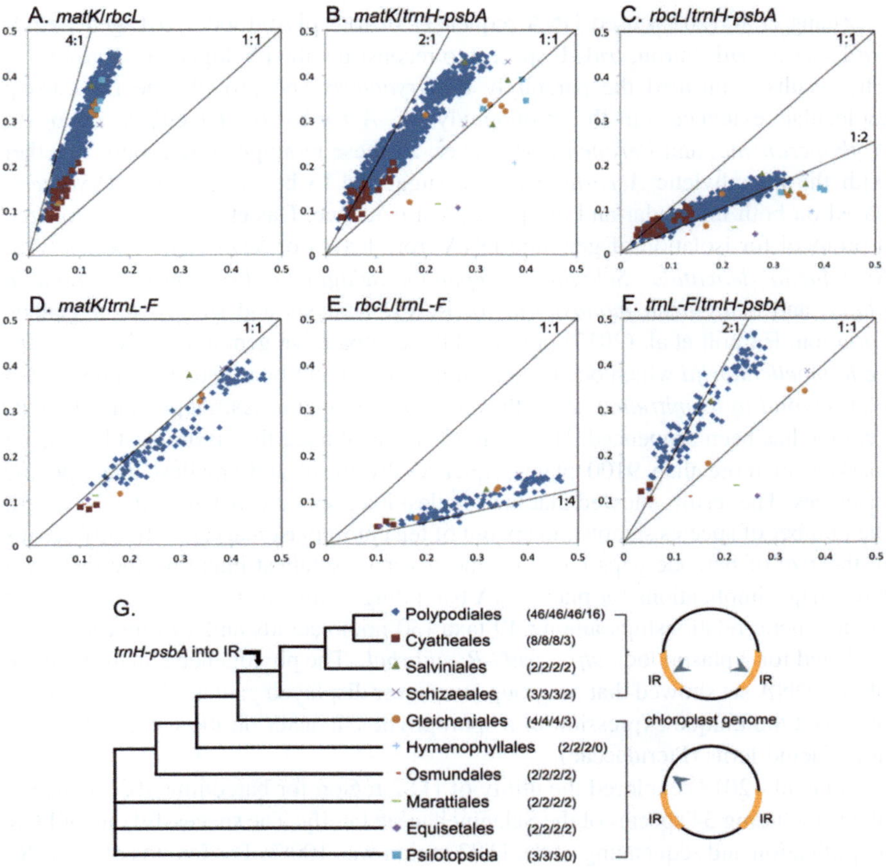


**Fig. 3.1** Large-scale loci comparisons across ferns. (Li et al. 2011: https://doi.org/10.1371/journal.pone.0026597.g001). The x- and y-axes depict the p-distances calculated for each species pair within each fern order (Psilotales and Ophioglossales are combined here into class Psilotopsida). All loci comparisons are presented as y-axis vs x-axis: (**a**) *matK* vs *rbcL*, (**b**) *matK* vs *trnH-psbA*, (**c**) *rbcL* vs *trnH-psbA*, (**d**) *matK* vs *trnL-F*, (**e**) *rbcL* vs *trnL-F*, and (**f**) *trnL-F* vs *trnH-psbA*. The lines in each panel (labeled with the ratios 1:1, 1:2, 1:4, 4:1, or 2:1) are not regression lines, but are drawn to guide the eye in interpreting the results. (**g**) Phylogenetic relationships among fern orders, taxonomic symbols, and number of species compared per order for each locus (*matK*, *rbcL*, *trnH-psbA*, and *trnL-F*, respectively). The arrowheads point to the locations of *trnH-psbA* in the plastid genome (Li et al. 2011). Copyright: © 2011 Li et al. This is an open-access article distributed under the terms of the Creative Commons Attribution License, which permits unrestricted use, distribution, and reproduction in any medium, provided the original author and source are credited

angiosperm lineages. *Selaginella* differs in post-transcriptional gene regulation, including small RNA regulation of repetitive elements, an absence of the trans-acting small interfering RNA pathway, and extensive RNA editing of organellar genes. Peredo et al. (2011) explained the use of dominant markers as molecular tools to identify fern diversity in natural populations.

Zhang et al. (2012) used DNA sequences of four plastid loci (*rbc*L gene, *rps*4-*trn*S spacer, *trn*L intron, *trn*L-F spacer) to reconstruct the phylogeny of *Dryopteris*. The results confirmed the paraphyly of *Dryopteris* and provide the first strong molecular evidence on the monophyly of *Acrophorus*, *Diacalpe*, *Dryopsis*, *Nothoperanema*, and *Peranema*. However, all these monophyletic groups together with the paraphyletic *Acrorumohra* are suggested to be merged into *Dryopteris* based on both molecular and morphological evidence. Das et al. (2012) optimized a protocol for isolation of genomic DNA from leaves of *Selaginella* species, viz., *Selaginella delicatula*, *Selaginella repanda*, *Selaginella bryopteris*, *Selaginella plana*, and *Selaginella monospora*, for RAPD analysis and studied their genetic variation. Harholt et al. (2012) presented the comparative genomic study of *Selaginella moellendorffii* which occupied an important evolutionary position among land plants with *Physcomitrella patens*, the only non-vascular terrestrial plant in which its genome has been sequenced. Hao et al. (2012) evaluated the chloroplast barcoding markers in more than 9100 plant sequences by mean and smallest inter-specific distances. The results showed that the smallest inter-specific distances decrease with the number of species sampled in six out of ten chloroplast markers. The differences in the size of barcode gaps based on mean versus smallest inter-specific distances have major implications for plant DNA barcoding. Johnson et al. (2012) studied the phylogenetic relationships among 49 taxa (30 notholaenids and 19 outgroup taxa) analyzed for 3 plastid loci, *atpA*, *trnG-R*, and *rbcL*. The phylogenetic analysis of the plastid DNA set showed that outgroup taxa never displayed gametophytic farina and observed the unique expression of a sporophytic character on the gametophytes of notholaenid ferns (Pteridaceae).

Gu et al. (2013) explored the utility of ITS2 region for barcoding 103 medicinal plants including 34 species of the Selaginellaceae family. The successful rate of PCR amplification and sequencing of the ITS2 region was 100%. The lengths of the ITS2 regions ranged from 145 to 189 bp, with an average length of 162 bp; the mean GC content was 56%, with a range of 46 to 67%. There was significant divergence between the inter-specific and intra-specific genetic distances of the ITS2 regions, while the barcoding gap was more obvious. Chen et al. (2013) performed the discriminating power of 3 barcodes (*matK*, *rbcL*, and *trnL-F*) on 16 vittarioid sporophytes. *trnL-F* showed the highest primer universality and discriminatory ability scores, whereas PCR success rates were very low for *matK* and *rbcL* regions (10.8% and 41.3%, respectively). BLAST analyses showed that all the sampled field gametophytes could be successfully identified to species level. Blanca et al. (2013) made an attempt to identify a cryptic fern (*Pityrogramma trifoliata*) distribution in the Western Andes of Peru through DNA sequencing. They identified that the collected sporophytes pointed to four different genera, two in Pteridaceae (*Anogramma* and *Pityrogramma*) and the others in Aspleniaceae (*Asplenium*) and Cystopteridaceae (*Cystopteris*). The results resolved that sequences of DNA (*rbcL* and *trnG-R*) pointed to *Pityrogramma trifoliata* of Pteridaceae. Chao et al. (2014) analyzed the molecular phylogeny and biogeography of the fern genus *Pteris* (Pteridaceae) using cpDNA *rbc*L and *mat*K. Shao et al. (2015) studied the molecular phylogeny of the cliff ferns (Woodsiaceae: Polypodiales) and proposed an

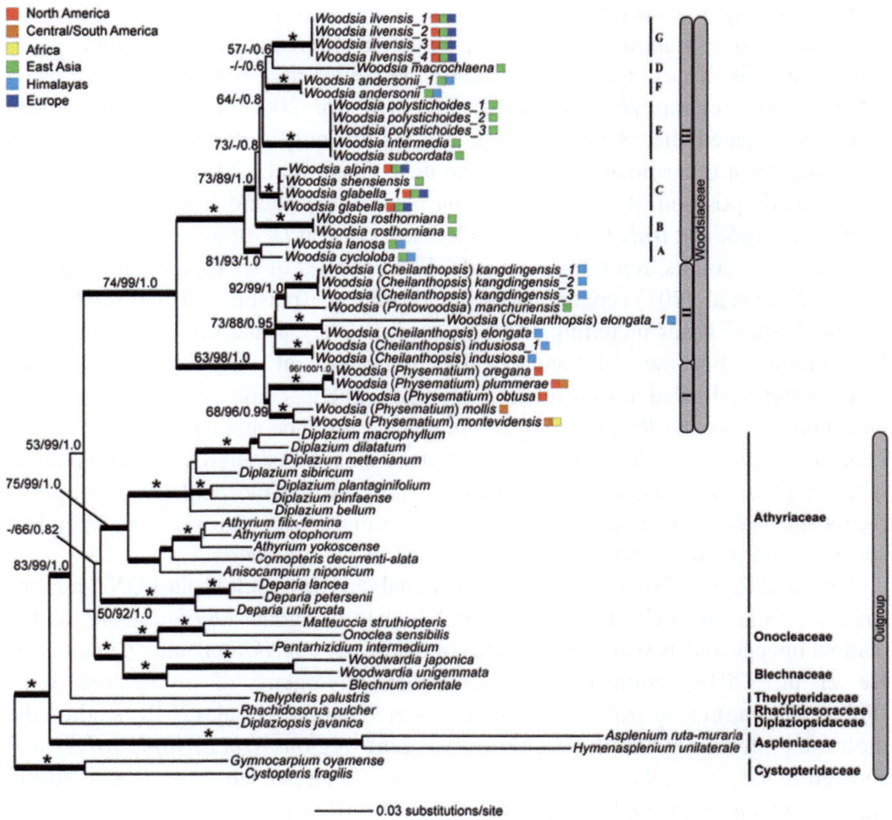


**Fig. 3.2** Phylogram of Woodsiaceae obtained from the maximum likelihood analysis of the combined dataset. (Source: Shao et al. 2015). Numbers on branches are support values [maximum parsimony bootstrap values ($BS_{MP}$)/maximum likelihood bootstrap values ($BS_{ML}$)/Bayesian inference posterior probability values ($PP_{BI}$)]. Bold branches indicate $BS_{MP}$, $BS_{ML}$ $\geq$60%, and $PP_{BI}$ $\geq$0.95. Asterisk indicates $BS_{MP}$, $BS_{ML}$ = 100%, and $PP_{BI}$ = 1.0. Dash (−) indicates nodes with $BS_{MP}$ or $BS_{ML}$ <50%. I to III mark the three major clades recognized in Woodsiaceae, and A to G represent the seven subclades in clade III. Other lineages in eupolypods II are marked with family names. Colored squares indicate the geographical distributions of each species. https://doi.org/10.1371/journal.pone.0136318.g002. Copyright: © 2015 Shao et al. This is an open-access article distributed under the terms of the Creative Commons Attribution License, which permits unrestricted use, distribution, and reproduction in any medium, provided the original author and source are credited

intrageneric classification using chloroplast genes (*atp*A, *mat*K, *rbc*L) and one non-coding region (*trn*L-F) (Fig. 3.2).

Wang et al. (2016) applied DNA barcodes to identify closely related species of Chinese *Adiantum* (Pteridaceae). Liu (2016) analyzed the phylogenetic relationships of pteridophytes occurring in China and reconstructed the DNA sequences using three plastid genes, viz., *atpA*, *atpB*, and *rbcL*. Zhou et al. (2016) analyzed the DNA sequences of 1 plastid (*rbcL*) and 1 nuclear (ITS) locus from 394 accessions representing approximately 200 species of *Selaginella* worldwide and inferred a phylogeny using maximum likelihood, Bayesian inference, and maximum parsimony methods. Klaus et al. (2017) studied the phylogeny of Selaginellaceae using DNA marker regions *rbcL* and *ITS1–5.8S-ITS2* for 200 species. Node density analyses revealed that Selaginellaceae has significantly older median and mean node ages than other putative ancient families. Chen et al. (2017) investigated the phylogenetic position of *Dryopolystichum* using three plastid DNA markers, viz., *rbcL*, *rps4-trnS*, and *trnL-F*. They provided new data on *Dryopolystichum* including spore number counts, reproductive mode, SEM images of spore, and chromosome counts. Nitta et al. (2017) created DNA barcode library (plastid *rbcL* and *trnH-psbA*) for the 2 island floras including 145 fern species. They found that fern sporophyte communities become phylogenetically clustered at high elevations and gametophytes showed no correlation of phylogenetic community structure with elevation. Shu et al. (2017) employed the genetic distance among nine members of *Diplopterygium* using DNA barcoding. *D. simulans* is considered to be synonymous with *D. glaucum* or *D. giganteum* based mainly on the morphology of its pinna rachis and blade. The molecular results indicate that *D. simulans* is an independent species rather than a synonymy of *D. glaucum* or *D. giganteum.*

Patel and Reddy (2018) phylogenetically analyzed three chloroplast DNA regions (*trnL-F*, *rbcL*, and *psbA-trnH*) to unambiguously designate *Ophioglossum* as the distinct lineage and as sister to the clade containing *O. parvifolium* and *O. nudicaule*. He et al. (2018) compared the molecular phylogeny of selligueoid ferns (Polypodiaceae) using *trn*L-F intergenic spacer. Shalimov et al. (2019) studied the molecular phylogeny of three chloroplast gene regions (*rbcL*, *atpI*, *psbA*) and showed that *Selaginella dianzhongensis* forms an independent branch with strong support which is distantly related to *S. amblyphlla* and *S. kurzii*, but sister to *S. bodinieri* which is quite different in habitat of erect or ascending stem. Zhang et al. (2020) analyzed phylogenetically using three chloroplast markers (*rbcL*, *atpI*, and *psbA*) and revealed that *Selaginella subvaginata* is a distinct species among the anisosporophyllous species clade in Selaginellaceae. Nitta et al. (2020) surveyed 176 species representing 69 genera and 22 families of pteridophytes. Plastid *rbcL* was selected as a DNA barcode marker and obtained for >95% of pteridophyte taxa. Combined molecular and morphological analyses revealed two previously undescribed taxa that appear to be of hybrid origin. Dong et al. (2020) analyzed the sequences of multiple chloroplast and nuclear regions of *Gymnosphaera*. They identified four clades within *Gymnosphaera* in mainland Asia, viz., *G. denticulata*, *G. gigantea*, *G. podophylla*, and *G. salletii*. The new species *G. saxicola*, which is special for its saxicolous habitat, was resolved as sister to *G. austroyunnanensis* in the *G. salletii* clade. The newly discovered *G. bachmaensis*, which is characterized specially by the spathulate frond, was positioned in the *G. podophylla* clade, being sister to *G. bonii*. Park et al. (2021) surveyed more places to find new independent gametophyte populations of *Antrophyum obovatum* using the *rbcL* gene sequence-based DNA barcoding.

## 3.4 Conclusion

A phenotypic marker in comparison to a DNA marker does not reveal the proper identification of morphologically similar pteridophytes. Therefore, the current trend on DNA barcoding-based molecular markers helps in the proper structuring of a species beyond any level of ambiguity. Among the different kinds of DNA markers used for analyzing the molecular phylogeny of pteridophytes, chloroplast gene *rbc*L is used by many of the research workers for finding the evolutionary relationship among similar plants. There is also a challenge for studying variation among plants despite having many advanced molecular techniques available to assess genetic variation. The level of polymorphism that different methodologies reveal is also important. If it reveals too little variation, then it may not be possible to discriminate taxa. If the variation found is too high, then the relationship between the taxa is concealed. Merging of these profiles will certainly help in developing a comprehensive understanding of a species.

## References

Arvind K, Ravikanth G, Uma Shankar R et al (2007) DNA barcoding: an exercise in futility or utility? Curr Sci 92(9):1213–1216

Banks JA, Nishiyama T, Hasebe M et al (2011) The *Selaginella* genome identifies genetic changes associated with the evolution of vascular plants. Science 332:960–963

Baracaldo PS (2004) Phylogenetics and biogeography of the Neotropical fern genera *Jamesonia* and *Eriosorus* (Pteridaceae). Am J Bot 91(2):274–284

Beck JB, Windham MD, Pryer KM (2011) Do asexual polyploid lineages lead short evolutionary lives? A case study from the fern genus *Astrolepis*. Evolution 65(11):3217–3229

Blanca L, Rothfels CJ, Arakaki M et al (2013) Revealing a cryptic fern distribution through DNA Sequencing: *Pityrogramma trifoliata* in the Western Andes of Peru. Am Fern J 103(1):40–48

CBOL Plant Working Group (2016) A DNA barcode for land plants. Proc Natl Acad Sci 106: 12794–12797

Chao Y, Rouhan G, Amoroso VB et al (2014) Molecular phylogeny and biogeography of the fern genus Pteris (Pteridaceae). Ann Bot 114:109–124

Chen CW, Huang YM, Kuo LY, et al. (2013) *trnL-F* is a powerful marker for DNA identification of field vittarioid gametophytes (Pteridaceae). Ann Bot 111(4):663–673

Chen C, Sundue M, Kuo L et al (2017) Phylogenetic analyses place the monotypic *Dryopolystichum* within Lomariopsidaceae. Phyto Keys 78:83–107

Dai Z, Ma SC, Wang GL et al (2006) A new glucoside from *Selaginella sinensis*. J Asian Nat Prod Res 8(6):529–533

Das S, Maumita Bandyopadhya Y, Bera S (2012) Optimization of protocol for isolation of genomic DNA from leaves of *Selaginella* species suitable for RAPD analysis and study of their genetic variation. Am Fern J 102(1):47–54

Dong S, Zuo Z, Xiao Y et al (2020) Contribution to *Gymnosphaera* (Cyatheaceae) in mainland Asia: Two new species, reinstatement of *Cyathea bonii*, and their phylogenetic positions. J Syst Evol. https://doi.org/10.1111/jse.12679

Dubuisson JY (1997) *rbc*L sequences: a promising tool for the molecular systematics of the fern genus *Trichomanes* (Hymenophyllaceae)?. Mol Phylogenet Evol 8(2):128–138

Ebaçh MC, Holdrege C (2005) DNA barcoding is not a substitute for taxonomy. Nature 434:697

Ebihara A, Nitta JH, Ito M (2010) Molecular species identification with rich floristic sampling: DNA barcoding the pteridophyte flora of Japan. PLoS One 5(12):e15136

Gao L, Yi X, Yang Y et al (2009) Complete chloroplast genome sequence of a tree fern *Alsophila spinulosa*: insights into evolutionary changes in fern chloroplast genomes. BMC Evol Biol 11(9):130–134

Gastony GJ, Ungerer MC (1997) Molecular systematics and a revised taxonomy of the onocleoid ferns (Dryopteridaceae: Onocleeae). Am J Bot 84(6):840–849

Green BR (1971) Isolation and base composition of DNA's of primitive land plants I. Ferns and Fern allies. BBA Sect Nucleic Acids Protein Synth 254(3):402–406

Groot GA, During HJ, Maas JW et al (2011) Use of *rbc*L and *trn*L-F as a two-locus DNA barcode for identification of NW-European ferns: an ecological perspective. PLoS One 6(1):e16371

Grusz AL, Windham MD, Pryer KM (2009) Deciphering the origins of apomictic polyploids in the *Cheilanthes yavapensis* complex (Pteridaceae). Am J Bot 96(9):1636–1645

Gu W, Song J, Cao Y et al (2013) Application of the ITS2 region for barcoding medicinal plants of Selaginellaceae in Pteridophyta. PLoS One 8(6):1–8

Guillon J (2007) Molecular phylogeny of horsetails (*Equisetum*) including chloroplast *atp*B sequences. J Plant Res 120(4):569–574

Hao DC, Xiao PG, Dong YPJ et al (2012) Evaluation of the chloroplast barcoding markers by mean and smallest interspecific distances. Pak J Bot 44(4):1271–1274

Harholt J, Sørensen I, Fangel J et al (2012) The glycosyltransferase repertoire of the spikemoss *Selaginella moellendorffii* and a comparative study of its cell wall. PLoS One 7(5):1–15

Hasebe M, Omori T, Nakazawa M, et al. (1994) *rbc*L gene sequences provide evidence for the evolutionary lineages of leptosporangiate ferns. Proc Natl Acad Sci U S A 91:5730–5734

Hasebe M, Wolf PG, Pryer KM et al (1995) Fern phylogeny based on *rbc*L nucleotide sequences. Am Fern J 85:134–181

Haufler CH, Grammar WA, Hennipman E et al (2003) Systematics of the ant-fern genus *Lecanopteris* (Polypodiaceae): testing phylogenetic hypotheses with DNA sequences. Syst Bot 28:217–227

Hauk WD, Parks CR, Chase MW (1996) A comparison between *trn*L-F intergenic spacer and *rbc*L DNA sequence data: an example from Ophioglossaceae. Am J Bot 83:126

Hauk WD, Parks CR, Chase MW (2003) Phylogenetic studies of Ophioglossaceae: evidence from *rbc*L and *trn*L-F plastid DNA sequences and morphology. Mol Phylogenet Evol 28:131–151

He L, Schneider H, Hovenkamp P et al (2018) A molecular phylogeny of selligueoid ferns (Polypodiaceae): implications for a natural delimitation despite homoplasy and rapid radiation. Taxon 67(2):237–249

Hebert PDN, Cywinska A, Ball SL et al (2003) Biological identifications through DNA barcodes. Proc Biol Sci 270(1512):313–321

Heede CJV, Viane RL, Chase MW (2003) Phylogenetic analysis of *Asplenium* subgenus *Ceterach* (Pteridophyta: Aspleniaceae) based on plastid and nuclear ribosomal ITS DNA sequences. Am J Bot 90(3):481–495

Hennequin S, Ebihara A, Ito M et al (2003) Molecular systematics of the fern genus *Hymenophyllum* (Hymenophyllaceae) based on chloroplastic coding and noncoding regions. Mol Phylogenet Evol 27:283–301

Hillis DM, Moritz C (1990) Molecular systematics. Sinauer, Sunderland, MA

Hollingsworth ML, Clark AA, Forrest LL et al (2009) Selecting barcoding loci for plants: evaluation of seven candidate loci with species-level sampling in three divergent groups of land plants. Mol Ecol Resour 9:439–457

Jimenez A (2011) Microsatellites: A powerful genetic marker for fern research. In: Fernandez H, Kumar A, Revilla MA (eds) Working with ferns: issues and applications. Springer, New York, pp 207–220

Johnson AK, Rothfels CJ, Windham MD et al (2012) Unique expression of a sporophytic character on the gametophytes of notholaenid ferns (Pteridaceae). Am J Bot 99(6):1118–1124

Klaus KV, Schulz C, Bauer DS et al (2017) Historical biogeography of the ancient lycophyte genus *Selaginella*: early adaptation to xeric habitats on Pangea. Cladistics 33:469–480

Kolukisaoglu HU, Marx S, Wiegmann C et al (1995) Divergence of the phytochrome gene family predates angiosperm evolution and suggests that *Selaginella* and *Equisetum* arose prior to *Psilotum*. J Mol Evol 41:329–337

Korall P, Kenrick P (2002) Phylogenetic relationships in Selaginellaceae based on *rbc*L sequences. Am J Bot 89:506–517

Korall P, Kenrick P (2004) The phylogenetic history of Selaginellaceae based on DNA sequences from the plastid and nucleus: extreme substitution rates and rate heterogeneity. Mol Phylogenet Evol 31:852–864

Korall P, Taylor WA (2006) Megaspore morphology in the Selaginellaceae in a phylogenetic context: a study of the megaspore surface and wall structure using scanning electron microscopy. Grana 45:22–60

Korall P, Pryer KM, Metzgar JS et al (2006) Tree ferns: monophyletic groups and their relationships as revealed by four protein-coding plastid loci. Mol Phylogenet Evol 39:830–845

Korall P, Conant DS, Metzgar JS et al (2007) A molecular phylogeny of scaly tree ferns (Cyatheaceae). Am J Bot 94(5):873–886

Kress WJ (2017) Plant DNA barcodes: applications today and in the future. J Syst Evol 55(4): 291–307

Kress WJ, Wurdack KJ, Zimmer EA et al (2005) Use of DNA barcodes to identify flowering plants. Proc Natl Acad Sci U S A 102:8369–8374

Kress WJ, Erickson DL, Jones FA et al (2009) Plant DNA barcodes and a community phylogeny of a tropical forest dynamics plot in Panama. Proc Natl Acad Sci U S A 106:18627–18632

Kumar A, Fernandez H (2019) Recent advances in fern research: a brief review. Indian Fern J 36: 110–132

Labiak PH, Sundue M, Rouhan G (2010) Molecular phylogeny, character evolution, and biogeography of the grammitid fern genus *Lellingeria* (Polypodiaceae). Am J Bot 97(8):1354–1364

Lahaye R, Van der Bank M, Bogarin D et al (2008) Barcoding the floras of biodiversity hotspots. Proc Natl Acad Sci 105:2923–2928

Lasalita-Zapico F, Aguilar CH, Lon B et al (2011) Phenotypic variations of extant ferns in Mt. Hamiguitan range, Mindanao Island, Philippines. World Appl Sci J 12(7):979–987

Lehtonen S, Tuomisto H, Rouhan G et al (2010) Phylogenetics and classification of the pantropical fern family Lindsaeaceae. Bot J Linn Soc 163:305–359

Li FW, Kuo LY, Huang YM et al (2010) Tissue-direct PCR, a rapid and extraction-free method for barcoding of ferns. Mol Ecol Resour 10(1):92–95

Li FW, Kuo LY, Rothfels CJ, et al. (2011) *rbcL* and *matK* earn two thumbs up as the core DNA barcode for ferns. PLoS One 6(10):1–7

Liu H (2016) Embracing the pteridophyte classification of Ren-Chang Ching using a generic phylogeny of Chinese ferns and lycophytes. J Syst Evol 54(4):307–335

Liu H, Zhang X, Wang W et al (2007) Molecular phylogeny of the fern family Dryopteridaceae inferred from chloroplast *rbc*L and *atp*B genes. Int J Plant Sci 168(9):1311–1323

Lu JM, Li DZ, Gao LM et al (2005) Paraphyly of *Cyrtomium* (Dryopteridaceae): evidence from *rbc*L and *trn*L-F sequence data. J Plant Res 118:129–135

Lu J, Barrington DS, Li D (2007) Molecular phylogeny of the polystichoid ferns in Asia based on *rbc*L sequences. Syst Bot 32(1):26–33

Ma X, Xie C, Liu C et al (2010) Species identification of medicinal pteridophytes by a DNA barcode marker, the chloroplast *psb*A-*trn*H intergenic region. Biol Pharm Bull 33(11): 1919–1924

Madeira PT, Pemberton RW, Center TD (2008) A molecular phylogeny of the genus *Lygodium* (Schizaeaceae) with special reference to the biological control and host range testing of *Lygodium microphyllum*. Biol Control 45:308–318

Mirialili MH, Fakhr-Tabatabaei SM, Alizadeh H et al (2009) Genetic and withaferin analysis of Iranian natural populations of *Withania somnifera* and *Withania coagulans* by RAPD and HPTLC. Nat Prod Commun 4(3):337–346

Nagalingam NS, Nowak MD, Pryer KM (2008) Assessing phylogenetic relationships in extant heterosporous ferns (Salviniales), with a focus on *Pilularia* and *Salvinia*. Bot J Linn Soc 157: 673–685

Nitta JH (2008) Exploring the utility of three plastid loci for biocoding the filmy ferns (Hymenophyllaceae) of Moorea. Taxon 57(3):725–736

Nitta JH, Meyer JY, Taputuarai R et al (2017) Life cycle matters: DNA barcoding reveals contrasting community structure between fern sporophytes and gametophytes. Ecol Monogr 87(2):278–296

Nitta JH, Ebihara A, Smith AR (2020) A taxonomic and molecular survey of the pteridophytes of the Nectandra Cloud Forest Reserve, Costa Rica. PLoS One 15(11):e0241231

Park SH, Kim JS, Kim HT (2021) A small number of gametophytes with gametangia and stunted sporophytes of *Antrophyum obovatum* Baker (Pteridaceae): The suppression of functional sporophyte production by prezygotic and postzygotic sterility. Plan Theory 10:170

Patel M, Reddy MN (2018) Discovery of the world's smallest terrestrial pteridophyte. Sci Rep 8: 5911

Peredo EL, Revilla A, Mendez M et al (2011) Diversity in natural fern populations: dominant markers as genetic tools. In: Fernandez H, Kumar A, Revilla MA (eds) Working with ferns: issues and applications. Springer, New York, pp 220–236

Perrie LR, Brownsey PJ, Lockhart PJ et al (2003) Biogeography of temperate Australasian *Polystichum* ferns as inferred from chloroplast sequence and AFLP. J Biogeogr 30:1729–1736

Pertoldi C, Bijlsma R, Loeschcke V (2007) Conservation genetics in a globally challenging environment: present problems, paradoxes and future challenges. Biodivers Conserv 16: 4147–4163

Pillai SV, Sundaresan P, Harisankar, et al. (2000) Molecular characterization of germplasm in tropical tuber crops. DAE-BRNS Symposium, Mumbai

Pryer KM, Smith AR, Skog JE (1995) Phylogenetic Relationships of extant ferns based on evidence from morphology and *rbcL* sequences. Am Fern J 85(4):205–282

Pryer KM, Schuettpelz E, Wolf PG et al (2004) Phylogeny and evolution of ferns (Monilophytes) with a focus on the early leptosporangiate divergences. Am J Bot 91:1582–1598

Pryer KM, Schuettpelz E, Huiet L et al (2010) DNA barcoding exposes a case of mistaken identity in the fern horticultural trade. Mol Ecol Resour 10(6):979–985

Ranker TA, Geiger JM, Kennedy SC et al (2003) Molecular phylogenetics and evolution of the endemic Hawaiian genus *Adenophorus* (Grammitidaceae). Mol Phylogenet Evol 26(3):337–347

Ranker TA, Smith AR, Parris BS et al (2004) Phylogeny and evolution of grammitid ferns (Grammitidaceae): a case of rampant morphological homoplasy. Taxon 53:415–428

Rouhan G, Dubuisson JY, Rakotondrainibe F et al (2004) Molecular phylogeny of the fern genus *Elaphoglossum* (Elaphoglossaceae) based on chloroplast non-coding DNA sequences: contributions of species from the Indian Ocean area. Mol Phylogenet Evol 33:745–763

Schneider H, Schuettpelz E (2006) Identifying fern gametophytes using DNA sequences. Mol Ecol Notes 6(4):989–991

Schneider H, Russell SJ, Cox CJ et al (2004a) Chloroplast phylogeny of asplenioid ferns based on *rbc*L and *trn*L-F spacer sequences (Polypodiidae, Aspleniaceae) and its implications for the biogeography of these ferns. Syst Bot 29:260–274

Schneider H, Smith AR, Cranfill R et al (2004b) Unraveling the phylogeny of polygrammoid ferns (Polypodiaceae and Grammitidaceae): exploring aspects of the diversification of epiphytic plants. Mol Phylogenet Evol 31:1041–1063

Schneider H, Smith AR, Pryer KM (2009) Is morphology really at odds with molecules in estimating fern phylogeny? Syst Bot 34(3):455–475

Schuettpelz E, Pryer KM (2007) Fern phylogeny inferred from 400 leptosporangiate species and three plastid genes. Taxon 56:1037–1050

Shalimov AP, Zhu YM, Zhang MH, et al. (2019) *Selaginella dianzhongensis* (Selaginellaceae), a new spikemoss from China. Phyto Keys 118: 75–87

Shao Y, Wei R, Zhang X et al (2015) Molecular phylogeny of the cliff ferns (Woodsiaceae: Polypodiales) with a proposed infrageneric classification. PLoS One 10(9):e0136318

Shu JP, Shang H, Jin DM et al (2017) Re-establishment of species from synonymies based on DNA barcoding and phylogenetic analyses using *Diplopterygium simulans* (Gleicheniaceae) as an example. PLoS One 12(3):e0163604

Skog JE, Zimmer EA, Mickel JT (2002) Additional support for two subgenera of *Anemia* (Schizaeaceae) from data for the chloroplast intergenic spacer region *trn*L-F and morphology. Am Fern J 92:119–130

Skog JE, Mickel JT, Moran RC et al (2004) Molecular studies of representative species in the fern genus *Elaphoglossum* (Dryopteridaceae) based on cpDNA sequences *rbc*L, *trn*L-F and *rps*4-*trn*S. Int J Plant Sci 165(6):1063–1075

Small RL, Lickey EB, Shaw J et al (2005) Amplification of noncoding chloroplast DNA phylogenetic studies in lycophytes and monilophytes with a comparative example of relative phylogenetic utility from Ophioglossaceae. Mol Phylogenet Evol 36:509–522

Smith AR, Pryer KM, Schuettpelz E et al (2006) A classification of extant ferns. Taxon 55:705–731

Soltis DE, Soltis PS (1999) Polyploidy: recurrent formation and genome evolution. Trends Ecol Evol 14:348–352

Song J, Yao H, Li Y et al (2009) Authentication of the family Polygonaceae in Chinese pharmacopoeia by DNA barcoding technique. J Ethnopharmacol 124(3):434–439

Su Y, Wang T, Zheng B et al (2004) Population genetic structure and phylogeographical pattern of a relict tree fern, *Alsophila spinulosa* (Cyatheaceae), inferred from cpDNA *atp*B-*rbc*L intergenic spacers. Theor Appl Genet 109(7):1459–1467

Su YJ, Wang T, Zheng B et al (2005) Genetic differentiation of relictual populations of *Alsophila spinulosa* in southern China inferred from cpDNA *trn*L-F noncoding sequences. Mol Phylogenet Evol 34(2):323–333

Taberlet P, Coissac E, Pompanon F et al (2006) Power and limitations of the chloroplast *trn*L (UAA) intron for plant DNA barcoding. Nucleic Acids Res 35:e14

Ting W, Ying-Juan SU, Bo Z et al (2003) Phylogenetic analysis of the chloroplast *trn*L intron and *trn*L-*trn*F intergenic spacer sequences of the Cyatheaceae plants from China. J Trop Subtrop Bot 11(2):137–142

Tsuda Y, Kimura M, Kata, et al. (2009) Genetic structure of *Cerasus jamasakura*, a Japanese flowering cherry, revealed by nuclear SSRs: implications for conservation. J Plant Res 122:367–375

Wang T, Su Y, Li X et al (2004) Genetic structure and variation in the relict populations of *Alsophila spinulosa* from southern China based on RAPD markers and cpDNA *atp*B-*rbc*L sequence data. Hereditas 140:8–17

Wang F, Lu JM, Wen J et al (2016) Applying DNA barcodes to identify closely related species of ferns: a case study of the Chinese *Adiantum* (Pteridaceae). PLoS One 11(9):e0160611

Winter P, Kahl G (1995) Molecular marker technologies for plant improvement. World J Microbiol Biotechnol 11:438–448

Wolf PG, Soltis PS, Soltis DE (1994) Phylogenetic relationships of dennstaedtioid ferns: evidence from *rbcL* sequences. Mol Phylogenet Evol 3(4):383–392

Wolf PG, Sipes SD, White MR et al (1999) Phylogenetic relationships of the enigmatic fern families Hymenophyllopsidaceae and Lophosoriaceae: evidence from *rbc*L nucleotide sequences. Plant Syst Evol 219:263–270

Wolf PG, Der JP, Duffy AM et al (2011) The evolution of chloroplast genes and genomes in ferns. Plant Mol Biol 76(3–5):251–261

Yatabe Y, Masuyama S, Darnaedi D et al (2001) Molecular systematics of the *Asplenium nidus* complex from Mt. Halimun National Park, Indonesia: evidence for reproductive isolation among three sympatric *rbc*L sequence types. Am J Bot 88(8):1517–1522

Zhang GM, Zhang XC, Chen ZD (2005) Phylogeny of cryptogrammoid ferns and related taxa based on *rbc*L sequences. Nord J Bot 23:485–493

Zhang L, Zhang L, Dong S et al (2012) Molecular circumscription and major evolutionary lineages of the fern genus *Dryopteris* (Dryopteridaceae). BMC Evol Biol 12:180

Zhang X, Shalimov AP, Kang J, et al. (2020) *Selaginella subvaginata* (Selaginellaceae), a new spikemoss from China. J Species Res 9(3): 221–232

Zhou X, Rothfels CJ, Zhang L et al (2016) A large-scale phylogeny of the lycophyte genus *Selaginella* (Selaginellaceae: Lycopodiopsida) based on plastid and nuclear loci. Cladistics 32:360–389

**Dr. N. Janakiraman** is working as an Assistant Professor in the Department of Botany, The Madura College (Autonomous), Madurai, from June 2019. He finished his M.Phil. and Ph.D. degrees from Manonmaniam Sundaranar University, Tirunelveli. He secured first rank in M.Phil. (Gold Medal) at university level. He served as a Junior Research Fellow in the UGC-sponsored major research project at St. Xavier's College (Autonomous), Palayamkottai, from August 2012 to December 2013. He published 17 research articles in Scopus-indexed journals and 20 research articles in peer-reviewed national and international journals. Further, he contributed five chapters in books and deposited three sequences in GenBank

**Dr. M. Narayani** attained her Master's Degree in Biotechnology and awarded with Doctoral Degree from Manonmaniam Sundaranar University, Tamil Nadu, India. Dr. M. Narayani has published 23 research articles in national and international journals. She has submitted 11 genes in GenBank. She worked as an Assistant Lecturer in DMI-St. Joseph University College of Agriculture Science and Technology, Tanzania, during 2015–2016. She worked as a Dean and Senior Lecturer in School of Health Sciences, MOSA University, Zambia, during the period of May 2020 to January 2021. And now she is a Vice-Chancellor at Harvest University, Zambia, Africa, from February 2021

**Dr. M. Johnson, M** .Sc., Ph.D., F.I.F.S., completed his Doctorate in Botany (Plant Tissue Culture) from Manonmaniam Sundaranar University, Tirunelveli District, Tamil Nadu, and is working as Assistant Professor of Botany, Director for Centre for Plant Biotechnology, St. Xavier's College (Autonomous), Palayamkottai, Tamil Nadu, India. He has 22 years of research experience and 18 years of teaching experience. He has published 220 papers in the national and international journals, 27 papers in the proceedings, and 38 book chapters. He is elected as Fellow of the Indian Fern Society 2020 for his contribution in Pteridology. He is a recipient of DST-SERB Young Scientist and Prof. S.S. Bir Gold Medal 2017 and Certificate of Merit 2014 prestigious awards by the Indian Fern Society, India. Eleven scholars have completed their doctoral degree under his guidance

# Plastogenomics Provides a Tool to Study Phylogenetic Interrelationships of Monilophytes: A Review

4

Ashwani Kumar, Priti Giri, and Prem Lal Uniyal

**Abstract**

Green plants and algae have plastids of which chloroplasts are most common. Chloroplasts contain their own genomes, which are relatively stable. Approximately 120 genes are in plastomes and mostly encode proteins used in photosynthesis, protein synthesis, and DNA replication. They provide valuable resources to study the plastogenomics, which helps in understanding the process of evolution in one of the oldest living plants and their interrelationship with angiosperms. Comparing the components of plastomes with those of whole genomes or proteome sequences provides important insights into plant phylogeny and evolution. In the recent years, phylogenomics based on chloroplast genomes has demonstrated numerous advantages in plant phylogenetics. Chloroplast phylogenomics can provide a framework for assessing the impact of reticulate evolution in the early evolution of ferns, especially now the more nuclear data is becoming available. In this review, an attempt is made to study plastid genomes and comparisons of gene content, gene arrangement, gene expression, and gene sequences between extant among various fern species.

**Keywords**

Plastome · Phylogenomics · Chloroplast · Monilophytes · Ferns

A. Kumar (✉)
Former Head of the Department of Botany and Biotechnology, University of Rajasthan, Jaipur, Rajasthan, India

P. Giri · P. L. Uniyal
Department of Botany, University of Delhi, Delhi, India





J. Marimuthu et al. (eds.), *Ferns*, https://doi.org/10.1007/978-981-16-6170-9_4

## 4.1 Introduction

Ferns are the closest sister group to all seed plants, yet little is known about their genomes other than that they are generally colossal. Tree ferns first occurred in the Late Triassic. Ferns as a whole include lineages that diverged from one another prior to the divergence of the major seed plant clades. In a broad sense, ferns include four main clades such as psilotoids (whisk ferns) + ophioglossoids, equisetoids (horsetails), marattioids, and leptosporangiates (Sessa et al. 2014). As our ability to infer evolutionary trees has improved, classifications aimed at recognizing natural groups have become increasingly predictive and stable (IPPG 2016). Most broad analyses of green plant relationships based on nuclear gene sequence data have relied largely on 18S/26S rDNA sequences (Soltis et al. 1999) although recent analyses have employed numerous nuclear genes (review Kumar et al., this volume). Most studies that have used nuclear markers were based on only a single locus and/or were at relatively shallow phylogenetic depths, with the goal of understanding reticulate patterns of evolution caused by hybridization and allopolyploidy (Chen et al. 2014). Next-generation sequencing has provided a wealth of plastid genome sequence data from an increasingly diverse set of green plants (Viridiplantae) (Ruhfel et al. 2014).

## 4.2 Methodology

Chloroplasts have their own DNA and ribosomes because of which they are able to synthesize some of their proteins and replicate independent of the nucleus. Plastids also contain 70s ribosomes that translate the mRNA produced from these genes.

Chloroplast genome sequences are most valuable for understanding plant evolution and phylogeny (Daniell et al. 2016). The utilities of whole chloroplast genomes, or plastomes, in fern phylogenetics have been documented in earlier studies (Wolf et al. 2003; Der 2010; Lu et al. 2015).

### 4.2.1 Plastid Genome Sequencing

The availability of next-generation sequencing (NGS) platforms (Alanazi et al. 2021) and bioinformatic tools (Langmead and Nellore 2018) has a great impact on understanding of plastogenomics or chloroplast (cp) genomics and its application in biotechnology (Daniell et al. 2016). In studies conducted before the availability of high-throughput methods, isolated chloroplasts were used for the amplification of the entire chloroplast genome by rolling circle amplification (Bausher et al. 2006). Eleven chloroplast genes encode *ndh* subunits, which are involved in photosynthesis. The ndh proteins assemble into the photosystem I complex to mediate cyclic electron transport in chloroplasts (Munekage et al. 2004) and facilitate chlororespiration (Peltier and Cournac 2002).

Plastomes of pteridophytes (spore-bearing vascular plants) are mined from NCBI organelle genome database. Plant plastomes possess a quadripartite structure composed of large single-copy (LSC) and small single-copy (SSC) regions divided by two parts of inverted repeat (IR) (Olejniczak et al. 2016). Olejniczak et al. (2016) reported that plastome size of higher plants is usually around 150,000 bp in length and comprises approximately 120–130 genes, among which about 75 genes encode proteins of photosystems I and II, as well as for other proteins, involved in photosynthesis (Daniell et al. 2016), while other genes encode ribosomal RNA and proteins and transfer RNA. IRs are usually regarded as the most stable part of the plastome (Olejniczak et al. 2016). Logacheva et al. (2017) suggested that IRs typically range in size from 15 to 30 kbp and contain a core set of genes consisting of four rRNA genes (*4.5S*, *5S*, *16S*, and *23S rRNA*) and five tRNA genes (*trnAUG*, *trnI-GAU*, *trnN-GUU*, *trnR-ACG*, and *trnV-GAC*). Plastome structure, gene contents, and GC contents are analyzed by the in-house developed Python code (see Kwon et al. 2020). Intronic features including presence/absence, length, and intron phases are analyzed manually in the annotated information in NCBI. Peng et al. (2020) sequenced complete chloroplast genome of the fern *Asplenium tenerum* (Aspleniaceae). The complete plastid genome of *A. tenerum* (GenBank accession no. MT700551) is 154,628 bp in length with an overall GC content of 40.82%. The genome displays a typical quadripartite structure consisting a small single-copy region (SSC; 21,374 b p), a large single-copy region (LSC; 81,205 bp), and a pair of identical inverted repeats (IR; 26,117 bp). Peng et al. (2020) suggested that the genome encoded a non-redundant gene set similar to that of the other Aspleniaceae plastomes, including 84 protein-coding genes, 8 rRNA genes, and 34 tRNA genes. Nine protein-coding genes (*ndhA*, *rpl2*, *rpl16*, *petD*, *petB*, *rpoC1*, *atpF*, *rps16*, and *ndhB*) were disrupted by one intron, and three genes (*clpP*, *rps12*, and *ycf3*) by two, including the trans-spliced *rps12* gene. Logacheva et al. (2017) characterized plastid genomes of three species of *Dryopteris*, using sequencing of chloroplast DNA-enriched samples, and performed comparative analysis with available plastomes of Polypodiales, the most species-rich group of ferns. Logacheva et al. (2017) determined the marked conservation of gene content and relative evolution rate of genes and intergenic spacers in the IRs of Polypodiales. Faster evolution of the four intergenic regions had been demonstrated (*trnA-or*f42, *rrn16-rps12*, *rps7-psbA*, and *ycf2-trnN*).

The chloroplast *translation initiation factor 1* (*infA*) is a homolog of the essential gene *infA* in *Escherichia coli* (Millen et al. 2001). This gene initiates translation in collaboration with two nuclear-encoded initiation factors to mediate interactions between mRNA, ribosomes, and initiator tRNA-Met (Millen et al. 2001).

Another evolutionary event is horizontal gene transfer (HGT) affecting the plastome structure (Logacheva et al. 2017). HGT between nucleus, mitochondria, and plastids has been shown to occur at a high rate and contributed significantly to the plant genome evolution by relocating and refashioning of the genes and consequently contributing to genetic diversity.

### 4.2.2 Simple Sequence Repeats (SSRs)

Microsatellites, or simple sequence repeats (SSRs), are ubiquitous throughout both the coding and non-coding regions of all eukaryotic genomes (Powell et al. 1996). Recently, they have been found and characterized within protein-coding genes and their untranslated regions (UTRs). Several studies on possible SSR functions have been undertaken (Li et al. 2004).

SSRs derived from EST (expressed sequence tag) libraries (EST-SSRs) show a higher rate of interspecies transferability than genomic SSRs, which reside in the non-coding region of the genome (Zwenger et al. 2010). Most of the SSR variation are functionally neutral, but the variations in coding regions of SSR may have functional significance, including chromosomal organization, DNA structure, protein binding, gene transcription, and translation (Li et al. 2002, 2004) which provides the basis for rapid evolution (Parisod et al. 2010).

Li et al. (2004) reviewed the SSR distributions within expressed sequence tags (ESTs) and genes including protein coding, 3′-UTRs and 5′-UTRs, and introns and discussed the consequences of SSR repeat-number changes in those regions of both prokaryotes and eukaryotes. Li et al. (2004) reported that substantial data indicates that SSR expansions and/or contractions in protein-coding regions can lead to a gain or loss of gene function via frameshift mutation or expanded toxic mRNA. They further said that SSR variations in 5′-UTRs could regulate gene expression by affecting transcription and translation and the SSR expansions in the 3′-UTRs cause transcription slippage and produce expanded mRNA, which can be accumulated as nuclear foci and which can disrupt splicing and, possibly, disrupt other cellular function.

### 4.2.3 Transcriptional Regulation

The regulation of transcription, that is, the synthesis of messenger RNA from a genomic DNA template, plays a crucial role in plant development. According to Lang et al. (2010), transcriptional regulation is primarily achieved by transcription-associated proteins (TAPs, comprising transcription factors [TFs] and other transcriptional regulators [TRs]), which control gene regulatory networks. Evolutionary retention of duplicated genes encoding transcription-associated proteins (TAPs, comprising transcription factors and other transcriptional regulators) has been hypothesized to be positively correlated with increasing morphological complexity and paleopolyploidizations, especially within the plant kingdom.

### 4.2.4 Posttranscriptional RNA Processing

Over the past 23 years, it has been well documented that RNAs transcribed from most eukaryotic genes can undergo a variety of posttranscriptional RNA processing events (splicing, capping, polyadenylation) that are required to convert RNA precursors into mature RNA species (Yoshinaga et al. 1996, see review Gott and Emeson (2000). Gott and Emeson (2000) suggested that the term RNA editing describes numerous cellular processes that result in the modification of RNA sequences differing from that designated by their DNA (or RNA) templates. The RNA sequence revisions, which include both the insertion and deletion of nucleotides and the conversion of one base to another, involve a wide range of largely unrelated mechanisms (Gott and Emeson 2000). Such mechanisms affect mRNAs, tRNAs, rRNAs, and 7 SLRNA (Ben-Shlomo et al. 1999) which in turn can alter the function or coding potential of the modified transcripts (see review Du et al. 2020).

The majority of the RNA-editing events that have been identified thus far involve changes in mRNA sequences and result in the production of altered protein products. Creation of new start and stop codons by uridine insertion and cytidine to uridine (C-to-U) conversions has been observed in plant organelles (Gott and Emeson 2000). Stop codons are also subject to removal by U-to-C changes in plants, most frequently in hornworts (Yoshinaga et al. 1996). "Silent" codon changes are also observed, but more often editing creates codons for highly conserved or functionally essential amino acids (Gott and Emeson 2000).

Labiak and Karol (2017) used next-generation sequencing methods to study changes in gene composition, plastome architecture, and putative RNA-editing sites. Although the rapid development of high-throughput sequencing technology has led to an explosion of plastome sequences, annotation remains a significant bottleneck for plastomes. In the absence of cDNA, the annotation of RNA editing in plastomes must be done manually. However, as compared to manual annotation, Robison and Wolf (2019) developed a tool ReFernment which offers a greater speed and accuracy for annotating RNA-editing sites. This software should be especially useful for researchers generating large numbers of plastome sequences for taxa with high levels of RNA editing. Likewise, Qu et al. (2019) introduced Plastid Genome Annotator (PGA), a standalone command line tool that can perform rapid, accurate, and flexible batch annotation of newly generated target plastomes based on well-annotated reference plastomes. PGA accurately identifies gene and intron boundaries as well as intron loss. PGA uses reference plastomes as the query and unannotated target plastomes as the subject to locate genes, which Qu et al. (2019) referred to as the reverse query-subject basic local alignment search tool (BLAST) (Altschul et al. 1990) search approach. BLAST was proposed by Altschul et al. (1990) as a new approach to rapid sequence comparison, and it directly approximates alignments that optimize a measure of local similarity, the maximal segment pair (MSP) score.

### 4.2.5 Plastid RNA Editing

The term RNA editing was first coined by Benne and colleagues (Benne et al. 1986) to describe the insertion of uridines into the cytochrome oxidase subunit II mRNA in kinetoplasts of *Trypanosoma brucei* and *Crithidia fasciculata.* Gott and Emeson (2000) said that RNA editing can be broadly defined as any site-specific alteration in an RNA sequence that could have been copied from the template, excluding changes due to processes such as RNA splicing and polyadenylation. They further reported that changes in gene expression attributed to editing have been described in organisms from unicellular protozoa to man and can affect the mRNAs, tRNAs, and rRNAs present in all cellular compartments. However, very little is known about intrageneric variation in frequency of plant RNA editing, and almost no study has been conducted in ferns as reported by Fauskee et al. (2021).

Recent studies of plant RNA editing have demonstrated that the number of editing sites can vary widely among large taxonomic groups. Grosche et al. (2012) suggested that RNA editing is a posttranscriptional process that results in modifications of ribonucleotides at specific locations. Thus, RNA editing acts upon transcripts from mitochondrial, nuclear, and chloroplast genomes.

Replacing Cytidine-to-Uridine (C-to-U) during RNA editing, is a process for converting a specific nucleotide of RNA in organellar genomes e.g. mitochondria, and plastids, throughout land plants but U to C is used less frequently. (Shikanai 2006; Du et al. 2020). In most cases, RNA editing alters translated amino acids or creates new start codons. Spike moss genus *Selaginella* (lycophytes), has the highest frequency of RNA editing. It has been used as a model to test the effects of extreme RNA editing on phylogenetic reconstruction (Du et al. 2020). They predicted the C-to-U RNA-editing sites in coding regions of 18 *Selaginella* plastomes and reconstructed the phylogenetic relationships within *Selaginella* based on 3 data set pairs consisting of plastome or RNA-edited coding sequences, first and second codon positions, and translated amino acid sequences, respectively. Du et al. (2020) reported that the numbers of RNA-editing sites in plastomes were highly correlated with the GC content of first and second codon positions, but not correlated with the GC content of plastomes as a whole.

Contrast phylogenetic analyses showed that there were substantial differences (e.g., the placement of clade B in *Selaginella*) between the phylogenies generated by the plastome and RNA-edited data sets. This empirical study provides evidence that extreme C-to-U RNA editing in the coding regions of organellar genomes alters the sequences used for phylogenetic reconstruction and might even confound phylogenetic reconstruction.

Shikanai (2006) wrote further that to specify the site of editing, the cis-element adjacent to the editing site functions as a binding site for the trans-acting factor. Genetic approaches using *Arabidopsis thaliana* have clarified that a member of the protein family with pentatricopeptide repeat (PPR) motifs is essential for RNA editing to generate a translational initiation codon of the chloroplast *ndhD* gene. The PPR motif is a highly degenerate unit of 35 amino acids and appears as tandem repeats in proteins that are involved in RNA maturation steps in mitochondria and plastids. The *Arabidopsis* genome encodes approximately 450 members of the PPR family, some of which possibly function as *trans*-acting factors binding the *cis*-elements of the RNA-editing sites to facilitate access of an unidentified RNA-editing enzyme (Shikanai 2006).

RNA editing can alter individual nucleotides in primary transcripts, which can cause the amino acids encoded by edited RNA to deviate from the ones predicted from the DNA template (see Jiang et al. 2012). Technique of bioinformatics is used to analyze the effect of editing events on protein secondary and three-dimensional structures. Example of cotton is given here to show comparison of RNA posttranscriptional editing in vascular plants and seed plants. Jiang et al. (2012) found that 21 editing sites in cotton chloroplast transcripts can affect protein secondary structures and 7 editing sites can alter three-dimensional protein structures. These results imply that 24 editing sites in seed plant cotton (all these editing sites were C-to-U conversion) may play an important role in their protein structures and functions (Jiang et al. 2012). As reported by Chen et al. (2011), C-to-U changes are most common in seed plants.

Grosche et al. (2012) further reported that in chloroplasts, single-nucleotide conversions in mRNAs via RNA editing occur at different frequencies across the plant kingdom. These range from several hundred edited sites in some mosses and ferns to lower frequencies in seed plants and the complete lack of RNA editing in the liverwort *Marchantia polymorpha.* RNA-editing sites have imbalanced distribution in genes, and most of them may function by changing protein structure or interaction (Chen et al. 2011). Grosche et al. (2012) said that analyses of the C-to-U conversions and the genomic context in which the editing sites are embedded provide evidence in favor of the hypothesis that chloroplast RNA editing evolved to compensate mutations in the first land plants. It was concluded that RNA-editing sites can be rapidly gained or lost throughout evolution but start or stop codons are relatively stable.

Ruiz-Ruano et al. (2019) reported that gene content of *Vandenboschia speciosa* (Hymenophyllaceae) of fern order Hymenophyllales plastome was similar to that in most ferns but an important number of genes required U-to-C RNA editing for proper protein translation and two genes showed start codons alternative to the canonical AUG (AUA).

## 4.3 Plastome

In addition to photosynthesis, chloroplast play vital roles in other aspects of plant physiology and development, including the synthesis of nucleotides, fatty acids, amino acids, vitamins, phytohormones, metabolites, and the assimilation of sulfur and nitrogen (Neuhaus and Emes 2000; Daniell et al. 2016).

Plastids (chloroplasts) possess their own genetic information and, consequently, express heritable traits (Bock 2007). Wicke et al. (2011) as well as few reported instances of gene duplication or horizontal gene transfer. The plastid genome provides a wealth of phylogenetically informative data that are relatively easy to obtain and use (Soltis and Soltis 1998).

Plastid genomes are structurally highly conserved coding and considerable diverse non-coding spacer regions. Because of their very low level of recombination, they are valuable sources of genetic markers for phylogenetic analyses (Daniell et al. 2006; Gao et al. 2009; Xu et al. 2019). Wolf et al. (2010) examined for the first time the structure of the plastome across fern phylogeny. According to Grewe et al.

(2013), the order of these plastid genes has remained consistent for most species, such that large syntenic tracks can be easily identified between genomes (Grewe et al. 2013).

The size of photosynthetic land plant plastid chromosomes ranges from 120 to 160 kb. The plastome in photosynthetic plants comprises 70 (gymnosperms) to 88 (liverworts) protein-coding genes and 33 (most eudicots) to 35 (liverworts) structural RNA genes (Bock 2007), totaling 100–120 unique genes. An almost universal feature of the circular chloroplast genome is a large inverted repeat sequence, some 10–25 kilobase pairs (kb) in size, which separates the remainder of the molecule into single-copy regions of 80 kb and 20 kb (Shim et al. 2021). The chloroplast genome includes 120–130 genes, primarily participating in photosynthesis, transcription, and translation.

Wicke et al. (2011) grouped plastid genes into four groups such as those involved in primary and secondary photosynthesis pathways, genes involved in sulfate transport and lipid acid synthesis, genes involved in transcription and translation, and a number of structural RNA genes. Wicke et al. (2011) demonstrated only a few functional gene gains, and more frequent gene losses have been inferred for land plants; the plastid Ndh complex is one example of multiple independent gene losses. Zhu et al. (2016) reported that the plastid genome (plastome) of nearly all land plants has a highly conserved quadripartite structure composed of two copies of an inverted repeat (IR) and two single-copy (SC) regions, termed the large single-copy (LSC) and small single-copy (SSC) regions. Intronic SSRs can affect gene transcription, mRNA splicing, or export to cytoplasm (Li et al. 2004).

## 4.4 Plastogenomics

Plastomes have now been extensively used for exploring phylogenetic relationships and understanding evolutionary processes of plants (Lehtonen 2011). Sequence data from the plastid genome have transformed plant systematics and contributed greatly to the current view of plant relationships (Ruhfel et al. 2014). Phylogenetic and biotechnological investigations are allowing novel insights and expanding the scope of plastome research (Ruhlman and Jansen 2014). Structural changes in the cp genome, such as gene rearrangements (Tangphatsornruang et al. 2010), gene/intron losses or duplications (Guisinger et al. 2011), and small inversions (Yi and Kim 2012), are well known at the genus, family, or ordinal levels of seed plants. However, the cp genome studies in ferns are limited to just a few lineages.

Liu et al. (2021) reported that most fern plastomes consist of four parts, including a pair of large inverted repeats (IRs), a large single-copy (LSC) region, and a small single-copy (SSC) region. Almost all fern IRs contains a core gene set of four ribosomal RNAs (16S, 23S, 4.5S, and 5S) and several tRNA genes (trnA-UGC, trnI-GAU, trnN-GUU, and trnR-ACG). IR regions are responsible for variations in chloroplast genome size and rearrangement and thus promoting genomic evolution (Raubeson et al. 2007; Gao et al. 2013; Grewe et al. 2013; Daniell et al. 2016; Logacheva et al. 2017). In chloroplast genome variables are present in LCS and SSC regions, while expansion and contraction were noted in the IR region (Li et al. 2016; Asaf et al. 2017). Logacheva et al. (2017) reported that IRs of Polypodiales plastomes are dynamic and are regulated by gene loss, duplication, and putative lateral transfer from mitochondria.

Molecular phylogenetic investigations have revolutionized our understanding of fern phylogeny (Rothfels et al. 2015). According to Rothfels et al. (2015), deep divergences in fern phylogeny have been recorded mainly in 12 studies almost exclusively on plastid data (Table 4.1). These studies relied almost exclusively on data from a single linkage group, the plastid genome, which is maternally inherited in ferns (Gastony and Yatskievych 1992; Guillon and Raquin 2000).

Lu et al. (2015) reported phylogeny of ferns based on plastome sequence data (Fig. 4.1). All the 132 complete plastome sequences in the NCBI RefSeq collection from GenBank as of 10 Feb 2019 were downloaded and analyzed by Liu et al. (2020). Sequenced species mainly included tree ferns (Cyatheales) and polypod ferns (Polypodiales), which contain most of the extant fern diversity. These include the inversion of a 3 kb region that is shared by *Equisetum* L. and other ferns (Gao et al. 2009) and the loss of *chlB*, *chlL*, and *chlN* in *Psilotum* Sw. and *Tmesipteris* Bernh (Grewe et al. 2013; Zhong et al. 2014).

## 4.5 Phylogeny of Ferns Based on Plastome Sequence Data

Studies on chloroplast genomes of ferns and lycophytes are relatively few in comparison with those on seed plants. Lu et al. (2015) suggested a basic phylogenetic framework of extant ferns (Fig. 4.1). However, the relationships among a few key nodes remain unresolved or poorly supported.

Pteridophytes are free-sporing vascular plants comprising two classes Lycopodiopsida (lycophytes) and Polypodiopsida (ferns) which form distinct evolutionary lineages in the tracheophyte phylogenetic tree (Shmakov 2016).

### 4.5.1 Lycopodiopsida (Lycophytes)

Only three orders are currently recognized within Lycopodiopsida, including Lycopodiales, Isoetales, and Selaginellales (Shmakov 2016). Order Lycopodiales includes 1 family and 16 genera, whereas orders Isoetales and Selaginellales each contain a single genus *Selaginella* (Zhou and Zhang 2015; Weststrand and Korall 2016a, b). Complete plastome sequences have been made available for all the three important orders: (1) Lycopodiales, *Huperzia* species; (Guo et al. 2016); (2) Isoetales, *Isoetes flaccida* (Karol et al. 2010); and (3) Selaginellales, *Selaginella* species (Smith 2009).

*Huperzia serrata* (Lycopodiaceae) chloroplast genome is reported to contain 120 unique genes, including 86 coding genes, 4 rRNA genes, and 30 tRNA genes (Guo et al. 2016). Wolf et al. (2005) studied completed genome of *Huperzia lucidula* (Lycopodiaceae) which is 154,373 bp, containing inverted repeats of 15,314 bp

**Table 4.1** Summary of main studies of deep fern phylogeny (Rothfels et al. 2015)

| Study | Phylogenetic depth | Ferns sampled | Characters used (bp) | Data types included |
|---|---|---|---|---|
| Pryer et al. (1995) | All ferns | 50 | 1206 (+77 morph. chars.) | Plastid (*rbcL*); Morphology |
| Hasebe et al. (1995) | All ferns | 107 | 1206 bp | Plastid (*rbcL*) |
| Pryer et al. (2001) | Vascular plants | 21 | 4072 (+136 morph. chars.) | Plastid (three loci); Nucleus (*18S*); Morphology |
| Pryer et al. (2004) | All ferns | 53 | 5049 | Plastid (three loci); Nucleus (*18S*) |
| Wikström and Pryer (2005) | All ferns | 20 | 5697 (+138 morph. chars.) | Plastid (three loci); Nucleus (*18S*); Mitochondrion (*atp1*); Morphology |
| Schuettpelz et al. (2006) | All ferns | 52 | 6113 | Plastid (four loci); Nucleus (*18S*) |
| Schuettpelz and Pryer (2007) | Leptosporangiate ferns | 400 | 4092 | Plastid (three loci) |
| Qiu et al. (2007) | Land plants | 36 | 14553 | Plastid (seven loci); Nucleus (*18S*); Mitochondrion (two loci) |
| Rai and Graham (2010) | Land plants | 34 | ~10,000[a] | Plastid (17 loci) |
| Kuo et al. (2011) | All ferns | 78 | 3876 | Plastid (three loci) |
| Lehtonen (2011) | All ferns | 2656 | 4406[b] | Plastid (four loci)[b] |
| Rothfels et al. (2012) | Polypodiales | 81 | 6596 | Plastid (five loci) |
| Grewe et al. (2013) | Vascular plants | 9 | 32547 | Plastid (49 loci) |
| Wickett et al. (2014) | Green plants | 6 | 1,701,170 | Nucleus (852 loci) |

Table 4.1 Rothfels et al. (2015) reported summary of main studies of deep fern phylogeny
[a]The total alignment was 36,139 base pairs long, but much of that length was due to single-taxon regions where the sequences were staggered due to uncertain homology
[b]Fewer than 10% of included taxa had all four loci; 54% were represented by only a single locus.
Source: Rothfels, C.J., et al. (2015). The evolutionary history of ferns inferred from 25 low-copy nuclear genes. American Journal of Botany, 102: 1089–1107. doi: 10.3732/ajb.1500089. Reproduced with license number 5087470694801 dated 14 June 2021

each, a large single-copy region of 104,088 bp, and a small single-copy region of 19,657 bp. Tsuji et al. (2007) determined complete nucleotide sequence of the chloroplast genome of *Selaginella uncinata*, a lycophyte belonging to the basal

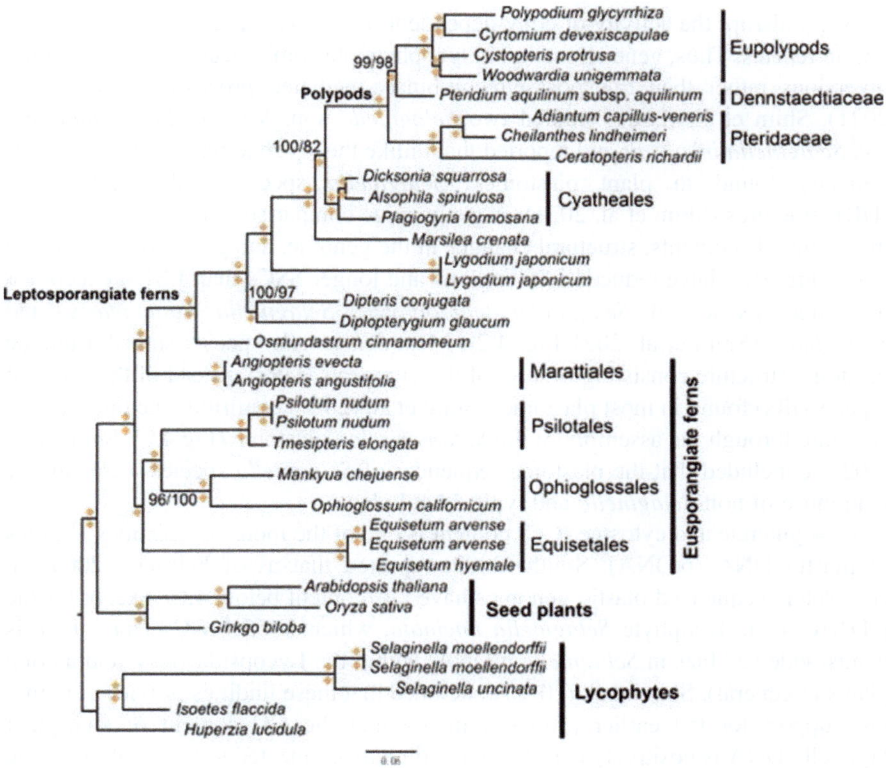


**Fig. 4.1** Lu et al. (2015) reported phylogeny of ferns based on plastome sequence data. Source: Lu, J.M. et al. (2015). Chloroplast phylogenomics resolves key relationships in ferns. Jnl of Systematics Evolution, 53: 448–457. doi: 10.1111/jse.12180. Reproduced with license number 5085221071437 dated 10 June 2021

lineage of the vascular plants. The circular double-stranded DNA is 144,170 bp, with an inverted repeat of 25,578 bp separated by a large single-copy region (LSC) of 77,706 bp and a small single-copy region (SSC) of 40,886 bp. Tsuji et al. (2007) showed that the gene order and arrangement are almost identical between the plastomes of *Huperzia lucidula* (Lycopodiaceae) and bryophytes, but the plastome of *S. uncinata* is considerably different from those of bryophytes. Several new pieces of evidence for monilophyte monophyly of *Isoetes flaccida*, a heterosporous lycophyte, have been recorded.

*Selaginella* with 700 species is distributed in a diverse range of habitats which include deserts, tropical rainforests, and alpine and arctic regions. *Selaginella* plastomes have the highest GC content and fewest genes and introns of any photosynthetic land plant. Uniquely, the canonical inverted repeat was converted into a direct repeat (DR) via large-scale inversion in some *Selaginella* species. Ancestral reconstruction identified additional putative transitions between an inverted and DR orientation in *Selaginella* and *Isoetes* plastomes. ADR orientation does not disrupt the activity of copy-dependent repair to suppress substitution rates within repeats. Thus, gene relocation in lycophyte plastomes occurs via overlapping inversions rather than transposase/recombinase-mediated processes (Shim et al. 2021). Shim et al. (2021) studied two *Selaginella* spp. *Selaginella stauntoniana* and *Selaginella involvens* and reported that unlike the inverted repeat (IR) structures typically found in plant plastomes, *Selaginella* species had direct repeat (DR) structures (Shim et al. 2021). A genus-wide comparison of genomic features, including GC contents, structural changes in the genome, and gene losses, revealed that *Selaginella* have reduced LSC regions and longer SSCs than LSCs, except for the three species of *Selaginella lepidophylla*, *Selaginella hainanensis*, and *S. uncinata* (Shim et al. 2021 Fig. 4.2). Most *Selaginella* species shared a unique plastome structure consisting of a set of direct repeats (DRs) instead of the inverted repeats (IRs) found in most plastomes. Shim et al. (2021) confirmed the unusual DR structure through the assembly of the *S. tamariscina* plastome (Fig.4.2). Shim et al. (2021) concluded that the plastome sequences of *Selaginella* species were smaller than those of non-*Selaginella* and typical land plants.

Low guanine and cytosine (GC) content is one of the more conspicuous features of plastid DNA (ptDNA). Smith (2009) reported that as of February 2009, all completely sequenced plastid genomes have GC content below 43% except for the ptDNA of the lycophyte *Selaginella uncinata*, which is 55% GC. Thus, there is genus-wide GC bias in *Selaginella* ptDNA, within the Lycopsida class (and among plants in general). Shim et al. (2021) concluded that these findings provide convincing support for the earlier proposed theory that the GC content of land-plant organelle DNA is positively correlated and directly connected to levels of organelle RNA editing.

### 4.5.2 Polypodiopsida (Ferns)

Recently, the complete cp genome sequences of four orders of eusporangiate ferns were analyzed, and the data aided in understanding the evolutionary history of eusporangiate ferns (Grewe et al. 2013; Karol et al. 2010). Seed plant lineages usually show small ranges of variation in both GC contents and effective numbers of codons (ENCs). Kim et al. (2014) reported that GC contents and the effective numbers of codons (ENCs) values of the early diverged leptosporangiate ferns showed intermediate levels between eusporangiate and core leptosporangiate ferns. The core leptosporangiate ferns show higher ENCs than early diverged leptosporangiate ferns. The cp gene sequences clearly indicated that the cp genome similarity between *O. cinnamomea* (Osmundales) and eusporangiate ferns is symplesiomorphies, rather than synapomorphies. Therefore, Kim et al. (2014) are in agreement with the view that Osmundales is a distinct early diverged lineage in the leptosporangiate ferns. In Pteridaceae, all subfamilies accepted by Christenhusz et al. (2011) were found to be monophyletic, although the monophyly of Cheilanthoideae had poor support. By contrast, numerous pteridoid genera, including *Adiantum*, were not monophyletic.

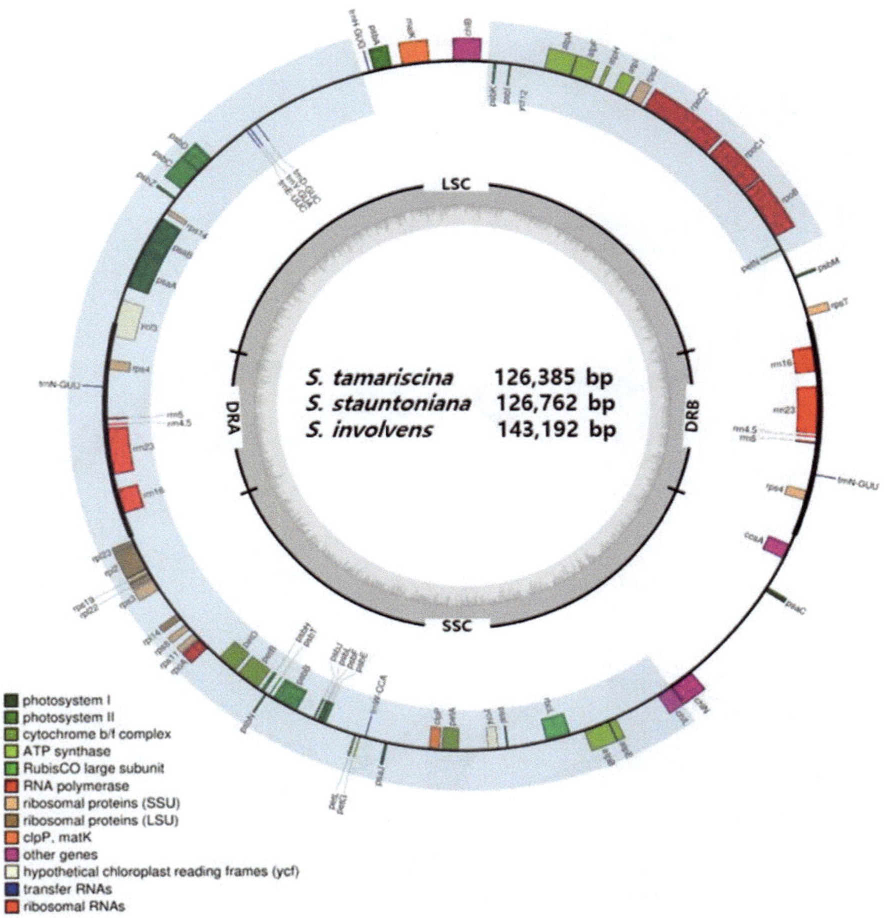


**Fig. 4.2** Shim et al. (2021) presented map of complete plastid genomes of the *Selaginella tamariscina*, *Selaginella stauntoniana*, and *Selaginella involvens.* Shaded areas indicate regions involved in the inversion event. Source: Shim, Hyeonah et al. (2021) "Plastid Genomes of the Early Vascular Plant Genus *Selaginella* Have Unusual Direct Repeat Structures and Drastically Reduced Gene Numbers" *Int. J. Mol. Sci.* 22, no. 2: 641. doi: 10.3390/ijms22020641. An open-access article distributed under the terms of the Creative Commons CC BY license

#### 4.5.2.1 Eusporangiate

Grewe et al. (2013) sequenced the plastid genomes from three early diverging species: *Equisetum hyemale* (Equisetales), *Ophioglossum californicum* (Ophioglossales), and *Psilotum nudum* (Psilotales). A comparison of fern plastid genomes showed that some lineages have retained inverted repeat (IR) boundaries originating from the common ancestor of land plants, while other lineages have experienced multiple IR changes including expansions and inversions (Grewe et al. 2013; Fig. 4.3).

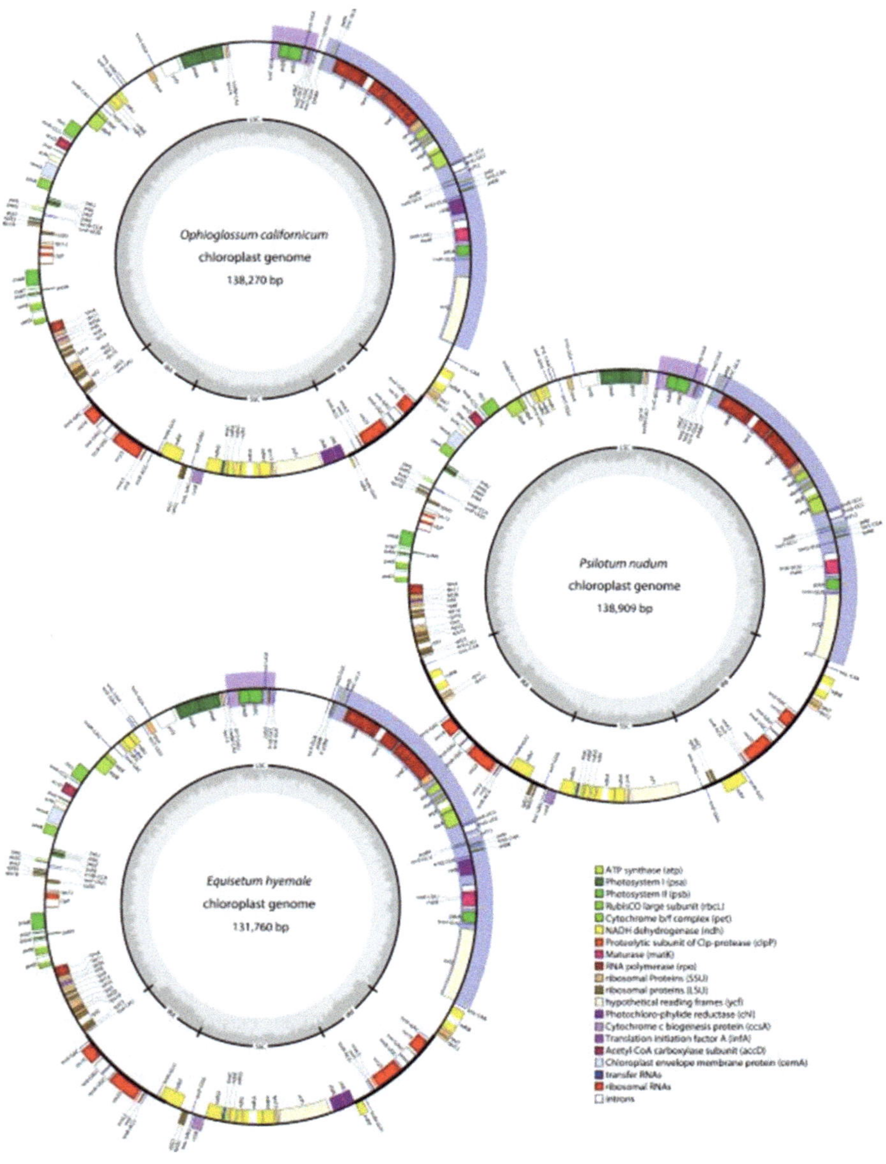


**Fig. 4.3** Grewe et al. (2013) presented plastome maps for newly sequenced monilophytes. Boxes on the inside and outside of the outer circle represent genes transcribed clockwise and anti-clockwise, respectively. The inner circle displays the GC content represented by dark gray bars. The location of the IRs is marked on the inner circle and represented by a thicker black line in the outer circle. The large euphyllophyte LSC inversion and the small monilophyte LSC inversion are highlighted on the outer circle by blue and purple bars, respectively. Source: Grewe et al. (2013). Complete plastid genomes from *Ophioglossum californicum*, *Psilotum nudum*, and *Equisetum hyemale* reveal an ancestral land plant genome structure and resolve the position of Equisetales among monilophytes. *BMC Evol Biol* 13, 8. doi: 10.1186/1471-2148-13-8. This is an open-access article distributed under the terms of the Creative Commons CC BY license

#### 4.5.2.2 Polypodiales

**Leptosporangiate Ferns**

Leptosporangiate ferns account for 80% of nonflowering vascular plants (including gymnosperms and lycophytes) (Schneider et al. 2004; Schuettpelz and Pryer 2009; Rai and Graham 2010), and the leptosporangiate order Polypodiales is by far the largest fern order, with more than 7000 extant species. The unique chloroplast genomic rearrangement of core leptosporangiate ferns (Salviniales, Cyatheales, and Polypodiales) and Schizaeales can be explained by an expansion of the IRs and "two inversions" (Wolf et al. 2003) which mainly affect the orientation and gene content of the IRs (Hasebe and Iwatsuki 1992).

Fan et al. (2021) studied complete chloroplast genomes of *Athyrium brevifrons* Nakai ex Kitagawa, *D. crassirhizoma* Nakai, *Dryopteris goeringiana* (Kunze) Koidz, and *Polystichum tripteron* (Kunze) Presl. Simple sequence repeats (SSRs), nucleotide diversity analysis, and RNA editing were investigated in all four species (Fan et al. 2021). Genome comparison analysis revealed that single-copy regions were more highly conserved than IR regions. IR boundary expansion and contraction varied among the four ferns (Fan et al. 2021; Figs.4.4 and 4.5). The genome size ranged from 149,468 (*D. crassirhizoma*) to 151,341 bp (*A. brevifrons*). The chloroplast genomes had a circular assembly and exhibited a typical quadripartite structure, including one LSC region (82,384–82,799 bp), one SSC region (21,600–21,708 bp), and two IR regions (22,040–22,682 bp).

The complete chloroplast genomes of *Dryopteris goeringiana* (Kunze) Koidz, *D. crassirhizoma* Nakai, *Athyrium brevifrons* Nakai ex Kitagawa, and *Polystichum tripteron* (Kunze) Presl were sequenced by Fan et al. (2020). Simple sequence repeats (SSRs), nucleotide diversity analysis, and RNA editing were investigated in all four species. They also demonstrated that ferns have a higher G + C content

**Fig. 4.4** Fan et al. (2021) presented morphological characteristics of *D. crassirhizoma*, *D. goeringiana*, *P. tripteron*, and *A. brevifrons*. (Source: Fan, R, Ma, W, Liu, S, Huang, Q. Integrated analysis of three newly sequenced fern chloroplast genomes: Genome structure and comparative analysis. *Ecol Evol.* 2021; 11: 4550–4563. doi: https://doi.org/10.1002/ece3.7350). This is an open-access article distributed under the terms of the Creative Commons CC BY license

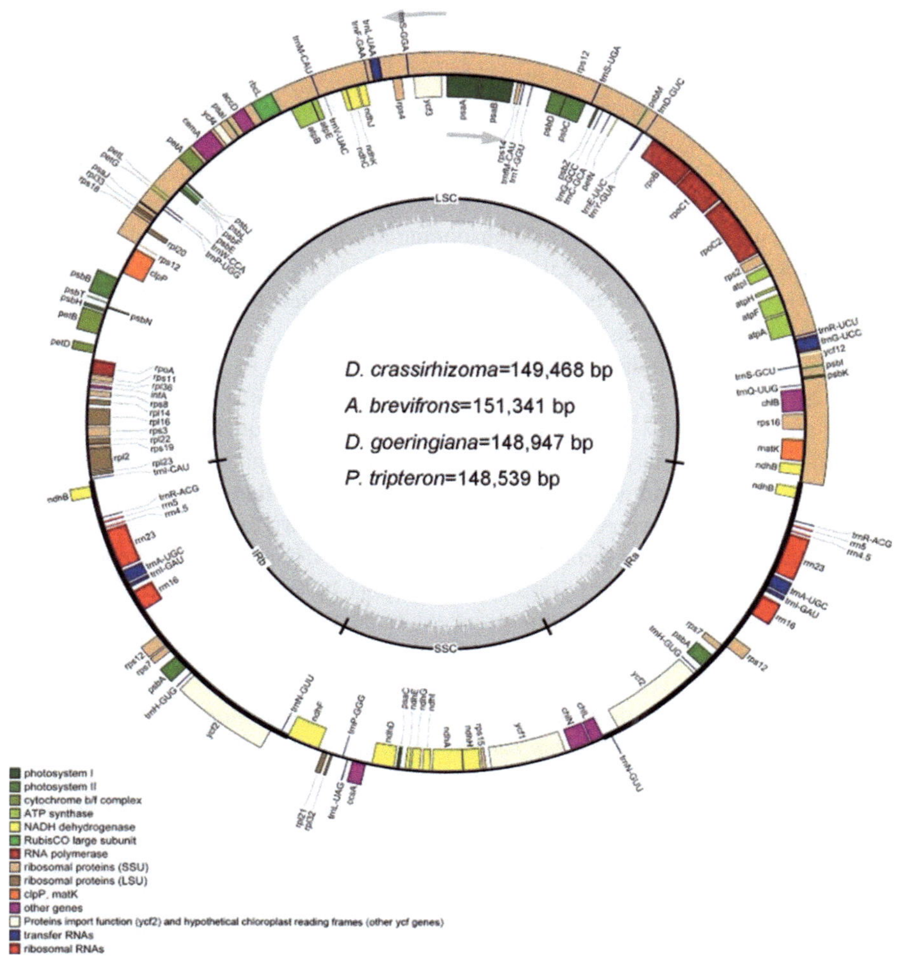


**Fig. 4.5** Fan et al. (2021) depicted the chloroplast genome maps of *D. crassirhizoma*, *D. goeringiana*, A. *brevifrons*, and *P. tripteron*. Genes drawn inside the circle are transcribed clockwise, and those outside the circle are transcribed counterclockwise. The light gray inner circle corresponds to the A + T content and the dark gray to the G + C content. Genes belonging to different functional groups are shown in different colors. (Source: Fan R, Ma W, Liu S, Huang Q. Integrated analysis of three newly sequenced fern chloroplast genomes: Genome structure and comparative analysis. Ecol Evol. 2021 Mar 18;11(9):4550–4563. doi: https://doi.org/10.1002/ece3.7350). This is an open-access article distributed under the terms of the Creative Commons CC BY license

and a higher number of C-to-U RNA editing events than other plants. Fan et al. (2021), speculated that *D. crassirhizoma*, *D. goeringiana*, *D. decipiens*, *P. tripteron*, and *C. devexiscapulae* are closely related. *D. crassirhizoma* and *D. goeringiana* are closely related to *D. decipiens*. *P. tripteron* was identified as a sister species of *C. devexiscapulae*. Interestingly, *D. decipiens* and *C. devexiscapulae* (Fig. 4.5) were found to be clustered into one branch in a study by Wei et al. (2017). The genomes of *Azolla* and *Salvinia* offer a new opportunity to examine the evolution of plant genes and gene families across all Viridiplantae (land plants plus green algae) (Li et al. 2018).

Li et al. (2018) reported on the genomes of *Azolla filiculoides* and *Salvinia cucullata* (Salviniales) and presented evidence for episodic whole-genome duplication in ferns – one at the base of "core leptosporangiates" and one specific to *Azolla*. One fern-specific gene seems to have been derived from bacteria through horizontal gene transfer (Fig. 4.6). This gene provides insect resistance. The relatively small genome (0.75 Gb; Obermayer et al. 2002) of *Azolla* is exceptional among ferns, a group that is notorious for genomes as large as 148 Gb (Hidalgo et al. 2017) and averaging 12 Gb (Sessa and Der 2016). *Azolla* is one of the fastest-growing plants on the planet, with demonstrated potential to be a significant carbon sink (Li et al. 2018).

*Azolla* is also remarkable in harboring an obligate, $N_2$-fixing cyanobacterium, *Nostoc azollae*, within specialized leaf cavities. Furthermore, the *Azolla* genome lacks genes that are common to arbuscular mycorrhizal and root nodule symbioses. Li et al. (2018) identified several putative transporter genes specific to *Azolla*-cyanobacterial symbiosis. These genomic resources will help in exploring the biotechnological potential of *Azolla* and address fundamental questions in the evolution of plant life (Li et al. 2018).

#### Polypodiaceae

Polypods are the lineage of most derived ferns that diversified in the Cretaceous period, displaying an ecologically opportunistic response to the diversification of angiosperms· (Schneider et al. 2004). They suggested that plastomes of polypods have undergone multiple complex genomic reconfigurations during fern evolution, and thus, their plastomes differ substantially from the plastomes of basal ferns (Psilotales, Ophioglossales, Marattiales, and Equisetales). Earlier plastome evolution among Polypodiaceae is considered relatively static compared with that in lineages other than polypods (Wolf et al. 2010), but Liu et al. (2021) suggested that the plastomes of Polypodiaceae are dynamic molecules, rather than constituting static genomes as previously thought. They further indicated that dispersed repeats flanking insertion sequences contribute to the repair mechanism induced by double-strand breaks and are probably a major driver of structural evolution in the plastomes of Polypodiaceae, e.g., *Neolepisorus fortunei*, *Neolepisorus ovatus*, and *Phymatosorus cuspidatus* (Liu et al. 2021; Fig. 4.7).

## 4.6 Analysis of SSRs in Plastomes

Genomic microsatellites (simple sequence repeats; SSRs), iterations of 1–6 bp nucleotide motifs, have been detected in the genomes of every organism (Guichoux et al. 2011). Nevertheless, SSRs are usually just considered as evolutionarily neutral DNA markers (e.g. Schlötterer and Wiehe 1999). SSR genetic and evolutionary

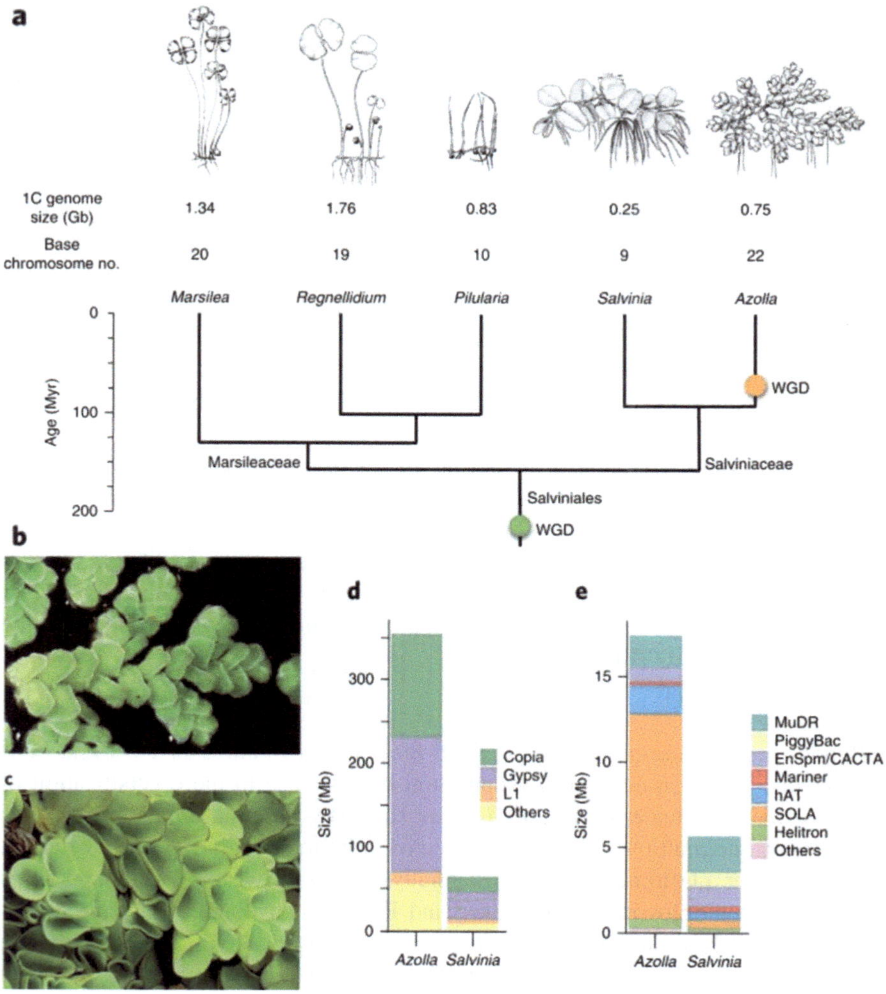


**Fig. 4.6** Genome size evolution in Salviniales (Li et al. 2018). (**a**) Members of Salviniales have smaller genome sizes than other ferns (averaging 1C = 12 Gb). Two whole-genome duplication (WGD) events identified in this study were mapped onto the phylogeny, with divergence time estimates obtained from Testo and Sundue. (**b, c**) Whole genomes were assembled from *A. filiculoides* (**b**) and *S. cucullata* (**c**). (**d, e**) The genome of *S. cucullata* has substantially reduced levels of RNA (**d**) and DNA (**e**) transposons compared to *A. filiculoides*. Image in panel **c** courtesy of P.-F. Source: Li, FW., Brouwer, P., Carretero-Paulet, L. et al. Fern genomes elucidate land plant evolution and cyanobacterial symbioses. *Nature Plants* **4,** 460–472 (2018). doi: 10.1038/s41477-018-0188-8. This is an open-access article distributed under the terms of the Creative Commons CC BY license

mechanisms remain controversial. Li et al. (2002) presented SSR putative functions/effects. Xu et al. (2019) investigated the phylogenetic relatedness among the plastid genomes of 30 species. The ferns of Leptosporangiatidae, Psilophytinae, and

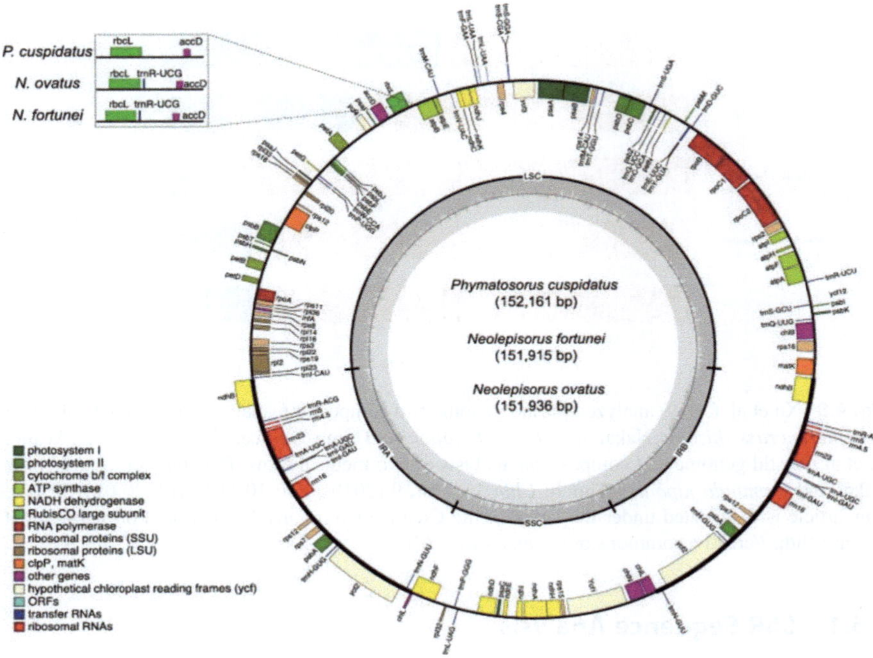


**Fig. 4.7** Liu et al. (2021) presented plastome gene maps of *Neolepisorus fortunei*, *Neolepisorus ovatus*, and *Phymatosorus cuspidatus*. The plastome map represents all three species since their gene numbers, orders, and names are the same, except that *N. fortunei* has lost the trnR-UCG gene. Genes located outside and within the black circle are transcribed in the clockwise and counterclockwise directions, respectively. Different colors represent genes belonging to different functional groups Source: Liu et al. (2021). Comparative genomic analysis of Polypodiaceae chloroplasts reveals fine structural features and dynamic insertion sequences. BMC Plant Biol. 2021 Jan 7;21(1): 31. doi: https://doi.org/10.1186/s12870-020-02800-x. This article is licensed under a Creative Commons Attribution 4.0 International License. http://creativecommons.org/licenses/by/4.0/. The Creative Commons Public Domain

Equisetinae were grouped into three separate clades, respectively. The two Eusporangiate ferns were not grouped in one clade, with Mankyua chejuensis B.Y. Sun was closer to Psilophytinae and the other one *Angiopteris evecta* (G. Forst.) Hoffm. (Marattiaceae) was identified as a sister genus to Leptosporangiatidae. The Leptosporangiatidae ferns formed two clades: *Osmuda* and the other clade contained the other species. This tree also indicated that the moss *Physcomitrella patens* (Hedw.) Bruch & Schimp was grouped in one clade with Leptosporangiatidae, Psilophytinae, Equisetinae, and Eusporangiate (Xu et al. 2019; Fig 4.8).

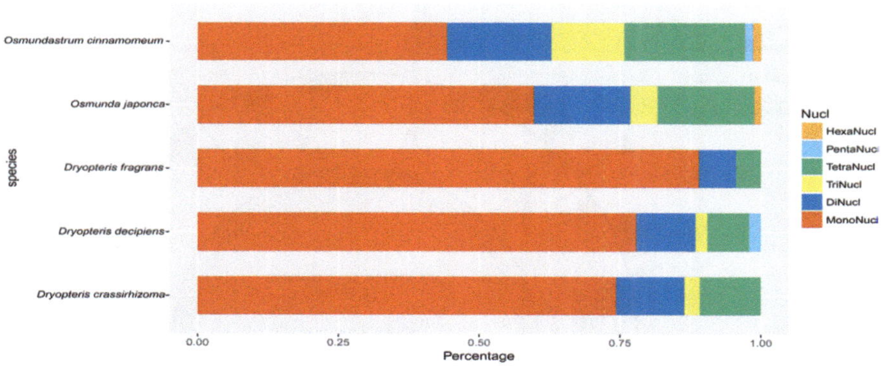


**Fig. 4.8** Xu et al. (2019) analyzed plastid genome and composition analysis of two medical ferns: *Dryopteris crassirhizoma* Nakai and *Osmunda japonica* Thunb. Source: Xu, L., Xing, Y., Wang, B. et al. Plastid genome and composition analysis of two medical ferns: *Dryopteris crassirhizoma* Nakai and *Osmunda japonica* Thunb. *Chin Med* **14,** 9 (2019). doi: 10.1186/s13020-019-0230-4. This article is distributed under the terms of the Creative Commons Attribution 4.0 International License (http://creativecommons.org/licenses/by/4.0/)

### 4.6.1 SSR Sequence Analysis

A total of 74 (1024) bp and 82 SSR (1191) bp loci that were 1024 bp and 1191 bp long, respectively, were detected in the *D. crassirhizoma* and *O. japonica* plastid genomes. The number of mono-repeats was dominant in the plastid genomes of both species. Compared with *D. fragrans*, *D. decipiens*, and *O. cinnamomeum* (Xu et al. 2019; Fig. 4.8), they found that the three species from *Dryopteris* (Dryopteridaceae) had more SSR mono-repeats than the two species from *Osmunda*.

Liu et al. (2021) recorded the total number of SSRs in 12 Polypodiaceae species, which ranged from 38 to 51. Four kinds of SSRs were detected: mononucleotides (62.8–88.3%), dinucleotides (8.7–20.9%), trinucleotides (6.6–22.5%), and tetranucleotides (0–4.4%). However, tetranucleotide repeats were discovered in only the plastomes of *L. clathratus*, *L. hemionitideus*, *L. hederaceum*, *S. yakushimensis*, and *D. roosii* (Fig. 4.9). SSRs were much more frequently located in the LSC region (48.0–71.1%) than in IR (10.5–36.0%) and SSC regions (9.3–22.0%).

The "Monilophyte" clade comprising ferns, horsetails, and whisk ferns receives unequivocal support from molecular data as the sister clade to seed plants. However, the branching order of its earliest emerging lineages, the Equisetales (horsetails), the Marattiales, the Ophioglossales/Psilotales, and the large group of leptosporangiate ferns, has remained dubious (Knie et al. 2015). However, Knie et al. (2015) ultimately obtained a well-supported molecular phylogeny is placing Marattiales as sister to leptosporangiate ferns and horsetails as sister to all remaining monilophytes.

Gao et al. (2009) studied complete chloroplast genome sequence of a tree fern *Alsophila spinulosa*. This provided insights into evolutionary changes in fern

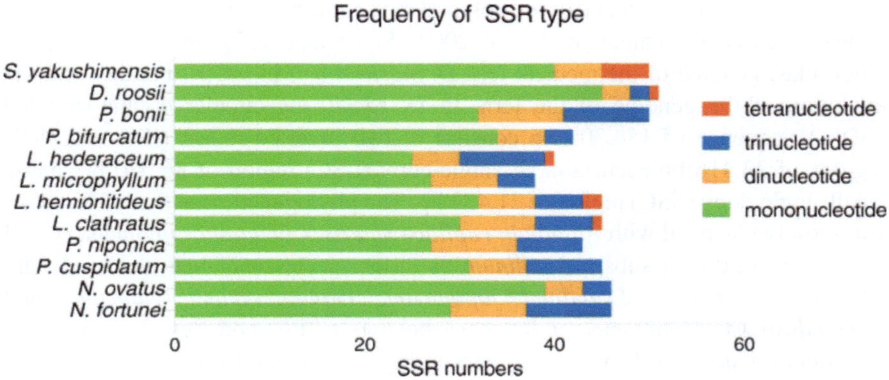


**Fig. 4.9** Source: Liu et al. (2021) Comparative genomic analysis of Polypodiaceae chloroplasts reveals fine structural features and dynamic insertion sequences. *BMC Plant Biol* 21, 31 (2021). The Creative Commons Public Domain Dedication waiver (http://creativecommons.org/publicdomain/zero/1.0/) applies to the data made available in this article, unless otherwise stated in a credit line to the data

chloroplast genomes. Gao et al. (2009) reported that the *Alsophila* cp genome is 156,661 base pairs (bp) in size and has a typical quadripartite structure with the large single-copy (LSC, 86,308 bp) and small single-copy (SSC, 21,623 bp) regions separated by two copies of an inverted repeat (IRs, 24,365 bp each). Thick black lines on the inner cycle indicate the inverted repeats (IRA and IRB) which separate the genome into the large single-copy (LSC) and small single-copy (SSC) regions. Gao et al. (2013) determined the complete chloroplast genome sequences of *Lygodium japonicum* (Lygodiaceae), a member of schizaeoid ferns (Schizaeales), and *Marsilea crenata* (Marsileaceae), a representative of heterosporous ferns (Salviniales). Wolf et al. (2003) determined the complete nucleotide sequence of the chloroplast genome of the leptosporangiate fern, *Adiantum capillus-veneris* L. (Pteridaceae).

Xu et al. (2019) carried out a SSR sequence analysis. A total of 74 and 82 SSR loci that were 1024 bp and 1191 bp long, respectively, were detected in the *D. crassirhizoma* and *O. japonica* plastid genomes. The number of mono-repeats was dominant in the plastid genomes of both species. There were 54 and 62 SSRs located in the LSC, 14 and 12 located in the IR, and 8 and 6 located in the SSC in the *D. crassirhizoma* and *O. japonica* plastid genome, respectively. Compared with *D. fragrans*, *D. decipiens*, and *O. cinnamomeum*, Xu et al. (2019) found that the three species from *Dryopteris* (Dryopteridaceae) had more SSR mono-repeats than the two species from *Osmunda*. The position of *Osmunda* may indicate that *Osmunda* diverged early in the lineage of leptosporangiate ferns (Kim et al. 2014).

Liu et al. (2020) have sequenced the complete plastid genome of a scaly tree fern *Alsophila spinulosa* (ab. *Alsophila*) (Cyatheaceae). In addition to tree ferns, heterosporous and polypod ferns are the other two main lineages within the "core leptosporangiates" (Li et al. 2016). They confirmed that two major rearrangements distinguish higher leptosporangiate ferns from basal fern lineages. The *Alsophila* cp genome is very similar to that of the polypod fern *Adiantum* in terms of gene content, gene order, and GC content (Wolf et al. 2003). Sun et al. (2017) studied the complete chloroplast genome of the medical fern *Drynaria roosii* in order to understand the evolution of the genome of the fern. In *D. roosii*, the circular double-stranded cpDNA sequence of 154,305 bp consists of two inverted repeat (IRA and IRB) regions of 23,416 bp each, a large single-copy (LSC) region of 86,040 bp, and a small single-copy (SSC) region of 21,433 bp. The phylogenetic position of *D. roosii* was closely clustered with *Adiantum capillus-veneris*, *Cheilanthes lindheimeri*, and *Pteridium aquilinum* subsp. *Aquilinum* as sister species and then clustered with *Alsophila spinulosa*, *Lygodium japonicum*, *Diplopterygium glaucum*, and *Osmundastrum cinnamomeum*. *D. roosii* belongs to Polypodiales. The complete chloroplast genome of *D. roosii* provides utility information for ferns' evolutionary and genomic studies.

## 4.7 Genes Translocated into the Plastid Inverted Repeat

Li et al. (2016) reported that plant chloroplast genomes (plastomes) are characterized by an inverted repeat (IR) region and two larger single-copy (SC) regions, and they further suggested that patterns of molecular evolution in the IR and SC regions differ, most notably by a reduced rate of nucleotide substitution in the IR compared to the SC region. Rates of molecular evolution vary dramatically among organismal lineages and across genomes (Bromham and Penny 2003), and understanding what causes this rate variation is a fundamental topic in evolutionary biology (Lanfear et al. 2010). Li et al. (2016) demonstrated that when genes are translocated into the IR, their nucleotide substitution rates dropped significantly (two- to threefold). They further reported that this deceleration is not shared with other nontranslocated chloroplast genes. They concluded that in addition to rate deceleration, GC content increases following translocation, indicating that the IR affects both substitution rates and GC content (Fig. 4.10). It may be finally concluded that without the knowledge of genome structure, or modeling for possible hidden rate shifts, the evolutionary inferences could be grossly misleading (Rothfels and Schuettpelz 2014).

Plastid ribosomes are ubiquitous organelles in plant cells and play a vital role in the biosynthesis of proteins. In higher plants, plastid ribosomes contain approximately 60 ribosomal proteins that are encoded in both the plastid and the nuclear genetic compartments (Eneas-Filho et al. 1981). The plastid ribosomal protein S12 encoded by the *rps*12 gene is a highly conserved protein located in the functional center of the 30S subunit of the ribosome (Yamaguchi and Subramanian 2003).

The plastomes of polypods have undergone multiple complex genomic reconfigurations during fern evolution, and thus, their plastomes differ substantially from the plastomes of basal ferns (Psilotales, Ophioglossales, Marattiales, and Equisetales).

The variation in the exon location and intron content of the *rps*12 gene in fern plastomes provides a unique opportunity to explore the effect of gene structure on

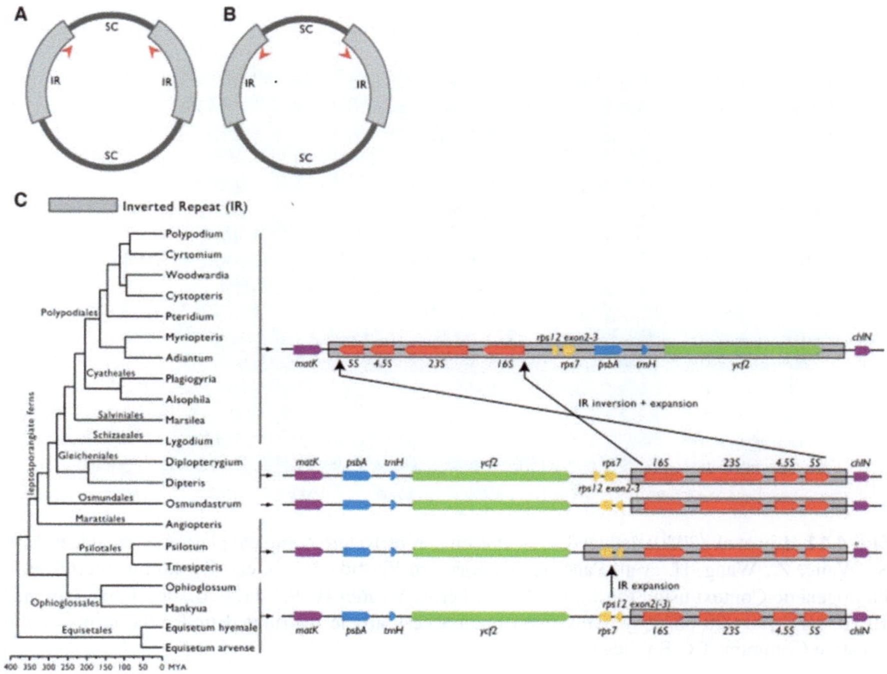


**Fig. 4.10** Li et al. (2016) showed chloroplast genome structure. (**a**) The typical plant chloroplast genome (plastome) comprises a pair of inverted repeat (IR) regions separating a single-copy (SC) region. The red arrowheads indicate the direction in which the rRNA genes in the IR are transcribed. (**b**) In some fern chloroplast genomes, the direction of transcription in the IR is inverted. (**c**) Genome rearrangements have resulted in changes in IR gene content. The phylogeny on the left shows the relationships among the sampled fern chloroplast genomes; a part of their genome organization is shown on the right. The tree topology and divergence times are derived from Rothfels et al. (2015). Gene lengths are not to scale, gene arrow tips indicate the direction of transcription, and a few genes are omitted for clarity. "*" indicates that *chlN* is not always present. Note that *rps12* in Ophioglossales, Psilotales, and Equisetales lacks the second intron, and therefore there is no exon 3. Source: Li, F. W., Kuo, L. Y., Pryer, K. M., & Rothfels, C. J. (2016). Genes Translocated into the Plastid Inverted Repeat Show Decelerated Substitution Rates and Elevated GC Content. *Genome biology and evolution*, *8*(8), 2452–2458. doi: 10.1093/gbe/evw167. Reproduced with license number 5101790146234 dated 4 July 2021

sequence evolution. Liu et al. (2020) reconstructed the phylogeny of ferns and inferred the patterns and rates of plastid *rps*12 gene evolution in a phylogenetic context (Figs. 4.11 and 4.12). In most ferns, the first exon of *rps*12 is located in the LSC, whereas the second and third exons reside in the IRs. The plastomes of polypods have undergone multiple complex genomic reconfigurations during fern evolution, and thus, their plastomes differ substantially from the plastomes of basal ferns (Psilotales, Ophioglossales, Marattiales, and Equisetales).

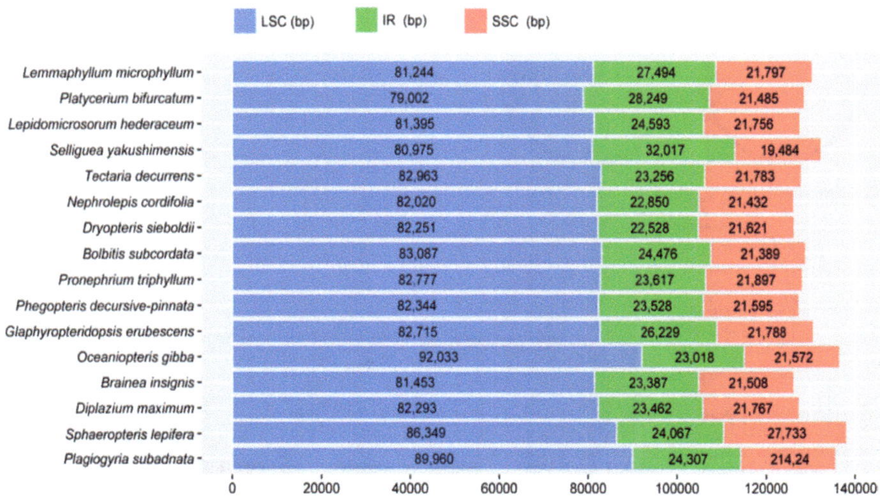


**Fig. 4.11** Liu et al. (2020) depicted sizes of each part of 16 fern complete plastome sequences. Liu, S., Wang, Z., Wang, H. et al. Patterns and Rates of Plastid *rps*12 Gene Evolution Inferred in a Phylogenetic Context using Plastomic Data of Ferns. *Sci Rep* **10,** 9394 (2020). doi: https://doi.org/10.1038/s41598-020-66219-y. This is an open-access article distributed under the terms of the Creative Commons CC BY license

## 4.8 Plant Transcriptome Evolution

Plant genomes encode many lineage-specific, unique transcription factors (Wilhelmsson et al. 2017). Wilhelmsson et al. (2017) further stated that expansion of transcription-associated proteins (TAPs) (total number of TAP genes per genome) comprising transcription factors and transcriptional regulators has been found to coincide with the evolution of morphological complexity (Lang et al. 2010). Lang et al. (2010) further reiterated that evolutionary retention of duplicated genes encoding transcription-associated proteins (TAPs) has been hypothesized to be positively correlated with increasing morphological complexity and paleopolyploidizations, especially within the plant kingdom. Both, the emergence and expansion of TAP families during land plant evolution, suggest a clear trend of increasing transcriptional complexity along with morphological complexity (Lang et al. 2010). Using phylogenetic comparative (PC) analyses, Lang et al. (2010) defined the timeline of TAP loss, gain, and expansion among Viridiplantae.

The evolution of plant transcription-associated proteins, including transcription factors (TFs, binding in sequence-specific manner to *cis*-regulatory elements to enhance or repress transcription) and transcriptional regulators (TRs, acting as part of the transcription core complex, via unspecific binding, protein-protein interaction, or chromatin modification) of ferns were exclusively represented by the *Pteridium aquilinum* transcriptome (Wilhelmsson et al. 2017). Szövényi et al. (2021) compared patterns of various levels of genome and epigenomic organization found in seed-free

**Fig. 4.12** Liu et al. (2020) depicted phylogram showing intron losses and the distribution of the *rps*12 gene in fern plastomes. The absence of the *rps*12 intron is indicated with a red line; the dashed line denotes that all the exons of *rps*12 were present in only one copy. Source: Liu, S., Wang, Z., Wang, H. et al. (2020). Patterns and Rates of Plastid *rps*12 Gene Evolution Inferred in a Phylogenetic Context using Plastomic Data of Ferns. *Sci Rep* **10,** 9394. doi: https://doi.org/10.1038/s41598-020-66219-y. This is an open-access article distributed under the terms of the Creative Commons CC BY license

plants to those of seed plants, e.g., some genomic features appear to be fundamentally different. For instance, hornworts, *Selaginella*, and most liverworts are devoid of whole-genome duplication, in stark contrast to other land plants. However, model systems are crucial to further our understanding about how changes in genes translate into evolutionary novelties (Szövényi et al. 2021).

The finding that the transcriptional regulator Polycomb group EZ (PcG_EZ) was lost in ferns is corroborated here by whole-genome data (Li et al. 2018). Conversely, the transcription factor ULTRAPETALA, which originated at the base of euphyllophytes and is present in *P. aquilinum*, was apparently secondarily lost in Salviniales. Conversely, the transcription factor ULTRAPETALA, which originated at the base of euphyllophytes and is present in *P. aquilinum*, was apparently secondarily lost in Salviniales (Li et al. 2018).

## 4.9 Discussion

Plastid genome data are beneficial in resolving species definitions because organelle-based "barcodes" can be established for a species and then used to unmask interspecies phylogenetic relationships (Yang et al. 2013). Sequence data from the plastid genome have transformed plant systematics and contributed greatly to the current view of plant relationships. The plastid genome provides a wealth of phylogenetically informative data that are relatively easy to obtain and use (Olmstead and Palmer 1994; Soltis and Soltis 1998). Next-generation sequencing has provided a wealth of plastid genome sequence data from an increasingly diverse set of green plants (*Viridiplantae*) (Rothfels et al. 2015).

Grewe et al. (2013) reported that phylogenetic affinities were revealed by mapping rare genomic structural changes in a phylogenetic context. Chloroplast genomes have stable maternal heredity which excludes recombination and are remarkably conserved in land plants. This has made them a valuable and ideal resource for species identification, plant phylogenetics, population genetics, and genetic engineering taxon-rich phylogenetic analyses (Nock et al. 2014). Rai and Graham (2010) also examined the utility of a large plastid-based data set in inferring backbone relationships for monilophytes and found it highly congruent with earlier multigene studies, corroborating clades in common across studies (Knie et al. 2015). However, the relationships among major fern lineages, especially the placement of Equisetales, remain enigmatic (Grewe et al. 2013; see also Wickett et al. 2014).

The ferns of Psilophytinae formed a sister clade to Equisetinae with strong support, which was different from a previous study (Borgstrom et al. 2011). Pryer et al. (2004) based on phylogenetic analyses confirmed that (1) Osmundaceae are sister to the rest of the leptosporangiates, (2) resolved a diverse set of ferns formerly thought to be a subsequent grade as possibly monophyletic (Dipteridaceae, Matoniaceae, Gleicheniaceae, Hymenophyllaceae), and (3) placed schizaeoid ferns as sister to a large clade of "core leptosporangiates" that includes heterosporous ferns, tree ferns, and polypods. Whole plastome sequences have been used to study evolution (Kim et al. 2014; Labiak and Karol 2017). Szövényi et al. (2021) reported that during the past few years, several high-quality genomes have been published from Charophyte algae, bryophytes, lycophytes, and ferns. Szövényi et al. (2021) compared patterns of various levels of genome and epigenomic organization found in seed-free plants to those of seed plants. They reported that in stark contrast to other land plants, *Selaginella* and most liverworts are devoid of whole-genome duplication.

The phylogenetic relationships among these four basal fern orders are the most debated topics in fern phylogeny. Pryer et al. (2001) suggested that maximum likelihood analysis showed unambiguously that horsetails and ferns together are the closest relatives to seed plants. However, this refutes the prevailing view that horsetails and ferns are transitional evolutionary grades between bryophytes and seed plants and has important implications for our understanding of the development and evolution of plants. Although lycophytes were abundant and dominant in land flora during the Carboniferous era (Kenrick and Crane 1997), the class

Lycopodiopsida diverged shortly after land plants evolved to acquire vascular tissues (Banks et al. 2011). Lycopods are similar to ferns in this regard, but ferns are the sister group of the seed plants (gymnosperms plus angiosperms), whereas the lycopods are sister to all other vascular plants (ferns plus the seed plants) (Christenhusz and Chase 2014). Psilotales (which include *Psilotum* and *Tmesipteris*) are closer to ferns (Hendy and Penny 1989; Pryer et al. 2001; Qiu et al. 2006, 2007; Korall et al. 2010; Zhong et al. 2011, 2014).

Clark et al. (2016) suggested that genome size was correlated with chromosome number across all ferns despite some substantial variation in both traits. Marchant et al. (2019) reported a single ancient polyploidy event and spread of repeat elements in the evolutionary history of C-Fern (*Ceratopteris richardii*) based on both genomic and cytogenetic data. According to Shim et al. (2021), plastid genomes (plastomes) are typically 120–160 kb long, with a quadripartite architecture comprising one long single-copy (LSC) region and a short single-copy (SSC) region separated by two inverted repeats (IRA and IRB). Shim et al. (2021) reported that the early vascular plants in the genus *Selaginella* are valuable resources for deciphering plant evolution. Comparative analyses of 19 lycophytes by Shim et al. (2021) revealed unique phylogenetic relationships between *Selaginella* species and related lycophytes. The changes were reflected by structural rearrangements involving two rounds of large inversions that resulted in dynamic changes between IR and DR blocks in the plastome sequence. Furthermore, lycopods present other uncommon characteristics, e.g., a small genome size, drastic reductions in gene and intron numbers, a high GC content, and extensive RNA editing. Their findings suggest that *Selaginella* plastomes have undergone unique evolutionary events yielding genomic features unparalleled in other lycophytes, ferns, or seed plants. Banks et al. (2011) reported that the transition from a gametophyte to a sporophyte dominated life cycle required far fewer new genes than the transition from a non-seed vascular to a flowering plant.

Fan et al. (2021) observed vast differences in SSRs among *D. fragrans*, *D. crassirhizoma*, and *D. goeringiana*, although they belonged to the same genus. They speculated that the types of SSRs are associated more with the surrounding environment than with the genus. This could explain the considerable differences among SSRs within the same species (Fan et al. 2021). Evolutionary significance has been shown in comparative studies between ferns and angiosperms. Palmer and Stein (1982) reported that *Osmunda* chloroplast genome was found to be remarkably similar in size, conformation, physical organization, and map positions of known genes, to chloroplast DNA from a number of angiosperms. Gene probes from tobacco, corn, and spinach were used to map the positions of six genes on the *Osmunda cinnamomea* chloroplast chromosome. Palmer and Stein (1982) made comparative studies on gene probes from tobacco, corn, and spinach to map the positions of six genes on the chloroplast chromosome. The 16S and 23S ribosomal RNAs are encoded by duplicate genes which lie within the inverted repeat. Genes for the large subunit of ribulose-1,5-bisphosphate carboxylase, a photosystem II polypeptide, and the alpha and beta subunits of chloroplast coupling factor are located in three different segments of the large single-copy region. The major difference between chloroplast DNA from this fern and angiosperms is that the inverted repeat is smaller in *Osmunda* (8–13 kb) than in angiosperms (22–25 kb).

Zhong et al. (2014) studied chloroplast genomes of a tree fern (*Dicksonia squarrosa*) (Cyatheaceae that includes the *Alsophila* genus) and a "fern ally" (*Tmesipteris elongata*). Gao et al. (2009) reported that *Alsophila* cp genome shows a high degree of synteny with that of *Adiantum*, but differs considerably from two basal ferns (*Angiopteris evecta* and *Psilotum nudum*). Availability of cp genome sequence from other tree ferns will facilitate interpretation of the evolutionary changes of fern cp genomes. Complete cp genome sequences of *Angiopteris yunnanensis* Hieron. (Marattiaceae) (Liu et al. 2019), *Angiopteris evecta* (G. Forst.) Hoffm. (Roper et al. 2007), and *Angiopteris angustifolia* C. Presl (Zhu et al. 2016) have been published. Comparative analyses confirmed the conservatism of plastid genome sequences among the species of *Angiopteris* and the distant related marattioid genus *Christensenia* (Jiang et al. 2019; Liu et al. 2019).

Haufler (1987, 2002 and 2014) suggested that ferns underwent multiple cycles of polyploidy. They reported whole-genome duplications (WGDs) accompanied by subsequent diploidization involving gene silencing, but without apparent chromosome loss, hence high chromosome numbers were retained leading to polyploidy. The high chromosome count of horsetails could then indeed be caused by a few paleopolyploidies of which a large fraction of genetic material has been retained (Haufler 1987). Vanneste et al. (2015) demonstrated that horsetails underwent an independent paleopolyploidy during the Late Cretaceous prior to the diversification of the genus but did not experience any recent polyploidizations that could account for their high chromosome number. This hypothesis has been provided by observations that polyploidy contributes to *c.* 31% of speciation events in ferns compared with *c.* 15% in angiosperms (Wood et al. 2009). This could also explain the high chromosome counts of homosporous ferns. Alternatively, the ancestor of all vascular plants could have exhibited a relatively high chromosome number (Soltis and Soltis 1987) so that a single paleopolyploidy could have resulted in the very high chromosome number in horsetails. It is emphasized that both theories are not mutually exclusive. The evolution of fern genomes has been considered paradoxical owing to the conservation of high chromosome numbers in taxa with demonstrated diploid gene expression (Haufler 2014).

WGD is often proposed as a driver of species diversification (Landis et al. 2018). Nakazato et al. (2006) suggested that abundance of gene duplicates is a potential mechanism for the past polyploidization in *Ceratopteris richardii*. They constructed a high-resolution genetic linkage map of the homosporous fern model species, *C. richardii* ($n = 39$). Single genome duplication isolates an individual from its parental species and forces the nascent polyploid to overcome numerical inferiority and parental competition if it is to survive (Levin 1975). The concurrent duplication of all nuclear genes is accompanied by widespread changes in gene expression (Adams and Wendel 2005) and often chromosomal rearrangements (Levin 2002; Gaeta et al. 2007). Yet despite the potential for ecological and genomic havoc, polyploidy is remarkably frequent, especially among plants. By some accounts, 20–40% (Stebbins 1971) of extant flowering plant species are neopolyploids, and as many as 70% are thought to have some polyploid ancestry (De Bodt et al. 2005; Cui et al. 2006).

Clark et al. (2016) reported that genome size was correlated with chromosome number across all ferns despite some substantial variation in both traits. They observed a trend toward conservation of the amount of DNA per chromosome, although Osmundaceae and Psilotaceae have substantially larger chromosomes.

Plastid genomes display remarkable organizational stability over evolutionary time (Robison et al. 2018). The chloroplast (cp) (plastid genomes – plastomes) has small size, high copy number, conservation, and extensive characterization at the molecular level (Raubeson and Jansen 2005). Plastomes contain high proportions of protein-coding genes compared with plant nuclear genomes, with many of these genes being essential to photosynthesis (Wicke et al. 2011). Chloroplast genomes have a typically circular structure with one large single-copy (LSC) region, one short single-copy (SSC) region, and two inverted repeat (IR) regions, ranging from 120 to 170 kb in length (Downie and Palmer 1992). Bromham and Penny (2003) reported that plant chloroplast genomes (plastomes) are characterized by an inverted repeat (IR) region and two larger single-copy (SC) regions. Patterns of molecular evolution in the IR and SC regions differ, most notably by a reduced rate of nucleotide substitution in the IR compared to the SC region. Xu et al. (2015) suggested that gain and loss of genes, gene content duplication, and gene order rearrangements appear to be phylogenetically and species informative.

Transcriptome sequencing or gene sequence resources as the 1000 Plants Project (*The 1000 Plants Project* http://www.onekp.com) are available, but genes alone are insufficient to answer the most pressing questions in fern and land plant genome evolution. In contrast, other phyloplastomic studies tend to support grouping Equisetales and Ophioglossales + Psilotales together as a monophyletic group and sister to the remaining ferns. Given that plastid genes generally evolve more slowly than the nuclear genes, these topological differences may be due to different numbers of phylogenetically informative sites contained within the diverse molecular data.

The relationships between the different horsetail species are now well resolved with *Equisetum bogotense* basal to both the subgenera *Hippochaete* and *Equisetum* that each contain seven species (Guillon 2004, 2007), while the Equisetopsida are most likely sister to both the whisk ferns (Psilotales) and ophioglossoid ferns (Ophioglossales; Grewe et al. 2013). Both molecular dating studies (Des Marais et al. 2003; Pryer et al. 2004) and fossil evidence (Stewart and Rothwell 1993) indicate that extant horsetails diverged in the Early Cenozoic, not long after the Cretaceous-Paleogene boundary ~66 Mya. The phylogenetic relationships among many ferns have been studied through different methods, and at the broadest level, Borgstrom et al. (2011) results were congruent with previous studies (Des Marais et al. 2003; Grewe et al. 2013); however, the phylogenetic evolution of ferns poses several unanswered answers. Liu et al. (2021) carried out whole-chloroplast genome comparison among Polypodiaceae. The three newly obtained plastomes of Polypodiaceae (*N. ovatus, N. fortunei, and P. cuspidatus*) were compared with nine previously published plastomes representing three subfamilies of Polypodiaceae, i.e.,

Microsoroideae, Platycerioideae, and Drynarioideae. The Polypodiaceae plastomes appeared to be structurally similar to each other, showing a typical quadripartite structure consisting of two IRs separated by LSC and SSC. Gao et al. (2018) have shown that repetitive structures with a higher GC content contribute to increasing the thermal stability of the *Dryopteris fragrans* plastome and maintaining its structure in the face of thermal changes during millions of years of evolution including Cretaceous period when Angiosperms made their appearance (Schneider et al. 2004). Thus, speculate that these repeating structures with a high GC content may be one of the molecular foundations of the adaptation of Polypodiaceae to the environment, which also provides new insights for understanding the environmental adaptation mechanism of plants.

## 4.10 Conclusion

Almost all recognized extant fern families and nearly all monilophyte genera at the early diverging nodes have now been sampled in published molecular phylogenetic studies, with few exceptions. Further analyses of the fern chloroplast genomes should provide new insights into the plastid genome evolution. Phylogenomics based on chloroplast genomes has shown many advantages in plant phylogenetics in recent years. With more nuclear data becoming available recently, chloroplast phylogenomics can provide a framework for testing the impact of reticulate evolution in the early evolution of ferns. The examination of plastid genomic features, such as gene content, gene order, intron gain/loss, genome size, nucleotide composition, and codon usage, may also offer independent tests of hypotheses derived from analysis of DNA sequence data. Further plastome sequencing of marattioid ferns and early diverging leptosporangiate ferns will likely be necessary to solidify the sister relationship between these two lineages, but the position of *Equisetum* is unlikely to be resolvable with more plastome data. However, Grewe et al. (2013) concluded that Equisetopsida is sister to Psilotopsida.

## 4.11 Prospects

There are structural variations, gene contents, and GC contents of the chloroplast genomes from green algae to flowering plants (Kwon et al. 2020). The diversity of plastome structures in ferns is insufficiently explored (Logacheva et al. 2016). They further suggested that the ancient HGT DNA transfer from mitochondrial to plastid genome occurred in a common ancestor of ferns which is an evolutionary event that can affect plastome structure. Wilhelmsson et al. (2017) updated rule sets for domain-based classification of transcription-associated proteins (TAPs), comprising transcription factors and transcriptional regulators. Kwon et al. (2020) also presented and corrected some false annotations on the introns in protein-coding and tRNA genes in the genome database, which might be confirmed by the chloroplast transcriptome analysis in the future.

RNA-editing sites should be corrected when plastid or mitochondrial genes and can be used for phylogenetic studies, particularly in those lineages with abundant organellar RNA-editing sites, e.g., hornworts, quillworts, spike mosses, and some seed plants (Du et al. 2020). Expansion or reduction or deletion of IRs resulted in the length variation of the plastomes. Ribosomal RNA genes, rrn, were located in the IRs so that they were present in a duplicate except of the species that had lost one of the IRs. The plastid introns are long compared with the nuclear introns, which might be related with the spliceosome nuclear introns and self-splicing group II plastid introns (Kwon et al. 2020). There were many annotation artifacts in the intron positions in the NCBI database. Fauskee et al. (2021) reported that RNA-editing sites among the three species *Adiantum* (Pteridaceae), *A. shastense*, *A. aleuticum*, and *A. capillus-veneris* showed a higher degree of conservation, with reverse (U-to-C) editing sites than forward (C-to-U) sites and sites involving start and stop codons were highly conserved. In contrast to this, in seed plants, RNA editing most commonly involves C-to-U changes (Chen et al. 2011). Further studies are needed to study the role of RNA editing in plant evolution.

**Acknowledgments** This chapter is dedicated to my (AK) teacher Dr. K. K. Sharma for introducing the subject of pteridophytes now "monilophytes" during master's study at the University of Rajasthan, Jaipur. We are also thankful to Professor S.K. Verma and Professor S.P. Khullar noted pteridologists for the support and during the preparation of this MS. Authors acknowledge with thanks the authors of publications quoted here:

- Table 4.1: Rothfels, C.J., et al. (2015). The evolutionary history of ferns inferred from 25 low-copy nuclear genes. American Journal of Botany, 102: 1089-1107. https://doi.org/10.3732/ajb.1500089. Reproduced with license number 5087470694801 dated 14 June 2021.
- Figure 4.1: Lu, J.-M. et al. (2015). Chloroplast phylogenomics resolves key relationships in ferns. Jnl of Systematics Evolution, 53: 448–457. https://doi.org/10.1111/jse.12180. Reproduced with license number 5085221071437 dated 10 June 2021
- Figure 4.2: Shim, Hyeonah et al. (2021) "Plastid Genomes of the Early Vascular Plant Genus *Selaginella* Have Unusual Direct Repeat Structures and Drastically Reduced Gene Numbers" *Int. J. Mol. Sci.* 22, no. 2: 641. https://doi.org/10.3390/ijms22020641. An open-access article distributed under the terms of the Creative Commons CC BY license
- Figure 4.3: Grewe et al. (2013). Complete plastid genomes from *Ophioglossum californicum*, *Psilotum nudum*, and *Equisetum hyemale* reveal an ancestral land plant genome structure and resolve the position of Equisetales among monilophytes. *BMC Evol Biol* **13,** 8. https://doi.org/10.1186/1471-2148-13-8
- Figure 4.4: Fan R, Ma W, Liu S, Huang Q. Integrated analysis of three newly sequenced fern chloroplast genomes: Genome structure and comparative analysis. Ecol Evol. 2021 Mar 18;11 (9):4550–4563. doi: https://doi.org/10.1002/ece3.7350. PMID: 33976830; PMCID: PMC8093657
- Figure 4.5: Fan R, Ma W, Liu S, Huang Q. Integrated analysis of three newly sequenced fern chloroplast genomes: Genome structure and comparative analysis. Ecol Evol. 2021 Mar 18;11 (9):4550–4563. doi: https://doi.org/10.1002/ece3.7350. This is an open-access article distributed under the terms of the Creative Commons CC BY license
- Figure 4.6: Li, FW., Brouwer, P., Carretero-Paulet, L. et al. Fern genomes elucidate land plant evolution and cyanobacterial symbioses. *Nature Plants* **4,** 460–472 (2018). https://doi.org/10.1038/s41477-018-0188-8. This is an open-access article distributed under the terms of the Creative Commons CC BY license
- Figure 4.7: Liu S, et al. (2021). Comparative genomic analysis of Polypodiaceae chloroplasts reveals fine structural features and dynamic insertion sequences. BMC Plant Biol. 2021 Jan 7;21

(1):31. doi: https://doi.org/10.1186/s12870-020-02800-x. This article is licensed under a Creative Commons Attribution 4.0 International License. http://creativecommons.org/licenses/by/4.0/. The Creative Commons Public Domain

- Figure 4.8: Xu, L., Xing, Y., Wang, B. et al. Plastid genome and composition analysis of two medical ferns: *Dryopteris crassirhizoma* Nakai and *Osmunda japonica* Thunb. *Chin Med* **14,** 9 (2019). https://doi.org/10.1186/s13020-019-0230-4. This article is distributed under the terms of the Creative Commons Attribution 4.0 International License (http://creativecommons.org/licenses/by/4.0/)
- Figure 4.9: Liu et al. (2021). Comparative genomic analysis of Polypodiaceae chloroplasts reveals fine structural features and dynamic insertion sequences. *BMC Plant Biol* 21, 31 (2021). The Creative Commons Public Domain Dedication waiver (http://creativecommons.org/publicdomain/zero/1.0/)
- Figure 4.10: Li, F. W., Kuo, L. Y., Pryer, K. M., & Rothfels, C. J. (2016). Genes Translocated into the Plastid Inverted Repeat Show Decelerated Substitution Rates and Elevated GC Content. *Genome biology and evolution*, *8*(8), 2452–2458. https://doi.org/10.1093/gbe/evw167. Reproduced with license number 5101790146234 dated 4 July 2021
- Figure 4.11: Liu et al. (2020) depicted sizes of each part of 16 fern complete plastome sequences. Liu, S., Wang, Z., Wang, H. et al. Patterns and Rates of Plastid *rps*12 Gene Evolution Inferred in a Phylogenetic Context using Plastomic Data of Ferns. *Sci Rep* **10,** 9394 (2020). https://doi.org/10.1038/s41598-020-66219-y. This is an open-access article distributed under the terms of the Creative Commons CC BY license
- Figure 4.12: Liu, S., Wang, Z., Wang, H. et al. (2020). Patterns and Rates of Plastid *rps*12 Gene Evolution Inferred in a Phylogenetic Context using Plastomic Data of Ferns. *Sci Rep* **10,** 9394. https://doi.org/10.1038/s41598-020-66219-y. This is an open-access article distributed under the terms of the Creative Commons CC BY license

AK is thankful to his wife Professor Dr. Vijay Rani Kumar Retd Professor of Botany for critically examining the MS and valuable suggestions.

## References

Adams KL, Wendel JF (2005) Polyploidy and genome evolution in plants. Curr Opin Plant Biol 8(2):135–141

Alanazi KM, Ali MA, Kim SY, Rahman MO, Farah MA, Alhemaid F, Lee J (2021) The cp genome characterization of Adenium obesum: gene content, repeat organization and phylogeny. Saudi J Biol Sci 28(7):3768-3775. https://doi.org/10.1016/j.sjbs.2021.03.048

Altschul SF, Gish W, Miller W, Myers EW, Lipman DJ (1990) Basic local alignment search tool. J Mol Biol 215(3):403–410. https://doi.org/10.1016/S0022-2836(05)80360-2

Asaf S, Khan AL, Khan MA, Waqas M, Kang SM, Yun BW, Lee IJJSR (2017) Chloroplast genomes of Arabidopsis halleri ssp. gemmifera and Arabidopsis lyrata ssp. petraea: Structures and comparative analysis. Sci Rep 7(1):7556

Banks JA, Nishiyama T, Hasebe M, Bowman JL, Gribskov M, Depamphilis CW, Albert VA, Aono N, Aoyama T, Ambrose BA et al (2011) The Selaginella genome identifies genetic changes associated with the evolution of vascular plants. Science 332:960–963

Bausher MG, Singh ND, Lee SB, Jansen RK, Daniell H (2006) The complete chloroplast genome sequence of *Citrus sinensis* (L.) Osbeck var 'Ridge Pineapple': organization and phylogenetic relationships to other angiosperms. BMC Plant Biol 6:21

Benne R, Burg JVD, Brakenhoff JPJ, Sloof P, Boom JHV et al (1986) Major transcript of the frameshifted coxII gene from trypanosome mitochondria contains four nucleotides that are not encoded in the DNA. Cell 46:819–826

Ben-Shlomo H, Levitan A, Shay NE, Goncharov I, Michaeli S (1999) RNA editing associated with the generation of two distinct conformations of the trypanosomatid Leptomonas collosoma 7SL RNA. J Biol Chem 274(36):25642–25650. https://doi.org/10.1074/jbc.274.36.25642

Bock R (2007) Structure, function, and inheritance of plastid genomes. In: Bock R (ed) Cell and molecular biology of plastids. Springer, Berlin, pp 29–63

Borgstrom E, Lundin S, Lundeberg J (2011) Large scale library generation for high throughput sequencing. PLoS One 6(4):e19119

Bromham L, Penny D (2003) The modern molecular clock. Nat Rev Genet 4:216–224

Chen H, Deng L, Jiang Y, Lu P, Yu J (2011) RNA editing sites exist in protein-coding genes in the chloroplast genome of Cycas taitungensis. J Integr Plant Biol 53(12):961–970. https://doi.org/10.1111/j.1744-7909.2011.01082.x

Chen CW, Nagan LT, Hidayat A (2014) Biodiversity, evolution and biogeography of plants. Blumea 59(1):49–58

Christenhusz MJ, Chase MW (2014) Trends and concepts in fern classification. Ann Bot 113(4): 571–594. https://doi.org/10.1093/aob/mct299

Christenhusz MJM, Zhang X, Schneider H (2011) A linear sequence of extant lycophytes and ferns. Phytotaxa 19:7–54

Clark J, Hidalgo O, Pellicer J, Liu H, Marquardt J, Robert Y, Christenhusz M, Zhang S, Gibby M, Leitch IJ, Schneider H (2016) Genome evolution of ferns: evidence for relative stasis of genome size across the fern phylogeny. New Phytol 210(3):1072–1082. https://doi.org/10.1111/nph.13833

Cui L, Wall PK, Leebens-Mack JH et al (2006) Widespread genome duplications throughout the history of flowering plants. Genome Res 16:738–749

Daniell H, Lin CS, Yu M et al (2016) Chloroplast genomes: diversity, evolution, and applications in genetic engineering. Genome Biol 17:134. https://doi.org/10.1186/s13059-016-1004-2

De Bodt S, Maere S, Van de Peer Y (2005) Genome duplication and the origin of angiosperms. Trends Ecol Evol 20:591–597

Der JP (2010) Genome perspectives on evolution in bracken fern. PhD thesis, Utah: Utah State University

Des Marais DL, Smith AR, Britton DM, Pryer KM (2003) Phylogenetic relationships and evolution of extant horsetails, equisetum, based on cp DNA sequence data (*rbcL* and *trnL-F*). Int J Plant Sci 164:737–751

Downie SR, Palmer JD (1992) The use of chloroplast DNA rearrangements in reconstructing plant phylogeny. In: Soltis PS, Soltis DE, Doyle JJ (eds) Molecular systematics of plants. Chapman and Hall, New York, NY, pp 14–35

Du XY, Lu JM, Li DZ (2020) Extreme plastid RNA editing may confound phylogenetic reconstruction: a case study of *Selaginella* (lycophytes). Plant Divers 42(5):356–361. https://doi.org/10.1016/j.pld.2020.06.009

Eneas-Filho J, Hartley MR, Mache R (1981) Pea chloroplast ribosomal proteins: characterization and site of synthesis. Mol Gen Genet 184:484–488. https://doi.org/10.1007/BF00352527

Fan R, Ma W, Liu S, Huang Q (2021) Integrated analysis of three newly sequenced fern chloroplast genomes: genome structure and comparative analysis. Ecol Evol 11(9):4550–4563. https://doi.org/10.1002/ece3.7350

Fauskee BD, Sigel EM, Pryer KM, Grusz AL (2021) Variation in frequency of plastid RNA editing within *Adiantum* implies rapid evolution in fern plastomes. Am J Bot 108(5):820–827

Gaeta RT, Pires JC, Iniguez-Luy F, Leon E, Osborn TC (2007) Genomic changes in resynthesized Brassica napus and their effect on gene expression and phenotype. Plant Cell 19:3403–3417

Gao L, Yi X, Yang Y, Su Y, Wang T (2009) Complete chloroplast genome sequence of a tree fern *Alsophila spinulosa*: insights into evolutionary changes in fern cp genomes. BMC Evol Biol 9: 130

Gao L, Wang B, Wang ZW, Zhou Y, Su YJ, Wang T (2013) Plastome sequences of *Lygodium japonicum* and *Marsilea crenata* reveal the genome organization transformation from basal ferns to core leptosporangiates. Genome Biol Evol 5:1403–1407. https://doi.org/10.1093/gbe/evt099

Gastony GJ, Yatskievych G (1992) Maternal inheritance of the chloroplast and mitochondrial genomes in cheilanthoid ferns. Am J Bot 79:716–722

Gott JM, Emeson RB (2000) Functions and mechanisms of RNA editing. Annum Rev Genet 34: 499–531

Grewe F, Guo W, Gubbels EA, Hansen AK, Mower JP (2013) Complete plastid genomes from Ophioglossum californicum, Psilotum nudum, and Equisetum hyemale reveal an ancestral land plant genome structure and resolve the position of Equisetales among monilophytes. BMC Evol Biol 13:8. https://doi.org/10.1186/1471-2148-13-8

Grosche C, Funk HT, Maier UG, Zauner S (2012) The chloroplast genome of Pellia endiviifolia: gene content, RNA-editing pattern, and the origin of chloroplast editing. Genome Biol Evol 4(12):1349–1357. https://doi.org/10.1093/gbe/evs114

Guichoux E, Lagache L, Wagner S, Chaumeil P, Léger P, Lepais O et al (2011) Current trends in microsatellite genotyping. Mol Ecol Resour 11:591–611

Guillon JM (2004) Phylogeny of horsetails (equisetum) based on the chloroplast rps4 gene and adjacent noncoding sequences. Syst Bot 29:251–259

Guillon JM (2007) Molecular phylogeny of horsetails (equisetum) including chloroplast atpB sequences. J Plant Res 120:569–574

Guillon JM, Raquin C (2000) Maternal inheritance of chloroplasts in the horsetail Equisetum variegatum (Schleich.). Curr Genet 37:53–56

Guisinger MM, Kuehl JV, Boore JL, Jansen RK (2011) Extreme reconfiguration of plastid genomes in the angiosperm family Geraniaceae: rearrangements, repeats, and codon usage. Mol Biol Evol 28:583–600

Guo ZY, Zhang HR, Shrestha N, Zhang XC (2016) Complete chloroplast genome of a valuable medicinal plant, *Huperzia serrata* (Lycopodiaceae), and comparison with its congener. Appl in plant Sci 4(11):apps.1600071. https://doi.org/10.3732/apps.1600071

Gao R, Wang W, Huang Q, Fan R, Wang X, Feng P et al (2018) Complete chloroplast genome sequence of Dryopteris fragrans (L.) Schott and the repeat structures against the thermal environment. Sci Rep 8:1–11

Hasebe M, Iwatsuki K (1992) Gene localization on the chloroplast DNA of the maiden hair fern; *Adiantum capillus-veneris*. Bot Mag (Tokyo) 105:413–419

Hasebe M, Wolf PG, Pryer KM, Ueda K, Ito M, Sano R, Gastony GJ et al (1995) Fern phylogeny based on rbcL nucleotide sequences. Am Fern J 85:134–181

Haufler CH (1987) Electrophoresis is modifying our concepts of evolution in homosporous pteridophytes. Am J Bot 74:953–966

Haufler CH (2002) Homospory 2002: an odyssey of progress in pteridophyte genetics and evolutionary biology. Bio Sci 52:1081–1093

Haufler CH (2014) Ever since Klekowski: testing a set of radical hypotheses revives the genetics of ferns and lycophytes. Am J Bot 101:2036–2042

Hendy MD, Penny DA (1989) Framework for the quantitative study of evolutionary trees. Syst Zool 38:297–309

Hidalgo O, Pellicer J, Christenhusz M, Schneider H, Leitch AR, Leitch IJ (2017) Is there an upper limit to genome size. Trends Plant Sci 22:567–573

IPPG (2016) A community-derived classification for extant lycophytes and ferns. J Syst and Evol 54(6):563–603. https://doi.org/10.1111/jse.12229

Jiang Y, Fan SL, Song MZ, Yu JN, Yu SX (2012) Identification of RNA editing sites in cotton (Gossypium hirsutum) chloroplasts and editing events that affect secondary and three-dimensional protein structures. Genet Mol Res 11(2):987–1001. https://doi.org/10.4238/2012

Jiang Q, Schneider H, Liu H (2019) Complete chloroplast genome of *Angiopteris yunnanensis* (Marattiaceae). Mitochondrial DNA B Resour 4(2):3912–3913. https://doi.org/10.1080/23802359.2019.1688111

Karol KG, Arumuganathan K, Boore JL, Duffy AM, Everett KD, Hall JD, Hansen SK, Kuehl JV, Mandoli DF, Mishler BD, Olmstead RG, Renzaglia KS, Wolf PG (2010) Complete plastome sequences of Equisetum arvense and Isoetes flaccida: implications for phylogeny and plastid genome evolution of early land plant lineages. BMC Evol Biol 10:321. https://doi.org/10.1186/1471-2148-10-321

Kenrick P, Crane PR (1997) The origin and early evolution of plants on land. Nature 389:33–39

Kim HT, Chung MG, Kim KJ (2014) Chloroplast genome evolution in early diverged leptosporangiate ferns. Mol Cells 37(5):372–382. https://doi.org/10.14348/molcells.2014.2296

Knie N, Fischer S, Grewe F, Polsakiewicz M, Knoop V (2015) Horsetails are the sister group to all other monilophytes and Marattiales are sister to leptosporangiate ferns. Mol Phylogenet Evol 90:140–149. https://doi.org/10.1016/j.ympev.2015.05.008

Korall P, Schuettpelz E, Pryer KM (2010) Abrupt deceleration of molecular evolution linked to the origin of arborescence in ferns. Evolution 64:2786–2792

Kuo L-Y, Li F-W, Chiou W-L, Wang C-N (2011) First insights into fern matK phylogeny. Mol Phylogenet Evol 59:556–566

Kwon EC, Kim JH, Kim NS (2020) Comprehensive genomic analyses with 115 plastomes from algae to seed plants: structure, gene contents, GC contents, and introns. Genes Genomics 42(5): 553–570. https://doi.org/10.1007/s13258-020-00923-x

Labiak PH, Karol KG (2017) Plastome sequences of an ancient fern lineage reveal remarkable changes in gene content and architecture. Am J Bot 104:1008–1018. https://doi.org/10.3732/ajb.1700135

Landis JB, Soltis DE, Li Z, Marx HE, Barker MS, Tank DC, Soltis PS (2018) Impact of whole-genome duplication events on diversification rates in angiosperms. Am J Bot 105:348–363. https://doi.org/10.1002/ajb2.1060

Lanfear R, Welch JJ, Bromham L (2010) Watching the clock: studying variation in rates of molecular evolution between species. Trends Ecol Evol 25:495–503

Lang D, Weiche B, Timmerhaus G, Richardt S, Riaño-Pachón DM, Corrêa LG, Reski R, Mueller-Roeber B, Rensing SA (2010) Genome-wide phylogenetic comparative analysis of plant transcriptional regulation: a timeline of loss, gain, expansion, and correlation with complexity. Genome Biol Evol 2:488–503. https://doi.org/10.1093/gbe/evq032

Langmead B, Nellore A (2018) Cloud computing for genomic data analysis and collaboration. Nat Rev Genet 19:208–219. https://doi.org/10.1038/nrg.2017.113

Lehtonen S (2011) Towards resolving the complete Fern tree of life. PLoS One 6(10):e24851. https://doi.org/10.1371/journal.pone.0024851

Levin DA (1975) Minority cytotype exclusion in local plant populations. Taxon 24:35–43

Levin DA (2002) The role of chromosomal change in plant evolution. Oxford University Press, New York

Li YC, Korol AB, Fahima T, Beiles A, Nevo E (2002) Microsatellites: genomic distribution, putative functions and mutational mechanisms: a review. Mol Ecol 11:2453–2465

Li YC, Korol AB, Fahima T, Nevo E (2004) Microsatellites within genes: structure, function, and evolution. Mol Biol Evol 6:991–1007. https://doi.org/10.1093/molbev/msh073

Li FW, Kuo LY, Pryer KM, Rothfels CJ (2016) Genes translocated into the plastid inverted repeat show decelerated substitution rates and elevated GC content. Genome Biol Evol 8:2452–2458. https://doi.org/10.1093/gbe/evw167

Li FW, Brouwer P, Carretero-Paulet L et al (2018) Fern genomes elucidate land plant evolution and cyanobacterial symbioses. Nat Plants 4:460–472. https://doi.org/10.1038/s41477-018-0188-8

Liu HM, Schneider H, Yu Y, Fujiwara T, Khine PK (2019) Towards the conservation of the Mesozoic relict fern *Christensenia*—a fern species with extremely small populations in China. J Plant Res 132(5):601–616

Liu S, Wang Z, Wang H et al (2020) Patterns and rates of plastid *rps*12 gene evolution inferred in a phylogenetic context using Plastomic data of ferns. Sci Rep 10:9394. https://doi.org/10.1038/s41598-020-66219-y

Liu S, Wang Z, Su Y et al (2021) Comparative genomic analysis of Polypodiaceae chloroplasts reveals fine structural features and dynamic insertion sequences. BMC Plant Biol 21:31. https://doi.org/10.1186/s12870-020-02800-x

Logacheva MD, Krinitsina AA, Belenikin MS et al (2017) Comparative analysis of inverted repeats of polypod fern (Polypodiales) plastomes reveals two hypervariable regions. BMC Plant Biol 17:255. https://doi.org/10.1186/s12870-017-1195-z

Lu J-M, Zhang N, Du X-Y, Wen J, Li D-Z (2015) Chloroplast phylogenomics resolves key relationships in ferns. J Syst Evol 53:448–457

Marchant DB, Sessa EB, Wolf PG, Heo K, Barbazuk WB, Soltis PS, Soltis DE (2019) The C-Fern (*Ceratopteris richardii*) genome: insights into plant genome evolution with the first partial homosporous fern genome assembly. Sci Rep 9(1):18181. https://doi.org/10.1038/s41598-019-53968-8

Millen RS, Olmstead RG, Adams KL, Palmer JD, Lao NT, Heggie L et al (2001) Many parallel losses of infA from chloroplast DNA during angiosperm evolution with multiple independent transfers to the nucleus. Plant Cell 13:645–658

Munekage Y, Hashimoto M, Miyake C, Tomizawa K-I, Endo T, Tasaka M, Shikanai T (2004) Cyclic electron flow around photosystem I is essential for photosynthesis. Nature 429:579–582

Nakazato T, Jung MK, Housworth EA, Rieseberg LH, Gastony GJ (2006) Genetic map-based analysis of genome structure in the homosporous fern Ceratopteris richardii. Genetics 173(3): 1585–1597. https://doi.org/10.1534/genetics.106.055624

Neuhaus HE, Emes MJ (2000) Nonphoto synthetic metabolism in plastids. Annu Rev Plant Physiol Plant Mol Biol 51:111–140

Nock CJ, Baten A, King GJ (2014) Complete chloroplast genome of Macadamia integrifolia confirms the position of the Gondwanan early-diverging eudicot family Proteaceae. BMC Genomics 15(S9):S13. https://doi.org/10.1186/1471-2164-15-S9-S13

Obermayer R, Leitch IJ, Hanson L, Bennett MD (2002) Nuclear DNA C-values in 30 species double the familial representation in pteridophytes. Ann Bot 90:209–217

Olejniczak SA, Lojewska E, Kowalczyk T, Sakowicz T (2016) Chloroplasts: state of research and practical applications of plastome sequencing. Planta 244(3):517–527

Olmstead RG, Palmer JD (1994) Chloroplast DNA systematics – a review of methods and data analysis. Am J Bot 81:1205–1224. https://doi.org/10.2307/2445483

Palmer JD, Stein DB (1982) Chloroplast DNA from the fern *Osmunda cinnamomea*: physical organization, gene localization and comparison to angiosperm. Curr Genet 5:165–170. https://doi.org/10.1007/BF00391801

Parisod C, Holderegger R, Brochmann C (2010) Evolutionary consequences of autopolyploidy. New Phytol 186:5–17

Peltier G, Cournac L (2002) Chlororespiration. Annu Rev Plant Biol 53:523–550

Peng Y, Zhang S, Shanshan D, Peng T (2020) The complete chloroplast genome of the fern *Asplenium tenerum* (Aspleniaceae). Mitochondrial DNA B 5(3):3743–3374. https://doi.org/10.1080/23802359.2020.1831985

Powell W, Michele M, Chaz A, Michael H, Julie V, Scott T et al (1996) The comparison of RFLP, RAPD, AFLP and SSR (microsatellite) markers for germplasm analysis. Mol Breed 2:225–238. https://doi.org/10.1007/BF00564200

Pryer KM, Smith AR, Skog JE (1995) Phylogenetic relationships of extant pteridophytes based on evidence from morphology and rbcL sequences. Am Fern J 85:205–282

Pryer KM, Schneider H, Smith AR, Cranfill R, Wolf PG, Hunt JS, Sipes SD (2001) Horsetails and ferns are a monophyletic group and the closest living relatives to seed plants. Nature 409:618–622

Pryer KM, Schuettpelz E, Wolf PG, Schneider H, Smith AR, Cranfill R (2004) Phylogeny and evolution of ferns (Monilophytes) with a focus on the early leptosporangiate divergences. Am J Bot 91:1582–1598. https://doi.org/10.3732/ajb.91.10.1582

Qiu YL et al (2006) The deepest divergences in land plants inferred from phylogenomic evidence. Proc Natl Acad Sci U S A 103:15511–15516

Qiu YL et al (2007) A non-flowering land plant phylogeny inferred from nucleotide sequences of seven chloroplast, mitochondrial and nuclear genes. Int J Plant Sci 168:691–708

Qu XJ, Moore MJ, Li DZ, Yi TS (2019) PGA: a software package for rapid, accurate, and flexible batch annotation of plastomes. Plant Methods. 15:50. https://doi.org/10.1186/s13007-019-0435-7

Rai HS, Graham SW (2010) Utility of a large, multigene plastid data set in inferring higher-order relationships in ferns and relatives (monilophytes). Am J Bot 97:1444–1456. https://doi.org/10.3732/ajb.0900305

Raubeson LA, Jansen RK (2005) Chloroplast genomes of plants. In: Henry RJ (ed) Plant diversity and evolution: genotypic and phenotypic variation in higher plants. CABI Publishing, London, pp 45–68

Raubeson LA, Peery R, Chumley TW, Dziubek C, Fourcade HM, Boore JL, Jansen RK (2007) Comparative chloroplast genomics: analyses including new sequences from the angiosperms Nuphar advena and Ranunculus macranthus. BMC Genomics 8(1):174

Robison TA, Wolf PG (2019) Refernment: an R package for annotating RNA editing in plastid genomes. Appl Plant Sci. 7(2):e01216. https://doi.org/10.1002/aps3.1216

Robison TA, Grusz AL, Wolf PG, Mower JP, Fauskee BD, Sosa K, Schuettpelz E (2018) Mobile elements shape Plastome evolution in ferns. Genome Biol and Evol 10(10):2558–2571. https://doi.org/10.1093/gbe/evy189

Roper JM, Kellon Hansen S, Wolf PG, Karol KG, Mandoli DF, Everett KDE, Kuehl J, Boore JL (2007) The complete plastid genome sequence of Angiopteris evecta (G. Forst.) Hoffm. (Marattiaceae). Am Fern J 97(2):95–106

Rothfels CJ, Schuettpelz E (2014) Accelerated rate of molecular evolution for vittarioid ferns is strong and not driven by selection. Syst Biol 63:31–54

Rothfels CJ, Sundue MA, Kuo LY, Larsson A, Kato M, Schuettpelz E, Pryer KM (2012) Overcoming deep roots, fast rates, and short internodes to resolve the ancient rapid radiation of eupolypod II ferns. Syst Biol 61:490–509

Rothfels CJ, Li F-W, Sigel EM, Huiet L, Larsson A, Burge DO, Ruhsam M, Deyholos M, Soltis DE, Stewart CN, Jr Shaw SW, Pokorny L, Chen T, de Pamphilis C, DeGironimo L, Chen L, Wei X, Sun X, Korall P, Stevenson DW, Graham SW, Wong GK-S, Pryer KM (2015) The evolutionary history of ferns inferred from 25 low-copy nuclear genes. Am J Bot 102:1089–1107. https://doi.org/10.3732/ajb.1500089

Ruhfel BR, Gitzendanner MA, Soltis PS et al (2014) From algae to angiosperms–inferring the phylogeny of green plants (*Viridiplantae*) from 360 plastid genomes. BMC Evol Biol 14:23. https://doi.org/10.1186/1471-2148-14-23

Ruhlman TA, Jansen RK (2014) The plastid genomes of flowering plants. Methods Mol Biol 1132: 3–38. https://doi.org/10.1007/978-1-62703-995-6-1

Ruiz-Ruano FJ, Navarro-Domínguez B, Camacho JPM, Garrido-Ramos MA (2019) Full plastome sequence of the fern Vandenboschia speciosa (Hymenophyllales): structural singularities and evolutionary insights. J Plant Res 132(1):3–17. https://doi.org/10.1007/s10265-018-1077-y

Sessa EB, Banks JA, Barker MS et al (2014) Between Two Fern Genomes. GigaSci 3:15. https://doi.org/10.1186/2047-217X-3-15

Schlötterer C, Wiehe T (1999) Microsatellites, a neutral marker to infer selective sweeps. In: Goldstein DB, Schlötterer C (eds) Microsatellites: evolution and applications. Oxford University Press, Oxford, pp 238–247

Schneider H, Schuettpelz E, Pryer KM, Cranfill R, Magallón S, Lupia R (2004) Ferns diversified in the shadow of angiosperms. Nature 428:553–557

Schuettpelz E, Pryer KM (2007) Fern phylogeny inferred from 400 leptosporangiate species and three plastid genes. Taxon 56:1037–1050

Schuettpelz E, Pryer KM (2009) Evidence for a Cenozoic radiation of ferns in an angiosperm-dominated canopy. Proc Natl Acad Sci U S A 106:11200–11205

Schuettpelz E, Korall P, Pryer KM (2006) Plastid atpA data provide improved support for deep relationships among ferns. Taxon 55:897–906

Sessa EB, Der JP (2016) Evolutionary genomics of ferns and lycophytes. Adv Bot Res 78:215–254

Shikanai T (2006) RNA editing in plant organelles: machinery, physiological function and evolution. Cell Mol Life Sci 63(6):698–708. https://doi.org/10.1007/s00018-005-5449-9

Shim H, Lee HJ, Lee J, Lee HO, Kim JH, Yang TJ, Kim NS (2021) Plastid genomes of the early vascular plant genus *Selaginella* have unusual direct repeat structures and drastically reduced gene numbers. Int J Mol Sci 22(2):641. https://doi.org/10.3390/ijms22020641

Shmakov AA (2016) Community-derived classification for extant lycophytes and ferns. J Syst Evol 54:563–603

Smith DR (2009) Unparalleled GC content in the plastid DNA of Selaginella. Plant Mol Biol 71: 627–639

Soltis DE, Soltis PS (1987) Polyploidy and breeding systems in homosporous pteridophyta - a reevaluation. Am Nat 130:219–232

Soltis DE, Soltis PM (1998) Choosing an approach and an appropriate gene for phylogenetic analysis. In: Soltis DE, Soltis PM, Doyle J (eds) Molecular systematics of plants II. Kluwer, Boston, pp 1–42

Soltis PS, Soltis DE, Chase MW (1999) Angiosperm phylogeny inferred from multiple genes as a tool for comparative biology. Nature 402:402–404

Stebbins GL (1971) Chromosomal evolution in higher plants. Edward Arnold, London

Stewart WN, Rothwell GW (1993) Paleobotany and the evolution of plants. Cambridge University Press, Cambridge

Sun M, Li J, Li D, Shi L (2017) Complete chloroplast genome sequence of the medical fern *Drynaria roosii* and its phylogenetic analysis. Mitochondrial DNA B Resour 2(1):7–8. https://doi.org/10.1080/23802359.2016.1275835

Szövényi P, Gunadi A, Li FW (2021) Charting the genomic landscape of seed-free plants. Nat Plants 7:554–565. https://doi.org/10.1038/s41477-021-00888-z

Tangphatsornruang S, Sangsrakru D, Chanprasert J, Uthaipaisanwong P, Yoocha T, Jomchai N, Tragoonrung S (2010) The chloroplast genome sequence of mungbean (Vigna radiata) determined by high-throughput pyrosequencing: structural organization and phylogenetic relationships. DNA Res 17:11–22

Tsuji S, Ueda K, Nishiyama T, Hasebe M, Yoshikawa S, Konagaya A, Nishiuchi T, Yamaguchi K (2007) The chloroplast genome from a lycophyte (microphyllophyte), *Selaginella uncinata*, has a unique inversion, transpositions and many gene losses. J Plant Res 120:281–290

Vanneste K, Sterck L, Myburg AA, Van de Peer Y, Mizrachi E (2015) Horsetails are ancient polyploids: evidence from *Equisetum giganteum*. Plant Cell 27(6):1567–1578. https://doi.org/10.1105/tpc.15.00157

Wei R, Yan YH, Harris AJ, Kang JS, Shen H, Xiang QP, Zhang XC (2017) Plastid Phylogenomics resolve deep relationships among Eupolypod II ferns with rapid radiation and rate heterogeneity. Genome Biol Evol 9(6):1646. https://doi.org/10.1093/gbe/evx107

Weststrand S, Korall P (2016a) Phylogeny of Selaginellaceae: there is value in morphology after all! Am J Bot 103:2136–2159

Weststrand S, Korall P (2016b) A subgeneric classification of Selaginella (Selaginellaceae). Am J Bot 103:2160–2169

Wicke S, Schneeweiss GM, dePamphilis CW et al (2011) The evolution of the plastid chromosome in land plants: gene content, gene order, gene function. Plant Mol Biol 76:273–297. https://doi.org/10.1007/s11103-011-9762-4

Wickett NJ, Mirarab S, Nguyen N, Warnow T, Carpenter E, Matasci N, Ayyampalayam S, Barker MS, Burleigh JG, Gitzendanner MA, Ruhfel BR, Wafula E, Der JP, Graham SW, Mathews S, Melkonian M, Soltis DE, Soltis PS, Miles NW, Rothfels CJ, Pokorny L, Shaw AJ, DeGironimo L, Stevenson DW, Surek B, Villarreal JC, Roure B, Philippe H, dePamphilis CW, Chen T, Deyholos MK, Baucom RS, Kutchan TM, Augustin MM, Wang J, Zhang Y, Tian Z, Yan Z, Wu X, Sun X, Wong GK, Leebens-Mack J (2014) Phylotranscriptomic analysis of the origin and early diversification of land plants. Proc Natl Acad Sci U S A 111(45):E4859–E4868. https://doi.org/10.1073/pnas.1323926111

Wikström N, Pryer KM (2005) Incongruence between primary sequence data and the distribution of a mitochondrial atp1 group II intron among ferns and horsetails. Mol Phylogenet Evol 36:484–493

Wilhelmsson P, Mühlich C, Ullrich KK, Rensing SA (2017) Comprehensive genome-wide classification reveals that many plant-specific transcription factors evolved in Streptophyte algae. Genome Biol Evol 9(12):3384–3397. https://doi.org/10.1093/gbe/evx258

Wolf PG, Rowe CA, Sinclair RB, Hasebe M (2003) Complete nucleotide sequence of the chloroplast genome from a Leptosporangiate Fern, *Adiantum capillus-veneris* L. DNA Res 10(2): 59–65. https://doi.org/10.1093/dnares/10.2.59

Wolf PG, Karol KG, Mandoli DF, Kuehl J, Arumuganathan K, Ellis MW, Mishler BD et al (2005) The first complete chloroplast genome sequence of a lycophyte, *Huperzia lucidula* (Lycopodiaceae). Gene 350:117–128

Wolf PG, Roper JM, Duffy AM (2010) The evolution of chloroplast genome structure in ferns. Genome 53:731–738. https://doi.org/10.1139/G10-061

Wood TE, Takebayashi N, Barker MS, Mayrose I, Greenspoon PB, Rieseberg LH (2009) The frequency of polyploid speciation in vascular plants. Proc Natl Acad Sci U S A 106:13875–13879

Xu JH, Liu Q, Hu W, Wang T, Xue Q, Messing J (2015) Dynamics of chloroplast genomes in green plants. Genomics 106(4):221–231. https://doi.org/10.1016/j.ygeno.2015.07.004

Xu L, Xing Y, Wang B et al (2019) Plastid genome and composition analysis of two medical ferns: *Dryopteris crassirhizoma* Nakai and *Osmunda japonica* Thunb. Chin Med 14:9. https://doi.org/10.1186/s13020-019-0230-4

Yamaguchi K, Subramanian AR (2003) Proteomic identification of all plastid-specific ribosomal proteins in higher plant chloroplast 30S ribosomal subunit. Eur J Biochem 270:190–205. https://doi.org/10.1046/j.1432-1033.2003.03359.x

Yang JB, Yang SX, Li HT, Yang J, Li DZ (2013) Comparative chloroplast genomes of camellia species. PLoS One 8(8):e73053. https://doi.org/10.1371/journal.pone.0073053

Yi DK, Kim KJ (2012) Complete chloroplast genome sequences of important oilseed crop *Sesamum indicum* L. PLoS One 7:e35872

Yoshinaga K, Iinuma H, Masuzawa T, Uedal K (1996) Extensive RNA editing of U to C in addition to C to U substitution in the *rbcL* transcripts of hornwort chloroplasts and the origin of RNA editing in green plants. Nucleic Acids Res 24:1008–1014

Zhong B et al (2011) Systematic error in seed plant phylogenomics. Genome Biol Evol 3:1340–1348

Zhong B, Fong R, Collins LJ, McLenachan PA, Penny D (2014) Two new fern chloroplasts and decelerated evolution linked to the long generation time in tree ferns. Genome Biol Evol 6: 1166–1173

Zhou X-M, Zhang L (2015) A classification of Selaginella (Selaginellaceae) based on molecular (chloroplast and nuclear), macromorphological, and spore features. Taxon 64:1117–1140

Zhu A, Guo W, Gupta S, Fan W, Mower JP (2016) Evolutionary dynamics of the plastid inverted repeat: the effects of expansion, contraction, and loss on substitution rates. New Phytol 209: 1747–1756. https://doi.org/10.1111/nph.13743

Zwenger SR, Reinsvold RE, Basu C (2010) Analysis and functional annotation of expressed sequence tags from the diesel tree (*Copaifera officinalis*). Plant Biotechnol 27:385–391

**Ashwani Kumar** Alexander von Humboldt Fellow (Germany), MSc (Botany) (Gold Medalist) (1967)), PhD (1971), postdoc AvH Fellow (Germany) and JSPS fellow (Japan) Selected in the Indian Administrative Services (IAS: IPS) (1972) but he opted career in botany. 37 years of teaching & research experience. Guided 39 PhD. research students and published 320 research papers and 26 books. Elected fellow FBS, FPSI, FIMPS and FIFS. Editor in Chief IJLS. V. Puri medal (2008) CEE award (2015). President of CHEC (India), Indian Botanical Society, Society for Promotion of Plant Sciences. Professor Emeritus, Department of University of Rajasthan, Jaipur. Presently, Chair Professor at JVWU, Jaipur and Adj. Professor Modi University, Sikar.

**Priti Giri** is a research scholar in the Department of Botany, University of Delhi. She has completed M.Sc. and M.Phil. from the Department of Botany University of Delhi, Delhi. Her area of specialization is conservation and phytochemical and biochemical screening of ethnomedicinally important ferns (pteridophytes), and she has published many research paper and book chapters on her subject.

**Prem Lal Uniyal** is presently working as a professor in the Department of Botany, University of Delhi, and has been teaching for the last 27 years. He is specialized in the archegoniates, biodiversity assessment, conservation biology, development and reproductive biology of ferns, conifers, Podostemaceae, and biological activity of lichens.

He is a member of several scientific bodies such as the International Association of Bryologists, American Bryological and Lichenological Society, International Society of Plant Reproductive Biology, International Society of Plant Morphologists, Orchid Society of India, Indian Fern Society, etc. He has done numerous plant collection expeditions in the Himalaya and Western Ghats and conservation programs with indigenous societies in the Himalaya.

He has authored 4 books and published 105 research papers, 35 review articles, and 25 popular articles. He has supervised 16 Ph.D. scholars.

# Molecular Markers in Pteridophytes 5

Shaiesh Morajkar, C. Suneetha, T. G. Harish Kumar, and Smitha Hegde

**Abstract**

Identification and classification of lifeforms is an elementary step in knowledge generation and biodiversity management. Traditionally, taxonomical studies of variation and discrimination of species were established based on their morphological and anatomical characteristics. The advancement of molecular techniques has generated information on the genetic basis for diversity and variability in organisms. For several years, biologists have used various DNA-based fingerprinting techniques such as plastid and nuclear SSRs, RAPD, AFLP, DNA sequencing, and classical taxonomy tools to study population dynamics, species delimitation, hybridization, and phylogenetics. *Albeit* on ad hoc basis, the techniques are used to identify specimen with variations or atypical morphological characters. The need for their application in pteridophyte identification and discrimination is more pressing than any other plant group because pteridophytes have limited stark variations in morphological characters and high species diversity. Molecular marker studies have a wide range of applications in pteridophytes, such as evolutionary studies, homology identification, diversity analysis, breeding for trait improvement, and detection of adulteration in compounded forms of herbal extracts.

S. Morajkar
Goa State Wetland Authority, O/o Goa State Biodiversity Board, Bardez, Saligao, Goa, India

C. Suneetha · T. G. Harish Kumar
Department of Horticulture, University of Agricultural Sciences, Bangalore, India

S. Hegde (✉)
Nitte (Deemed to be University), Nitte University Center for Science Education and Research, Deralakatte, Mangalore, India
e-mail: smitha.hegde@nitte.edu.in



J. Marimuthu et al. (eds.), *Ferns*, https://doi.org/10.1007/978-981-16-6170-9_5

**Keywords**

DNA barcoding · Molecular marker · Pteridophytes

## 5.1 Introduction

Biological diversity is often described as the variations in diverse forms of life on earth consequential to the evolutionary process of millions of years. According to the Convention on Biological Diversity (CBD), biodiversity is the variability among living organisms from all sources, including terrestrial, marine, aquatic ecosystems, and ecological complexes. It encompasses diversity within species and between species. It also includes various ecosystems (CBD 1992). Biodiversity represents a wealth of systematic ecological data that aids to apprehend the natural world and its origins. Identifying and monitoring biological diversity is an enormous and potentially infinite assignment considering its variability in time, space, levels, and spectra. Biological diversity is dynamic and constantly evolving and changing in response to biotic and abiotic fluctuations; other environmental influences must document its status quo in time and space. Subsequently, it is crucial to monitor changes and identify and assess their impacts. Such impact studies may necessitate intervention measures to protect the future conservation of biological diversity and its sustainable use.

Species identification is a rudimentary step in the evaluation and discovery of biodiversity. It is critical in natural habitats to assess the biodiversity before changes occur, as the extant diversity is subject to extinction. It is predominantly relevant to tropical ecosystems, where the levels of endemism tend to be higher than in temperate regions, increasing the risks of species becoming globally extinct. The loss of biodiversity is an ongoing global issue that results in species extinction and threatens the ecosystem services and resources it provides. The threatened status and rate of extinction of plant species is far greater than any other groups of species (Pimm and Joppa 2015). In a review of studies on species' biodiversity by Pimm et al. (2014), more than 290,000 land plant species are expected to reach 400,000 species when disputes over currently unknown species are resolved.

Knowledge of the distribution of biodiversity is a requirement of conservation planning (Grantham et al. 2008) and identifying sites for their biodiversity value (Kukkala and Moilanen 2013). The limited availability of conservation resources means that measures cannot be applied to all areas, and instead, certain areas have to be prioritized for conservation action.

## 5.2 The Need for Molecular Information in Pteridophytes

Pteridophytes are the only nonflowering primitive vascular plant group having global distribution (Chapman 2009). This group of lower vascular plants that complete their life cycle in two phases, viz., sporophytic and gametophytic, are generally neglected (Shukla and Chakravarty 2012). Lack of conservation efforts of this group of plants has led to the addition of several fern species to the IUCN red list (García Criado et al. 2017), while some of them are on the verge of extinction in a recent decade.

In their sporophytic phase, ferns have distinct features such as foliage pattern, shape, size, exine of spores, sporangia patterns in mature fronds, the rhizomes, and chromosome number. Depending on these morphological characteristics, pteridophyte flora has been taxonomically treated, identified, and classified by numerous classical pteridologists for decades. Kholia and Fraser-Jenkins (2011) opine that conflicting species classification and identification makes scientific publication worthless. Discrimination and classification of fern sporophyte are difficult and problematic, but the identification of gametophytic stages is impossible using mere morphological data. Therefore, it is essential to supplement the classical taxonomic approach with molecular systematics to improve species discrimination and classification and hasten species identification and augment discovery to aid conservation strategy planning.

## 5.3 Molecular Markers: A Solution

Accurate identification and documentation of plant genetic resources is the primary step toward its conservation and sustainable use. The differences that discern one plant from another are encoded in its genetic material. The advent of DNA-based markers has transformed species identification techniques (Botstein et al. 1980). Thottappilly et al. (2000) refer to molecular markers as naturally occurring polymorphism within proteins and nucleic acids that are measurably different. Over the years, the spectacular development in molecular genetics has provided researchers and plant genetic resource conservationists an array of new techniques for simple and reliable species identification. Molecular-based techniques have effectively been used to study the extent and distribution of variation in species, which have also answered archetypal evolutionary and taxonomic questions (Smith et al. 2008).

Numerous molecular markers and techniques are available (Table 5.1). The choice of the molecular marker is based on the objectives of the study and its distinct advantages and limitations. According to Adhikari et al. (2017), a good molecular marker should be moderate to highly polymorphic, co-dominantly inherited, unambiguous assignment of alleles, frequently occurring and distributed throughout the genome; selectively neutral in behavior and highly reproducible, the data should be easily exchangeable between laboratories, with affordable cost for marker development and assay. There are two major types of molecular markers, namely, biochemical or protein-based molecular markers and DNA-based molecular markers.

**Table 5.1** Consolidated list of pteridophytes subjected to various marker studies and their applications

| Species name | Molecular marker | Application | Reference |
|---|---|---|---|
| *Biochemical or protein-based markers* | | | |
| *Anemia hirsute*<br>*Anemia raddiana* | Monoterpenes: Alpha-pinene, beta-pinene, limonene, 3-carene, and myrcene | Taxonomic and evolutionary studies | Hanover (1966a, b); Squillace (1971); Strauss and Critchfield (1982), Sturgeon (1979), Latta et al. (2000) |
| *Anemia tomentosa* var. *anthriscifolia* | Essential oil profile | Similarities between species | Santos et al. (2013) |
| *Asplenium csikii* | Allozymes | Genetic variation studies with high polymorphism and co-dominant expression | Vogel et al. (1999) |
| *Botrychium campestre* | Starch-gel enzyme electrophoresis | Genetic relationships | Farrar and Gilman (2017) |
| Polyploid pteridophytes | Isozyme electrophoresis | Pteridophyte polyploid complexes | Werth (1989) |
| *Culcita macrocarpa* | Isozyme electrophoresis | Genetic diversity-related evolutionary studies | Luis and Santiago (2007) |
| *Woodwardia radicans* | Isozyme electrophoresis | Genetic diversity-related evolutionary studies | Luis and Santiago (2007) |
| *Tectaria coadunata* | Isoperoxidase banding | Morphologically and genetically distinctiveness | Johnson et al. (2010a, b) |
| *Tectaria wightii* | Isoperoxidase banding | Morphologically and genetically distinctiveness | Johnson et al. (2010a, b) |
| *Tectaria paradoxa* | Isoperoxidase banding | Morphologically and genetically distinctiveness | Johnson et al. (2010a, b) |
| *Adiantum* spp. | Isoperoxidase banding | Detection of adulterants in herbal medicines | Johnson et al. (2010a, b) |
| *Pronephrium triphyllum*<br>*Sphaerostephanos unitus* | Isozyme profile | Genetic uniformity of in vitro and in vivo ferns | Marimuthu and Manickam (2011) |
| *Thelypteris ciliata* | Isoperoxidase banding | Confirm intra- and inter-population's genetic difference | Johnson et al. (2012) |

(continued)

**Table 5.1** (continued)

| Species name | Molecular marker | Application | Reference |
|---|---|---|---|
| *DNA-based molecular markers* | | | |
| *C. richardii* | A high-resolution genetic linkage map using restriction fragment length polymorphism (RFLP) | Test the hypothesis of paleopolyploidy in homosporous ferns | Nakazato et al. (2006) |
| *Asplenium viride* | Diversity arrays technology (DArT) | Phylogeographic and substrate specificity patterns | James et al. (2008) |
| *Cibotium x heleniae* | Randomly amplified polymorphic DNA (RAPD) | Conformation of true hybrid origin | Timothy and Motley (2001) |
| *Angiopteris chauliodonta* | Randomly amplified polymorphic DNA (RAPD) | Conformation of 18 survivors of this endemic fern on Pitcairn Islands | Naomi et al. (2004) |
| *Marsilea* spp. | Randomly amplified polymorphic DNA (RAPD) | Morphological plasticity within species | Priyanka and Nilima (2014) |
| *Ceratopteris thalictroides* | Randomly amplified polymorphic DNA (RAPD) | Genetic diversity | Yuan-Huo et al. (2007) |
| *Archangiopteris itoi* | Randomly amplified polymorphic DNA (RAPD) | Genetic variation | Tsai-Wen et al. (2009) |
| *Athyrium distentifolium* | Simple sequence repeats (SSRs) | Potentially transferable polymorphic markers suitable for biodiversity | Woodhead et al. (2003) |
| *Huperzia serrata* (Thunb.) Trevis | Simple sequence repeats (SSRs) | Comprehensive transcriptome characterization | Luo et al. (2010) |
| *Pteridium aquilinum* | Simple sequence repeats (SSRs) | Population genetic variation | Gichira et al. (2016) |
| *Isoetes sinensis* | Simple sequence repeats (SSRs) | Population genetic variation | Gichira et al. (2016) |
| *Neottopteris nidus* | Simple sequence repeats (SSRs) | Population genetic variation | Jia et al. (2016) |
| *Alsophila gigantea* | Simple sequence repeats (SSRs) | Survey population genetic variation and local adaptation | Xiao et al. (2017) |
| *Brainea insignis* | Simple sequence repeats (SSRs) | Expressed sequence tag-simple sequence repeat (EST-SSR) | Liu et al. (2017) |

(continued)

**Table 5.1** (continued)

| Species name | Molecular marker | Application | Reference |
|---|---|---|---|
| *Dryopteris fragrans* | Simple sequence repeats (SSRs) | Genetic diversity | Gao et al. (2018) |
| *A. hispidulum,*<br>*A. incisum,*<br>*A. raddianum*<br>*A. zollingeri* | Inter-simple sequence repeat (ISSR) | Genetic variation | Sankar and Moore (2001)<br>Helena et al. (2005) |
| *Isoetes orientalis*<br>*I. sinensis* | Inter-simple sequence repeat (ISSR) | Genetic diversity | Jin-Ming and Qing-Feng (2006) |
| *Ceratopteris thalictroides* | Inter-simple sequence repeat (ISSR) | Genetic diversity | Yuan-Huo et al. (2007) |
| *Alsophila spinulosa* | Inter-simple sequence repeat (ISSR) | Relative impact of sexual and asexual reproduction | Wang et al. (2012a, b) |
| *Adiantum incisum* | Inter-simple sequence repeat (ISSR) | Genetic variation | Chelliah et al. (2014) |
| *Asplenium nidus* L | Inter-simple sequence repeat (ISSR) | Genetic diversity | Abirami et al. (2018) |
| *Nephrolepis exaltata*<br>*Polystichum acrostichoides*<br>*Pteris acanthoneura*<br>*Nephrolepis furcans*<br>*Adiantum caudatum*<br>*Adiantum capillus-veneris* | Inter-simple sequence repeat (ISSR) | Genetic diversity | Animasaun et al. (2018) |
| *Sticherus flabellatus* | Amplified fragment length polymorphism (AFLP) | Characterize the genetic diversity | Keiper and McConchie (2000) |
| *C. richardii* | Amplified fragment length polymorphism (AFLP) | Linkage map studies | Nakazato et al. (2006) |
| *Serpocaulon* spp. | Amplified fragment length polymorphism (AFLP) | Phylogeny | Kreier et al. (2008) |
| *Asplenium cimmeriorum*<br>*A. gracillimum* | Amplified fragment length polymorphism (AFLP) | DNA fingerprinting | Leon et al. (2010) |
| *Ceratopteris pteridoides* | Amplified fragment length polymorphism (AFLP) | Genetic diversity | Chen et al. (2010) |
| *Matteuccia struthiopteris* | Amplified fragment length polymorphism (AFLP) | Genetic variation | Koenemann et al. (2011) |
| *Sphaeropteris brunoniana* | Amplified fragment length polymorphism (AFLP) | Population genetics | Zi-Juan and Kai-Yun (2011) |

(continued)

**Table 5.1** (continued)

| Species name | Molecular marker | Application | Reference |
|---|---|---|---|
| *Marsilea quadrifolia* | Amplified fragment length polymorphism (AFLP) | Conservation studies | Bruni et al. (2013) |
| *Marsilea quadrifolia* L | Amplified fragment length polymorphism (AFLP) | Genetic diversity | Agnieszka et al. (2014) |
| *Asplenium paleaceum* | Amplified fragment length polymorphism (AFLP) | Genetic relationship | Daniel et al. (2014) |
| *Ceratopteris pteridoides* | Amplified fragment length polymorphism (AFLP) | Genetic structure and diversity | Xiang et al. (2018) |
| *Isoetes yunguiensis* | Amplified fragment length polymorphism (AFLP) | Genetic diversity | Xiang et al. (2018) |
| *Osmunda lancea* | Cleaved amplified polymorphic sequences (CAPS) | Polymorphisms among individuals | Kakugawa and Ootsuki (2014) |
| *Sequence-based DNA studies* | | | |
| *Asplenium ceterach* | cpDNA | Physical mapping | Trewick et al. (2002) |
| *Adiantum capillus-veneris* L | cpDNA | The complete nucleotide sequence | Wolf et al. (2003) |
| *Elaphoglossum* spp. | cpDNA | Phylogenetic relationships | Judith et al. (2004) |
| *Lepisorus clathratus* | cpDNA | Phylogeography | Wang et al. (2011) |
| *Pteridium*, *Cystopteris*, *Ceratopteris*, (*Polypodium*), Cyatheales (*Plagiogyria*), and Gleicheniales (*Dipteris)* | cpDNA | Analyze genomes | Wolf et al. (2015) |
| *Polystichum braunii* | cpDNA | Phylogenetic studies | Stacy and David (2017) |
| *Adiantum hispidulum* | cpDNA | Plastome sequencing | Maria and Logacheva (2017). |
| *Dryopteris fragrans* | cpDNA | Complete nucleotide sequence | Gao et al. (2018) |
| *Alsophila podophyllum* | cpDNA | Genome sequencing | Liu et al. (2018) |
| *Isoetes yunguiensis* | cpDNA | Whole-genome next-generation sequencing | Feng et al. (2019) |

(continued)

**Table 5.1** (continued)

| Species name | Molecular marker | Application | Reference |
|---|---|---|---|
| *Ophioglossum californicum* *Psilotum nudum* | mtDNA | Generation of complete mitogenomes | Guo et al. (2017) |
| *Deparia lancea* | mtDNA | Mitogenome inheritance | Yaung et al. (2018) |
| *Alsophila*, *Gymnosphaera*, and *Sphaeropteris* | Single nucleotide polymorphism (SNPs) | Phylogenetic reconstruction | Dong et al. (2019) |
| *Botrychium lunaria* | Single nucleotide polymorphism (SNPs) | Phylogenetic resolution | Vinciane et al. (2020) |
| *DNA barcoding* | | | |
| *Cheilanthes wrightii* | rbcL, atpA, and trnG-R | Molecular identification | Pryer et al. (2010) |
| *Cheilanthes marginata* | trnH-psbA and trnL-F | Plastid barcoding | Li F-W et al. (2011) |
| *Selaginella* spp. | ITS2 | Phylogenetic analysis | Gu et al. (2013) |
| *Abrodictyum*, *Crepidomanes*, and *Didymoglossum* | rbcL nucleotide sequences | Phylogenetic analysis | Varma and Madusoodanan (2014) |
| *Adiantum aleuticum C.A.Paris* | *atp*A, *mat*K, *rbc*L, and *trn*L-F | Morphological variation, geographic distribution pattern, and evolutionary history | Shao et al. (2015) |
| *Stenochlaena palustris* | rbcL | Molecular identification | Morajkar et al. (2015) |
| *A. pedatum* sensu stricto | | Phylogenetic analysis | Williams et al. (2016) |
| *Selaginella* spp. | rbcL gene | Evolutionary divergences | Nair et al. (2017) |
| *O. parvifolium* and *O. nudicaule Azolla caroliniana* | trnL-F, rbcL, and psbA-trnH | Phylogenetic analysis | Patel and Reddy (2018) |
| *Ophioglossum gujaratense* *O. polyphyllum*, *O. parvifolium*, and *O. nudicaule* | rbcL, trnH-psbA, trnF-trnE, and trnL-trnF | Distinctness of the new taxon | Patel and Reddy (2018) |
| *P. vittata* and *P. vittata* | rbcL | Phylogenetic analysis | Morajkar and Hegde (2021) |

### 5.3.1 Biochemical or Protein-Based Molecular Markers

These markers are limited to plants that produce a suitable range of biochemical products (e.g., proteins, enzymes, secondary metabolites, etc.), which can be easily analyzed and distinguished between different varieties (Joshi et al. 1999). As the name suggests, protein-based molecular markers are based on protein polymorphism and depend on the migration properties of proteins in a gel resulting in separation by electrophoresis, which is revealed by histochemical stains specific to the enzyme to be assayed. The mobility changes in enzymes reflect changes in the encoding DNA sequences (Sharma and Gupta 2013).

#### 5.3.1.1 Monoterpenes

These are biochemical markers of a subgroup of the terpenoid substances found in plants' resins and essential oils. The concentrations of different monoterpenes, such as alpha-pinene, beta-pinene, limonene, 3-carene, and myrcene, are determined by gas chromatography and are useful as genetic markers (Hanover 1966a, b, 1992; Squillace 1971; Strauss and Critchfield 1982) and have been applied primarily to taxonomic and evolutionary studies (Sturgeon 1979; Latta et al. 2000). A study was conducted to evaluate the chemical composition of essential oils from two new fern species, *Anemia hirsuta* and *Anemia raddiana*. The essential oils were subjected to gas chromatography/mass spectrometry for identification. Thirteen compounds were detected in *Anemia hirsuta* (92.3% of the total oil), and the sesquiterpene ß-caryophyllene was one of the major compounds (48.7%). The chemical profile of essential oils obtained from *A. hirsuta* and *A. raddiana* did not show similarities. Likewise, extremely low similarities (< 6.2%) were observed between these species and three previously published profiles of *Anemia tomentosa* var. *anthriscifolia* (Santos et al. 2013).

#### 5.3.1.2 Allozyme Markers

Allozymes represent allelic forms of enzymes that differ structurally but not functionally. Due to their structural differences, they can be distinguished through electrophoresis. Allozymes were extensively used in estimating genetic variation in plant species due to their high polymorphism and co-dominant expression. Vogel et al. (1999) studied the genetic structure of isolated populations of the rock fern *Asplenium csikii*, growing on perpendicular walls of natural rocks using allozymes. All genetic variations observed in this taxon were divided between localities; no allozyme variation was found within the site, and a single multilocus phenotype colonized each site. Thus, allozymes are not very discriminatory in ferns.

#### 5.3.1.3 Isozyme Markers

Isozymes represent enzymes from homologous genes (genes that have diverged over time) that vary in amino acid sequence but catalyze the same reaction. Isozymes with different amino acid composition generally have a different charge and/or size, which results in mobility differences in gel electrophoresis. Isozymes find application in breeding programs in flowering plants, e.g., *Brassica* (Sammour et al. 2021).

However, in nonflowering plants, it is used for species characterization (Calò et al. 1989; Chase et al. 1991) and identification (Prasad et al. 2003) and to derive genetic relationships (Farrar and Gilman 2017). Numerous pteridophyte species are of polyploid origin. The identity of diploid ancestors is well documented in few cases. Studies of pteridophyte polyploid complexes using isozyme electrophoresis have demonstrated the value of this technique for elucidating the ancestry of polyploids (Werth 1989).

Their northern limits using isozymes are reported by genetic variation and distribution studies of clonal ferns of *Culcita macrocarpa* and *Woodwardia radicans* in the northwestern Iberian Peninsula. Despite their high chromosome numbers, both species were isozymic diploids. The very low (*W. radicans*, H T, = 0.012) or zero (*C. macrocarpa*) genetic diversity was detected. It was attributed to genetic drift associated with reducing populations in the last glaciation and founder effects in the subsequent Holocene expansion (Luis and Santiago 2007).

Isozyme studies have indicated the existence of diversity at the molecular level in *Tectaria coadunata* (J. Smith) C.Chr., *Tectaria wightii* (Clarke) Ching, and *Tectaria paradoxa* (Fee) Sledge. The isoperoxidase banding profiles confirmed the classification based on morphology, confirming that all the three species are morphologically and genetically distinct despite being cytologically uniform (Johnson et al. (2010a, b)). An interesting application of isozyme studies for detecting contaminants in crude herbal drugs is reported for six different species of *Adiantum* from the Western Ghats, South India. The molecular markers could successfully identify plant biomass to the species levels, confirming their presence in the crude drugs tested. The isoperoxidase analysis showed the identity of the selected 6 species, and the number of bands varied from 6 to 18. The study revealed the presence of interspecific variation in the isozyme pattern for peroxidase. These unique bands can distinguish and characterize the species and differentiate the original crude drug from the adulterant (Johnson et al. (2010a, b)). Genetic uniformity of in vitro raises sporophytes/plants using isozymes is reported by Marimuthu and Manickam (2011). A protocol for conserving two endangered ferns of southern Western Ghats of India using in vitro spore culture was undertaken. In addition, they conducted spore germination, gametophyte development, changes in the reproductive phases, and sporophyte formation of the medicinally important ferns *Pronephrium triphyllum* (Sw.) Holttum and *Sphaerostephanos unitus* (L.) (Holttum) in vitro. Cytological and isoperoxidase analysis confirmed the genetic uniformity between mother plants and in vitro raised sporophytes/plants.

Diverse populations of *Thelypteris ciliata* (Wall. ex Benth.) Holttum (*T. ciliata*) collected from different localities on Tirunelveli Hills (Kakachi, Kothayar Upper Dam) India and their genetic variations were studied using isoperoxidase isolation. They confirm the intra- and inter-population's genetic difference using isozymes as markers *T. ciliate* (Johnson et al. 2012).

### 5.3.2 DNA-Based Molecular Markers

These types of markers are used to evaluate DNA polymorphism. The versatility of molecular markers has found its prominence and application in various fields like taxonomy, plant breeding, genetic engineering, etc. (Joshi et al. 1999). There are a wide range of DNA-based molecular markers, namely, restriction fragment length polymorphism (RFLP) (Botstein et al. 1980); sequence characterized amplified regions (SCAR) (McDermott et al. 1994; Paran and Michelmore 1993); the variable number of tandem repeats (VNTR) (Nakamura et al. 1987; Jeffreys et al. 1990); single-strand conformation polymorphism (SSCP) (Orita et al. 1989; Hayashi 1992; Hayashi and Yandell 1993); fluorescence-based PCR-SSCP (F-SSCP) (Makino et al. 1992); random amplified polymorphic DNA (RAPD) (Williams et al. 1990); DNA amplification fingerprinting (DAF) (Caetano-Anolles et al. 1991); expressed sequence tags (EST) (Adam et al. 1991); microsatellites, also known as simple sequence repeats (SSR) (Akkaya et al. 1992), short tandem repeats (STR) (Hamada et al. 1982), or simple sequence length polymorphisms (SSLP) (Dietrich et al. 1992); cleaved amplified polymorphic sequences (CAPS) also called PCR-RFLP (Akopyanz et al. 1992; Konieczny and Ausube 1993); microsatellite-primed PCR (MP-PCR) (Meyer et al. 1993; Weising et al. 1995); inter-simple sequence repeat (ISSR) (Meyer et al. 1993; Zietkiewicz et al. 1994); single nucleotide polymorphisms (SNPs) (Brookes 1999; Cho et al. 1999); randomly amplified microsatellite polymorphisms (RAMP) (Wu et al. 1994); single primer amplification reactions (SPAR) (Gupta et al. 1994); amplified fragment length polymorphism (AFLP) (Vos et al. 1995); inverse PCR (IPCR) (Rohde 1996); random amplified microsatellite (RAM) (Hantula et al. 1996); anchored simple sequence repeats (ASSR) (Wang et al. 1998); diversity arrays technology (DArT) (Jaccoud et al. 2001); sequence-related amplified polymorphism (SRAP) (Li and Quiros 2001); target region amplification polymorphism (TRAP) (Hu and Vick 2003); gene sequencing (regions/loci of chloroplast DNA (cpDNA), mitochondrial DNA (mtDNA), and nuclear ribosomal DNA (nrDNA) or internal transcribed spacer (ITS) region of rDNA); DNA barcoding (Hebert et al. 2004); and start codon targeted polymorphism (SCoT) (Collard and Mackill 2009).

Out of these aforementioned molecular markers, a few of the important and most frequently used in pteridophyte genetic diversity studies and systematic, each with its advantages and disadvantages, are discussed below. The molecular markers can be classified into different groups based on (1) mode of transmission (biparental nuclear inheritance, maternal nuclear inheritance, maternal organelle inheritance, or paternal organelle inheritance), (2) mode of gene action (dominant or co-dominant markers), and (3) method of analysis (hybridization-based, PCR-based, or sequence-based markers). For better illustration, the molecular markers have been classified according to the third criteria, i.e., method of analysis.

### 5.3.3 Hybridization-Based DNA Molecular Markers

#### 5.3.3.1 Restriction Fragment Length Polymorphism (RFLP)

RFLP's are simple markers inherited naturally in Mendelian pattern. This technique is based on restriction enzymes that reveal a difference between DNA fragment sizes between individual samples. Chemically or radioactively labelled DNA probe of known sequence (from a cDNA or genomic library) is hybridized with a southern blot of restriction digested DNA samples, resulting in differential DNA fragment profile. The differential DNA profile is generated due to point mutations within the restriction enzyme recognition site or DNA rearrangements like insertion/deletion, translocation, inversion, and transposition (Schlotterer and Tautz 1992). Restriction fragment length polymorphism (RFLP) markers are used to determine the number of gene copies and their distribution in the genome. Yaung et al. (2018) determined the maternity of each sporophyte by barcoding its connected gametophytes using a genetic marker (i.e., ndhF PCR-RFLP). A high-resolution genetic linkage map of the *C. richardii*, a homosporous fern species, was undertaken using a population of doubled haploid lines (DHLs). Their goal was to test the hypothesis of paleopolyploidy in homosporous ferns and provide permanent genetic resources for pteridophytes, the neglected lineage of vascular plants (Nakazato et al. 2006).

**Advantages:** RFLPs are co-dominant and locus-specific and have high genomic abundance, reproducibility, and reliability.

**Limitations:** The expensive technique is time-consuming and hazardous. It requires a large amount of high-quality DNA and prior sequence information for probe generation. Additionally, the risk of exposure to radioactivity in the case of radiolabelled probes is high.

#### 5.3.3.2 Diversity Arrays Technology (DArT)

DArT is a microarray hybridization-based technique that permits simultaneous screening of thousands of polymorphic loci without prior sequence information. DArT markers are polymorphic segments of genomic DNA present in a particular genomic representation. They are identified through differential hybridization on a diversity "genotyping array" developed specifically for this purpose. It is based on the use of microarrays that detect a single base change and insertions and deletions and differences in DNA methylation depending on the enzyme used to generate the fragments. The initial step of DArT analysis involves assembling a "discovery array" from a metagenome representing the germplasm of interest subjected to a reduction in repetitive DNA level (Kilian et al. 2005). A discovery array is then used to identify polymorphic DArT markers assembled into a "genotyping array." Diversity arrays generally detect polymorphisms due to single base pair changes (SNPs) at the restriction sites of endonucleases and insertion-deletions/rearrangements within restriction fragments (Jaccoud et al. 2001). James et al. (2008) study the utility of diversity arrays technology (DArT) in evolutionary studies of non-model organisms like *Asplenium viride*. Phylogenetic analyses of DArT genotypes reveal phylogeographic and substrate specificity patterns in *A. viride*.

**Advantages:** Diversity arrays technology is reproducible and cost-effective and allows simultaneous typing of several hundred polymorphic loci spread over a genome without any previous sequence information about these loci (Jaccoud et al. 2001; Wenzl et al. 2004).

**Limitations:** DArT is the dominant markers.

### 5.3.4 PCR-Based DNA Molecular Markers

#### 5.3.4.1 Random Amplified Polymorphic DNA (RAPD)

RAPD amplifies genomic DNA with short single PCR primers of arbitrary nucleotide sequence (Williams et al. 1990), which detect polymorphisms in the absence of specific nucleotide sequence information. The polymorphisms function as genetic markers. RAPDs can be used to construct genetic maps. Timothy and Motley (2001) indicated that randomly amplified polymorphic DNA (RAPD) markers provide data consistent with the conclusion based on morphological characters that the recently named taxon *Cibotium xheleniae* is of true hybrid origin. Naomi et al. (2004) identified related populations of *Angiopteris chauliodonta*, endemic to remote Pitcairn Island, by RAPD analysis. A field survey indicated that only 18 individuals of *Archangiopteris itoi*, an endemic fern in Taiwan, are surviving in the wild. The genetic variation of the population was assessed by RAPD fingerprinting. RAPD data indicated that some variation existed within the population, suggesting that materials could re-establish the populations (Tsai-Wen et al. 2009). According to Priyanka and Nilima (2014), *Marsilea* shows pronounced morphological plasticity within species, and as such, it is difficult to distinguish species depending on traditional morphology only. Genomic analysis through RAPD markers has been employed for the first time to study inter-population differences between the same plants growing in different habitats. The results so obtained indicate a sufficient level of genetic diversity.

**Advantages:** This low-cost technique is quick, simple, and efficient. A small amount (10–100 ng/reaction) of DNA is sufficient for a reaction (Karp et al. 1997). RAPD primer designing does not need prior sequence information.

**Limitation:** Reproducibility and low-resolution power (Karp et al. 1997).

#### 5.3.4.2 Microsatellites, or Simple Sequence Repeats (SSRs)

SSRs are polymorphic loci in DNA consisting of simple repeating units of one to six base pairs in length found at high frequency in the nuclear genome (Beckmann and Weber 1992). Examples of mono-, di-, tri-, tetra-, penta-, and hexa-nucleotide repeats are (A)11, (GT)12, (ATT)9, (ATCG)8, (TAATC)6, and (TGTGCA)5, respectively (Muhammad and Muhammad 2014). Tri-nucleotide repeats are most abundant in plants (Beckmann and Weber 1992; Kantety et al. 2002; Chen et al. 2006). The polymorphism is caused by frequent mutation of SSRs due to slippage and proofreading errors during DNA replication, which changes the number of repeats and the length of repeats (Eisen 1999) detected through PCR using locus-specific flanking region primers. The most common method to detect microsatellites is to design PCR primers unique to one locus in the genome and base pair on either side of the repeat sequence. Therefore, a single pair of PCR primers can be used for other individuals in the species and result in different-sized products for each of the different lengths of microsatellites. The utility of EST-simple sequence repeats (EST-SSRs) was evaluated in the fern *Athyrium distentifolium*. EST-derived SSRs provide robust, informative, and potentially transferable polymorphic markers suitable for biodiversity research (Woodhead et al. 2003). Simple sequence repeat (SSR) markers have developed for *Huperzia serrata* (Thunb.) Trevis. (Luo et al. 2010). The study was conceived and developed by Joshua et al. (2011); is the first comprehensive transcriptome characterization for a fern gametophyte, including an assessment of transcriptome coverage, gene family and functional representation, SSR identification, etc., It includes a comparative analysis of gene sets across land plants. In bracken fern (*Pteridium aquilinum*), to date, simple sequence repeat (SSR) markers have also been developed in several fern species, such as *Isoetes sinensis* Palmer (Gichira et al. 2016) and *Neottopteris nidus* (L.) J. Sm. ex Hook. (Jia et al. 2016). Alsophila gigantea is a large frond tree fern in the family Cyatheaceae. The plant is a vulnerable species that prefers a specific subtropical montane climate. Little work has been done so far to access the population genetic variation in *A. gigantea*. The novel polymorphic SSR markers characterized here will be used to survey population genetic variation and local adaptation in *A. gigantea*, which helps design effective conservation strategies (Xiao et al. 2017). *Brainea insignis* (Aspleniaceae) is an endangered tree fern in China whose wild populations have been seriously damaged due to overexploitation. Liu et al. (2017) developed expressed sequence tag-simple sequence repeat (EST-SSR) primers for *B. insignis* and investigated genetic diversity and provide resources for future conservation studies. The number and percent of repeat structures are extremely high in ferns (Gao et al. 2018). Simple sequence repeats (SSRs) and long repeat structure pairs (30–55 bp) were identified in *Dryopteris fragrans* (L.) Schott.

**Advantages:** These are co-dominant markers, often presenting high levels of inter- and intra-specific polymorphism with exhaustive genome coverage, and are highly reproducible.

**Limitations:** Primer development flanking microsatellite loci is often a tedious and costly process that needs prior sequence information.

#### 5.3.4.3 Inter-Simple Sequence Repeat (ISSR)

Inter-simple sequence repeat (ISSR) analysis involves amplifying the DNA segment at an amplifiable distance between two identical microsatellite repeat regions oriented in the opposite direction. ISSRs are usually 16–25 bp long as primers in a single primer PCR reaction targeting multiple genomic loci to amplify different sizes of inter-SSR sequences. The microsatellite repeats used as a primer can be di-, tri-, tetra-, or penta-nucleotides. ISSR segregate mostly as dominant markers following simple Mendelian inheritance. However, they have also shown to segregate as co-dominant markers in some cases (Sankar and Moore 2001). Helena et al. (2005) analyzed the level and pattern of genetic variation in four species of the fern genus *Adiantum* L., *A. hispidulum* Sw., *A. incisum* Forssk., *A. raddianum* C. Presl, and *A. zollingeri* Mett. ex Kuhn, originating from South India, using the ISSR fingerprinting method.

The populations of *Adiantum* possessed a considerable level of genetic variation, the diversity indices ranging from 0.284 to 0.464. ISSR markers were used to measure the genetic diversity of two rare and endangered fern species in China. Seventy-two individuals from four natural populations of *Isoetes orientalis* and *I. sinensis* were used. UPGMA cluster analysis showed that there was distinct genetic differentiation between populations of *I. sinensis*. The probable causes that have contributed to the current pattern of genetic structure were discussed, and the suggestions on the future protection of the *Isoetes* species were also given (Jin-Ming and Qing-Feng (2006). The pattern and levels of the genetic variation within and among populations of the endangered fern *Ceratopteris pteridoides* (Hook.) Hieron in China were investigated by Yuan-Huo et al. (2007) using inter-simple sequence repeat (ISSR) markers. Yuan-huo et al. (2008) studied the genetic diversity of the endangered aquatic fern *Ceratopteris thalictroides* from five regions of China. He analyzed the diversity within and among 13 populations using random amplification of polymorphic DNA (RAPD) markers and inter-simple sequence repeat (ISSR) markers. The results indicate that *C. thalictroides* possess an intermediate level of genetic diversity at the species level. They have a low level of genetic differentiation among populations.

Using the diploid homosporous fern *Alsophila spinulosa* as an example species, a study was conducted by Wang et al. (2012a; b) to assess the relative impact of sexual and asexual reproduction on the level and structure of population genetic variation using inter-simple sequence repeat analysis. The study involved 140 individuals collected from 7 populations (HSG, LCH, BPC, MPG, GX, LD, and ZHG) in China. A terrestrial fern in India was undertaken by the genetic variation and patterns of the population structure within the five populations of *Adiantum incisum*. Inter-simple sequence repeat (ISSR) markers are used to measure genetic diversity. The percentage of polymorphism showed a superior genotype that could be used for the conservation of species. The ISSR has proved to be a helpful marker for genotype identification prediction within a closed group of interspecific population in the study area (Chelliah et al. 2014).

Inter-simple sequence repeat (ISSR) analysis was used to study the genetic diversity of the species *Asplenium nidus* L., one of the most important terrestrial ferns in Andaman and Nicobar Islands ornamental value. The genetic variability studies with ISSR marker grouped the accessions of *A. nidus* L. on the basis their geographical locations. It would help in the rapid identification of polymorphism and assist in germplasm collection, conservation, and domestication programs (Abirami et al. 2018). A genetic diversity study conducted by Animasaun et al. (2018) in Nigeria parks and gardens for six ornamental ferns using inter-simple sequence repeat (ISSR) markers concludes that ISSR markers are effective in the assessment of the genetic study of the ferns. The genetic diversity information provided could be utilized to select, improve, and conserve the ornamental plants.

**Advantages**: The use of ISSRs is quick and simple, and the use of high annealing temperature due to long primer length leads to higher stringency and reproducibility. ISSR markers usually show high polymorphism (Kojima et al. 1998), and no requirement of prior information about genomic sequence (Bornet and Branchard 2001) adds to its value.

**Limitations:** As ISSR is a multilocus technique, disadvantages include the possible non-homology of similar-sized fragments. The ISSR primers sometimes have less specificity to the genome to be scanned, leading to ambiguous fingerprints. The poor quality of genomic DNA used in the technique also leads to poor reproducibility of ISSR results.

#### 5.3.4.4 Amplified Fragment Length Polymorphism (AFLP)

AFLP analysis technique is based on selective PCR utilizing directed PCR amplification of the restriction fragments of a total digest of genomic DNA (Vos et al. 1995). It combines the specificity of restriction analysis with PCR technology's flexibility by ligating primer recognition sequences (adapters) to the restricted DNA (Lynch and Walsh 1998). Restriction digestion of highly purified DNA with tetracutter (MseI) and hexacutters (EcoR1) is followed by ligation of adapters to all restricted fragments. Firstly, the pre-selective PCR (pre-amplification) in which amplification of the restriction fragments with EcoRI primer + AC and MseI primer + CC is carried out followed by the second PCR, (the selective PCR) is run and the fragments are viewed on a gel and analyzed by automated sequencing machine.

Amplified fragment length polymorphisms (AFLPs) are used to characterize the genetic diversity within and among natural populations of *Sticherus flabellatus* (Keiper and McConchie 2000), whereas isozyme markers and amplified fragment length polymorphisms (AFLPs) are employed to increase the saturation of the linkage map of the homosporous fern species *C. richardii* using a large population of doubled haploid lines (Nakazato et al. 2006). Kreier et al. (2008) used amplified fragment length polymorphism (AFLP) data to study and infer the phylogeny of epiphytic fern genus *Serpocaulon*. The diversification of species complexes resulted in phylogenetic trees with similar topologies and some notable differences.

Genetic diversity and structure of eight populations *Ceratopteris pteridoides* (Hook.) Hieron collected from the mid-lower reaches of Yangtze River were investigated using amplified fragment length polymorphisms (Chen et al. 2010). Leon et al. (2010) demonstrated parallel polyploid speciation within the ferns *Asplenium cimmeriorum* and *A. gracillimum* using AFLP DNA fingerprinting technology. Koenemann et al. (2011) use AFLP data to investigate the genetic variation of the fiddlehead fern at two geographic scales to infer the historical biogeography of the species *Matteuccia struthiopteris*, which segregates globally into minimally divergent (0.3%) Eurasian and American lineages.

Amplified fragment length polymorphism (AFLP) was employed to analyze the population genetics of relict tree fern *Sphaeropteris brunoniana* (Zi-Juan and Kai-Yun 2011). Manuela et al. (2011) compared population genetic structure and genetic diversity derived from AFLP markers of two epiphytic fern species differing in their ability to colonize secondary habitats. Bruni et al. (2013) investigated the conservation status of *Marsilea quadrifolia*, an endangered fern found in paddy fields, irrigation ditches, and ponds. Finally, a DNA analysis using the AFLP approach was employed to identify the most suitable genetic pool for plant reintroduction efforts.

The study was conducted by Agnieszka et al. (2014) to evaluate the genetic diversity of selected European populations of *Marsilea quadrifolia* L. and to assess the applicability of those genetic resources of *Marsilea quadrifolia* L. using AFLP markers. Daniel et al. (2014) investigated species boundaries and relationships in the *Asplenium paleaceum* (Aspleniaceae) species complex from eastern Australia, using AFLP fingerprinting. According to Xiang et al. (2018), mating system has important implications for populations' genetic structure and diversity especially threatened and endangered species. The mating system of the endangered aquatic fern *Ceratopteris pteridoides* in China was investigated using AFLP markers and a selfing test. In another mating system study, the genetic diversity and genetic structure of the endangered endemic aquatic lycophyte *Isoetes yunguiensis* in China were investigated using AFLP markers (Xiang et al. 2018).

**Advantages:** AFLP analysis is applicable to all species and is highly reliable and reproducible (Jones et al. 1997). The analysis does not require prior sequence information and can work with partially degraded DNA samples.

**Limitations:** Requires 300–1000 ng of DNA, free of inhibitors that interfere with restriction enzymes, per reaction. Nuclear and chloroplast sequences occasionally fail to reveal variability in closely related plant species.

#### 5.3.4.5 Cleaved Amplified Polymorphic Sequences (CAPS)

CAP markers rely on differences in restriction enzymes' digestion patterns of PCR fragments caused by nucleotide polymorphism between samples. Single base changes like SNPs and insertions/deletions in PCR amplicons modify restriction endonuclease recognition sites and generate locus-specific polymorphism. In CAPS analysis, the amplification of target DNA with sequence-specific primers is followed by digestion of amplified products with one or more restriction enzymes (Michael and Amasino 1998). The digested products are then separated by gel electrophoresis on agarose or polyacrylamide gels. Cleaved amplified polymorphic sequence (CAPS) markers can be developed based on various sequences in nucleotide databases. For example, expressed sequence tag (EST) library in a leptosporangiate fern, *Osmunda lancea*, has been used to develop primers in exons, which have lower nucleotide substitution rates introns or intergenic spacer regions (Kakugawa and Ootsuki 2014).

**Advantages:** Co-dominance and faster than hybridization techniques (Matsumoto and Tsumura 2004).

**Limitations:** Uncertain occurrence of mutation in the restriction endonuclease recognition sites often limits the ability of CAPS in detecting polymorphism.

## 5.4 Sequence-Based DNA Studies

DNA sequencing is the determination of the nucleotide arrangement of the bases A (adenine), G (guanine), C (cytosine), and T (thymine) content in a target molecule of DNA (Arif et al. 2010). The dye terminator sequencing technique is the standard method in automated sequencing analysis (Olsvik et al. 1993). The technique uses four dideoxynucleotide chain terminators labelled with fluorescent dyes, each with a different fluorescence emission wavelength. The labelling of the chain terminator ddNTPs allows sequencing in a single reaction. The first partial genome sequence assembly of *Ceratopteris richardii*, a homosporous fern, was carried out by Marchant et al. (2019). Their efforts to unravel the expansive genomes of ferns are the first stepping stone in understanding the evolutionary genomics of land plants.

### 5.4.1 Next-Generation Sequencing Techniques (NGS)

The advent of a new technique, i.e., next-generation sequencing techniques (NGS), has significantly decreased the time and cost needed for producing large quantities of sequence data. These technologies have increased the feasibility of analyzing genomes and transcriptomes, identifying new genes and genetic markers, and investigating genome homology, differential gene expression, and epistasis. A variety of NGS technologies are available, including Roche 454 pyrosequencing (Ronaghi et al. 1996), Illumina sequencing-by-synthesis (Turcatti et al. 2008), Ion Torrent sequencing-by-synthesis (Rothberg et al. 2011), Life Technologies SOLiD sequencing (Pandey et al., 2008), and Pacific Biosciences single-molecule real-time sequencing (Eid et al. 2009). These present technologies trade off in reading length, read depth, the total amount of data produced, run time, and cost. Longer reads are desirable for sequencing across repetitive regions of the genome and can also distinguish homologous regions from one another.

The concept of NGS technology is similar to CE sequencing. DNA polymerase catalyzes the fluorescently labelled deoxyribonucleotide triphosphates (dNTPs) into a DNA template strand. It is followed by sequential cycles of DNA synthesis. At the point of incorporation, during every cycle, the nucleotides are identified by fluorophore excitation. The principle difference is that NGS extends this process across millions of fragments in a massively parallel fashion instead of a single gene locus. NGS (Illumina sequencing-by-synthesis) workflows include four basic steps. The first step is the library preparation, which is prepared by random fragmentation of the DNA, followed by 5′ and 3′ adapter ligation. "Tagmentation" combining the fragmentation and ligation reactions into a single step, greatly increasing the efficiency of the library preparation process, is undertaken. Adapter-ligated fragments are then amplified in PCR and gel purified. The generation of cluster follows library preparation. The library is mounted into a flow cell. Here fragments are immobilized on an array of surface-bound oligos complementary to the library adapters. The fragments obtained are amplified by bridge amplification into distinct, clonal clusters.

When the cluster is complete, the templates are ready for the next step, i.e., sequencing. The technology uses a proprietary reversible terminator-based method that detects single bases as incorporated into DNA template strands. As all four reversible terminator-bound dNTPs are available during each sequencing cycle, the natural competition reduces bias and greatly reduces raw error rates compared to other technologies (Ross et al. 2013; Bentley et al. 2008). The result is an accurate base-by-base sequencing that virtually eliminates sequence-specific errors, even within repetitive sequence regions and homopolymers. The final step is data analysis. The freshly identified sequence reads are aligned to a reference genome. Following alignment, many analysis variations include single nucleotide polymorphism (SNP), insertion-deletion (indel) identification, read counting for RNA methods, phylogenetic or metagenomic analysis, etc. These techniques are robust population-genetic studies based on complete genomes rather than just short sequences of a single gene. A comparative genomic approach is productive for identifying functional loci related to morphological, geographical, and physiological variation.

Thus, next-generation sequencing technology will enable us to understand the process of plant evolution better. Few researchers have used NGS technologies to analyze genomes of several species in pteridophytes. Six Polypodiales ferns (*Pteridium*, *Cystopteris, Ceratopteris*, *Polypodium*), Cyatheales (*Plagiogyria*), and Gleicheniales (*Dipteris*) nuclear genome were explored for repetitive sequences including DNA transposons, retrotransposons, ribosomal DNA and simple repeats, and protein-coding genes. They have used the data for drawing a comparative account of variations and similarities in repeat structure between seed plants and ferns (Wolf et al. 2015).

**Advantages:** The technique is automated and robust and has high accuracy (>98%).

**Limitations:** Next-generation sequencing requires sophisticated bioinformatics systems, fast data processing, and large data storage capabilities. Additionally, genome sequencing can be costly, and every laboratory may not afford the costs.

## 5.4.2 Gene Sequencing

Identification of easily amplifiable and relatively rapidly evolving unambiguously alignable DNA regions that provide sufficient suitable variation within a short sequence segment from the nuclear genome is a necessary step for preliminary phylogenetic studies. Several genes (e.g., mitochondria and chloroplast) have been utilized for various molecular systematic studies, particularly the chloroplast genome that has been extensively surveyed to reconstruct plant phylogeny (Olmstead and Palmer 1994).

#### 5.4.2.1 cpDNA

The chloroplast DNA (cpDNA) structural mutations have been identified by Raubeson and Stein (1995) through physical mapping within leptosporangiate ferns. Chloroplast DNA sequences obtained from 331 *Asplenium ceterach* plants representing 143 populations throughout Europe and outlying sites in North Africa and the near East (Trewick et al. 2002) were used. Wolf et al. (2003) determined the complete nucleotide sequence of the chloroplast genome of the leptosporangiate fern, *Adiantum capillus-veneris* L. (Pteridaceae). Phylogenetic relationships among 48 species representing the 9 sections within the fern genus *Elaphoglossum* were investigated using cpDNA sequence data from rbcL, trnL-F, and rps4-trnS (Judith et al. 2004). Wolf et al. (2010) examine the structure of the plastome across fern phylogeny for the first time. The plastid genome (plastome) is a rich source of phylogenetic and other comparative data in plants. Most land plants possess a plastome of a similar structure. However, in a major group of plants, the ferns, a unique plastome structure has evolved. The gene order in ferns has been explained by a series of genomic inversions relative to the plastome organization of seed plants.

Wang et al. (2011) employed sequences of two chloroplast DNA regions, rps4-trnS and trnL-trnF, to reconstruct phylogeography of the Sino-Himalayan fern *Lepisorus clathratus*; Lestari et al. (2014) analyzed a molecular phylogeny of *Adiantum* collected from the Lesser Sunda Islands to reveal its phylogenetic relationships. Two cpDNA regions (rbcL and trnL-F) were chosen as markers. Stacy and David (2017) used the chloroplast markers rbcL, rps4-trnS, and trnL-F and the nuclear markers pgiC and gapCp to demonstrate that *Polystichum braunii* is a single allotetraploid with a minimum of two origins. A comparative analysis with available plastomes of Polypodiales, the most species-rich group of ferns, was characterized by Maria and Logacheva (2017). They also conducted the plastid genomes analysis of three species of *Dryopteris*, one of the largest fern genera, using sequencing of chloroplast DNA-enriched samples. They also sequenced the plastome of *Adiantum hispidulum* (Pteridaceae).

Liu et al. (2018) completely sequenced the chloroplast genome of the tree fern *Alsophila podophyllum*. Gao et al. (2018) sequenced the complete nucleotide sequence of *Dryopteris fragrans* (L.) Schott. using chloroplast (cp) genome. De novo assembly of whole-genome next-generation sequencing and the chloroplast genome sequencing of *Isoetes yunguiensis*, a rare and endangered fern endemic to Yunnan-Guizhou Plateau of China, was completed (Feng et al. 2019). Wei et al. (2017) propose that 14 chloroplast markers are particularly phylogenetically informative for eupolypods II at both the familial and generic levels. They demonstrate the power of a character-rich plastome dataset and high-throughput sequencing to resolve the recalcitrant lineages, which have undergone rapid evolutionary radiation and dramatic changes in substitution rates.

#### 5.4.2.2 mtDNA

The "Monilophyte" clade comprising ferns, horsetails, and whisk ferns receives unequivocal support from molecular data as the sister clade to seed plants. Knie et al. (2015) investigated the mitochondrial nad2 and rpl2 genes as two new, intron-containing loci for a wide sampling of taxa. Guo et al. (2017) generated complete mitogenomes from the adder's tongue fern (*Ophioglossum californicum*) and the whisk fern (*Psilotum nudum*). Sequencing of fern mitogenomes could shed light on the major evolutionary transitions that established mitogenomic diversity among extant lineages. Yaung et al. (2018) developed a new and efficient method to examine plastome and mitogenome inheritance in a fern species, *Deparia lancea* (Athyriaceae, Aspleniineae, Polypodiales). They found that plastid and mitochondrial DNAs were transmitted from only the maternal parentage to the next generation.

#### 5.4.2.3 Single Nucleotide Polymorphism (SNPs)

SNPs are advanced DNA based molecular markers to detect single nucleotide mutations (A, T, G, or C) among the same species. SNP is the most abundant marker system in both animal and plant genomes. SNPs may occur in the coding, non-coding, and intergenic regions of the genome. SNPs located in the non-coding region are the most frequent type of variations found in DNA. Their discovery and insertions/deletions have formed the basis of most differences between alleles. Thus, SNPs can be explained as any polymorphism between two genomes based on a single nucleotide exchange. SNPs are consequences of transition or transversion event and are highly abundant; their density differs substantially in different genome regions (Weising et al. 1995). In brief, the SNP and detection method protocol includes preparation of sample reactions using template and primer, performing SNP shot reactions by PCR, and the post-extension conditions of the product. Automated electrophoresis of the samples is essential for analyzing the data for SNP after sequencing. SNPs are co-dominant and many loci that can be assessed (Foster et al. 2010).

The evolutionarily conserved nature of SNPs makes them less subject to the problem of homoplasy (Brumfield et al. 2003). Kao et al. (2019) report a species phylogeny for notholaenid ferns. They use four low-copy nuclear loci and three plastid loci. A total of 61 individuals were sampled (49 notholaenids and 12 outgroup taxa), including 31 out of 37 recognized notholaenid species. The presence of farina and their biochemical constitutions were mapped, and a consensus phylogenetic tree was drawn. The characters were homoplastic and had complex evolutionary histories. Hybridization among recognized species of the notholaenid clade appears to be relatively rare compared to that observed in other well-studied fern genera.

Dong et al. (2019) generated transcriptome sequences of five species in three genera of Cyatheaceae (*Alsophila*, *Gymnosphaera*, and *Sphaeropteris*). They used them to search for single-copy nuclear loci for phylogenetic reconstruction. Phylogenetic analyses based on multispecies coalescent and, alternatively, concatenation models yielded congruent results with high resolution.

Improved phylogenetic resolution among early-branching ferns was reported by Vinciane et al. (2020). They present an early phylogenetically branching fern with an extremely large genome of 27.5 Gbp as a reference transcriptome for *Botrychium lunaria.* Based on this, an additional 11 transcriptomes of the same species were identified. They report unexpected variation in population-level heterozygosity.

## 5.5 DNA Barcoding

DNA barcode sequences are short sequences relative to a genome, and they can be obtained reasonably quickly and cheaply (Kress et al. 2005). DNA barcoding shows sequence diversity in short, standardized gene regions to aid species identification and discovery in large life assemblages (Ratnasingham and Hebert 2007). DNA barcode has been applied successfully in animals using COX gene loci (Hebert et al. 2004), with excellent results. It find application in molecular identification, phylogenetic studies, and molecular discrimination of various species. There have been disputes over the best gene loci for representing plant DNA barcode. Various genes have advantages over one another as a plant DNA barcode. Recently, plant DNA barcode researchers proposed two chloroplast genes, rbcL and matK, taken together as appropriate for the barcoding of plants (CBOL 2009). High efficiency, reproducibility, reliability, sequence information, and discrimination of various species based on one gene give DNA barcoding a very distinct advantage, while low resolution still limits in some cases its use (Zuo et al. 2011). DNA barcoding has been used for molecular identification of plant species such as *Stenochlaena palustris* (Morajkar et al. 2015).

### 5.5.1 Species Identification

To test the efficacy of DNA barcoding, Ma et al. (2010) initiated the analysis using five DNA sequence markers (rbcL, psbA-trnH intergenic region, rpoB, matK, and rpoC1) using six chloroplast genomic sequences from GenBank. They found psbA-trnH intergenic region the best candidate for the availability of universal primers.

The Japanese pteridophyte flora comprising 733 taxa, including subspecies and varieties, were used to test the utility of 2 plastid DNA barcode regions (rbcL and trnH-psbA) to develop an identification system for native gametophytes. The overall success of DNA barcodes for species identification in the Japanese pteridophyte flora was observed (Ebihara and Ito 2010).

DNA barcoding approaches are useful to the horticultural community for keeping plants in the trade accurately identified, e.g., cheilanthoid ferns. Pryer et al. (2010) used plastid rbcL, atpA, and trnG-R sequence data to demonstrate a fern marketed as *Cheilanthes wrightii*. It is endemic to the Southwestern USA and Northern Mexico. In the horticultural trade, *Cheilanthes distans* is endemic to Australia and adjacent islands. DNA sequence data is linked to vouchered plant material deposited in public and private databases. They provide an important resource for conservation and fostering collaborations between the academic and horticultural communities. We strongly advocate the barcoding approach as a valuable new technology available to the horticulture industry to help correct plant identification errors in international trade.

de Groot et al. (2011) evaluated the potential of a combination of rbcL with a non-coding plastid marker, trnL-F, to obtain DNA identifications for fern species. A regional approach was adopted by creating a reference database of trusted rbcL and trnL-F sequences for the wild occurring homosporous ferns of NW Europe. Their results suggest a high potential for combining rbcL and trnL-F as a two-locus cpDNA barcode to identify fern species.

DNA barcoding has revolutionized our understanding of fern ecology, especially because the accurate identification of the independent but cryptic gametophyte phase of the fern's life history, an endeavor previously impossible, is now feasible. Li et al. (2011) assessed the discriminatory power of the core plant DNA barcode (rbcL and matK), as well as alternatively proposed fern barcodes (trnH-psbA and trnL-F), across all major fern lineages. They present plastid barcode data for two genera in the hyper-diverse polypod clade, *Deparia* (Woodsiaceae) and the *Cheilanthes marginata* group (currently being segregated as a new genus of Pteridaceae), to further evaluate the resolving power of these loci (Li et al. 2011).

Gu et al. (2013) reported the ITS2 barcode could effectively identify medicinal plants of Selaginellaceae. The results provide a scientific basis for identifying plants of the family Selaginellaceae and the reasonable development of these resources. This study may broaden the application of DNA barcoding in the medicinal plant field and benefit phylogenetic investigations. DNA barcoding is a novel molecular identification method that aids in identifying traditional Chinese Materia Medica using traditional identification techniques. However, a study is needed to assess the stability and accuracy of DNA barcoding. ITS2 and psbA-trnH exhibited an average intraspecific divergence of 0.001 and 0, respectively, as well as an average interspecific divergence of 0.0331 and 0.0161 (Dianyun et al. 2013).

Five plastid regions (rbcL, matK, trnH-psbA, trnL-F, and rps4-trnS) and eight nuclear regions (ITS, pgiC, LEAFY, gapC, ITS2, IBR3_2, DET1, and SQD1_1) were screened and evaluated in the fern genus *Adiantum* from China and neighboring areas (Wang et al. 2016). To compare patterns of community structure between fern sporophytes and gametophytes, Nitta et al. (2017) surveyed the ferns of the islands of Moorea and Tahiti (French Polynesia). They first constructed a DNA barcode library (plastid rbcL and trnH-psbA) for the 2 island floras, including 145 fern species, and then used these DNA barcodes to identify more than 1300 field-collected gametophytes from 25 plots spanning an elevational gradient from 200 to 2000 m. The phylogenetic analysis based on rbcL sequences showed partially resolved clades within Pteridaceae, including few ecotypes of *P. vittata* and *P. vittata* L. nano sequence. In addition to sporophyte morphology, molecular evidence helped report a new variant of *P. vittata* L. The above investigation was undertaken to verify the claim of *P. vittata* L. as *P. vittata* L. nano, a new variant. Molecular data and phylogenetic analysis and morphological considerations were used as evidence (Morajkar and Hegde 2021).

### 5.5.2 Phylogeny

Complete 18S ribosomal RNA sequence data from representatives of all extant pteridophyte lineages and RNA sequences from different seed plants were used to infer a molecular phylogeny of vascular plants that included all major land plant lineages. The molecular data indicate that lycopsids are monophyletic and are the earliest diverging group within the vascular land plants. In contrast, *Psilotum nudum* is more closely related to the seed plants than to other pteridophyte lineages (Kranz and Volkar 1996). Shepherd et al. (2007) sequenced the chloroplast trnL-trnF locus for all of the Blechnaceae species indigenous to New Zealand, plus several non-indigenous species. The monophyletic Pteridaceae accounts for roughly 10% of extant fern diversity and occupies an unusually broad range of ecological niches, including terrestrial, epiphytic, xeric-adapted rupestral, and even aquatic species. Schuettpelz et al. (2007) present the results of the first broad-scale and multi-gene phylogenetic analyses of these ferns and determine the affinities of several previously unsampled genera. Korall et al. (2007) investigated the phylogenetic relationship of tree ferns based on DNA sequence data from five plastid loci (rbcL, rbcL-accD IGS, rbcL-atpB IGS, trnG-trnR, and trnL-trnF). As a case study exploring the utility of selected molecular data for species identification within this integrative taxonomic framework, chloroplast DNA from three regions (rbcL, trnSGG, trnH-psbA) were sequenced for all species of filmy ferns (Hymenophyllaceae) known from Moorea. The relative utility of each of these regions for phylogenetic analysis and DNA-based identification was inferred by estimating support for phylogenetic trees reconstructed from each region and calculating intraspecific and interspecific distance values (uncorrected p) between taxa for each region. All three of these regions were potentially useful for phylogenetic studies at the appropriate taxonomic level (Nitta 2008).

Molecular systematics of the filmy ferns (Hymenophyllaceae) of South India is done based on rbcL nucleotide sequences (Varma and Madusoodanan 2014). Phylogenetic analysis was obtained showing the trichomanoid lineage of South India comprising three strongly supported clades corresponding to the three genera recognized by Ebihara et al. (2006), viz., *Abrodictyum*, *Crepidomanes*, and *Didymoglossum* (Varma and Madusoodanan 2014). A molecular phylogeny of *Pteris* is presented for 135 species, based on cpDNA rbcL and matK and using maximum likelihood, maximum parsimony, and Bayesian inference approaches. The inferred phylogeny was used to assess the biogeographical history of *Pteris* and to reconstruct the evolution of one ecological and four morphological characters commonly used for infrageneric classifications (Yi-Shan et al. 2014).

Morphological diversity and inadequate taxon sampling make the phylogenetic relationships among the genera of *Woodsiaceae* remain unclear. Shao et al. (2015) carry out a comprehensive phylogenetic analysis of *Woodsiaceae* using molecular evidence from four chloroplast DNA marker loci (*atp*A, *mat*K, *rbc*L, and *trn*L-F) and covering over half the currently recognized species. They present a new subgeneric classification of the redefined *Woodsia* based on phylogenetic and ancestral state reconstructions to reflect better the morphological variation, geographic distribution pattern, and evolutionary history of the genus. *Adiantum* populations of growing on serpentine barrens in Maryland and Pennsylvania are identified as *A. aleuticum* (Rupr.) C.A.Paris (=*A. pedatum* L. *var. aleuticum* Rupr.). The species, and the allopolyploid *A. viridimontanum* C.A.Paris, are known from Canada serpentine and in New England. A phylogenetic analysis using two plastid markers showed that the Maryland and Pennsylvania populations are *A. pedatum* sensu stricto rather than *A. aleuticum* or *A. viridimontanum* (Williams et al. 2016). Blechnaceae, a leptosporangiate fern family, typically divided among seven to nine genera, is nested within eupolypods II. It comprises 200–250 species. Despite recent molecular studies, the family lacks a modern taxonomic update. de André et al. (2017) have assembled the broadest dataset from three plastid regions (rbcL, rps4-trnS, trnL-trnF) with taxonomic sampling focused on major diversity centers based on broad sampling from the two centers of diversity, viz., the Neotropics and Australasia/Oceania. They outline and explain a plan to resolve the polyphyly of *Blechnum* by recognizing additional, monophyletic, segregate genera.

The research study was undertaken with three major objectives: primarily, to screen the fern flora and spatial distribution and conduct an index-based diversity assessment of the fern flora in the Kudremukh National Park (KNP) of Western Ghats; secondly, to identify the edaphic factors responsible for fern distribution; and most importantly, to contribute to the knowledge of DNA barcodes of the pteridophytes in the Indian context (Morajkar 2018). *Ophioglossum L.*, commonly known as "adder's tongue fern," has been of great interest due to the highest number of chromosomes. Phylogenetic analysis of three chloroplast DNA (cpDNA) regions (trnL-F, rbcL, and psbA-trnH) unambiguously designates this adder's tongue fern (*O. malviae* sp. nov.) as the distinct lineage and is sister to the clade containing *O. parvifolium* and *O. nudicaule. Azolla caroliniana*—an aquatic fern (average size, 0.5–1.5 cm), is the smallest fern on the earth (Patel and Reddy 2018).

A new species, *Ophioglossum gujaratense*, is described by Patil et al. (2018) from Gujarat state (India). A comparative account of morphologically similar species, viz., *O. gujaratense*, *O. polyphyllum*, *O. parvifolium*, and *O. nudicaule*, is provided. The distinctness of the new taxon has been confirmed using molecular data from chloroplast genome markers, viz., rbcL, trnH-psbA, trnF-trnE, and trnL-trnF. Liu et al. (2019) focus on the epiphytic genus *Scleroglossum*, emphasizing the occurrences in Hainan and Yunnan of mainland China. Their integrative results show that the Yunnan accessions are distinct from those in Hainan in phenotypic and genotypic variation. The study contributes to the crucial assessment of the plant diversity in Yunnan. It illustrates the importance of integrating collection efforts and DNA barcoding approaches to enable effective assessment of the epiphytic diversity of Yunnan. Nitta et al. (2020) combine classical taxonomic and molecular phylogenetic (DNA barcoding) approaches to catalogue the diversity of pteridophytes (ferns and lycophytes) of the Nectandra Cloud Forest Reserve, Costa Rica. A plastid loci rbcL was selected as a DNA barcode marker and obtained for >95% of pteridophyte taxa in this site. Combined molecular and morphological analyses revealed a hybrid that was previously undescribed taxa.

### 5.5.3 Quality Control

Pteridophytes are an important group of plants used in traditional herbal medicine. There is no simple and straightforward way to differentiate various species of this group by morphological traits. DNA barcoding is considered a novel method to discriminate species with universal primers and indicate sufficient variation. Ma et al. (2010) tested the first five DNA sequence markers (psbA-trnH intergenic region, rbcL, rpoB, rpoC1, and matK) for the effectiveness of DNA barcoding. They were the first to analyze six chloroplast genomic sequences from GenBank and found that the psbA-trnH intergenic region is the best candidate for the generation of universal primers.

Ma et al. (2010) also amplified the psbA-trnH region from 79 samples of medicinal pteridophyte plants representing 51 species from 24 families, including all the authentic pteridophyte species listed in the Chinese pharmacopoeia (2005 version) and some commonly used adulterants. DNA barcoding is a simple, low-cost method to identify medicinal plant materials in various forms. Implementing this method for the quality control of medicinal plant products in Mexico is expected to bring several benefits to the herbal industry and the consumers. The authentication of medicinal plants by DNA barcoding is expected to guarantee high-quality herbal medicinal products. It is envisaged that the quality of the herbal medicines without adulterants and side effects will encourage the national herbalist sector (Marcial-Quino et al. 2015). Indian traditional medicine has invited a lot of attention in recent times. Medicinal plants have become a major source of natural drugs. New molecular approaches toward identifying these medicinal plants are becoming inevitable to avoid confusions in identification. In this regard, DNA barcoding using the rbcL gene is suggested as a marker for pteridophyte identification. Molecular characterization of *Selaginella* spp. was performed using polymerase chain reaction (PCR) for the rbcL gene. A phylogenetic tree was built using Molecular Evolutionary Genetics Analysis (MEGA5) software for the obtained sequences. The evolutionary divergences between its closely related species were also performed to disclose its amendments during evolution (Nair et al. 2017).

## 5.6 Conclusion

Molecular techniques generate a dynamic volume of information on the genome of pteridophytes and their diversity, homology, and genetic variations. Among them, plastid DNA-based markers are popular in systematics. Whole-genome sequencing and organellar DNA sequencing reports are limited. It is observed that DNA-based markers are a preferred method of genetic studies in pteridophytes as compared to probe-based marker techniques. Perhaps because the probe-based molecular markers are popular tools of plant breeding programs, viz., selection of elite parents with desired traits, studies on trait inheritance patterns, etc. Limited studies are available on marker-based fern breeding programs. However, the majority of marker-based studies in pteridophytes are on homology studies and phylogenetic analysis and, more recently, resolving taxonomical issues in pteridophytes using molecular marker tools. The simplicity of gametophyte morphology and complexity of sporophytic phenotypes perplexed taxonomist for ages. They were challenged with the cryptic species often seen in pteridophytes due to natural polyploidy, apospory, apogamy, and hybridization events. Molecular markers are gaining popularity in pteridophytes due to their instant resolving capacity for species identification irrespective of the stages of life cycles of the pteridophyte, i.e., independent of gametophytic or sporophytic tissue. Detection of intentional or unintentional adulteration in powdered herbal medicines will always be a challenge to the medicinal plant industry producers and consumers. Few molecular markers have shown the potential and promise of detecting contaminants in herbal drugs produced using pteridophytes.

## References

Abirami K, Baskaran V, Simhachalam P et al (2018) Molecular characterization of Asplenium nidus L. – a potential ornamental fern species from Andaman. Agric Sci Digest 38(4):265–269

Adam MD, Kelley JM, Gocayne JD et al (1991) Complementary DNA sequencing: expressed sequence tags and human genome project. Science 252:1651–1656

Adhikari S, Saha S, Biswas A et al (2017) Application of molecular markers in plant genome analysis: a review. Nucleus 60:283–297

Agnieszka J, Katarzyna G, Jerzy B et al (2014) A study of the genetic variation of the aquatic fern Marsilea quadrifolia L. preserved in botanical collections in Poland and originated from natural populations in Europe. Flora: Morphol Distrib Funct Ecol Plants 209(11):655–665

Animasaun DA, Oyedeji S, Ingalhalli RS et al (2018) Polymorphism and genetic deiversity assessment of some ornamental fern by microsatellite (ISSR) markers. J Appl Hortic 20(3):1–5. https://doi.org/10.37855/jah.2018.v20i03.41

Akkaya MS, Bhagwat AA, Cretan PB (1992) Length polymorphisms of simple sequence repeat DNA in soybean. Genetics 132:1131–1139

Akopyanz N, Bukanov NO, Westblom TU (1992) PCR-based RFLP analysis of DNA sequence diversity in the gastric pathogen helicobacter pylori. Nucleic Acid Res 20:6221–6225

Arif IA, Bakir MA, Khan HA et al (2010) A brief review of molecular techniques to assess plant diversity. Int J Mol Sci 11:2079–2096

Beckmann JS, Weber JL (1992) Survey of human and rat microsatellites. Genomics 12:627–631

Bentley DR, Balasubramanian S, Swerdlow HP et al (2008) Accurate whole human genome sequencing using reversible terminator chemistry. Nature 456(7218):53–59

Bornet B, Branchard M (2001) Nonanchored inter simple sequence repeat (ISSR) markers: reproducible and specific tools for genome fingerprinting. Plant Mol Biol Rep 19:209–215

Botstein D, White RL, Skolnick MH et al (1980) Construction of a genetic map in man using restriction fragment length polymorphisms. Am J Hum Genet 32:314–331

Brookes A (1999) The essence of SNPs. Gene 234:177–186

Brumfield RT, Beerli P, Nickerson DA et al (2003) The utility of single nucleotide polymorphisms in inferences of population history. Trends Ecol Evol (Amst.) 18:249–256

Bruni I, Gentili R, De Mattia F et al (2013) A multi-level analysis to evaluate the extinction risk of and conservation strategy for the aquatic fern Marsilea quadrifolia L. in Europe. Aquat Bot 111:35–42. https://doi.org/10.1016/j.aquabot.2013.08.005

Caetano-Anolles G, Bassam BJ, Gresshoff PM (1991) DNA amplification fingerprinting using very short arbitrary oligonucleotide primers. Biotechnology 9:553–557

Calò A, Costacurta G, Paludetti G et al (1989) The use of isozyme markers to characterize grape cultivars. Riv Vitic Enol 1:15–22
CBD (1992) The convention on biological diversity. United Nations
CBOL Plant Working Group (2009) A DNA barcode for land plants. Proc Natl Acad Sci U S A. 106:12794–12797
Chapman AD (2009) Numbers of living species in Australia and the world, Second edn. Canberra, Australian Biological resource study
Chase CD, Ortega VM, Vallejos CE (1991) DNA restriction fragment length polymorphisms correlate with isozyme diversity in Phaseolus vulgaris L. Theor Appl Genet 81:806–811
Chelliah DA, De John BA, Selvin Samuel A (2014) Genetic variations of Adiantum incisum Forssk. Revealed by ISSR markers in the Western Ghats of Tamil Nadu, India. J Res Biol 4 (8):1604–1610
Chen CX, Zhou P, Choi YA et al (2006) Mining and characterizing microsatellites from citrus ESTs. Theor Appl Genet 112:1248–1257
Chen YY, Han QX, Cheng Y, Li ZZ, Li W (2010) Genetic variation and clonal diversity of the endangered aquatic fern Ceratopteris pteridoides as revealed by AFLP analysis. Biochem Syst Ecol 38(6):1129–1136. https://doi.org/10.1016/j.bse.2010.12.016
Cho RJ, Mindrinos M, Richards DR (1999) Genome-wide mapping with biallelic markers in Arabidopsis thaliana. Nat Genet 23:203–207
Collard BCY, Mackill DJ (2009) Start codon targeted (SCoT) polymorphism: a simple, novel DNA marker technique for generating gene-targeted markers in plants. Plant Mol Biol Rep 27:86–93
Daniel J, Ohlsen A, Leon R et al (2014) Investigation of species boundaries and relationships in the Asplenium paleaceum complex (Aspleniaceae) using AFLP fingerprinting and chloroplast and nuclear DNA sequences. Aust Syst Bot 27(6):378–394
de André G, ThaisA VD et al (2017) Molecular phylogeny of the fern family Blechnaceae (Polypodiales) with a revised genuslevel treatment. Cladistics 33:429–446. https://doi.org/10.1111/cla.12173
de Groot GA, During HJ, Maas JW et al (2011) Use of rbcL and trnL-F as a two-locus DNA barcode for identification of NW-European ferns: an ecological perspective. PLoS One 6(1):e16371. https://doi.org/10.1371/journal.pone.0016371
Dianyun H, Jingyuan S, Shi Li, et al (2013) Stability and accuracy assessment of identification of traditional Chinese Materia Medica using DNA barcoding: a case study on Flos Lonicerae Japonicae. BioMed Res Int 1–08. https://doi.org/10.1155/2013/549037
Dietrich W, Katz H, Lincoln SE (1992) A genetic map of the mouse suitable for typing intraspecific crosses. Genetics 131:423–447
Dong S, Xiao Y, Kong H, Chao F, Yan Y-H, Kang M (2019) Nuclear loci developed from multiple transcriptomes yield high resolution in phylogeny of scaly tree ferns (Cyatheaceae) from China and Vietnam. Mol Phylogenet Evol:106567. https://doi.org/10.1016/j.ympev.2019.106567
Ebihara JHN, Ito M (2010) Molecular species identification with rich floristic sampling: DNA barcoding the Pteridophyte Flora of Japan. PLoS One 5(12):15136
Ebihara A, Dubuisson JY, Iwatsuki K et al (2006) A taxonomic revision of Hymenophyllaceae. Blumea 51:221–280
Eid J, Fehr A, Gray J et al (2009) Real-time DNA sequencing from single polymerase molecules. Science 323:133–138
Eisen JA (1999) Mechanistic basis for microsatellite instability. In: Goldstein DB, Schlotterer C (eds) Microsatellites: evolution and applications. Oxford University Press, Oxford
Farrar DR, Gilman AV (2017) Relationships in the Botrychium campestre(Ophioglossaceae) complex. Brittonia 69(3):265–275
Feng T, Qun-Ying X, Liang Q et al (2019) The complete chloroplast genome of Isoetes yunguiensis (Isoetaceae), a rare and endangered fern endemic to China. Mitochondrial DNA B 4 (1):563–564. https://doi.org/10.1080/23802359.2018.1555015
Foster JT, Allan GJ, Chan AP et al (2010) Single nucleotide polymorphisms for assessing genetic diversity in castor bean (Ricinus communis). BMC Plant Biol 10:13–23

Gao R, Wang W, Huang Q, Fan R, Wang X, Feng P et al (2018) Complete chloroplast genome sequence of Dryopteris fragrans (L.) Schott and the repeat structures against the thermal environment. Sci Rep 8(1):1–11

García Criado M, Väre H, Nieto A et al (2017) European red list of Lycopods and ferns. Brussels, Belgium; IUCN. iv + 59

Gichira AW, Long ZC, Wang QF et al (2016) Development of expressed sequence tag-based microsatellite markers for the critically endangered Isoëtes sinensis (Isoetaceae) based on transcriptome analysis. Genet Mol Res 15:15038497

Grantham HS, Moilanen A, Wilson KA et al (2008) Diminishing return on investment for biodiversity data in conservation planning. Conserv Lett 1:190–198

Gu W, Song J, Cao Y et al (2013) Application of the ITS2 region for barcoding medicinal plants of Selaginellaceae in Pteridophyta. https://doi.org/10.1371/journal.pone.0067818

Guo W, Zhu A, Fan W (2017) Complete mitochondrial genomes from the ferns Ophioglossum californicum and Psilotum nudum are highly repetitive with the largest organellar introns. New Phytol 213:391–403

Gupta M, Chyi YS, Romero-Severson J et al (1994) Amplification of DNA markers from evolutionary diverse genomes using single primers of simple-sequence repeats. Theor Appl Genet 89:998–1006

Hamada H, Petrino MG, Kakunaga T (1982) A novel repeat element with Z-DNA-forming potential is widely found in evolutionarily diverse eukaryotic genomes. Proc Natl Acad Sci U S A 79:6465–6469

Hanover JW (1966a) Environmental variation in the monoterpenes of Pinus monticola dougl. Phytochemistry 5(4):713–717

Hanover JW (1966b) Gene control of monoterpene levels in Pinus monticola dougl. Heredity 21:73–84

Hantula J, Dusabenygasani M, Hamelin RC (1996) Random amplified microsatellites (RAMS) - a novel method for characterizing genetic variation within fungi. Eur J For Pathol 26:15–166

Hayashi K (1992) PCR-SSCP: a method for detection of mutations. Genet Anal Tech Appl 3:73–79

Hayashi K, Yandell DW (1993) How sensitive is PCR-SSCP? Hum Mutat 2:338–346

Hebert PDN, Cywinska A, Ball S et al (2004) Biological identifications through DNA barcodes. Proc R Soc London B Biol Sci 270:313–321

Helena K, de John B, Doublet J et al (2005) Four tropical, closely related fern species belonging to the genus Adiantum L. are genetically distinct as revealed by ISSR fingerprinting. Genetica 125:283–291

Hu J, Vick BA (2003) TRAP (target region amplification polymorphism), a novel marker technique for plant genotyping. Plant Mol Biol Report 21(3):289–294

Jaccoud D, Peng K, Feinstein D et al (2001) Diversity arrays: a solid-state technology for sequence information independent genotyping. Nucleic Acids Res 29:25

James KE, Schneider H, Ansell SW et al (2008) Diversity arrays technology (DArT) for pan-genomic evolutionary studies of non-model organisms. PLoS ONE 3(2):e1682. https://doi.org/10.1371/journal.pone.0001682

Jeffreys AJ, Neumann R, Wilson V (1990) Repeat unit sequence variation in minisatellites: a novel source of DNA polymorphism for studying variation and mutation by single molecule analysis. Cell 60:473–485

Jia XP, Deng YM, Sun XB et al (2016) De novo assembly of the transcriptome of Neottopteris nidus using Illumina paired-end sequencing and development of EST-SSR markers. Mol Breed 36:94

Jin-Ming C, Qing-Feng W (2006) Genetic diversity in the rare and endangered Fern Isoetes orientalis. J Wuhan Bot Res 24(6):569–573

Johnson M, Irudaya Raj V, Rajkumar D (2010a) Isozymic variation studies on the selected species of Tectaria from India. J Chem Pharm Res 2(5):334–338

Johnson M, Irudaya Raj V, Rajkumar D et al (2010b) Isozyme markers for the crude drugs of maiden hair ferns from the Western Ghats, South India. Natural products: an Indian J :6

Johnson M, Irudayaraj V, Rajkumar D et al (2012) Isoperoxidase analysis on Thelypteris ciliata (Wall. ex Benth.) Holttum (Thelypteridaceae). Asian Pac J Trop Biomed 2:S27–S29. https://doi.org/10.1016/S2221-1691(12)60124-5

Jones CJ, Edwards KJ, Castaglione S et al (1997) Reproducibility testing of RAPD, AFLP, and SSR markers in plants by a network of European laboratories. Mol Breed 3:381–390

Joshi SP, Prabhakar K, Ranjekar PK (1999) Molecular markers in plant genome analysis. Curr Sci 77(2):230–240

Joshua PD, Michael SB, Norman J (2011) De novo characterization of the gametophyte transcriptome in bracken fern, Pteridium aquilinum. Genomics 12:99

Judith ES, John TM, Robbin C et al (2004) Molecular studies of representative species in the Fern genus Elaphoglossum (Dryopteridaceae) based on cpDNA sequences rbcL, trnL-F, and rps4-trnS. Int J Plant Sci 165(6). https://doi.org/10.1086/423877

Kakugawa YY, Ootsuki R (2014) Development and analysis of CAPS markers in ferns. In: Shavrukov Y (ed) Cleaved amplified polymorphic sequences (CAPS) markers in plant biology. New York, NOVA Publisher

Kantety RV, Rota ML, Matthews DE (2002) Data mining for simple sequence repeats in expressed sequence tags from barley, maize, rice, sorghum and wheat. Plant Mol Biol 48:501–510

Kao T-T, Pryer K, Freund F, Windham M, Rothfels C (2019) Low-copy nuclear sequence data confirm complex patterns of farina evolution in notholaenid ferns (Pteridaceae). Mol Phylogenet Evol 138. https://doi.org/10.1016/j.ympev.2019.05.016

Karp A, Kresovich S, Bhat KV et al (1997) Molecular tools in plant genetic resources conservation: a guide to the technologies. International plant genetic resources institute, Rome. IPGRI technical bulletin no. 2

Pryer KM, Eric S, Huiet L et al (2010) DNA barcoding exposes a case of mistaken identity in the fern horticultural trade. Mol Ecol Resour 10:979–985. https://doi.org/10.1111/j.1755-0998.2010.02858.x

Keiper FJ, McConchie R (2000) An analysis of genetic variation in natural populations of Sticherus flabellatus [R. Br. (St John)] using amplified fragment length polymorphism (AFLP) markers. Mol Ecol 9(5):571–581

Kholia BS, Fraser-Jenkins CR (2011) Misidentification makes scientific publications worthless-save our taxonomy and taxonomists. Curr Sci 100(4):458–461

Kilian A, Huttner E, Wenzl P et al (2005) The fast and the cheap: SNP and DArT-based whole genome proWling for crop improvement. In: Tuberosa R, Phillips RL, Gale M (eds) Proceedings of the international congress "in the wake of the double helix: from the green revolution to the gene revolution", 2003. Bologna, Avenue Media

Knie N, Fischer S, Grewe F et al (2015) Horsetails are the sister group to all other monilophytes and Marattiales are sister to leptosporangiate ferns. Mol Phylogenet Evol 90:140–149

Koenemann D, Maisonpierre J, Barrington D (2011) Broad-scale integrity and local divergence in the fiddlehead Fern Matteuccia struthiopteris (L.) Todaro (Onocleaceae). Am Fern J 101 (4):213–230

Kojima T, Nagaoka T, Noda K (1998) Genetic linkage map of ISSR and RAPD markers in einkorn wheat in relation to that of RFLP markers. Theor Appl Genet 96:37–45

Konieczny A, Ausube FM (1993) A procedure for mapping Arabidopsis mutations using co-dominant ecotype-specific PCR based markers. Plant J 4:403–410

Korall P, Conant D, Jordan M (2007) A molecular phylogeny of scaly tree ferns (Cyatheaceae). Am J Bot 94:873–886. https://doi.org/10.3732/ajb.94.5.873

Kranz HD, Volkar AR (1996) Molecular evolution of pteridophytes and their relationship to seed plants: evidence from complete 18S rRNA gene sequences. Plant Syst Evol 202:1–11

Kreier HP, Rex M, Weising K et al (2008) Inferring the diversification of the epiphytic fern genus Serpocaulon (Polypodiaceae) in South America using chloroplast sequences and amplified fragment length polymorphisms. Plant Syst Evol 1:274

Kress WJ, Wurdack KJ, Zimmer EA (2005) Use of DNA barcodes to identify flowering plants. Proc Natl Acad Sci U S A 102:8369–8374

Kukkala AS, Moilanen A (2013) Core concepts of spatial prioritization in systematic conservation planning. Biol Rev 88:443–464

Latta RG, Linhart YB, Lundquist L et al (2000) Patterns of monoterpene variation within individual trees in ponderosa pine. J Chem Ecol 26(6):1341–1357

Leon RP, Shepard LD et al (2010) Parallel polyploid speciation: distinct sympatric gene-pools of recurrently derived Allo-octoploid Asplenium ferns. Mol Ecol. https://doi.org/10.1111/j.1365-294X.2010.04705.x

Lestari WS, Adjie B, Jaruvantanaphan T et al (2014) Molecular phylogeny of maidenhair fern genus Adiantum (Pteridaceae) from lesser Sunda Islands Indonesia based on rbcL and trnL-F. Reinwardtia 14(1):143–156

Li G, Quiros CF (2001) Sequence-related amplified polymorphism (SRAP), a new marker system based on a simple PCR reaction: its application to mapping and gene tagging in brassica. Theor Appl Genet 103:455–461

Li F-W, Kuo L-Y, Rothfels CJ et al (2011) rbcL and matK earn two thumbs up as the Core DNA barcode for ferns. PLoS One 6(10):e26597. https://doi.org/10.1371/journal.pone.002659

Liu H, Yan Z, Xu H, Li C, Fan Q, Liao W, Liao B (2017) Development and Characterization of EST-SSR Markers Via Transcriptome Sequencing in Brainea insignis (Aspleniaceae S. L.). Appl Plant Sci. 5:1700067. https://doi.org/10.3732/apps.1700067

Liu S, Ping J, Wang Z et al (2018) Complete chloroplast genome of the tree fern Alsophila podophylla (Cyatheaceae). Mitochondrial DNA B 3(1):48–49. https://doi.org/10.1080/23802359.2017.1419095

Liu H-M, Shen J-Y, Liang Z-L et al (2019) Two out of one: revising the diversity of the epiphytic fern genus Scleroglossum (Polypodiaceae, Grammitidoideae) in southern China. In: Cai J, Yu W-B, Zhang T, Li D-Z (eds) Revealing of the plant diversity in China's biodiversity hotspots, vol 130. PhytoKeys, London, pp 115–133

Luis Q, Santiago P, Emilia Pet al (2007) Allozyme variation in the sympatric ferns Culcita macrocarpa and Woodwardia radicans at the northern extreme of their ranges. Plant Syst Evol 263:135–144

Luo HM, Li Y, Sun C et al (2010) Comparison of 454-ESTs from Huperzia serrata and Phlegmariurus carinatus reveals putative genes involved in lycopodium alkaloid biosynthesis and developmental regulation. BMC Plant Biol 10:209

Lynch M, Walsh B (1998) Genetics and analysis of quantitative traits. Sinauer Associates, Sunderland, MA

Ma XY, Xie CX, Liu C et al (2010) Species identification of medicinal pteridophytes by a DNA barcode marker, the chloroplast psbA-trnH intergenic region. Biol Pharm Bull 33 (11):1919–1924

Makino A, Sakashita H, Hidema J (1992) Distinctive responses of ribulose-1,5-bisphosphate carboxylase and carbonic anhydrase in wheat leaves to nitrogen nutrition and their possible relationships to CO2-transfer resistance. Plant Physiol 100:1737–1743

Manuela W, Marcus K, Peter H (2011) High gene flow in epiphytic ferns despite habitat loss and fragmentation. Conserv Genet 12:1411–1420

Marchant DB, Sessa EB, Wolf PG et al (2019) The C-Fern (Ceratopteris richardii) genome: insights into plant genome evolution with the first partial homosporous fern genome assembly. Sci Rep 9:18181. https://doi.org/10.1038/s41598-019-53968-8

Marcial-Quino J, Mendoza-Espinoza AJ, Sierra-Palacios E et al (2015) DNA barcoding: an alternative for the identification of the medicinal plants employed in Mexico. J Plant Sci 10 (4):116–127

Maria D, Logacheva AA (2017) A comparative analysis of inverted repeats of polypod fern (Polypodiales) plastomes reveals two hypervariable regions. BMC Plant Biol 17(2):255

Marimuthu J, Manickam VS (2011) Ex-situ conservation of two threatened ferns of the Western Ghats through in vitro spore culture. J Threat Taxa 3(7):1919–1928

Matsumoto A, Tsumura Y (2004) Evaluation of cleaved amplified polymorphic sequence markers for *Chamaecyparis obtusa* based on expressed sequence tag information from *Cryptomeria japonica*. Theor Appl Genet 110:80–91

McDermott JM, Brandle U, Dutly F (1994) Genetic variation in powdery mildew of barley: development of RAPD, SCAR and VNTR markers. Phytopathology 84:1316–1321

Meyer W, Michell TG, Freedman EZ et al (1993) Hybridization probes for conventional DNA fingerprinting used as single primers in polymerase chain reaction to distinguish strain of Cryptococcus neoformans. J Clin Microbiol 31:2274–2280

Michael SD, Amasino RM (1998) A robust method for detecting single-nucleotide changes as polymorphic markers by PCR. Plant J 14:381–385

Morajkar S (2018) Biodiversity studies of Ferns of Western Ghats using Molecular Markers. PhD thesis submitted to Mangalore University

Morajkar S, Hegde S (2021) DNA barcoding identifies a potential new ecotype of Chinese brake Fern, Pteris vittata L. Nano. Proc Natl Acad Sci India Sect B Biol Sci. https://doi.org/10.1007/s40011-021-01231-4

Morajkar S, Sajeev S, Hegde S (2015) A DNA barcode for Stenochlaena palustris: an edible medicinal fern. Indian Fern J 32:124–131

Muhammad I, Muhammad I (2014) Molecular markers in plants for analysis of genetic diversity: a review. Eur Acad Res 2(1):1513–1540

Nair A, Fathima M, Tilton F (2017) DNA barcoding of medicinal pteridophyte Selaginella delicatula from Western Ghats of Kerala. Int J Curr Res 9(5):50887–50890

Nakamura Y, Leppert M, O'Connell P (1987) Variable number tandem repeat (VNTR) markers for human gene mapping. Science 235:1616–1622

Nakazato T, Jung MK, Housworth EA et al (2006) Genetic map-based analysis of genome structure in the homosporous fern Ceratopteris richardii. Genetics 173(3):1585–1597

Naomi K, Steve W, Noeleen S (2004) Conservation genetics and ecology of Angiopteris chauliodonta Copel. (Marattiaceae), a critically endangered fern from Pitcairn Island, south Central Pacific Ocean. Biol Conserv 117(3):309–319

Nitta JH (2008) Exploring the utility of three plastid loci for biocoding the filmy ferns (Hymenophyllaceae) of Moorea. Taxon 57(3):725–736

Nitta JH, Meyer J, Taputuarai R et al (2017) Life cycle matters: DNA barcoding reveals contrasting community structure between fern sporophytes and gametophytes. Ecol Monogr 87:278–296

Nitta JH, Atsushi E, Smith AR (2020) Taxonomic and molecular survey of the pteridophytes of the Nectandra cloud Forest reserve, Costa Rica. PLoS One 10:1371

Olmstead RG, Palmer JD (1994) Chloroplast DNA systematics: a review of methods and data analysis. Am J Bot 81(9):1205–1224

Olsvik O, Wahlberg J, Petterson B (1993) Use of automated sequencing of polymerase chain reaction-generated amplicons to identify three types of cholera toxin subunit B in vibrio choleraeO1 strains. J Clin Microbiol 31:22–25

Orita M, Iwahana H, Kanazawa H (1989) Detection of polymorphisms of human DNA by gel electrophoresis as single-strand conformation polymorphisms. Proc Natl Acad Sci U S A 86:2766–2770

Pandey V, Nutter RC, Prediger E (2008) Applied biosystems SOLiD™ system: ligation-based sequencing. In: Michal J (ed) Next generation genome sequencing: towards personalized medicine. Betz-Druck GmbH, Darmstadt, pp 29–42. https://doi.org/10.1002/9783527625130.ch3

Paran I, Michelmore RW (1993) Development of reliable PCR-based markers linked to downy mildew resistance genes in lettuce. Theor Appl Genet 85:985–993

Patel M, Reddy MN (2018) Discovery of the World's smallest terrestrial Pteridophyte. Sci Rep 8:5911

Patil SM, Kachhiyapatel RN, Patel RS et al (2018) Phytotaxa 351(4):273–280

Pimm SL, Joppa LN (2015) How many plant species are there, where are they, and at what rate are they going extinct? Ann Mo Bot Gard 100:170–176

Pimm SL, Jenkins CN, Abell R et al (2014) The biodiversity of species and their rates of extinction, distribution, and protection. Science 344:987–997

Prasad P, Eswara Reddy NP, Anandam RJ et al (2003) Isozymes variability among fusarium udum resistant cultivars of pigeonpea (Cajanuscajan (L.) (Millsp)). Acta Physiol Plant 25:221–228

Priyanka S, Nilima B (2014) Assessment of genetic diversity among Marsilea populations of Hadauti plateau as revealed by RAPD markers. Int J Pure App Biosci 2(2):254–261

Ratnasingham S, Hebert PDN (2007) BOLD: the barcode of life data system. Molecular Ecology Notes 7:355–364

Raubeson LA, Stein DB (1995) Insights into fern evolution from mapping chloroplast genomes. American Fern J 85(4):193–204. https://doi.org/10.2307/1547809

Rohde W (1996) Inverse sequence-tagged repeat (ISTR) analysis, a novel and universal PCR-based technique for genome analysis in the plant and animal kingdom. J Genet Breed 50:249–261

Ronaghi M, Karamohamed S, Pettersson B (1996) Real-time DNA sequencing using detection of pyrophosphate release. Anal Biochem 242:84–89

Ross MG, Russ C, Costello M (2013) Characterizing and measuring bias in sequence data. Genome Biol 14(5):R51

Rothberg JM, Hinz W, Rearick TM et al (2011) An integrated semiconductor device enabling non-optical genome sequencing. Nature 475:348–352

Sammour RH, Karam MA, Morsi YS et al (2021) Population structure and phylogenetic relationships in Brassica rapa L. subspecies by using isozyme markers. Braz J Biol 81 (3):601–610

Sankar AA, Moore GA (2001) Evaluation of inter-simple sequence repeat analysis for mapping in Citms and extension of genetic linkage map. Theor Appl Genet 102:206–214

Santos M, Fernandes C, Tietbohl L et al (2013) Chemical composition of essential oils from two Fern species of anemia. Am Fern J 103(4):215–224

Schlotterer C, Tautz D (1992) Slippage synthesis of simple sequence DNA. Nucleic Acids Res 20:211–215

Schuettpelz E, Schneider H, Huiet L, Windham MD, Pryer KM (2007) A molecular phylogeny of the fern family Pteridaceae: assessing overall relationships and the affinities of previously unsampled genera. Mol Phylogenet Evol 44(3):1172–1185

Shao Y, Wei R, Zhang X et al (2015) Molecular phylogeny of the cliff ferns (Woodsiaceae: Polypodiales) with a proposed Infrageneric classification. PLoS One 9:8–10

Sharma U, Gupta PD (2013) A study on population genetics and molecular markers with its applications in conservation of wildlife. Int J Pharm Bio Sci 4(3):287–298

Shepherd LD, Perrie LR, Parris BS, Brownsey PJ (2007) A molecular phylogeny for the New Zealand Blechnaceae ferns from analyses of chloroplast trn L-trn F DNA sequences. N Z J Bot 45(1):67–80

Shukla G, Chakravarty S (2012) Fern diversity and biomass at Chilapatta reserve forest of West Bengal Terai Duars in sub-humid tropical foothills of Indian eastern Himalayas. J For Res 23 (4):609–613

Smith AR, Pryer KM, Schuettpelz E et al (2008) Fern classification. Biology and evolution of ferns and lycophytes. Cambridge University Press, Cambridge, pp 417–467

Squillace AE (1971) Inheritance of monoterpene composition in cortical oleoresin of slash pine. For Sci 17:381–387

Stacy AJ, David SB (2017) Two Beringian origins for the Allotetraploid Fern Polystichum braunii (Dryopteridaceae). Syst Bot 42(1):6–16

Strauss SH, Critchfield WB (1982) Inheritance of β-pinene in xylem oleoresin of Knobcone x Monterey pine hybrids. For Sci 28(4):687–696

Sturgeon KB (1979) Monoterpene variation in ponderosa pine xylem resin related to western pine beetle predation. Evolution 33:803–814

Thottappilly G, Magonouna HD, Omit Ogun OG (2000) The use of DNA markers for rapid improvement of crops in Africa. Afr Crop Sci J 8(1):99–108

Timothy J, Motley CWM (2001) Utility of RAPD markers in evaluating the status of the Hawaiian tree Fern Cibotium xheleniae L. Pac Sci 55(2):145–155
Trewick SA, Morgan M, Richards SS (2002) Polyploidy, phylogeography and Pleistocene refugia of the rockfern Asplenium ceterach: evidence from chloroplast DNA. Mol Ecol 11:2003–2012
Tsai-Wen H, Shan-Juei M, Chiang T-Y et al (2009) Low RAPD polymorphism in Archangiopteris itoi, a rare and endemic fern in Taiwan. Bot Bull Acad Sin. 41:15–18. 2000
Turcatti G, Romieu A, Fedurco M (2008) A new class of cleavable fluorescent nucleotides: synthesis and optimization as reversible terminators for DNA sequencing by synthesis. Nucleic Acids Res 36:e25
Varma PKV, Madusoodanan PV (2014) Molecular systematics of the filmy ferns (hymenophyllaceae) of South India. Indian Fern J 31:65–83
Vinciane M, Dauphin B, Grant J (2020) A transcriptome for the early-branching fern Botrychium 2 lunaria enables fine-grained resolution of population 3 structure. https://doi.org/10.1101/2020.02.17.952283
Vogel J, Rumsey FJ et al (1999) Genetic structure, reproductive biology and ecology of isolated populations of Asplenium csikii (Aspleniaceae, Pteridophyta). Heredity 83:604–612
Vos P, Hogers R, Bleeker M et al (1995) AFLP: a new technique for DNA fingerprinting. Nucleic Acids Res 23(21):4407–4414
Wang G, Mahalingan R, Knap HT (1998) (C-A) and (GA) anchored simple sequence repeats (ASSRs) generated polymorphism in soybean, Glycine max (L.) Merr. Theor Appl Genet 96:1086–1096
Wang L, Wu ZQ, Bystriakova N et al (2011) Phylogeography of the Sino-Himalayan Fern Lepisorus clathratus on "the roof of the world". PLoS One 6(9):e25896. https://doi.org/10.1371/journal.pone.0025896
Wang T, Yingjuan S, Li Y (2012a) Population genetic variation in the tree Fern Alsophila spinulosa (Cyatheaceae): effects of reproductive strategy. PLoS One 7(7):1–7
Wang T, Su Y, Li Y (2012b) Population genetic variation in the tree Fern Alsophila spinulosa (Cyatheaceae): effects of reproductive strategy. PLoS One 7(7):1–7
Wang FH, Lu JM, Wen J et al (2016) Applying DNA barcodes to identify closely related species of ferns: a case study of the Chinese Adiantum (Pteridaceae). PLoS One 11(9):e0160611. https://doi.org/10.1371/journal.pone.0160611
Wei R, Yan YH, Harris AJ, Kang JS, Shen H, Xiang QP, Zhang XC (2017) Plastid phylogenomics resolve deep relationships among eupolypod II ferns with rapid radiation and rate heterogeneity. Genome Biol Evol 9(6):1646–1657
Weising K, Nybom H, Wolff K (1995) DNA fingerprinting in plants and fungi. CRC Press, Boca Raton
Wenzl P, Carling J, Kurn D (2004) Diversity arrays technology (Dart) for whole genome profiling of barley. Proc Natl Acad Sci U S A 101(26):9915–9920
Werth RC (1989) The use of isozyme data for inferring ancestry of polyploid pteridophytes. Biochem Syst Ecol 17(2):117–130
Williams JGK, Kubelik AR, Livak KJ (1990) DNA polymorphisms amplified by arbitrary primers are useful as genetic markers. Nucleic Acids Res 18:6531–6535
Williams E, Theis Z, Hoess C (2016) Identifying a cryptic Adiantum population through DNA barcoding. Am Fern J 106(2):135–142
Wolf PG, Carol A, Rowe R et al (2003) Complete nucleotide sequence of the chloroplast genome from a Leptosporangiate Fern, Adiantum capillus-veneris L. DNA Res 10(2):59–65
Wolf PG, Jessie M, Roper JM (2010) The evolution of chloroplast genome structure in ferns. Genome 53(9):731–738
Wolf PG, Sessa EB, Marchant DB et al (2015) An exploration into fern genome space. Genome Biol Evol. 7(9):2533–2544. https://doi.org/10.1093/gbe/evv163
Woodhead M, Russell J, Squirrell J, Hollingsworth PM, Cardle L, Ramsay L et al (2003) Development of EST-SSRs from the alpine Lady-fern, Athyrium distentifolium. Mol Ecol Notes 3(2):287–290

Wu KS, Jones R, Danneberger L (1994) Detection of microsatellite polymorphisms without cloning. Nucleic Acids Res 22:3257–3258

Xiang D, Hong L, Wei GA et al (2018) Estimation of mating system in the endangered aquatic Fern Ceratopteris pteridoides in China based on AFLP molecular marker and Selfing test: implications for conservation. Not Bot Horti Agrobo 46(2):688–699

Xiao XR, Zhen W, Wang T et al (2017) Novel nuclear SSR markers in the large frond tree fern Alsophila gigantea and its congeneric species Alsophila spinulosa. Acta Sci Nat Univ Sunyatseni 56(1):115–120

Yaung KL, Te-Yen T, Fay-Wei L et al (2018) Organelle genome inheritance in Deparia ferns (Athyriaceae, Aspleniineae, Polypodiales). Front Plant Sci 9:486. https://doi.org/10.3389/fpls.2018.00486

Yi-Shan C, Germinal R, Victor A et al (2014) Molecular phylogeny and biogeography of the fern genus Pteris (Pteridaceae). Ann Bot. 114:1. https://doi.org/10.1093/aob/mcu086

Yuan-Huo D, Jin-Ming C, Robert WG et al (2007) Gene flow in populations of the endangered aquatic fern Ceratopteris pteridoides in China as revealed by ISSR markers. Aquat Bot 87:69–74

Yuan-huo D, Jin-ming C, Gituruwahiti R et al (2008) Genetic variation in the endangered aquatic fern Ceratopteris thalictroides (Parkeriaceae) in China: implications from RAPD and ISSR data. Bot J Linn Soc 157:657–671

Zietkiewicz E, Rafalski A, Labuda D (1994) Genome fingerprinting by simple sequence repeat (SSR) anchored polymerase chain reaction amplification. Genomics 20:176–183

Zi-Juan W, Kai-Yun G (2011) High genetic diversity and low genetic differentiation in the relict tree fern Sphaeropteris brunoniana (Cyatheaceae) revealed by amplified fragment length polymorphism (AFLP). Bot Stud 52:231–238

Zuo Y, Chen Z, Kondo K et al (2011) DNA barcoding of Panax species. Planta Med 77:182–187

**Shaiesh Morajkar** His expertise is in the utility of molecular markers in plant taxonomic, diversity, and breeding studies. Along with the molecular diversity studies, his research interests include studies of stress resistance in crops, plant-environment association, hyperspectral remote sensing, soil sciences, and wetland ecosystems. He has got several peer-reviewed publications in his field of interest.

**C. Suneetha** is working as Assistant Professor at the Department of Horticulture, College of Agriculture, University of Agricultural Sciences, GKVK, Bangalore. She holds a M.Sc. in Horticulture and her Ph.D., Biotechnology. She is specialized in the field of Floriculture, Horticulture, and Plant Biotechnology. She has over 15 years of research experience in the area of tissue culture, molecular markers, and plant molecular biology. She has completed a research project entitled "DNA Barcoding studies in Ferns of South India for Genetic Diversity Studies" funded by SERB, DST, GOI, New Delhi, as Principal Investigator. She also worked as Co PI in university-funded projects. She has got several peer-reviewed publications to her credit.

**Harish Kumar T. G.** Ph.D. Scholar at the Department of Horticulture, College of Agriculture, University of Agricultural Sciences, GKVK, Bangalore. He holds a M.Sc. Horticulture in Vegetable Science degree to his credit. He is specialized in the field of Vegetable Science, Horticulture, Floriculture, and Plant Breeding and Biotechnology. His research interests are molecular markers and plant molecular biology. He has got several peer-reviewed publications.

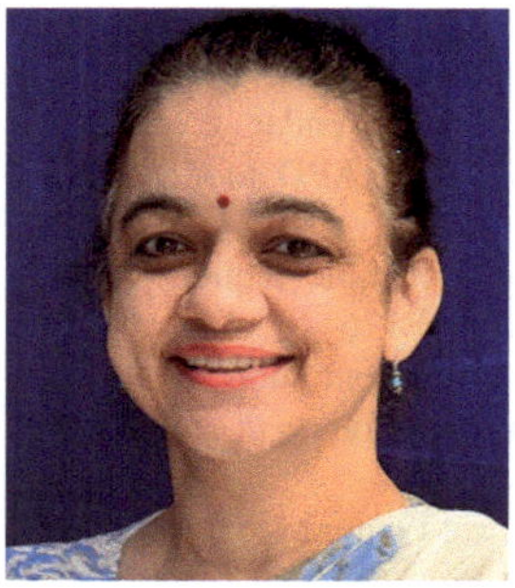

**Smitha Hegde** NUCSER, Nitte DU, Deralakatte, Mangalore, has 30 years of experience in the pteridology. She obtained her M.Sc. from the University of Bombay and Ph.D. Biosciences from St. Aloysius College, Mangalore University, India. Her doctoral thesis was on tissue culture propagation of *Drynaria quercifolia* L. She has conducted pioneering work in biodiversity and distribution studies, developmental studies, phytoremediation, plant tissue culture, and molecular markers related to pteridophytes of the region. She has over 74 publications including book chapters and journals, edited 3 books, 2 published patents, and successfully guided 112 post-graduate research project students and 10 doctoral students. She is currently the PI of a state-funded project on plant tissue culture and heavy metal remediation using pteridophytes.

# The Crucial Role of *Ceratopteris richardii* in Understanding the Evolution of the *WOX* Gene Family

6

Christopher E. Youngstrom, Erin E. Irish, and Chi-Lien Cheng

**Abstract**

While continuing to be developed as a fern model, *Ceratopteris richardii* has demonstrated its value in elucidating evolutionary developmental questions in land plants, in addition to other aspects of plant biology. Plants are unique in that they have the ability to produce new organs throughout their life due to persistent proliferation of tissue-specific stem cell populations. The family of WUSCHEL-related homeobox (WOX) genes are important transcription factors for stem cell maintenance. WOX genes are found in all land plants and some extant algae. This chapter concentrates on the current understanding of the evolution of the WOX gene family in controlling cell proliferation of various meristem domains in distantly related land plant models, *Physcomitrella patens, C. richardii*, and *Arabidopsis thaliana*.

**Keywords**

*Ceratopteris richardii* · Evo-devo · Plant evolution · WOX · WUSHEL-related homeobox

## 6.1 Introduction

### 6.1.1 *Ceratopteris richardii* as a Fern Model

The advantages of choosing *Ceratopteris richardii* as a fern model were detailed in two seminal papers by Hickok et al. (1987, 1995). It is worth listing these desirable features here again 35 years later. Like 90% of ferns, *C. richardii* is homosporous;

C. E. Youngstrom · E. E. Irish · C.-L. Cheng (✉)
Department of Biology, The University of Iowa, Iowa City, IA, USA
e-mail: christopher-youngstrom@uiowa.edu; erin-irish@uiowa.edu; chi-lien-cheng@uiowa.edu





J. Marimuthu et al. (eds.), *Ferns*, https://doi.org/10.1007/978-981-16-6170-9_6

unlike most ferns, *C. richardii* grows as an annual with a short life cycle of ~120 days from spore to spore. A single mature *C. richardii* sporophyte can produce more than a million spores per month. In parallel, *C. richardii* sporophytes can be vegetatively propagated from bulbils (leaf buds) often seen on the leaf margins. The relatively small stature of sporophytes allows them to be cultured conveniently in laboratory settings. The single-cell-layered gametophytes are tiny (~1 mm diameter for hermaphrodites and even smaller for males) and can be easily grown on defined media. *C. richardii* gametophytes take 10–12 days to become sexually mature, at which time, intra- or inter-gametophyte fertilization can be achieved. In addition to these favorable features, *C. richardii*, being a diploid fern, is suitable for classical genetic analysis.

Because spores can be mutagenized and phenotypes screened in the haploid gametophytes, mutants resistant to herbicides, insensitive to hormones ABA and antheridiogen, and defective in sperm movement, among others, have been isolated and characterized (reviewed in Hickok 1987, 1995; Atallah and Banks 2015; McAdam et al. 2016). A genetic linkage map with more than 700 molecular markers was constructed (Nakazato et al. 2006), which can facilitate mutant analysis and the identification of quantitative trait loci (Nakazato et al. 2007). Although *C. richardii* offers an excellent system for genetic analysis, the large genome size (11.25 Gb) hampers forward genetic approaches to identify the mutated genes. An obvious solution for this problem is to take the reverse genetic approach. RNAi uptake by spores (Stout et al. 2003) and DNA bombardment into gametophytes (Rutherford et al. 2004) were able to cause systemic gene suppression. Now with new and necessary resources and technologies developed as described below, it is possible to carry out reverse genetic analysis in this model fern in a routine and tractable manner.

### 6.1.2 Further Development of *C. richardii* as a Fern Model

Recently, transcriptomes of *C. richardii* gametophytes (Atallah et al. 2018) and sporophytes (Geng et al. 2021) have been published. They can be a valuable resource for identifying the homologs for genes of interest in *C. richardii*. The publication of a partial genome assembly of *C. richardii* (Marchant et al. 2019) provides additional information. Furthermore, the sequenced and annotated genomes of two heterosporous ferns with exceptionally small genomes, *Azolla filiculoides* and *Salvinia cucullata* (Li et al. 2018), further serve as valuable fern reference genomes. In addition to these genome resources, efficient and tractable transformation of *C. richardii* has been made possible using *Agrobacterium*-mediated transformation of gametophytes directly (Bui et al. 2015) or by bombardment of callus generated from sporophytes (Plackett et al. 2014).

### 6.1.3 A Fern Model Facilitates Evolutionary Development Study

The fern clade consists of over 10,000 extant species (Smith et al. 2006; PPG1. 2016), second in diversity only to angiosperms among vascular plants. Ferns are located favorably in the land pant phylogeny: sister to seed plants yet possessing free-living gametophytes characteristic of earlier diverging, nonvascular plants. The One Thousand Plant Transcriptomes Initiative (1KP) provides a robust phylogenomic framework for green plant evolution (1KP 2019). This facilitates phylogenetically informed experiments to investigate evo-devo questions, but more model organisms at key branches in the phylogeny are urgently needed. A well-described life cycle is essential for the fern to serve as an evo-devo model. Excellent studies of defining *C. richardii* developmental stages were made for the gametophyte (reviewed by Banks 1999; Bartz and Gola 2018) as well as the embryo (Johnson and Renzaglia 2008), leaf (Hill 2001), and root (Hou and Blancaflor 2009; Hou and Hill 2002) of the sporophyte. Recently, two additional, more comprehensive studies of the life cycle of *C. richardii* were published (Conway and Di Stilio 2020; Aragón-Raygoza et al. 2020). With multiple reverse genetic methodologies firmly established in this model fern, functional studies of genes important in the evolution of plant development, including *AINTEGUMENTA* in apogamy (Bui et al. 2017), *LEAFY* in the control of apical stem cell (Plackett et al. 2018), and a *WUSCHEL-related homeobox* (*WOX*) gene in stem cell proliferation (Youngstrom et al. 2019), have been carried out. This chapter focuses on the evo-devo study of *WOX* genes in land plants represented by *Arabidopsis thaliana*, *C. richardii*, and *Physcomitrella patens*.

## 6.2 Discovery of *WUSCHEL* and the Stem Cell Maintenance Model

The WOX proteins are homeodomain transcription factors. The founding member of the *WOX* gene family *WUSCHEL* (*WUS*) was identified in *A. thaliana* by isolating and characterizing loss-of-function mutants that are defective in shoot and floral meristem development (Laux et al. 1996) and followed by the cloning and characterization of the *WUS* gene (Mayer and Schoorman 1998). Because stem cells in the shoot apical meristem (SAM) seem to be mis-specified in the *wus* mutants and *WUS* is expressed in the organizing center (OC) underneath the stem cells but not in the stem cells themselves, Mayer and Schoorman (1998) proposed a model in which WUS functions to specify stem cells in a non-cell autonomous manner. This model eventually led to the CLAVATA-WUS feedback loop model of shoot meristem maintenance (Schoof et al. 2000) after incorporating the discovery of the regulatory function of the CLV1 and CLV3 receptor/ligand pair in the SAM (Clark et al. 1997; Fletcher et al. 1999; Brand et al. 2000). One key feature of the model predicts that the WUS protein moves from the OC to the stem cells where it maintains their identity; the cell-to-cell movement of WUS was later demonstrated (Yadav et al. 2011; for a recent review, see Fuchs and Lohmann 2020).

Other CLAVATA-WUS-like gene regulatory pathways for stem cell maintenance have been found to act in the root apical meristem (RAM) (Sarkar et al. 2007; Stahl et al. 2009) and in the cambium (Hirakawa et al. 2010) by involving different members of the *WUS*-like genes and *CLAVATA* gene pairs.

## 6.3 The *WOX* Gene Family

There are 15 *WOX* genes in *A. thaliana* (Haecker et al. 2004), and their functions in this dicot model remain the best investigated among all plant species. Phylogenic trees of the WOX proteins group them into three clades, Clades I, II, and III (Zhang et al. 2010). Clade I, also named the WUS-like, or modern clade (consisting of AtWUS and AtWOX1-7) contain members from seed plants and ferns; Clade II, the intermediate clade (consisting of AtWOX8, 9, 11, and 12), are present in vascular plants; and Clade III, the ancient clade (consisting of AtWOX10, 13, and 14), are present in all land plants (van der Graaff et al. 2009; Nardmann et al. 2009; Lian et al. 2014; Segatto et al. 2016). A more recent WOX phylogenic tree was built by including an expanded sample of species spanning Viridiplantae and Rhodophyta with improved sequence alignment methods. This tree encompasses three ancient superclades, termed T1WOX, T2WOX, and T3WOX (Wu et al. 2019). The overall topology of the tree is not altered; three land plant WOX clades are present as in previous trees. The major difference concerns the lycophyte and fern WOX proteins grouped previously in the intermediate clade. In this new tree, they are grouped, although weakly supported, to separate branches of the T3WOX clade. Note that in this tree, there are no T2WOX proteins in the extant lycophytes and ferns; the authors consider them lost from these seed-free vascular plants. The fern *C. richardii* contains five members of the *WOX* gene family: the ancestral clade *CrWOX13A* and *CrWOX13B* (T1WOX clade), the intermediate clade *CrWOXA* and *CrWOXB* (now in the T3WOX clade), and the modern clade *CrWUL* (T3WOX clade) (Nardmann and Werr 2012).

In addition to *C. richardii*, the WOX proteins from two heterosporous ferns, *Azolla filiculoides* and *Salvinia cucullata*, were included in the recent tree. *A. filiculoides* has eight and *S. cucullata* has six WOX family members, both numbers higher than that in *C. richardii*. More *WOX* genes in *A. filiculoides* than in *S. cucullata* is likely due to a whole genome duplication event in the *A. filiculoides* lineage after the divergence of the two heterosporous ferns (Li et al. 2018). Because the complete genome sequence and annotation of *C. richardii* is not available, it is possible that additional yet unidentified WOX members are present in this model fern.

### 6.3.1 The T1WOX Proteins

T1WOX proteins exist in all land plants examined and are the only type of WOX proteins found in the bryophytes. The *PpWOX13*-like genes were first identified in

*P. patens* among genes whose expression level peaks in the first 25 h of protoplast reprogramming in a global gene expression analysis (Xiao et al. 2012). Expression and functional analysis by Sakakibara et al. (2014) demonstrated that two *PpWOX13*-like genes, *PpWOX13A* and *PpWOX13B*, are expressed in all tissues of gametophytes and sporophytes but more strongly in the apical cells. These genes function redundantly for zygote development and for reprogramming of detached leaves and protoplasts into stem cells. The authors attributed the requirement of *PpWOX13*-like genes in reprogramming stem cells to the upregulation of cell wall loosening factors by these WOX proteins. Cell loosening factors change the osmolarity and cause cell expansion which is considered a sine qua non condition for cell division. As T1WOX genes are found in many chlorophytes (Wu et al. 2019), the *PpWOX13*-like genes' function in regulating cell wall loosening enzymes is likely to be ancestral. This notion is further supported by results of experiments testing whether the *A. thaliana* T1WOX proteins still retain this ancient function. The two *A. thaliana* T1WOX genes, *AtWOX13* and *AtWOX14*, are expressed in similar but not identical patterns in seedlings and in developing flowers and fruits (Deveaux et al. 2008; Romera-Branchat et al. 2013); however, these *Arabidopsis* T1WOX proteins are not required for root meristem (RAM) regeneration, after wounding, suggesting that they are not involved in regulating osmolarity in *A. thaliana* (Sakakibara et al. 2014). Because these *PpWOX13*-like genes are expressed in all cells examined, their function in the stem cells are unlikely to be non-cell autonomous. Given the early divergence of *P. patens* in land plants, this feature of the T1WOX genes in the moss likely represents an ancestral feature of the gene family. Consistent to this view, the *A. thaliana* T1WOX proteins AtWOX13 and AtWOX14 have never been shown to be mobile.

The analysis of the *C. richardii* T1WOX genes is limited. *CrWOX13A* is expressed quite high in root tips, gametophytes, and young sporophytes based on RT-PCR results, but *CrWOX13B* expression was not detected (Nardmann and Werr 2012). Protein comparison showed that all T1WOX proteins have a conserved motif (WOX13 motif) N-terminal to the homeodomain and a subset of the land plant T1WOX proteins also contain a conserved motif (ESExE motif) C-terminal to the homeodomain. Additionally, angiosperm T1WOX proteins contain a conserved motif (YxDpI motif) located between the WOX13 motif and the homeodomain (Zhang et al. 2017). Apart from sequence homology, there is little understanding as to the roles of each conserved motif, and it is unknown whether these motifs are conserved among fern and lycophyte T1WOX proteins.

### 6.3.2 The T2WOX Proteins

The *A. thaliana* T2WOX clades consists of *AtWOX8, 9, 11*, and *12*. AtWOX8 and AtWOX9 act redundantly with the T3WOX protein AtWOX2 to determine the apical-basal polarity of the zygote (see Chandler and Werr 2019 for a review). In addition to embryo development, AtWOX9 is also involved in post-embryonic meristematic pluripotency in vegetative SAM, leaf primordia, and floral meristems and in emerging floral organs and in roots by preventing cells from exiting the cell cycle prematurely and entering differentiation (Wu et al. 2005). AtWOX9 acts downstream of cytokinin signaling and upstream of sugar sensing/signaling in promoting SAM establishment (Skylar et al. 2010). During adventitious root formation from *A. thaliana* leaf explants, the expression of *AtWOX11* is induced by auxin and marks the transition from a procambium cell to a root founder cell, with AtWOX11 and AtWOX12 acting redundantly in this process (Liu et al. 2014). Together, they activate T3WOX genes *AtWOX5* and *AtWOX7* to form root primordium (Hu and Xu 2016).

Although T2WOX genes are thought to have been lost from ferns and lycophytes (Wu et al. 2019), evidence of their phylogenic positions has poor support (Nardmann and Werr 2013; Wu et al. 2019). RT-PCR showed that *CrWOXA* is preferentially expressed in the root tips when compared with gametophytes and young sporophytes but *CrWOXB* is expressed highly in all three types of tissues. Whereas *CrWOXB* is expressed broadly in the roots, the cellular expression pattern of *CrWOXA* gene is more specific (Nardmann and Werr 2012). In *C. richardii*, the root arises from a swollen root apical mother cell (RAMC) in the hypodermal cell layer. The lateral root mother cells (LRMC) originate in the root endodermis, usually starting from the fifth root (Hou et al. 2004). Both RAMC and LRMC divide four times to form the tetrahedral root apical cell and four daughters with the most distal one fated as the root cap initial (Hou and Blancaflor 2009; Hou and Hill 2002). *CrWOXA* is specifically expressed in the RAMC and the LRMC with expression ceasing after the completion of the fourth cell division and then is restricted to the distal merophyte that acquires root cap identity. These dynamic changes of *CrWOXA* expression, although not tested, reflect positional cues for establishing the RAMC and LRMC fate and the polarity for root development (Nardmann and Werr 2012). The cellular expression of *CrWOXA* in fern RAMC and LRMC is reminiscent of the expression and function of the *A. thaliana* T2WOX genes *AtWOX11* and *AtWOX12* in the transition from a procambium cell to a root founder cell during adventitious root formation. Indeed, normally only one stem-borne root is formed at each node, but auxin treatment stimulates multiple RAMC formation at a single node. These auxin-induced RAMCs correlate with where *CrWOXA* and *CrWUL* are expressed and the eventual development of supernumerary roots (Yu et al. 2020). The authors further demonstrated, using a tobacco transient expression system, that CrWOXA binds to promoter regions of *CrWUL* to stimulate its transcription. In planta, whether the CrWOXA protein activates the transcription of *CrWUL* in RAMC/LRMC to form the merophytes and whether *CrWOXA* continues to express in the root cap cells are unknown (also see CrWUL in 3.3).

So far, *CrWOXB* is the only fern WOX gene whose functional analysis in the fern has been published (Youngstrom et al. 2019). This gene is expressed in actively dividing cells in both gametophytes and sporophytes. In the gametophyte, it is expressed in the notch meristem of the hermaphrodite, while in males, its expression in the meristematic cells is most apparent before those cells give rise to antheridia. In sporophytes, they are expressed in all areas with active cell division, including SAM, root primordia, vascular bundle, and leaf primordia. Interestingly, the *CrWOXB* transcript is not detectable in the shoot apical cell or root apical cell but is present in the leaf apical cell. The expression pattern is consistent with the phenotypes of the *CrWOXB* RNAi knockdown lines. Gametophytes are smaller due to fewer cell divisions in the notch meristem region, and sporophytes have fewer vegetative leaves, fewer sporophyll pinna, and fewer roots and lateral roots. These phenotypes can be explained as due to fewer pluripotent cells available for differentiation into organs, thereby delaying development. The expression pattern and the function of *CrWOXB* are similar to the post-embryonic function of the *A. thaliana* T2WOX gene *AtWOX9* which prevents cells from exiting the cell cycle and differentiating prematurely (Wu et al. 2005).

The AtWOX8/9 orthologs of T2WOX proteins in gymnosperms and dicots contain three conserved motifs outside of the homeodomain, two at the C-terminus (VFIN WOX8 motif and LQxG WOX8 motif) and one N-terminal to the homeodomain (WOX9 motif), whereas rice does not have the VFIN WOX8 motif (Zhang et al. 2016). The WOX11/12 orthologs within seed plants only have the two conserved domains at the C-terminus (Zhang et al. 2016). CrWOXA and CrWOXB appear to have a rudimentary LQxG WOX8 motif but lack the other two motifs. The T2WOX-specific WUS box conserved in *Arabidopsis* T2WOX proteins is not found in CrWOXA and CrWOXB (Youngstrom et al. 2019). How each of these motifs lends itself in T2WOX gene function remains unclear.

### 6.3.3 The T3WOX Proteins

The T3WOX or the WUS-like WOX clade is only found in lycophytes, ferns, and seed plants and is greatly expanded in the seed plants where they function embryonically and/or post-embryonically. In the embryo, in addition to working together with the T2WOX genes *ATWOX8* and *AtWOX9* to determine embryo polarity, *AtWOX2* acts together with *AtWOX1*, *AtWOX3,* and *AtWOX5* as a module for properly setting up the embryonic SAM (Zhang et al. 2017). In addition to embryo development, these and other T3WOX genes function in a variety of primordia throughout the plant. *AtWUS*, *AtWOX5*, and *AtWOX4* play major roles in stem cell maintenance at the SAM, the RAM, and the cambium, respectively. *AtWUS* and *AtWOX5* are transcribed in the OC of the SAM and the quiescent center of the RAM, respectively, and their proteins move to the central zone and columella stem cells to maintain the stem cell fate in the SAM and RAM, respectively. Whereas the SAM maintenance requires a mobile AtWUS (reviewed by Fuchs and Lohmann 2020), the presence of mobile AtWOX5 in the columella stem cells is not necessary to maintain their stemness (Berckmans et al. 2020). Interestingly, *AtWOX4* is the only T3WOX that does not complement the *AtWUS* function in loss-of-function mutants, suggesting that the AtWOX4 protein is not mobile or is unable to regulate transcription in a similar way as AtWUS (at least not in the SAM). The cell autonomy of AtWOX4 in the cambium stem cells may be explained by a recent functional characterization of vascular cambium development during radial expansion of the hypocotyl. This study showed that *AtWOX4* is transcribed in cambium stem cells where a dividing stem cell is capable of giving rise to xylem inwardly and phloem outwardly (Shi et al. 2019).

Apart from poorly classified CrWOXA and CrWOXB, CrWUL is more supported as a T3WOX protein in *C. richardii*. The initial analysis of *CrWUL* was described by Nardmann and Werr (2012, 2013). RT-PCR results showed that *CrWUL* is strongly expressed in root tips and gametophytes but only weakly in young sporophytes. In sporophyte tissues, in situ hybridization showed that *CrWUL* is transiently expressed in the developing leaf vasculature. During root and lateral root initiation, *CrWUL* marks the three proximal merophytes but not the distal root cap initial cell, the root apical cell, or the lateral root apical cell. This expression pattern suggests that *CrWUL* is associated with a pluripotent cell fate. Taken together with the dynamic expression pattern of *CrWOXA* during root and lateral root initiation (see Sect. 6.3.2), the authors proposed that in this process, CrWOXA acts as the positioning cues for RAMC and LRMC identity and after cell divisions, CrWUL maintains the pluripotency of the three proximal merophytes. This notion is in consistent with CrWOXA binding to the *CrWUL* promoter to stimulate transcription (Yu et al. 2020).

So far, there is no functional study of *CrWUL* in *C. richardii*; however, its possible function in the SAM and RAM of *A. thaliana* has been tested (Zhang et al. 2017). When the full-length CrWUL peptide is expressed in the OC under the control of *AtWUS* promoter, it does not move nor does it complement the SAM phenotype of *wus-1*. When the CrWUL peptide is shortened by removing some regions between the homeodomain and the WUS motif, the resulting protein moves to the stem cells in the central zone and complements the SAM phenotype of *wus-1*. A similar observation is made in the *A. thaliana* RAM; only the shortened, but not the full-length, CrWUL when expressed in the quiescent center under the control of *AtWOX5* promoter moves to the distal columella stem cells and complements the *wox5-1* phenotype. In addition, the full-length gymnosperm WUS, having a length similar to that of AtWUS, but much shorter than CrWUL, is able to move and complement the *wus-1* and *wox5-1* phenotypes when under the control of their respective promoters. When the full-length CrWUL is expressed under the control of *CLV3* promoter, which is specifically expressed in the central zone, it complements the SAM phenotype of *wus-1*. The authors thus interpreted the results as the consequences of a two-step innovation occurring during WUS/WOX evolution. Step 1 was the acquisition of stem cell maintenance activity in T3WOX genes, which originated in the last common ancestor of lycophytes and seed plants. Step 2 is intercellular mobility that emerged in the last common ancestor of seed plants. Determining whether *CrWUL* indeed functions in the *C. richardii* SAM will be necessary to lend supports for this hypothesis.

Members of the T3WOX clade, in addition to the homeodomain, variably contain up to three conserved motifs. The most conserved is the WUS box which is present in all seed plant T3WOX proteins (Haecker et al. 2004; Dolzblasz et al. 2016) and is reported to interact with TOPLESS-type co-repressors (Kieffer et al. 2006). The WOX1 subclade of T3WOX proteins have three additional conserved motifs, motif 1, motif 2, and STF domain (Zhang et al. 2016). WOX2 and WOX4 were also among the analyzed T3WOX in Dolzblasz et al. (2016). They found that WOX2 orthologs in angiosperms have an additional WOX2 motif at the N-terminus that is absent from WOX2 orthologs in gymnosperms. WOX4 orthologs have two additional conserved domains, a N-terminal PEST domain and a C-terminal acidic domain. The latter, a putative transcriptional activation domain, is also present in the WUS/WOX5 subclade. As found in the WUS/WOX5 orthologs, the fern T3WOX protein CrWUL, in addition to the homeodomain, contains a conserved WUS box, an EAR motif, and an acidic domain, additional to the homeodomain (Zhang et al. 2016). Of the heterosporous fern *A. filiculoides* and *S. cucullata,* the former has two T3WOX proteins that group closely to the CrWUL, while the latter has none (Wu et al. 2019). These two *A. filiculoides* proteins do not have a recognizable WUS box but do have the EAR motif and an acidic domain (Youngstrom et al. unpublished data). It would be interesting to investigate whether the two *A. filiculoides* WOX proteins function similarly to CrWUL in *C. richardii*.

## 6.4 Conclusion

Although present in all land plants, the function and evolution of the T1WOX genes are the least understood. In addition to the apical cells, *PpWOX13*-like genes are broadly expressed, but it is unclear what regulatory role they have in these cells. The biological functions of their orthologs in *A. thaliana* seem to have much evolved as they function in floral and fruit organs which the moss does not have. Understanding the T1WOX gene function in *C. richardii* will bridge the disparate functions discovered in the two distant species. *CrWOXA* and *CrWOXB*, interestingly, seem to have conserved biological functions with T2WOX genes in *A. thaliana*, however are difficult to place confidently in phylogenic trees. Further investigation of the cellular function of the *CrWOXB* such as its interplay with cytokinin and sugar observed in its homolog *AtWOX9* in *A. thaliana* could reveal more evolutionary conservation.

The T3WOX clade contains the well-studied WOX genes. Much of the interest stems from the SAM and RAM maintenance properties of *AtWUS* and *AtWOX5*. This property, at least in the SAM, requires the mobility of AtWUS. The major difference between the SAM and RAM of *A. thaliana* and *C. richardii* is that the *A. thaliana* SAM contain multiple stem cells to generate different cell layers to produce future tissue types, while the *C. richardii* SAM contain a single stem cell, the apical cell, that produces all tissue types. How an apical cell establishes and maintains its identity is unknown. Detailed analysis of *CrWOXB* in the fern showed that it is not expressed in the apical cells of SAM and RAM and has a general role in cell proliferation of both gametophyte and sporophyte generations. Similar analysis of the remaining *C. richardii WOX* genes will shed light on our understanding of the single apical cell meristems of the fern.

## References


Aragón-Raygoza A, Vasco A, Blilou I, Herrera-Estrella L, Cruz-Ramírez A (2020) Development and cell cycle dynamics of the root apical meristem in the fern Ceratopteris richardii. *bioRxiv*, 2020.2008.2027.271049

Atallah NM, Banks JA (2015) Reproduction and the pheromonal regulation of sex type in fern gametophytes. Front Plant Sci 6:100

Atallah NM, Vitek O, Gaiti F, Tanurdzic M, Banks JA (2018) Sex determination in Ceratopteris richardii is accompanied by transcriptome changes that drive epigenetic reprogramming of the young gametophyte. G3 Genes Genomes Genet 8:2205–2214

Banks JA (1999) Gametophyte development in ferns. Annu Rev Plant Physiol Plant Mol Biol 50: 163–186

Bartz M, Gola EM (2018) Meristem development and activity in gametophytes of the model fern, Ceratopteris richardii. Dev Biol 444:107–115

Berckmans B, Kirschner G, Gerlitz N, Stadler R, Simon R (2020) CLE40 Signaling regulates root stem cell fate. Plant Physiol 182:1776–1792

Brand U, Fletcher JC, Hobe M, Meyerowitz EM, Simon R (2000) Dependence of stem cell fate in Arabidopsis on a feedback loop regulated by CLV3 activity. Science 289:617–619

Bui LT, Cordle AR, Irish EE, Cheng C-L (2015) Transient and stable transformation of Ceratopteris richardii gametophytes. BMC Res Notes 8:214

Bui LT, Pandzic D, Youngstrom CE, Wallace S, Irish EE, Szovenyi P, Cheng C-L (2017) A fern AINTEGUMENTA gene mirrors BABY BOOM in promoting apogamy in Ceratopteris richardii. Plant J 90:122–132

Chandler JW, Werr W (2019) In: Grossniklaus U (ed) Current topics in developmental biology, vol 131. Academic Press, Cambridge, MA, pp 545–564

Clark SE, Williams RW, Meyerowitz EM (1997) The CLAVATA1Gene encodes a putative receptor kinase that controls shoot and floral meristem size in Arabidopsis. Cell 89:575–585

Conway SJ, Di Stilio VS (2020) An ontogenetic framework for functional studies in the model fern Ceratopteris richardii. Dev Biol 457:20–29

Deveaux Y, Toffano-Nioche C, Claisse G, Thareau V, Morin H, Laufs P, Moreau H, Kreis M, Lecharny A (2008) Genes of the most conserved WOX clade in plants affect root and flower development in Arabidopsis. BMC Evol Biol 8:291

Dolzblasz A, Nardmann J, Clerici E, Causier B, van der Graaff E, Chen J, Davies B, Werr W, Laux T (2016) Stem cell regulation by Arabidopsis WOX genes. Mol Plant 9:1028–1039

Fletcher JC, Brand U, Running MP, Simon R, Meyerowitz EM (1999) Signaling of cell fate decisions by CLAVATA3 in Arabidopsis shoot meristems. Science 283:1911–1914

Fuchs M, Lohmann JU (2020) Aiming for the top: non-cell autonomous control of shoot stem cells in Arabidopsis. J Plant Res 133:297–309

Geng Y, Cai C, McAdam S, Banks JA, Wisecaver J, Zhou Y (2021) A De novo transcriptome assembly of Ceratopteris richardii provides insights into the evolutionary dynamics of complex gene families in land plants. Genome Biol Evol 13:evabo42

Haecker A, Gross-Hardt R, Geiges B, Sarkar A, Brueninger H, Herrmann M, Laux T (2004) Expression dynamics of WOX genes mark cell fate decisions during early embryonic patterning in Arabidopsis thaliana. Development 131:657–668

Hickok LG, Warne TR, Slocum MK (1987) Ceratopteris richardii: applications for experimental plant biology. Am J Bot 74:1304–1316

Hickok LG, Warne TR, Fribourg R (1995) The biology of the Fern Ceratopteris and its use as a model system. Int J Plant Sci 156:332–345

Hill J (2001) Meristem development at the Sporophyll pinna apex in Ceratopteris richardii. Int J Plant Sci 162:235–247

Hirakawa Y, Kondo Y, Fukuda H (2010) TDIF peptide signaling regulates vascular stem cell proliferation via the *WOX4* Homeobox gene in *Arabidopsis*. Plant Cell 22:2618–2629

Hou G, Blancaflor EB (2009) Fern Root Development. Annu Plant Rev Root Dev 37:192–208

Hou G, Hill JP (2002) Heteroblastic root development in *Ceratopteris richardii* (Parkeriaceae). Int J Plant Sci 163:341–351

Hou G, Hill JP, Blancaflor EB (2004) Developmental anatomy and auxin response of lateral root formation in Ceratopteris richardii. J Exp Bot 55:685–693

Hu X, Xu L (2016) Transcription factors WOX11/12 directly activate *WOX5/7* to promote root primordia initiation and organogenesis. Plant Physiol 172:2363–2373

Johnson GP, Renzaglia KS (2008) Embryology of Ceratopteris richardii (Pteridaceae, tribe Ceratopterideae), with emphasis on placental development. J Plant Res 121:581–592

Kieffer M, Stern Y, Cook H, Clerici E, Maulbetsch C, Laux T, Davies B (2006) Analysis of the transcription factor WUSCHEL and its functional homologue in antirrhinum reveals a potential mechanism for their roles in meristem maintenance. Plant Cell 18:560–573

Laux T, Mayer KF, Berger J, Jürgens G (1996) The WUSCHEL gene is required for shoot and floral meristem integrity in Arabidopsis. Development 122:87–96

Li F-W et al (2018) Fern genomes elucidate land plant evolution and cyanobacterial symbioses. Nat Plants 4:460–472

Lian G, Ding Z, Wang Q, Zhang D, Xu J (2014) Origins and evolution of WUSCHEL-related Homeobox protein family in plant kingdom. Sci World J 2014:534140

Liu, J. *et al.* (2014) *WOX11* and *12* are involved in the first-step cell fate transition during de novo root organogenesis in *Arabidopsis*. Plant Cell 26, 1081–1093

Marchant DB, Sessa E, Wolf P, Heo K, Barbazuk B, Soltis P, Soltis D (2019) The C-Fern (Ceratopteris richardii) genome: insights into plant genome evolution with the first partial homosporous fern genome assembly. Sci Rep 9:18181

Mayer RC, Schoorman FD (1998) Differentiating antecedents of organizational commitment: a test of march and Simon's model. J Organ Behav 19:15–28

McAdam S, Brodribb T, Banks JA, Hedrich R, Atallah N, Cai C, Geringer M, Lind C, Nichols D, Stachowski K, Geiger D, Sussmilch F (2016) Abscisic acid controlled sex before transpiration in vascular plants. Proc Natl Acad Sci U S A 113:12862–12867

Nakazato T, Jung M-K, Housworth EA, Rieseberg LH, Gastony GJ (2006) Genetic map-based analysis of genome structure in the Homosporous Fern Ceratopteris richardii. Genetics 173: 1585–1597

Nakazato T, Jung M-K, Housworth EA, Rieseberg LH, Gastony GJA (2007) Genomewide study of reproductive barriers between allopatric populations of a Homosporous Fern, Ceratopteris richardii. Genetics 177:1141–1150

Nardmann J, Werr W (2012) The invention of WUS-like stem cell-promoting functions in plants predates leptosporangiate ferns. Plant Mol Biol 78:123–134

Nardmann J, Werr W (2013) Symplesiomorphies in the WUSCHEL clade suggest that the last common ancestor of seed plants contained at least four independent stem cell niches. New Phytol 199:1081–1092

Nardmann J, Reisewitz P, Werr W (2009) Discrete shoot and root stem cell-promoting WUS/WOX5 functions are an evolutionary innovation of angiosperms. Mol Biol Evol 26: 1745–1755

One Thousand Plant Transcriptome Initiative (1KP) (2019) One thousand plant transcriptomes and the phylogenomics of green plants. Nature 574:679–685

Plackett ARG, Huang L, Sanders HL, Langdale JA (2014) High-efficiency stable transformation of the model Fern species Ceratopteris richardii via microparticle bombardment. Plant Physiol 165: 3–14

Plackett A, Conway S, Hazelton K, Rabbinowitsch E, Langdale J, Stilio V (2018) LEAFY maintains apical stem cell activity during shoot development in the fern Ceratopteris richardii. eLife 7:e39625

PPG I (2016) A community-derived classification for extant lycophytes and ferns. J Syst Evol 54: 563–603

Romera-Branchat M, Ripoll JJ, Yanofsky MF, Pelaz S (2013) The WOX13 homeobox gene promotes replum formation in the Arabidopsis thaliana fruit. Plant J 73:37–49

Rutherford G, Tanurdzic M, Hasebe M, Banks JA (2004) A systemic gene silencing method suitable for high throughput, reverse genetic analyses of gene function in fern gametophytes. BMC Plant Biol 4:6–6

Sakakibara K, Reisewitz P, Aoyama T, Friedrich T, Ando S, Sato Y, Tamada Y, Nishiyama T, Hiwatashi Y, Kurata T, Ishikawa M, Deguchi H, Rensing S, Werr W, Murata T, Hasebe M, Laux T (2014) WOX13-like genes are required for reprogramming of leaf and protoplast cells into stem cells in the moss Physcomitrella patens. Development 141:1660–1670

Sarkar A, Luijten M, Miyahima S, Lenhard M, Hashimoto T, Nakajima K, Scheres B, Heidstra R, Lauz T (2007) Conserved factors regulate signalling in Arabidopsis thaliana shoot and root stem cell organizers. Nature 446:811–814

Schoof H, Lenhard M, Haecker A, Mayer K, Jurgens G, Lauz T (2000) The stem cell population of Arabidopsis shoot meristems in maintained by a regulatory loop between the CLAVATA and WUSCHEL genes. Cell 100:635–644

Segatto ALA, Thompson CE, Freitas LB (2016) Contribution of WUSCHEL-related homeobox (WOX) genes to identify the phylogenetic relationships among petunia species. Genet Mol Biol 39:658–664

Shi D, Lebovka I, López-Salmerón V, Sanchez P, Greb T (2019) Bifacial cambium stem cells generate xylem and phloem during radial plant growth. Development 146:dev171355

Skylar A, Hong F, Chory J, Weigel D, Wu X (2010) STIMPY mediates cytokinin signaling during shoot meristem establishment in Arabidopsis seedlings. Development 137:541–549

Smith A, Pryer K, Schuettpelz E, Korall P, Schneider H, Wolf P (2006) A classification for extant ferns. Taxon 55:705–731

Stahl Y, Wink RH, Ingram GC, Simon R (2009) A signaling module controlling the stem cell niche in Arabidopsis root meristems. Curr Biol 19:909–914

Stout SC, Clark GB, Archer-Evans S, Roux SJ (2003) Rapid and efficient suppression of gene expression in a single-cell model system, Ceratopteris richardii. Plant Physiol 131:1165–1168

van der Graaff E, Laux T, Rensing SA (2009) The WUS homeobox-containing (WOX) protein family. Genome Biol 10:248

Wu X, Dabi T, Weigel D (2005) Requirement of Homeobox gene STIMPY/WOX9 for Arabidopsis meristem growth and maintenance. Curr Biol 15:436–440

Wu C-C, Li F-W, Kramer EM (2019) Large-scale phylogenomic analysis suggests three ancient superclades of the WUSCHEL-RELATED HOMEOBOX transcription factor family in plants. PLoS One 14:e0223521

Xiao L, Zhang L, Yang G, Zhu H, He Y (2012) Transcriptome of protoplasts reprogrammed into stem cells in Physcomitrella patens. PLoS One 7:e35961

Yadav R, Perales M, Gruel J, Girke T, Jonsson H, Reddy GV (2011) WUSCHEL protein movement mediates stem cell homeostasis in the Arabidopsis shoot apex. Genes Dev 25:2025–2030

Youngstrom CE, Geadelmann LF, Irish EE, Cheng C-L (2019) A fern WUSCHEL-RELATED HOMEOBOX gene functions in both gametophyte and sporophyte generations. BMC Plant Biol 19:416

Yu J, Zhang Y, Liu W, Wang H, Wen S, Zhang Y, Xu L (2020) Molecular evolution of auxin-mediated root initiation in plants. Mol Biol Evol 37:1387–1393

Zhang X, Zong J, Liu J, Yin J, Zhang D (2010) Genome-wide analysis of WOX gene family in rice, sorghum, maize, Arabidopsis and poplar. J Integr Plant Biol 52:1016–1026

Zhang Y, Jiao Y, Jiao H, Zhao H, Zhu Y-X (2016) Two-step functional innovation of the stem-cell factors WUS/WOX5 during plant evolution. Mol Biol Evol 34:640–653

Zhang Z, Tucker E, Hermann M, Laux T (2017) A molecular framework for the embryonic initiation of shoot meristem stem cells. Dev Cell 40:264–277.e264

**Christopher E. Youngstrom** is an aspiring plant biologist with great interests in plant cell communication and evo-devo of transcription factor families. I am a PhD candidate at the University of Iowa Department of Integrated Biology in the laboratory of Dr. Chi-Lien Cheng, where I have investigated the function of *WUSCHEL* transcription factor family, among other projects, in the fern *Ceratopteris richardii*.

**Erin Irish** is a plant developmental geneticist with a long-standing interest in plant life cycles. Her PhD at Indiana University was followed by postdocs at Penn and Yale University, before starting her own lab at the University of Iowa. In addition to work on alternation of generation in ferns, she studies vegetative phase change in maize, leveraged by her use of meristem culture to "rejuvenate" adult plants, leading to the finding that jasmonic acid promotes juvenility in maize. She has also conducted research on maize inflorescence development, studied the tassel seed mutations, and discovered the wandering carpel mutation which affects zygomorphy.

**Chi-Lien Cheng** is a plant biologist with broad interests. After receiving my PhD in plant physiology from the University of Connecticut, I obtained my postdoctoral training in Harvard Genetics Department and Massachusetts General Hospital. There, I cloned the nitrate reductase gene from barley and subsequently mapped the two *Arabidopsis* nitrate reductase genes. I continued to investigate the regulation of nitrate reductase genes after establishing my laboratory at the University of Iowa where I am now a professor in the Biology Department. Fifteen years ago, my laboratory changed the course of our research to investigating sexual and asexual alternation of generations using the model fern *Ceratopteris richardii*.

# Domestication of the Floating Fern Symbiosis *Azolla*

7

H. Schluepmann, I. Bigot, N. Rijken, A. Correas Grifoll, P. A. N. M. Gudde, L. W. Dijkhuizen, and E. Güngör

**Abstract**

Ferns from the *Azolla* genus are highly productive without nitrogen fertilizer because filamentous cyanobacteria, *Nostoc azollae*, associated with the shoot stem cells, invade leaf cavities for $N_2$ fixation and reproductive structures for generational transfer. Previously used as nitrogen biofertilizer, their domestication is now considered for circular economy including the sustainable production of plant protein. The symbiosis recently transgressed into molecular research. Sequences from metagenomes of several species are available to study the contribution of the microbiome components to the symbiosis traits. A first assembly and annotation of the reference genome *A. filiculoides* was released; it allowed reconstruction of tannin biosynthesis, which determines *Azolla* biomass quality as a feed. Here, we begin with describing novel research areas required to integrate agrosystem development with domestication. We next describe first achievements to control the life cycle of the symbiosis in relation to dissemination, storage, and pre-breeding. We then identify key traits of the symbiosis that will need to be considered to achieve yield stability and discuss these traits with the little mechanistic insight available thus far. We conclude that for rapid breeding, the next vital development will be genome editing of fern host and cyanobacterial symbiont and describe our first steps toward this end.

**Keywords**

*Azolla* · Dinitrogen fixation · Ferns · Life cycle control · Neo-domestication · *Nostoc azollae* · Metagenome editing · Symbiosis

H. Schluepmann (✉) · I. Bigot · N. Rijken · A. C. Grifoll · P. A. N. M. Gudde · L. W. Dijkhuizen · E. Güngör
Molecular Plant Physiology, Biology Department, Utrecht University, Utrecht, The Netherlands
e-mail: h.schlupmann@uu.nl; i.j.bigot@students.uu.nl; N.rijken@students.uu.nl; p.a.n.m.gudde@students.uu.nl; l.w.dijkhuizen@uu.nl; e.gungor@uu.nl





J. Marimuthu et al. (eds.), *Ferns*, https://doi.org/10.1007/978-981-16-6170-9_7

## Abbreviations

| | |
|---|---|
| AP2 | APETALA2 |
| GA | Gibberellic Acid |
| GAMYB | Gibberellin- and Abscisic acid-regulated MYB |
| GFP | Green fluorescent protein |
| HiC | High-range interaction capture |
| Ipt | Isopentenyl synthase |
| MAG | Metagenome-Assembled Genome |
| MIKC$^{C}$ | Transcription factor with MADS domain, I region, K domain, and C terminal domain of the C-type |
| RFP | Red fluorescent protein |
| STM | SHOOT MERISTEMLESS |
| WOX | WUSCHEL-Related Homeobox transcription factor |

## 7.1 *Azolla*: A Model Aquatic Symbiosis that Requires Conceptually Novel Research in Ecology, Physiology, Development, and Genetics

*Azolla* ferns have been used as biofertilizer in diverse wetland agrosystems and have tremendous potential for sustainable plant protein or substrate production in tropical to temperate regions. They are wild strains, however, that require intensive labor input for their use, at times implying cheap labor, or child labor. A major challenge constitutes their domestication to reduce input labor: we currently lack industrialized methods to disseminate, store, and breed seed-free plants in general and more specifically these symbiotic heterosporous ferns. We also have a very scant understanding of fundamental processes relating to the development of heterosporous ferns or the physiology of phototrophic nitrogen-fixing symbionts.

We herewith aim to provide a timely review of what is needed for *Azolla* domestication based on recent advances in the control of the life cycle and the molecular genetics of these and other ferns. The recent advances are critically important for characterization of the biodiversity in the genus *Azolla*, for development of nursery practices and pre-breeding, which are required to eventually stabilize yield and tune biomass quality to specific applications by breeding. More advances will be needed, however, to achieve domestication; some will require conceptually novel research in biology and ecology, the latter because the ferns colonized the water surface. To begin with, we will introduce the topic of *Azolla* domestication by briefly presenting a few of these conceptually novel research areas to motivate readers from different fields to join the challenge.

### 7.1.1 Azolla Ferns Form Thick Floating Mats, Not Reefs, that Fixate $CO_2$ and $N_2$

Unlike most other floating plants, *Azolla* ferns are peculiar in that they form thick mats. Figure 7.1a shows 6-month-old cultures of *A. filiculoides*: the mats contain a significant amount of floating senescent plant and root biomass that does not degrade rapidly in the darkness the canopy generates. The potential of massive $CO_2$

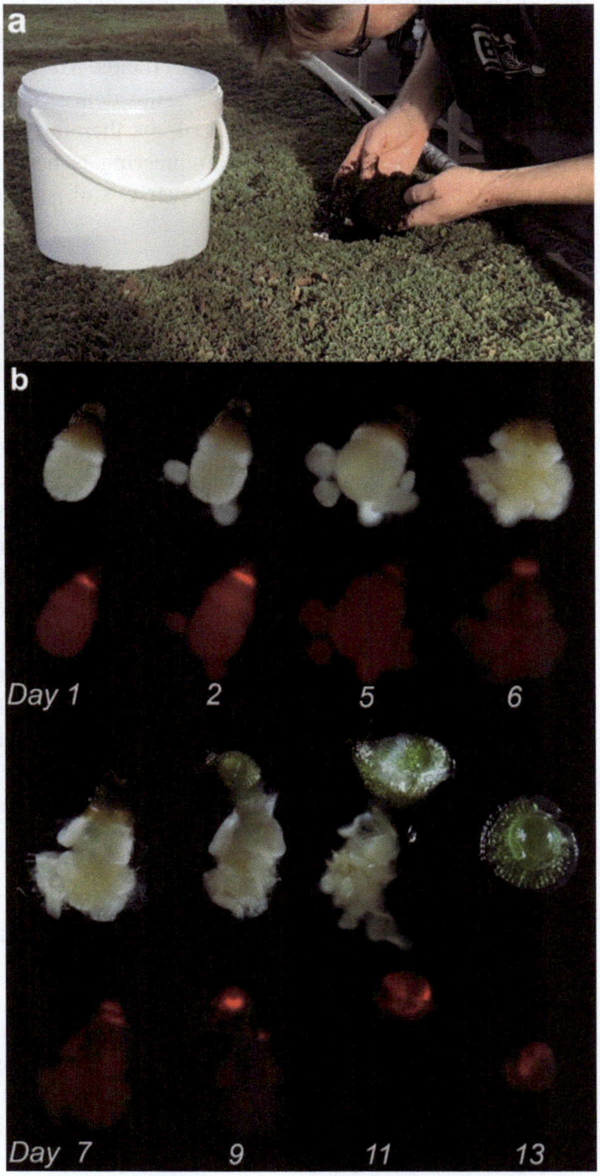


**Fig. 7.1** Novel research areas important for integrated agrosystem development and domestication of *Azolla* ferns. (**a**) The true growth habit of *Azolla* ferns: mats with a substantial amount of floating senescent plant and root biomass that does not degrade rapidly in the darkness the canopy generates. Six-month-old cultures of *A. filiculoides* maintained in the glasshouses of Radboud University by Bas van de Riet from B-WARE in Nijmegen. (**b**) The coordinated development of symbionts and host for the establishment of the floating sporophyte. "Sporeling" and *N. azollae* akinetes germinate within 2 weeks after transfer to 20 °C and light. Images were taken using a Zeiss Axio Zoom V16 binocular with a 280× magnification, using white light to document megasporocarp development (above row), or using RFP filter fluorescence settings (bottom row) to reveal *N. azollae*. All images result from focus stacking (Helicon Focus)

drawdown when the mats sink is illustrated in sediments from the North Pole dating from the Eocene (Speelman et al. 2009). Drawdown over geological time periods may not be of much concern for the present realities. Nevertheless, the potential of *Azolla* mats has often been overlooked, or their development was deemed negative, even if nutrient remobilization was measured in mesocosms, as, for example, in Pinero-Rodríguez et al. (2021). Mats are of particular interest, however: they generate anoxia which in turn is required for the mobilization of phosphate precipitates (Gu et al. 2019; Wang et al. 2019); this represents an opportunity that should be exploited as phosphate will become a limiting resource for agriculture (Geissler et al. 2019). Mats causing anoxia in flooded wetlands further reduce $CO_2$ emissions from respiration of drained wetlands. Miller et al. (2008) pointed out that the mats also function to reverse soil subsidence in subsiding deltas such as the Sacramento-San Joaquin Delta. In ecology, therefore, research on the possible role of *Azolla* mats for biogeological engineering is needed to circularize the use of P-fertilizer before it runs out or reverse soil subsidence in subsiding deltas worldwide.

### 7.1.2 Azolla Ferns Exhibit Nitrogen Autotrophy at Astonishing Productivities

Thus far, the production of plant protein by nitrogen-fixating crops such as the legumes is N-fertilized because $N_2$ fixation rates that depend on carbon supply are limiting production (Salvagiotti et al. 2008). In contrast, *Azolla* fern biomass productivity, even at its peak, was not limited by $N_2$ fixation when tested under small-scale production conditions: it was driven by light energy while yielding some 50 ton/h/a dry weight biomass containing roughly 20% w/w protein (Brouwer et al. 2017, 2018). Research on the physiological mechanisms maintaining such high productivities and nitrogen fixation is needed to engineer staple crops with organelles that fixate $N_2$ using light energy; chloroplasts likely were derived from cyanobacteria that did not fixate $N_2$ (Nowack and Weber 2018). This idea is not novel, but molecular research on the specific case of the double photosynthesis in the leaves of *Azolla* is untouched, except for data collected some 40 years ago without molecular genetic insight (Peters et al. 1976; Ray et al. 1979; Tyagi et al. 1981; Calvert and Peters 1981).

### 7.1.3 Azolla Ferns and Symbionts Coordinate their Development So as to Colonize the Water Surface

*Azolla* ferns have adapted to the floating habit; they even evolved a mechanism to ditch their anchor, which in this case means to rapidly abscise their roots (Uheda and Kitoh 1994). They therefore are equipped to deal with sudden increases in water level by responding to submergence when respiration is inhibited. The developmental response does not require transcription and is mediated by oxidative stress including reactive oxygen, nitrogen, and sulfur species; these species were proposed to trigger cell wall loosening, thus releasing the roots at the abscission zone in less than 1 h (Gurung et al. 2012; Cohen et al. 2015; Yamasaki et al. 2019).

The coordination of development between symbiont and fern is not understood at the mechanistic level. First insights on possible metabolites involved in the coordination were gained when testing the effects of deoxyanthocyanins synthesized by the fern on the development of the filamentous cyanobacteria *Nostoc punctiforme*, the facultative symbionts of cycads (Cohen et al. 2002).

Many ultrastructural observations lead to the conclusion that trichomes from the host could guide migration of the motile *N. azollae* into leaf cavities and sporocarps when the organs form in the shoot apical meristem (Calvert and Peters 1981; Calvert et al. 1985; Perkins and Peters 1993; Peters and Perkins 1993). Two types of trichomes seem to co-exist at the sites of attraction: the branched and unbranched type. Active membrane networks and other ultrastructure feature suggest that these trichomes are secreting substances. In a leaf cavity, only a single branched trichome was observed in contrast to the two-celled unbranched trichomes which are numerous. Sporophytes still form the trichomes inside the leaf cavity in the absence of the *N. azollae* (Forni et al. 1991). Glandular trichome metabolism in ferns is not studied; we expect differences in the types of trichomes as described for trichomes from the Solanaceae (Schuurink and Tissier 2020). To find out what substances may be secreted capitalizing on recent developments, we could begin with profiling gene expression and metabolites of cavity packets generated as in (Uheda 1986). More insight, however, would be gained from analysis of the thick glandular trichomes at the fern shoot apices, but these are more challenging to harvest for single cell profiling.

An understanding of the dormant stages in the life cycle of the symbiosis is lacking, yet these stages are critical for storage and dissemination of *Azolla* strains. It will be important, for example, to find out how the *N. azollae* (and possibly other low-abundance bacteria) are induced to form resting stages and how these resting stages are spurred into life again when the sporeling grows into the inoculation chamber under the indusium cap. *A. filiculoides* sporeling and *N. azollae* akinetes germinate within 2 weeks after transfer of megasporocarp/massulae clumps from 4 °C to 20 °C and to light (about 80 μmol/m$^2$/s photosynthetic active radiation; Fig. 7.1b). Under these conditions, the akinetes are seen as a red fluorescence under the indusium cap until day 6; at day 7, when the sporeling shoot apex reaches the indusium chamber, a clear separation of the red fluorescent signal is seen between indusium and shoot apex; from day 9 onward, when the shoot apex has emerged from the megaspore and the indusium cap is ditched, most of the red fluorescence from the *N. azollae* is seen at the sporeling shoot apex (Fig. 7.1b).

Understanding coordinated development of the host with the bacteria in the *Azolla* Ark will be key to engineering yield stability by efficient dissemination with inoculum that may also be stored long term. Such inoculum could be the fertilized megasporocarps, which in the case of *A. filiculoides* can be stored over several years at 4 °C as well as frozen (Brouwer et al. 2014).

### 7.1.4 Breeding Heterosporous Ferns and the Microbiome Associated with the Azolla Symbiosis

Combined molecular genetics and phylogenetic approaches will reveal mechanisms ruling the fern life cycle because of the likely conservation of at least some of the mechanism components with those found in well-studied plant lineages. Knowledge of these mechanisms will be required to control the production of dissemination and storage stages of ferns in general. The knowledge will further provide insight into the origins of life cycle regulators in seed plants, as the last common ancestor of ferns and seed plants had a homosporous life cycle.

The symbiosis is currently a wild plant with extremely promising yield potential, but we have yet to engineer yield stability. The most cost-effective approach to achieve yield stability in the case of seed plants has been by breeding (Cox et al. 1988; Bailey-Serres et al. 2019). Classical breeding by selection and inbreeding is not desirable, however, because it reduces the genetic diversity of the cultured populations (Fu 2015).

We know little of the genetic diversity of the symbiotic microflora and how to engineer the key traits associated with them including the further amelioration of $N_2$ fixation. The cyanobiont and fern genomes have co-evolved, and additional bacteria are persistently associated with the *Azolla* symbiosis (Dijkhuizen et al. 2018; Li et al. 2018). Co-evolution of the genomes implies that fitness of the symbiosis is determined by its metagenome and that, therefore, not only the fern but also associated microflora need to be considered when breeding from high yield potential to high yield. This is conceptually novel.

### 7.1.5 Engineering Sustainability and Resilience of Primary Production through Rapid Domestication of Botanic Diversity

Taken together, *Azolla* ferns represent a botanic oddity, but their applications are aligned with improving the sustainability of our primary production systems while increasing their resilience and yield. We lack much of the foundational knowledge to develop these applications. Nevertheless, recent advances in molecular genetics of cyanobacteria and ferns, including *Azolla* ferns, allow now to consider domestication of fern symbioses.

In the following, we will attempt at drafting the biotechnology approaches that already have been undertaken and those still needed toward *Azolla* fern domestication. These represent a change in perspective where, more generally, we aim to exploit botanic diversity in rapid breeding/domestication schemes of plant/microbe associations that are adapted to thrive in specialized environments. Proof-of-concept rapid breeding/domestication have recently been reported for a wild tomato, *Solanum pimpinellifolium*, for example (Zsögön et al. 2018). We have yet to embrace the concept for plant-microbe symbioses. When complemented with more rapid and systematic agrosystem engineering, including paludiculture, the method will extend arable regions and widen crop diversity so as to achieve yield resilience (Massawe et al. 2016).

We will first review how the life cycle of *Azolla* ferns and the cyanobionts may be controlled as this is key to storage, dissemination, and breeding of strains for any one application. We secondly will describe what information is available to link traits with genes from the different components of the metagenome and provide a first impression of the pangenome diversity in the cyanobiont. Thirdly, we will concentrate on what traits could be tackled for engineering initially, given priorities for sustainable primary production with *Azolla* and commercial viability. Lastly, we will look at emerging advances in engineering the genomes of the fern and cyanobiont. In an outlook, we provide the rationale for the challenge of wanting to develop genetic tools for the heterosporous fern and the obligate cyanobacterial symbiont.

## 7.2 Domestication of the *Azolla* Symbiosis: What Bases Are Available?

### 7.2.1 Controlling the Symbiosis Life Cycle for Strain Storage and Dissemination and for Breeding

A sketch of the life cycle of *Azolla* ferns is provided in Fig. 7.2a using *A. filiculoides*, for example; *A. filiculoides* can complete its life cycle in 2 to 3 months. Sporophytes (Fig. 7.2a-*j*, *a*) transit into sexual reproduction when their shoot apical meristems form sporocarp initials in pairs (Dijkhuizen et al. 2020, Fig. 7.1). Such pair usually consists of a megasporocarp (mega) with a single megasporangium (Fig. 7.2a-*b*) and a microsporocarp (micro) with the degenerating megasporangium initial and many de-repressed microsporangia (Fig. 7.2a-*e*, arrow). In the meristem region of the shoot apex, *N. azollae* are recruited into the closing indusium cap (Fig. 7.2a-*b*, arrow). Pairs of sporocarps are found on the branch next to the branch point where roots are formed. When they mature, megasporocarps will form a megagametophyte fuelled by the megaspore (Fig. 7.2a-*d*). We do not know whether the gametophyte development occurs before or after the megasporocarp detaches from the sporophyte; eventually archegonia are formed by the megagametophyte (Fig. 7.2a-*d*, arrow). Mature microsporocarps rupture to release massulae containing several microspores (Fig. 7.2a-*f*); we do not know when the much reduced microgametophytes (Fig. 7.2a-*g*) develop from the microspores and when the microgametophytes generate the flagellate gametes, but this must happen before or when massulae attach to the megasporocarps with their glochidia (Fig. 7.2a-*h*). The flagellate gametes swim toward the archegonia and fertilize the egg cells which starts the outgrowth of the diploid "sporeling," still fuelled by the megaspore reserves (Fig. 7.2a-*i*, *j*, *k*).

The *Azolla* life cycle can be completed within 2 to 3 months for *A. filiculoides*. Sporocarps are the natural storage and dissemination stages which in the case of *A. filiculoides* may be stored up to 4 years at 4 °C or indefinitely, when first dried and then cryopreserved at −80 °C (Brouwer et al. 2014). In contrast, sporophytes cannot

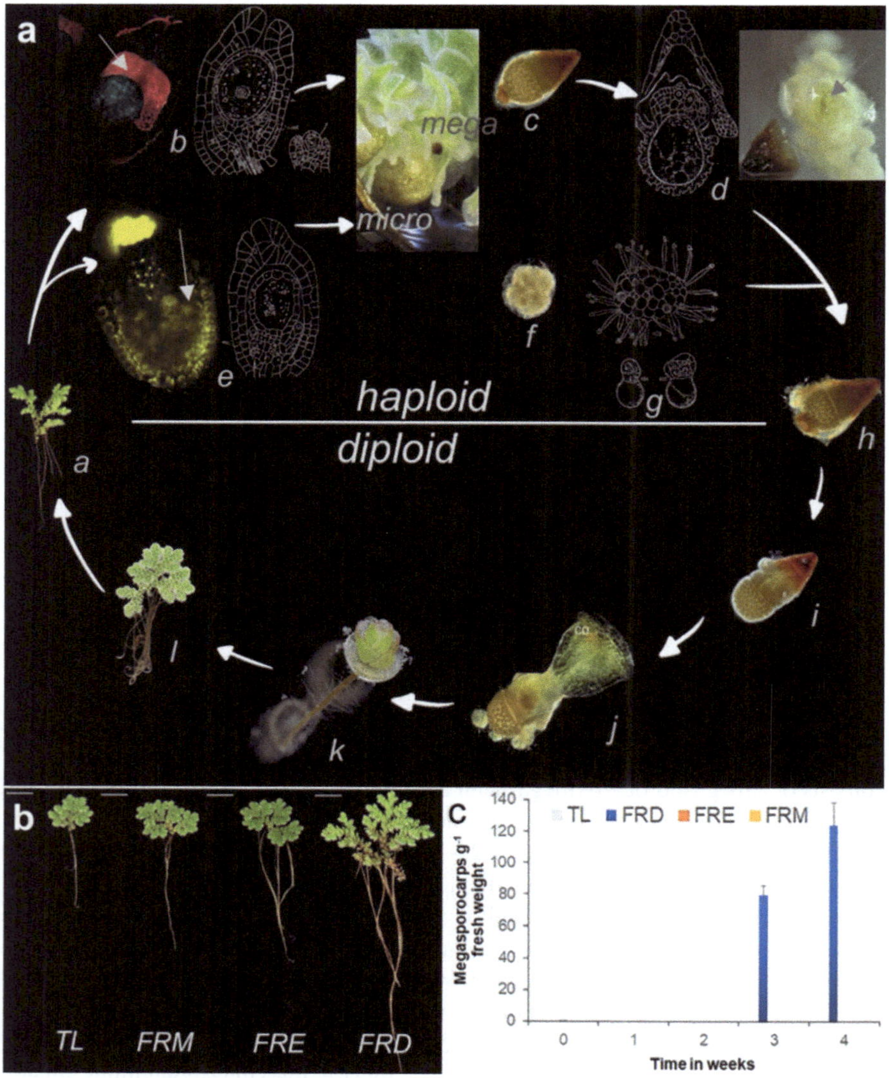


**Fig. 7.2** The *A. filiculoides* symbiosis heterosporous life cycle and repression of sexual reproduction by red light. Graphic insets are from Coulter (1910). **(a)** Sporophytes (*i*–*a*) transit into sexual reproduction when they form sporocarp initials in pairs: a megasporocarp (mega) with a single megasporangium (*b*) and a microsporocarp (micro) with the degenerating megasporangium initial and many de-repressed microsporangia (*e*, arrow). *N. azollae* are recruited into the closing indusium cap (*b*, arrow). When they mature, megasporocarps will form a megagametophyte (*d*) which develops archegonia with egg cells (*d*, arrow). Mature microsporocarps rupture to release massulae (*f*) attaching to megasporocarps with their glochidia (*h*); microspores inside massulae germinate into microgametophytes, generating the flagellate gametes that fertilize the egg cells: this starts the outgrowth of the diploid "sporeling" (*i, j, k*). *haploid*, life stages with haploid tissues; *diploid*, life stages with diploid tissues. **(b)** Sporophyte habit after 4 weeks with a 16 h light period with tube light (*TL*, which contains no far-red light), tube light with far-red light during either the entire day length (*FRD*), 4 h in the morning (*FRM*), or 4 h in the afternoon (*FRE*). **(c)** Sporocarps produced under the conditions in **(b)**. Shown are averages with standard deviation, from three replicate

be stored; they are comparatively fragile and heavy due to their high water content. Industrialized inoculation of production fields via sporocarps, therefore, needs to be considered in the future and will rely on methods to produce them in high quantity and quality.

The induction of the haploid phase change in fern sporophytes occurs when sporangia initials are formed. This phase change was shown to depend on photoperiod, temperature, or sugars in the medium in many ferns, including members of the heterosporous Salviniaceae (Labouriau 1958; Sussex and Steeves 1958; Harvey and Caponetti 1974). Analogies were made from the onset with the transition to flowering in seed plants. Unlike in seed plants, however, the formation of the meristems that developed sporangia was reported to be surprisingly plastic: it occurred at many locations in the sporophytes and even in gametophytes (Labouriau 1958). In the case of *A. filiculoides*, induction of sporocarps was dependent on far-red light; furthermore, numbers of sporocarps increased with the density of the fern mat (Dijkhuizen et al. 2020). Far-red light was required during the entire light period for complete elongation and three-dimensional growth of the sporophytes (Fig. 7.2b); it was also required for the induction of sporocarps (Fig. 7.2c); neither end of day nor morning or afternoon far-red light sufficed (Fig. 7.2b, c). We therefore conclude that the typically red-light-dominated spectrum of open fields represses entry into sexual reproduction in *A. filiculoides*. Light quality did not, in our hands, induce sexual reproduction in sporophytes from the species of the Anzali Lagoon or from *A. pinnata* (Dijkhuizen et al. 2020). Environmental cues inducing sporocarp formation, therefore, differed for other *Azolla* species. Environmental cues are, however, predicted to feed into a conserved phase transition signaling network with component regulators known from seed plants: the topologies of phylogenetic trees obtained with regulators such as the MIKC$^C$, AP2, or GAMYB and the conserved regulons such as the GAMYB/microRNA319 substantiate the prediction (Ambrose and Vasco 2016; Dijkhuizen et al. 2020). Furthermore, in situ hybridization with *MIKC$^C$* and *AP2* probes strongly light up the sporangia initials and show that these regulators are active during sporangia initiation (Hasebe et al. 1998; Ambrose and Vasco 2016). Together, the preliminary work suggests that control of flowering in seed plants may well have its origins in the diploid to haploid phase transition of the common ancestor of ferns and seed plants. If confirmed, much of the networks of control known from seed plants could be conserved, and this knowledge could be exploited to more quickly control the phase transitions in all *Azolla* fern species, as well as other ferns.

What induces spore germination and the development of gametophytes in *Azolla* is not known and difficult to study by comparison with homosporous ferns because the gametophytes are much reduced in size and develop inside closed structures (Fig. 7.2a-*c, f*). As a result, we expect regulation of spore germination and

**Fig. 7.2** (continued) measurements of three independently grown cultures for each condition. The cultures were reset once per week to a fixed density gametophyte development to be taken over by the sporocarp to some extent. Ripe megasporocarps, for example, exhibit a distinct break between indusium cap and body due to the growth of the mega-gametophyte (Fig. 7.2a-*h*). In *A. filiculoides* and *A. caroliniana*, these sporocarps will clump with massulae. This behavior was exploited to clean sporocarp clumps away from debris. In the case of *A. caroliniana*, the sporophyte biomass was dried before sieving of the clumps and setup of in vitro germination assays (Singh et al. 1990). Germination of dried sporocarps from *A. caroliniana* was improved with GA (Singh et al. 1990). Unlike *A. caroliniana*, mats of the temperate *A. filiculoides* did not yield viable sporocarps after drying. The *A. filiculoides* sporocarps were harvested by showering sporophytes over sieves instead. Then, gentle agitation of the harvest in water lead to clumps of megasporocarps and massulae which are picked up manually for storage or germination assays.

Mats of *A. filiculoides* grown with 0.3% w/v NaCl could be harvested for such sporocarp clumps, but when these clumps were germinated at RT with 16 h light/day in distilled water, they only yielded few viable sporelings, in contrast to clumps obtained from sporophytes grown without NaCl (Fig. 7.3a). Upon closer inspection, the clumps from sporophytes raised on NaCl had many very immature sporelings arrested in their development, presumably because of insufficient nutrient reserve in the megaspore (not shown). Germination rates of clumps were variable in our hands. They were, however, unaffected by dipping (12 h) into surfactant (Silvet L-77 at 0.02% w/v) or *Agrobacterium* suspensions. In contrast, dipping into sugars caused infections which affected germination rates, even if the clumps had been rinsed with abundant distilled water after exposure to the sugars and if the sporelings were transferred to IRRI medium as soon as they emerged (Fig. 7.3b).

Whether compounds in the part of the floating *Azolla* mat made of dead sporophytes and roots improve germination of sporelings is unknown. Given that this mat part floats, it retains the naturally sinking detached sporocarps close to the water surface, and thus to light, thereby increasing likelihood of germination and establishment of the sporelings. The mat structure and biochemical content, we expect therefore, have been selected for the ferns' after-life traits that facilitate re-establishment of generations of ferns. Analysis of the lipids and waxes in *Azolla* ferns reveals that they not only are characteristic but particularly abundant in hydrophobic very long chains (Nierop et al. 2018) which promotes buoyancy. Moreover, the very-long-chain lipids and waxes are recalcitrant to biodegradation since they are found in the fossil sediment layers dating from the Eocene *Azolla* (Speelman et al. 2009).

The industrialized collection of sporocarps still needs to be invented and may need genetic alteration of fern architecture and abscission of the megasporocarps for evenly induced development and detachment of many more megasporocarps, in analogy with seed shattering (Di Vittori et al. 2019).

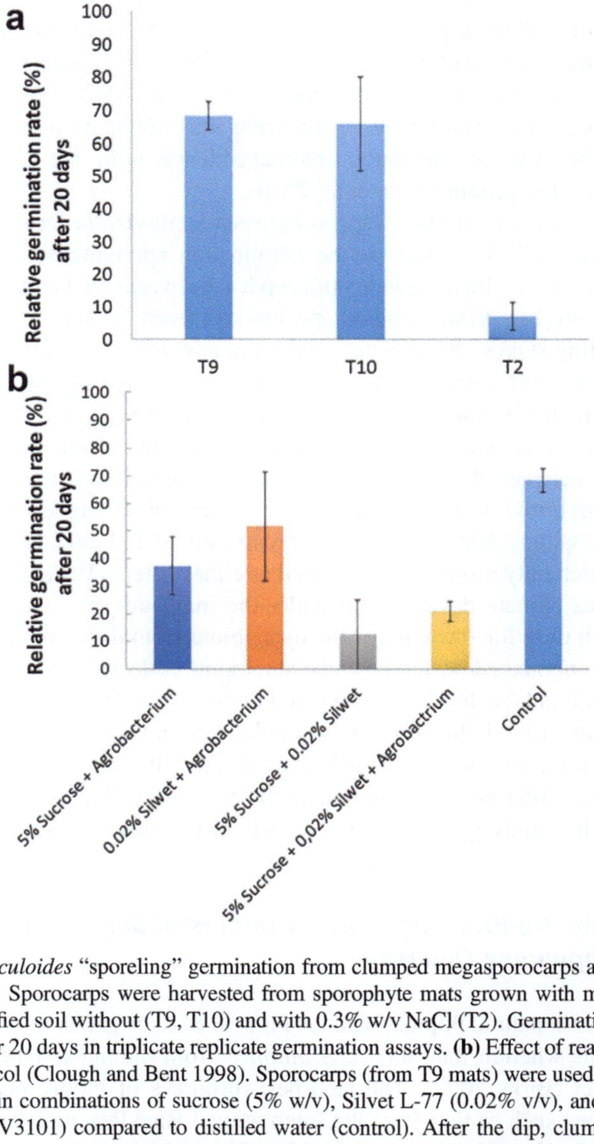


**Fig. 7.3** *A. filiculoides* "sporeling" germination from clumped megasporocarps and massulae. **(a)** Effect of NaCl. Sporocarps were harvested from sporophyte mats grown with micronutrient and phosphate-fortified soil without (T9, T10) and with 0.3% w/v NaCl (T2). Germination of the clumps was scored after 20 days in triplicate replicate germination assays. **(b)** Effect of reagents used in the floral dip protocol (Clough and Bent 1998). Sporocarps (from T9 mats) were used to test overnight (12 h) dipping in combinations of sucrose (5% w/v), Silvet L-77 (0.02% v/v), and *Agrobacterium tumefaciens* (GV3101) compared to distilled water (control). After the dip, clumps were cleaned with distilled water and transferred to distilled water for sporulation at RT with 16 h light/day. Sporelings emerging to the surface were transferred to IRRI medium. Shown are averages with standard error of three replicate assays.

### 7.2.2 Life Cycle of Cyanobacteria inside the Fern: The Apical Colony and Germinating Akinetes Matter

Key to propagation of the *N. azollae* is its apical colony in the shoot apex and the consequent movement of its motile filaments, the hormogonia, to organ initials such as the leaf cavities and the sporocarps. How the movement of the hormogonia and the differentiation of cell types are controlled by the fern host is not known. A role for host trichomes that are of the secretory style is likely and for deoxyanthocyanins which may regulate hormogonia formation (Cohen et al. 2002). A critical handicap for biotechnology is the fact that *N. azollae* cells may not be propagated outside the ferns, likely because of metabolic interdependence with the host, given the cyanobionts' eroded genome (Ran et al. 2010).

*N. azollae*, however, can be swapped between some *Azolla* fern species. A key stage for "swapping" *N. azollae* is the germinating sporeling of a "decapitated" megasporocarp, from which the indusium cap has been removed and, then, to which a new indusium cap from another species has been added. The germinating *N. azollae* resting stages, the akinetes, under the new indusium cap will apparently invade the developing sporeling shoot apex to generate hybrid fern cyanobacterial symbioses (Lin and Watanabe 1988; Plazinski et al. 1989). The hybrids are viable suggesting that the *N. azollae* functions and mechanisms of recognition/immunity are conserved between *Azolla* species. In contrast, hybrids of *Azolla* fern species (sexual fern hybrids) were not generally viable: for example, when crossing *A. microphylla* with *A. filiculoides*, only 8 crosses out of 114 gave viable fern sexual hybrids, of which only a single grew as well as either parent (Watanabe et al. 1993).

Synchronous akinete development under the indusium cap is a biotechnology opportunity not only for "swapping" the cyanobiont from different species but also for generating engineered bacteria using cyanobionts collected from under indusium caps. If we could induce the akinetes into cell divisions while they are still trapped in the mucilage under the indusium cap, we could attempt conjugation with *E. coli* by soaking indusium caps and then transfer the soaked indusium caps to "decapitated" megasporocarps. Alternatively, to engineer the *N. azollae*, *E. coli* conjugation must be carried out in situ using sporophytes so as to target the colony at the shoot apex.

### 7.2.3 Azolla *Pre-Breeding: Current Understanding of the Azolla Metagenome Diversity*

To exploit traits in natural populations, the ferns need to be preserved along with the wetland regions wherein they thrive. *A. nilotica* from the upper Nile regions was reported extinct (Birks 2002). The next-best resources of biodiversity are *Azolla* strain collections such as that from the International Rice Research Institute (IRRI) biofertilizer collection (Watanabe et al. 1992). The IRRI collection of *Azolla* strains was recently transferred to the University of the Philippines (Philippines) and the Dr. Cecilia Koo Botanic Conservation Center (Taiwan). The IRRI collection was conserved over several decades by sub-culturing which is labor-intensive and prone to mislabelling and could alter the microbial consortium associated with the original strains. Nevertheless, the first sample of sequenced accessions from the IRRI collection has shown that the ferns are indeed diverse and that also the diversity of their *N. azollae* symbiont was maintained along with other low-abundance symbionts (Dijkhuizen et al. 2018; Li et al. 2018). To improve on these collections, recent knowhow of how to induce sporocarps and store them at −80 °C by cryopreservation should be extended beyond *A. filiculoides* and deployed for germplasm conservation (Brouwer et al. 2014; Dijkhuizen et al. 2020).

To make the most of the natural biodiversity of the *Azolla* symbioses, an improved characterization of the taxonomy inside the *Azolla* genus is required; it will likely uncover novel species or natural hybrids as recent studies have shown (Madeira et al. 2019; Dijkhuizen et al. 2020). Sequencing approaches that go beyond PCR amplifications between conserved regions in the chloroplast will allow more resolution of the analyses. Analyses of amplicons from the ITS gene regions are flawed by the intra-genomic variations in the many ITS repeats within the nuclear genome (Dijkhuizen et al. 2020). Nevertheless, the ITS1 region was used to infer that some *A. caroliniana* strains in China are hybrids: the PCR-amplified ITS1 sequence consensus from the "hybrid" when submitted to BLAST search of the NCBI database (Johnson et al. 2008) returned sequence regions more identical to *A. cristata* and to *A. caroliniana* (Madeira et al. 2019). This could be an artifact when assembling the PCR consensus, however. We suggest that whole-genome sequences should be used to reconstruct the diversity of ITS1 in each genome accurately, if we decide to continue using the ITS1 sequences to infer phylogeny for historical reasons.

### 7.2.4 Rapid Breeding Using Molecular Genetics

Accelerated breeding using molecular breeding approaches requires a high-quality inventory of genes in the symbiosis (Kole et al. 2015). Genome sequencing data is available for *Azolla* symbioses from six species as well as the non-symbiotic and related water fern *Salvinia cucullata* (Dijkhuizen et al. 2018; Li et al. 2018), mostly short-read technology. Assemblies of the reads from these species have not been released. In addition, genome assemblies from key species are missing including a species representing the second branch of the *Rhizosperma* section of the *Azolla* genus, for example, a (sub)tropical *A. pinnata*. The data has not been exploited beyond the figure of the co-evolution of fern host and cyanobiont (Li et al. 2018).

Gene expression for simultaneous profiling of all transcripts in the symbiosis was shown to be possible by rRNA depletion of both plant and gram-negative bacterial origins. At 20 M pair-end reads per sample sequencing depth, the assays were sensitive enough to extract differential gene expression in sporophytes for the fern host nucleus, the chloroplast, and the cyanobiont, but not for the associated bacteria (Dijkhuizen et al. 2020). Pooling six replicate samples of sporophyte revealed that increasing sequencing depth sixfold would only allow to profile the most highly expressed genes in the associated bacteria.

The *A. filiculoides* reference genome ($2n = 44$ chromosomes, predicted size 750 Mb) is not yet a chromosome-scale assembly. Long-read sequencing (PacBio RS II N50 14 kb) led to the first assembly of the *A. filiculoides* genome with some 4700 scaffolds covering some 650 Mb non-haplotype resolved sequences assigned as belonging to "Streptophyta" (Li et al. 2018). Annotation yielded some 32 k genes based on de novo prediction and sporophyte RNA sequencing. Consequently, many genes uniquely expressed in specialized reproductive tissues are missing or have an erroneous annotation. To date, the miRNA loci discovered in Dijkhuizen et al. (2020) have not been included in the annotation of the *A. filiculoides* genome on fernbase.org. Functional annotation was exclusively done using homology predictions. Key resources to improve annotations are intron/exon annotation predictors (Keilwagen et al. 2019), more diverse RNA sequencing data, and the One KP resource (Leebens-Mack et al. 2019). To improve contiguity, HiC and optical mapping have been carried out (unpublished); we will ideally want whole chromosome assemblies and haplotype resolution of at least one cross to predict segregation. More urgently, the future assembly and annotation quality should be such that it can be shared and uploaded onto the NCBI and other databases as a more accessible resource.

### 7.2.5 Metagenomes from Six Species Have Provided Insight into Microbial Assemblages Associated with the Symbiosis

Breeding of the *Azolla* symbioses needs to consider the important traits conveyed by its microbiome. Our initial analysis comparing rRNA sequences in short reads of DNA from six species of *Azolla* with those in scaffolds from *A. filiculoides* long-read assemblies revealed near-complete bacterial genomes persistently associated with the symbioses. The two Rhizobiales genomes persistent in several *Azolla* species were found enriched in leaf juice DNA from *A. filiculoides*. The two Rhizobiales when combined were capable of denitrification (Dijkhuizen et al. 2018); more recent gene expression data suggests that the bacteria express specific sugar transporters at high levels and therefore may be thriving on imported sugars (data not shown).

Metagenome-assembled genomes (MAGs) of the *N. azollae* from the six sequenced *Azolla* species had over 95% identity with the *N. azollae* assembly from the *A. filiculoides* strain Stockholm, suggesting that they represent a single operational taxonomic unit (Ran et al. 2010; Dijkhuizen et al. 2020). This is surprising given the relative isolation of the *N. azollae* inside the fern host: crosses between the different fern species did not generally lead to viable offspring (Watanabe et al. 1993). The observation further suggests that the genome erosion caused by active transposons may have occurred early on during the evolution of the symbiosis. Alignment of the *N. azollae* MAGs from the different species suggested that assembly breaks often aligned (Fig. 7.4); manual inspection revealed that more than half of the breaks (Fig. 7.4 *red bars*) coincide with transposase genes as annotated in the reference *N. azollae 0708* genome (NCBI microbial genomes database). The data suggest that some 60% of the transposon coordinates date back to a common ancestor of the *N. azollae* strains inside the diverse fern species tested. We conclude, firstly, that the *N. azollae* are monophyletic and, secondly, that the evolution of their genomes might be constrained by the symbiosis. Early erosion would be consistent with theory proposing that the loss of functions no longer required presents benefits (McCutcheon and Moran 2010).

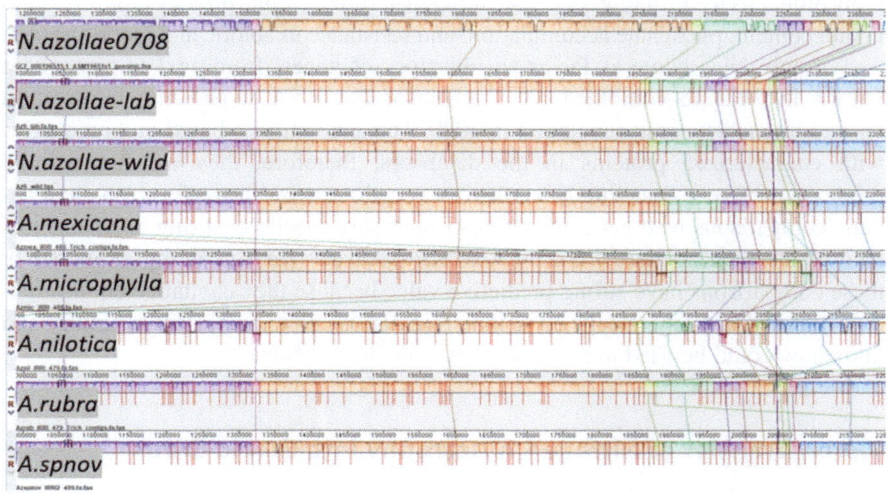


**Fig. 7.4** Alignment of contig breaks in metagenome-assembled genomes (MAG) of *N. azollae* from different species of *Azolla*. MAG were sorted and aligned with MAUVE (Rissman et al. 2009). Sequences were assembled as described in Dijkhuizen et al. (2020) which yielded 341–366 contigs in the section *Azolla* assemblies. Sequences from *N. azollae* in *A. filiculoides* were from either ferns originating in Stockholm (*N. azollae* 0708) or ferns originating from Utrecht, laboratory cultured or collected in the wild (*N. azollae*_Lab and *N. azollae_wild*, respectively). Mauve colored "syntenic" regions, the *red bars* represent contig breaks in the MAG

Genome erosion by transposon mutagenesis is more generally found in the vertically transmitted species of *Burkholderia* bacteria associated with leaf nodules of angiosperm plants from the Primulaceae and Rubiaceae, yet these symbioses are of polyphyletic origin (Pinto-Carbó et al. 2018). Given erosion in all *N. azollae* genomes and presence of associated Rhizobiales bacteria in all *Azolla* ferns, we further suspect that, as in the case of the aphid/*Buchnera aphidicola* symbiosis, some tasks within the symbiosis have been transferred to associated bacteria in the leaf cavity (Dijkhuizen et al. 2018). This could have occurred by horizontal gene transfer as in the case of the aphid di-symbiontic systems (Manzano-Marín et al. 2020). A more in-depth analysis of the bacteria within the leaf cavity could shed light on as of yet undiscovered functions these Rhizobiales bacteria have for the *Azolla* host, for *N. azollae* cyanobionts, and for each other.

## 7.3 Biotechnology for Sustainable Development with *Azolla* Ferns: What Traits Could be Tackled in Priority?

Methods in rapid domestication require that we identify the relationships between traits and gene expression in a broader sense. This requires insight in molecular mechanisms underpinning key traits of the *Azolla* symbiosis. Traits of particular relevance toward the use of *Azolla* ferns for sustainable primary production include high productivity in the absence of nitrogen fertilizer, high protein biomass, high tannin content, phosphate accumulation, and resistance to abiotic and biotic stresses.

We only have come as far as an inventory of genes: from next to zero genes known in 2013 to combined pro- and eukaryote RNA sequencing information backed by metagenome annotation in 2020. The annotation is based on homologies with characterized proteins in the databases; therefore, they are not definite predictions of gene function. Nevertheless, such inventory allows for deep phylogenetic comparisons with gene inventories from many species covering all lineages of land plants such as those from the One KP database (Ka-Shu Wong et al. 2020); these comparisons have strong predictive value because they capitalize on information gathered in many lineages. We therefore discuss how the present inventory and knowhow could be used to tackle the key traits listed above.

### 7.3.1 Transfer of Nitrogen Autotrophy into Staple Crops

Transfer of nitrogen autotrophy into staple crops requires knowing which processes in the fern regulate interactions with the $N_2$-fixating phototrophic symbionts so as to eventually align light-driven nitrogen fixation of the endophyte with high growth. This feature is unique for *Azolla*. In most other documented symbioses, including those with facultative symbionts from the Nostocales, $N_2$ fixation is fuelled with photosynthate derived from the plant and therefore presumed less efficient.

Engineering chloroplasts for $N_2$ fixation may be the first alternative to consider (nitroplast) when engineering nitrogen autotrophy of a crop plant since the coordinated development of a bacterium and vertical transmission during the life cycle of the crop plant are not needed. The nitroplast idea has been around for quite some time, but the endeavor is possibly still out of experimental reach, possibly because of the complex regulation needed to obtain the nitrogenase assembled. Very recent studies report a breakthrough with the expression of a whole recombinant *Nif*-operon from filamentous cyanobacteria: this led to nitrogenase complex formation. More work is needed to obtain functional complexes (Thiel 2019). Key for the engineering of a nitroplast will be the protection of nitrogenase from $O_2$ as well as the metabolic integration of $N_2$ fixation with reducing power and ATP production; this may now become feasible with improved efficiencies for chloroplast transformation and synthetic biology tools (Kwak et al. 2019).

Obviously, the common ancestor of the Archaeplastida (red algae, glaucophyte, green algae, and land plants) already possessed the machinery to phagocyte and maintain cyanobacteria (Nowack and Weber 2018); the machinery for phagocytosis likely was lost in the plant lineage when thicker cell walls became predominant. Genomic approaches indicate that chloroplasts are derived from an early-branching cyanobacterium related to *Gloeomargarita lithophora*; *G. lithophora* thrives in fresh water, is coccal, and lacks $N_2$ fixation (Ponce-Toledo et al. 2017). We conclude, therefore, that a machinery to specifically maintain $N_2$-fixating filamentous cyanobacteria, for example, may have evolved after chloroplasts were acquired; this occurred in hornworts, mosses, the *Azolla* ferns, the gymnosperm cycads, or plants from the *Gunnera* genus (Angiosperm). Nevertheless, functions of the many nuclear genes unique to all Archaeplastida (Karpowicz et al. 2011) should be taken into account when engineering chloroplasts with $N_2$-fixating capability. Alternatively, the specific metabolism of the recently discovered UCYN cyanobacteria from the *Braarudosphaera bigelowii* marine haptophyte algae should be used as a basis of understanding (Landa et al. 2021). Given the slow progress on turning chloroplasts into nitroplasts, stimulation of crop symbioses with phototrophic $N_2$-fixating bacteria remains an attractive alternative.

The conserved set of symbiosis genes from the plants in symbioses with Rhizobiales or arbuscular mycorrhiza has been invoked for the facultative interactions of Nostocaceae with many plants, including rice and *Gunnera* species, but not in the case of *Azolla* (Li et al. 2018; Alvarez et al. 2020). These symbioses are thus very likely polyphyletic. This is corroborated by a phylogenetic analysis of the Nostocales engaging in plant symbioses (Warshan et al. 2018). The later study revealed a list of genes common to all the *Nostoc* strains able to enter a symbiotic relation with plants. Once we understand how facultative phototrophic $N_2$-fixating bacteria may be recruited into crop plants, we can move on to reconstituting obligate symbioses such as those from *Azolla* and the UCYN-A/haptophyte; the latter showed that $N_2$ fixation is viable even in N-rich coastal or high-latitude water (Mills et al. 2020). In the meantime, we are using the genomes of the different *N. azollae* in the species of *Azolla* ferns to understand which genes were inactivated first, given that these genomes eroded mostly by transposon activity (Ran et al. 2010).

### 7.3.2 Controlling Formation of Megasporocarps

The number of megasporocarps limits sexual reproduction of *Azolla* ferns in a nursery setting or when reproductive structures, the megasporocarps/massulae clumps, are used for dissemination. Sporocarps are made as a result of the *Azolla* fern transition from the diploid to the haploid phase; these structures are likely homologous to sori of the homosporous ferns. The process of sori induction in the homosporous fern *Ceratopteris richardii* is characterized by specific and high expression of the *Cr*MADS1 transcription factor (TF), from a TF clade sister to that of the floral homeotic genes and AtSOC1 (Hasebe et al. 1998; Dijkhuizen et al. 2020). *Azolla* ferns contain two homologues of *Cr*MADS1, consistent with the whole-genome duplication at the base of the *Azolla* fern evolution, which are induced upon sporocarp induction; we have yet to find out whether these *Azolla* homologues are responsible for sporocarp specification. Sporangia in ferns further exhibit specific EuAP2 expression as shown in the case of *C. richardii* (Zumajo-Cardona et al. 2021).

In *Azolla*, once sporocarp initials are induced, sporangia initials develop. In megasporocaprs, the most advanced sporangium will develop into a megasporangium, while the remaining sporangia are stopped in their development. Alternatively in microsporocarps, the most advanced sporangium initial degenerates which permits development if many microsporangia inside the growing microsporocarp (Coulter 1910; Fig. 7.2a). To regulate the number of mega- to microsporocarps generated on the sporophytes, we need more insight into the mechanisms triggering megasporangium degeneration in the early development of *Azolla* sporangia. The mechanisms may be related to those of the nucellus (megasporangium) degradation in seed plants. In some seed plants, these mechanisms were linked to ethylene signalling and de-repression by the polycomb protein FERTILIZATION-INDEPENDENT SEED (Ingram 2017).

### 7.3.3 Quality of the Protein-Rich Biomass for Applications in the Food/Feed Sector

*Azolla* biomass is a protein-rich feed without any further modification; it is suited for the on-farm recycling of mineral nutrients where it does not need to be stored since it can be produced continuously (Leterme et al. 2009, 2010; Brouwer et al. 2018). *Azolla* biomass used in feed mixes has maximum inclusion rates that are somewhat lower than those for soymeal. This is likely due to the fact that *Azolla* fern sporophytes accumulate proanthocyanidins (PA, also known as condensed tannins; Güngör et al. 2021). PA likely function as antifeedants blocking enzymes of the digestive system and thus protecting sporophytes from grazers. PA extract together with proteins and they cause an unappetizing brown color upon oxidation. PA are regarded as the main bottleneck of *Azolla* protein extraction. On the other hand, PA could be beneficial anti-bloating agents in feed formulations (Kelln et al. 2020).

Manipulating the PA biosynthesis pathway would, therefore, be a first possible target for *Azolla* pre-breeding. PA have been reported to result from spontaneous polymerizations of *trans*- and *cis*-flavan-3-ols such as catechin and epicatechin (Dixon et al. 2005). However, new insights show that the polymerization might still be regulated by enzymes such as leucoanthocyanidin reductase (LAR). LAR is traditionally known as the enzyme converting leucoanthocyanidins into *trans*-flavan-3-ols such as catechin (Fig. 7.5). Consistently, LAR from *A. filiculoides* was shown to produce catechin from leucocyanidin in vitro (Güngör et al. 2021). A knock-out of the *LAR* gene in *Medicago truncatula*, however, caused accumulation of PA with a high degree of polymerization (DP) (Liu et al. 2016). LAR, therefore, was proposed to regulate the ratio of starter units (epicatechin) and extension units (epicatechin carbocation): knocking out LAR leads to a significant reduction in starter units causing the extension units to polymerize onto the few remaining starter units forming very large PAs. If there are more starter units, more smaller PAs are formed. Knocking out *Azfi*LAR might thus not directly lead to a reduction of the PA content in *Azolla* but in longer PA polymers. To manipulate PA accumulation instead, a transcription factor (TF) should be targeted that regulates expression of the PA biosynthesis pathway enzymes coordinately. Such a TF has not been characterized yet in *Azolla*, but phylogenetic analyses of MYB TF indicate that fern class VIII MYB may be involved (Güngör et al. 2021).

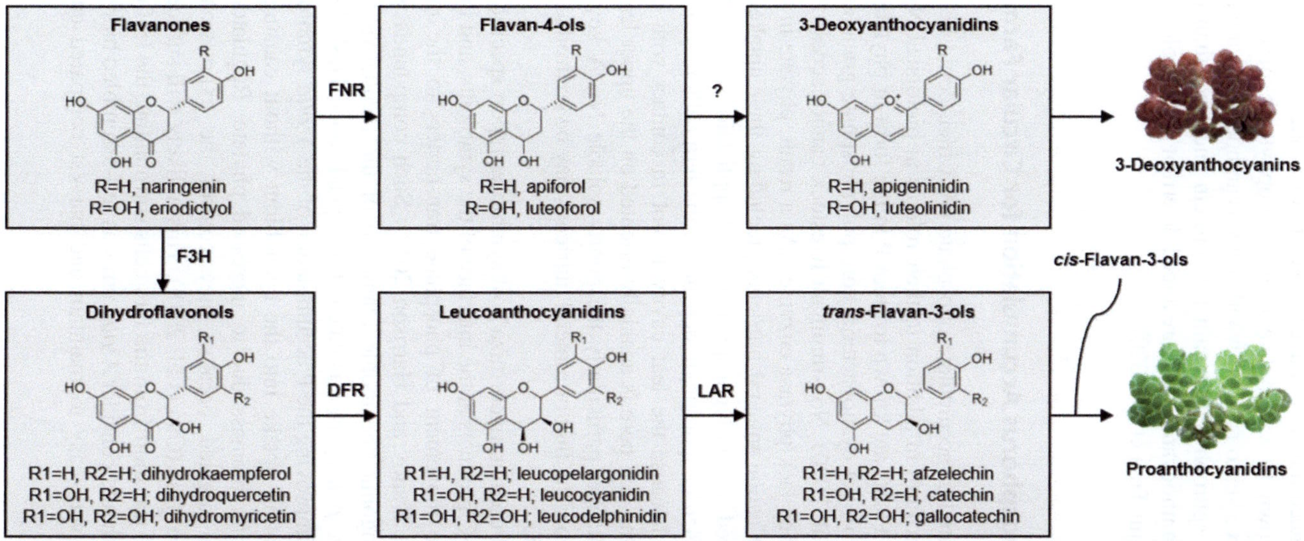


**Fig. 7.5** Partial flavonoid biosynthesis pathway in *A. filiculoides* (Güngör et al. 2021). DFR, dihydroflavonol 4-reductase; F3H, flavanone 3-hydroxylase; FNR, flavanone 4-reductase; LAR, leucoanthocyanidin reductase. Proanthocyanidins accumulate in green ferns with a degree of polymerization consistent with a function as antifeed. The activity of the *Azolla* LAR was shown in vitro, yet LAR enzymes from some plants also control the extent to which proanthocyanidins condensate. Under abiotic stress, *Azolla* ferns are known to accumulate 3-deoxyanthocyanin which are made from the same precursor naringenin as proanthocyanidins. Enzymes for the biosynthesis of the 3-deoxyanthocyanins are unknown

Alternatively, reducing dihydroflavonol 4-reductase (DFR) activity may reduce the intermediates supplying the PA biosynthesis (Fig. 7.5). Knocking out DFR, however, could lead to accumulation of alternative products such as the red-colored deoxyanthocyanins, known to accumulate when sporophytes experience abiotic stress (Fig. 7.5). We have yet to find out whether deoxyanthocyanins are beneficial colorants. Stressed red sporophytes contain less protein and lipid than the green unstressed ones. Deoxyanthocyanins also are a feed deterrent for tadpoles and snails (Cohen et al. 2002; Nham Tran et al. 2020).

### 7.3.4 Breeding Phosphorus Accumulation for Circular Farming

*Azolla* ferns were reported to accumulate phosphate (P) which could be useful to circularize P nutrient on farm in addition to their role as in biological $N_2$ fixation (Temmink et al. 2018). Plants are known to store P in the form of phytate (inositol hexakisphosphate, InsP6), yet cyanobacteria store polyphosphate granules (poly-P bodies; Sanz-Luque et al. 2020). Non-ruminant livestock cannot access P nutrient from phytate because they lack phytase enzymes. As a result, phytate in their guts binds essential divalent cation mineral nutrients precluding their uptake. Phytate, therefore, is less preferred over polyphosphate in feed applications. *N. azollae* from the leaf pockets of the *Azolla* sporophyte depend on the fern cells to provide them with mineral nutrients because the leaf cavity is not in contact with the water surrounding the ferns; the leaf pore is adaxially oriented on the upper leaf lobe so as to promote gas exchange, particularly the poorly soluble $N_2$. A key research question, therefore, is how the flux of mineral nutrients may be controlled from the host to the cyanobacteria.

Recent studies on arbuscular mycorrhiza/host symbioses report specific accumulation of phospholipase C products at the interface of the symbionts, and these have been reported to cause movement of phosphate transporters to the membrane (Nakamura et al. 2005; Ivanov and Harrison 2019). Such compounds were also found to induce hormogonia (the motile filaments) of the facultative symbiotic Yaku-1 strain of *Nostoc punctiforme* isolated from cycad roots (Hashidoko et al. 2019). Taken together, improving the P accumulation of the *Azolla* symbiosis could be achieved by stimulating uptake into the cyanobiont without causing mineral deficiencies of the *Azolla* biomass due to increased phytate. Regulators of the Pho-regulon could be targeted because they control the expression of the polyphosphate kinase enzyme (Gao et al. 2018). This unlikely will suffice without the further increase of P transport from the fern cells surrounding the leaf cavities.

More generally, we should consider *Azolla* ferns as self-assembled bioreactors of cyanobacteria that are very easy to maintain and harvest compared cultures of cyanobacteria or algae.

### 7.3.5 Breeding Yield Stability of Floating Mats in Aquaculture to Counter Soil Subsidence, Grazing by Specialist Insects, and Tannase Activity

Populations of *Azolla* are effectively controlled by grazers, the most notorious being *Stenopelmus rufinasus* (McConnachie et al. 2004). *S. rufinasus* has become ubiquitous because it has been actively deployed in regions where *Azolla* ferns are deemed invasive weeds; the weevil prefers *Azolla* species with lower PA content, *A. filiculoides* and *A. cristata*, compared to *A. pinnata* (Madeira et al. 2016; Güngör et al. 2021). The populations are furthermore affected by aphids; aphid feeding precludes *A. filiculoides* from reaching reproductive maturity (most notably the water lily aphid *Rhopalosiphum nymphaeae* (Hance et al. 1994)). Fungal infections with *Sclerotium rolfsii*, *Rhizoctonia* sp., *Acremonium* sp., *Aspergillus* sp., *Curvularia* sp., *Botryodiplodia*, and *Fusarium thapsinum* have been described (Shahjahan et al. 1980; Natural and Mendoza 1991; Dey et al. 2017), but also unidentified fungi affect *Azolla* mat growth. Much work is needed in this area to obtain yield stability in an open aquaculture setting, for example, when flooding land on low-land regions.

We distinguish the rapid production mode from the slow growth mode that leads to the buildup of a thick mat complete with a large zone of dead fern/root material. In the former mode, fern mats are harvested to 60% of the surface weekly which avoids formation of senescing ferns and permits disease evasion. In the latter mode, disease and pests need to be controlled. Engineering resistance to biotic stress represents the biggest breeding/biotechnology challenge in the domestication of the *Azolla* ferns because of their aquatic habitat; in these habitats, spray applications of systemic insecticides or immune stimulating substances are restricted. To this date, we fail to understand how grazers evade the effects of tannins as well as other not yet discovered anti-grazing compounds from the *Azolla* biomass.

## 7.4 Development of *Azolla* Breeding Tools and Current Limitations

### 7.4.1 Breeding by Genome Editing to Counter Inbreeding and Maximize on-Farm Crop Diversity

Breeding by selection and inbreeding is not desirable because it reduces the genetic diversity of the cultured populations (Fu 2015). Genetic diversity of receptors is required, for example, to maintain crop resistance to a variety of biotic stresses (Wu et al. 2018). Molecular breeding and genome editing represent the better alternatives.

Technologies to edit genomes seamlessly exist today; they mostly employ RNA-guided DNA binding nickases such as Cas9 and result in a very high frequency of edited cells in spite of their dependence on the target cell repair mechanisms (Atkins and Voytas 2020). Recently, the RNA-guided insertion of a transposon could be achieved in *E. coli* (Strecker et al. 2019): this means that editing may now become possible beyond a mere few bases even if the method is not yet seamless. The high efficacy of editing is key to its success: then only few meristem cells need to be targeted to engineer a new cell lineage. Alternatively, a genome-edited vegetatively growing lineage still needs to be induced into producing natural stem cells, including spores or gametes, for the production of homozygous gene-edited offspring.

Access to stem cells remains a critical bottleneck of all genome editing in practice, even with the latest genome editing techniques. Induction of shoot meristems from vegetative cells using WUSCHEL, SHOOT MERISTEMLESS (STM), and the bacterial isopentenyl synthase (ipt) for cytokinin biosynthesis proved an effective alternative to targeting natural or cultured stem cells in several Solanaceae, *Vitis vinifera*, and *Arabidopsis* (Maher et al. 2020). We know little of the components required for meristem induction in ferns (Ambrose and Vasco 2016). The only *Ceratopteris richardii* (Cer) WUSCHEL-related homeobox (WOX) from the modern WUSHEL clade of WOX proteins is not expressed in the shoot apical meristem. In contrast, *Cer*WOXB from the intermediate clade of WOX proteins is expressed, and its knockdown by RNAi causes a reduction of the sporophyte and meristem size (Youngstrom et al. 2019). Gene constructs that induced meristems in seed plants may thus not necessarily work in ferns. In addition, Maher et al. (2020) used constructs containing viral replicons such that the incoming T-DNA may be replicated and possibly trafficked between cells; viral replicons have never been tested in ferns.

### 7.4.2 Transient Transformation of the Fern Host and how to Progress from There

Tissue culture proved possible with the surface-sterilized *A. filiculoides* fern host devoid of cyanobacteria. When maintained in the dark, such sterilized sporophytes on C-medium stopped the outgrowth of leaves and branches; after 8 weeks, they produced cauliflower-like structures (Fig. 7.6a). C-medium was made of agar-solidified half-strength MS medium (Murashige and Skoog 1962) with 6% w/v sucrose, 1 mg/L 2,4-dichlorophenoxyacetic acid, and 1.5 mg/L 6-benzylaminopurine. Upon transfer to low light and R-medium for 2 weeks and then to liquid IRRI medium as in Brouwer et al. (2017), outgrowth resumed (Fig. 7.6b). R-medium was made of agar-solidified (0.7% w/v) half-strength MS medium with 150 mg/L ascorbic acid and 1 mg/L gibberellic acid at pH 6.0.

We have not yet learned how to re-inoculate ferns without *N. azollae* with their symbiont. Also we have not yet been able to induce sporocarp formation in ferns without *N. azollae*. Development of a transformation/genome editing method that does not require tissue culture, therefore, was the focus of initial studies.

For genome editing to be successful in *Azolla*, the following steps need to be achieved: (1) transfer of the DNA (RNA or ribozyme) into a cell, (2) movement to the nucleus, (3) transient expression of genes required for meristem activation and

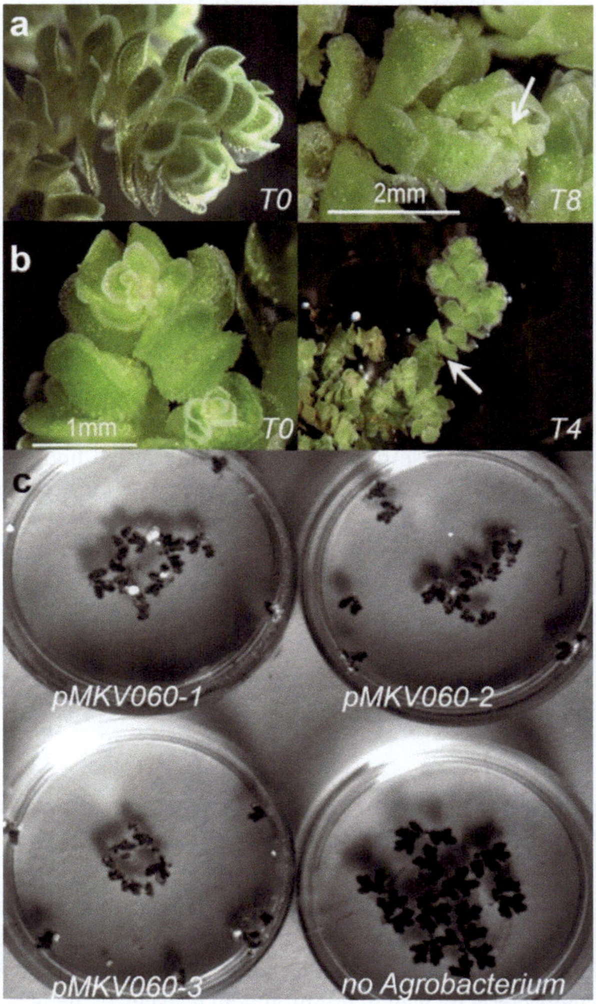


**Fig. 7.6** Results toward developing transformation and genome editing for the *Azolla* symbiosis. **(a)** Growth of sterilized *Azolla* sporophytes on agar-solidified half-strength Murashige and Skoog (MS, Murashige and Skoog 1962) medium with 6% w/v sucrose, 1 mg/L 2,4-dichlorophenoxyacetic acid, and 1.5 mg/L 6-benzylaminopurine. T0, at the beginning; T8, after 8 weeks in the dark. Cauliflower-like structures at the shoot apex (arrow) are typically observed after 8 weeks of dark incubation. **(b)** Outgrowth of new shoot apices (arrow) from the cauliflower-like structures. Ferns with cauliflower-like structures, obtained as in **(a)**, were transferred in low light and to agar-solidified (0.7% w/v) half-strength MS medium with 150 mg/L ascorbic acid and 1 mg/L gibberellic acid at pH 6.0. After 2 weeks, they were further transferred to liquid medium as in Brouwer et al. (2017). T0, shoot apex at the beginning; T4, representative situation after 4 weeks. **(c)** Transient expression of the luciferase gene 24 h after co-cultivation with *A. tumefaciens* GV3101 containing pMKV060 (Maher et al. 2020), compared to no *Agrobacterium* control. Luminescence is seen as white spots overlaid on the black and white image of the ferns. We tested that *Agrobacterium* with this construct do not express the luciferase unlike those with the pMM131 construct (Maher et al. 2020) which we also tested (data not shown)

catalytic insertion into the plant chromosome, (4) DNA insertion/editing in a meristematic cell, and (5) generation of reproductive organs/cells to obtain a homozygous offspring.

We first tested *Agrobacterium*-mediated transfer of DNA into cells of whole, cut, or dissected fern after dipping procedures. We used the previously tested plasmid pMKV060 (https://www.addgene.org/133315/) containing a transfer-DNA (T-DNA) encoding the Bean yellow dwarf viral (BeYDV) replicon and the firefly *LUCIFERASE* expression cassette and pMKV057 (https://www.addgene.org/133312/) containing the same T-DNA with in addition *STM* and *ipt* expression cassettes (Maher et al. 2020). The luciferase reporter was active in sporophyte leaf cells exposed to *Agrobacterium* with either of these plasmids, while it was inactive when the ferns were not exposed to the *Agrobacterium* (Fig. 7.6c). Moreover, the reporter was not active in these *Agrobacterium* strains, unlike those containing the pMM131 plasmid with the *LUCIFERASE* cloned behind the CaMV promoter (Maher et al. 2020). Luciferase reporter activity only lasted until 4 days after the co-cultivation, however. Also, the luminescence was never located in the shoot tips, unless we manually dissected these before co-cultivation (Fig. 7.6c). We conclude that *Agrobacterium* is able to transfer T-DNA into cells of *Azolla* ferns and that T-DNA is moved to the nucleus and then expressed. The T-DNA likely was silenced/degraded and unlikely incorporated into the fern genome. In addition, shoot meristem cells were not exposed to the *Agrobacterium* during dipping, regardless of whether Silvet L-77 and sucrose were added during the dipping procedure as in Clough and Bent (1998).

### 7.4.3 Transient Transformation for the *N. Azollae* and how to Progress from there

Shoot apices of the *A. filiculoides* sporophytes have proven impermeable to the outside solutions. It may be possible to dissect shoot apices to expose the apical *N. azollae* colony, apply a DNA transfer technique, and, possibly, then select for the engineered cyanobacteria by way of erythromycin resistance: the ferns but not wild-type *N. azollae* resist 60 mg/L erythromycin when grown on nitrogen medium (Forni et al. 1991).

Preliminary, tri-parental conjugation assays using leaf juice from sporophytes revealed that GFP was transiently expressed in *N. azollae* for up to 24 h after conjugation with a cargo containing the GFP expression cassette. This occurred at low frequencies, however, as the cells were not dividing and possibly no longer viable outside the fern leaf cavity (data not shown).

The process needs optimizing so as to achieve high rates of incorporation of the incoming cargo. One possible alternative to achieve this is to catalyze the insertion of the cargo into the chromosome or plasmids that are expected to be present in multiple copies per cyanobacterial cell. Recent characterization of CRISPR-associated transposases (CAST) and their optimization in *E. coli* reveal that this may be possible by way of RNA-guided transposition (Strecker et al. 2019). The

CAST were found in many filamentous cyanobacteria, including *Anabaena cylindrica* PCC 7122, and, therefore, they likely will function. The engineered version of CAST, however, should be tested in *Anabaena* first and then in *N. azollae*.

To move *Azolla* symbioses from high yield potential to high yield stability, we conclude, is a tall order that will require building a community of scientists able to communicate so as to transgress discipline boundaries and organization from molecular to ecosystem level. Many new developments fostering fundamental research on symbioses and their development as model organisms will facilitate developments with *Azolla* ferns; let us thus embrace the challenge together.

**Acknowledgments** We would like to thank Bas van de Riet and Prof. Fons Smolders for allowing us to repeatedly collect sporocarps from their cultures. NWO-ALW EPS grant (ALWGS.2016.020) supported LWD; NWO-TTW grant (Project 16294) supported EG. We acknowledge funding from the Gordon and Betty Moore Foundation's Symbiosis in Aquatic Systems Initiative.

## References

Alvarez C, Navarro JA, Molina-Heredia FP, Mariscal V (2020) Endophytic colonization of rice (*Oryza sativa* L.) by the symbiotic strain *Nostoc punctiforme* PCC 73102. Mol Plant-Microbe Interact 33:1040–1045. https://doi.org/10.1094/MPMI-01-20-0015-SC

Ambrose BA, Vasco A (2016) Bringing the multicellular fern meristem into focus. New Phytol 210: 790–793. https://doi.org/10.1111/nph.13825

Atkins PA, Voytas DF (2020) Overcoming bottlenecks in plant gene editing. Curr Opin Plant Biol 54:79–84. https://doi.org/10.1016/j.pbi.2020.01.002

Bailey-Serres J, Parker JE, Ainsworth EA, Oldroyd GE, Schroeder JI (2019) Genetic strategies for improving crop yields. Nature 575:109–118. https://doi.org/10.1038/s41586-019-1679-0

Birks HH (2002) The recent extinction of *Azolla nilotica* in the Nile Delta, Egypt. Acta Palaeobot 42:203–213

Brouwer P, Bräutigam A, Külahoglu C, Tazelaar AO, Kurz S, Nierop KG, van der Werf A, Weber AP, Schluepmann H (2014) *Azolla* domestication towards a biobased economy? New Phytol 202:1069–1082. https://doi.org/10.1111/nph.12708

Brouwer P, Bräutigam A, Buijs VA, Tazelaar AO, van der Werf A, Schlüter U, Reichart GJ, Bolger A, Usadel B, Weber AP, Schluepmann H (2017) Metabolic adaptation, a specialized leaf organ structure and vascular responses to diurnal $N_2$-fixation by *Nostoc azollae* sustain the astonishing productivity of *Azolla* ferns without nitrogen fertilizer. Front Plant Sci 8:442. https://doi.org/10.3389/fpls.2017.00442

Brouwer P, Schluepmann H, Nierop KG, Elderson J, Bijl PK, van der Meer I, de Visser W, Reichart GJ, Smeekens S, van der Werf A (2018) Growing *Azolla* to produce sustainable protein feed: the effect of differing species and $CO_2$ concentrations on biomass productivity and chemical composition. J Sci Food Agric 98:4759–4768. https://doi.org/10.1002/jsfa.9016

Calvert HE, Pence MK, Peters GA (1985) Ultrastructural ontogeny of leaf cavity trichomes in Azolla implies a functional role in metabolite exchange. Protoplasma 129:10–27. https://doi.org/10.1007/BF01282301

Clough SJ, Bent AF (1998) Floral dip: a simplified method for *Agrobacterium*-mediated transformation of *Arabidopsis thaliana*. Plant J 16:735–743. https://doi.org/10.1046/j.1365-313X.1998.00343.x

Cohen MF, Sakihama Y, Takagi YC, Ichiba T, Yamasaki H (2002) Synergistic effect of deoxyanthocyanins from symbiotic fern *Azolla* spp. on *hrmA* gene induction in the cyanobacterium *Nostoc punctiforme*. Mol Plant-Microbe Interact 15:875–882. https://doi.org/10.1094/MPMI.2002.15.9.875

Cohen MF, Gurung S, Birarda G, Holman HY, Yamasaki H (2015) Bimodal effect of hydrogen peroxide and oxidative events in nitrite-induced rapid root abscission by the water fern *Azolla pinnata*. Front Plant Sci 6:518. https://doi.org/10.3389/fpls.2015.00518

Coulter JM (1910) A textbook of botany for colleges and universities. American Book Company, Atlanta

Cox TS, Shroyer JP, Ben-Hui L, Sears RG, Martin TJ (1988) Genetic improvement in agronomic traits of hard red winter wheat cultivars 1919 to 1987. Crop Sci 28:756–760. https://doi.org/10.2135/cropsci1988.0011183x002800050006x

Dey S, Hore M, Biswas J, Biswas M, Mandal BK, Das P, Gupta S (2017) A new record of brown rot disease in water fern *Azolla microphylla* (Azollaceae): loss of important bio-resource. Fern Gaz 20(6):245–254. https://ebps.org.uk/wp-content/uploads/2017/11/FGV20P6M21-AUQjPl5u7L.pdf

Di Vittori V, Gioia T, Rodriguez M, Bellucci E, Bitocchi E, Nanni L, Attene G, Rau D, Papa R (2019) Convergent evolution of the seed shattering trait. Genes (Basel) 10:68. https://doi.org/10.3390/genes10010068

Dijkhuizen LW, Brouwer P, Bolhuis H, Reichart GJ, Koppers N, Huettel B, Bolger AM, Li FW, Cheng S, Liu X, Wong GK (2018) Is there foul play in the leaf pocket? The metagenome of floating fern *Azolla* reveals endophytes that do not fix $N_2$ but may denitrify. New Phytol 217: 453–466. https://doi.org/10.1111/nph.14843

Dijkhuizen LW, Tabatabaei BE, Brouwer P, Rijken N, Mehta RS, Güngör E, Schluepmann H (2020) The *Azolla* fern symbiosis sexual reproduction requires far-red light and involves responsive CMADS1 homologue, miR319-controlled GAMYB, miR529 in the fern, and transporters in the symbiont. Front Plant Sci (in press) bioRxiv. https://www.biorxiv.org/content/10.1101/2020.09.09.289736v2

Dixon RA, Xie DY, Sharma SB (2005) Proanthocyanidins - a final frontier in flavonoid research? New Phytol 165:9–28. https://doi.org/10.1111/j.1469-8137.2004.01217.x

Forni CI, Tel-Or E, Bar E, Grilli-Caiola M (1991) Effects of antibiotic treatments on *Azolla, anabaena* and *Arthrobacter*. In: Polsinelli M, Materassi R, Vincenzini M (eds) Nitrogen fixation, Developments in plant and soil sciences, vol 48. Springer, Dordrecht

Fu YB (2015) Understanding crop genetic diversity under modern plant breeding. Theor Appl Genet 128:2131–2142. https://doi.org/10.1007/s00122-015-2585-y

Gao F, Wu H, Zeng M, Huang M, Feng G (2018) Overproduction, purification, and characterization of nanosized polyphosphate bodies from *Synechococcus sp*. PCC 7002. Microb Cell Factories 17:1–2. https://doi.org/10.1186/s12934-018-0870-6

Geissler B, Mew MC, Steiner G (2019) Phosphate supply security for importing countries: developments and the current situation. Sci Total Environ 677:511–523. https://doi.org/10.1016/j.scitotenv.2019.04.356

Gu S, Gruau G, Dupas R, Petitjean P, Lic Q, Pinay G (2019) Respective roles of Fe-oxyhydroxide dissolution, pH changes and sediment inputs in dissolved phosphorus release from wetland soils under anoxic conditions. Geoderma 338:365–374. https://doi.org/10.1016/j.geoderma.2018.12.034

Güngör E, Brouwer P, Dijkhuizen LW, Shaffar DC, Nierop KG, de Vos RC, Sastre Torano J, van Der Meer IM, Schluepmann H (2021) *Azolla* ferns testify: seed plants and ferns share a common ancestor for leucoanthocyanidin reductase enzymes. New Phytol 229:1118–1132. https://doi.org/10.1111/nph.16896

Gurung S, Cohen MF, Fukuto J, Yamasaki H (2012) Polyamine-induced rapid root abscission in *Azolla pinnata*. J Amino Acids 2012:1–9. https://doi.org/10.1155/2012/493209

Hance T, Nibelle D, Lebrun PH, Van Impe G, Van Hove C (1994) Selection of *Azolla* forms resistant to the water lily aphid, *Rhopalosiphum nymphaeae* life history of *Rhopalosiphum nymphaeae*. Entomol Exp Appl 70:19–25. https://doi.org/10.1111/j.1570-7458.1994.tb01754.x

Harvey WH, Caponetti JD (1974) *In vitro* studies on the induction of sporogenous tissue on leaves of cinnamon fern. III. The role of reduced nitrogen. Can J Bot 52:2611–2614. https://doi.org/10.1139/b74-339

Hasebe M, Wen CK, Kato M, Banks JA (1998) Characterization of *MADS* homeotic genes in the fern *Ceratopteris richardii*. Proc Natl Acad Sci U S A 95:6222–6227. https://doi.org/10.1073/pnas.95.11.6222

Hashidoko Y, Nishizuka H, Tanaka M, Murata K, Murai Y, Hashimoto M (2019) Isolation and characterization of 1-palmitoyl-2-linoleoyl-sn-glycerol as a hormogonium-inducing factor (HIF) from the coralloid roots of *Cycas revoluta* (Cycadaceae). Sci Rep 9:1–2. https://doi.org/10.1038/s41598-019-39647-8

Calvert H, Peters G (1981) The *Azolla-Anabaena azollae* relationship: IX. Morphological analysis of leaf cavity hair populations. New Phytol 89:327–335. https://doi.org/10.1111/j.1469-8137.1981.tb07493.x

Ingram GC (2017) Dying to live: cell elimination as a developmental strategy in angiosperm seeds. J Exp Bot 68:785–796. https://doi.org/10.1093/jxb/erw364

Ivanov S, Harrison MJ (2019) Accumulation of phosphoinositides in distinct regions of the periarbuscular membrane. New Phytol 221:2213–2227. https://doi.org/10.1111/nph.15553

Johnson M, Zaretskaya I, Raytselis Y, Merezhuk Y, McGinnis S, Madden TL (2008) NCBI BLAST: a better web interface. Nucleic Acids Res 36:W5–W9. https://doi.org/10.1093/nar/gkn201

Karpowicz SJ, Prochnik SE, Grossman AR, Merchant SS (2011) The greenCut2 resource, a phylogenomically derived inventory of proteins specific to the plant lineage. J Biol Chem 286:21427–21439. https://doi.org/10.1074/jbc.M111.233734

Ka-Shu Wong G, Soltis DE, Leebens-Mack J et al (2020) Sequencing and analyzing the transcriptomes of a thousand species across the tree of life for green plants. Annu Rev Plant Biol 71:741–765. https://doi.org/10.1146/annurev-arplant-042916

Keilwagen J, Hartung F, Grau J (2019) GeMoMa: homology-based gene prediction utilizing intron position conservation and RNA-seq data. In: Gene prediction. Humana, New York, pp 161–177. https://doi.org/10.1007/978-1-4939-9173-0_9

Kelln B, Penner GB, Acharya SN, McAllister TA, Lardner HA (2020) Impact of condensed tannin containing legumes on ruminal fermentation, nutrition and performance in ruminants: a review. Can J Anim Sci. https://doi.org/10.1139/cjas-2020-0096

Kole C, Muthamilarasan M, Henry R, Edwards D, Sharma R, Abberton M, Batley J, Bentley A, Blakeney M, Bryant J, Cai H (2015) Application of genomics-assisted breeding for generation of climate resilient crops: Progress and prospects. Front Plant Sci 6:563. https://doi.org/10.3389/fpls.2015.00563

Kwak SY, Lew TT, Sweeney CJ, Koman VB, Wong MH, Bohmert-Tatarev K, Snell KD, Seo JS, Chua NH, Strano MS (2019) Chloroplast-selective gene delivery and expression in planta using chitosan-complexed single-walled carbon nanotube carriers. Nat Nanotechnol 14:447–455. https://doi.org/10.1038/s41565-019-0375-4

Labouriau LFG (1958) Studies on the Initiation of Sporangia in Ferns. Dissertation (Ph.D.), California Institute of Technology. https://doi.org/10.7907/3BNE-T071. https://resolver.caltech.edu/CaltechETD:etd-10082004-105932

Landa M, Turk-Kubo KA, Cornejo-Castillo FM, Henke BA, Zehr JP (2021) Critical role of light in the growth and activity of the marine $N_2$-fixing UCYN-A symbiosis. Front Microbiol 12:1024. https://doi.org/10.3389/fmicb.2021.666739

Leebens-Mack JH, Barker MS, Carpenter EJ et al (2019) One thousand plant transcriptomes and the phylogenomics of green plants. Nature 574:679–685. https://doi.org/10.1038/s41586-019-1693-2

Leterme P, Londono AM, Munoz JE, Súarez J, Bedoya CA, Souffrant WB, Buldgen A (2009) Nutritional value of aquatic ferns (*Azolla filiculoides* lam. And *Salvinia molesta* Mitchell) in pigs. Anim Feed Sci Technol 149:135–148. https://doi.org/10.1016/j.anifeedsci.2008.04.013

Leterme P, Londoño AM, Ordoñez DC, Rosales A, Estrada F, Bindelle J, Buldgen A (2010) Nutritional value and intake of aquatic ferns (*Azolla filiculoides* lam. And *Salvinia molesta* Mitchell.) in sows. Anim Feed Sci Technol 155:55–64. https://doi.org/10.1016/j.anifeedsci.2009.10.002

Li F-W, Brouwer P, Carretero-Paulet L et al (2018) Fern genomes elucidate land plant evolution and cyanobacterial symbioses. Nat Plants 4:460–472. https://doi.org/10.1038/s41477-018-0188-8

Lin C, Watanabe I (1988) A new method for obtaining *anabaena*-free *Azolla*. New Phytol 108:341–344. https://doi.org/10.1111/j.1469-8137.1988.tb04172.x

Liu C, Wang X, Shulaev V, Dixon RA (2016) A role for leucoanthocyanidin reductase in the extension of proanthocyanidins. Nat Plants 2:1–7. https://doi.org/10.1038/nplants.2016.182

Madeira PT, Hill MP, Dray FA, Coetzeeb JA, Paterson ID, Tippinga PW (2016) Molecular identification of *Azolla* invasions in Africa: the *Azolla* specialist, *Stenopelmus rufinasus* proves to be an excellent taxonomist. South Afr J Bot 105:299–305. https://doi.org/10.1016/j.sajb.2016.03.007

Madeira PT, Dray FA, Tipping PW (2019) Molecular identification of *Azolla* in the Yangtze River watershed, China. Aquat Bot 159:103149. https://doi.org/10.1016/j.aquabot.2019.103149

Maher MF, Nasti RA, Vollbrecht M, Starker CG, Clark MD, Voytas DF (2020) Plant gene editing through *de novo* induction of meristems. Nat Biotechnol 38:84–89. https://doi.org/10.1038/s41587-019-0337-2

Manzano-Marín A, Coeur d'acier A, Clamens AL, Orvain C, Cruaud C, Barbe V, Jousselin E (2020) Serial horizontal transfer of vitamin-biosynthetic genes enables the establishment of new nutritional symbionts in aphids' di-symbiotic systems. ISME J 14:259–273. https://doi.org/10.1038/s41396-019-0533-6

Massawe F, Mayes S, Cheng A (2016) Crop diversity: an unexploited treasure trove for food security. Trends Plant Sci 21:365–368. https://doi.org/10.1016/j.tplants.2016.02.006

McConnachie AJ, Hill MP, Byrne MJ (2004) Field assessment of a frond-feeding weevil, a successful biological control agent of red waterfern, *Azolla filiculoides*, in southern Africa. Biol Control 29:326–331. https://doi.org/10.1016/j.biocontrol.2003.08.010

McCutcheon JP, Moran NA (2010) Functional convergence in reduced genomes of bacterial symbionts spanning 200 my of evolution. Genome Biol Evol 2:708–718. https://doi.org/10.1093/gbe/evq055

Miller R, Fram M, Fujii R, Wheeler G (2008) Subsidence reversal in a re-established wetland in the Sacramento-san Joaquin Delta, California, USA. San Franc Estuary Watershed Sci 6(3). https://doi.org/10.15447/sfews.2008v6iss3art1

Mills MM, Turk-Kubo KA, van Dijken GL, Henke BA, Harding K, Wilson ST, Arrigo KR, Zehr JP (2020) Unusual marine cyanobacteria/haptophyte symbiosis relies on $N_2$ fixation even in N-rich environments. ISME J 14:2395–2406. https://doi.org/10.1038/s41396-020-0691-6

Murashige T, Skoog F (1962) A revised medium for rapid growth and bio assays with tobacco tissue cultures. Physiol Plant 15:473–497. https://doi.org/10.1111/j.1399-3054.1962.tb08052.x

Nakamura Y, Awai K, Masuda T, Yoshioka Y, Takamiya KI, Ohta H (2005) A novel phosphatidylcholine-hydrolyzing phospholipase C induced by phosphate starvation in Arabidopsis. J Biol Chem 280:7469–7476. https://doi.org/10.1074/jbc.M408799200

Natural MP, Mendoza RL (1991) New fungal pathogens of Azolla. UPLB-National Crop Protection Center Library College, Laguna

Nham Tran TL, Miranda AF, Abeynayake SW, Mouradov A (2020) Differential production of phenolics, lipids, carbohydrates and proteins in stressed and unstressed aquatic plants, *Azolla filiculoides* and *Azolla pinnata*. Biology 9:342. https://doi.org/10.3390/biology9100342

Nierop KG, Brouwer P, Dekker R, Schluepmann H, Reichart GJ (2018) ω20-Hydroxy and ω9, ω10-dihydroxy biomarker lipids in ferns from the Salviniaceae family. Org Geochem 125:229–242. https://doi.org/10.1016/j.btre.2019.e00368

Nowack ECM, Weber APM (2018) Genomics-informed insights into endosymbiotic organelle evolution in photosynthetic eukaryotes. Annu Rev Plant Biol 69:51–84. https://doi.org/10.1146/annurev-arplant-042817-040209

Perkins SK, Peters GA (1993) The *Azolla–anabaena* symbiosis: endophyte continuity in the *Azolla* life-cycle is facilitated by epidermal trichomes : I. partitioning of the endophytic *anabaena* into developing sporocarps. New Phytol 123:53–64. https://doi.org/10.1111/j.1469-8137.1993.tb04531.x

Peters GA, Perkins SK (1993) The *Azolla–anabaena* symbiosis: endophyte continuity in the *Azolla* life-cycle is facilitated by epidermal trichomes : II. Re-establishment of the symbiosis following gametogenesis and embryogenesis. New Phytol 123:65–75. https://doi.org/10.1111/j.1469-8137.1993.tb04532.x

Peters G, Evans W, Toia R (1976) *Azolla-Anabaena azollae* relationship: IV. Photosynthetically driven, Nitrogenase-catalyzed $H_2$ production. Plant Physiol 58:119–126. https://doi.org/10.1104/pp.58.2.119

Pinero-Rodríguez MJ, Fernández-Zamudio R, Arribas R, Gomez-Mestre I, Díaz-Paniagua C (2021) The invasive aquatic fern *Azolla filiculoides* negatively impacts water quality, aquatic vegetation and amphibian larvae in Mediterranean environments. Biol Invasions 23:755–769. https://doi.org/10.1007/s10530-020-02402-6

Pinto-Carbó M, Gademann K, Eberl L, Carlier A (2018) Leaf nodule symbiosis: function and transmission of obligate bacterial endophytes. Curr Opin Plant Biol 44:23–31. https://doi.org/10.1016/j.pbi.2018.01.001

Plazinski J, Zheng Q, Taylor R, Rolfe BG, Gunning BE (1989) Use of DNA/DNA hybridization techniques to authenticate the production of new *Azolla-anabaena* symbiotic associations. FEMS Microbiol Lett 65:199–203. https://doi.org/10.1111/j.1574-6968.1989.tb03622.x

Ponce-Toledo RI, Deschamps P, López-García P, Zivanovic Y, Benzerara K, Moreira D (2017) An early-branching freshwater cyanobacterium at the origin of plastids. Curr Biol 27:386–391. https://doi.org/10.1016/j.cub.2016.11.056

Ran L, Larsson J, Vigil-Stenman T, Nylander JA, Ininbergs K, Zheng WW, Lapidus A, Lowry S, Haselkorn R, Bergman B (2010) Genome erosion in a nitrogen-fixing vertically transmitted endosymbiotic multicellular cyanobacterium. PLoS One 5:e11486. https://doi.org/10.1371/journal.pone.0011486

Ray T, Mayne B, Toia R, Peters G (1979) *Azolla-anabaena* relationship: VIII. Photosynthetic characterization of the association and individual partners. Plant Physiol 64:791–795. https://doi.org/10.1104/pp.64.5.791

Rissman AI, Mau B, Biehl BS, Darling AE, Glasner JD, Perna NT (2009) Reordering contigs of draft genomes using the Mauve aligner. Bioinformatics 25:2071–2073. https://doi.org/10.1093/bioinformatics/btp356

Salvagiotti F, Cassman KG, Specht JE, Walters DT, Weiss A, Dobermann A (2008) Nitrogen uptake, fixation and response to fertilizer N in soybeans: a review. F Crop Res 108:1–13. https://doi.org/10.1016/j.fcr.2008.03.001

Sanz-Luque E, Bhaya D, Grossman AR (2020) Polyphosphate: a multifunctional metabolite in cyanobacteria and algae. Front Plant Sci 11:938. https://doi.org/10.3389/fpls.2020.00938

Schuurink R, Tissier A (2020) Glandular trichomes: micro-organs with model status? New Phytol 225:2251–2266. https://doi.org/10.1111/nph.16283

Shahjahan AKM, Miah SA, Nahar MA, Majid MA (1980) Fungi attack Azolla in Bangladesh. Int Rice Res Newslett 5:17–18

Singh PK, Bisoyi RN, Singh RP (1990) Collection and germination of sporocarps of *Azolla caroliniana*. Ann Bot 66:51–56. https://doi.org/10.1093/oxfordjournals.aob.a087999

Speelman EN, van Kempen MM, Barke J, Brinkhuis H, Reichart GJ, Smolders AJ, Roelofs JG, Sangiorgi F, de Leeuw JW, Lotter AF, Sinninghe Damste JS (2009) The Eocene Arctic *Azolla* bloom: environmental conditions, productivity and carbon drawdown. Geobiology 7:155–170. https://doi.org/10.1111/j.1472-4669.2009.00195.x

Strecker J, Ladha A, Gardner Z, Schmid-Burgk JL, Makarova KS, Koonin EV, Zhang F (2019) RNA-guided DNA insertion with CRISPR-associated transposases. Science 365:48–53. https://doi.org/10.1126/science.aax9181

Sussex IM, Steeves TA (1958) Experiments on the control of fertility of Fern leaves in sterile culture. Bot Gaz 119:203–208. https://doi.org/10.1086/335980

Temmink RJ, Harpenslager SF, Smolders AJ, van Dijk G, Peters RC, Lamers LP, van Kempen MM (2018) *Azolla* along a phosphorus gradient: biphasic growth response linked to diazotroph traits and phosphorus-induced iron chlorosis. Sci Rep 8:1–8. https://doi.org/10.1038/s41598-018-22760-5

Thiel T (2019) Organization and regulation of cyanobacterial nif gene clusters: Implications for nitrogenase expression in plant cells. FEMS Microbiol Lett 366:fnz077. https://doi.org/10.1093/femsle/fnz077

Tyagi VV, Ray TB, Mayne BC, Peters GA (1981) The *Azolla-Anabaena azollae* relationship : XI. Phycobiliproteins in the action spectrum for nitrogenase-catalyzed acetylene reduction. Plant Physiol 68:1479–1484. https://doi.org/10.1104/pp.68.6.1479

Uheda E (1986) Isolation of empty packets from *anabaena*-free *Azolla*. Plant Cell Physiol 27:1187–1190. https://doi.org/10.1093/oxfordjournals.pcp.a077203

Uheda E, Kitoh S (1994) Rapid shedding of roots from *Azolla filiculoides* plants in response to inhibitors of respiration. Plant Cell Physiol 35:37–43. https://doi.org/10.1093/oxfordjournals.pcp.a078568

Wang H, Wang F, Wang C, Han Y (2019) Effects of floating *Azolla* on phosphorus fluxes and recovery from former agricultural lands in wetland microcosms. Soil Sci plant Nutr 65:90-9 AND (2019) Statement of retraction: effects of floating *Azolla* on phosphorus fluxes and recovery from former agricultural lands in wetland microcosms. https://doi.org/10.1080/00380768.2018.1536387

Warshan D, Liaimer A, Pederson E, Kim SY, Shapiro N, Woyke T, Altermark B, Pawlowski K, Weyman PD, Dupont CL, Rasmussen U (2018) Genomic changes associated with the evolutionary transitions of *Nostoc* to a plant symbiont. Mol Biol Evol 35:1160–1175. https://doi.org/10.1093/molbev/msy029

Watanabe I, Roger PS, Ladha JK, Van Hove C (1992) Biofertilizer germplasm collections at IRRI. International Rice Research Institute, Manila

Watanabe I, Lapis-Tenorio MT, Ventura TS, Padre BC Jr (1993) Sexual hybrids of *Azolla filiculoides* with *a. microphylla*. Soil Sci Plant Nutr 39:669–676. https://doi.org/10.1080/00380768.1993.10419184

Wu CH, Derevnina L, Kamoun S. Receptor networks underpin plant immunity. Science. 2018 Jun 22;360(6395):1300–1. https://doi.org/10.1126/science.aat2623

Yamasaki H, Ogura MP, Kingjoe KA, Cohen MF (2019) D-cysteine-induced rapid root abscission in the water fern *Azolla pinnata*: implications for the linkage between d-amino acid and reactive sulfur species (RSS) in plant environmental responses. Antioxidants 8:411. https://doi.org/10.3390/antiox8090411

Youngstrom CE, Geadelmann LF, Irish EE, Cheng CL (2019) A fern *WUSCHEL-RELATED HOMEOBOX* gene functions in both gametophyte and sporophyte generations. BMC Plant Biol 19:1–3. https://doi.org/10.1186/s12870-019-1991-8

Zsögön A, Čermák T, Naves ER, Notini MM, Edel KH, Weinl S, Freschi L, Voytas DF, Kudla J, Peres LE (2018) *De novo* domestication of wild tomato using genome editing. Nat Biotechnol 36:1211–1216. https://doi.org/10.1038/nbt.4272

Zumajo-Cardona C, Pabón-Mora N, Ambrose BA (2021) The evolution of *euAPETALA2* genes in vascular plants: from plesiomorphic roles in sporangia to acquired functions in ovules and fruits. Mol Biol Evol. https://doi.org/10.1093/molbev/msab027

**H. Schluepmann** is a well-traveled plant biologist with a great respect for the achievements of young scientists. "After a Masters researching homologous recombination in the lab of Jerzy Paszkowski at ETH (Zürich, Switzerland), I completed my PhD on pollen tube cell wall biosynthesis in the laboratory of Tony Bacic (Melbourne, Australia). Since 2004, I am Assistant Professor at the Utrecht University Molecular Plant Physiology where I uncovered the regulatory role of trehalose metabolism in *Arabidopsis thaliana* before developing the innovation-driven *Azolla* research. I am very much hands-on, both in the laboratory and in bio-informatics, in spite of the visionary ambitions in sustainability of my research."

**I. Bigot** was a Master's Bio Inspired Innovation intern at the *Azolla* laboratory in 2019–2020 (Molecular Plant Physiology Department of Utrecht University). Ivan explored dissemination and storage of *Azolla filiculoides*, specifically the induction and the maturity of spores, fertilization, and sporeling establishment, and worked on T-DNA delivery in both sporophytes and spores. He is currently working at the INRA plum trees.

**N. Rijken** was a Master's Environmental and Plant Biology intern at the *Azolla* laboratory in 2018–2019 (Molecular Plant Physiology Department of Utrecht University); he currently is a research assistant at Koppert Biological Systems. Niels explored the effect of far-red light on, and analyzed miRNA loci of, *Azolla filiculoides*.

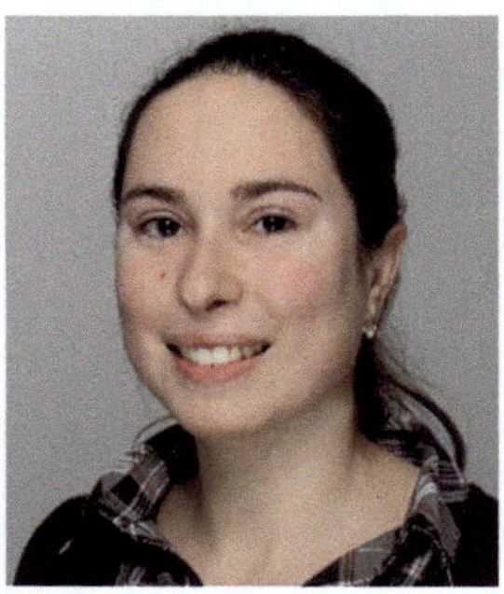

**A. Correas Grifoll** was a Master's Science and Business intern at the *Azolla* laboratory in 2018–2019 (Molecular Plant Physiology Department of Utrecht University). Using confocal microscopy, Anna explored the diversity of natural fluorophores synthesized during the development of reproductive structures of the *A. filiculoides* symbiosis. She currently is working as Marketing Associate at OPTI Medical Systems.

**P. Gudde** was a Master's Bio Inspired Innovation intern at the *Azolla* laboratory in 2019–2020 (Molecular Plant Physiology Department of Utrecht University). Peter analyzed the genomes of *Nostoc azollae* from different *Azolla* species. He is now working in the field of science communication.

**L.W. Dijkhuizen** is a Ph.D. candidate in the *Azolla* laboratory of Henriette Schluepmann. Laura studies the microbes associated with ferns of the *Azolla* genus using metagenomics techniques. In other *Azolla* projects, she supports colleagues with computational infrastructure and analyses.

**E. Güngör** is a Ph.D. candidate in the *Azolla* laboratory of Henriette Schluepmann. He obtained a bachelor in Medical Biology and a master in Bio Inspired Innovation. After practical research in methane microbiology, he moved on to investigating *Azolla* secondary metabolism in relation to biomass quality.

# Ferns, a Source of Phytoecdysones, and their Applications in Pestiferous Insect Management

# 8

Kitherian Sahayaraj

**Abstract**

Steroidal compounds which are closely related structurally to ecdysone are grouped as ecdysteroids. Ecdysteroids identified in plants and animals are named phytoecdysteroids and zooecdysteroids, respectively. Phytoecdysteroids are distributed in a large number of land plants (6%). Phytoecdysteroids have been recorded from 27 families of pteridophytes, 10 families of gymnosperms, and 74 families of angiosperms. Ecdysteroids are not found in all fern families, but in Polypodiaceae; it is very common. Ferns likes *Pteridium aquilinum*, *Polypodium vulgare*, *Schizaea dichotoma*, *Cheilanthes farinosa*, *Cheilanthes tenuifolia*, *Microsorum scolopendria*, *Adiantum pedatum*, *Dryopteris nipponensis*, *Adiantum raddianum*, *Asplenium aethiopicum*, *Cyclosorus interruptus*, *Dicranopteris linearis*, *Diplazium polypodioides* etc. more commonly synthesize Ecdysteroids. Chemically, phytoecdysteroids are triterpenoids (triterpene saponins, phytosterols, and phytoecdysteroids). Approximately, more than 300 different phytoecdysteroids have been identified. Phytoecdysteroids have the same structural features as ecdysteroids found at much higher concentrations in insects or other arthropods. Zooecdysteroids, ecdysone, and 20-hydroxyecdysone (20E) are integral to the growth and reproductive functions in arthropods. Literature reveals that phytoecdysteroids could exert deterrent or antifeedant (repellents) or toxicant or develop interfering effects or affect oviposition. It showed insecticidal (*Helicoverpa armigera*, *Spodoptera litura*, *Helopeltis theivora*) and miticidal (*Oligonychus coffeae*) activities. Monophagous insects which feed on ecdysteroids would rather starve to death than the oligophagous insects, but suffer developmental defects on the ingestion of larger amounts. Very few nonsteroidal compounds like halofenozide, methoxyfenozide,

K. Sahayaraj (✉)
Crop Protection Research Centre, Department of Zoology, St. Xavier's College, Palayamkottai, Tamil Nadu, India





J. Marimuthu et al. (eds.), *Ferns*, https://doi.org/10.1007/978-981-16-6170-9_8

tebufenozide, and chromafenozide have also shown insecticidal activity against lepidopteran pests. Halofenozide has broad-spectrum activity against lepidopteran and coleopteran pests. Both tebufenozide and methoxyfenozide exhibit selective contact and ovicidal activity. A prominent disadvantage is that phytoecdysteroids of few plants would deter nonadapted insect predators (beneficial insects).

**Keywords**

Ferns · Phytoecdysteroids · Insect pest · Deterrency · Toxicant · Morphogenesis · Oviposition

## 8.1 Introduction

Insects cause severe damage to numerous economically important crops, and their control relies on pesticides. Besides the beneficial actions of pesticides/insecticides in agricultural production, pesticides could have impacts on the environment; contaminate food, air, soil, and water, and also affect various beneficial microbes, plants, domestic and wild animals, and human beings. Pesticides are also classified based on their persistence in the environment and categorized as unstable (spread out up to 12 weeks), moderately persistent (spread out in 1–18 months), and highly persistent (spread out 75–100% within 2–3 years). Biotic stress includes infection and infestations of microbes and micro- and macro-organism including insect pests. Therefore, the new approach to agriculture and horticulture focuses on the introduction of new, reliable, less harmful to living beings, and easily degradable green pesticides, which significantly reduce the risk of pests [pestiferous insects, other arthropods, phytopathogens] and make the use of pesticides largely unnecessary.

Green pest control is becoming increasingly popular due to concerns about the negative impacts of pesticides on the environment (Kristina et al. 2020). It includes botanical pesticides, microbial insecticides, semio-chemicals, natural enemies, etc. Plants produce enormous chemicals to perform their daily life and also to protect themselves from the biotic and abiotic stress. We have been using rotenone (*Derris* sp.), carboxin, fluoroacetate, nicotine, neem (*Azadirachta indica*), microbial pesticide *Bacillus thuringiensis*, and pyrethrins in pest control. The importance of botanicals has been highlighted by various researchers (Guleria and Tiku 2009; Oguh et al. 2019; Reddy and Chowdary 2021).

Initially, plants that have phytoecdysones are suggested to utilize pest control. Of the various plants suggested in pest control, ferns play a vital role because of their specific chemical nature, i.e., phytoecdysones. Ferns are the second largest lineage of vascular plants, yet our understanding of their interactions with herbivory is very limited (Cooper-Driver 1978) because of the natural chemical molecules. Hemiptera, Lepidoptera, and Coleoptera are the main insect orders which affect fern species (Jones and Firn 1978a; b; Fuentes-Jacques et al. 2021).

Of the various primary and secondary metabolites of ferns, phytoecdysone is a special secondary metabolite produced by ferns. Phytoecdysteroids are found in ferns and affect a wide range of insects at very low concentrations (Das et al. 2020). The phytoecdysteroids (PEs) comprise a large group of biologically active plant steroids, which have structures similar to those of insect-molting hormones. The ecdysteroids derived from plants are called phytoecdysteroids; ecdysteroids derived from animals are known as zooecdysteroids.

Over 200 ecdysteroids occurring in plants are yet described; the most common are 20-hydroxyecdysone and polypodine B. Their concentrations vary over a wide range (up to 1023 M), and about 6% of the plants contain high amounts of ecdysteroids. Major phytoecdysones include 20-hydroxyecdysone, ecdysone, 2-deoxy-20-hydroxyecdysone, and 2-deoxyecdysone, whereas inokosterone, makisterone A, makisterone C, and 2-deoxyecdysone are considered as minor ones. These were recorded from rhizomes, fronds, and leaves.

Phytoecdysones have been screened from ferns belonging to various countries. Ecdysteroids are not found in all fern families, but in some (e.g., Polypodiaceae), almost all investigated species contain these molecules, sometimes at very high concentrations. These ferns can be used as safe biological alternatives to pesticides. In Japan, a total of 170 species, 22 varieties, and 1 form have been found to show the insect molting hormone activity (Hikino et al. 1973). Latter, *Diplazium donianum*, makisterone A, makisterone D, and an unidentified stereoisomer of makisterone B have been isolated (Hiroshi et al. 1976). *Christella parasitica* (L.) Khun (Dennstaedtiaçeae: Pteridophyceae), endemic to Western Ghats of Tamil Nadu, contains two major phytoecdysteroids α- and β-ecdysteroids (Balasubramanian et al. 2008). In India, the HPLC analysis of *Pteridium aquilinum* (L) Kuhn (Pteridophyceae) has both ∝ and β ecdysones (Selvaraj et al. 2005). Five ecdysteroids, namely, (22R,24R,25S,26S)-2β,3β,14α,20R-tetrahydroxy-26-α-methoxy-6-oxo-stigmast-7-ene-22,26-lactone; (22R,24R,25S)-2β,3β,14α,20R,26S-pentahydroxy-6-oxo-stigmast-7-ene-22,26-lactone; (22R, 25S)-2β,3β,14α,20R,24S-pentahydroxy-6,26-dioxo-stigmast-7-ene-22,26-lactone; (22R, 25S)-2β, 3β, 14α, 20R, 24S,26S-hexahydroxy-6-oxo-stigmast-7-ene-22,26-lactone; and capitasterone, were isolated from *Diplopterygium rufopilosum* (Hu et al. 2014). Major and minor phytoecdysones recorded from various fern species are listed in Table 8.1.

Relatively little is known about insect defense mechanisms in ferns. The most generally accepted hypothesis is that they protect against attacks by phytophagous insects, by either disturbing development or reducing food intake. An account is given of the method by which crude filicin and the "Filixsäure" were prepared from a commercial extract of the rhizome of male fern (*Aspidium filix-mas*) and of laboratory tests on the toxicity of crude filicin to a pollinator *Aphis rumicis [Aphis fabae]*. Results revealed that the product will not affect the life traits or behavior of the honeybee tested. In another case, the product was also tested on the human vector *Culex fatigans [Culex quinquefasciatus]*, with an acetone solution of "Filixsäure" proving four times as toxic as one of crude filicin, of which "Filixsäure" is therefore considered to be one of the important toxic principles. Tests on houseflies (*Musca*

**Table 8.1** Major and minor phytoecdysones recorded from various fern species distributed worldwide

| Fern genus | Fern species | Major and minor phytoecdysones | Reference |
|---|---|---|---|
| Athyriaceae | *Diplazium esculentum* | ecdysteroids | Masato et al. (2021) |
| Azollaceae | *Azolla imbricata* | Ecdysteroids | |
| Adiantaceae | *Adiantum aethiopicum* *Adiantum capillus-veneris* *Adiantum cunninghamii* *Adiantum flabellulatum* *Adiantum hispidulum* *Adiantum philippense* | Ecdysteroids | |
| Blechnaceae | *Brainea insignis* | Braines-teroside A [14-deoxy-14a-15a-epoxyponasteroside A], brai-nesteroside B [14-deoxy-14,15-didehydroponasteroside A], brainesteroside C [25-deoxycalonysterone-3-O-b-D-glucopyrano-side], brainesteroside D [25-deoxyecdysone-3-O-b-D-gluco-pyranoside], brainesteroside E [5-epi-ponasteroside A] | Wu et al. (2010) |
| Cyatheaceae | *Cyathea cooperi* | Ecdysteroids | Lafont et al. (2011) |
| Dryopteridaceae | *Acrophorus stipellatu* | Ecdysteroids | |
| Polypodiaceae | *Microsorum grossum* | 20-Hydroxyecdysone | Snogan et al. (2007) |
| | *Crypsinus hastatus* | β-sitosterol, 20-hydroxyecdysone, β-ecdysone | Yi-Ru (2007) |
| | *Cheilanthes farinosa* | Ecdysteroids | Josephrajkumar et al. (2000) |
| | *Neocheiropteris multiflorins* | Ecdysterone, β-sitosterol | Aulakh et al. (2019) |
| | *Lepidogrammitis drymoglossoides* | Ponasteroside B | Yao et al. 2014 |
| | *Lepisorus contortus* | β-sitosterol | Yang et al. 2011 |
| | *Phymatosorus membranifolum* | Ecdysone, 20-hydroxyecdysone, 2-deoxy-20-hydroxyecdysone, 2-deoxyecdysone | Aulakh et al. (2019) |
| | *Phymatosorus scolopendria* | Ecdysone, amarasterone A, 20-hydroxyecdysone, | Snogan et al. (2007) |

(continued)

**Table 8.1** (continued)

| Fern genus | Fern species | Major and minor phytoecdysones | Reference |
|---|---|---|---|
| | | amarasterone A, 25-deoxyecdysone, 22-glucoside | |
| | *Pteridium aquilinum* | Ecdysteroids, ∝ and β ecdysones | Sahayaraj et al. (2007) |
| | | Ecdysone, ecdysterone | Dreier (1987) |
| | *Polypodium leucotomos* | Ecdysone, ecdysterone | Garcia et al. (2006) |
| | *Polypodium vulgare* | Ecdysone-20E, abutasterone, polypodine B | Lafont et al. (2011) |
| | | 5-hydroxyecdysone, 20-deoxyshidasterone and polypodine B 2-b-D-glucoside | Simon et al. (2011) |
| | *Microsorum insigne* | Phytoecdysteroids | Varangkana et al. (2016) |
| | *Microsorum scolopendria* | 20-deoxymakisterone A, 25-epimer of amarasterone A, and 25-deoxyecdysone 22-glucoside; inokosterone, makisterone A, makisterone C, and 2-deoxyecdysone | Snogan et al. (2007) |
| | *Microsorum maximum* | 20-Hydroxyecdysone, makisterone A, inokosterone, 2-deoxy-20-hydroxyecdysone, and ecdysone | Ho et al. (2007) |
| | *Microsorum membranifolium* | 20-Hydroxyecdysone, ecdysone, 2-deoxy-20-hydroxyecdysone, 2-deoxyecdysone, makisterone C | Ho et al. (2007) |
| | *Microsorum punctatum* | 20-Hydroxyecdysone | Ho et al. (2007) |
| | *Microsorum commutatum* | Ecdysone | Ho et al. (2007) |
| | *Cheilanthes tenuifolia* | Ecdysone analogues, cheilanthones A and B | Faux et al. (1970) |
| | *Blechnum minus* | Deoxyecrustecdysone, deoxyecdysone, and α-ecdysone | Chong et al. (1970) |
| | *Pteridium aquilinum* (L.) Kuhn. | Alpha-ecdysone and 20-hydroxyecdysone | Macek and Vaněk (1994) |
| | *Blechnum vulcanicum* | 2-Deoxy-3-epiecdysone and ecdysone, a-ecdysone (11), and 2-deoxy-a-ecdysone | Russell et al. (1981) |
| | *Podocarpus nakaii* | Ponasterone | Nakanishi et al. (1996) |
| | *Cheilanthes tenuifolia* | Cheilanthones A and B | Faux et al. (1970) |
| | *Costa rica* | Ponasterone A | Tanaka et al. (1981) |

(continued)

**Table 8.1** (continued)

| Fern genus | Fern species | Major and minor phytoecdysones | Reference |
|---|---|---|---|
| | *Chenopodium album* | Polypodine | DDinan (1992) |
| | *Pteris inaequalis* | P-sister01 P-2 dehydroxy D glucoside | Murakami et al. (1978) |
| Pteridaceae | *Acrostichum aureum* | Ecdysteroids | Vetter (2018) |
| Schizaeaceae | *Anemia phyllitidis* | Ecdysteroids | Lafont et al. (2011) |
| Sinopteridaceae | *Aspidotis densa* | Ecdysteroids | Dreier (1987) |
| | *Athyrium aphanoneuron* | Ecdysteroids | Lafont et al. (2011) |
| | *Athyrium arisanense* | Ecdysteroids | Yen et al. (1971) |
| | *Athyrium atkinsonii* | Ecdysteroids | Vetter (2018) |
| | *Athyrium crenulata serratum* | Ecdysteroids | Lafont et al. (2011) |
| | *Athyrium decurrenti-alatum* | Ecdysteroids | |
| | Athyrium yokoscense | 24-Epi-pterosterone | Ohta et al. (1996) |
| Thelypteridaceae | *Chingia sakayensis* | β-sitosterol | Sutoyo et al. (2007) |

*domestica*, L.) were made with a spray containing 0.2% crude filicin and 0.05% pyrethrins. The tests were made according to the Peet-Grady method, and the percentage mortality given by the spray was 52.0, as compared with 53.1 by the official test insecticide (Gupta et al. 2018). These three tested reveal phytoecdysones of ferns have potentiality against various types of insects.

## 8.2 Feeding Detergency and Food Consumption

There was a good association which is observed between the insects and various fern species. Feeding and gustatory responses to phytoecdysones were investigated by researchers. In this part, selective examples were included. Jones and Firn (1978a; b) studied the effect of ecdysteroids on the feeding behavior of seven insect species. Results showed that four species were deterred at or above 60 mg ecdysteroid/kg fresh weight of the diet and an insect at 6 mg ecdysteroid/kg fresh weight of the diet or above. However, two species were not affected at the higher ecdysteroid concentrations. Latter, in 1983, Ottosson and Anderson studied the impact of mature frond extracts of six fern species (*Dryopteris filix-mas*, *D. dilatata*, *D. borreri*,

*Phyllitis scolopendrium*, *Polystichum setiferum*, and *Polypodium vulgare*) against *Spodoptera littoralis.* Authors found that tannins of these plants decreased feeding, survivorship, and growth rates of *Spodoptera* larvae.

The antifeedant activity of the crude ethyl acetate extract of *Lygodium microphyllum* was evaluated against the third instar larvae of *Spodoptera litura* in the laboratory at different concentrations, i.e., 1, 3, 5, and 7%. The mean percent feeding was 39.54%, and percent protection due to treatment was 32.15% with an extract concentration of 3%. Furthermore, adult emergence (percent survival) was low (42.4%) when compared to the control. In addition, anti-ovipositional and ovicidal activities were exhibited by the leaf extract (Ragupathy et al. 2004). Markham et al. (2006a; b) found that saponins of sensitive fern (*Onoclea sensibilis*) and glade fern (*Athyrium pycnocarpon*) caused a greatest decrease in the larval growth of corn earworm, *Helicoverpa zea*, and fall armyworm, *Spodoptera frugiperda*.

In 2008, Sahayaraj and his co-workers tested antifeedant and insecticidal activities of the ethanolic root extract of *Pedalium murex* (Linn) (Pedaliaceae) at 0.1, 0.2, 0.4, and 0.8% against third, fourth, and fifth instar larvae of *Spodoptera litura* (Fab.) by leaf-dip method. Sahayaraj et al. (2008) results revelas that *P. murex* extracts reduced the food consumption index, growth rate, approximate digestibility, and efficiency of conversion of ingested and/or digested food into the body of *S. litura* which indicate the antifeedant activity of the tested plant. This might be due to the presence of steroids, terpenoids, phenolics, saponins, tannins, and flavonoids (e.g. Phenol, 2-(5,6-dimethyl pyrazinyl) methyl (molecular weight 214); O-Terphenyl-13C (molecular weight 230); and 3,3A, 4,9B-Tetrahydro2H-Furo(3,2-C)(1) Benzopyran).

Methanol extract of *Pronephrium megacuspe* showed antifeedant activity to the third instar larvae of *Plutella xylostella*, *Ostrinia furnacalis*, and *Pieris rapae*. The fern possessed 7-hydroxy-5-methoxy-6,8-dimethyl-flavanone, 2′,4′-dihydroxy-3-′-methyl-6′-methoxychalcone, 5,7-dihydroxy-6,8-dimethylflavanone, and 5-hydroxy-6,7-dimethoxyflavone (Suqing et al. 2013). Protein extracts from fern and moss species were compared with those from a lepidopteran-susceptible soybean (*Glycine max*) cultivar (Cobb) in bioassays for insect resistance. Under laboratory conditions, the protein extracts of the few ferns like ebony spleenwort (*Asplenium platyneuron*), sensitive fern (*Onoclea sensibilis*), and glade fern (*Athyrium pycnocarpon*) decrease the larval growth of corn earworm (*Helicoverpa zea*) and fall armyworm (*Spodoptera frugiperda*) (Markham et al. 2006a, b). The feeding inhibition was due to a specialized sensory perception rather than to a poisoning effect. Results show that the larvae were avoiding food treated with 20E $2\times1024$ M (feeding ratio of treated substrate vs control: 4%). At the same dosage, feeding was much less affected in *Mamestra brassicae* (feeding ratio: 50%), while higher doses of 20E yielded a maximal feeding ratio of 40% (Jr Descoins and Marion-Poll 1999). But some of the ecdysteroids might still act as nontoxic feeding deterrents. Antifeedant index was calculated for aqueous extracts of *Adiantum raddianum*, *Asplenium aethiopicum*, *Cyclosorus interruptus*, *Dicranopteris linearis*, *Diplazium polypodioides*, and *Pteridium aquilinum* against red spider mite

*Oligonychus coffeae* Nietner and tea mosquito bug *Helopeltis theivora* Waterhouse. Results reveal that at higher concentration (5%), AFI was high at 24 h and gradually decreased at 72 h observation (Pandian et al. 2017). These observations reveal that ingested phytoecdysones differently affect the endocrine organs at different levels and inhibit feeding behavior (biting), oral intake of the food, and then physiologically alter the hemolymph quality and quantity which leads to the starvation and death.

## 8.3 Mortality

*Athyrium filix-femina*, *Diyopleris austriaca*, and *D. filix-mas* treated leaves were covered with an aqueous extract of the host plant (*Alnus glutinosa* or *Salix* spp., respectively) or fern leaves. Mortality of young larvae (1-day old) of *A. alni* was significantly higher when provided with treated or untreated fern leaves and treated food plant leaves compared to untreated food plant leaves. Mortality of older larvae (6-day old) was significantly lower than that of young larvae. Young larvae did not feed on fern leaves at all, while older larvae ingested material from treated fern leaves, however at low rates. *Phratora vitellinae* individuals showed abnormal elytra and irregular melanization patterns when they had taken up fern substances. In both species, the larvae produced less defensive secretion if they had ingested an extract of *Athyriumfilix-femina* (Michael and Sisrun 2005). Aqueous extracts of *Pteridium aquilinum* (Dennstaedtiaceae) on mortality of *Myzus persicae* and *Ascia monuste orseis* were evaluated (Gehardt et al. 2012). Results showed greater effectiveness of the commercial product followed by aqueous extract of leaves of *P. aquilinum*, resulting in higher mortality of aphids and caterpillar (Gehardt et al. 2012). *Adiantum raddianum, Asplenium aethiopicum, Cyclosorus interruptus, Dicranopteris linearis, Diplazium polypodioides*, and *Pteridium aquilinum* were evaluated against the two major pests of tea, red spider mite *Oligonychus coffeae* Nietner and tea mosquito bug *Helopeltis theivora* Waterhouse. The extracts of *P. aquilinum* and *D. linearis* showed good contact toxicity at a 5% concentration to *O. coffeae* (Pandian et al. 2017).

*Oryctes rhinoceros* Linn. is one of the most serious pests of coconuts and other palms. Following the bioassay -guided method, a larvicidal compound 22-hydroxyhopane has been isolated for the first time from methanol extract of leaves of *Adiantum* latifolium Lam. against the pest ($LC_{50}$ value 20.81 μg/g). It is a hopanoid triterpene with a molecular mass of 442.42 g/mol. The compound exhibited antibacterial activity against symbiotic gut bacteria, caused histolysis of midgut tissues, and inhibited secretion of digestive enzymes such as protease, amy lase, and trehalase resulting in weight loss of larvae. Enzyme immunoassay showed an elevation of 20-hydroxyecdysone level in hemolymph causing disruption of meta morphosis of larvae (Kumar et al. 2019).

*Diplazium esculentum, Christella parasitica,* and *Blechnum orientale* ethanol extract (1%, 5%, 10%, 15%, and 20%) has been tested on the larva of Diamond back moth *P. xylostella,* a notorious pest of cabbage in the lab condition. Ferns caused dose-dependent mortality at 20% concentration. The maximum larval mortality was 36.66% after 24 h of treatment in case of *D. esculentum* which increased to 73.33% after 72 h which is well above the untreated control as 0.66% after 24 h and 13.33% after 72 h in the latter case, indicating *D. esculentum* to be the most toxic to larvae of *P. xylostella*. The efficacy of *D. esculentum* altered the feeding behavior of *P. xylostella*, reduced the larval and pupal weight, prolonged the pupation period, caused malformation in pupae and adult under in vitro condition (Murasing et al. 2019).

Very recently the insecticidal activity of crude ethanol extracts of the fern species *Dicksonia sellowiana* and *Nephrolepis cordifolia* was evaluated against *Oncopeltus fasciatus* (Hemiptera) under laboratory conditions. The *N. cordifolia* extract was responsible for 63% ($p < 0.0001$) of insect mortality around 16 days after treatment, whereas the D. sellowiana exhibited 50% ($p < 0.0001$) on the 21st day post-treatment. The extracts also caused delays in insect molting and metamorphosis. Both the fern extracts exhibited 18% similarity in the terpenoid profile and 0% for phenolic substances. They also showed potential for research on selective biodegradable substances for use as green insecticides (de Souza et al. 2020).

## 8.4 Physiological Activity

The immunological impact of the Crude Ecdysone Fractions (CEF) of *Christella parasitica* (L.) H.lev., *Pteridium aquilinum* (L.) Kuhn, and *Hemionitis arifolia* (Burm) T. Moore was tested against two major polyphagous pests, *Helicoverpa armigera* Hubner and *Spodoptera litura* (Fabr.) (Lepidoptera: Noctuidae) (Sahayaraj et al. 2007). Hemolymph volume in proportion to animal weight (Hemosomic index-HI) of both the experimental and control categories was increased as the exposure time increased. Prohemocytes (PR), plasmatocytes (PL), granular cells (GR), spherule cells (SP), oenocytoids (OE), and adipohemocytes (AD) provide immunity to these animals. Proportion of plasmatocytes population was higher in control *H. armigera*, followed by spherule cells, granular cells, Prohemocytes, oenocytoids, and adipohemocytes (Sahayaraj et al. 2007). But in the experimental categories, highest Prohemocytes count was recorded for *P. aquilinum* treatment followed by *H. arifolia* and *C. parasitica* at 24 h. Plasmatocytes level was decreased in all the three ferns treated groundnut leaves fed *H. armigera*. For instance, highest reduction was observed in *P. aquilinum* followed by *H. arifolia* (Table 8.2) and *C. parasitica*. Even though PL count of the control gradually decreased from 24 to 72 h of exposure, its population increased at 96 h of observation. The reduction in plasmatocytes count was more pronounced in the experimental categories and it was extended up to 72 h of observation. Similar trend was also observed in granular cells. However, in *C. parasitica* treated category, the granular cells count was gradually decreased up to 96 h. Same trend was observed for oenocytoid levels from 24 h up to 96 h of observation. This increasing trend was more pronounced in the experimental

**Table 8.2** Proportions of various hemocytes (%) observed in *Helicovera armigera* treated with phytoecdysone fractions of *Christella parasitica* (L.) H.lev. (CP), *Pteridium aquilinum* (L.) Khun (PA), and *Hemionitis arifolia* (Burm) T. Moore (HA) in relation to the control category. Prohemocytes (PR), Plasmatocytes (PL), Granular cells (GR), Spherule cells (SP), Oenocytoids (OE), and Adipohemocytes (AD)

| Time (h) | Extracts | Prohemocytes | Plasmatocytes | Granular cells | Spherule cells | Oenocytoids |
|---|---|---|---|---|---|---|
| 24 | CP | 14.3 | -7.6 | 3.1 | 4.5 | 11.1 |
| | PA | 66.7 | -16.5 | 3.1 | 8.7 | 100.0 |
| | HA | 54.8 | -12.3 | 2.3 | 7.3 | 11.0 |
| 48 | CP | 18.2 | -5.1 | -1.5 | 2.4 | 70.0 |
| | PA | 68.2 | -17.5 | -5.3 | 9.0 | 20.0 |
| | HA | 59.1 | -13.8 | -3.8 | 5.9 | 100.0 |
| 72 | CP | 08.3 | -2.8 | 4.0 | 0.9 | 6.7 |
| | PA | 08.3 | -9.1 | -10.5 | 11.9 | 37.5 |
| | HA | 16.7 | -8.2 | -2.4 | 4.1 | 33.3 |
| 96 | CP | 37.5 | -6.4 | -4.2 | 7.3 | -14.3 |
| | PA | 25.0 | -9.8 | -6.8 | 13.7 | -14.3 |
| | HA | 31.3 | -10.1 | -2.0 | 9.3 | 0 |

After Sahayaraj et al. (2007)

categories. For instance, the order of increase was 31.2, 31.4, 35.5, and 39.0% from 24 to 96 h for *P. aquilinum* (Sahayaraj et al. 2007). Similar trend was also recorded for *S. litura.*

## 8.5 Acaricidal Activity

The acaricidal activity was observed for the aqueous extracts of *Adiantum raddianum, Asplenium aethiopicum, Cyclosorus interruptus, Dicranopteris linearis, Diplazium polypodioides*, and *Pteridium aquilinum* to red spider mite *Oligonychus coffeae* Nietner (Pandian et al. 2017). The order of toxicity is *P. aquilinum* > *D. linearis* > *C. interruptus* > *A. raddianum* > *D. polypodioides* > *A. aethiopicum* (Pandian et al. 2017). Eagle fern *(Pteridium aquilinum* (L.) Kuhn) was assessed against the adult female of *Tetranychus urticae*, one of the most important pests of crops and greenhouse plants. The aqueous extract of blackberry did not have any acaricidal effect on the pest, and this effect varied from 9.2% to 49.8% for other aqueous extracts. However, the blackberry ethanolic extract had the highest oviposition inhibitory effect in comparison with other extracts (94% at 1 mg/cm$^2$) (Salehi et al. 2019). Recently Taha and Baioumy Ali (2020) investigated the acaricidal activity of ethanolic and methanolic extracts of *Adiantum capillus-veneris* at different concentrations (1%, 2%, 3%, and 4%) on semi-fed females of *Argas persicus* collected from the field. The results indicated that all the tested concentrations caused a significant decrease in the percentages of mobile ticks and a significant increase in the percentages of dead ones when compared with the control.

## 8.6 Fern Protein for GM Cotton Crop Production

*Bacillus thuringiensis* (Bt) crystalline proteins (Cry) such as Cry1Ab, Cry1Fa, and Cry2Ab or vegetative insecticidal proteins (Vip) such as Vip3Aa have been very effective against lepidopteran pests in transgenic crops worldwide (Liu et al. 2019). Authors further stress that though *Bt* diversity remains a major source for insecticidal proteins, it has become challenging to find new *Bt* proteins with an optimal activity spectrum and potency that are suitable to replace those active in current commercial products. Alternative sources, including plants, that produce insecticidal proteins with activity against lepidopteran species, have not been explored extensively. To date, only two major groups of plant-derived proteins—proteinase inhibitors and carbohydrate-binding lectins—have been shown to possess insect inhibiting activity, but their use in transgenic crop plants has had only limited success, especially against lepidopteran pests of major crops.

Ferns are considered as an important candidate in this line of research; hence a potential protein (Tma12) isolated from an edible fern, *Tectaria macrodonta* (Fee). Authors utilized this insecticidal protein to produce genetically modified (GM) crop plants and transgenic cotton lines that express Tma12 at ~0.01%. It is effective against whitefly *Bemisia tabaci* with median lethal concentration = 1.49 μg/ml in in vitro feeding assays (Shukla et al. 2016). Further it was reported that transgenic cotton lines that express Tma12 at ~0.01% of total soluble leaf protein were resistant to whitefly infestation in contained field trials, with no detectable yield penalty. The transgenic cotton lines were also protected from whitefly-borne cotton leaf curl viral disease. Rats fed Tma12 showed no detectable histological or biochemical changes, and this, together with the predicted absence of allergenic domains in Tma12, indicates that Tma12 might be well suited for deployment in GM crops to control whitefly and the viruses it carries. Recently a low molecular weight protein of ~90 kDa was isolated and identified from *Adiantum pedatum* fronds and the protein showed potential insecticidal activity against European corn borer (ECB, *Ostrinia nubilalis*) (Liu et al. 2019). Using this protein, transgenic soybean plants expressing IPD083Aa were evaluated for leaf disk feeding protection against neonate of corn earworm (CEW, *Helicoverpa zea*), fall armyworm (FAW, *Spodoptera frugiperda*), soybean looper (SBL, *Chrysodeixis includens*), and velvetbean caterpillar (VBC, *Anticarsia gemmatalis*), which are primary insect pests of soybean (*Glycine max*). After the field trial it has been utilized for farmers. Furthermore, Tma12, isolated from the fern *T. macrodonta,* has been proposed as a next-generation insecticidal protein because transgenic cotton expressing Tma12 exhibits resistance against whitefly and viral diseases (Yadav et al. 2019).

## 8.7 Life Traits

The sweet potato hornworm, *Agrius convolvuli*, and the cabbage armyworm, *Mamestra brassicae*, were reared from hatching on artificial diets supplemented with ecdysone or 20-hydroxyecdysone. The effects of the two ecdysteroids on larval development were investigated. In *A. convolvuli*, an extra larval molt was induced by 400 ppm ecdysone and a maximum of 4 additional larval ecdyses were induced by 800 ppm ecdysone. Conversely, 20-hydroxyecdysone did not affect larval development although much larger amounts of 20-hydroxyecdysone were present in the hemolymph than was found in the hemolymph of larvae-fed ecdysone. In M brassicae, neither ecdysone nor 20-hydroxyecdysone affected larval development. Although larvae were fed on large amounts of ecdysteroids, the ecdysteroid level in the hemolymph remained as low as that in control larvae. These results suggest that ecdysone, but not 20-hydroxyecdysone, can induce multiple additional larval ecdyses in *A. convolvuli*, as it does in *Bombyx mori*, but that the sensitivities to ingested ecdysteroids differ widely among the species (Tanaka and Naya 1995).

*Cheilanthes farinosa* Kaulf. (Polypodiaceae: Pteridophyta) incorporated into a semi-synthetic diet significantly extended the larval period, reduced pupal weight, and adversely affected pupation of the cotton bollworm, *Helicoverpa armigera* (Hübner) (Noctuidae: Lepidoptera) (Josephrajkumar et al. 2000). *Pteridium aquilinum* (L) Chloroform (CE) and ethanol extracts (EE) and also the hexane (HF) and aqueous fractions (AF) were tested against *Helicoverpa armigera and Spodoptera litura*. Results reveal that *Helicoverpa armigera* showed more susceptibility to the fern extracts than *Spodoptera litura*. The different extracts of the same plant showed differences in their toxic as well as growth disrupting responses. The order of concentration required for 50% mortality ($LD_{50}$) of *H. armigera* was 0.198%, 0.112%, and 0.198% for CE, CEF, and EE. As observed in *H. armigera*, *S. litura* also recorded the highest lethal activity in EE (0.141%) followed by 0.128% of CEF and 0.160% of CE (Selvaraj et al. 2005).

Larvicidal and pupicidal effects of *Dryopteris filix-mas* root and rhizome's ethanolic extract were studied against the third instar larvae of *Corcyra cephalonica* (Staint.). *D. filix-mas* extract 0.20% (v/w) caused 100% larval mortality. The plant extracts reduce pupation percent, pupal death, and adult emergence indicating absolute toxicity to the pest. However, further detailed studies are essential under pots, controlled-field cage and field studies (Shukla and Tiwari 2011).

Methanol extracts of *Cyclosorus interruptus* (Willd.) H. Itô (Theylepteridaceae), *Christella dentata* (Willd.) (Forssk.) Brownsey et Jermy (Theylepteridaceae), and *Nephrolepis cordifolia* (L.) Presl revealed the presence of secondary metabolites like alkaloids, steroids, tannins, flavonoids, cardiac glycosides, and phenolic compounds. Synthesized AgNP reduced developmental period, pupal weight, percentage of pupation, and adult emergence and also caused larval, pupal, and adult deformities that confirm the insecticidal activity against third instar larvae of Spodoptera litura (Fab.) (Lepidoptera: Noctuidae) (Selvaraj et al. 2018).

Phytoecdysone fractions of the ferns *Cyclosorus interruptus, Christella dentata,* and *Nephrolepis cordifolia* are used for silver nanoparticles (AgNPs) formulation. It was tested against *Spodoptera litura* (Fab.) (Lepidoptera: Noctuidae) third instar larvae. The crude extract and Fe-AgNP formulations exerted their influence on the developmental period, pupation rate, pupal weight, and adult emergence which were reduced significantly. They also caused larval, pupal, and adult deformities that confirm the insecticidal activity of the plant (Xavier et al. 2016). They conduced that the absorption of these compounds by insects is slow and limited, their excretion rapid, and absorbed ecdysterone is rapidly catabolized into compounds with little or no molting hormone activity.

## 8.8 Field Applications

In India, during 2013, the field efficacy of three fern extracts was evaluated for their pesticidal property against *Helicoverpa armigera* (Hubner) and *Spodoptera litura* (Fab.) under groundnut field condition. The fern extracts were effective in controlling the pests at field level. The order of pesticidal efficacy was *P. aquilinum* > *H. arifolia* > *C. parasitica*. Highest population reduction was recorded in *P. aquilinum* treated plots followed by *H. arifolia* at 49 and 64 days after seedling (DASE). The highest yield was recorded in *P. aquilinum* (1400 kg/ha) followed by *H. arifolia* (1370 kg/ha) and *C. parasitica* (1250 kg/ha). The cost benefit ratio was 1:1.82, 1:1.79, and 1:1.63 for *P. aquilinum, H. arifolia,* and *C. parasitica,* respectively (Sahayaraj and Selvaraj 2013). Under field conditions, the aqueous extract of *D. linearis* and *P. aquilinum* showed a 50% reduction in the population of red spider mite and caused no phytotoxic effect to tea leaves. But their insecticidal activity was less pronounced against *H. theivora* (Pandian et al. 2017).

Alternative substances to manage thrips population (*Thrips tabaci* Lind.) on onion in an organic agriculture system were evaluated. The experiments were carried out in Ituporanga, Santa Catarina State, Brazil. Plants of onion cv. Crioula were transplanted into two fields on 9 September 1996 and 13 August 1997 and harvested respectively on 10 January 1997 and 11 December 1997. Randomized blocks with four replicates were used in both experiments. In 1996 the treatments included the anaerobic liquid biofertilizer at 50%, aerobic liquid biofertilizer at 5%, manganese sulfate at 1%, hydroalcoholic extract of propolis at 0.2%, macerate of herbs at 2% and 4%, tobacco extract (*Nicotiana tabacum*) 2 L/ha + neutral detergent at 1%, and no spray. In 1997, the treatments were: macerate of herbs 5% and 10%; sulfur waterable powder 0.25% + propolis hydroalcoholic extract 0.2% + extract of fern (*Pteridium aquilinum*) 3%; anaerobic liquid biofertilizer 50%; extract of wormseed goosefoot (*Chenopodium ambrosioides*) 10%; aerobic liquid biofertilizer 5%; extract of *Ateleia glazioviana* 0.5%; extract of fern (*Pteridium aquilinum*) 10%; extract of chinaberry (*Melia zedarach*) 10%; extract of camomile (*Matricaria chamomilla*) 5%; and control without spray. The products were applied using a $CO_2$ sprayer with constant pressure. The different treatments did not cause significant reduction in the thrips population nor any increase in yield in comparison to control treatment, without spray (Gonçalves et al. 2004).

*Pteridium auquilinum* and *Ricinus communis* were utilized for the protection of three vegetable crops *Lactuca sativa, Solanum nigrum,* and *Raphanus sativus* of BIONATURE (Bafoussam, Cameroon). Four types of preparations were made and bioefficacy tests consisted of contact treatments. Direct observations on the physical aspect of the plant, agronomic measurements, and pathology monitoring were carried out. The results show that the fermentation of the fern is complete after 5 days versus 8 days for the castor. The aqueous extracts of fern have insect repellent, insecticidal and fungicidal properties. Diluted maceration of fern (88.75%) was more efficient than the pure maceration (68%) (Mala et al. 2019). Authors concluded that manure and castor oil have insecticidal and repellent properties. The monitoring of pathologies after treatment reveals that castor oil is more effective than fern manure for the control of gray rots and tip burn attacks.

## 8.9 Storage Pests

Using surface film method and filter paper disk methods, the rhizome of *Drynaria quercifolia* (J. Smith) was evaluated for pesticidal and pest repellency activities against *Tribolium castaneum* (Herbst), a harmful pest of stored grain and flour-based products in tropical and subtropical region. Chloroform soluble fraction of ethanol extract of rhizome of *D. quercifolia* showed significant pesticidal activity (0.88 to 1.77 mg/cm$^2$) and significant pest repellency activity (0.94–0.23 mg/cm$^2$). However, no pesticidal and pest repellency activity was found for petroleum ether, ethyl acetate, and methanol soluble fractions of ethanol extract as well as for 3,4-dihydroxybenzoic acid (Khan et al. 2014). Authors conclude that the chloroform soluble fraction of rhizome of *D. quercifolia* is useful in controlling stored grain and flour-based product pests.

Both laboratory and field observations reveal the potentiality of ferns against many pestiferous insects, which suggest that these botanicals can be considered as an import Biointensive Integrated Pest Management (BIPM) component. However, the following points should be incorporated, while selecting the ferns.

## 8.10 Recommendations

- More number of ferns, their extracts, and/or bioactive compounds can be screened against pestiferous insects of economic importance.
- Screening can be carried out under controlled field cage and then recommended for field application.
- Synergistic activity of ferns and other botanical/microbial pesticides can be studied.
- Biosafety of the fern extracts and/or their bioactive principles or products can be screened against laboratory model organism like silkworm/drosophila, pollinators, and natural enemies (predators/parasitoids).
- Farmers friendly commercial products can be brought out.

## References


Aulakh MK, Kaur NAVDEEP, Saggoo MIS (2019) Bioactive phytoconstituents of pteridophytes–a review. Indian Fern J 36:37–79

Balasubramanian R, Selvaraj P, Sahayaraj K (2008) Partial purification and characterization of phytoecdysone from *Chrystella parasitica* (L.) and screening its pesticidal properties on lepidopteran pests. J Biopest 1(2):201–205

Chong YK, Galbraith MN, Horn DHS (1970) Isolation of deoxyecrustecdysone, deoxyecdysone, and α-ecdysone from the fern *Blechnum minus*. J Chem Soc D Chem Commun 18:1217–1218

Cooper-Driver GA (1978) Insect-Fern associations. Entomol Exp Appl 24:110–116

Das N, Siddhartha KM, Anusha B, Eunüs SA, Anupam B (2020) The phytochemical, biological, and medicinal attributes of phytoecdysteroids: an updated review. Acta Pharm Sin B. https://doi.org/10.1016/j.apsb.2020.10.012

de Souza, J. B., Lima, G. P., Feder, M. D., Lancellotti, IR., Santos MG. (2020) *Dicksonia sellowiana* hook. And *Nepholepis cordifolia* (L.) Lellinger extracts as potential green pesticides: insecticidal activity. Res Soc Dev, 9(8), e120985182-e120985182

Dinan L (1992) The analysis of phytoecdysteroids in single (pre-flowering stage) specimens of fat hen, *Chenopodium album*. Phytochem Anal 3(3):132–138. https://doi.org/10.1002/pca.2800030309

Dreier SI (1987) Occurrence and activity of ecdysterone (insect moulting hormone) in plants. Ph.D. Thesis, University of British Columbia

Faux A, Galbraith MN, Horn DHS, Middleton EJ, Thomson JA (1970) The structures of two ecdysone analogues, cheilanthones A and B, from the fern *Cheilanthes tenuifolia*. J Chem Soc D 4:243–244

Fuentes-Jacques LJ, Hanson-Snortum P, Hernández-Ortiz V, Díaz-Castelazo C, Mehltreter K (2021) A global review and network analysis of phytophagous insect interactions with ferns and lycophytes. Plant Ecol. https://doi.org/10.21203/rs.3.rs-301036/v1

Garcia F, Pivel JP, Guerrero A, Brieva A, Martinez-Alcazar MP, Caamano-Somoza M, Gonzalez S (2006) Phenolic components and antioxidant activity of Fernblock (R), an aqueous extract of the aerial parts of the fern *Polypodium leucotomos*. Methods Find Exp Clin Pharmacol 28 (3):157–160

Gehardt A, Putzke MTL, Lovatto PB (2012) Insecticide activity of botanical extracts from three wild plant species of Rio Grande do Sul, Brazil, on *Myzus persicae* (Hemiptera: Aphididae) and *Ascia monuste orseis* (Lepidoptera: Pieridae). Cad Pesqui Sér Biol 24(2):55–64

Gonçalves PAS, Werner H, Debarba JF (2004) Evaluation of biofertilizers, plant extracts, and some alternative substances to manage onion thrips in organic agriculture system. Hortic Bras 22 (3):659–662

Guleria S, Tiku AK (2009) Botanicals in pest management: current status and future perspectives. In: Integrated pest management: innovation-development process. Springer, Dordrecht, pp 317–329

Gupta MK, Marangmei L, Sarma AK (2018) Efficacy of extracts of some ferns against *Plutella xylostella*. Indian J Entomol 80(4):1667–1671

Hikino H, Okuyama T, Jin H, Takemoto T (1973) Screening of Japanese ferns for phytoecdysones. I. Chem Pharm Bull 21(10):2292–2302

Hiroshi H, Kiminori M, Toru O, Tsunematsu T, Kun-Ying Y (1976) Phytoecdysones from *Diplazium donianvm*. Steroids 28(5):649–654

Ho R, Teai T, Loquet D, Bianchini J-P, Girault J-P, Lafont R, Raharivelomanana P (2007) Phytoecdysteroids in the genus Microsorum (Polypodiaceae) of French Polynesia. Nat Prod Commun 2(8):1934578X0700200803

Hu J, Shi X, Mao X (2014) Ecdysteroids from the ethanol extract of *Diplopterygium rufopilosum*. Phytochem Lett 8:73–76

Jones CG, Firn RD (1978a) The role of phytoecdysteroids in bracken fern, *Pteridium aquilinum* (L.) Kuhn as a defense against phytophagous insect attack. J Chem Ecol 460(4):117–138

Jones CG, Firn RD (1978b) The role of phytoecdysteroids in bracken fern, *Pteridium aquilinum* (L.) Kuhn as a defense against phytophagous insect attack. J Chem Ecol 4(2):117–138

Josephrajkumar A, Subrahmanyam B, Devakumar C (2000) Growth-regulatory activity of silver fern extract on the cotton bollworm, *Helicoverpa armigera* (Hübner). Int J Trop Insect Sci 20 (4):295–302

Jr Descoins C, Marion-Poll F (1999) Electrophysiological responses of gustatory sensilla of *Mamestra brassicae* (Lepidoptera, Noctuidae) larvae to three ecdysteroids: ecdysone, 20-hydroxyecdysone and ponasterone A. J Insect Physiol 45(10):871–876

Khan A, Islam MH, Islam ME (2014) Pesticidal and pest repellency activities of rhizomes of *Drynaria quercifolia* (J. Smith) against *Tribolium castaneum* (Herbst). Biol Res. 47:51. https://doi.org/10.1186/0717-6287-47-51

Kristina KG, Stenberg JA, Lankinen Å (2020) Making sense of integrated Pest management (IPM) in the light of evolution. Evol Appl 13(8):1791–1805

Kumar RP, Dinesh Babu KV, D. A. Evans DA. (2019) Isolation, characterization and mode of action of a larvicidal compound, 22-hydroxyhopane from *Adiantum latifolium* lam. Against *Oryctes rhinoceros* Linn. Pestic Biochem Physiol 153:161–170

Lafont R, Raimana H, Phila R, Laurie D (2011) Ecdysteroids in ferns: distribution, diversity, biosynthesis, and functions. In: Working with ferns. Springer, New York, NY, pp 305–319

Liu L, Schepers E, Lum A, Rice J, Yalpani N, Gerber R, Jiménez-Juárez N, Haile F, Pascual A, Barry J, Qi X (2019) Identification and evaluations of novel insecticidal proteins from plants of the class Polypodiopsida for crop protection against key lepidopteran pests. Toxins 11(7):383

Macek T, Vaněk T (1994) Pteridium aquilinum (L.) Kuhn (bracken fern): in vitro culture and the production of ecdysteroids. In: Medicinal and aromatic plants VI. Springer, Berlin, pp 299–315

Mala C, Kekeunou S, Djoukouo N, Denis Z, Bassala J, Nukenine E (2019) Biopesticide potentialities of eagle Fern (*Pteridium aquilinum*) and ricin (*Ricinus communis*) in the protection of vegetables crops. J Exp Agric Int 35(6):1–14. https://doi.org/10.9734/jeai/2019/v35i630222

Markham K, Chalk T, Neal Stewart Jr C (2006a) Evaluation of fern and moss protein based defenses against phytophagous insects. Int J Plant Sci 167(1):111–117

Markham K, Chalk T, Stewart CN Jr (2006b) Evaluation of fern and moss protein-based defenses against phytophagous insects. Int J Plant Sci 167(1):111–117

Masato W, Tomoka M, Hari Prasad D (2021) Phenolic compounds and ecdysteroids of *Diplazium esculentum* (Retz) Sw (Athyriaceae) from Japan and their chemotaxonomic significance. Biochem Syst Ecol. 94:104211. https://doi.org/10.1016/j.bse.2020.104211

Michael Schmitt, Sisrun B (2005) Leaf beetles (Coleóptera: Chrysomelidae) suffer from feeding on fern leaves. Bonner Zool Beitr 54(4):261–269

Murakami T, Minoru K, Satoshi T, Nobutoshi T, Yasuhisa S, Chiuming C (1978) Weitere Inhaltsstoffe aus Pteris inaequalis Baker var. aequata (MIQ.) TAGAWA. Chem Pharm Bull 26(2):643–645

Murasing J, Vignesh M, Gupta MK (2019) Bio-efficacy study on alcoholic extracts of fern plants against diamondback moth, Plutella xylostella L. in cabbage. Pharm Innov J 8(7):05–07

Nakanishi K, Koreeda M, Sasaki S, Chang ML, Hsu HY (1996) Insect hormones. The structure of ponasterone A, insect-moulting hormone from the leaves of *Podocarpus nakaii* Hay. Chem Commun (Camb) 24:915–917

Oguh CE, Okpaka CO, Ubani CS, Okekeaji U, Joseph PS, Amadi EU (2019) Natural pesticides (biopesticides) and uses in pest management-a critical review. Asian J Biotechnol Genet Eng 2:1–18

Ohta S, Guo J-R, Hiraga Y, Suga T (1996) 24-epi-pterosterone: a novel phytoecdysone from the roots of *Athyrium yokoscense*. Phytochemistry 41(3):745–747

Ottosson JG, Anderson JM (1983) Seasonal and interspecific variation in the biochemical composition of some British fern species and their effects on *Spodoptera littoralis* larvae. Biol J Linn Soc 19(4):305–320

Pandian P, Balakrishnan R, Kodakkadal KS, Bastian SK (2017) Efficacy of certain common ferns against red spider mite *Oligonychus coffeae* and tea mosquito bug *Helopeltis theivora* infesting tea. Plant Prot Sci 53(4):232–242

Ragupathy E, Jesudasan RWA, Jeyasankar A, Subramanian K (2004) Efficacy of ethyl acetate extract of a fern, *Lygodium microphyllum* against the Asian army worm, *Spodoptera litura* (Fabr.). J Appl Zool Res 15(1):4–8

Reddy DS, Chowdary NM (2021) Botanical biopesticide combination concept—a viable option for pest management in organic farming. Egypt J Biol Pest Control 3(1):1–10

Russell GB, Greenwood DR, Lane GA, Blunt JW, Munro MH (1981) 2-Deoxy-3-epiecdysone from the fern Blechnum vulcanicum. Phytochemistry 20(10):2407–2410

Sahayaraj K, Selvaraj P (2013) Field evaluation of three fern extracts on *Helicoverpa armigera* (Hubner) and *Spodoptera litura* (Fab.) incidence and infestation and groundnut production. Legume Res Int J 36:84–86

Sahayaraj K, Selvaraj P, Balasubramanian R (2007) Cell mediated immune response of *Helicoverpa armigera* Hubner and *Spodoptera litura* Fab. to fern phytoecdysone. J Entomol 4(4):289–298

Sahayaraj K, Venkateshwari M, Balasubramanian R (2008) Insecticidal and antifeedant effect of *Pedalium murex* Linn. Root and on *Spodoptera litura* (fab.) (Lepidoptera: Noctuidae). J Agric Technol 4(2):73–80

Salehi AS, Mohammadi SM, Hadizadeh A (2019) Biological effects of *Sambucus ebulus* L., *Urtica dioica* L., *Rubus fructicosus* Boiss. and *Pteridium aquilinum* (L.) Kuhn extracts against two-spotted spider mite. Iran J Med Aromat Plants 35(4):565–576

Selvaraj P, De Britto AJ, Sahayaraj K (2005) Phytoecdysone of *Pteridium aquilinum* (L) Kuhn (Dennstaedtiaceae) and its pesticidal property on two major pests. Arch Phytopathol Plant Protect 38(2):99–105

Selvaraj P, Jayaseeli PJR, Ansilin M (2018) Pest control potential of *Cyclosorous interruptus, Christella dentata* and *Nephrolepis cordifolia* on the biology of *Spodoptera litura* (fab.). J Biopest 11(1):76–81

Shukla S, Tiwari SK (2011) Insecticidal activity of *Dryopteris filix-mas* (Linn.) schott ethanolic extract against Corcyra cephalonica Staint. (Lepidoptera: Pyralidae). J Biopest 4(2):138

Shukla AK, Upadhyay SK, Mishra M, Saurabh S, Singh R, Singh H, Thakur N, Rai P, Pandey P, Hans AL, Srivastava S (2016) Expression of an insecticidal fern protein in cotton protects against whitefly. Nat Biotechnol 34(10):1046–1051

Simon A, Vanyolos A, Beni Z, Dekany M, Toth G, Bathori M (2011) Ecdysteroids from *Polypodium vulgare* L. Steroids 76:1419e24

Snogan E, Vahirua-Lechat I, Ho R, Bertho G, Girault JP, Ortiga S, Lafont R (2007) Ecdysteroids from the medicinal fern *Microsorum scolopendria* (Burm. f.). Phytochem Anal 18(5):441–450

Suqing H, Hanhong X, Yongqing T (2013) Insecticidal activities and active ingredients of *Pronephrium megacuspe*. China Agrice Sci 45(1):77–84

Sutoyo S, Indrayanto G, Zaini NC (2007) Chemical constituents of the fern Chingia sakayensis (Zeiller) Holtt. Nat Prod Commun 2(5):1934578X0700200513

Taha MA, Baioumy Ali AA (2020) The first study for the Acaricidal activity of alcoholic extracts of Adiantum capillus-veneris and Funaria hygrometric against Argas persicus. Egypt Acad J Biol Sci F Toxicol Pest Control 12(2):203–217

Tanaka Y, Naya S (1995) Dietary effect of ecdysone and 20-hydroxyecdysone on larval development of two lepidopteran species. Appl Entomol Zool 30(2):285–294

Tanaka N, Murakami T, Saiki Y, Chiuming C (1981) Chemical and chemotaxonomical studies of ferns. XXXVII. Chemical studies on the constituents of costa Rican ferns (2). Chem Pharm Bull 29(12):3455–3463

Varangkana J, Somnuk P, Namfon T, Sahanat P (2016) Screening of some species of Thai Microsoroid ferns for phytoecdysteroid. SDU Res J 9(3):81–97

Vetter J (2018) Secondary metabolites of ferns. In: Current advances in fern research. Springer, Cham, pp 305–327

Wu P, Xie H, Tao W, Miao S, Wei X (2010) Phytoecdysteroids from the rhizomes of *Brainea insignis*. Phytochemistry 71((8–9):975–981

Xavier AGS, Selvaraj P, Nida J (2016) Impact of phytoecdysone fractions of the ferns *Cyclosorous interruptus, Christella dentata* and *Nephrolepis cordifolia* on the biology of *Spodoptera litura* (Fab.). J Biopest. 9(2):125–134

Yadav SK, Archana S, R. (2019) Insecticidal fern protein Tma12 is possibly a lytic polysaccharide monooxygenase. Planta 249:1987–1996. https://doi.org/10.1007/s00425-019-03135-0

Yang JH, Kondratyuk TP, Jermihov KC, Marler LE, Qiu X, Choi Y, Zhang HJ (2011) Bioactive compounds from the fern Lepisorus contortus. J Nat Prod 74(2):129–136

Yao JN, Li ZF, Lou HY, Huang L, Liang GY, Cao PX (2014) A new ecdysteroidal glycoside from *Lepidogrammitis drymoglossoides* (Bak.) Ching. J Carbohydr Chem 33:206e11

Yen KY, Yang LL, Tom WM (1971) Studies on physiological active Phytohormones of ferns in Taiwan I. J Taiwan Pharm Assoc 23(1):83–92

Yi-Ru C (2007) Studies on the chemical constituents of *Crysinus hastatus*(Thunb.)Copel. Ph.D. Thesis. China Medical University, Taiwan

**Kitherian Sahayaraj**, M.Sc., M.Phil., Ph.D., D.Sc., focuses his studies mainly on repellent, insecticidal, ovicidal, and ovipositional activities of terrestrial as well as seaweeds recorded for lepidopteran, hemipteran, and coleopteran pests. Terrestrial and seaweed extracts and entomopathogenic fungi either alone or along with reduviids were field evaluated for cotton and groundnut production. Biosafety of botanical pesticides against few non-target earthworm, silkworm, and reduviids was studied. Sahayaraj has published more than 160 peer-reviewed scientific papers and 15 book chapters, edited 13 books, and authored 13 research project consensus reports for national and international organizations. He has graduated 47 M.Sc. and 20 Ph.D. students and mentored 5 postdoctoral/visiting scientists.

# Part II

# Propagation of Ferns

# Micropropagation of Pteridophytes

# 9

C. Suneetha and Smitha Hegde

**Abstract**

Pteridophytes are haplodiplontic plants that do not produce seeds but haploid spores. Pteridophytes, in general, prefer shaded, moist areas, although their habitat varies considerably. They find different uses such as ornamentals, foliage plants, medicine, and food, and few are good biofertilizers. These plants play an important role in the ecological systems of forests and grasslands and have even successfully colonized urban niches. Their exquisite foliage patterns and resilient growing make them popular ornamental plants globally. Pteridophytes can be propagated by vegetative, sexual, and tissue culture methods. The most common method of propagation is the division of the rhizome. Other means of vegetative propagation are bulbils, aerial growths, stolons, gemmae, stipules, layering tubers, cuttings, and root buds. Pteridophytes are also raised through the spore, with a fusion of gametes in the sexual propagation method. Micropropagation has emerged as a powerful tool for the rapid propagation of rare and endangered plants. Due to environmental changes, the pteridophyte population has gone down drastically in the wild. Thus, it would be appropriate to employ scientific methods such as tissue culture to conserve and propagate these pteridophytes. A much superior and uniform quality of ferns can be produced independent of the season by this technology.

C. Suneetha
Department of Horticulture, University of Agricultural Sciences, Bangalore, Karnataka, India

S. Hegde (✉)
Nitte (Deemed to be University), Nitte University Centre for Science Education and Research, Deralakatte, Mangalore, Karnataka, India
e-mail: smitha.hegde@nitte.edu.in





J. Marimuthu et al. (eds.), *Ferns*, https://doi.org/10.1007/978-981-16-6170-9_9

**Keywords**

Pteridophytes · Spores · Media · Phytohormones · Regeneration · Spermatophyte · Gametophyte

## Abbreviations

| | |
|---|---|
| MS | Murashige and Skoog |
| LS | Linsmaier and Skoog |
| SH | Schenk and Hildebrandt |
| P and T | Parker and Thompson |
| B5 | Gamborg |
| IAA | Indole-3-acetic acid. |
| NAA | 1-Naphthaleneacetic acid |
| KIN | Kinetin |
| BAP | 6-Benzylaminopurine |
| BA | Benzyl adenine |
| 2ip | Isopentenyl adenine |
| 2,4-D | 2,4-Dichlorophenoxyacetic acid |
| $GA_3$ | Gibberellic acid |
| ABA | Abscisic acid |
| GGB | Green globular bodies |
| SFS | Synthetic seeds using fern spores |
| GM | Genetically modified |
| DNA | Deoxyribonucleic acid |
| RNA | Ribonucleic acid |
| GUS | Beta-*glucuronidase* |

## 9.1 Introduction

Pteridophyta (*pteron, feather*; *phyton, plants*) are primitive vascular plants, known as "vascular cryptogams." They are spore-bearing plants that include ferns and fern allies. The pteridophytes are considered the first group of cryptogams showing well-developed tracheids, xylem, and phloem bearing conducting system (Dudani et al. 2013). They form a distinct element of the earth's vegetation. They are significant from the evolutionary point of view as they demonstrate the evolution of the vascular system and indicate the emergence of seed habitat in the plants (Dixit 2000).

Pteridophytes have an independent sporophytic phase and gametophytic phase in their life cycle. Their life cycle shows the alteration of generations, with sporophyte as the dominant phenotype. The sporophyte in pteridophytes is generally differentiated into root, stem, and leaves. The primary root is short lived and is usually replaced by adventitious roots. The stem is often branched, and the branches seldom arise in the axil of the leaves. The fern leaf is called a "frond," and the small are termed "pinnae." The leaves are simple (*Equisetum*), simple and sessile (*Lycopodium*, *Selaginella*), or pinnately compound (*Adiantum*, *Dryopteris*). The vascular system consists of the xylem (composed of tracheids, while true fibers, and vessels absent) and phloem (consisting of only sieve tubes) (Lucas et al. 2013). Reproduction in pteridophytes is via spores. Some "heterosporous" pteridophytes give rise to two types of spores: large, female spores called megaspores and small male spores called microspores. Pteridophytes that bear only one kind of nonsexual reproductive spores are referred to as "homosporous" pteridophytes (Dudani et al. 2013).

They occur in a diverse variety of habitats like terrestrial (*Equisetum*, *Selaginella*), aquatic (*Azolla*, *Marsilea*), epiphytic (*Lepisorus*, *Drynaria*), or lithophytic (*Psilotum*, *Adiantum*). Pteridophytes grow convivially in moist tropical forests and temperate forests. Their distribution across different ecogeographical regions and threatened habitats and from sea level to the highest mountains is intriguing (Dixit 2000). Pteridophytes continue to occupy a crucial position in the evolutionary history of the plant kingdom despite their replacement by the spermatophytes in the modern-day flora.

## 9.2 Threats and Need for Conservation of Pteridophytes

Pteridophytes are moisture and shade-loving plants dependent on microclimatic conditions of a region for their successful survival. Any disturbance in its microclimatic conditions can hinder the growth and evolutionary processes in these plant populations leading to its decline in nature. Factors like climate change, increasing urbanization, industrialization, the encroachment of forest lands, unplanned developmental activities, and overexploitation of natural resources are known to pose a major threat to the existence and survival of these groups of plants. It is imperative to develop in situ and ex situ conservation methods to conserve the diminishing biodiversity of pteridophytes in the environment. The in situ strategy emphasizes the protection of the ecosystem for its biodiversity conservation. The establishment of networks of protected areas selected for high conservation is the basic tenet of in situ conservation. National parks, wildlife sanctuaries, and biosphere reserve contribute to in situ conservation. The in situ conservation is very beneficial as it allows the evolution of the species to continue within the area of natural occurrence. The ex situ conservation includes botanical gardens or conservatories, germplasm banks, DNA banks, and seed banks. It also involves using techniques such as plant tissue culture; cryopreservation; the introduction of disease, pest, and stress tolerance traits using genetic transformation; and ecological restoration of rare plant species and their populations (Kapai et al. 2010).

## 9.3 Need for Tissue Culture in Pteridophytes

Pteridophytes are plants with a variety of medicinal properties, and many of the ferns are consumed as food. They can be grown in soil, as indoor and outdoor plants in pots and baskets, and as epiphytes on other plants. Their aesthetic appeal and their exquisite foliage patterns make them popular plants for landscaping. Chemical analysis of sun-dried *Azolla* showed crude protein, trace minerals, and vitamins. Its rich protein content makes it suitable as unconventional livestock feed (Anitha et al. 2016). Pteridophytes have a great affinity for heavy metals. The two species identified as a hyperaccumulator of arsenic were *Pteris vittata* and *Pityrogramma calomelanos*; these ferns have an extraordinary capacity to accumulate 2 and 2.3% arsenic in its biomass, respectively (Klopper 2011).

Pteridophytes are conventionally propagated using both sexual and asexual vegetative methods. The sexual method involves raising plants from spores. This method is considered more advantageous over the vegetative mode of propagation for economic and transportation reasons. However, the production of plants from spores depends on several factors such as viability and storage of spores, media and soil surfaces, sterilization of spores and soil, size of the spore, the density of spore sowing temperature, pH range, and gametophyte-sporophyte interaction (Kaur 1991). Raising plants from spores is a method, which requires skill and is also comparatively slow. In contrast, the vegetative method incorporates propagation of vegetative organs, such as proliferous frond tip, bulbils, offsets, aerial growths, stolon and tubers, gemmae, root buds, and stipules (Shukla and Khare 2014). These methods of vegetative propagation are limited and mostly seasonal. The application of tissue culture techniques and other molecular techniques of biotechnology has brought about a significant change in the commercialization of ferns. Micropropagation is now being used for the large-scale production of ornamental plants. According to an analysis by Pierik (1991), 157 million plants, which are 74% of the total production of micropropagated plants, were ornamental species. The fern *Nephrolepis* heads the list with 17.8 million plants.

The propagation of plants using tissue culture has progressed rapidly in the last quarter of this century. Ferns are some of the earliest mass multiplied plants for commercial purposes (Murashige 1974; Burr 1976). But today, if one were to appraise the progress tissue culture has made in the branches of angiosperms and gymnosperms, the work in pteridophytes appears to be limited. Only a few scattered reports of propagation of ferns by tissue culture are available. The tissue culture studies on ferns can be broadly reviewed as work done on spore germination/gametophytic tissue culture and sporophytic tissue culture.

## 9.4 Tissue Culture of Pteridophytes

Plant tissue culture broadly refers to the growth of plant cells, tissues, organs, and plantlets on an artificial medium under aseptic and controlled environmental conditions. The concept of tissue culture resulted from Gottlieb Haberlandt's experiments to understand the properties and potentialities of a cell as an elementary organism (Haberlandt 1902). Tissue culture methods are employed as an essential aid in conventional methods of plant improvement. They are also used to propagate genetically manipulated superior clones and ex situ conservation of valuable germplasm (Razdan 2003). Producing pathogen-free plants in the synthesis of many commercially important secondary metabolites has changed the conventional production method. The technique has made landmark contributions in the screening and production of phytochemicals regularly used in pharmaceutical industries.

### 9.4.1 Spores/Gametophytic Culture

Pteridophyte tissue culture began with the culture of spores on artificial substrates first under partially sterile conditions such as sterile sandbanks and then under fully axenic conditions on agar plated petri dishes. Spore germination using tissue culture methods facilitates contamination-free cultures (Fig. 9.1a, b). It is free from cross-contamination by spores of other species and infection of bacteria, fungi, algae, and mosses (Debergh 1994).

Different tissue culture experiments were carried out in the following species for regeneration using spores as explants, namely, *Osmunda regalis* (Morini 2000), *Pteridium aquilinum* (Zhang et al. 2001), *Osmunda japonica* (Yi-Fa 2003), *Osmunda cinnamomea* (Huang et al. 2004), *Cyathea spinulosa* Wall. ex Hook (Shukla and Khare 2005a; b), *Blechnum spicant* L (Menéndez et al. 2006), *Cyathea delgadii* Sternb. and *Blechnum brasiliense* Desv. (Hiendlmeyer and Randi 2007), *Pronephrium triphyllum* (Sw.) Holttum and *Sphaerostephanos unitus* (L.) (Holttum) (Marimuthu and Manickam 2011), *Dipteris wallichii* (R.Br.) T. Moore (Bharati et al. 2013), and *Asplenium nidus* (Haddad and Bayerly 2014). In the above experiments, the spores were cultured either for the production of gametophytes or to carry out fundamental studies in different fields, e.g., photomorphologists used the spores to study the effect of light of various wavelengths on the first division of the spore cell. The subsequent division of the cells results in the formation of the protonemata, which later forms the prothalli of the fern gametophytes.

Gametophytes are also produced for the production of sexual sporophytes (Fig. 9.1d). The production of sexual sporophytes is of primary interest to fern collectors and nurseries as they are the source of variety. To environmentalists, they are a means to maintain biodiversity, i.e., pteridophytes of ecological value are tissue cultured for conservation. While spores continue to form the primary source of explants in pteridophyte tissue culture, some of the other gametophytic tissues used are green sporangia in *Dicranopteris linearis* (Henson 1979) and gemmae of gametophytic tissue in *Trichomanes speciosum* (Raine and Sheffield 1997); Janssens and Sepelie (1989) have used homogenized prothalli tissue of *Blechnum* species as explants. In *O. cinnamomea*, gametophyte size is negatively related to the population density, which significantly affects gametophytes' sexual expression (Huang et al. 2004). Experiments were carried out on gametophytes and very young fronds of sporophytes with Murashige and Skoog (1962) medium. The experiment describes

**Fig. 9.1** (**a**) Opening of the trilete spore for germination. (**b**) In vitro germination of gametophytes on sterile filter paper discs. (**c**) Aposporous gametophytes developing on the sporophytic tissue in vitro. (**d**) Mature sporophyte with roots and circinate fronds at the end of 8 weeks in culture in vitro

15 years of in vitro experiments on 16 species of tree ferns belonging to genera *Blechnum*, *Cibotium*, *Cyathea*, and *Dicksonia*. Genus *Cyathea* was represented by *C. dealbata* (G. Forest) Sw., *C. australis* (R.Br.) Domin., *C. capensis* (L.f.) Sm., *C. cooperi* (F. Muell.) Domin, *C. brownii* Domin, *C. robertsiana* (F. Muell.) Domin., *C. dregei* Kunze, *C. leichhardtiana* (F. Muell.) Copel., *C. smithii* Hook.f., *C. schanschin* Mart., and *Cyathea* sp. In genus *Dicksonia*, only two species, *D. fibrosa* Colenso and *D. sellowiana* Hook, were used. Taxon *Blechnum* was presented by *B. brasiliense* Desv. and *Cibotium* by *C. glaucum* (Sm.) Hook. and Am. and *C. schiedei* Schltdl. and Cham. All the above species showed various responses to culture conditions depending on the stages of development. The time required for spore germination differed between species ranging from a few weeks to 16 weeks. Prothallus formations showed various types of growth depending on the

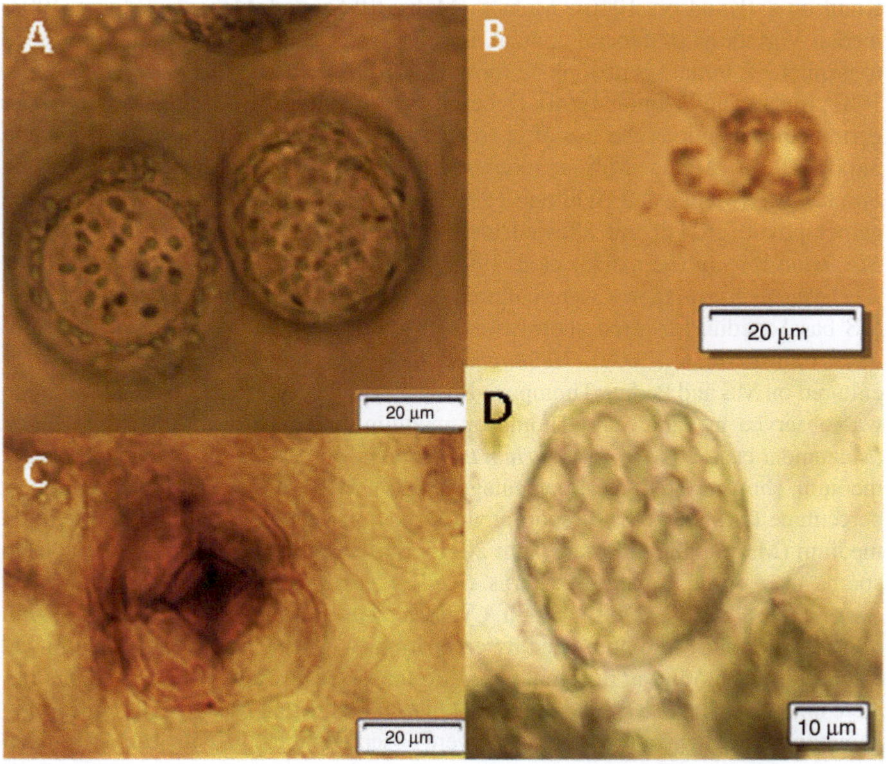


**Fig. 9.2** Reproductive structures on mature gametophytes of ferns. (**a**) Dorsal view of antheridia with distinct opercular cell. (**b**) Release of active, coiled, and multiflagellate antherozoids. (**c**) Dorsal view of opened archegonium. (**d**) Lateral view of mature antherozoids with spherical coiled antherozoids

marginal meristems. For all test species, long-term in vitro cultures of gametophyte were established. Mature gametophyte developed functional antheridia and archegonia (Fig. 9.2).

Spontaneous fertilization helps to establish the culture of young sporophytes (Goller and Rybczyński 2007). Marimuthu and Manickam (2011) reported gametophytic development in two medicinally important ferns, *Pronephrium triphyllum* (Sw.) Holttum and *Sphaerostephanos unitus* (L.) (Holttum). *Anemia rotundifolia*, a rare fern species, showed that the explant of gametophyte can be used to multiply the plants (gametophytes and sporophytes). The study demonstrated that explants produced secondary regenerates, which matured into cordate gametophytes and bore only archegonia (Singh et al. 2012).

#### 9.4.1.1 Media and Methods

Spores are generally cultured on an agar plate medium. Among the various media used for spore germination, the most popular ones are Knop's (Knop 1865),

Knudson's (Knudson 1946), Mohr's (Mohr 1956), and Moore's (Moore 1903) media. Variations in mineral content, nitrogen source, and temperature are needed to promote or initiate gametophyte formation from spores in some species (Whittier 1981; Melan and Whittier 1990). Debergh (1994) in his key protocol for fern spore germination mentions the use of charcoal in the 1/2 MS medium (Murashige and Skoog 1962). Media with ammonium nitrogen are useful for germination in *Botrychium* (Melan and Whittier 1990). Nutrient-deficient media decreased the gametophytic growth but allowed antheridia formation in a shorter period than *Blechnum spicant* (Fernández et al. 1996a, b).

*Drynaria roosii* spores were cultured on Knop's medium and 1/2 MS medium. MS basal medium favored spore germination but inhibited gametophyte development (Yin-Li et al. 2009). *Helminthostachys zeylanica* (L.) Hook spores were cultured on MS and Parker-Thompson Basic C fern medium. Maximum germination was observed in Parker and Thompson medium with IAA (1.41 μM) supplement (Mazumder et al. 2010). *Pronephrium triphyllum*, spores sown on Knop's basal agar medium, showed the highest percentage ($38.3 \pm 1.13$) of germination. The highest percentage ($52.3 \pm 1.43$) of sporophyte formation was observed in Knop's liquid medium (Marimuthu and Manickam 2011). Surface sterilization of freshly isolated spores from fronds of the tree ferns *Cyathea australis* (R. Br.) Domin, *Cyathea cooperi* (Hook. ex F. Muell.) Domin, *Cyathea cunninghamii* Hook.f., and *Dicksonia antarctica* Labill. were optimized by comparing two sterilizing agents (sodium and calcium hypochlorite) at different exposure times. Best results (0% contamination) were obtained with 5% (v/v) commercial bleach for 15 min. Best results concerning gametophyte growth and development were observed on media with 1/2 MS mineral salts and 2% sucrose supplemented or not with 208 μM of $NaH_2PO_4$ and on Dyer mineral salt medium. In terms of sporophyte development, the best results were obtained on a culture medium supplemented with 208 μM of $NaH_2PO_4$ (Reis Moura et al. 2012). An in vitro method for micropropagation of *D. wallichii* Moore was carried out using Murashige and Skoog medium for spore germination and cell proliferation. It was observed that supplements of phytohormones such as auxins (indole-3-acetic acid) were favorable for sporophyte development from gametophytic prothallus. Further growth and development of the prothallus and sporophyte were seen on various hormone concentrations and combinations. The best results were observed in 0.85 μ mol IAA + 0.92 μmol KIN (Bharati et al. 2013). *Diplazium proliferum* is indigenous to the Mascarene region and is considered a rare species in Mauritius. MS culture media supplemented with BAP did not significantly increase the growth rate of both gametophytes and sporophytes of *D. proliferum* (Golamaully et al. 2015). The *Polypodium aureum* sporangia were seeded in a different culture, namely, media agar, MS, SH, White, and B5. The medium White presented the highest average for the germination speed index, whereas the agar media, MS, SH, and B5, had the same germination speeds (Santos Alves et al. 2019).

Reports of the effect of plant growth regulators on gametophytic culture are few (Camloh et al. 1996). *Osmunda regalis* spores germinated on gelled Hoagland and Arnon growth medium produced normal gametophytes and sporophytes. Sporophytes cultivated on both MS and Hoagland and Arnon growth media consistently were susceptible to tissue vitrification, even when different cytokinins were used (Morini 2000). Spores of *Osmunda japonica* cultured on MS medium produced sporophytes with a spray of 2,4-D, GA3 (144.35 μM), and 1% $KH_2PO_4$ (Yi-Fa 2003)· Effects of sucrose and irradiance were studied on germination and early gametophyte growth of the endangered tree fern *Dicksonia sellowiana* Hook (Renner and Randi 2004). The effect of the gibberellins $GA_3$ at a dose of 2.88 μM favored a significant increase in the percentage of female gametophytes in *Blechnum spicant* L (Menéndez et al. 2006). Spores of *Asplenium nidus* were used as explants to study different growth hormones in vitro; their response on ferns in different stages of propagation. Consequently, the hormone concentration that results in initiation and proliferation for prothalli, differentiation and rooting was determined (Haddad and Bayerly 2014).

Carbohydrates were considered very critical for spores (Whittier 1964). Later the role of the osmotic potential of the media was proved to be of greater significance (Whittier 1975). In *Pteridium* ferns raised on different sugars, it was observed that a steady rise in organ development (33%) was observed in the sucrose containing medium as compared to sorbitol. Sorbitol is very poorly metabolized by ferns but is preferable to mannitol, inhibiting development (Hirsch 1975). However, the presence of sucrose is known to encourage the subsequent growth of gametophytes and sometimes even apogamy (Whittier 1962). The culture conditions cannot be generalized as one set of culture conditions cannot be effective for all species. For example, spores of some species of ferns such as *Botrychium* and *Ophioglossum* require darkness for germination (Whittier 1981). In contrast, Dyer (1979) has found that spores of most ferns require exposure to light for germination and growth. The ability to germinate in the dark may vary with specimen age, exposure to different temperature, and plant hormone treatment (Page 1979; Dyer 1979; Miller 1968). For all the species examined to date, sugar is necessary to grow spores in the dark as these are non-photosynthetic; gametophytes require sugar in the medium (Whittier and Moyroud 1993). Ford and Fay (1990) state that the percentage of dark germination is low and subsequent growth is abnormal or retarded. It is believed that a period of 3–96 h in the dark after sowing is required for imbibition before the spores become light receptive (Jarvis and Wilkins 1973). Dyer (1979) and Douglas and Sheffield (1992) have reported extensive studies on conditions favorable for in vitro spore germination and gametophyte development.

The effect of light on spore germination in two ornamental ferns native to the Atlantic forest *Cyathea delgadii*, a tree fern and *Blechnum brasiliense*, a sub-arborescent fern was conducted under natural conditions by Hiendlmeyer and Randi (2007). Results showed that germination of *Cyathea delgadii* spores at 22% light reached 76% and mean germination time was 19.7 days; at 5% light, germination reached 83.5%, and mean germination time was 20.16 days. Germination of *Blechnum brasiliense* at 22% light reached 76%, and mean germination time was 9.06 days; at 5% light, germination reached 84%, and mean germination time was 13.18 days. Studies using spores of leptosporangiate ferns have shown that the best germination occurs at a slightly acidic or neutral pH (Miller 1968). Promotion of gametophyte development by strongly acidic conditions is unusual in ferns, but in

*Ophioglossum palmatum*, a reduction in the pH increased germination percentage; the highest germination percentage was observed at pH 4.0 (Higuchi and Amaki 1989). Ethylene has been reported to affect spore germination (Warner and Hickok 1987), growth of gametophytes (Miller et al. 1970), and production of apogamous buds (Elmore and Whittier 1973). The germination time for fern spores varies from a few days to a year. Smith and Yee (1975) reported that *Nephrolepis* spores germinate in 3–4 days in culture, while in Ophioglossaceae, germination is known to occur at the end of 6 months of incubation in the dark (Whittier 1981). Certain *Helminthostachys* species are known to take 8 months to germinate (Whittier 1987). In vitro gametophyte production and its use in the phytoremediation of lead and mercury are reported in *Pityrogramma calomelanos* (Pulukkunadu Thekkeveedu and Hegde 2020).

#### 9.4.1.2 Challenges

Some of the problems of pteridophyte in vitro spore germination and culture are (a) high-density population of gametophytes in limited space, (b) drying of the media, (c) discoloration of the media because of the browning at the base of the gametophytes, and (d) mass sporophyte formation. They have been well discussed and analyzed with solutions by Ford and Fay (1990). There are several problems like collection, sterilization, and breaking dormancy of spores to use them in micropropagation. Suppose there is a serious risk of contamination from other sources, e.g., spore from fungi and other fern species (found in surrounding air and on the surface of the frond). In that case, the spore must be cleaned before sowing (Hoshizaki and Moran 2001). The spore propagated ferns produce the healthiest seedlings. Alternatively, sowing spores directly like plant seeds is challenging as fern spores are very small, making their handling difficult in field conditions. It also requires continuous and strict environmental control. In vitro culture of spores, gametophytes, and sporophytes attempted in recent times reduces labor in addition to facilitation of year-round production of plants. However, in vitro culture requires expensive facilities and equipment that increase production costs substantially. Further, as knowledge and training in plant tissue culture are required, access to farmers and field use is still limited (Jang et al. 2020a, b).

### 9.4.2 Sporophytic Tissue Culture

Some of the significant reports of tissue culture of sporophytic tissues are by Murashige (1974). Green globular bodies (GGBs) are produced when rhizome segments prepared from mature plants of *Adiantum*, *Asplenium*, *Nephrolepis*, *Pteris*, and *Rumohra* were cultured on modified Murashige and Skoog medium supplemented with benzyl adenine (Amaki and Higuch 1991). *Adiantum raddianum*, *Asplenium nidus*, *Nephrolepis cordifolia*, *Pteris ensiformis*, and *Rumohra adiantiformis* are used for the production of callus and regeneration of sporophytes from rhizome segments. Camloh et al. (1994), Camloh and Gogala (1991), and Camloh and Zel (1995) have carried out extensive tissue culture studies on *Platycerium coronarium* and *Platycerium bifurcatum*. Kwa et al. (1995a, 1995b, 1995c, 1997) worked on the role of ethylene in the production of sporophytes from sporophytic tissues, the activity of ammonium, nitrate uptake, and nitrate reductase activity of callus cultures and also studied the effect of $CO_2$ enrichment on photoautotrophic callus and its morphogenetic plasticity in *Platycerium coronarium*. Fernández et al. (1991a, b) and Fernández et al. (1997a, b) report very interesting findings of sporophyte regeneration from rhizome homogenate of *Asplenium nidus* and *Pteris ensiformis*.

Explants of sporophytic tissue are prepared from tissues such as the young circinate fronds in *Drymoglossum piloselloides* (Kwa et al. 1988), whole or section of fronds in A splenium nidus-avis (FernÃndez et al. 1997b), leaflet primordium in *Cyathea gigantea* (Padhya 1987), pinnae in *Nephrolepis exaltata* (Bryne and Caponetti 1992a, b), leaf margin meristems in *Ceratopteris* (Gottlieb 1972), rhizome tips in *Adiantum cuneatum* (Murashige 1974), rhizome segments in *Adiantum raddianum* (Amaki and Higuch 1991), and lateral buds of rhizome in *Matteuccia struthiopteris* (Dykeman and Cumming 1985). Other reports include root apices in *Cyclosorus dentatus* (Mehra and Palta 1971), shoot apex/whole tips in *Platycerium stemaria* (Cooke 1979) or macerated as in *Davallia bullata* (Cooke 1979), shoot segments in *Marsilea quadrifolia* (Breznovitis and Mohay 1987), runner tips in *Nephrolepis cordifolia* (Torres 1989), stolon segments in Nephrolepis exaltata, and stolon tips in *Nephrolepis exaltata* (Bryne and Caponetti 1992a, b) as explants. A variety of scales were used in *Platycerium bifurcatum* (Dolinsek and Camloh 1997), while an embryo was used in *Todea barbara* (DeMaggio and Wetmore 1961).

Goller and Rybczyński (2007) investigated 16 species of fern for long-term gametophyte in vitro cultures. Mature gametophyte develops antheridia and archegonia. The spontaneous fertilization helped to establish the culture of young sporophytes. The study conducted by Marimuthu and Manickam (2011) reports spore germination, gametophyte development, changes in the reproductive phases, and sporophyte formation of the medicinally important ferns *Pronephrium triphyllum* (Sw.) Holttum and *Sphaerostephanos unitus* (L.) (Holttum) from mature spores. A new micropropagation protocol for leather leaf fern (*Rumohra adiantiformis* (G. Forst.) Ching) was successfully established using rhizomes as the donor explant, following appropriate sterilization (Winarto and Teixeira 2012). Shastri et al. (2012) attempted to standardize the in vitro protocol for the mass multiplication and conservation of a threatened tree fern *Cyathea spinulosa* Wall. ex. Hook employing leaf primordium explants excised from in vitro raised sporophytes through spore culture. Nair et al. (2013) report the successful in vitro culture of a circinate part of young leaves of *D. esculentum*. Aseptic cultures were raised at the morphogenic level of callus, axillary shoot, multiple shoots, and rooted plants for mass multiplication of five ornamental economically important ferns (*Nephrolepis biserrata* (Sw.) Schott., *N. cordifolia* cv. 'Duffii' (L.) Presl., *N. exaltata* cv. *bostoniensis* (L.) Schott., *Pteris vittata* L., and *Cyclosorus dentatus* Link.) and three threatened ferns, namely, *Cyathea spinulosa* Wall. ex. Hook, *Pityrogramma calomelanos* (L.) Link., and *Microsorum punctatum* (L.) Schott., through in vitro techniques (Shukla and Khare 2014).

The study by Nair et al. (2014) reports spore germination, gametophyte development, changes in the reproductive phases, and sporophyte formation of the fern *Diplazium esculentum* (Retz) SW. An efficient micropropagation and in vitro conservation method via direct and indirect organogenesis from seed and leaf explants of *Ramonda serbica* and *Ramonda nathaliae*, respectively, was established (Gashi et al. 2015). Aldea et al. (2017) developed a viable experimental system for in vitro multiplication of *Asplenium trichomanes*, *Polypodium vulgare*, and *Asplenium filix-femina* to be adapted to other fern species of biotechnological and conservative interest. An efficient in vitro propagation method for *Selaginella martensii* using shoot tip cultures with Murashige and Skoog (MS) medium was reported by Park et al. 2020. *Selaginella pulvinate* was propagated in vitro using frond tips as explants (Yu et al. 2021).

#### 9.4.2.1 Contamination

Contamination is controlled by using tissues of sporophytes formed in vitro using germinated spores (Fernández et al. 1996a, b). However, there are a few reports of axenic cultures of sporophytic explants from field-grown ferns, e.g., *Drynaria quercifolia* (Hegde 1998) and *Nephrolepis* (Torres 1989). In ferns, in vitro infection of bacteria, fungi, algae, and mosses is a major challenge when explants are prepared from field-grown sporophytes (Agretious et al. 1996). Particularly establishing sterile cultures from rhizome explants is very difficult. The contamination percentage is high in young circinate fronds' explants due to their spiral and pubescent nature. Hairy explants are usually difficult to decontaminate because of the ineffective penetrating activity of the surface sterilant (George 1993). Among various surface sterilants, the most common sterilant used for ferns is sodium hypochlorite. The concentration used and the length of exposure vary greatly depending on the type of plant material. A surfactant such as a tween or an antiseptic soap solution is normally included when some tissues required more exacting sterilization conditions. Other sterilants used are calcium hypochlorite, mercuric chloride, and hydrogen peroxide (Fay 1994).

The main cause of the low effectiveness of in vitro propagation has been the lethal browning of *Matteuccia struthiopteris* (L.) Tod. fern rhizome tissue after isolation. In vivo pretreatment of creeping rhizomes of ostrich fern induced their rejuvenation, manifested by the development of numerous buds (Zenkteler 2006).

#### 9.4.2.2 Media and Methods

Most often used media for sporophyte tissue culture are modified Murashige and Skoog (Murashige and Skoog 1962), Fern Multiplication Medium (Murashige 1974), and modified Knudson's (Knudson 1946) medium. The use of growth regulators is minimal. Some of the cytokinins reported are kinetin, BAP (Padhya 1995), and 2-iP (Hedge and D'souza 1997). Certain additives such as adenine sulfate and coconut milk are also used. For example, adenine sulfate is necessary for shoot tip cultures, as reported in *Platycerium stemaria* (Hennen and Sheehan 1978). Coconut milk is found to encourage embryo growth in *Todea barbara* (DeMaggio and Wetmore 1961). Sporophytes are produced from cultured tissues by either axillary shoot proliferation (Murashige 1974) or adventitious shoots formed directly or from callus (Amaki and Higuch 1991). Direct adventitious shoot formation is considered the simplest and most practical method for large-scale production of true to type plants. While working with *Drynaria quercifolia*, where direct plants could be induced on rhizome explants, our experience was simple and productive. Plantlet differentiation from callus is a relatively new method for ferns. The induction of callus in ferns was considered a rare and spontaneous occurrence rather than a controllable event (Partanen 1972). However, callus induction and differentiation have been simple for several ferns, e.g., *Ampelopteris prolifera* (Mehra and Sulklyan 1969) and *Cyclosorus dentatus* (Mehra and Palta 1971). In *Drynaria quercifolia*, morphogenetic callus formed sporophytes in cultures containing 2-iP (Hedge and D'souza 1997). During the development of sporophytes from callus, we observed interesting juvenile characteristics in vitro (Hegde et al. 2006). Another method popular in the horticulture industry is the regeneration of plants from sporophytic tissue homogenate (Kyte 1987). It is simple, practical, and economically feasible for large-scale production of popular ornamental ferns such as *Nephrolepis exaltata*.

*Nephrolepis biserrata* runner tips embedded into the medium with 1/2-strength MS produced buds and shoots (Aimee 2000). The rare fern *Diplazium cognatum* (Hieron.) was multiplied through indirect organogenesis in vitro (Manickam et al. 2003). The work of Ambrósio and Melo (2004) reports the effect of different pH and sucrose concentrations on in vitro propagation of *Nephrolepis biserrata* fronds aseptically obtained from stolon segment and cultured. An attempt was made to standardize the protocol for the shoot regeneration via caulogenesis in *Pteris vittata* L. employing leaf primordium explants. Calli were induced on Murashige and Skoog (MS) and Parker and Thompson (P and T) media supplemented with different combinations of growth regulators (Shukla and Khare 2012a; b). To conserve and multiply the aquatic fern *Marsilea quadrifolia* L., in a long-term in vitro procedure, Rolli et al. (2015) suggest that the in vitro hormone-free micropropagation could be useful in the development of ex situ conservation programs of *Marsilea quadrifolia*, even to reintroduce the plants in their natural environment possibly.

#### 9.4.2.3 Challenges

The prolonged culture of pteridophyte tissues produces certain aberrations due to ethylene production in the culture containers. Ethylene has been reported to inhibit the regeneration of plants of *Platycerium coronarium* from fronds and rhizome pieces (Kwa et al. 1995a). Variations in ploidy, apospory (Fig. 9.1c), and apogamy cause ambiguity in ferns, such as gamma radiation-induced in vitro hormetic apogamy observed in the fern *Pityrogramma calomelanos*. There are many factors to be considered, micropropagation of ferns like explant (maturity and size), surface sterilization (types of sterilant used and duration), media (different media with sugar or without and different hormone conditions), culture condition (light, humidity, and temperature), and duration (months to years). Some propagation protocols are varying with species and specific to genera also.

### 9.4.3 Suspension Culture of Pteridophytes

Cell suspension cultures are potential systems for the mass propagation of plants. Few reports are available on *Platycerium* species of plants. Pteridophyte in vitro suspension cultures demonstrate gametophytic and sporophytic tissues (Kwa et al. 1997; Camloh 2006). The gametophyte and sporophyte callus of *Platycerium coronarium* was established on kinetin containing MS media and subcultured onto phytohormone-free MS media in cell suspensions to obtain morphogenesis. Leaf cell suspension cultures of *Platycerium bifurcatum* were successfully regenerated into sporophytes on a growth regulator-free medium with activated charcoal into basal MS medium (Teng 1997). The role of activated charcoal in cell suspension culture is elucidated by the improvement of regeneration of sporophytes, prevention of gametophytic clusters and nodule-like bud clusters of sporophytes, and prevention of hyperhydricity in regenerated sporophytes. Teng and Teng (1997) demonstrated two phase culture mechanisms for large-scale production of *Platycerium bifurcatum*, viz., aposporous gametophytes or directly from cell suspensions. Single cells aggregate of 100 cells gave rise to sporophytes indirectly via formation of aposporous gametophyte, while aggregation of 500–1000 cells regenerated sporophytes directly.

### 9.4.4 Cryopreservation of Pteridophytes

Cryopreservation is an important component of pteridophyte conservation. Ex situ conservation for pteridophytes is a form of spore banks, gametophyte, and sporophyte in botanical gardens and parks. In vitro conservation methods include maintenance of cultures and cryopreservation (Ballesteros and Pence 2018). The methods of cryopreservation are similar to those exercised for seed plants. Sporophytic and gametophytic tissues can be used for preservation. However, spores are a preferred choice of explant. Several pteridophyte sporophyte and gametophytes have demonstrated high regeneration potentials in vitro. They exhibit a natural potential for vegetative propagation from small fragments of sporophytic or gametophytic tissues. However, the spore is the preferred tissue for cryopreservation as they are small in size and simple in organization and have a natural desiccation tolerance (Pence 2008). Among various methods used for cryopreservation, vis-à-viz, encapsulation-dehydration method (Fabre and Dereuddre 1990); drying without encapsulation along with a 7-day preculture on abscisic acid (ABA) (Pence 2000), the encapsulation-dehydration method with preculture on ABA provided the best results for long-term banking of pteridophyte tissues. Regeneration of cryopreserved tissue has shown encouraging results. Sixty percent or more germination was seen in spores of most species stored for $\geq$20 years. However, the percent germination was variable and depended on the species. Various aspects of cryopreservation in pteridophytes are exhaustively detailed (Ballesteros and Pence 2018).

A simple and reliable method for in vitro multiplication and long-term preservation using liquid nitrogen was developed for gametophytes of *Osmunda regalis*. The application of the in vitro and cryopreservation methods made it possible to improve the number and time of *O. regalis* sporophyte production. The protocols suggest new avenues for the mass propagation, germplasm conservation, and resource management of the species (Makowski et al. 2016). Bracken fern synthetic seeds were prepared using bracken spores and alginate matrix. Spore germination, gametophyte, sporophyte growth and development from synthetic seeds using fern spores (SFS) was conducted. The spore density influenced the gametophyte and sporophyte numbers. Even 7-year-old SFS prepared using cold (4 °C) long-term storage spores could effectively form sporophytes. Developing synthetic seeds is an economically feasible solution for ensuring efficient transport and handling small-sized fern spores; furthermore, this SFS technology provides the basis for fern seedling culture and fern spore industrialization (Jang et al. 2020a, b). Fern spores are haploid plant germplasm and microscopic that can regenerate full plants via germination. Conventional storage (i.e., dry at −20 °C) is found damaging in few species. Therefore, the use of cryopreservation has prospected as an option for long-term ex situ conservation. The cryopreservation process for pteridophyte spores is used in the Millennium Seed Bank of Royal Botanic Gardens, Kew. They have included accounts of the methods of harvesting and cleaning spores in the study (Nebot et al. 2021).

## 9.5 Pteridophytes as Experimental Organisms

Pteridophytes are considered ideal biological systems for research by scientists from numerous branches of science. They are interested in taxonomists, morphologists, photomorphologists, anatomists, plant ecologists, plant physiologists, cytogeneticists, and biologists in reproductive studies. However, most of the experimentation is carried out on the gametophytes for various reasons, viz., they are small, thalloid, heart-shaped short-lived structures. The gametophytes are suited to study the detailed analysis of cell and tissue growth. They are used to study the mechanisms that regulate organ growth, cell division process, genetic variations, etc. The organization of the pteridophyte gametophyte is simple. The cell number in the prothallus is also small, making the cellular analysis of the tissue growth much easier than in the tissues of flowering plants. Gametophytes are useful for both research and teaching related to genetics, physiology, morphology, and systematics. Pteridophyte spores and gametophytes have also been used for toxicity assessments. Pteridophytes have been used to study antibiotic resistance. Advantage has been taken of pteridophytes as outstanding experimental material for applying newly developed techniques in photobiology. Fern spores and protonemata are excellent systems for studies of cell cycle regulation because the progression of the cell cycle, i.e., cell growth, division, and differentiation, can be directly observed under the microscope.

## 9.6 Genetic Transformation of Pteridophytes

Pteridophytes are important members of the plant kingdom. However, they lag behind other taxa regarding our understanding of their genetics, genomics, and molecular biology. Stable genetic transformation is accomplished in a few species outside angiosperms and gymnosperms, especially among bryophytes, but rare in the Pteridophyta. In pteridophytes, the study of genetic programs underpinning developmental processes has been hampered by large genome size, a lack of available mutants, and an inability to create stable transgenic lines.

Overexpression and knockdown studies of individual genes in a wide variety of pteridophyte species will accelerate our learning about their biology and develop novel products from them. Furthermore, a facile transformation system will accelerate functional genomics and systems biology in pteridophytes. For example, knockdown analysis of genes involves interesting biosynthetic pathways that can greatly facilitate gene and biochemical discoveries. A stable transformation protocol for pteridophytes using Agrobacterium-mediated transformation (Bui et al. 2015; Muthukumar et al. 2013) and particle bombardment method (Andrew et al. 2015; Plackett et al. 2015) in *Ceratopteris richardii* and *Pteris vittata* was successfully demonstrated. Bui et al. (2018) have elaborated the concept of gametophyte transformation. Pteridophyte gametophytes are dominant haploid systems that are excellent models of non-vascular, non-seed plant systems. They report a transient and stable *Agrobacterium*-mediated transformation method for gametophytes in the model fern of *Ceratopteris richardii. Marsilea vestita* was used by Klink and Wolniak (2001) for gene expression suppression studies in *C. richardii* using RNA interference. Microprojectile bombardment studies were demonstrated in gametophytes of *Adiantum capillus-veneris* (Kawai-Toyooka et al. 2004) and *C. richardii* (Rutherford et al. 2004).

## 9.7 Pteridophytes and Transgenics

Phytophagous insects rarely infest pteridophytes and mosses in the wild. Insect infestation on pteridophytes is 30-fold lower than on flowering plants. Whitefly (*Bemisia tabaci*) damages field crops by sucking sap and transmitting viral diseases. None of the insecticidal proteins used in genetically modified (GM) crop plants to date are effective against whitefly. Shukla et al. (2016) report on the identification of a protein (Tma12) from an edible fern, *Tectaria macrodonta* (Fee) C. Chr., that is insecticidal to whitefly (median lethal concentration = 1.49 mg/ml in vitro feeding assays) and interferes with its life cycle at sublethal doses. Transgenic cotton lines that express Tma12 at ~0.01% of total soluble leaf protein were resistant to whitefly infestation in contained field trials, with no detectable yield penalty. The transgenic cotton lines were also protected from whitefly-borne cotton leaf curl viral disease. Rats fed Tma12 showed no detectable histological or biochemical changes. Together with the predicted absence of allergenic domains in Tma12, the findings indicate that

Tma12 might be well suited for deployment in GM crops to control whitefly and the viruses it carries.

The protoporphyrin IX magnesium chelatase (CrChlI) under the control of P35S plus a construct of P35S-driven GUS reporter has used study gene suppression in *C. richardii.* The function of the arsenite transporter in *Pteris vittata*, known for its hyperaccumulation of arsenic, was studied using RNA interference by Indriolo et al. (2010). *Agrobacterium*-mediated transient transformation in *C. richardii* was also demonstrated (Bui et al. 2015). Singh et al. (2020) reported *A. rhizogenes*-induced hairy root formation in *Selaginella bryopteris*, a medicinally and commercially important plant. The experiment was targeted toward the production of secondary metabolites in vitro. *Ceratopteris richardii* has long been proposed as a model pteridophyte and has recently become tractable due to stable transgenesis and increasing genomic resources, allowing researchers to test detailed questions about gene function in a pteridophyte for the first time (Stephanie and Conway 2020).

## 9.8 Summary

Pteridophytes are an important group of vascular plants with high economic importance and aesthetic value. A rich diversity of pteridophytes are distributed worldwide, but factors like climate change, increasing urbanization, industrialization, and encroachment of forest lands pose a major threat to the survival of these plants. It is vitally important to conserve the diminishing biodiversity of pteridophytes by either in situ or ex situ conservation methods.

The application of biotechnology, especially tissue culture techniques to pteridophytes, has resulted in some major changes in the propagation of pteridophytes. The conventional method of propagating pteridophytes in nurseries faced inherent dangers of moss and algal invasion and contamination with other pteridophytes. The pteridophytes also showed generally slow and uneven growth. Tissue culture solved all these problems. The production of many healthy pteridophytes for international trade that meets the standards of plant health restrictions is today a reality. *Nephrolepis*, one of the most popular house plants, is produced in thousands in a relatively short period. With this method, the production of a million plants per year is within reach of a modest laboratory. Endangered pteridophytes, rare pteridophytes, and pteridophytes that are difficult to grow are today micropropagated and restored to their natural habitat using this technique. The technique could be further applied to the culture of pteridophyte tissues with medicinal properties. Tissue culture could also be utilized to produce secondary metabolites and extraction of active principles from them (Tables 9.1 and 9.2).

**Table 9.1** Trend of in vitro propagation of pteridophytes before 2000

| Species name | Source of explant | Media | Results | References |
|---|---|---|---|---|
| *Acrostichum aureum* | Spores | Klekowski Jr (1969) | Gametophytes with sex organs | Lloyd (1980) |
| *Acrostichum aureum* *Acrostichum danaeifolium* | Spores | Klekowski Jr (1969) | Germination and gametophyte | Lloyd and Buckley (1986) |
| *Adiantum capillus-veneris* | Coiled up young leaves | Pais and Casal (Pais and Casal 1987) | Direct adventitious shoots. Rooted plants | Pais and Casal (1987) |
| *Adiantum capillus-veneris* | Spores | 1/10 MS (1962) | Germination and early gametophyte development | Uchida and Furuya (1997) |
| *Adiantum cuneatum* | Rhizome tips | Murashige (1974) | Direct adventitious and auxiliary shoot proliferation | Murashige (1974) |
| *Adiantum pedatum* | Shoot apex | Wetmore (1954) | Shoot growth one plant per explant | Wetmore (1954) |
| *Adiantum raddianum* | Rhizome segments | MS (1962) with variation in strength | Globular bodies from callus – Sexual sporophytes | Amaki and Higuch (1991) |
| *Adiantum tenerum* | Prothalli | 1/2 MS (1962) | Prothallus ex vitro homogenized gametophytic tissue – Sexual sporophytes | Knauss (1976) |
| *Adiantum trapeziforme* | Rhizome | B5 – Gamborg media with 2,4-D KIN and BAP | Callus subcultured on KIN, BAP | Padhya (1995) |
| *Alsophila australis* | Rhizome tip | Modified Murashige (1974) | Direct adventitious shoots | Harper (1976) |
| *Alsophila australis* | Rhizome tip | Murashige (1974) | Direct adventitious shoots | Murashige (1974) |
| *Ampelopteris prolifera* | Germinating spores and leaves | Steeves et al. (1955) + 2,4-D | Apogamous sporophytes from prothalli Aposporous gametophytes from leaves | Mehra and Sulklyan (1969) |

(continued)

**Table 9.1** (continued)

| Species name | Source of explant | Media | Results | References |
|---|---|---|---|---|
| *Anemia phyllitidis* | Spores | Knop's (1865); modified Miller and Miller (1961) | Prothalli with sex organs | Reynoids (1979) |
| *Anemia phyllitidis* | Spores | Moore's (1903) | Gametophytes grown and immobilized in polyurethane foam | Douglas and Sheffield (1990) |
| *Anemia phyllitidis* | Spores | Moore's (1903) | Gametophytes grown in airlift fermenter | Sheffield et al. (1997) |
| *Arthromeris tenuicauda* | Spores | Modified Dyer's | Notched prothalli development | Roy and Choudhury (1989) |
| *Asplenium nidus* | Rhizome segments | Modified MS (1962) | Colloid with green globular bodies proliferated plants | Amaki and Higuch (1991) |
| *Asplenium nidus* | Homogenized rhizome tissue | MS (1962) | Sporophyte and gametophytes | Fernandez et al. (1991a, b) |
| *Asplenium nidus-avis* | Section of frond | Modified MS (1962) | Direct sporophyte and gametophyte formation of saprophytic plants | Fernandez et al. (1991a, b) |
| *Asplenium ruta-muraria* | Germinated spores | Mohr (1956) | Formation of sex organs and presence of antheridiogen system | Schneller and Hess (1995) |
| *Blechnum brasiliense* (cv. *Crispum*) | Homogenized prothalli tissue | Modified Knudson's (1946) | Prothallus, sexual sporophytes | Janssens and Sepelie (1989) |
| *Blechnum laevigatum* | Spores | Steeves et al. (1955) | Gametophytes with sex organs | Duran and Anton (1996) |
| *Blechnum punctulatum* | Homogenized prothalli tissue | Modified Knudson's (1946) | Prothalli sexual sporophytes | Janssens and Sepelie (1989) |
| *Blechnum spicant* | Rhizome | MS (1962) with BAP BAP + NA A | Green globular bodies, sporophytes, apospory | Fernandez et al. (1996a,b) |
| *Blechnum spicant* | Spores | 1/2 MS, 1/4 MS, Knop's (1865), Knudson's (1946), Klekowski Jr (1969) | Prothalli with antheridia | Fernandez et al. (1996a,b) |

(continued)

**Table 9.1** (continued)

| Species name | Source of explant | Media | Results | References |
|---|---|---|---|---|
| *Botrychium dissectum* | Leaves of sporophyte | Modified Knudson's (1946) | Gametophytes by apospory | Whittier (1978) |
| *Botrychium* spp. | Gametophytes | Modified Knudson's (1946) | Prothalli with sex organs | Whittier (1972, 1976, 1981) |
| *Ceratopteris thalictroides* | Gametophytes from spores | Knop's (1865), MS (1962) with KIN, BAP, NAA, and 2,4-D | Gametophyte callus induction and regeneration of both gametophytes and sporophytes | Cheema and Sharma (1991) |
| *Ceratopteris thalictroides* | Pinnae, leaf callus | MS (1962) with 2,4-D, NAA, KIN, and BAP | Multiple shoots, rooted and sporophytes formed | Cheema and Sharma (1994) |
| *Ceratopteris* spp. | Leaf margin's meristems | Modified Knudson's (1946) | Sporophytes from callus regeneration | Gottlieb (Gottlieb 1972) |
| *Cheilanthes alabamensis* | Spores | Whittier (1964) | Gametophytes and then apogamous sporophytes | Whittier (1965) |
| *Cheilanthes tomentosa* | Spores | Whittier (1964) | Gametophytes and then apogamous sporophytes | Whittier (1965) |
| *Christella parasitica* | Spores | Modified Dyer's (1979) | Germination | Roy and Choudhury (1989) |
| *Cyathea dregei* | Spores | MS (1962) | Gametophyte tissue. Macerated for ex vitro sporophytes | Finni (1987) |
| *Cyathea gigantea* | Leaflet primordia apical meristem | 1/2 Knudson's (1946) Dilute White's (White 1954) | Fronds developed. Rooted plantlet | Padhya (1987) |
| *Cyclosorus dentatus* | 1. Sporophyte root apices.<br>2. Callus.<br>3. Gametophytes with sex organs. | Steeves et al. (1955) + 2,4-D | Suspension culture sporophytes | Mehra and Palta (1971) |
| *Cyrtomium falcatum* | Gametophytic tissue | 1/2 MS (1962) | Sporophytes regenerated from gametophytic homogenate | Knauss (1976) |

(continued)

**Table 9.1** (continued)

| Species name | Source of explant | Media | Results | References |
|---|---|---|---|---|
| *Davallia bullata* | Sporophytes from homogenated | LS (1965) | Multiple shoots from plated tissue | Cooke (1979) |
| *Dicranopteris linearis* | Green sporangia | 1/2 MS (1962) | Prothalli and sporophytes | Henson (1979) |
| *Diphasiastrum habereri* | Spores | Modified Moore's (1903) | Gametophytes | Whittier and Britton (1995) |
| *Drymoglossum piloselloides* | Strips from young fronds | MS (1962) | Direct aposporous gametophytes | Kwa et al. (1988) |
| *Drynaria quercifolia* | Spores rhizome | Knop's (1865) Knudson's (1946) | Gametophytes. Sporophytes formed directly on the rhizome as well as from the callus | Hedge and D'souza (1997) |
| *Drynaria quercifolia* | Rhizome | MS medium | Sporophyte production | Hegde and D'Souza (2000) |
| *Dryopteris affinis* sp. *affinis* | Gametophyte callus | MS (1962) with BAP and NAA | Sporophyte and gametophyte regeneration | Fernandz et al. (1996a, b) |
| *Dryopteris filix-mas* | Gametophytes | Modified Moore's (1903) | Haploid apogamous, sporophytes from callus | Breznovitis and Mohay (1987) |
| *Equisetum arvense* | Spores | Murashige and Skoog (1962) liquid medium | Gametophytes and globular cell mass | Kuriyama and Maeda (1999) |
| *Helminthostachys* spp. | Spores | Modified Moore's (1903) | Early gametophytes | Whittier (1987) |
| *Lycopodiella inundata* | Vegetative apices | CM-1 to CM-5 media | Development of nodular callus | Atmane et al. (2000) |
| *Lygodium japonicum* | Spores | Dyer's (1979) | Germination | Roy and Choudhury (1989) |
| *Lygodium japonicum* | Leaves from in vitro sporophytes | Maeda et al. (1990) | Protoplasts are isolated and cultured Gametophytes regenerated | Maeda et al. (1990) |
| *Marsilea quadrifolia* | Sporophyte shoot segments | Modified MS (1962) and Moore's (1903) | Direct sporophytes with rhizomes | Breznovits and Mohay (1987) |

(continued)

**Table 9.1** (continued)

| Species name | Source of explant | Media | Results | References |
|---|---|---|---|---|
| *Matteuccia struthiopteris* | Shoot apices | Modified Knudson's (1946) | Shoot multiplication | Hicks and Von Aderkas (1986) |
| *Matteuccia struthiopteris* | Lateral buds of rhizomes | Dykeman and Cumming (1985) | Shoot proliferation. Rooted shoots | Dykeman and Cumming (1985) |
| *Matteuccia struthiopteris* | Rhizome pieces + meristems | Modified Knudson's (1946) | Aposporous | Von Aderkas (1986) |
| *Matteuccia struthiopteris* | Spores | Modified Knudson's (1946) | Gametophytes, apogamy | Von Aderkas (1984) |
| *Matteuccia struthiopteris* | Young | Hirsch (1975) | Sporophytic and gametophytic regenerants from callus | Hirsch (1975) |
| *Microlepia strigosa* | Rhizome tip | Modified Murashige (1974) | Direct adventitious shoots | Harper (1976) |
| *Microlepia strigosa* | Rhizome tip | Modified Murashige (1974) | Direct adventitious shoots | Murashige (1974) |
| *Nephrolepis cordifolia* | Runner tips | MS with variation in salt strength | Callus producing green globular bodies | Higuchi et al. (1987) |
| *Nephrolepis cordifolia* | Leaves root runners | Steeves et al. (1955) | Sporophytic callus gametophyte regeneration apogamous roots and aposporous gametophytes | Sulklyan and Mehra (1977) |
| *Nephrolepis cordifolia* | Rhizome segments | MS (1962) with variation in salt strength | Green globular bodies from callus plantlets | Amaki and Higuch (1991) |
| *Nephrolepis exaltata* | Stolon segments (1–2 cm) | B5 (1968) modified | Numerous direct adventitious shoots | Harper (1976) |
| *Nephrolepis exaltata* | Runner tips | Modified Murashige (1974) | Direct adventitious shoots | Harper (1976) |
| *Nephrolepis exaltata* | Runner tips | Murashige (1974) | Direct adventitious shoots | Murashige (1974) |
| *Nephrolepis exaltata* | Sporophyte shoot segments | Modified MS (1962) | Direct sporophytes | Breznovitis and Mohay (1987) |

(continued)

**Table 9.1** (continued)

| Species name | Source of explant | Media | Results | References |
|---|---|---|---|---|
| *Nephrolepis exaltata* | Runner tips | Burr (1975) | Direct adventitious shoot | Burr (1975; 1976) |
| *Nephrolepis exaltata* | Runner tips | Modified B5 (1968) | Direct adventitious shoots | Padhya and Mehta (1982) |
| *Nephrolepis exaltata* | Runner tips | Modified B5 (1962) | Rooted shoots transferred to soils | Camloh et al. (1989) |
| *Nephrolepis exaltata* cv. bostoniensis | Runner tips | Modified MS (1962) | Direct adventitious shoot formation | Loescher and Albrecht (1979) |
| *Nephrolepis exaltata* cv. Bostoniensis | Runner tips | 1/2 MS (1962) | Shoot proliferation plants | Brogan and Naess (1987) |
| *Nephrolepis exaltata* cv. Bostoniensis | Rhizome, stolon tips, and pinnae | Murashige (1974) | Callus induction and regeneration of sporophyte plantlets | Bryne and Caponetti (1992a) |
| *Nephrolepis exaltata* cv. Bostoniensis | Runner tips | Ben-Jaacov and Dax (1981) | Multiple shoot formation | Leffring and Soede (1982) |
| *Nephrolepis falcata* | Runner tips (2 cm) | Modified Murashige (1974) | Shoot proliferation shoots rooted | Beck and Caponettii (1983) |
| *Nephrolepis hirsutula* | Spores | Knudson's (1946) | Increased rate of growth of gametophytes | Smith and Yee (1975) |
| *Nephrolepis multiflora* | Spores | Knudson's (1946) with kinetin and BAP | Sporophytes, apogamy | Sara et al. (1998) |
| *Nephrolepis* spp. | Rhizome tips | Miller and Murashige (1976) | Shoot proliferation shoots rooted | Peterson (1979) |
| *Nephrolepis* spp. | Stolon tips | 1/2 MS (1962) | Axillary and direct adventitious shoots | Henson (1979) |
| *Notholaena 'sun-stuff'* | Mature sori on leaflets | Rogers and Banister (1992) | Plantlet formation | Rogers and Banister (1992) |
| *Ophioglossum palmatum* | Spores | Modified Moore's (1903) | Germination and early development of gametophytes | Whittier and Moyroud (1993) |
| *Ophioglossum* spp. | Spores | Modified Knudson's (1946) | Germination of gametophytes | Whittier (1981) |

(continued)

**Table 9.1** (continued)

| Species name | Source of explant | Media | Results | References |
|---|---|---|---|---|
| *Osmunda cinnamomea* | Prothalli | Whittier and Steeves (1960) | Apogamous sporophytes from prothalli | Whittier and Moyroud (1993) |
| *Osmunda cinnamomea* | Leaves | Caponetti and Steeves (1970) | Morphogenesis | Caponetti (1972) |
| *Osmunda cinnamomea* | Spores | Morel and Wetmore (1951) | Callus growth | Morel and Wetmore (1951) |
| *Osmunda regalis* | Gametophytes from germinated spores | Modified Moore's (1963) | Direct apogamous sporophyte formation | Breznovitis and Mohay (1987) |
| *Osmunda regalis* | Gametophytes from germinated spores | Knop's (1865) Knudson's (1946) 1/4 MS (1962) | Gemmation | Fernandez et al. (1997a) |
| *Pellaea glabella* var. *occidentalis* | Gametophytes from germinated spores | Knudson's (1946) medium with variation in sucrose | Haploid apogamous sporophytes on gametophytes | Rigby (1973) |
| *Pellaea rotundifolia* | Prothalli from spores | Modified Knudson's (1946) | Prothallus tissue homogenized and sexual sporophytes induced | Janssens and Sepelie (1989) |
| *Pilularia globulifera* | Sporophyte shoot segments | Modified Moore's (1903) | Sporophytes with rhizomes induced directly | Breznovitis and Mohay (1987) |
| *Pityrogramma calomelanos* | Spores | Modified dyer's (1979) | Sexual sporophytes | Singh and Roy (1989) |
| *Platycerium bifurcatum* | Gametophytic tissue homogenate | 1/2 MS (1962) | Regeneration of sporophytes for homogenate | Knauss (1976) |
| *Platycerium bifurcatum* | Leaves from in vitro grown plants | MS modified by Hennen and Sheehan (1978) | Direct adventitious shoots, rooted | Camloh and Gogala (1991) |
| *Platycerium bifurcatum* | Gametophytes germinated from spores | MS modified by Hennen and Sheehan (1978) | Sporophytes established on the soil | Camloh and Gogala (1992) |
| *Platycerium bifurcatum* | Spore germination and early gametophyte development | MS modified by Hennen and Sheehan (1978) | Gametophytes | Camloh (1993) |

(continued)

**Table 9.1** (continued)

| Species name | Source of explant | Media | Results | References |
|---|---|---|---|---|
| *Platycerium bifurcatum* | Leaf | MS modified by Hennen and Sheehan (1978) | Sporophytes regenerated and established on the soil | Camloh et al. (1994) |
| *Platycerium bifurcatum* | Scales | MS (1962) modified | Gametophyte and sporophyte regeneration | Dolinsek and Camloh (1997) |
| *Platycerium bifurcatum* | Leaves of young sporophytes | MS modified by Hennen and Sheehan (1978) | Protoplasts are isolated and cultured. First divisions observed | Camloh and Zel (1995) |
| *Platycerium bifurcatum* | 1. Spores for germination.<br>2. Leaves of young sporophytes for protoplast culture. | Knop's (1865), with the addition of jasmonic acid | 1. Jasmonic acid promotes early gametophyte development.<br>2. It has a stimulatory effect on protoplast divisions. | Camloh et al. (1996) |
| *Platycerium bifurcatum* | Homogenized gametophytic tissue | LS (1965) | Sporophytes regenerated from planted tissue | Cooke (1979) |
| *Platycerium coronarium* | Rhizome and fronds | MS (1962) with NAA | Sporophyte generation | Kwa et al. (1995a) |
| *Platycerium coronarium* | Gametophytic callus | MS (1962) with KIN | Regeneration of gametophytes and sporophytes | Kwa et al. (1997) |
| *Platycerium stemaria* | Shoot tips | Modified MS by Hennen and Sheehan (1978) with IAA and adenine sulfate | Adventitious and axillary | Hennen and Sheehan (1978) |
| *Platycerium stemaria* | Shoot tips | LS (1965) | Sporophytes induced directly and through callus | Cooke (1979) |
| *Polypodium cambricum* | Rhizome, frond, petiole, or root tip | MS media (1962) | Green globular bodies (GGB) | Bertrand et al. (1999) |
| *Pronephrium triphyllum* | Spores | Knop's agar and liquid medium | Germination | Marimuthu and Manickam (2011) |
| *Pteridium aquilinum* | Spores | Moore's (1903) | The gametophyte is grown in an airlift fermenter | Sheffield et al. (1997) |

(continued)

**Table 9.1** (continued)

| Species name | Source of explant | Media | Results | References |
|---|---|---|---|---|
| *Pteridium aquilinum* | Spores | Modified Knudson's (1946) | Apogamous sporophytes | Whittier (1964) |
| *Pteridium aquilinum* | Gametophyte | Whittier and Pratt (1971) | Apogamous buds on gametophytes | Elmore and Whittier (1973) |
| *Pteridium aquilinum* | Spores | Moore's (1903) | Gametophytes used for streptomycin-resistant line isolation | Breznovitis and Sheffield (1990) |
| *Pteridium aquilinum* (*bracken*) | Spores | Moore's (1903) | Gametophytes grown and immobilized polyurethane foam | Douglas and Sheffield (1990) |
| *Pteridium argyraea* | Rhizome tips | Murashige (1974) | Direct adventitious shoots | Murashige (1974) |
| *Pteris cretica* | Leaf rachis | Bristow (1962) | Gametophytes and sporophytes from callus | Breznovitis and Mohay (1987) |
| *Pteris cretica* 'Albolineata' | Gametophytes | Modified Moore's (1903) | Haploid apogamous sporophytes from gametophytic callus | Breznovitis and Mohay (1987) |
| *Pteris ensiformis* | Gametophytes homogenized | 1/2 MS (1962) | Sporophytes regenerated | Knauss (1976) |
| *Pteris ensiformis* | Rhizome homogenized | MS (1962) | Sporophytes regenerated, apospory | Fernandez et al. (1996a, b) |
| *Pteris ensiformis* 'Victoriae' | Rhizome segments | MS (1962) | Callus produced rapidly growing globular bodies. Plantlets | Amaki and Higuch (1991) |
| *Pteris henryi* | Gametophytes from germinated spores | Modified Moore's (1903) | Haploid | Breznovitis and Mohay (1987) |
| *Pteris vittata* | Rhizome segments | Dilute White's (White 1954) | Adventitious shoots (diploid) from callus. Rooted | Padhya (1987) |
| *Pteris vittata* | Spores | Modified Miller and Miller (1961) | Gametophytes | Beri and Bir (1993) |

(continued)

**Table 9.1** (continued)

| Species name | Source of explant | Media | Results | References |
|---|---|---|---|---|
| *Pteris vittata* | Rhizome pieces | Kshirsagar and Mehta (1978) | Callus formed a suspension culture. Sporophytic plantlets from callus | Kshirsagar and Mehta (1978) |
| *Pteris vittata* | Rhizome segments with apex | Dilute White's (White 1954) | Multiple shots. Rooted plantlets | Paadhya (1987) |
| *Rumohra (Arachniodes) adiantiformis* 'Florida' | Rhizome segments | MS (1962) | Callus produced green globular bodies Plantlets | Amaki and Higuch (1991) |
| *Salvinia* | Shoot apices | Hunter and Vergnano (1953) | Plantlets | Seilheimer (1974) |
| *Selaginella willdenowii* (peacock fern) | Shoot apices | Wetmore (1954) | Shoot growth | Wetmore (1954) |
| *Thelypteris dentata* | Spores | Knop's (1865) | Germination | Seilheimer (1978) |
| *Thelypteris ovata* | Prothalli from spores | Klekowski Jr (1969) | Development of sex organs and presence of antheridiogen system | Nester-Hudson et al. (1997) |
| *Todea barbara* (king fern) | Embryos from prothalli | 10% coconut milk | Embryos growth, young sporophytes | DeMaggio and Wetmore (1961) |
| *Trichipteris corcovadensis* | Spores | Dyer (1979) | Germination | Esteves and Gil (1985) |
| *Trichomanes speciosum* | Gemmae | Moore's (1903) and Hoagland and Arnon's (1950) | Germination of secondary gemmae on gametophytic tissue | Raine and Sheffield et al. (1997) |
| *Vittaria* spp. | Gametophytes | Knudson's (1946) | Sporophytes from callus | Caponetti et al. (1982) |
| *Woodwardia fimbriata* | Rhizome tip | Modified Murashige (1974) | Direct adventitious shoots | Harper (1976) |

**Table 9.2** Recent advances in the propagation of pteridophytes in vitro after 2000

| Species name | Source of explant | Media | Results | References |
|---|---|---|---|---|
| *Anemia rotundifolia* | Apical Explants of gametophyte | Parker and Thompson media | Gametophytes and sporophytes | Singh et al. (2012) |
| *Arachniodes aristata* | Spore | Knop's (1865) and MS (1962) media | Germination and prothalli growth | Cho et al. (2017) |
| *Asplenium nidus* | Spores | 1/4-strength MS medium (1962) | Shoots and roots | Khan et al. (2008) |
| *Asplenium nidus* | Spores | MS medium | Initiation and proliferation of prothalli | Haddad and Bayerly (2014) |
| *Athyrium sheareri* | Spore | Knop's (1865), MS medium (1962) + activated charcoal | Formation and growth of sporophytes | Jang et al. (2019) |
| *Athyrium yokoscense* | Spores | Modified MS medium (1962) | Cadmium-tolerant sporophytes | Yosihara et al. (2005) |
| *Blechnum spicant* L | Spores | MS supplemented with 2% sucrose (w/v) and 0.1 and 1 mg/l $GA_{4+7}$ | $GA_{4+7}$ at a dose of 1 mg/l favored a significant increase in the percentage of female gametophytes | Menéndez et al. (2006) |
| *Ceratopteris pteroides* | Spores | Knop's, Knudson's, Moore's, and MS medium | Multiple bud formation and plant regeneration | Cheema (2005) |
| *Cibotium barometz* | Spores | 1/4 MS medium + activated charcoal | Sporophyte multiplication | Xue et al. (2010) |
| *Cheilanthes viridis* (Forrsk.) Swartz | Spores | Knudson C and Knop's media | A higher percentage of germination | Manickam et al. (2003) |
| *Coniogramme japonica* (Thunb.) Diels | Prothallus | Murashige and Skoog (1962) Knop's medium | Growth of prothallus | Park et al. (2019) |
| *Cyathea atrovirens* | *Spore* | Meyer et al. (1955) | Gametophyte development | Vargas and Droste (2014) |
| *Cyathea spinulosa* wall. Ex hook. | *Spores* | Parker and Thompson nutrient medium | Large-scale propagation was possible through spores | Shukla and Khare (2005a, b) |
| *Cyathea spinulosa* | Leaf primordium in vitro raised sporophyte | Parker and Thompson basal media + BAP + NAA | Mass multiplication | Shukla and Khare (2014) |

(continued)

**Table 9.2** (continued)

| Species name | Source of explant | Media | Results | References |
|---|---|---|---|---|
| *Cyathea spinulosa* | Leaf primordium | Parker and Thompson media | Callus, multiple shoots, rooting | Shukla and Khare (2012a, b) |
| *Cyclosorus dentatus* link. | Leaf primordia | Parker and Thompson basal media + BAP + IAA | Mass multiplication | Shukla and Khare (2014) |
| *Dicksonia sellowiana* | Spores | Dyer and MS medium supplemented with 0 to 5% sucrose | The germination was lower with the addition of sucrose | Renner and Randi (2004) |
| *Diplazium cognatum* (Hieron.) | Crozier explants | 1/2 MS agar medium supplemented with 3% sucrose | Semi-friable callus developed | Manickam et al. (2003) |
| *Diplazium esculentum* | Crosiers | 1/2-strength Murashige and Skoog (MS) medium | Good morphogenic response | Nair et al. (2013) |
| *Diplazium esculentum* | Spores | Knop's basal agar medium with BAP | Multiplication of prothallus | Nair et al. (2014) |
| *Diplazium proliferum* | Spores | Murashige and Skoog (1962) media | Germination of spores | Golamaully et al. (2015) |
| *Dipteris wallichii* | Spore | Murashige and Skoog (1962) with IAA + KIN | Differentiation of prothallus from spores | Bharati et al. (2013) |
| *Drynaria fortune* | Spores | 1/2-strength MS medium (1962) | Antheridia and archegonia in the gametophytes | Chang et al. (2007) |
| *Drynaria quercifolia* (L.) | Spores | MS-Z4 medium Parker and Thompson | Germination and proliferation | Mazumder et al. (2011) |
| *Drynaria quercifolia* | Leafy structures and rhizome tips | Knop's medium | Leafy sporophyte | Hegde et al. (2006) |
| *Drynaria roosii* | Spores | Knop's medium and 1/2 MS medium | MS basal medium favored spore germination but inhibited gametophyte development | Yin-Li et al. (2009) |
| *Dryopteris affinis* | Green sporangia (sori) | 1/2 MS (1962) | Embryo | Soare et al. (2010) |

(continued)

**Table 9.2** (continued)

| Species name | Source of explant | Media | Results | References |
|---|---|---|---|---|
| *Dryopteris nipponensis* Koidz | Spore and prothallus | Knop's and Murashige and Skoog (MS) media | Spore germination and prothallus development | Jang et al. (2017) |
| Elaphoglossum nilgiricum | Spores | KC basal solid medium | Spore germination | Shibila and Johnson (2016) |
| *Helminthostachys zeylanica* (L.) hook | Spores | MS medium, Parker-Thompson basic C medium | Spore germination | Mazumder et al. (2010) |
| *Lemmaphyllum microphyllum* C. Presl | Spore | Murashige and Skoog (MS) basal medium +1/2, 1/4, or 1/8 Knop's medium, activated charcoal | Increase in fresh weight of the gametophyte | Jang et al. (2020a, b) |
| *Marsilea minuta* | Spores | Knop's, Knudson's, Moore's, MS medium | Multiple bud formation and plant regeneration with Knop's medium | Cheema (2005) |
| *Matteuccia struthiopteris* (L.) Tod. | Creeping rhizomes | Murashige Fern multiplication medium | Regeneration of micro-plantlets | Zenkteler (2006) |
| *Microlepia strigosa* | Spore | Knop's and Murashige and Skoog (MS) media | Spore germination and prothallus development; antheridium formation | Cho et al. (2017) |
| *Microsorum punctatum* (L.) | Leaf primordia | Parker and Thompson basal media | In vitro studies for mass multiplication | Shukla and Khare (2014) |
| *N. exaltata* cv. Bostoniensis (L.) | Leaf primordia (0.5–1 cm) | Murashige and Skoog (1962) 2,4-D + BAP | In vitro studies for mass multiplication | Shukla and Khare (2014) |
| *Nephrolepis biserrata* | Stolon explants | Parker and Thompson basal media 2,4-D + NAA | In vitro studies for mass multiplication | Shukla and Khare (2014) |
| *Nephrolepis cordifolia* cv. 'Duffii' | Leaf primordia | Parker and Thompson basal media 2,4-D + NAA | In vitro studies for mass multiplication | Shukla and Khare (2014) |
| *Osmunda japonica* | Spores | MS medium | Sporophyte production | Yi-Fa (2003) |
| *Osmunda regalis* | Spores | Murashige and Skoog (MS) basal medium | Sporophytes produced | Makowski et al. (2016) |

(continued)

**Table 9.2** (continued)

| Species name | Source of explant | Media | Results | References |
|---|---|---|---|---|
| *Osmunda regalis* L | Spores | Knop's + 1/2 or 1/4 MS (1962) | Gametophyte proliferation Production of sporophytes | Makowski et al. (2016) |
| *Pityrogramma calomelanos* | Crozier and frond explants | 1/4-strength MS medium | Apospory and apogamy and sporophyte | Martin et al. (2006) |
| *Pityrogramma calomelanos* (L.) | Leaf primordium of in vitro raised sporophyte | Parker and Thompson basal media | In vitro studies for mass multiplication | Shukla and Khare (2014) |
| *Platycerium bifurcatum* | Leaf | Modified MS medium | Aposporous gametophytes. | Ambrozic Dolinsek et al. (2002) |
| *Platycerium bifurcatum* | Juvenile leaf explants | MS medium (1962) | Multiple sporophytes, GGB regeneration | Liao and Wu (2011) |
| *Polypodium aureum* | Spore | MS, SH, White, and B5 | The medium White presented the highest average for the germination speed index, whereas the agar media, MS, SH, and B5, had the same germination speeds | Santos Alves et al. (2019) |
| *Pronephrium triphyllum* (Sw.) | Spores | Knop's basal agar medium | The highest percentage of germination | Marimuthu and Manickam (2011) |
| *Pteridium aquilinum* | Spores | MS medium (1962) | Spore germination | Zhang et al. (2001) |
| *Pteris multifida* | Spores | MS (1962) medium | Prothallus | Jianmin (2004) |
| *Pteris Tripartita* Sw. | Spores | 1/2-strength MS medium | Spore-derived gametophytes were formed | Ravi et al. (2015a, b) |
| *Pteris vittata* | Spores | 1/2-strength MS | Sporophyte formation | Aimee (2000) |
| *Pteris vittata* | Leaf primordium segments | MS medium (1962) | Callus induction | Shukla and Khare (2005a, b) |

(continued)

**Table 9.2** (continued)

| Species name | Source of explant | Media | Results | References |
|---|---|---|---|---|
| *Pteris vittata* | Spores | Diluted MS (Murashige and Skoog 1962) medium | Germination of gametophyte and sporophyte | Zheng et al. (2008) |
| *Pteris vittata* | Leaf primordium | MS media (1962) and Parker and Thompson media | Callus Shoot differentiation | Shukla and Khare (2012a, b) |
| *Pteris vittata* (L.) | Leaf primordia | MS (1962) IAA+ BAP | Mass multiplication | Shukla and Khare (2014) |
| *Ramonda serbica* and *Ramonda nathaliae* | Spores | Jungnickel/ Gliemeroth basal medium with BAP and IAA | The highest number of shoots and multiplication rate | Gashi et al. (2015) |
| *Rumohra adiantiformis* | Rhizomes | Murashige and Skoog (MS) medium | Regeneration of rhizome | Winarto and Teixeira (2012) |
| *Marsilea quadrifolia* L. | Sporocarps | 1/2-strength murashige and Skoog (MS) | After 6 weeks, sporophytes were formed | Rolli et al. (2015) |
| *Selaginella martensii* | Shoot tip | Murashige and Skoog (1962) | Node formed two new shoot tips | Park et al. (2020) |
| *Selaginella pulvinata* (hook. & Grev.) maxim | Frond tips | 1/2-strength MS medium supplemented with BAP | Shoot induction, adventitious shoot proliferation, and plantlet growth | Yu et al. (2021) |
| *Sphaerostephanos unitus* | Spores | Knop's agar and liquid medium | Germination | Marimuthu and Manickam (2011) |

## References

Agretious TK, Martin KP, Hariharan M (1996) *In vitro* clonal multiplication of *Alpinia calcarata* Rosc. Phytomorphology 46:133–138

Aimee JA (2000) Micropropagation of ferns: *Pteris vittata* spores and *Nephrolepis biserrata* runners. An undergraduate thesis submitted to the Institute of Biology, College of Science, University of the Philippines in partial fulfillment for the B.S. Biology degree

Aldea F, Banciu C, Brezeanu A et al. (2017) *In vitro* micropropagation of fern species (Pteridophyta) of biotechnological interest, for *ex situ* conservation. Oltenia J Stud Nat Sci32 (2):27–35

Amaki W, Higuch HA (1991) Possible propagation system of Nephrolepis, Asplenium, Pteris, Adiantum and Rumohra through tissue culture. Acta Hortic 300:237–243

Ambrósio ST, Melo NF (2004) Interaction between sucrose and pH during *in vitro* culture of *Nephrolepis biserrata* (Sw.) Schott (Pteridophyta). Acta Botanica Brasilica 18(4):809–813

Ambrozic Dolinsek J, Camloh M, Bohanec B, Zel J (2002) Apospory in leaf culture of staghorn fern (*Platycerium bifurcatum*). Plant Cell Rep 20:791–796

Andrew RGP, Ester HR et al (2015) Protocol: genetic transformation of the fern *Ceratopteris richardii* through microparticle bombardment. Plant Methods 11:37

Anitha KC, Rajeshwari YB, Prasanna SB et al (2016) Nutritive evaluation of azolla as livestock feed. J Exp Bio and Agri Sci 4(6):670–674

Atmane N, Blervacq AS, Michaux-Ferriere N et al (2000) Histological analysis of indirect somatic embryogenesis in the marsh clubmoss *Lycopodiella inundata* (L.) Holub (Pteridophytes). Plant Sci 156:159–167

Ballesteros D, Pence VC (2018) Fern conservation: spore, gametophyte, and sporophyte *ex situ* storage, *in vitro* culture, and cryopreservation. In: Fernández H (ed) Current advances in fern research. Springer, Cham, pp 227–249

Beck MJ, Caponetti JD (1983) The effects of kinetin and naphthaleneacetic acid in vitro shoot multiplication and rooting in the fishtail fern. Am J Bot 70:1–7

Ben-Jaacov J, Dax E (1981) In vitro propagation of *Grevillea rosmarinifolia*. HortScience 16:309–310

Beri A, Bir SS (1993) Germination of stored spores of *Pteris vittata* (L.). Am Fern J 83(3):73–78

Bertrand AM, Abuerne MA, Fernández H et al. R (1999) *In vitro* organogenesis of *Polypodium cambricum*. Plant Cell Tiss Org 57: 65–69

Bharati SK, Manabendra D, Behari MP (2013) "*In vitro* propagation of *Dipteris wallichii* (R. BR.) T. Moore: a hope for conservation of an endangered pteridophyte. Int Res J Pharm 4: 215–219

Borgan AK, Naess SK (1987) Homogeneity and plant quality of *in vitro* propagated *Nephrolepis exaltata*. Acta Hortic 212:433–438

Breznovitis A, Mohay J (1987) *In vitro* problems related to propagation of different fern species. Acta Hortic 212: 427–431

Breznovitis A, Sheffield E (1990) Isolation and investigation of streptomycin resistant lines of the fern *Pteridium aquilinum*. In Vitro Cell Dev Biol 26:301–304

Bristow JM (1962) The controlled *in vitro* differentiation of callus derived from a fern *Pteris cretica* L. into gametophytic and sporophytic tissues. Dev Biol 4:361

Bryne TE, Caponetti JD (1992a) Morphogenesis in three cultivars of Boston fern. II. Callus production from stolon tips and plantlet differentiation from callus. Am Fern J 82:1–11

Bryne TE, Caponetti JD (1992b) Morphogenesis in three cultivars of Boston fern. III. Callus production and plantlet differentiation from cell suspensions. Am Fern J 82:12–22

Bui LT, Cordle AR, Irish EE et al (2015) Transient and stable transformation of *Ceratopteris richardii* gametophytes. BMC Res Notes 8:214–223

Bui LT, Long H, Irish EE et al (2018) The power of gametophyte transformation. In: Fernández H (ed) Current advances in Fern research. Springer, Cham

Burr RW (1975) Mass propagation of the Boston fern through tissue culture. Proc Int Plant Prop Soc 25:122

Burr RW (1976) Mass propagation of ferns through tissue culture. In Vitro 12:309

Camloh M (1993) Spore germination and early gametophyte development of *Platycerium bifurcatum*. Am Fern J 83(3):79–85

Camloh M (2006) *In vitro* culture of the fern *Platycerium bifurcatum* as a tool for developmental and physiological studies. In: da Silva JAT (ed) Floriculture, ornamental and plant biotechnology, Advances and topical issues, vol 1. Global Science Book, Kagawa ken, pp 163–170

Camloh M, Gogala N (1991) *Platycerium bifurcatum* – adventitious buds and root formation without growth regulators *in vitro*. Acta Hortic 289: 89–90

Camloh M, Gogala N (1992) *In vitro* culture of *Platycerium bifurcatum* gametophytes. Sci Hortic 51: 343–346

Camloh M, Zel J (1995) Isolation and first divisions of *Platycerium bifurcatum* protoplasts. Acta Pharma 45:301–304

Camloh M, Gogala N, Ruzic R (1989) The micropropagation of *Nephrolepis exaltata*. Biol Vestn 37(3):23–32

Camloh M, Gogala N, Janko R (1994) Plant regeneration from leaf explants of the fern *Platycerium bifurcatum in* vitro. Sci Hortic 56:257–266

Camloh M, Ravishankar M, Zel J (1996) Jasmonic acid promotes divisions of fern protoplasts, elongation of rhizoids and early development of gametophytes. Physiol Plant 97:659–664

Caponetti J (1972) Morphogenetic studies on excised leaves of *Osmunda cinnamomea*: developmental capabilities of excised leaf primordial apices in sterile culture. Bot Gazette. 133(3): 31–335

Caponetti JD, Steeves (1970) Morphogenetic studies on excised leaves of *Osmunda cinnamomea*: histological studies of leaf development in sterile nutrient culture. Can J Bot 48(6):1005–1016. https://doi.org/10.1139/b70-143

Caponetti JD, Whitten M, Beck MJ (1982) Axenic culture and induction of callus and sporophytes of the *Appalachian vittaria* gametophyte. Am Fern J 72(2):36–39

Chang HC, Agrawal DC, Kuo CL et al.., (2007) *In vitro* culture of *Drynaria fortune* - a fern species source of Chinese medicine "Gu-Sui-Bu". In Vitro Cell Dev Biol Plant 43: 133–139

Cheema HK (2005) Multiple bud formation and plant regeneration in aquatic fern *in vitro*. Plant Biotech 89:290–296

Cheema HK, Sharma MB (1991) *In vitro* induction and differentiation of gametophytic callus in *Ceratopteris thalictroides*. Indian Fern J 8:74–77

Cheema HK, Sharma MB (1994) Induction of multiple shoots from adventitious buds and leaf callus in *Ceratopteris thalictroides*. Indian Fern J 8:74–77

Cho JS, Han JH, Lee CH (2017) Effects of medium components and composition on mass propagation *Arachniodes aristata* (G. frost) Tindale. Hortic Sci Technol 35:131–114

Cooke RC (1979) Homogenization as an aid in tissue culture propagation of Platycerium and Davallia. HortScience 14(1):21–22

Debergh P (1994) *In vitro* culture of ornamentals. In: Thorpe AT (ed) Vasil IK. Plant Cell, Tissue and Organ Culture, Dordrecht, pp 561–573

DeMaggio AE, Wetmore RH (1961) Morphogenetic studies on the fern *Todea Barbara* experimental embryology. Am J Bot 48:551–565

Dixit RD (2000) Conspectus of Pteridophytic diversity in India. Indian Fern J 17:77–91

Dolinsek JA, Camloh J (1997) Gametophytic and sporophytic regeneration from bud scales of the fern Platycerium (Cav.) C Chr *In vitro*. Ann Bot 80:23–28

Douglas GE, Sheffield E (1990) A new technique for the culture of fern gametophytes. Plant Cell Rep 8:632–634

Douglas GE, Sheffield E (1992) The investigation of existing and novel artificial growth systems for the production of fern gametophytes. In: Fern horticulture: past, present and future perspectives. Edinburgh University Press, Edinburgh, pp 183–187

Dudani SN, Mahesh MK, Mukri V et al (2013) An appraisal and conservation strategies for the pteridophytes of uttara kannada. CES Tech Rep 129

Duran M, Anton AM (1996) Erect gametophytes in *Blechnum laevigatum* in axenic cultures. Phytomorphology 46(1):59–64

Dyer AF (1979) The culture of fern gametophytes for experimental investigation. In: Dyer AF (ed) The experimental biology of ferns. Academic Press, London, pp 253–305

Dykeman BW, Cumming BG (1985) *In vitro* propagation of the ostrich fern (*Matteuccia struthiopteris*). Can J Plant Sci 65(4): 1025–1032

Elmore HW, Whittier DP (1973) The role of ethylene in the induction of apogamous buds in Pteridium gametophytes. Planta (Berl) 111:85–90

Esteves LM, Gil MF (1985) Germination and morphology of spores of *Trichipteris corcovadensis*. Am Fern J 75:92–102

Fabre J, Dereuddre J (1990) Encapsulation-dehydration: a new approach to cryopreservation of solanum shoot tips. Cryo Letters 11:413–426

Fay MF (1994) In what situations is *in vitro* culture appropriate to plant conservations. Biodivers Conserv 3:176–183

Fernández GE, Bertrand A, Sanchez TR (1991a) Micropropagation of *Asplenium nidus-avis* (L.). Acta Hortic 289:113–114
Fernández H, Bertrand AM, Sanchez TR (1991b) Micropropagation and phase change in *Blechnum spicant* and *Pteris ensiformis*. Plant Cell Tissue Cult 44:261–265
Fernández H, Bertrand AM, Sánchez-Tamés R (1996a) Influence of tissue culture conditions on apogamy in *Dryopteris affinis sp. affinis*. Plant Cell Tissue Organ Cult 45:93–97
Fernández H, Bertrand AM, Sachez TR (1996b) Micropropagation and phase change in *Blechnum spicant* and *Pieris ensiformis*. Plant Cell Tissue Organ Cult 44:261–265
Fernández H, Bertrand AM, Sanchez TR (1997a) Gemmation in cultured gametophytes of *Osmunda regalis*. Plant Cell Rep 16:358–362
Fernández H, Bertrand AM, Sánchez TR (1997b) Plantlet regeneration in *Asplenium nidus* (L.) and *Pteris ensiformis* (L.) by homogenisation of BA treated rhizomes. Sci Hortic 68:243–247
Finni JF, Van staden J (1987) Multiplication of the tree fern *Cyathea dregei*. HortScience 22: 665
Ford MV, Fay MF (1990) Growth of ferns from spores in axenic culture. In: Pollard JW, Walker JM (eds) Methods in molecular biology, Plant cell and tissue culture, vol 6. Humana Press, Totowa, NJ, pp 171–180
Gamborg OLR, Miller A, Ojima K (1968) Nutrient requirements of suspension cultures of soybean root cells. Exp Cell Res 50:148–151
Gashi B, Abdullai K, Sota V et al (2015) Micropropagation and *in vitro* conservation of the rare and threatened plants *Ramonda serbica* and *Ramonda nathaliae*. Physiol Mol Biol Plants 21:123–136
George EF (Ed.) (1993) Micropropagation in practice. Ferns and club mosses. In: Plant propagation by tissue culture. Part I. The Technology Exegetic Ltd., Edington; pp. 909–916
Golamaully ZM, Vishwakalyan B, Nadeem Z et al. (2015) *In vitro* culture as an aid to conservation of indigenous ferns: *Diplazium proliferum*. Int J Plant Biol 6(1): 33–36
Goller K, Rybczyński JJ (2007) Gametophyte and sporophyte of tree ferns *in vitro* culture. Acta Soc Bot Pol 76:193–199
Gottlieb JE (1972) Control of marginal leaf meristem growth in the aquatic fern Ceratopteris. Bot Gaz 133:299–304
Haberlandt G (1902) Kulturversuche mit isolierten Pflanzenzellen. Sitzungsber Akad Wiss Wien Math-Naturwiss Kl Abt J 111:69–92
Haddad S and Bayerly R (2014) *In vitro* propagation of ferns (*Asplenium nidus*) via spores culture. Jordan J Agric Sci 10(1):145–152
Harper KL (1976) Asexual multiplication of leptosporangiate ferns through tissue culture. M.S. thesis University of California-Riverside
Hedge S, D'souza L (1997) *In vitro* studies of *Drynaria quercifolia*– An endangered fern. In: Ravishankar GA, Venkataraman LV (eds) Recent advances in Biotechnological applications of plant tissue culture and cell culture. Oxford & IBH Publishing Co Pvt. Ltd., New Delhi, pp 113–115
Hegde S (1998) PhD thesis, Mangalore University, Mangalore
Hegde S, D'Souza L (2000) *In vitro* sporophyte production from rhizome of *Drynaria quercifolia*. Indian Fern J 17:72–76
Hegde S, Menon VP, Noronha R et al (2006) Callus culture and an unconventional pattern of sporophyte regeneration in *Drynaria quercifolia*– a medicinal fern. In Vitro Cell Dev Biol Plant 42:508–513
Hennen GR, Sheehan TJ (1978) *In vitro* propagation of *Platycerium stemaria*. HortScience 13(3): 245
Henson N (1979) Aseptic culture of ferns at Kew. Plantsman 1(2):74–80
Hicks G, Von Aderkas P (1986) A tissue culture of the ostrich fern *Matteuccia struthiopteris* (L). Todaro. Plant Cell Tissue Organ Cult 5:199–204
Hiendlmeyer R, Randi AM (2007) Response of spores and young gametophytes of *Cyathea delgadii* Sternb. (Cyatheaceae) and *Blechnum brasiliense* Desv. (Blechnaceae) to different light levels. Acta Bot Bras 21. https://doi.org/10.1590/S0102-33062007000400015

Higuchi H, Amaki W (1989) Effects of 6-benzylaminopurine on the organogenesis of *Asplenium nidus* (L.) through *in vitro* propagation. Sci Hortic 37:351–359

Higuchi H, Amaki W, Suzuki S (1987) *In vitro* propagation of *Nephrolepis cordifolia* Prsel. Sci Hortic 32: 105–113

Hirsch AM (1975) The effect of sucrose on the differentiation of excised fern leaf tissue into either gametophytes or sporophytes. Plant Physiol 56:390–393

Hoagland DR, Arnon DI (1950) The water culture method of growing plants without soil. Calif Agric Exp Station Circ 347

Hoshizaki BJ, Moran R (2001) Fern grower's manual, 2nd edn. Timber Press, Portland, OR

Huang YM, Chou HM, Chiou WL (2004) Density affects gametophyte growth and sexual expression of *Osmunda cinnamomea* (Osmundaceae: Pteridophyta). Ann Bot 94:229–232

Hunter JG, Vergnano O (1953) Trace element toxicities in oat plant. Ann Appl Biol 40:761–777

Indriolo E, Na G, Ellis D et al (2010) A vacuolar arsenite transporter necessary for arsenic tolerance in the arsenic hyperaccumulating fern *Pteris vittata* is missing in flowering plants. Plant Cell 22: 2045–2057

Jang BK, Cho JS, Lee KC et al (2017) Culture conditions affecting spore germination, prothallus propagation and sporophyte formation of *Dryopteris nipponensis* Koidz. Hortic Sci Techol 35: 480–489

Jang BK, Cho JS, Park K (2019) A methodology for large scale *Athyrium sheareri* gametophyte proliferation and sporophyte production using tissue culture. In Vitro Cell Dev Biol Plant 55: 519–526

Jang BK, Cho JS, Park K (2020a) Practical methodology for gametophyte proliferation and sporophyte production in green penny fern (*Lemmaphyllum microphyllum* C. Presl) using mechanical fragmentation. In Vitro Cell Dev Biol Plant 56:318–324

Jang BK, Cho JS, Lee CH (2020b) Synthetic seed technology development and production studies for storage, transport, and industrialization of bracken spores. Plan Theory 9(9):1079

Janssens J, Sepelie M (1989) *In vitro* multiplication of *Blechnum* spp. and *Pellaea rotundifolia* (Forst.) hook by homogenization. Sci Hortic 38: 161–164

Jarvis SJ, Wilkins MB (1973) Photoresponses of *Matteuccia struthiopteris* (L.) Todaro. I. Germination. J Exp Bot 24:1149–1157

Jianmin Y (2004) Culture and observation of the gametophyte of *Pteris multifida* Poir. J Hubei Univ (Natural Science Edition)

Kapai VY, Kapoor P, Rao IU (2010) *In vitro* propagation for conservation of rare and threatened plants of India–a review. Int J Biol Technol 1: 1–14

Kaur S (1991) Perspectives in pteridology: present and future. In: Bhardwaj TN, Gena, CB (eds.). Today and Tomorrow's Printers and Publishers, New Delhi, India. Aspects of Plant Sciences. 13: 83–89

Kawai-Toyooka H, Kuramoto C, Orui K et al (2004) DNA interference: a simple and efficient gene-silencing system for high-throughput functional analysis in the fern Adiantum. Plant Cell Physiol 45:1648–1657

Khan S, Raziq M, Kayani HA (2008) *In vitro* propagation of bird's nest (*Asplenium nidus*) from spores. Pak J Bot 40: 91–97

Klekowski EJ Jr (1969) Reproductive biology of the Pteridophyta, II. Theoretical considerations. Bot J Linn Soc London 62:347–359

Klink VP, Wolniak SM (2001) Centrin is necessary for the formation of the motile apparatus in spermatids of Marsilea. Mol Biol Cell 12:761–776

Klopper R (2011) The use of ferns in phytoremediation. Pteridoforum 96:1–5

Knauss JF (1976) A partial tissue culture method of pathogen-free propagation of selected ferns from spores. Proc Fla State Hortic Soc 89:363–365

Knop W (1865) Quantitative Untersuchungen über die Ernahrungsprozesse der Pflanzen. Landwirtsch Vers Stn 7:93–107

Knops N (1885) Quantitative Untersuchungan uber die Ernahrungsprozesse der pflangen L and wirtsch Vers, stn. 7:93–107

Knudson L (1946) A nutrient solution for the germination of orchid seed. Bullet Amer Orchid Soc 15:214–217

Kshirsagar MK, Mehta AR (1978) *In vitro* studies in ferns: growth and differentiation in rhizome callus of *Pteris vittata*. Phyomorphy 28: 50–58

Kuriyama A, Maeda M (1999) Direct production of sporophytic plants from spores of *Equisetum arvense*. Plant Cell Tissue Organ 58:77–79

Kwa SH, Wee YC, Loh CS (1988) Production of aposporous gametophytes from *Drymoglossum piloselloides* (L.) Presl. Frond strips. Plant Cell Rep 7:142–143

Kwa SH, Wee YC, Kumar PP (1995a) Role of ethylene in the production of sporophytes from *Platycerium coronarium* (Koenig) Desv. Frond and rhizome pieces cultured *in vitro*. J Plant Growth Regul 14:183–189

Kwa SH, Wee YC, Lim TM et al (1995b) Establishment and physiological analysis of photoautotrophic callus cultures of the fern *Platycerium coronarium* (Koenig) Desv. Under $CO_2$ enrichment. J Exp Bot 46(291):1535–1542

Kwa SH, Wee YC, Lim TM et al (1995c) Ammonium and nitrate uptake and nitrate reductase activity of photoautotrophic callus cultures of the fern *Platycerium coronarium* (Koenig). Desv. In vitro Cell Dev Biol 31:211–214

Kwa SH, Chin WY, Meng LT, Kumar P (1997) Morphogenetic plasticity of callus reinitiated from cell suspension cultures of the fern *Platycerium coronarium*. Plant Cell Tissue Organ Cult 48(1):37–44

Kyte L (1987) Plants from test tubes: an introduction to micropropagation. Timber Press, Portland, OR, p 160

Leffring L, Soede AC (1982) Influence of salt concentration on the homogeneity of Nephrolepis plants from tissue culture. In: Fujiwara A (ed.) Plant tissue culture 1982. Proc. 5th Int. Congr. Plant tissue and cell culture, Maruzen, Tokyo, pp. 499–500

Liao Y, Wu Y (2011) In vitro propagation of *Platycerium bifurcatum* (Cav.) C. Chr. Via green globular body initiation. Bot Stud 52:455–463

Lloyd RM (1980) Reproductive biology and gametophyte morphology of new world populations of *Acrostichum aureum*. Am Fern J 70(3):99–110

Lloyd RM, Buckley DP (1986) Effect of salinity on gametophyte growth of *Acrostichum aureum* and *A. danaeifolium*. Fern Gaz 13(2):97–102

Loescher WH, Albrecht CN (1979) Development in vitro of *Nephrolepis exaltata cv. Bostoniensis* runner tissues. Physiol Plant 47:250–254

Lucas WJ, Groover A, Lichtenberger R et al (2013) The plant vascular system: evolution, development and functions f. J Integr Plant Biol 55(4):294–388

Maeda M, Sugimoto Y, Nakamura M et al (1990) Division and gametophytic tissue formation from protoplasts of young sporophytes in fern *Lygodium japonicum*. Plant Cell Rep 2:179–182

Makowski D, Tomiczak K, Rybczyński JJ et al (2016) Integration of tissue culture and cryopreservation methods for propagation and conservation of the fern *Osmunda regalis* L. Acta Physiol Plant 38:19

Manickam VS, Vallinayagam S, Johnson M et al (2003) Micropropagation and conservation of rare and endangered ferns of the southern Western Ghats through *in vitro* culture. In: Chandra S, Srivastava M (eds) Pteridology in the new millennium. Springer, Dordrecht. https://doi.org/10.1007/978-94-017-2811-934

Marimuthu J, Manickam VS (2011) *Ex situ* conservation of two threatened ferns of the Western Ghats through *in vitro* spore culture. J Threat Taxa 3:1919–1928

Martin KP, Sini S, Zhang CL et al (2006) Efficient induction of apospory and apogamy *in vitro* in silver fern (*Pityrogramma calomelanos* L.). Plant Cell Rep 25:1300–1307

Mazumder PB, Sharma GD, Choudhury MD (2010) *In vitro* propagation of *Helminthostachys zeylanica* (L.) hook-a rare medicinal fern. J Sci Tech Biol Environ Sci 5: 129–133

Mazumder PB, Mazumder B, Choudhury MD et al.. (2011) *In vitro* propagation of *Drynaria quercifolia* (L.) a medicinal fern. Assam Univ J Sci Tech Biol Environ Sci 7:79–83

Mehra PN, Palta HK (1971) *In vitro* controlled differentiation of the root callus of *Cyclosorus dentatus*. Phytomorphology 21: 367–375
Mehra PN, Sulklyan DS (1969) *In vitro* on apogamy, apospory and controlled differentiation of rhizome segments of the fern *Ampelopteris prolifera* (Retz). Copel Bot J Linn Soc 62: 431–443
Melan MA, Whittier DP (1990) Effects of inorganic nitrogen sources on spore germination and gametophyte growth in *Botrychium dissectum*. Plant Cell Environ 13:477–482
Menéndez V, Revilla MA, Bernard P et al (2006) Gibberellins and antheridiogen on sex in *Blechnum spicant* L. Plant Cell Rep 25:1104–1110
Meyer BS, Anderson DB, Swanson CA (1955) Laboratory plant physiology. van Nostrad Co. Inc., Princeton, p 168
Miller JH (1968) Fern gametophytes as experimental material. Bot Rev 34:361–440
Miller J, Miller P (1961) The effect of different light conditions and sucrose on the growth and development of the gametophyte of the Fern, *Onoclea sensibilis*. Am J Bot 48(2):154–159
Miller LR, Murashige T (1976) Tissue culture propagation of tropical foliage plants. In Vitro Cell Dev Biol Plant 12:797–813
Miller PM, Sweet HC, Miller JH (1970) Growth regulation by ethylene in fern gametophytes. I. Effects on protonemal and rhizoidal growth and interaction with auxin. Am J Bot 57:212–217
Mohr H (1956) Die Beeinflussung der Keimung von Farnsporen durch Licht und andere Faktoren. Planta 46:534–551
Moore GT (1903) Methods for growing pure cultures of algae. J Appl Microsc 6:2309–3214
Morel G, Wetmore RM (1951) Fern callus tissue culture. Am J Bot 38:141–114
Morini S (2000) *In vitro* culture of *Osmunda* regalis fern. J Hortic Sci Biotech 75: 31–34
Murashige T (1974) Plant propagation through tissue culture. Ann Rev Plant Physiol 25:135–166
Murashige T, Skoog FA (1962) Revised medium for rapid growth and bioassays with tobacco tissue cultures. Physiol Plant 15:473–497
Muthukumar B, Joyce BL, Elless MP et al (2013) Stable transformation of ferns using spores as targets: *Pteris vittata* and *Ceratopteris thalictroides*. Plant Physiol 163:648–658
Nair AG, Pradeesh S, Nikhila GS et al. (2013)*In vitro* propagation of a rare medicinal fern of Western Ghats—*Diplazium esculentum* (Reytz.) Indian J Exp Biol 5111919–923
Nair AG, Pradeesh S, Nikhila GS et al (2014) Conservation of vegetable ferns (*Diplazium esculentum*) of western ghats through *in vitro* spore culture. J Aqua Biol Fish 2:43–50
Nebot A., Philpott VJ, Pajdo A (2021) Cryopreservation of fern spores and pollen. In: Wolkers Wolkers WF, Oldenhof H. (eds) Cryopreservation and freeze-drying protocols. Methods in molecular biology, Humana Press Totowa, NJ, 2180 :0716–07832015
Nester-Hudson JE, Ladas C, McClurd (1997) Gametophyte development and antheridiogen activity in *Thelypteris ovata var. lindheimeri*. Am Fern J 87(4):131–142
Padhya MA (1987) Mass propagation of ferns through tissue culture. Acta Hortic 212:645–649
Padhya MA (1995) Rapid propagation of *Adiantum trapeziformae* L. through tissue culture. Ind Fern J 12:20–23
Padhya MA, Mehta AR (1982) Propagation of fern *Nephrolepis* through tissue culture. Plant Cell Rep 1:261–263
Page CN (1979) Experimental aspects of fern ecology. In: Dyer AF (ed) In the experimental biology of ferns. Academic Press, London, pp 551–589
Pais SM, Casal M (1987) Propagation of the fern *Adiantum capillus-veneris* through tissue culture of the circinate part of young leaves. Acta Hortic 212:651–654
Park KT, Jang BK, Lee CH (2019) Culture condition for gametophyte and sporophyte mass propagation of bamboo fern (*Coniogramme japonica*) using tissue culture. J Plant Biotechnol 46:119–126
Park KT, Jang BK, Lee HM et al (2020) An efficient method for *in vitro* shoot-tip culture and sporophyte production using *Selaginella martensii* spring sporophyte. Plan Theory 9:235
Partanen CR (1972) Comparison of gametophytic callus and tumor tissues of *Pteridium aquilinum*. Bot Gaz 133:287–292

Pence VC (2000) Cryopreservation of *in vitro* grown fern gametophytes. Am Fern J 90:16–23
Pence VC (2008) Cryopreservation of bryophytes and ferns. In: Reed BM (ed) Plant cryopreservation: a practical guide. Springer, New York, NY
Peterson GL (1979) Review of the Folin phenol protein quantitation method of Lowry, Rosebrough, Farr and Randall. Anal Biochem 100(2):201–220
Pierik RLM (1991) Commercial aspects of micropropagation. In: Prakash J, Pierik RML (eds) Horticulture—new technologies and applications. Kluwer Academic Publishers, Dordrecht, pp 141–153
Plackett AR, Rabbinowitsch EH, Langdale JA (2015) Protocol: genetic transformation of the fern *Ceratopteris richardii* through microparticle bombardment. Plant Methods 11:37
Pulukkunadu Thekkeveedu R, Hegde S (2020) *In vitro* spore germination and phytoremediation of Hg and Pb using gametophytes of *Pityrogramma calomelanos*. Int J Phytoremediation 4:1–9
Raine CA, Sheffield E (1997) Establishment and maintenance of aseptic culture of *Trichomanes speciosum* gametophytes from gemmae. Am Fern J 87(3):87–92
Ravi BX, Varuvel GVA, Robert J (2015a) Apogamous sporophyte development through spore reproduction of a South Asia's critically endangered fern: *Pteris tripartita* Sw. Asian Pacific. J Reprod 4(2):135–139
Ravi BX, Geo Vigila A, Rajan K et al (2015b) Apogamous sporophyte development through spore reproduction of a South Asia's critically endangered fern: *Pteris tripartita* Sw. Asian Pac J Reprod 4(2):135–139
Razdan MK (2003) Introduction to plant tissue culture. Science Publishers, New York
Reis Moura I, Simões-Costa MC, Garcia JS et al. (2012). *In vitro* culture of tree fern spores from Cyatheaceae and Dicksoniaceae families. Acta Hortic 937. 455–461
Renner GDR, Randi AM (2004) Effects of sucrose and irradiance on germination and early gametophyte growth of the endangered tree fern *Dicksonia sellowiana* Dicksoniaceae. Acta Bot Bras 18:375–380
Reynoids LT (1979) Apical dominance in *Anemia phyllitidis* gametophytes. Am Fern J 63(4): 158–163
Rigby SJ (1973) Induction of apogamy in *Pellaea glabella var. occidentalis*. Amer Fern J 69(3): 92–95
Rogers SMD, Banister S (1992) Micropropagation of Notholaena 'Sun-Tuff' fern. HortScience 27: 1224–1225
Rolli E, Brunoni F, Marieschi M et al. (2015) *In vitro* micropropagation of the aquatic fern *Marsilea quadrifolia* L. and genetic stability assessment by RAPD markers. Plant Biosyst149(1)7–14
Roy S, Choudhury A (1989) Development of fern gametophytes on nitrogen-deficient medium. Indian Fern J 6:12–17
Rutherford G, Tanurdzic M, Hasebe M et al (2004) A systemic gene silencing method suitable for high throughput, reverse genetic analyses of gene function in fern gametophytes. BMC Plant Biol 4:6
Santos Alves PA, Alline MA, William Hiroshi ST (2019) Germination of spores of Amazonian ferns *Polypodium aureum* in different culture media. Ornament Hortic 25(3):218–224
Sara SC, Manickam VS, Antonisamy R (1998) Regeneration in kinetin -treated gametophytes of *Nephrolepis multiflora* (Roxb.) Jarret in Morton. Curr Sci 75:503–508
Schneller JJ, Hess A (1995) Antheridiogen system in the fern *Asplenium ruta-muraria*. Fern Gaz 15(2):64–70
Seilheimer AV (1974) Axenic culture of the aquatic fern, Salvinia. Am Fern J 64(1):13–14
Seilheimer AV (1978) Chlorophyll and lipid changes on germination in the no-green spores of *Thelypteris dentata*. Am Fern J 65(3):67–70
Shastri P, Shukla L, Khare PB (2012) *In vitro* mass multiplication of a threatened tree Fern, *Cyathea spinulosa* Wall. ex Hook, Int J Genet Eng Biotechnol 3(1): 15–23
Sheffield E, Douglas GE, Cove DJ (1997) Growth and development of fern gametophytes in an airlift fermenter. Plant Cell Rep 16:561–564

Shibila T, Johnson M (2016) *Ex situ* conservation of endangered fern *Elaphoglossum nilgiricum* (Krajina Ex Sledge and Bonner) T. Moore Ex Alston & Bonner using in vitro spore culture. Austin J Biotechnol Eng 3 (2): 1064–1067

Shukla PS, Khare PB (2005a) *In vitro* callusing of the leaf primordium of *Pteris vittata L.* Indian Fern J 21

Shukla SP, Khare PB (2005b) In-vitro mass multiplication of a threatened tree fern, *Cyathea spinulosa* wall. Ex hook. Indian Fern J 22:157–161

Shukla PS, Khare PB (2012a) *In vitro* shoot regeneration via caulogenesis in fern *Pteris vittata* L. J Environ Biol 33: 683–687

Shukla PS, Khare PB (2012b) *In vitro* mass multiplication of a threatened tree fern, *Cyathea spinulosa* Wall. ex Hook, Int J Genet Eng Biotechnol 3(1):15–23

Shukla PS, Khare PB (2014) *In vitro* conservation of some threatened and economically important ferns belonging to the Indian subcontinent. J Bot.1-8:152-156

Shukla JN, Kalsi M, Sethi A et al (2016) Reduced stability and intracellular transport of dsRNA contribute to poor RNAi response in lepidopteran insects. RNA Biol 13(7):656–669

Singh IP, Roy SK (1989) Effect of colchicine on growth and differentiation of sex organs on fern gametophyte. Indian Fern J 6:238–243

Singh AP, Johari D, Singh A et al (2012) Studies on regeneration of gametophytes and mass multiplication of *Anemia rotundifolia* Schrad (Pteridophyte). Int J Healthcare Pharm Res 1(1): 18–23

Singh RS, Jha VK, Chattopadhyay T et al (2020) First report of *agrobacterium rhizogenes*- induced hairy root formation in *Selaginella bryopteris*: a Pteridophyte recalcitrant to genetic transformation. Braz Arch Biol Technol 63:1–8

Smith CW, Yee RNS (1975) The effect of coconut milk on the germination and growth of spores of *Nephrolepis hirsutula*. Am Fern J 65:13–18

Soare LC, Visqiu E, Dobrescu CM (2010) *In vitro* culture and regeneration of the fern *Dryopteris affinis* species, growing in a protected area. Rom Biotechnol Lett 15:45–54

Steeves TA, Sussex IM, Partanen CR (1955) In vitro studies on abnormal growth of prothalli of the bracken fern. Am J Bot 42:232–244

Stephanie J, Conway VS (2020) An ontogenetic framework for functional studies in the model fern *Ceratopteris richardii*. Dev Biol 457(1):20–29

Sulklyan DS, Mehra PN (1977) *In vitro* morphogenetic studies in *Nephrolepis cordifolia.* Phytomorphology 27: 396–407

Teng WL (1997) Activated charcoal affects morphogenesis and enhances sporophyte regeneration during leaf cell suspension culture of *Platycerium bifurcatum.* Plant Cell Rep 17:77–83

Teng WL, Teng MC (1997) *In vitro* regeneration patterns of *Platycerium bifurcatum* leaf cell suspension culture. Plant Cell Rep 16: 820–824

Torres KC (1989) Tissue culture techniques for horticultural crops. Chapman and Hall, New York

Uchida K, Furuya M (1997) Control of the entry into S phase by phytochrome and blue light receptor in the first cell cycle of fern spores. Plant Cell Physiol 38(9):1075–1079

Vargas IB, Droste A (2014) *In vitro* propagation of *Cyathea atrovirens* (Cyatheaceae): spore storage and sterilization conditions. Int J Trop Biol 62 (1): 299–308

Von Aderkas P (1984) Promotion of apogamy in *Matteuccia struthiopteris*, the ostrich Fern. Am Fern J 74(1):1–6

Von Aderkas P (1986) Enhancement of apospory in liquid culture of *Matteuccia struthiopteris*. Ann Bot 57:505–510

Warner TR, Hickok LG (1987) (2-Chloroethyl) phosphonic acid promotes germination of immature spores of *Ceratopteris richardii* Brongn. Plant Physiol 83:723–725

Wetmore RH (1954) The use of *in vitro* cultures in the investigation of growth and differentiation in vascular plants. In: Abnormal and pathological plant growth. Brookhaven Symposia in Biology. 6

White PR (1954) The cultivation of animal and plant cells. The Ronald Press, New York

Whittier DP (1962) The origin and development of apogamous structures in the gametophyte of Pteridium in sterile culture. Phytomorphology 12:10–20

Whittier DP (1964) The influence of cultural conditions on the induction of apogamy in *Pteridium aquilinum* gametophytes. Am J Bot 51:730–736

Whittier DP (1965) Obligate apogamy in *Cheilanthes tomentosa* and *C. alabamensis*. Bot Gaz 126: 275–281

Whittier DP (1972) Gametophytes of *Botrychium dissectum* as grown in sterile culture. Bot Gaz 133:336–339

Whittier DP (1975) The influence of osmotic conditions on induced apogamy in Pteridium gametophytes. Phytomorphology 25:246–249

Whittier DP (1976) Tracheids, Apogamous leaves, and sporophytes in gametophytes of *Botrychium dissectum*. Bot Gaz 137(3):237–241

Whittier DP (1978) Apospory in haploid leaves of Botrychium. Phytomorphology 28:215–219

Whittier DP (1981) Spore germination and young gametophyte development of Botrychium and Ophioglossum in axenic culture. Am Fern J 71(1):13–19

Whittier DP (1987) Germination of Helminthostachys spores Amer. Fern J I 77(3):95–99

Whittier DP, Britton DM (1995) Gametophytes of *Diphasiastrum habereri*. Am Fern J 85(3):89–94

Whittier DP, Moyroud R (1993) The promotion of spore germination and gametophyte development in *Ophioglossum palmatum* by low pH. Am Ferm J 83(2):41–46

Whittier DP, Pratt LH (1971) The effect of light quality on the induction of apogamy in prothalli of *Pteridium aquilinum*. Planta 99:174–178

Whittier DP, Steeves TA (1960) The induction of apogamy in the bracken fern. Can J Bot 38:925–930

Winarto B, Teixeira DS (2012) Improved micropropagation protocol for leatherleaf fern (*Rumohra adiantiformis)* using rhizomes as donor explant. Sci Hortic 140:74–80

Xue L, Qing-Mei Y, Qi-Cheng ZH et al (2010) *In vitro* culture and plant regeneration of *Cibotium barometz* (L.). J Sm N Horticul 6:152–155

Yi-Fa H (2003) The adjustment and control of the progress of spore sprouting and sporophyte shaping of *Osmunda japonica*. J Hubei Inst Nat (Nat Sci) 21(1):65–67

Yin-Li Z, Hong-Hong D, Yang L (2009) Studies on the effect of different sterilization procedures, medium and different light intensity on spore propagation of *Drynaria roosii*. Acta Horticul Sini 36(5):711

Yosihara T, Tsunokawa K, Miyano Y et al (2005) Induction of callus from a metal hypotolerant fern, *Athyrium yokoscense*, and evaluation of its cadmium tolerance and accumulation capacity. Plant Cell Rep 23:579–585

Yu R, Ying C, Yanfei Pu FL et al (2021) *In vitro* propagation of resurrection plant *Selaginella pulvinata* using frond tips as explants. Hort Sci 56(3):313–317

Zenkteler E (2006) Micropropagation of *Matteucia struthiopteris* (L.) Tod. through meristem proliferation fromrhizomes. Biodiver Res Conserv 1(2):167–173

Zhang XY, Junyi JN, Chengsheng D (2001) A morphologic survey of spore development of *Pteridium aquilinum* under culture circumstances. J Gansu Agric Univ 2001

Zheng Y, Xu W, He Z et al (2008) Plant regeneration of the arsenic hyperaccumulator *Pteris vittata* L. from spores and identification of its tolerance and accumulation of arsenic and copper. Acta Physiol Plant 30:249–255

**C. Suneetha** is working as Assistant Professor at the Department of Horticulture, College of Agriculture, University of Agricultural Sciences, GKVK, Bangalore. She holds a M.Sc. in Horticulture and her Ph.D. in Biotechnology. She is specialized in the field of Floriculture, Horticulture, and Plant Biotechnology. She has over 15 years of research experience in the area of tissue culture, molecular markers, and plant molecular biology. She has completed a research project entitled "DNA Barcoding studies in Ferns of South India for Genetic Diversity Studies" funded by SERB, DST, GOI, New Delhi, as Principal Investigator. She also worked as Co PI in university-funded projects. She has got several peer-reviewed publications to her credit.

**Smitha Hegde** , NUCSER, Nitte DU, Deralakatte, Mangalore, has 30 years of experience in the pteridology. She obtained her M.Sc. from the University of Bombay and Ph.D. Biosciences from St. Aloysius College, Mangalore University, Mangalore. Her doctoral thesis was on tissue culture propagation of *Drynaria quercifolia* L. She has pioneered work on pteridophytes of the region on biodiversity and distribution studies, developmental studies, phytoremediation, plant tissue culture, and molecular markers. She has over 74 publications, 3 books, 2 published patents, and successfully guided 112 post-graduate research project students and 10 doctoral students. She is currently the PI of state-funded projects on plant tissue culture and heavy metal remediation using pteridophytes.

# In Vitro Propagation of *Histiopteris incisa* (Thunb.) J. Sm.: An Ornamental Fern

**10**

Vallinayagam Sambantham

**Abstract**

*Histiopteris incisa* (Thunb.) J. Sm. is commonly known as bat's wing fern which has been identified as metallophytes recently. The spores of *H. incisa* were cultured in vitro using Knop (KN) medium at pH 5.8. The spores showed high germination percentage (88%) in vitro, and antheridia and archegonia appeared on gametophytes after 60–70 and 70–80 days of culture, respectively. After fertilization (100–120 days of culture), the zygote developed into embryo and subsequently resulted in the emergence of the first sporophytic frond. The in vitro raised sporelings were successfully transferred to greenhouse condition for hardening. These micropropagated ferns were transferred to Kodaikanal Botanic Garden (KBG) for field establishment.

**Keywords**

Bat's wing fern · In vitro culture · Propagation

## 10.1 Introduction

The *Histiopteris incisa* (Thunb.) J. Sm. is popularly called as bat's wing fern which belongs to the family Dennstaedtiaceae. It is pervasive across tropical, subtropical, and various oceanic islands. It is a wild ornamental fern with attractive green foliage and hence worth introducing into commercial horticulture and indoor foliage decoration purposes. Recently, it has been identified as metallophytes which tolerate the different Cu and As concentrations in the soil; thus, it could be useful and effective for ecological restoration as an option to post-mining rehabilitation (Claveria et al.

V. Sambantham (✉)
Department of Plant Science, Manonmaniam Sundaranar University, Tirunelveli, Tamil Nadu, India





J. Marimuthu et al. (eds.), *Ferns*, https://doi.org/10.1007/978-981-16-6170-9_10

2019). In the last few decades, the in vitro culture methods have replaced the conventional methods of propagation in many fern species. Germination of spores in some species can be greatly increased by the use of in vitro methods, where low or no germination is achieved using conventional techniques, due to dormancy or specific germination requirements. In vitro culture methods are useful to investigate many basic aspects involved in gametophyte and sporophyte development (Vallinayagam et al. 2002; Manickam et al. 2003; Aspiras 2010; Vargas and Droste 2014), and also it is being used in an increasing number of botanic gardens for the propagation and conservation of wild plant species (Barnicoat et al. 2011; Baker et al. 2014). Therefore, in vitro spore culture and subsequent regeneration of sporophytes of the ferns will be very much helpful for mass cultivation. Hitherto, no other in vitro culture works have been reported in *Histiopteris incisa* species. In the present investigation, *Histiopteris incisa* has been selected for large-scale multiplication through in vitro spore culture and to achieve practical lab-to-land transfer through reestablishment of in vitro raised plantlets back to their native forest segments.

## 10.2 Materials and Methods

The fertile fronds with matured spores were carefully collected from the Kothayar hills (1200 m) of Tirunelveli, Tamil Nadu, India. The collected fronds were allowed to dry over white absorbent paper at room temperature for 24 h; the liberated spores were collected and passed through 60 μm nylon mesh to remove the sporangial wall materials. The clean spores were collected and stored in a refrigerator at 5 °C. The spores collected from the fertile fronds were mixed with sterile distilled water and then filtered through a sterile funnel with Whatman No. 1 filter paper (sterile) under aseptic condition. The filtrate solution were collected in a sterile beaker where the funnel with filter paper fixed. The spores collected on the filter paper were surface sterilized with 0.1% mercuric chloride ($HgCl_2$) solution for 5 min and then washed with sterile distilled water for 15 min using sterile distilled water. The surface-sterilized spores were cultured in plant growth regulators free Knop (1865) (KN) medium supplemented with 0.5% (w/v) agar and 1% sucrose, at pH 5.8. They were incubated at 25 °C $\pm$ 2 °C under dark condition for 4–5 days and subsequently transferred to light condition with 12 h photoperiod (2000 lux).

Germination percentage of the spores and their development pattern of prothalli were analyzed. Different stages of growth and development of thalli were microphotographed using a trinocular research microscope (Nikon, Japan). For acclimatization, the in vitro raised sporelings with well-developed roots (5–8 cm) were removed from culture tubes, washed in running tap water to remove the remnants of agar, and planted separately onto a 7-cm-diameter polycup filled with different potting mixtures: river sand, garden soil, and farm yard manure (1:1:1). The sporelings were kept in a mist chamber with a relative humidity of 70% under greenhouse condition. The sporelings were irrigated at 8 h intervals for 3–4 weeks, and establishment rate was recorded. Subsequently, they were transferred to community pots after 4 weeks of weaning process. The sporelings established in community pots were transferred to a shade net house for 3–4 weeks and then repotted in larger pots (20 cm diameter) with one plant in each pot. These micropropagated ferns were transferred to Kodaikanal Botanic Garden (KBG, 1200 m), Perumalmalai, Kodaikanal of Palni Hills (maintained by St. Xavier's College, Palayamkottai), for field establishment after successful acclimatization in the greenhouse condition.

## 10.3 Results and Discussion

Ferns reproduce themselves through spores which are unicellular microscopic structures. They are reproductive cells containing only half the normal complement of chromosomes. It supports sexual reproduction since it involves the union of sex cells. The collected spores of *H. incisa* were 40 × 35 μm in size and pale green in color. The spores germinated in vitro within 5–7 days of culture. The highest percentage (88%) of spore germination was observed in KN medium. Rhizoid initials appeared first, and the first cell of the filament was formed subsequently (Fig. 10.1a). Successive divisions of the first cell led to the formation of filamentous prothalli (Fig. 10.1b). The prothallial plate was formed after 30–40 days of culture. The thallus was dorsiventrally flat with a row of neatly arranged elongated pluricellular meristematic cells at the notch. Moreover, horizontal divisions resulted in the formation of cordate-shaped thalli. Dermal appendages were also formed all over the surface of the thalli. Antheridia and archegonia were formed on the thalli after 60–70 and 70–80 days of culture, respectively. Antheridia and archegonia were found interspread to each another from the posterior end to anterior cushion of the prothalli (Fig. 10.1c). The common feature of homosporous ferns is that their thallus differentiates antheridia earlier than archegonia (Dyer 1979; Raghavan 1989; Rachenmacher et al. 2014). The same kind of growth pattern of antheridia and archegonia formation was observed in *H. incisa* of the present investigation. Spermatozoids were released from the antheridia, which fertilized the archegonial egg cell in the gametophytes. After fertilization (100–120 days of culture), the zygote developed into embryo and subsequently resulted in the emergence of first sporophyte frond (Fig. 10.1d, e). Sprouting of frond with shoot and root initial was from the foot especially very near the apical notch.

Reduced nitrogen source is necessary for spore germination, and early growth of gametophytes in *Onoclea sensibilis* (Miller et al. 1970; Edwards and Miller 1972a, b), *Arthromeris tenuicauda* (Roy and Choudhry 1989), *Platycerium coronarium* and *P. grande* (Aspiras 2010), *Colysis latiloba* (Parajuli and Joshi 2013), *Dipteris wallichii* (Bharati et al. 2013), and *Pityrogramma calomelanos* (Sajeeva et al. 2018) were studied earlier. Our present experimental results are also consonance with the above earlier research works as it could respond well in low salt concentrated Knop medium.

Approximately 7-cm-long sporelings raised from the in vitro grown gametophytes were carefully removed from the culture vessel, washed thoroughly

**Fig. 10.1** In vitro spore culture of *Histiopteris incisa*. (**a**, **b**) Filamentous stage prothalli (bar 1 cm = 70 mm and 90 mm). (**c**) Cordate-shaped thallus with antheridia and archegonia formation (bar 1 cm = 220 mm). (**d**, **e**) Sporophytic emergence (bar 1 cm = 310 mm and 0.75 cm, respectively). (**f**) Four-week-old sporophytes under hardening. (**g**) Three-month-old hardened sporophytes. (**h**) Transferred sporophytes in beds of shade house at Kodaikanal Botanic Garden

in tap water to remove traces of agar sticking on to the roots, and implanted into 7-cm-diameter polycups (Fig. 10.1f) containing a mixture of sand/garden soil/ farmyard manure (1:2:1) covered with unperforated polybags and irrigated with 10× diluted MS liquid medium (without sucrose) once in a week in the greenhouse under misting (RH 80%). After 4–6 weeks, 115 plants were transferred to pots containing the same potting mixture and kept under misting again in the greenhouse (Fig. 10.1g). The micropropagated plants showed more than 67% establishment under the conditions of potting and hardening. Three months after transplantation in pots, the micropropagated plants showed normal growth and were free from morphological abnormalities. The micropropagated plants were transferred to Kodaikanal Botanic Garden (1100 m), Perumalmalai, Kodaikanal of Palni Hills, for field establishment. After 2–3 months of acclimatization at the garden, the plants were transferred to the beds inside the shade house. The plants were established well without loss and growing normally in the beds (Fig. 10.1h). The observation that 3 months after transplantation the micropropagated plants showed normal growth and morphology indicates the uniformity of the plants. These plants could also be successfully transferred to Kodaikanal Botanic Garden (1200 m), Kodaikanal, for reintroduction into selected forest habitats.

Success in micropropagation will not be completed unless the rooted plantlets are weaned and established in pots and thereafter in the field. The in vitro raised plants are the best semi-autotrophic showing abnormal morphology with delicate shoots, leaves without cuticle, stomata and root systems that are not well adapted for functioning in the external environment (Bozena 2001; Aygun and Dumanoglu 2015). Gradual decreasing of humidity of the in vitro culture raised plantlets played an important role in plant adaptation for ex vitro conditions. Therefore, these plants are usually passed through a transitional period of hardening or acclimatization to induce autotrophy and ensure their survival (George et al. 2008). The field transferred ferns of the present investigation showed normal growth and morphology. The knowledge of spore germination characteristics and the growth pattern of gametophyte development are important to understand the entire life cycle of *Histiopteris incisa* which provide information to support the cultivation of the species. These micropropagated plants could be used effectively as metallophytes in ecological restoration in post-mining rehabilitation.

**Acknowledgments** The author is thankful to the Ministry of Environment and Forests (MOEF), Govt. of India, for the financial support of the project during the period of research.

## References

Aspiras RA (2010) Sporophyte and gametophyte development of *Platycerium coronarium* (Koenig) Desv. and *P. grande* (fee) C. Presl. (Polypodiaceae) through *in vitro* propagation. Saudi J Biol Sci 17(1):13–22

Aygun A, Dumanoglu H (2015) *In vitro* shoot proliferation and *in vitro* and *ex vitro* formation of *Pyrus elaeagrifolia* Pallas front. Plant Sci 6:225–232

Baker KP, Lambdon E, Jones J et al (2014) Rescue, ecology, and conservation of a rediscovered island endemic fern (*Anogramma ascensionis*) *ex situ* methodologies and a road map for species reintroduction and habitat restoration. Bot J Lin Soc 174:461–477

Barnicoat HR, Cripps J, Kendon J et al (2011) Conservation *in vitro* of rare and threatened ferns - case studies of biodiversity hotspot and island species. In Vitro Cell Dev Biol 47:37–45

Bharati SK, Choudhury D, Manabendra et al (2013) *In vitro* propagation of *Dipteris wallichii* (R. BR.) T. Moore: a hope for conservation of an endangered pteridophyte. Intl Res J Pharm 4(3):215–219

Bozena B (2001) Morphological and physiological characteristics of micropropagated strawberry plants rooted *in vitro* and *ex vitro*. Sci Hortic 8(3):195–206

Claveria RJR, Perez TR, Perez REC et al (2019) The identification of indigenous Cu and As metallophytes in the Lepanto Cu-Au mine, Luzon, Philippines. Environ Monit Assess 191(3):185

Dyer AF (1979) The culture of fern gametophytes for experimental investigation. In: Dyer AF (ed) The experimental biology of ferns. Academic press, London, pp 253–305

Edwards ME, Miller JH (1972a) Growth regulation by ethylene in fern gametophytes. II. Inhibition of cell division. Am J Bot 59:450–457

Edwards ME, Miller JH (1972b) Growth regulation by ethylene in fern gametophytes. III. Inhibition of spore germination. Am J Bot 59:458–465

George EF, Hall MA, Klerk GJD (2008) Plant propagation by tissue culture - Vol 1. Springer, Dordrecht, pp 29–64

Knop N (1865) Quantitative Untersuchungan Uber die Ernahrungsprozesse der pflangen. Land Wirtsch Versuchssat 7:93–107

Manickam VS, Vallinayagam S, Johnson M (2003) Micropropagation and conservation of rare and endangered ferns of the southern western ghats through *in vitro* culture. In: Chandra S, Srivastava M (eds) Pteridology in new millennium (biology and taxonomy of ferns). Kluwer Academic Publishers, Dordrecht, pp 497–504

Miller PM, Sweet HC, Miller JH (1970) Growth regulation by ethylene in fern gametophytes. I. Effects on protonemal and rhizoidal growth and interaction with auxin. Am J Bot 57(2):212–217

Parajuli J, Joshi SD (2013) *In vitro* morphogenesis of *Colysis latiloba* (Ching.) Ching. Int J Sci Technol Res 2(3):247–253

Rachenmacher C, Schmitt JL, Droste A (2014) Spore germination and gametophytes development of *Cyathea atrovirens* (Langsd. & Fisch.) Domin (Cyatheaceae) under different pH conditions. Braz J Biol 70(4):1155–1160

Raghavan V (1989) Developmental biology of fern gametophytes. Cambridge University Press, Cambridge, pp 212–260

Roy S, Choudhry A (1989) Development of fern gametophytes on nitrogen-deficient medium. Indian Fern J 5:1–4

Sajeeva S, Melo JS, Hegde S (2018) Gamma-radiation induced *in vitro* hormetic apogamy in the fern *Pityrogramma calomelanos*. Biosystems 173:221–224

Vallinayagam S, Manickam VS, Seeni S (2002) Multiplication of *Diplazium cognatum* through *in vitro* culture. Phytomorphology 52(1):23–27

Vargas IB, Droste A (2014) *In vitro* propagation of *Cyathea atrovirens* (Cyatheaceae): spore storage and sterilization conditions. Rev Biol Trop 62(1):299–308

**Vallinayagam Sambantham** received Bachelor's (1995) degree from St. Xavier's College (Autonomous), Palayamkottai, and pursued Master's (1997) and Doctoral degree (2003) from Manonmaniam Sundaranar University, Tirunelveli, India. Then he did his Post-Doctoral Research at the Centre for Cellular and Molecular Biology (CSIR), Hyderabad. In his 20 years of Teaching cum Research career, he worked as Lecturer (BITS-Pilani Goa Campus, India), Research Professor (Korea University, Seoul, South Korea), Research Scientist (Biopolis-ASTAR and Singapore General Hospital, Singapore), Assistant Professor (Quest International University Perak, Ipoh, Malaysia), and Assistant Professor-UGC Plan XII (Bharathidasan University, Tiruchirappalli, India). Currently he is teaching at the Department of Plant Science, Manonmaniam Sundaranar University, Tirunelveli, India

# In Vitro Propagation of Two Epiphytic Ferns 11

Shibila Thangaiah and Johnson Marimuthu

**Abstract**

The present study was initiated to optimize the protocol for the mass multiplication of *Lepisorus nudus* (Hook.) Ching and *Elaphoglossum stelligerum* (Wall. ex Baker) T. Moore using in vitro spore culture. The surface-sterilized spores were inoculated onto different media devoid of sugar and plant growth regulators. The cultures were incubated at 25 °C ± 2 °C under 12 h photoperiod (1500 lux). KC basal medium (pH 5.7) showed the highest percentage of spore germination, viz., 78.51 ± 3.09% for *L. nudus* and 50.37 ± 3.31% for *E. stelligerum*. Among the two epiphytic ferns, *E. stelligerum* revealed the highest sporophyte proliferation (39.33 ± 2.78%) on KN basal medium, and *L. nudus* exhibited 36 ± 2.78% sporophyte proliferation on KC liquid medium. The in vitro spore-derived sporophytes of *E. stelligerum* and *L. nudus* showed 46.66 ± 3.33% and 42 ± 2.98% of re-establishment, respectively, at Kodaikanal Botanical Garden, Kodaikanal, Tamil Nadu, India.

**Keywords**

Ex situ conservation · Ontogeny · Mass propagation · Pteridophytes

S. Thangaiah
Sri Paramakalyani Centre for Excellence in Environmental Sciences, Manonmaniam Sundaranar University, Alwarkurichi, Tamil Nadu, India

J. Marimuthu (✉)
Department of Botany, Centre for Plant Biotechnology, St. Xavier's College (Autonomous), Palayamkottai, India


J. Marimuthu et al. (eds.), *Ferns*, https://doi.org/10.1007/978-981-16-6170-9_11

## Abbreviations

| | |
|---|---|
| ANOVA | Analysis of variance |
| $HgCl_2$ | Mercuric chloride |
| KBG | Kodaikanal Botanical Garden |
| KC | Knudson C |
| KN | Knops |
| Mi | Mitra |
| MS | Murashige and Skoog |

## 11.1 Introduction

Pteridophytes are a very ancient group of land plants, growing in varied climatic regions. Ferns also have medicinal uses (Homoeopathic, Ayurvedic, Tribal, and Unani medicines), and decoction or infusion of some ferns is used to ease labor pains. It provides food and is used as insecticides and ornamentations. Nowadays, many species of pteridophytes are listed under the threat due to natural habitat destruction, overexploitation, and biodiversity depletion (Shukla and Khare 2014).

An in vitro spore culture technique is the best and appropriate technique for the large-scale multiplication and re-establishment of economically, ecologically, medicinally important ferns and rare, threatened, and endangered ferns. Due to environmental factors, germination of the spores under natural condition is very difficult (Morini 2000). Additionally, in vitro spore culture techniques provide a platform to understand the developmental and reproductive biology of ferns (Sara et al. 1998; Sara 2001; Vallinayagam et al. 2002; Fernández and Revilla 2003; Irudayaraj et al. 2003; Johnson et al. 2005; Garcia and Ruiz 2006; Sara and Manickam 2007; Johnson et al. 2008; Zhang et al. 2008; Martinez 2010; Liliana et al. 2010; Liliana et al. 2013; Archana et al. 2014; Shukla and Khare 2014; Shibila and Johnson 2016).

Globally, in vitro spore culture technique is applied to propagate the commercially (Patricia et al. 2019), ecologically (Johnson and Manickam 2011; Shukla and Khare 2014), and medicinally important ferns (Mazumder et al. 2011; Archana et al. 2013; Gayathiri et al. 2018). A number of ferns are multiplied and conserved through in vitro spore culture, viz., *Thelypteris confluens*, *Athyrium nilgripes*, and *Cyathea crinita* (Sara 2001; Sara and Manickam 2005; Sara and Manickam 2007); *Diplazium cognatum* and *Hypodematium crenatum* (Vallinayagam et al. 2002; Manickam et al. 2003); *Cheilanthes viridis*, *Pronephrium triphyllum*, *Sphaerostephanos unitus*, and *Phlebodium aureum* (Johnson 2003; Manickam et al. 2003; Johnson and Manickam 2007; Johnson and Manickam 2011); *Metathelypteris flaccida* (Johnson and Manickam 2006a); *Pronephrium articulatum* (Johnson and Manickam 2006b); *Diplazium proliferum* (Golamaully et al. 2015); *Elaphoglossum nilgiricum* (Shibila and Johnson 2016); *Elaphoglossum stigmatolepsis* (Johnson and Shibila 2018);

*Rumohra adiantiformis* (Fonseka 2020); and *Selaginella tamariscina* (Park et al. 2021). Epiphytic plants are the major components of rainforest plant diversity (Barthlott et al. 2001). With this background, the present study was initiated to optimize the protocol for the mass multiplication of *Lepisorus nudus* (Hook.) Ching and *Elaphoglossum stelligerum* (Wall. ex Baker) T. using in vitro spore culture. In addition, the current exertion provides knowledge on the effect medium on the spore germination and the relative growth rate of gametophytes and sporophyte proliferation. The outcome of the present study will explain various developmental stages of *Lepisorus nudus* and *Elaphoglossum stelligerum*, viz., spore – protonema – cordate prothallus without sex organ – cordate prothallus without sex organ – sporophyte, and the results will help to understand the developmental biology and provide an outline for the ontogeny of the selected epiphytic and lithophytic ferns.

## 11.2 Materials and Methods

Mature fronds of *Lepisorus nudus* (Hook.) were collected from Kothayar, Kalakkad Mundanthurai Tiger Reserve, Tamil Nadu, India, and those of *Elaphoglossum stelligerum* (Wall. ex Baker) T. Moore were collected from Gurukula Botanical Sanctuary, Wayanad, Kerala, India, and Kodaikanal Botanical Garden, Eetipallum, Kodaikanal, Tamil Nadu, India. The collected fronds were washed in the running tap water for a few minutes to remove the surface contaminants. For the release of spores, the fronds were cut into small pieces and dried over white absorbent paper at room temperature (30 °C) for 24 h. Then, the liberated spores were filtered through 40 μm micro-sieves to remove the unwanted sporangial wall materials. The clean spores were collected and stored in refrigerator at 7 °C.

A solution of $HgCl_2$ (0.05, 0.1, 0.2, 0.3, and 0.4%) was employed to sterilize the spores for 5–15 min (Manickam et al. 2003; Archana et al. 2013; Shukla and Khare 2014; Lisha et al. 2016; Shibila and Johnson 2016; Johnson and Shibila 2018). Then, they were washed with sterile distilled water for 15–30 min. Once disinfected, the spores of both species were inoculated onto various basal media (liquid and semi-solid—0.5% agar), viz., Murashige and Skoog, Mitra, Knudson C, and Knops media using sterile Pasteur pipettes. Cultures were maintained at 25 °C ± 2 °C under 12 h photoperiod (1500 lux). The different developmental stages from spore germination, gametophyte formation, sex organ formation, and sporophyte proliferation all were observed and noted regularly. The frequency of spore germination, gametophyte multiplication, and sporophyte proliferations were calculated. The experiment was repeated thrice and calculated their average and standard deviations. ANOVA analyses were performed to know the significance of nutrient media on spore germination, gametophyte multiplication, and sporophyte proliferations using SPSS. Alpha subsets mean differences are significant at $p < 0.05$ level.

## 11.3 Results

### 11.3.1 Lepisorus Nudus

The developmental biology of *L. nudus* is presented in Fig. 11.1. The spores of *Lepisorus nudus*, sterilized with 0.1% mercuric chloride for 10 min, showed high rate of survival and reduced microbial contamination. Treatments of 0.2%–0.4% of $HgCl_2$ and prolonged exposure (more than 10 min with $HgCl_2$) of *L. nudus* spores resulted in high percentage of mortality and low percentage of spore germination. Spore germination was observed on Knudson C medium after 60 days, and germination pattern was vittaria type. The single cell divided transversely for outward appearance of two-celled stage. The second division occurred in the apical cell of two-celled stage resulting in a three-celled filamentous stage on the 68th day (Fig. 11.1d). The protonema cell is formed by a sequential division of a filament cell (Fig. 11.1e). Hairs were observed from some marginal cells of the thallus (Fig. 11.1e, f). The continuous cell division of the filaments produced a small spathulate thallus, and a nearly symmetrical cordate gametophyte is formed on the 87th day (Fig. 11.1g). Superficial hairs were located on both dorsal and ventral sides of the wing and midrib (Fig. 11.1f). On the 97th day, the female sex organs were detected (Fig. 11.1i) and the male sex organs after the 107th day of inoculation (Fig. 11.1h). Archegonia formed on the midrib of both ventral and dorsal sides. The proliferation of sporophytes was observed on the 125th day (Fig. 11.1j, k).

The influence of medium on spore germination and ontogeny of *L. nudus* is displayed in Fig. 11.2. Various media, viz., Knops, Knudson C, Mitra, and Murashige and Skoog medium (liquid and solid), were employed for the spore germination of *L. nudus* (Fig. 11.2). Among the tested media, KC basal medium showed the highest percentage of spore germination ($78.51 \pm 3.09\%$) at pH 5.7. Highest percentage of gametophyte multiplication ($90 \pm 3.33\%$) was recorded in KN liquid medium. *L. nudus* exhibited $36 \pm 2.78\%$ sporophyte proliferation on KC liquid medium. The re-establishment of in vitro spore-derived sporophytes of *L. nudus* was attained with $42 \pm 2.98\%$.

### 11.3.2 Elaphoglossum Stelligerum

The spore of *E. stelligerum* (Fig. 11.3a, b) sterilized with 0.1% mercuric chloride for 15 min showed high rate of survival and low frequency of microbial contamination. The spores sterilized above 0.1% $HgCl_2$ and prolonged exposure showed high percentage of spore mortality and low percentage of spore germination. The influence of medium on spore germination and ontogeny of *E. stelligerum* is displayed in Fig. 11.3c–l. Spores began to germinate after sowing on the 69th day. The protonema cell is formed by a sequential division of a filament cell (Fig. 11.3c, d). The length of filaments varied from 2 to 12 cells. Hairs were observed from some marginal cells of the thallus (Fig. 11.3c, d). The continuous cell division of the filaments produced a branch and grew to small spathulate stages (Fig. 11.3d, e). The

**Fig. 11.1** In vitro spore culture of *Lepisorus nudus*. **a**) Sori arrangement in *L. nudus*. (**b** and **c**) Spore morphology of *L. nudus*. (**d** and **E**) Gametophytes – filamentous stage and formation of cordate gametophytes. (**f**) Gametophyte multiplication. (**g**) Matured cordate gametophytes – microscopic view. (**h**) Antheridia – microscopic view. (**i**) Archegonia – microscopic view. (**j**) Emergence of sporophytes from gametophytes – microscopic view. (**k**) Sporophyte proliferation—in vitro condition. (**l**) In vitro cultured sporophytes established in KBG. (**m**) In vitro cultured sporophytes established in Wayanad. (**n**) 150-day-old plant of *L. nudus* at KBG

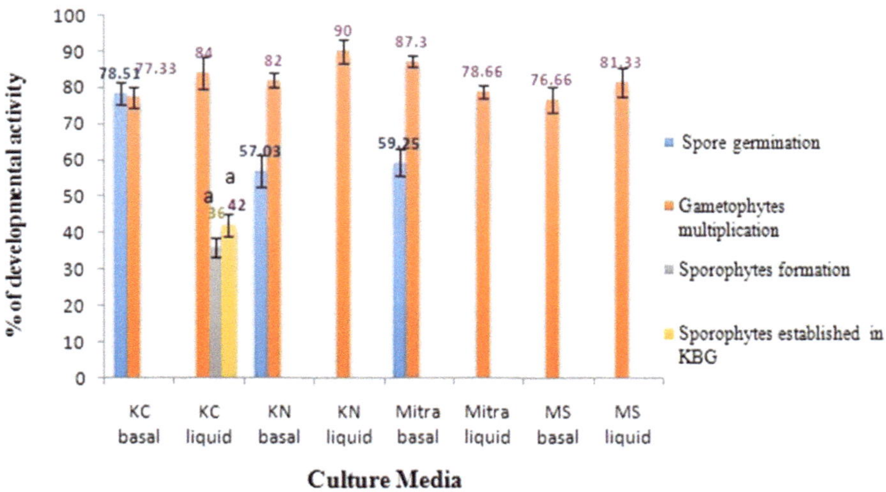


**Fig. 11.2** Effect of media on spore germination, gametophyte multiplication, and sporophyte formation of *L. nudus*. *Basal media means semi-solid media without sucrose and plant growth regulators

meristematic cell divided repeatedly and resulted in a broad spathulate thallus (Fig. 11.3d–f). A multicellular meristem later replaced the single meristematic cell (Fig. 11.3d–f), and a nearly symmetrical cordate gametophyte is formed (Fig. 11.3f). A wedge-shaped meristematic cell developed at the apex (Fig. 11.3f). A midrib at the central region developed after the cordate gametophyte is formed. Marginal and superficial unicellular hairs were abundant (Fig. 11.3d, e). Superficial hairs were located on both dorsal and ventral sides of the wing and midrib (Fig. 11.3f). After the 102nd day of inoculation, the appearance of male and female sex organs was noticed (Fig. 11.3g, h). Antheridia appeared on gametophyte margins in a crammed manner. Archegonia formed on the midrib of both ventral and dorsal sides. The proliferation of sporophytes was observed on the 136th day (Fig. 11.3i, j).

The effect of media on spore germination, gametophyte multiplication, and sporophyte formation of *E. stelligerum* is depicted in Fig. 11.4. The spore germination ($50.37 \pm 3.31\%$) was only recognized on KC basal medium at pH 5.7. The maximum percentage of gametophyte multiplication ($71.33 \pm 5.05\%$) was observed in Mitra liquid medium. *E. stelligerum* revealed the highest sporophyte proliferation ($39.33 \pm 2.78\%$) on KN basal medium. $46.66 \pm 3.33\%$ of in vitro spore-derived *E. stelligerum* sporophytes was re-established in KBG and Wayanad.

## 11.4 Discussion

Spores are significant reproductive structures in pteridophytes. Due to unfavorable environmental conditions, mostly the rare and endangered fern spores are unable to attain germination and development. The researchers identified plant tissue culture

**Fig. 11.3** In vitro spore culture of *Elaphoglossum stelligerum.* (**a** and **b**) Spore morphology of *E. stelligerum*. (**c** and **d**) Gametophytes – filamentous stage and formation of cordate gametophytes. (**e**) Gametophyte multiplication. (**f**) Matured cordate gametophytes – microscopic view. (**g**) Antheridia – microscopic view. (**h**) Archegonia – microscopic view. (**i**) Emergence of sporophytes from gametophytes – microscopic view. (**j**) Sporophyte proliferation—in vitro condition. (**k**) In vitro cultured sporophytes established in KBG. (**l**) In vitro cultured sporophytes established in Wayanad

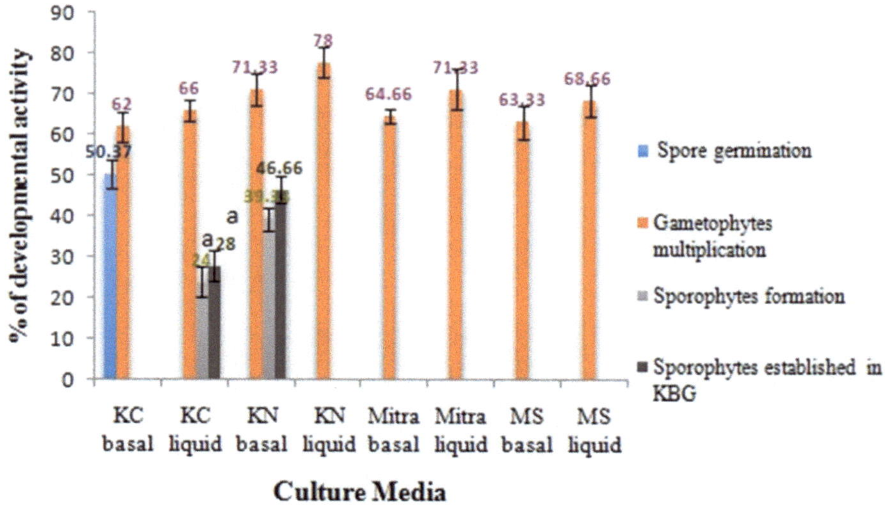


**Fig. 11.4** Effect of media on spore germination, gametophyte multiplication, and sporophyte formation of *E. stelligerum.* *Basal media means semi-solid media without sucrose and plant growth regulators

technique as an alternative tool to multiply the rare and endangered ferns (Sara 2001; Vallinayagam et al. 2002; Johnson 2003; Shibila and Johnson 2016; Espinosa-Leal et al. 2018). The spore germination of ferns are influenced and regulated by various factors, viz., age of spores, color of spores, influence of season, culture media, pH, sterilant, and sterilization. An optimized protocol was established for the large-scale multiplication of *Lepisorus nudus* and *Elaphoglossum stelligerum.*

For the large-scale propagation of selected ferns, mature spores were used as explants. A number of studies reported the spore as initial explants for the propagation of ferns (Johnson and Manickam 2007; Johnson et al. 2008; Khan et al. 2008; Johnson and Manickam 2011; Archana et al. 2014). The matured spores of *Lepisorus nudus* and *Elaphoglossum stelligerum* were employed for in vitro spore culture for the differentiation study and achieved varied frequency of spore germination. The in vitro spore culture has been used to assess the effects of sterilant, sterilization time, medium, sucrose, pH, and plant growth regulators on spore germination (Shibila and Johnson 2016; Ballesteros and Pence 2018). The spores of *L. nudus* sterilized with 0.1% (w/v) $HgCl_2$ for 10 min and the spores of *E. stelligerum* sterilized with 0.1% $HgCl_2$ for 15 min showed the highest percentage of spore viability and contamination-free cultures. Similarly, the spores of *Pronephrium triphyllum*, *Sphaerostephanos unitus*, *Cheilanthes viridis*, *Diplazium cognatum*, and *Elaphoglossum nilgiricum* were sterilized with 0.1% $HgCl_2$ for 5–15 min and obtained the contamination-free cultures (Manickam et al. 2003; Johnson and Manickam 2011; Shibila and Johnson 2016).

The influence of pH over fern spore germination was studied (Symonds et al. 2001; Ambrosio and De Melo 2004; Viviani and Randi 2008; Rechenmacher et al.

2010; Ravi et al. 2014). Most of the fern spores were germinated on various media with the pH around 5.0 to 6.0 (Johnson and Manickam 2011; Archana et al. 2013; Shibila and Johnson 2016). In accordance with previous observation, the spores of *Lepisorus nudus* and *Elaphoglossum stelligerum* were germinated on KC basal medium, KN basal medium, and Mitra basal medium with the pH 5.6 and 5.7.

The germination period for the studied ferns is different; however, the spores of lithophytic and epiphytic fern are relatively similar in KC basal medium. Lithophytic fern *Lepisorus nudus* and epiphytic fern *E. stelligerum* were germinated at the 60th and 69th days from the inoculation, respectively. Similarly, *Pronephrium triphyllum* and *Sphaerostephanos unitus* started germination after 38 and 35 days in Knops medium, respectively. The gametophytes of *P. triphyllum* and *S. unitus* cultured on Knops medium showed the sporophyte proliferation after 30 days and 180 days of inoculation, respectively (Johnson and Manickam 2011). The spores of *Alsophila odonelliana* began to germinate 9–12 days after inoculation (Bonomo et al. 2013). Manonmani and Sara (2014) observed male and female sex organ formation in *A. radiata* after 45–60 days of spore germination under various media (KC, KN, and brick pieces supplemented with KC medium). We observed the formation of sex organs after 3 months in lithophytic and epiphytic fern. *L. nudus* showed female sex organs after 37 days and male sex organs after 47 days after spore germination. In *E. stelligerum*, the female sex organs and male sex organs emerged on the 33rd day after spore germination.

The sporophytes of *Platycerium coronarium* and *Platycerium grande* developed on various media after 7 and 9 weeks correspondingly (Reyno 2010). Oh et al. (2013) visualized the significant morphological changes under microscope in MS medium after 20 days of culture. The in vitro spore culture of *Dryopteris affinis* emerged juvenile sporophytes after 45 days of inoculation (Liliana et al. 2010). But in our study, the sporophyte emergence of *L. nudus* and *E. stelligerum* were observed on the 125th and 136th days. Otherwise, the culture medium plays a significant role in the formation of sex organs, morphological changes of gametophytes, and sporophyte formation. Among the various media tested, KC medium produced the highest percentage of spore germination, gametophyte development, and sporophyte proliferations. Similar to that, Shibila and Johnson (2016) and Johnson and Manickam (2011) also observed the highest percentage of spore germination, gametophyte development, and sporophyte proliferations of *E. nilgiricum*, *Pronephrium triphyllum*, and *Sphaerostephanos unitus* using the KC medium. Gayathiri et al. (2018) also revealed the maximum germination (95%) in *Adiantum caudatum* on KC medium with vittaria type of germination. Similarly, the vittaria type of germination was observed for *L. nudus*.

## 11.5 Conclusion

The observed results suggested that the Knudson (KC) basal medium was the best for spore germination, gametophyte multiplication, and sporophyte proliferations of *L. nudus* and *E. stelligerum*. The time required for spore germination showed difference between species. The outcome of the study produced an alternate pathway for the large-scale multiplication of studied ferns *L. nudus* and *E. stelligerum* using spore as explants. This protocol will be used to propagate in large scale and conserve the ferns *L. nudus* and *E. stelligerum*, as well as apply the developed protocol to other species.

## References

Ambrosio ST, De Melo NF (2004) Interaction between sucrose and pH during *in vitro* culture of *Nephrolepis biserrata* (Sw.) Schott (Pteridophyta). Acta Bot Bras 18(4):809–813

Archana GN, Pradeesh S, Nikhila GS, Sangeetha G, Mini I, Swapna TS (2013) *In vitro* propagation of a rare medicinal fern of Western Ghats - *Diplazium esculentum* (Reytz). Indian J Exp Biol 51: 919–923

Archana GN, Pradeesh S, Nikhila GS, Sangeetha G, Mini II, Swapna TS (2014) Conservation of vegetable fern (*Diplazium esculentum*) of Western Ghats through *in vitro* spore culture. J Aquat Biol Fish 2(2):43–50

Ballesteros D, Pence VC (2018) Fern conservation: spore, gametophyte, and sporophyte ex situ storage, *in vitro* culture, and cryopreservation. In: Fernández H (ed) Current advances in Fern research. Springer, New York, pp 227–249

Barthlott W, Schmit-Neuerburg V, Nieder J, Engwald S (2001) Diversity and abundance of vascular epiphytes: a comparison of secondary vegetation and primary montane rain forest in the Venezuelan Andes. Plant Ecol 152:145–156

Bonomo MC, Martinez OG, Tanco ME, Cardozo R, Aviles (2013) Spore germination and gametophytes of *Alsophila odonelliana* (Cyatheaceae) in different sterile media. Int J Exp Bot 82:119–126

Espinosa-Leal CA, Puente-Garza CA, Garcia-Lara S (2018) *In vitro* plant tissue culture: means for production of biological active compounds. Planta 248(1):1–18

Fernández H, Revilla MA (2003) *In vitro* culture of ornamental ferns. Plant Cell Tissue Organ Cult 73:1–13

Fonseka DLCK (2020) Propagation of leather fern (*Rumohra adiantiformis*) from rhizomes by *in vitro* techniques. Int J Agric Biol Sci 4:113

Garcia BP, Ruiz AM (2006) Gametophyte morphology in three Mexican species of *Bolbitis* (Lomariopsidaceae). Am Fern J 96(4):115–126

Gayathiri M, Roselin RI, Sujatha S, Sara CS (2018) *In vitro* life cycle micro morphological studies and reproductive biology of an anti-diabetic fern. Int J Curr Microbiol Appl Sci 7(7):3590–3603

Golamaully ZM, Bhoyroo V, Nazurally N, Gopal V (2015) *In vitro* culture as an aid to conservation of indigenous ferns: *Diplazium proliferum* Int J Plant Biol 6(2) https://doi.org/10.4081/pb.2015.6020

Irudayaraj V, Manickam VS, Johnson M (2003) Incidence of apospory in *Pteris confusa* T. G. Walker. Phytomorphology 53:73–77

Johnson M (2003) In vitro studies on some rare and endangered ferns of Western Ghats, South India. Ph. D Thesis. Manonmaniam Sundaranar University, Tamil Nadu, India

Johnson M, Manickam VS (2006a) *In vitro* studies on normal and abnormal life cycle of *Metathelypteris flaccida* (Bl.) Ching. Ethiop J Sci Technol 4(1):37–44

Johnson M, Manickam VS (2006b) Adventitious proliferation of secondary and tertiary prothalli from the primary prothalli of *Pronephrium articulatum* (Houlst. & Moore) Holt. Ethiop J Sci Technol 3(2):93–96

Johnson M, Manickam VS (2007) Influence of sucrose on morphogenetic changes of four rare and endangered ferns of Western Ghats, South India. Biochem Indian J 1(4):18

Johnson M, Manickam VS (2011) *Ex situ* conservation of two threatened ferns of the Western Ghats through *in vitro* spore culture. J Threat Taxa 3(7):1919–1928

Johnson M, Shibila T (2018) In vitro spore culture and ontogeny of *Elaphoglossum stigmatolepsis* (fee.) Moore Phytomorphol 68(1&2):35–38

Johnson M, Manickam VS, Irudayaraj V (2005) *In vitro* studies on agamospours fern *Pteris gongalensis* T.G. Walker. Ethiop J Sci Technol 3:1–6

Johnson M, Manickam VS, Benniamin A, Irudayaraj V (2008) Conservation of endangered ferns of Western Ghats through micropropagation. In: Verma SC, Khuller SP, Cheema HK (eds) Perspective in pteridophytes. Bishen Singh Mahendra Pal Singh, Dehradun, India, pp 183–191

Khan S, Raziq M, Kayani HA (2008) *In vitro* propagation of bird's nest fern (*Asplenium nidus*) from spores. Pak J Bot 40(1):91–97

Liliana CS, Emilia V, Codruta MD (2010) *In vitro* culture and regeneration of the fern *Dryopteris affinis* species growing in a protected area. Rom Biotechnol Lett 5(1):45–54

Liliana CS, Codruta MD, Luminita B, Anca NS (2013) Research on the influence of two insecticides on the gametophyte of some leptosporangiate pteridophytes. Scientific Annals of Alexandru Ioan Cuza University of Iasi, New Series, Section II a. Vegetal Biol 59(2):5–12

Lisha DC, Prutha M, Shaiesh M, et al. (2016) In vitro studies of spore germination in Pityrogramma calomelanos. Proceedings of the National Seminar on 'Herbal Technology & Ethnobotany', pp. 51–54

Manickam VS, Vallinayagam S, Johnson M (2003) Micropropagation and conservation of rare and endangered ferns of Western Ghats through *in vitro* culture. In: Chandra S, Srivastava M (eds) Pteridology in the new millennium. Kluwer Academic Publishers, Dordrecht, Netherlands, pp 497–504

Manonmani R, Sara SC (2014) *In vitro* early ontogenic studies of *Actiniopteris radiata* (Swartz) link under different medium. Int J Pharm Biol Sci 5(1):483–491

Martinez OG (2010) Gametophytes and young sporophytes of four species of the fern genus *Pteris* (Pteridaceae) naturalized in the American continent. Rev Biol Trop 58(1):89–102

Mazumder PB, Mazumder B, Choudhury DM, Sharma GD (2011) In vitro propagation of *Drynaria quercifolia* (L.) J. Sm a medicinal fern. Assam Univ J Sci Technol Biol Environ Scis 7(1):79–83

Morini S (2000) *In vitro* culture of *Osmunda regalis* fern. J Hortic Sci Biotechnol 75:31–34

Oh MJ, Kim C, Na HR, Shin H, Liu JR, Choi HK, Kim SW (2013) High frequency sporophytes regeneration from the spore culture of the endangered aquatic fern *Isoetes coreana*. Am J Plant Sci 4:14–20

Park K, Jang BK, Lee HM et al (2021) Effective propagation of *Selaginella tamariscina* through optimized medium composition. Agron 11:578. https://doi.org/10.3390/agronomy11030578

Patricia ASA, Alline MA, William HST (2019) Germination of spores of Amazonian ferns *Polypodium aureum* in different culture media. Ornam Hortic 25(3). https://doi.org/10.1590/2447-536X.v25i3.1455

Ravi BX, Robert J, Gabriel M (2014) In vitro spore germination and gametophytic growth development of a critically endangered fern *Pteris tripartite* Sw. Afr J Biotechnol 13(23): 2350–2358

Rechenmacher C, Schmitt JL, Droste A (2010) Spore germination and gametophyte development of *Cyathea atrovirens* (Langsd. & Fisch.) Domin (Cyatheaceae) under different pH conditions. Braz J Biol 70(4):1155–1160

Reyno AA (2010) Sporophyte and gametophyte development of *Platycerium coronarium* (Koenig) Desv. and *Platycerium grande* (Fee) C. Presl. (Polypodiaceae) through *in vitro* propagation. Saudi J Biol Sci 17(1):13–22

Sara SC (2001) Conservation of selected rare and endangered ferns of the Western Ghats through micropropagation and restoration. PhD Thesis, Manonmaniam Sundaranar University, Tirunelveli, India

Sara CS, Manickam VS (2005) *In vitro* propagation of the critically endangered fern *Cyathea crinita* (Hook.) Copel. Asian J Microbiol Biotechnol Environ Sci 7:527-536

Sara CS, Manickam VS (2007) *In vitro* developmental ontogeny and life cycle of a rare fern species- *Thelypteris confluens* (Thunb.) Morton. Indian J Biotechnol 6:372–380

Sara SC, Manickam VS, Antonisamy R (1998) Regeneration in kinetin - treated gametophytes of *Nephrolepis multiflora* (Roxb.). Jarret Morton Curr Sci 75(5):503–508

Shibila T, Johnson M (2016) Ex situ conservation of *Elaphoglossum nilgiricum* (Krajina Ex Sledge and Bonner) T. Moore Ex Alston & Bonner using *in vitro* spore culture. Austin J Biotechnol Bioeng 3(2):1064–1067

Shukla SP, Khare PB (2014) *In vitro* conservation of some threatened and economically important ferns belonging to the Indian subcontinent. J Bot 2014:1–8

Symonds WL, Camphell LC, Clemens J (2001) Response of ornamental eucalyptus from acidic and alkaline habitats to potting medium pH. Sci Hortic 88:121–131

Vallinayagam S, Manickam VS, Seeni S (2002) Multiplication of *Diplazium cognatum* through *in vitro* culture. Phytomorphology 52:23–27

Viviani D, Randi AM (2008) Effects of pH, temperature and light intensity on spore germination and growth analysis of young sporophytes of *Polypodium lepidopterist* (Pteridophyta, Polypodiaceae). Rodriguesia 59(4):751–760. https://doi.org/10.1590/2175-7860200859406

Zhang KM, Shi L, Zhang XC, Jiang CD, Tim-Chun WL (2008) Gametophyte morphology and development of six Chinese species of *Pteris* (Pteridaceae). Am Fern J 98(1):33–41

**Shibila Thangaiah** obtained her PhD in Botany from Manonmaniam Sundaranar University, Tirunelveli District, Tamil Nadu, and is working as an Assistant Professor at Sri Paramakalyani Centre for Excellence in Environmental Sciences, Manonmaniam Sundaranar University, Alwarkurichi, Tamil Nadu. She has published more than two book chapters, 20 proceedings, and 25 research papers in international journals. She has attended 18 national and international conferences and seminars as well as received best paper presentation award in a national conference. She has 7 years of research experience in plant tissue culture and optimized the large-scale propagation protocol for four ferns.

**Johnson Marimuthu**, F.I.F.S., completed his Doctorate in Botany (Plant Tissue Culture) from Manonmaniam Sundaranar University, Tirunelveli District, Tamil Nadu, and is working as Assistant Professor of Botany and Director, Centre for Plant Biotechnology, St. Xavier's College (autonomous), Palayamkottai, Tamil Nadu. He has 21 years of research experience and 18 years of teaching experience. He has published 223 papers in reputed journals, 27 papers in proceedings, and 37 papers as book chapters. He is a recipient of the prestigious SERB Young Scientist—2013, Prof. S. S. Bir Gold Medal—2017, and Certificate of Merit 2014 by Indian Fern Society, India.

# In Vitro Gametophyte Development, Reproductive Biology, and Nitric Oxide Signaling in Ferns

12

Meenam Bhatia and Prem L. Uniyal

**Abstract**

We studied the reproductive biology in a few ferns and the effects of nitric oxide on the development of gametophyte and gametangial production. Longer time in initiation of germination indicates longer viability of the spore, and their gametophytes are comparatively healthier than those germinated earlier. The spore germination percentage in epiphytic ferns is relatively low (30%) and high in semi-aquatic species (95%). Gametophyte growth is faster in epiphytic species as compared to those of aquatic and terrestrial species. In aquatic species, the antheridia appeared much early as compared to those of in terrestrial and epiphytic species. The male gametophytes were found to be smaller, spathulate, and ameristic, whereas bisexual gametophytes were large, cordate, and meristic in composite populations. There was a sexual gap of 15–20 days in the expression of antheridia followed by formation of archegonia suggesting that the gametophytes escaped intragametophytic selfing for sporophyte production. In some case, no sporophyte was produced in isolated population indicating that the possibility of intragametophytic selfing is very rare. The percentage of sporophyte production in composite culture of epiphytic species was highly variable indicating the mixed mating system which provides flexibility in ferns and allows them to exercise both evolutionary and ecologically strategies. The intragametophytic mating is used to establish new populations through long-distance wind dispersal of single, minute spores and simultaneously maintaining significant genetic variation through intergametophytic mating. The germination of spores was found to be increased in *Ceratopteris thalictroides* by sodium nitroprusside (SNP – a nitric oxide donor) treatment at higher concentration (50 μM and

M. Bhatia
Department of Botany, Daulat Ram College, University of Delhi, Delhi, India

P. L. Uniyal (✉)
Department of Botany, University of Delhi, Delhi, India





J. Marimuthu et al. (eds.), *Ferns*, https://doi.org/10.1007/978-981-16-6170-9_12

100 μM) while inhibited at lower concentrations (5 μM and 25 μM). SNP and NO supplementation inhibited the vegetative growth of the thallus. At higher NO concentrations (50 μM and 100 μM), the gametophytes showed abnormal growth, and antheridia formation was promoted, but archegonia formation was inhibited.

**Keywords**

Gametophyte · Fern · Nitric oxide · Prothallus · Signaling · Sodium nitroprusside

## Abbreviations

| | |
|---|---|
| BA | 6-Benzylaminopurine |
| IAA | Indole-3-acetic acid |
| IBA | Indole-3-butyric acid |
| Kin | Kinetin |
| NO | Nitric oxide |
| SNP | Sodium nitroprusside |
| P&T | Parker & Thompson medium |

## 12.1 Introduction

Pteridophytes had been the dominant species of this planet about 280–230 million years ago (Sureshkumar and Ayyanar 2020), but at present they comprise of only 5% of the total vegetation of the world. The group includes over 12,000 species in the world, and India harbors 1100 species of ferns of which nearly 235 species are endemic (Chandra 2000; Dixit 2000); however, Fraser-Jenkins (2008) revisited the list and reported 47 species as endemic to India. Nearly 16% of global pteridophyte species are endangered and 22% are threatened (Brummitt et al. 2016). In India, 160 species are critically endangered, 82 are near threatened, and 113 are rare (Chandra et al. 2008). Fraser-Jenkins (2008) reported the high endemism is found in South India (27 sp.), followed by North Eastern India (7 sp.) and Western Himalayan ferns (2 sp.).

Pteridophytes constitute an important component of tropical and temperate biodiversity. They are ecologically important and have enormous potential as source of food, fodder, fiber, flavoring agents, aromatic oil, perfumes, dyes, and folk remedies. Besides their ornamental and medicinal values, aesthetic appeal, and exquisite foliage pattern which make them popular plants for landscaping, they have also been exploited as pollution indicator, insect repellents, antimicrobial agent, and phytoremediators (Shukla and Khare 2014). *Ceratopteris richardii* is recognized as an excellent model organism in developmental biology and physiological studies to identify and study mutants.

Epiphytic ferns grow in extreme environmental conditions which are very different from those of terrestrial species. Ferns found on bare bark are able to exploit microhabitats with low water availability; they drop their leaves under severe drought. Because of their habitat, the water availability is irregular; thus these plants are tending to endure drought stress. They develop different morphological, anatomical, and physiological adaptations necessary to conquer the epiphytic habit (Ranker and Haufler 2008). These include changes in nutrient and water uptake mechanisms (Watkins Jr et al. 2010), the evolution of desiccation tolerance in the gametophyte generation (Watkins Jr et al. 2007), alteration in sporophytic hydraulic and stomatal systems, and shifts in reproductive strategy (Watkins Jr et al. 2010; Pittermann et al. 2011). Epiphytic ferns show many water balance mechanisms like poikilohydry. Many species from the family Polypodiaceae like *Pyrrosia longifolia* shows a shift from C3 cycle to Crassulacean acid cycle for photosynthesis during transition from gametophyte to sporophyte generation (Martin et al. 2005). Successful epiphytic species would have required major modification in the gametophyte generation to include indeterminate growth, extreme stress tolerance, and an outcrossing breeding system.

In heterosporous taxa the gametophytes are endosporic, which is self-supportive in apt nutrition supply and ensures embryo development; however, in homosporous taxa the gametophytes are exosporic and dependent on the external environmental conditions. Pteridophytes display various other pathways of reproduction through apospory and apogamy (Ranker and Haufler 2008). In some ferns, the gametophytes also show vegetative propagation by means of dispersible gemmae. Gemma production has been commonly observed in epiphytic ferns. The fern *Vittaria appalachiana* is only known from its gametophytes (Chambers and Emery 2016). It multiplies by vegetative buds, or gemmae, and forms mats in dark, moist cavities and rock shelters in the Appalachian Mountains.

Ferns are long-lasting, stress tolerant (water, salt, heavy metals), fungal, and pest resistant and possess a number of useful phytochemicals. They play important role in restoration program. A large number of fern taxa in India have become threatened due to decreasing trend of their natural populations caused by changing land use pattern and overexploitation for commercial purposes. It will be pertinent to understand the diversity, distribution, ecology, reproductive biology, and present status in the context of threat of extinction and endangered situation and propose strategies for their conservation. In vitro culture of spores of ferns has been found beneficial for mass propagation, conservation of germplasm, and phytochemicals analysis. The program gives priority to the curation of plant collections and development of multiplication protocol, addressing reproductive constraints of populations. We studied the reproductive biology, mating system, and colonization potentiality in some fern species such as *Acrostichum aureum* L., *Ceratopteris thalictroides* (L.) Brongn., *Tectaria polymorpha* (Wall. ex Hook.) Copel., *Phymatosorus scolopendria* (Burm.f.) Pic.Serm., *P. membranifolium* (R.Br.) S.G. Lu, *Christella dentata* (Forssk.) Brownsey & Jermy, *Macrothelypteris torresiana* (Gaudich.) Ching, *Platycerium bifurcatum* (Cav.) C. Chr., *Phlebodium aureum (L.) J. Sm.*, and *Allantodia aspera* (Blume) Ching and studied the effects of nitric oxide in the growth and development of gametophyte and gametangial production in *Ceratopteris thalictroides*.

## 12.2 Materials and Methods

Mature fertile fronds with well-developed sori of the abovementioned fern species were collected from Wayanad, Athirappilly, Pooyamkutty, and Calicut Kerala in the month of October 2016. Mature sporophylls were packaged in brown paper packets and stored in a desiccator at room temperature for the release of the spores. The Parker's macro and Thompson's micronutrient culture medium (P&T) was used for the culture of spores (Srivastava and Uniyal 2014). Periodically the spore germination percentage, gametophyte growth, differentiation, and gametangial ontogeny were observed under Olympus CX21 microscope and photographed using Olympus camera EP-1. Gametophytes in stock culture were observed at regular intervals and were isolated in separate petri plates containing P&T medium before initiation of gametangia. The ratios of gametophyte bearing male, female, bisexual, or neuter conditions were recorded. The gametophytes were transferred in other plates in two sets (Set 1, 25 petri plates with single gametophyte in each isolate culture; Set 2, 5 petri plates with 25 gametophytes in each composite culture). After the initiation of the gametangia in stock cultures, watering of all the isolate and composite populations was done from above with sterile distilled water twice a week to facilitate fertilization. Percentage of sporophytes was recorded in both the abovementioned sets (Srivastava et al. 2007). Genetic load and fern species are generally estimated by isolating gametophytes and counting the proportion that does not produced viable sporophytes in laboratory cultures (Klekowski Jr 1970). We calculated isolate potential and sib potential following Peck et al. (1990). Isolate potential (the frequency of successful intragametophytic events) is the number of sporophytes produced by isolated gametophytes divided by the total number of isolated gametophytes for that species. Sib potential (the frequency of successful intergametophytic events) is the number of sporophytes produced by paired gametophytes divided by the total number of gametophytes in the paired treatment for each species.

## 12.3 Results and Discussion

### 12.3.1 In Vitro Spore Culture

In vitro spore culture technique is a very important tool for mass propagation and reintroduction and to study the reproductive biology of fern species. Dyer (1979) discussed the spore storage techniques, spore germination, composition of the medium, and cultural conditions like temperature, light quality, and intensity that influence the gametophyte development. Spore sterilization is necessary before sowing (Wu et al. 2009). Sheffield et al. (2001) suggested that early developments of gametophyte occur better in solid media than liquid media. Chang et al. (2007) reported the light quality, pH, and concentration of sugar in the medium significantly influence the spore germination, early gametophyte development, and development of gametangia. Fern spores are relatively resistant to extreme conditions of the environment and remain viable in active metabolic state for quite long period of time. Viability of the spore varies from species to species. Generally, spore retains viability for 2–3 months, but some may have over a year like *Onychium*, *Pityrogramma*, etc. In some taxa which produce green spores like *Osmunda*, the spore remains viable for few days. Several cases of long viability are reported by Ballesteros (2010).

The spores are grown in artificial medium, the events of germination pattern and growth forms of gametophyte are recorded, and reproductive attributes are studied to establish their mating system. Regenerated sporophytes of rare taxa are hardened and reintroduced in their natural habitat that helps in their conservation in wild. In culture conditions, though spore persists in a viable condition for long, they require adequate moisture, optimal temperature and pH, oxygen, and light. Generally, spore on germination produces a primary rhizoid followed by the formation of an elongated uniseriate germ filament. In some more primitive taxa, in place of a germ filament, a mass or a plate of cells was formed.

For in vitro germination, the sori containing fronds can be sterilized using sodium hypochlorite and sown on Parker's & Thompson's culture media. Two different set of experiments comprising of composite gametophyte culture and isolate gametophytes culture can be established to investigate spore germination percentage, gametophyte development, sexual expression, mating system, reproductive barriers, genetic load, and colonization and regeneration potentiality.

Nayar and Kaur (1971) recognized three distinct types of the homosporous ferns, namely, Polar, Equatorial, and Amorphous, on the basis of patterns of spore germination. Polar germination is further classified into three types: *Anemia*, *Osmunda*, and *Vittaria* types. Equatorial germination is classified into six types: *Cyathea*, *Christiopteris*, *Gleichenia*, *Hymenophyllum*, *Mecodium*, and *Trichomanes* types, on the basis of the plane and sequence of cell divisions. *Vittaria* type is the most common type of germination pattern in ferns. *Osmunda* type is the simplest type of Polar germination exhibited in Osmundaceae. *Anemia* type is characteristic of Anemiaceae and Lygodiaceae. Among Equatorial germination is the *Gleichenia* type, the simplest type which occurs in the Gleicheniaceae, Dipteridaceae, Loxogrammaceae, and many of the Polypodiaceae. A variation of this is found in *Christiopteris* (Polypodiaceae). The *Cyathea* type of spore germination is observed in Cyatheaceae, Loxsomaceae, and Cheiropleuriaceae. Hymenophyllaceae shows a characteristic of tripolar growth. *Trichomanes* type occurs in *Trichomanes*; *Mecodium* type is found in the genus *Mecodium*.

In our study, *Vittaria* type and polar type of spore germination pattern were recorded in most of the species. *Drynaria* type gametophyte development pattern is observed in all epiphytic species presently studied. Much variation is observed among the terrestrial species. *T. dentata* and *M. torresiana* showed *Drynaria* type, *T. polymorpha* showed *Aspidium* type, and *A. aspera* showed *Adiantum* type.

**Fig. 12.1** Development stages of *Ceratopteris thalictroides*. (**a**) Spore, (**b**) filamentous prothallus, (**c**) two-dimensional prothallus, (**d**) spathulate prothallus, (**e**) male prothallus with antheridia, (**f**) asymmetric prothallus with apical notch, (**g**) bisexual cordate prothallus, (**h**) antheridium, (**i**) archegonium, (**j**) isolate population, (**k**) composite population, (**l**) sporophyte ready for hardening

Aquatic species, *C. thalictroides* (Fig. 12.1a, b), and *A. aureum* showed *Ceratopteris* type. Following germination, the spore produced a rhizoid. In case of epiphytic species filament which appeared in 10–15 days after sowing, in aquatic species, it appeared much earlier (less than 10 days), while in terrestrial species, it delayed in *T. polymorpha* (25 days) and *A. aspera* (28 days). Some abnormally enlarged filaments and branched filaments were observed in the dense population as found in *T. polymorpha* and *P. scolopendria*.

On comparing epiphytic, aquatic, and terrestrial fern growth pattern, we found that spores of aquatic and epiphytic ferns germinate early than those of terrestrial. Epiphytic species showed early germination. In *Platycerium bifurcatum* (Bhatia et al. 2018) and *Phlebodium aureum*, the spore germination began in 3–5 days, and in *P. scolopendria* and *P. membranifolium*, it took 7–10 days after spore sowing confirming that the spores of these two taxa require specific substratum and favorable conditions for germination just after released from the plants in nature. In general, the typical germination time was from 3 to 10 days. Among the terrestrial species, germination begun in 2–21 days. *C. dentata* germination begun in 2–3 days, in *Macrothelypteris torresiana* 5–7 days, and in *Tectaria polymorpha* 10 days, and *Allantodia aspera* took 21 days. In *Allantodia aspera* and *T. polymorpha*, spore took longer time in germination, indicating that viability of the spore germination in these two species is longer, and for prolonged period, therefore, population of these are comparatively healthier than other investigated species. Aquatic species took less time in spore germination. In *Ceratopteris thalictroides*, germination begun in 2–3 days, and *Acrostichum aureum* took 4–6 days to germinate (Bhatia et al. 2017). This indicates that plant take advantage of the prevailing moist conditions of the habitat and undergo rapid growth phase to complete the life cycle.

A comparative study on spore germination percentage in all species revealed that the germination percentage of epiphytic ferns is relatively low. The percentage of germination is highest in aquatic fern *A. aureum* (95%) (Bhatia et al. 2017) and lowest in epiphytic fern *P. aureum* (23.03%). However *Allantodia aspera* and *M. torresiana* being terrestrial fern showed germination as low as 24.59% and 26.92%, respectively, while *P. bifurcatum* being epiphytic fern showed as high as 85.39%.

Chang et al. (2007) reported the high germination (63.3%) occurred on Murashige and Skoog's basal medium (half strength) supplemented with 2% sucrose at pH 7.7 under white light condition and red light, far red, blue, and white light caused maximum spore germination. Ranil et al. (2008) reported effective spore germination media, gametophyte morphology, and the successful raising of sporophytes from gametophytes in *Cyathea walkerae* Hook. Majumdar et al. (2010) obtained maximum germination at 0.023 μM IAA and maximum growth of gametophyte at 0.045 μM IAA, and maximum growth of sporophytes was observed in medium supplemented with 0.045 μM IAA+ 0.057 μM KN+ 0.023 μM IBA. Some protocols are optimized to obtain sporophytes in the laboratory in tree fern, *Platycerium* and *Asplenium* (Rybczyński and Mikuła 2011; Camloh and Ambroži-č-Dolinšek 2010; Marszał-Jagacka and Kromer 2011).

### 12.3.2 Gametophyte: Growth Forms and Morphology

The fern gametophytes are haploid, simple in structure and independent of the sporophyte plant, but it has great potential to yield important and significant insights in many areas of plant development research. It can serve as a model system for experimental studies especially morphological studies and genetics (Mikuła et al. 2015). Its small size (1–1/2 cm) offers many advantages to culture in a petri dish, and all aspects of its growth and development can be observed and manipulated in a nondestructive way by using simple tools. Another advantage of using gametophyte is its relatively short maturation period which allows a rapid evaluation of treatment effects and execution of cumulative sequence experiments in a much lesser time as compared to the higher plants.

The fern gametophytes show certifiable criteria for taxonomic and phyletic studies. The morphology of the gametophyte, events of spore germination to the complete development of gametophyte, and gametangia formation are significant to complete and update revision of the species. The significance of the morphological diversity of species to the evolution and systematics of ferns cannot be properly assessed without further knowledge of morphological variation and reproductive strategies of ferns especially of epiphytic species. Fernandez and Revilla (2003) documented the nutritional, environmental, and other factors required for growth and development in different stages of the life cycle of ferns and highlighted some protocols for enhancing their morphogenic and reproductive capacity. De Brum and Randi (2002, 2006) reported the effect of irradiation, temperature, and cryopreservation on spore germination, gametophyte, and sporophyte development. Gametophyte shows a wide range of responses to different light intensities. Light can affect in the growth of the fern from spore germination to sporophyte development in terms of the length of photoperiod or its wavelength. Red light promotes spore germination while far red inhibits (Sou et al. 2015). In addition to far red, blue and UV lights also inhibit germination, and this effect can be neutralized by red light. Manipulation of the sucrose concentration in the medium stimulates the multiplication of the gametophyte, but its absence may induce the production of gemmae (Goller and Rybczynski 2007). Addition of sucrose is also reported to induce apogamy in many ferns (Atallah and Banks 2015), or it may inhibit spore germination as reported in *Bolbitis costata* (Majumdar et al. 2010) or promote it (Sheffield et al. 2001).

The prothallus of majority of ferns is one cell thick that permits to trace growth as a function of increase in surface area. It enables the investigator to make a rapid assay of relative development and to establish precise growth curves for an individual. Wada (2007) reported cell lineages of a prothallus and growth at cellular level. The sequence of effects of chemical treatment or surgical treatment can be observed throughout the entire living organism without killing it. It provides a means of studying morphogenesis at biochemical level. We recorded the events of gametangial initiation, fertilization, and sporophyte development in some epiphytic ferns. The morphological variations of the gametophyte and reproductive strategies of the epiphytic ferns are assessed in vitro, so that these characters can significantly be used in systematic studies of ferns.

A great variation in gametophytes has been exhibited by ferns like standard cordate-thalloid shaped, filamentous, ribbon shaped, strap shaped, and tuberous (Pinson et al. 2017). Cordate-thalloid gametophyte is most common in ferns, which has a distinct meristem in the apical notch and a well-defined midrib. Ribbon-like gametophytes are narrow, elongated, and laterally branched and lack midrib and well-organized meristem like in Loxogrammaceae, Vittariaceae, some Hymenophyllaceae, and some Polypodiaceae. Here the meristem is described as discontinuous marginal meristem (Imaichi 2008). Strap-like gametophyte are intermediates between cordate and ribbon like and much elongated, unbranched, and interrupted midrib and cordate apex with a definite apical meristem like in Grammitidaceae, some Lomariopsidaceae, Elaphoglossaceae, and Polypodiaceae.

Tuberous and filamentous types are perennial and rare in ferns. Filamentous type have branched uniseriate filament with indefinite growth. These are found in *Schizaea* and in some *Hymenophyllaceae*. Tuberous type is subterranean, erect, and irregular in shape with uniseriate growing apex. Tuberous gametophytes are subterranean, cylindrical, or irregular in shape and very slow growing as found in Ophioglossaceae, *Actinostachys*, and *Lophidium* (Schizaeaceae), and *Stromatopleris* (Gleicheniaceae) possess tuberous gametophytes. Such diversity in the morphology of gametophytes enables them to grow indeterminately and branch to form clones of perennial gametophyte. Farrar (2003) documented the significance of gametophyte morphology in the classification of the ferns.

Nayar and Kaur (1971) described seven patterns of development in homosporous ferns, viz., *Adiantum* type, *Aspidium* type, *Ceratopteris* type, *Drynaria* type, *Kaulinia* type, *Marattia* type, and *Osmunda* type, on the basis of sequence of cell divisions during the stages of development, the region at which a meristem is established, and the resultant form of the adult thallus. In all, except the *Marattia* and *Osmunda* types, spore germination leads to a uniseriate, elongated, germ filament composed of barrel-shaped chlorophyllous cells and bearing one or more rhizoids at the basal end. Usually there is a sudden change which occurs in the plane of cell divisions, at a time when germ filament is two to ten0 cells long. However, it is often much delayed in some taxa, such as the Grammitidaceae, and results into more extensive uniseriate stage. Sometimes unfavorable conditions also delay this change. Some genera show no change in the plane of cell divisions, and the prothallus remains filamentous throughout their life as in *Schizaea* and *Trichomanes*.

In epiphytic species, two-dimensional prothallus was achieved in 15–20 days, in aquatic it occurred between 15 and 20 days, while in terrestrial species, it was formed in more than 25 days. Further divisions resulted in the development of spathulate thallus in 28–35 days. One or two terminal hairs on either side of the developing notch may be produced at the apex. The marginal cells in apical region of gametophytes usually became smaller and thick and transform into meristematic cells. Spathulate gametophyte developed in 21–35 days in epiphytic species with earliest emergence in *P. bifurcatum* (Bhatia et al. 2018) and *P. scolopendria* in 25 days, *P. aureum* in 28 days, and *M. nigrescens* in 35 days.

In terrestrial species, spathulate stage was achieved in more than 30 days. *M. torresiana* showed spathulate stage in 20 days, *C. dentata* in 30 days, and *A. aspera* and *T. polymorpha* in 40 and 42 days, respectively. Many unicellular hairs developed on the margins and a few superficial hairs at the spathulate stage in most of the species of our study. In aquatic species, spathulate growth was very fast developed much early on the 15th day in *C. thalictroides*, and *A. aureum* showed spathulate stage (Fig. 12.1c–e) in 20 days.

In epiphytic species cordate gametophyte was achieved much early in 35–42 days. The earliest development was observed in *P. bifurcatum* (35 days) and delayed in *M. nigrescens* showed last (42 days). However, terrestrial ferns showed wide range of days in development of cordate thallus in *C. dentata* and *M. torresiana* in 50 days and *T. polymorpha* and *A. aspera* in 60 days. In *C. thalictroides*, cordate thallus (Fig. 12.1f, g) developed in 28 days, followed by

*A. aureum* in 40 days. In *C. thalictroides*, *A. aureum*, and *A. aspera*, an asymmetric lopsided thalloid shape gametophyte was observed (Bhatia et al. 2017). *P. bifurcatum* (Bhatia et al. 2018), *P. scolopendria*, *P. membranifolium*, and *P. aureum* showed persistent gametophyte which grew for more than 2 years. In the terrestrial species, *C. dentata* and *M. torresiana* gametophyte remained for more than a year. However, they developed profuse clone forming branching in older gametophytes. Perhaps the long-lived gametophyte growth strategy adopted by ferns in the epiphytic habitat applies more generally, and the terrestrial ferns develop the competitive advantage over mosses in the terrestrial habitat.

De Brum and Randi (2002, 2006) and Chang et al. (2007) reported the effects of pH, light quality, presence of different sugars, and different concentrations of sucrose in the medium on spore germination, early gametophyte development, and change in the reproductive phase. Spore storage at low temperature has become a profitable tool for ex situ fern conservation (Li et al. 2010; Quintanilla et al. 2002).

### 12.3.3 Gametangial Sequence and Reproductive Biology

The main functions of gametophyte are sex determination and gamete production and develop a system for mating. Verma (2003) mentioned three types of mating system in homosporous ferns, each expressing different level of inbreeding. These are intragametophytic mating or intragametophytic selfing (mating between gametes developed from same gametophyte), intergametophytic selfing (mating between gametes from different gametophyte developed from same parent sporophyte), and intergametophytic crossing (mating between gametes derived from different sporophytes). The intragametophytic selfing results in a totally homozygous sporophyte. Development of both male and female gametes simultaneously and the gametophyte genotype which must be free from any genetic lethal defect are the two prerequisites for the success of this system. Intergametophytic crossing is equivalent to outbreeding in heterosporous ferns. The survival of the completely homozygous sporophytes may be hampered by deleterious recessive alleles because lethal recessive gene in homozygous condition causes death and removes the individual from the progeny during natural selection. Lethal or deleterious recessive alleles can be eliminated by natural selection in homozygous sporophytes over a long period of time, and alternatively "frequent occurrence of intragametophytic selfing will eliminate recessive lethal effect rapidly" (Sessa et al. 2016). Wubs et al. (2010) reported that these deleterious recessive genes usually expressed as abortive embryos which appear as swollen tissue within the archegonium or small and abnormal sporophytic tissue or abnormalities in the production of the first few fronds or roots.

Haufler (2002) explained the correlation between homozygosity and polyploidy through a model. Wood et al. (2009) reported that homosporous ferns have higher number chromosomes as compared to the heterosporous fern. There is an excellent correlation between the utilization of polyploidy and the evolution of taxa in the homosporous pteridophyta. Polyploidy provided the means to main inter-locus

**Table 12.1** Chronological changes in the sex ratio of a composite culture in *Platycerium bifurcatum* (Cav.) C. Chr

| Sample size | Days after sowing | Male | Female | Neuter | Bisexual |
|---|---|---|---|---|---|
| 20 | 28 | 0 | 0 | 20 | 0 |
| 20 | 35 | 2 | 0 | 18 | 0 |
| 20 | 42 | 9 | 0 | 11 | 0 |
| 20 | 49 | 5 | 1 | 13 | 1 |
| 20 | 56 | 10 | 1 | 4 | 5 |
| 20 | 63 | 12 | 2 | 4 | 0 |
| 20 | 70 | 9 | 0 | 1 | 8 |
| 20 | 77 | 4 | 0 | 1 | 15 |

**Table 12.2** Chronological changes in the sex ratio of a composite culture in *Ceratopteris thalictroides* (L.) Brongn

| Sample size | Days after sowing | Male | Female | Neuter | Bisexual |
|---|---|---|---|---|---|
| 20 | 15 | 10 | 0 | 6 | 4 |
| 20 | 20 | 0 | 2 | 0 | 18 |
| 20 | 25 | 4 | 8 | 2 | 4 |
| 20 | 30 | 2 | 4 | 2 | 12 |
| 20 | 35 | 3 | 6 | 1 | 10 |
| 20 | 40 | 2 | 1 | 0 | 17 |
| 20 | 45 | 6 | 0 | 4 | 10 |
| 20 | 50 | 13 | 4 | 0 | 3 |
| 20 | 55 | 2 | 0 | 0 | 18 |
| 20 | 60 | 11 | 0 | 1 | 9 |
| 20 | 65 | 0 | 0 | 0 | 20 |

heterozygosity and release of variability via occasional homoeologous chromosome pairing and recombination (Haufler 2014). Gametangial pattern in the present study showed that all the species studied were protandrous. Antheridia appeared in the terrestrial species with earliest in 30 days in *C. dentata* and in 35 days in *M. torresiana*. However, it took more time in *A. aspera* (42 days) and *T. polymorpha* (55 days). Antheridia appeared much early in the aquatic species with earliest in *C. thalictroides* (Fig. 12.1h) in 15 days and in *A. aureum* in 35 days. In epiphytic species such as *P. bifurcatum*, the antheridia appeared in 35 days (Table 12.1) followed by 40–50 days in *P. scolopendria*, *P. aureum* (Bhatia et al. 2017), and *P. membranifolium*. In the aquatic species, *C. thalictroides* archegonia appeared as early as in 20 days (Fig. 12.1i and Table 12.2) and in *A. aureum* in 40 days. Among the terrestrial species, *C. dentata* and *M. torresiana*, it was formed in 50 days, and *A. aspera* and *T. polymorpha* developed archegonia in 63 days and 65 days, respectively. In case of epiphytic species, archegonia developed in moderate time. In *P. bifurcatum* (Table 12.1) and *P. aureum*, it developed in 49 days and in *P. scolopendria* and *P. membranifolium* in 60–65 days.

**Table 12.3** Breeding behavior of different populations in *Platycerium bifurcatum* (Cav.) C. Chr

| Population | Number of sporophytes produced | Number of sporophytes produced | Percentage of sporophytes produced |
|---|---|---|---|
| Isolation | 25 | 4 | 16 |
| Composite | 75 | 15 | 20 |

**Table 12.4** Breeding behavior of different populations in *Ceratopteris thalictroides* (L.) Brongn

| Population | Number of gametophytes observed | Number of sporophytes produced | % sporophytes produced |
|---|---|---|---|
| Isolation | 25 | 15 | 60.00 |
| Composite | 75 | 55 | 73.33 |

We observed formation of dimorphic gametophytes in the composite population. Male gametophytes were small, spathulate, and ameristic, whereas bisexual gametophytes were large, cordate, and meristic. In *C. thalictroides*, such dimorphism (Fig. 12.1f, g) was prominent and well established under the influence of antheridiogen. Later, these early formed male prothalli may become bisexual as found in *A. aureum*, *C. dentata*, *M. torresiana*, and *A. aspera*. Bisexual gametophyte in *P. aureum* (Bhatia et al. 2017) and *P. bifurcatum* appeared in 70 days (Table 12.1), and in *P. membranifolium* and *P. scolopendria*, it appeared in 80 days. Among the terrestrial species, earliest bisexual gametophyte developed in *M. torresiana* in 50 days, in *C. dentata* and *T. polymorph* in 61–65 days, and in *A. aspera* in 84 days. In aquatic species, bisexual stage appears as early as in 15 days in *C. thalictroides* (Fig. 12.1g and Table 12.2) and in *A. aureum* in 42 days.

There was a sexual gap of 15–20 days in the expression of antheridia followed by the formation of archegonia in *P. membranifolium*, *P. scolopendria*, and *P. bifurcatum* (Bhatia et al. 2018). Therefore the mating between the gametangia originating from an individual gametophyte did not occur, which suggest that the gametophytes escaped intragametophytic selfing for sporophyte production. However, in case of *P. membranifolium*, there was no sporophyte in isolated population; thus the possibility of intragametophytic selfing is very rare. This indicates that though both the sexes developed in the same thallus, the lap in their emergence did not provide enough opportunity for their mating or they may be incompatible. However, there were numbers of antheridiate as well as archegoniate gametophytes in composite culture, which have imparted the gametangial fusion through intergametophytic selfing by the fusion of gametes which arise on two or more gametophytes of common parental origin. Thus intergametophytic mating was shown by *P. scolopendria*, *P. membranifolium*, and *P. bifurcatum* (Table 12.3).

In aquatic species *C. thalictroides* and *A. aureum* and terrestrial species *C. dentata*, *M. torresiana*, *T. polymorpha*, and *A. aspera*, the sexual gap between antheridia and archegonia was less than 15 days (Table 12.4). In epiphyte *P. aureum*, antheridia appeared with a gap of 5 days after the expression of archegonia. It is suggested that both the sexes occur on an individual gametophyte for a very short gap; hence these genera exhibited possibilities for gametic union between gametes developed either on the same gametophyte or two different gametophytes of common parental origin. The gametophytes remained bisexual with a slight gap of a day. Therefore, they performed reasonable intragametophytic selfing as well as intergametophytic selfing between the gametangia arisen from the common parent origin. As both the gametangia expressed with a short gap of period, therefore, they did not impart cent percent selfing and must have resulted in the reasonable rate of genetic load.

Wubs et al. (2010) studied species to colonization over longer distances and found a mixed mating system in *Asplenium scolopendrium* with outcrossing when possible and occasional selfing when needed. The resulting sporophyte, which is completely homozygous, shed large amounts of spores over time. Each year this creates a bed of gametophytes in the vicinity of the mother plant. Any unrelated spore, which arrives, is then selectively favored to reproduce and contribute its genes to the new population. Thus, while selfing facilitates initial colonization success, inbreeding depression promotes genetically diverse populations through outcrossing. The results provide further evidence against the overly simple dichotomous distinction of fern species as either selfing or outcrossing. Page (2002) discussed some important ecological strategies of ferns and allied plants and their underlying selection pressures, based on an extensive survey of tropical and temperate species. Khare (2003) described *Dryopteris cochleata* as a good colonizer in disturbed areas due to the presence of intragametophytic selfing, intergametophytic selfing, as well as intergametophytic crossing. Lott et al. (2003) suggested that the mating system of ferns shows enough plasticity to colonize a diversity of habitat. Outcrossing is density dependent; thus intragametophytic selfing would likely be the primary mode of reproduction in the pioneer population. As the density increases, opportunities for outcrossing would be greater as number of spores and resulting gametophytes increases. The presence of potential antheridial system supports this hypothesis, since they increase gametophyte density, particularly of males by inducing dark germination of spores. Khare et al. (2005) concluded that the species has less genetic diversity and less genetic load; thus it is a potential colonizer. Srivastava and Uniyal (2014) reported that *Polystichum lentum* is a good colonizer because sufficient numbers of sporophytes were developed by selfing but a few plants were found in the natural habitat of its occurrence.

Sex in an organism may be determined genetically or environmentally or by the interaction of the two. In organisms that exhibit environmental sex determination, individuals in poor quality habitat, low nutrient availability, high population density, or who are slow growing would expect to develop into males (DeSotot et al. 2008; Quintanilla et al. 2007). In many ferns antheridiogen, a pheromone secreted by bisexual gametophyte can control the sex expression in undifferentiated spores present in the in the surrounding. Though it is generally species specific, some of them show a broad range of plants to act upon (Jimenez et al. 2008). In many homosporous ferns including *C. richardii*, $A_{CE}$ has been shown experimentally to effect the gender decision of undifferentiated spores. The absence of antheridiogen $A_{CE}$ directs the spores toward hermaphrodite development and contains antheridia, archegonia, and a notch meristem (Tanurdzic and Banks 2004). The lack of $A_{CE}$ is thought to lead to expression of both female and male genes. In the presence of $A_{CE}$, spores tend toward male development through a process called induction, which is thought to be a consequence of the signal transduction system initiated by $A_{CE}$ that leads to the suppression of female genes (Strain et al. 2001). This signal transduction system is reactive to $A_{CE}$ within a narrow window of development, approximately 3–6 days from sowing, and $A_{CE}$ exposure outside of this window has no effect (Ganger and Sturey 2012). Not all spores grown in the presence of antheridiogen develop into males, or its absence leads to total bisexual gametophytes (Atallah and Banks 2015).

Relative position of antheridia and archegonia on the bisexual or unisexual prothalli in a population directly affects the breeding system of the species. This can vary among species, with the presence or absence of antheridiogen system and with density of the gametophyte population. Antheridiogen may act in two possible ways. It may act directly through signal transduction by suppressing female genes or alternatively may slow down the growth and development coupled with limited resources responsible for induction. $A_{CE}$ was reported to have a negative effect on growth of male gametophyte in a dosage-dependent manner in *Woodwardia radicans* (Quintanilla et al. 2007), *Onoclea sensibilis*, and *C. richardii*. $A_{CE}$ effected the growth of bisexual gametophyte as well, but the effect is not dosage dependent, and gametophytes remain receptive to $A_{CE}$ only prior to the decision to become bisexual (Ganger and Sturey 2012).

### 12.3.4 Sporophyte Production

Among the epiphytic species investigated in the present study, *P. bifurcatum*, *P. scolopendria*, and *P. membranifolium* showed intergametophytic mating system, whereas *P. aureum* showed mixed mating type. All terrestrial species studied, viz., *C. dentata*, *M. torresiana*, *A. aspera*, and *T. polymorpha*, exhibited mixed type of mating system showing both intra- and intergametophytic mating type. However, intergametophytic mating was predominant as expressed by high percentage of sporophyte production in composite culture than that of isolate culture. Aquatic species *C. thalictroides* and *A. aureum* exhibited mixed type of mating system. Majority of the ferns studied have potential of both intragametophytic (gametophytic selfing) and intergametophytic (sporophytic selfing) and exhibited mixed mating type.

In the present study, sporophytes were produced in earliest composite culture, in aquatic species *C. thalictroides* in 40 days (Fig. 12.1k and Table 12.4) followed by terrestrial *M. torresiana* in 65 days, *A. aureum* in 77 days, *C. dentata* in 85 days, *T. polymorpha* in 90 days, and *A. aspera* in 120 days. In epiphytes sporophyte emerged little late. In *P. aureum* first sporophyte appeared in 80 days in composite population and 88 days in isolate population. In *P. bifurcatum* sporophyte emerged in 120 days in composite and 145 days in isolated population (Table 12.3). In *P. scolopendria* sporophyte developed in 150 days in composite and 95 days in isolated cultures. In *P. membranifolium* sporophyte developed in 170 days in composite cultures and no sporophyte in isolated culture.

The sporophyte production in composite culture in epiphytic species revealed that *P. bifurcatum* showed minimum sporophyte production percentage (20%) (Table 12.3), followed by *P. scolopendria* (50.66%), *P. membranifolium* (54.66%), and *P. aureum* (80%). Among the terrestrial species, the percentage of sporophyte production in composite culture was minimum in *A. aspera* (29.33%) followed by *T. polymorpha* (36%) and *C. dentata* (44%) and maximum in *M. torresiana* (96%). Aquatic species shows high percentage of sporophyte production (*A. aureum* 68% and *C. thalictroides* 73.33%) (Fig. 12.11 and Table 12.4).

The species *M. torresiana* (96%), *A. aureum* (68%), and *C. thalictroides* (73.33%) with high sporophyte production showed mixed type of mating, i.e., both the intra- and intergametophytic selfing mode were favored in these species. Minimum percentage of sporophytes developed in isolated culture of *A. aureum* (25%), followed by C. *dentata* (32%), *A. aspera* (36%), *T. polymorpha* (56%), *C. thalictroides* (Fig. 12.1j) (60%), and *M. torresiana* (84%).

Majority of the ferns have potential of both intragametophytic (gametophytic selfing) and intergametophytic (sporophytic selfing) and exhibited mixed mating type. Among the epiphytic species, *Phlebodium aureum*, *P. scolopendria*, and *P. membranifolium* showed intergametophytic mating system, whereas *Platycerium bifurcatum* showed mixed mating type. All the terrestrial species studied, *Thelypteris dentata*, *Macrothelypteris torresiana*, *A. aspera*, and *Tectaria polymorpha*, exhibited mixed type of mating system. Aquatic species *Ceratopteris thalictroides* and *Acrostichum aureum* exhibited mixed type of mating system. Less genetic load seems to be found in *M. torresiana* and *C. thalictroides*. Mixed mating type provides flexibility in ferns and allows them to exercise both evolutionary and ecologically strategies. They use intragametophytic mating to establish new populations through long-distance wind dispersal of single, minute spores and simultaneously maintaining significant genetic variation through intergametophytic mating.

Makowski et al. (2016) advocated the application of the in vitro tissue culture and cryopreservation methods for propagation and conservation of the fern *Osmunda regalis*. They reported increased number and early development of sporophyte by integrating two techniques. Ravi (2016) induced polyembryony in *Pteris tripartita* in half strength MS basal media with 3 mg/L of BA and 30% sucrose and found similarity of these sporophytes to the normally developed sporophytes. Fernandez and Revilla (2003) reported that low concentration of ammonia in the media (Knops, Knudson, or 1/4 MS) was most effective for gametophytic growth of *Osmunda regalis*. Gemmae formation was also reported for the first time in this species with sucrose and in darkness.

### 12.3.5 Effects of Nitric Oxide (NO) in the Gametophyte Development and Gametangial Production in *Ceratopteris Thalictroides* L

The immense effects NO on plant cell functioning and plant development and response to changes in the environmental conditions have been studied. Nitric oxide (NO) and its role in physiology especially in regulating blood pressure and relieving heart conditions were first discovered in animal cell. The enzyme responsible for NO generation in animal organisms is nitric oxide synthase (NOS) with the participation of $O_2$and NADPH.

It is has also been found to be involved into the diverse biological activities in plants. Nitric oxide is a crucial signaling molecule with diverse physiological functions in plants. It plays an important role in plant growth and development, starting from germination to flowering, ripening of fruit and senescence of organs, respiratory metabolism, as well as plant response to abiotic and biotic stressors. NO play important role in the disruption of seed dormancy and seed germination (Arc et al. 2013) and root nodules (Boscari et al. 2013). Baudouin and Hancock (2014) discussed NO-dependent protein modification during germination of seeds. NO has a key role in the control of stomatal apertures (Gayatri et al. 2013).

In plants, nitric oxide can be produced by any of the four routes: (1) by nitric oxide synthase, (2) by plasma membrane-bound nitrate reductase, (3) by mitochondrial electron transport chain, or (4) by non-enzymatic reactions. SNP treatment (100 μM) is also reported to inhibit hypocotyl growth in potato, lettuce, and *Arabidopsis* (Beligni and Lamattina 2000). Nitric oxide has been found to effect photosynthesis, photorespiration, and photosynthetic electron transport chain directly.

Growth and development of gametophyte under the influence of sodium nitroprusside (SNP – a NO donor) were recorded in *Ceratopteris thalictroides*. In general germination of spores was found to be increased by SNP treatment at higher concentration (50 μM and 100 μM) while inhibited at lower concentrations (5 μM and 25 μM). The percentage of germination observed in first week was minimum at 5 μM NO (11.35 ± 18.19), and with the increase in NO concentration, the percentage germination increased. The percentage germination at 5 μM and 25 μM was less than that of controls. At 50 μM (91.14 ± 20.09) and 100 μM (94.72 ± 6.95), the percentage germination in first week was much higher than that of the control. The promotion was found to be higher at higher concentrations of the NO (50 μM and100 μM) (Table 12.5).

SNP supplementation in the growth medium inhibited the vegetative growth of the prothallus. The growth was found to be retarded by NO treatment. The growth of the prothallus was recorded in terms of number of prothallial cells and rhizoidal cells. The growth was found to be retarded with the change in concentration of NO as compared to the control. Lower concentrations (5 μM & 25 μM) were found to be more effective in inhibiting the growth. In the second week, the number of prothallial cells observed was very less in all treatments than that of the control (197.7 ± 26.80). The number of rhizoidal cells in control (9.7 ± 1.95) was much higher than that found in all NO treatments. With increase in NO concentration, the number of

**Table 12.5** Average number of prothallial cells and average number of rhizoids produced in different treatments in 2 weeks

| Treatment (μM) | Within in 2 weeks | | Within 6 weeks | |
|---|---|---|---|---|
| | Average no. of prothallial cells | Average no. of rhizoidal cells | Average no. of prothallial cells | Average no. of rhizoidal cells |
| Control | 197.7 ± 26.81 | 9.7 ± 1.95 | 558.4 ± 54.26 | 26.2 ± 5.71 |
| 5 NO | 42.2 ± 15.02 | 3.3 ± 0.82 | 288.3 ± 121.72 | 12.6 ± 5.13 |
| 5NO + MB | 66 ± 16.25 | 5.7 ± 1.25 | 551.5 ± 43.80 | 26.7 ± 2.58 |
| 25 NO | 65.3 ± 15.43 | 3.7 ± 0.78 | 364.7 ± 125.14 | 15.8 ± 5.22 |
| 25 NO +MB | 30.8 ± 10.21 | 3.9 ± 0.74 | 383.6 ± 25.15 | 15.5 ± 3.69 |
| 50 NO | 73.1 ± 36.22 | 3.5 ± 0.85 | 508.1 ± 83.27 | 25 ± 9.66 |
| 50 NO +MB | 147.6 ± 25.31 | 2.5 ± 0.53 | 325.9 ± 55.77 | 14 ± 3.62 |
| 100 NO | 62.5 ± 27.15 | 4.5 ± 0.97 | 378.5 ± 52.28 | 16.5 ± 2.84 |
| 100 NO +MB | 67.7 ± 30.71 | 5.6 ± 2.07 | 401.4 ± 87.69 | 14.1 ± 4.12 |

prothallial cells as well as rhizoids increased, indicating that growth inhibition is less effective at higher concentration of SNP (50 and 100 μM). With increase in NO concentration, the number of prothallial cells increased indicating that growth inhibition is less effective at higher concentration of SNP (NO). However, 50 μM SNP has no inhibitory effect on the number of prothallial cells; rather it was found to promote the number of rhizoids. The number of rhizoid (15.5 ± 6.36) was much higher than that of control (10.6 ± 2.17) (Table 12.5).

Two-dimensional stages reached in 15 days with all the treatments of NO, which was achieved in 7 days in control. However, there was a quantitative increase in the percentage of two-dimensional stage with the increase in NO concentration, but the number of two-dimensional prothalli decreases. Spathulate stage reached in 4 weeks in all the treatments. In bisexual prothalli, usually the basal cell was not divided and remained short and broad. In the fifth week, the thallus developed a notch in the apices and resulted into cordate type in all the treatments. The number of prothallial cells and rhizoids increased with the increase in concentration of SNP, but it was less than that of control. But 50 μM showed exceptionally higher values for the number of prothallial cells and rhizoids; however, it was less than that of control.

At higher NO concentrations (50 μM and 100 μM), growth of the prothalli was showing exceptional results. In some gametophytes the cells of the lower portion were highly elongated. Gametophytes were much asymmetrical and dissected. Some gametophytes were showing abnormal growth as the apical notch was lacking in them. Their middle portion grew continuously extending the two sides like drooping branch. In general, all NO treatments resulted in the browning and degeneration of cells after fifth week. At lower concentration (5 μM NO), it occurred after 5 weeks, but with the increase in concentration, it was found to be delayed. At 25 μM NO and 50 μM NO, it occurred in 6 weeks (Table 12.5), while in 100 μM NO, it took

7 weeks. The primary gametophyte was also found degenerating at its lower portion, and new secondary gametophyte was also developed.

In many homosporous fern species, the sexual generation is featured by sexual dimorphism, where male, female, and bisexual (hermaphroditic) prothalli are formed. Such sexual phenotypes are induced by antheridiogen, which is secreted by the young gametophytes especially at the cordate stage (Tanaka et al. 2014; Atallah and Banks 2015). Antheridiogen is released into the external environment, which induces the development of antheridia in the nearby asexual gametophytes (Strain et al. 2001). In our experiment, *Ceratopteris thalictroides* showed male prothalli, female prothalli, as well as bisexual prothalli.

Antheridia formation was promoted at higher concentrations of NO (50 μM and 100 μM). At these higher concentrations, majority of the prothallus, antheridia developed in large numbers in 2 weeks which took 3 weeks in control. Both the numbers of male prothalli and the numbers of antheridia per prothallus were increased with the concentration of NO. Although under controlled conditions (without NO), the formation of male and female prothalli was recorded; however, no female prothalli were formed under all the treatments of NO. Later a few bisexual prothalli were found at late spathulate or early quadrate stage. In general, development of bisexual stage was delayed by NO treatment. Bisexual prothallus appeared in 28 days in all concentration (5 μM and 25 μM and 100 μM) while concomitant with control at concentration at 50 μM (21 days). The quantitative increase was found in number of bisexual prothalli with the increase in concentration of NO. In 7 weeks (49 days), sporophyte was developed in cultures supplemented with 5 NO only, while in all other treatments, no sporophyte was observed in 7 weeks. In control setup, first sporophyte was produced after 40 days of spore sowing.

Nitric oxide shows a dose-dependent effect on the growth and development. Low micro-molar concentrations produced an increase in the rate of leaf expansion, whereas it does not show any promoting effect at higher concentrations by using NO donor sodium nitroprusside (SNP) as reported in wheat and pea seedlings (Tian and Lei 2006). Low micro-molar concentrations of SNP are found to induce root growth in maize. Treatment with SNP led to the inhibition of primary root length and led to a higher number of lateral roots of tomato seedlings (Correa-Aragunde et al. 2006) and adventitious root formation in cucumber (Pagnussat et al. 2002). SNP treatment enhances the rate of photosynthesis, chlorophyll content, transpiration rate, and stomatal conductance in cucumber seedlings (Fan et al. 2007). SNP has been found to decrease the level of enzymes that regulate photosynthesis in wheat. It is possible that ABA function as an antagonist of antheridiogen is also involved in the formation of antheridia (Romanenko et al. 2020).

## 12.4 Conclusions

The data on the reproductive biology of the species is an essential aspect which would help in propagation and multiplication of the threatened taxa. Information on spore germination studies is prerequisite for large-scale multiplication. Species showing low regeneration capacities are sorted and screened to find out the problems associated with their regeneration. The study is one step forward toward multiplying species with the help of biotechnology, and efforts will be directed toward their reintroduction to the natural habitats. From conservation point of view, it is important to identify the specific sites, besides the existing locations where reintroduction of the endangered/threatened species can be made through ecological niche modeling.

The data presented here are an important step toward a more thorough understanding of the capacity for intragametophytic mating in homosporous ferns and can serve as a guide for researchers wishing to know about the mating system (s) potentially in use by taxa of interest.

The fern gametophytes are haploid, simple in structure, and independent of the sporophyte plant, but it has great potential to yield important and significant insights in many areas of plant development research. It can serve as a model system for experimental studies especially morphological studies and genetics. Its small size (1–1/2 cm) offers many advantages to culture in a petri dish, and all aspects of its growth and development can be observed and manipulated in a nondestructive way using simple tools. Another advantage of using gametophyte is its relatively short maturation period which allows a rapid evaluation of treatment effects and execution of cumulative sequence experiments in a much lesser time as compared to the higher plants. The prothallus of majority of ferns is one cell thick that permits to trace growth as a function of increase in surface area. It enables the investigator to make a rapid assay of relative development and to establish precise growth curves for an individual and help in tracing cell lineages of prothalli and development at cellular level. The sequence of effects of chemical treatment or surgical treatment can be observed throughout the entire living organism without killing it. It provides a means of studying morphogenesis at biochemical level.

Nitric oxide plays a key role in improving early growth of plants. It can be can be used as a potential stress ameliorant in salt-affected soils. However, further studies are suggested to explore the NO-mediated biochemical mechanisms responsible for various aspects like growth and metabolism, sex expression, and different stress tolerance. It would be interesting to investigate the interaction between NO and plant hormones particularly IAA actively involved in growth activities like spore germination, gametophyte form, sex expression, and sporophyte development of plant species.

**Acknowledgments** We thank the Institute of Eminence, University of Delhi, for the partial financial support. We are grateful to the Malabar Botanical Garden and Institute for Plant Sciences, Kozhikode, Kerala, for the help in the collection and identification of some of the specimens. The help provided by the Botanical Survey of India, NRC, Dehradun, in confirming the identity of the specimens is gratefully acknowledged.

## References

Arc E, Galland M, Godin B, Cueff G, Rajjou L (2013) Nitric oxide implication in the control of seed dormancy and germination. Front Plant Sci 4:346. https://doi.org/10.3389/fpls.2013.00346

Atallah NM, Banks JA (2015) Reproduction and pheromonal regulation of sex type in fern gametophyte. Front Plant Sci 6:1–6

Ballesteros D (2010) Conservation of Fern spores. In: Fernández H et al (eds) Working with ferns: issues and applications. Springer, New York, NY, pp 165–172

Baudouin E, Hancock JT (2014) Nitric oxide signaling in plants. Front Plant Sci 4:553–556. https://doi.org/10.3389/fpls.2013.00553

Beligni MV, Lamattina L (2000) Nitric oxide stimulates seed germination and de-etiolation, and inhibits hypocotyl elongation, three light-inducible responses in plants. Planta 210(2):215–221

Bhatia M, Priti, Uniyal PL, Madhusoodanan PV (2017) Studies on the gametophyte development and reproductive behavior of *Acrostichum aureum* L. Int J Plant Reprod Biol 9(2):89–92

Bhatia M, Kumar B, Uniyal PL (2018) In vitro spore germination, gametophyte ontogeny and sporophyte regeneration in Platycerium bifurcatum (Cav.)C. Chr. Indian Fern J 35:54–66

Boscari A, Meilhoc E, Castella C, Bruand C, Puppo A, Brouquisse R (2013) Which role for nitric oxide in symbiotic N2-fixing nodules: toxic by-product or useful signaling/metabolic intermediate? Front Plant Sci 4:384. https://doi.org/10.3389/fpls.2013.00384

Brummitt N, Aletrari E, Syfert MM, Mulligan M (2016) Where are threatened ferns found? Global conservation priorities for pteridophytes. J Syst Evol 54:604–616

Camloh M, Ambrožič-dolinšek J (2010) *In vitro* regeneration systems of *Platycerium*. In: Fernández H, Kumar A, Revilla A (eds) Working with ferns: issues and applications. Springer, New York, NY, pp 111–125

Chambers SA, Emery NC (2016) Population differentiation and countergradient variation throughout the geographic range in the fern gametophyte *Vittaria appalachiana*. Am J Bot 103(1): 86–98

Chandra S (2000) The ferns of India (enumeration, synonyms and distribution). International Book Distributors, Dehradun, India

Chandra CR, Fraser J, Kumari A, Srivastava A (2008) A summary of the status of threatened Pteridophytes of India. Taiwania 53(2):170–209

Chang HC, Agrawal DC, Kho CL, Wen JL, Chen CC, Tsay HS (2007) In vitro culture of *Drynaria fortunei,* a fern species source of Chinese medicine "Gu-sui-Bu". In Vitro Cell Dev Biol Plant 43:133–139

Correa-Aragunde N, Graziano M, Chevalier C, Lamattina L (2006) Nitric oxide modulates the expression of cell cycle regulatory genes during lateral root formation in tomato. J Exp Bot 57(3):581–588

De-Brum FR, Randi AM (2002) High irradiance and temperature inhibit the germination of spores of the fern *Rumohra adiantiformis* (Forst.) Ching (Dryopteridaceae). Braz J Bot 25(4):391–396

De-Brum FR, Randi AM (2006) Germination of spores and growth of gametophytes and sporophytes of *Rumohra adiantiformis* (Forst.) Ching (Dryopteridaceae) after spore cryogenic storage. Rev Bras Bot 29(3):489–495

DeSotot L, Quintanilla LG, Méndez M (2008) Environmental sex determination in ferns: effects of nutrient availability and individual density in *Woodwardia radicans*. J Ecol 96:1319–1327

Dixit RD (2000) Conspectus of pteridophytic diversity in India. Indian Fern J 17:77–91

Dyer AF (1979) The culture of fern gametophytes for experimental investigations. In: Dyer AF (ed) The experimental biology of fern. Academic Press, New York, NY, pp 254–291

Farrar DR (2003) Gametophyte morphology and breeding systems in ferns. In: Chandra S. Srivastava M (eds) Pteridology in The New Millennium. Kluwer Academic Publishers, the Netherlands, pp 447–454

Fan H, Guo S, Jiao Y, Zhang R, Li J (2007) Effects of exogenous nitric oxide on growth, active oxygen species metabolism, and photosynthetic characteristics in cucumber seedlings under NaCl stress. Front Agric China 1(3):308–314

Fernandez H, Revilla M A (2003) *In vitro* culture of ornamental ferns. Plant Cell Tissue Organ Cult 73:1–13
Fraser-Jenkins CR (2008) Endemics and pseudo-endemics in relation to the distribution patterns of Indian pteridophytes. Taiwania 53(3):264–292
Ganger M, Sturey T (2012) Antheridiogen concentration and spore size predict gametophyte size in *Ceratopteris richardii*. Botany 90:175–179. https://doi.org/10.1139/B11-097
Gayatri G, Agurla S, Raghavendra AS (2013) Nitric oxide in guard cells as an important secondary messenger during stomatal closure. Front Plant Sci 4:425. https://doi.org/10.3389/fpls.2013.00425
Goller K, Rybczynski JJ (2007) Gametophyte and sporophyte of tree ferns *in vitro* culture. Acta Soc Bot Pol 76(3):193–199
Haufler CH (2002) Homospory 2002: an odyssey of progress in pteridophyte genetics and evolutionary biology. Bioscience 52(12):1081–1093
Haufler CH (2014) Ever since Klekowski: testing a set of radical hypotheses revives the genetics of ferns and lycophytes. Am J Bot 101(12):2036–2042
Imaichi R (2008) Meristem organization and organ diversity. In: Haufler CH, Ranker T (eds) Biology and evolution of ferns and lycophytes. Cambridge University Press, Cambridge, UK, pp 75–106
Jimenez A, Quintanilla LG, Pajaron S, Pangua E (2008) Reproductive and competitive interactions among gametophytes of the allotetraploid fern *Dryopteris corleyi* and its two diploid parents. Ann Bot 102:353–359
Khare PB (2003) Reproductive biology of *Dryopteris cochleata* C.Chr. in relation to its distributional pattern. In: Pandey AK, Dhakal MR (eds) Advances in plant reproductive biology. Narendra Publishing House, Delhi, India, pp 173–178
Khare PB, Behera SK, Srivastava R, Shukla SP (2005) Studies on reproductive biology of a threatened tree fern, *Cyathea spinulosa* Wall. ex. Hook. Curr Sci 89(1):173–177
Klekowski EJ Jr (1970) Reproductive biology of the pteridophyta IV: an experimental study of mating systems in *Ceratopteris thalictroides* (L.) Brongn. Bot J Linn Soc 63(2):153–169. https://doi.org/10.1111/j.1095-8339.1970.tb02547.x
Li Y, Zhang YL, Jiang CD, Wang T, Wang Q, Shi L (2010) Effect of storage temperature on spore viability and early gametophyte development of three vulnerable species of *Alsophila* (Cyatheaceae). Aust J Bot 58(2):89–96
Lott MS, Vollin JC, Pemberton RW, Austin DF (2003) Reproductive biology of the invasive ferns *Lygodium microphyllum* and *L. japonicum* (Schizaeaceae): implications for invasive potential. Am J Bot 90(8):1144–1152
Majumdar B, Chaudhury MD, Majumdar PB (2010) Effect of growth regulators on in vitro propagation of *Bolbitis costata* (wall ex. hook) C. Chr. Assam Univ J Sci Technol Biol Environ Sci 5(1):23–33
Makowski D, Tomiczak K, Rybczyński JJ, Mikuła A (2016) Integration of tissue culture and cryopreservation methods for propagation and conservation of the fern *Osmunda regalis* L. Acta Physiol Plant 38(19):1–12
Marszał-Jagacka J, Kromer K (2011) *In vitro* propagation of rare and endangered serpentine fern species. In: Hernández H, Kumar A, Revilla MA (eds) Working with ferns: issues and applications. Springer, New York, NY, pp 149–164
Martin SL, Davis R, Protti P, Lin T-C, Lin S-H, Martin CE (2005) The occurrence of Crassulacean acid metabolism in epiphytic ferns, with an emphasis on the Vittariaceae. Int J Plant Sci 166(4): 623–630
Mikuła A, Pozoga M, Tomiczak K, Rybczyński JJ (2015) Somatic embryogenesis in ferns: a new experimental system. Plant Cell Rep 34:783–794
Nayar BK, Kaur S (1971) Gametophytes of homosporous ferns. Bot Rev 37(3):295–396
Page CN (2002) Ecological strategies in fern evolution: a neopteridological overview. Rev Palaeobot Palynol 119:1–33

Pagnussat GC, Simontacchi M, Puntarulo S, Lamattina L (2002) Nitric oxide is required for root organogenesis. Plant Physiol 129(3):954–956

Peck JH, Peck CJ, Farrar DR (1990) Influences of life history attributes on formation of local and distant fern populations. Am Fern J 80:126–142

Pinson BJ, Chambers SM, Nitta JH, Kuo L-Y, Sessa EB (2017) The separation of generations: biology and biogeography of long-lived Sporophyteless Fern gametophytes. Int J Plant Sci 178(1):1–18

Pittermann J, Limm E, Rico C, Christman MA (2011) Structure function constraints of tracheid-based xylem: a comparison of conifers and ferns. New Phytol 192:449–461

Quintanilla LG, Amigo J, Pangua E, Pajaron S (2002) Effect of storage method on spore viability in five globally threatened fern species. Ann Bot 904:461–467

Quintanilla LG, De-Soto L, Méndez M (2007) Do antheridiogens act via gametophyte size? A study of *Woodwardia radicans* (Blechnaceae). Am J Bot 94(6):986–990

Ranil RHG, Pushpakumara DK, Wijesundara DSA (2008) Domestication of *Cyathea walkerae* hook. (Cyatheaceae). J Agric Sci 45:47–58

Ranker TA, Haufler CH (2008) Biology and evolution of ferns and lycophytes. Cambridge University Press, New York, NY

Ravi BK (2016) *In vitro* polyembryony induction in a critically endangered fern, *Pteris tripartita* Sw. Asian Pac J Reprod 5(4): 345–350

Romanenko KO, Babenko LM, Vasheka OV, Romanenko PO, Kosakivska IV (2020) *In vitro* phytohormonal regulation of fern gametophytes growth and development. Russ J Dev Biol 51: 71–83. https://doi.org/10.1134/S106236042002006X

Rybczyński JJ, Mikuła A (2011) Tree ferns biotechnology: from spores to sporophytes. In: Hernández H, Kumar A, Revilla MA (eds) Working with ferns: issues and applications. Springer, New York, NY, pp 135–147

Sessa EB, Testo WL, Watkins EJ Jr (2016) On the widespread capacity for, and functional significance of extreme inbreeding in ferns. New Phytol 211:1108–1119. https://doi.org/10.1111/nph.13985

Sheffield E, Douglas GE, Hearne SJ, Huxham S, Wynn M (2001) Enhancement of fern spore germination and gametophyte growth in artificial media. Am fern J 91(4):179–186

Shukla SP, Khare PB (2014) *In vitro* conservation of some threatened and economically important ferns belonging to the Indian subcontinent. J Bot 2014:1–8, https://doi.org/10.1155/2014/949028

Sou J, Chen S, Zhao Q, Shi L, Dai S (2015) Fern spore germination in response to environmental factors. Front Biol 10:358–376

Srivastava R, Uniyal PL (2014) Developmental studies in an ornamental fern *Polystichum lentum* (D. Don) Moore. Natl Acad Sci Lett 37:281–284

Srivastava R, Srivastava J, Behera SK, Khare PB (2007) *In vitro* studies on development of gametophyte, sex ontogeny and reproductive biology of a threatened fern, *Microsorum punctatum*. Indian J Biotechnol 7: 266–269

Strain E, Hass B, Banks JA (2001) Characterization of the mutations that feminize gametophytes of the fern *Ceratopteris*. Genetics 159(3):1271–1281

Sureshkumar J, Ayyanar M, Silambarasan R (2020) Pteridophyte species richness along elevation gradients in Kolli Hills of the Eastern Ghats, India. J Asia-Pac Biodivers 13(2020):92–106

Tanaka J, Yano K, Aya K, Hirano K, Takehara S, Koketsu E, Ordonio L, Park S, Nakajima M, Ueguchi-Tanaka M, Matsuoka M (2014) Antheridiogen determines sex in ferns via a spatio-temporally split gibberellin synthesis pathway. Science 346(6208):469–473

Tanurdzic M, Banks J (2004) Sex-determining mechanisms in land plants. Plant Cell 16:S61–S71

Tian X, Lei Y (2006) Nitric oxide treatment alleviates drought stress in wheat seedlings. Biol Plant 50(4):775–778

Verma SC (2003) Some aspects of reproductive biology of the gametophyte generation of homosporous ferns. In: Chandra S, Srivastava M (eds) Pteridology in the new millennium. Springer, Dordrecht, India, pp 455–484. https://doi.org/10.1007/978-94-017-2811-9_31

Wada M (2007) The fern as a model system to study photomorphogenesis. J Plant Res 120(1):3–16. https://doi.org/10.1007/s10265-006-0064-x

Watkins JE Jr, Mack MC, Sinclair TR, Mulkey SS (2007) Ecological and evolutionary consequences of desiccation tolerance in tropical fern gametophytes. New Phytol 176:708–717

Watkins JE Jr, Holbrook NM, Zwieniecki MA (2010) Hydraulic properties of fern sporophytes: consequences for ecological and evolutionary diversification. Am J Bot 97:2007–2019

Wood TE, Takebayashi N, Barker MS, Mayrose I, Greenspoon PB, Rieseberg LH (2009) The frequency of polyploid speciation in vascular plants. Proc Natl Acad Sci U S A 106:13875–13879

Wu H, Chen P-T, Yuan L-P, Chen L-Q (2009) An efficient method for surface sterilization and sowing fern spores *in vitro*. Am Fern J 99(3):226–230

Wubs G, de Groot A, During HJ, Vogel JC, Grundmann M, Bremer P, Schneider H (2010) Mixed mating system in the fern *Asplenium scolopendrium*: implications for colonization potential. Ann Bot 106:583–590

**Meenam Bhatia** is presently working as Associate Professor in the Department of Botany, Daulat Ram College (University of Delhi), Delhi. She did her Masters from Rajasthan University, Jaipur, Rajasthan, and PhD in Botany from the University of Delhi. Dr. Meenam has been teaching Archegoniates to undergraduate students for more than 15 years. Dr. Meenam has been working on pteridophytes for more than 7 years. Her main areas of research are gametophyte development and reproductive biology of ferns and conservation biology. She has done many syllabus-based excursions for plant collection in the Himalaya and Western Ghats.

**Prem L. Uniyal** is presently working as Professor in the Department of Botany, University of Delhi. He did his Masters from HNB Garhwal University, Srinagar, Uttarakhand, and PhD in Botany from the Panjab University, Chandigarh, and has been teaching since last 27 years. Dr. Uniyal is specialized in the Biology of Archegoniates. His areas of research are conservation biology, gametophyte development and reproductive biology of ferns, population biology of conifers, biology of Podostemaceae, and biological activity of lichens. His research group has discovered a few new taxa of mosses and *Ephedra* from India. He organized a few conservation programs with indigenous societies in Himachal Pradesh. He has published 4 books and more than 105 original research papers in peer-reviewed national and international journals; 35 review articles in proceedings, book, and scientific magazine; and 25 popular scientific articles. He has supervised 17 PhD scholars and 22 MPhil students.

# Apogamy, Apospory, Somatic Embryogenesis, and Vegetative Propagation in Ferns: A Review

13

Johnson Marimuthu, Helena Fernández, and Shibila Thangaiah

**Abstract**

The present review summarizes the different morphogenetic events of ferns, viz., apogamy, apospory, somatic embryogenesis, polyembryony, and vegetative propagation. In concrete, we provide information about the importance of culture conditions (mineral media, carbohydrates, phytohormones, and other physical factors such as light). In addition, we pay attention to other aspects affecting morphogenesis such as explants choice and manipulation, the creeping of rhizome, tissue homogenization, and/or wounding. Finally, a molecular approach on some of these processes is uncovered, to give evidence about new promising research lines, committed to deciphering the molecular clues operating behind them, and also to envisage important biotechnological repercussions.

**Keywords**

Apogamy · Apospory · Somatic embryogenesis · Polyembryony · Rhizome divisions · Vegetative bud

J. Marimuthu (✉)
Department of Botany, Centre for Plant Biotechnology, St. Xavier's College (Autonomous), Palayamkottai, India

H. Fernández
Department of Organisms and Systems Biology, Area of Plant Physiology, University of Oviedo, Oviedo, Spain
e-mail: fernandezelena@uniovi.es

S. Thangaiah
Sri Paramakalyani Centre for Excellence in Environmental Sciences, Manonmaniam Sundaranar University, Alwarkurichi, Tamil Nadu, India



J. Marimuthu et al. (eds.), *Ferns*, https://doi.org/10.1007/978-981-16-6170-9_13

## Abbreviations

| | |
|---|---|
| BA | 6-Benzylaminopurine |
| GGB | Green globular bodies |
| HXK | Hexokinase |
| IAA | Indole-3-acetic acid |
| IBA | Indole-3-butyric acid |
| Kin | Kinetin |
| MS | Murashige and Skoog mineral medium |
| NAA | Naphthalene acetic acid |
| PGR | Plant growth regulators |

## 13.1 Introduction

Ferns represent an excellent material for coping with morphogenesis (Fig. 13.1). Harrison quoted that "A world without ferns would be a great deal poorer one for experimental biology." Bulk of experimental investigations carried out to date dealt with the reproductive and developmental physiology of this plant group. In line with it, there are many contributions focusing on the ontogeny of ferns through in vitro spore culture and both generations of its vital cycle (Vallinayagam et al. 2002; Fernández and Revilla 2003; Manickam et al. 2003b; Johnson and Manickam 2006; Zenkteler 2006; Goller and Rybcynski 2007; Khan et al. 2008; Biradar et al. 2011; Johnson and Manickam 2011; Liao and Wu 2011; Mazumder et al. 2011; Shukla and Khare 2012; Winarta and Da Silva 2012; Archana et al. 2013; Ashraf et al. 2013; Bharati et al. 2013; Supriya et al. 2013; Archana et al. 2014; Baskaran et al. 2014; Haddad and Bayerly 2014; Shukla and Khare 2014; Vargas and Droste 2014; Aldea et al. 2016; Shibila and Johnson 2016; Gayathiri et al. 2018; Johnson and Shibila 2018; Cyaria and Anirudra 2019; Fonseka 2020; Park et al. 2021).

Typically, the life cycle of ferns (Fig. 13.1) encompasses the alternation of two generations, gametophyte and sporophyte, which are generally haploid and diploid, respectively; each one develops into a multicellular body. Sporogenesis takes place in the frond of sporophyte, and through the meiosis process, haploid spore production takes place. Then, mature spores are released from sporangia into the environment, and once germinated, a prothallus is formed, which passes by sequential aspects: filamentous, spatulate, and heart-shaped. In the gametophyte, the formation of male and female sexual organs, antheridia and archegonia, takes place, respectively, producing gametes, whose fusion arises the zygote, the first cell of the sporophyte generation. Apogamy and apospory are fascinating divergences from normal life cycle. Apogamy means sporophyte formation directly from the gametophyte, without sexual gametes fusion. On the other hand, apospory means

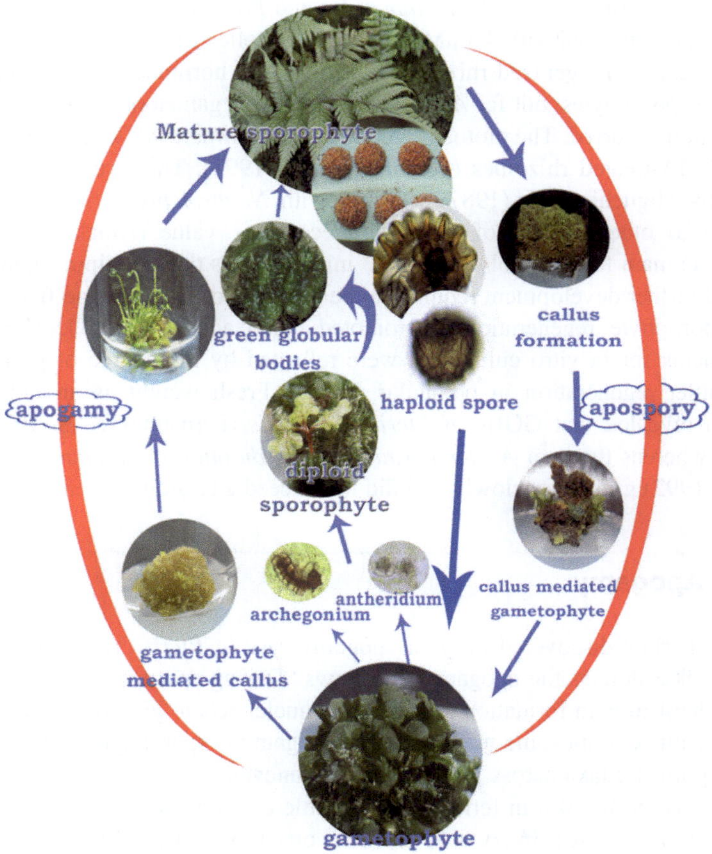


**Fig. 13.1** Life cycle in ferns and some morphogenetic events

gametophyte development from sporophytic tissues, without mediating spores and meiosis (Raghavan 1989).

As it can be seen in Fig. 13.1, both gametophyte and sporophyte allow to carrying out protocols on morphogenesis. It should be said that the gametophyte of the fern lends itself even more than that of seed plants, because it is a free-living structure, and of greater structural complexity. At this regard, it is worth mentioning the induction of dedifferentiated tissue or callus, usually obtained through the addition to the culture medium of developmental regulators, although it also originates in the absence of them, by the simple fragmentation of the tissue (Rivera et al. 2018). In addition, green globular bodies (GGB) are a frequent response, especially from the sporophyte rhizome, in culture media supplemented with cytokinins (Fernández et al. 1997) (Fig. 13.1).

In vitro conditions reveal the morphogenetic potential of both generations. There are numerous papers describing the fern multiplication system in vitro basing on culture of spores, gametophyte, and sporophytes (Rybczyński et al. 2018). The sporophyte and rhizomes of *Asplenium nidus* and *Pteris ensiformis* cultured on MS medium supplemented with 4.4 μM BA became swollen due to bud proliferation, as GGB. Indeed, homogenized rhizomes plated on MS hormone-free medium developed into sporophytes, but for *Asplenium nidus*, both gametophyte and sporophyte regeneration occurred. The protocol accounted approximately 500 sporophytes from 0.5 g of BA-treated rhizomes (Fernández et al. 1997). The term GGB was first coined by Higuchi et al. (1987) working with *Nephrolepis cordifolia* and then extended to other species of ferns with economic value (Amaki and Higuchi 1992). The main factor involved in GGB initiation was the cytokinin supplementation, and further development required to be transferred to a hormone-free medium, when sporophyte regeneration is promoted (Liao and Wu 2011). The various requirements for in vitro cultivation were reflected by the degree of proliferation and plantlet regeneration in particular species. Fresh weight as an indicator of variation revealed that GGBs of *Pteris ensiformis* (Fernández et al. 1996) grew rapidly, whereas those of *Asplenium nidus* and *Adiantum raddianum* (Amaki and Higuchi 1992) grew very slowly and did not exceed a twofold increase in weight.

## 13.2 Apogamy

In 1874, Farlow discovered the first apogamy case in *Pteris cretica*. Later on, in 1979, Walker defines the apogamy as follows "During sporogenesis, a compensatory mechanism with formation of restitution nuclei acts to give rise to diplospores with the same chromosome number as the apogamous sporophytes." The distribution of apomictic taxa across plant lineages is uneven, with estimations of 0.1% in angiosperms, up to 10% in ferns, and with little or no evidence of its existence in gymnosperms, mosses, liverworts, or hornworts (Dyer et al. 2012). Apomixis is especially frequent inside the Dryopteridaceae family, which, together with Pteridaceae, comprises around 70% of the reported apomictic fern species (Liu et al. 2012). At this regard, in vitro culture has been a powerful tool to confirm this process (Fig. 13.2).

In sexual ferns, sporogenesis takes place within the sporangium, and through meiosis, the haploid spores are produced. In the apomictic ferns, the unreduced diplospores are formed mediating the premeiotic endomitosis (PE) or meiotic first division restitution (MFDR) pathways (Manton 1950; Braithwaite 1964; Walker 1985). The diplospores of the apomictic ferns possess full chromosome set and have the potential to germinate (Amanda 2016).

The induction of apogamy is achieved in many ferns, by altering the culture conditions. Thus, the influential role of sugar (sucrose) on induction apogamous sporophyte was observed in *Pteridium aquilinum* (Whittier and Steeves 1960), *Platycerium bifurcatum* (Cav.) C. Chr (Dolinsek and Camloh 1997; Dolinsek et al. 2002), *Cheilanthes viridis* (Manickam et al. 2003b), *Pteris confusa* (Manickam et al. 2003a), *Dryopteris affinis* sp. *affinis* (Menendez et al. 2006), *Pityrogramma calomelanos* L. (Martin et al. 2006), *Ceratopteris richardii* (Cordle et al. 2007, 2011), and *Pityrogramma calomelanos* (L.) Link (Sajeev et al. 2018). Otherwise, the

**Fig. 13.2** Apogamy in *Dryopteris affinis* ssp. *affinis*, from spore cultures on MS medium in vitro. (**a**) Antheridia; (**b**) apogamous embryo near the apical notch; (**c**) apogamous embryo developing its fronds; (**d**) sporophytes emerging from the gametophyte clusters; (**e**) sporophytes growing inside a culture chamber; (**f**) sporophytes transferred to the soil, in the greenhouse. Bar = 1 mm

impact of plant growth regulators was studied since long (Whittier 1964). As examples, the role of ethylene was reported by Elmore and Whittier (1973), and later, among others, in *Dryopteris affinis* ssp. *affinis*, proliferation of apogamous sporophytes was induced on MS medium supplemented with 2% sucrose and the auxin NAA (Menendez et al. 2006). The combinational influence of sucrose and plant growth regulators was also noticed in *Ampelopteris prolifera* (Mehra and Sulklyan 1969), *Adiantum capillus-veneris* (Bhambie and Gupta 1994), *Nephrolepis multiflora* (Sara et al. 1998), *Sphaerostephanos unitus* (Johnson and Manickam 2011), and *Pteris tripartita* (Baskaran et al. 2015).

## 13.3 Molecular Advance on Apogamy

We assist recently to the use of ferns to resolve interesting problems in the plant world caused by abiotic and biotic stress. Concretely, the fern-ally *Selaginella* is one of the most primitive vascular resurrection plants, which can survive a desiccated state and recover when water becomes available, by morphological adaptations, hormonal regulation, antioxidant protection, and accumulation of osmolytes, which could serve to cope with drought in crops (Wang et al. 2010). Recently, a fascinating paper was published based upon the feature that ferns and mosses are rarely infested by phytophagous insect in the field, and in which an insecticidal protein from the fern species *Tectaria macrodonta* (Fee) C. Chr. was identified and expressed in transgenic cotton lines, conferring protection against whitefly, a sap-sucking pest (Shukla et al. 2016).

As mentioned above, ferns apomixis is an important mode of asexual reproduction, which has evolved several times independently within the group (Ekrt and Koutecky 2016). Performing molecular analyses in fern has been elusive, as they exhibit higher chromosome numbers and larger genomes than mosses and seed plants, which makes it difficult to get genomic and transcript sequence data. However, with the advent of the next-generation sequencing (NGS) technologies, it is possible to characterize the transcriptome in plants, which represents a small but information reach target compared to complete genome characterization (Ward et al. 2012). The variation in gene expression, induced by whatever environmental or inner conditions, should be examined also in non-model organisms, because these techniques have become more feasible as automation and efficiency have reduced costs. Therefore, other species than the current models can contribute to throw light and complete the puzzle that always represents to decipher a so complex process.

Until present, some transcriptome and proteome data sets have been published in ferns (Der et al. 2011; Salmi et al. 2010; Cordle et al. 2012; Valledor et al. 2014; Aya et al. 2015; Grossmann et al. 2017; Atallah et al. 2018; Chen et al. 2019; Wyder et al. 2020). The interest of ferns to deepen on reproduction in this plant group is exemplified in some of the aforementioned papers. Some of them refer to reproduction by sexual means and others to asexual by apogamy. The primer include the work of Atallah et al. (2018), who, using RNA-seq, identified genes differentially expressed as sex, being determined in young gametophytes of the model fern species *C. richardii*, when treated with the pheromone antheridiogen. Several of the genes are associated to chromatin remodeling and epigenetic regulation. In line with this research, Chen et al. (2019) carried out proteomic analyses in male and hermaphrodite gametophytes and reported proteins involved in photosynthesis and chaperone proteins overrepresented in hermaphrodites, whereas in male gametophytes, increased proteins involved in metabolism to keep their development under relatively nutritionally deficient conditions. Regarding apogamy, there are just a couple of papers so far. Cordle et al. (2012) investigating on the fern model *C. richardii* reported an apogamy library, where in silico expression analyses of the closest *Arabidopsis* homologs revealed many genes that display preferential expression in seed and flower tissues, structures that are absent in ferns. The same research team recognized recently an AINTEGUMENTA-like unigene, inducing the sporophyte formation without fertilization or apogamy. It mirrors *BABY BOOM* (*BBM*) gene, a transcription factor of AP2/ERF family, which in angiosperms are known to promote somatic embryogenesis (Bui et al. 2017).

Otherwise, the genotypic diversity of apogamous ferns has been studied by employing allozymes, simple sequence repeats (SSRs), inter-simple sequence repeats (ISSRs), and amplified fragment length polymorphisms (AFLPs). They revealed that apomictic lineages contain lower genotypic diversity than sexual ones (Schneller and Krattinger 2010; Ootsuki et al. 2011; Peredo et al. 2013; Grusz and Pryer 2015).

## 13.4 Apospory

Apospory means the capacity of sporophytic tissue to bear gametophytes (Fig. 13.3) without meiosis and spore formation (Raghavan 1989; Fernández et al. 1993; Menéndez et al. 2011). Apospory was first discovered by Druery (1884), as a natural phenomenon, in *Athyrium filix-femina* var. *clarissima* Jones. Aposporous gametophytes have the potency to grow as vegetatively, in clusters, and, as they are fertile, can induce polyploidy series.

To date, various aspects of the induction and subsequent development of aposporous gametophytes have been investigated, and the researchers identified the influential role of several factors such as detachment of organs from the mother plants, sporophytic tissue homogenization and the physiological isolation of sporophytic cells from their neighbors, condition of culture medium, juvenility of the plant material, and the culture media with either no sucrose or only with low concentration of it (Raghavan 1989; Macklis and Beresford 1991; Fernández et al. 1993; Materi et al. 1995; Cheema 1997; Menéndez et al. 2011).

The available literature confirmed that wounding significantly increased the potential for aposporous development (Hirsch 1976; Sheffield and Bell 1981; Von Aderkas 1986; Kwa et al. 1988, 1991; Raghavan 1989; Fernández et al. 1993; Dolinsek and Camloh 1997; Teng and Teng 1997; Dolinsek et al. 2002; Menéndez et al. 2011). In the natural condition, wounding is causing a mechanical and histological damage to the plants; however, with the self-support wound healing mechanism, plant recovers from wounds. Wound healing is not a simple process but a very complex process regulated by a variety of genes (Seo et al. 1997; Hara et al. 2000). The activation of genes leads to enhancement of certain enzyme and endogenous growth substances. The synthesis and movement of the growth substances to damaged sites accelerate the cell division and wound healing process (George 1994).

Wounding is the stimulus for the release of inhibitory substances, accelerates the synthesis of ethylene, and stimulates organogenesis. Due to wounding, ethylene

**Fig. 13.3** Regeneration features from homogenized sporophytes of *Asplenium nidus avis.* (**a**) Piece of tissue mostly browning; (**b**) cell division recovering from the fragmented tissue; (**c**, **d**) apospory evolution; (**e**) sporophyte evolving from one fragmented tissue; (**f**) sporophytes growing on soil. Bars = 1 mm

accumulation was recorded in the tissues (Abeles et al. 1992), which has been reported to be involved in various developmental pathways, including elongation of dark-grown filaments (Miller et al. 1970) and inhibition of rhizoids growth in gametophytes and regeneration of sporophytes (Kwa et al. 1995).

The reduction and profusion of sugar (carbohydrates) concentration can boost or suppress the expression of genes and interrelate with plant growth regulators and other nutrients; as a result, the biochemical, physiological, and molecular development at the cell, organ, and whole plant might be regulated or controlled (Koch 1996). Young et al. (1997) observed that high concentration of sugar enhances apogamy but inhibits apospory induction while low sucrose or glucose ($< / =$ 0.5%) supports apospory induction. They noted that basal medium (without sugar and PGR) failed to support the sporophyte proliferation and apospory induction; the prolonged cultures (explants) in the basal medium are turned to brown due to the lack of sufficient sugar on the agar plates. High sugar in the culture condition or tissues may trigger the overexpression of the HXK (hexokinase) signaling pathway and leads to cell division (Bolouri-Moghaddam et al. 2010). Thus, it is feasible that low sugar levels enhance the induction of apospory by retarding senescence.

The induction of aposporous gametophytes could be achieved either without or by the supplementation of plant growth substances. Apospory on PGR-free medium has been reported in *Microgramma vacciniifolia* (Hirsch 1975), *Asplenium nidus avis* (Fernández et al. 1993; Menéndez et al. 2011), and *Platycerium bifurcatum* (Dolinsek and Camloh 1997; Teng and Teng 1997; Dolinsek et al. 2002). Among the various PGR tested, combinations of cytokinins BAP and Kin reported higher efficiency on the induction of apospory than auxin and cytokinin (BAP) combination (Johnson 2003). The cytokinins Kin and BA in combination with the auxin NAA induced high frequency of gametophytes in *Pyrrosia piloselloides* (Kwa et al. 1988, 1990). In addition, ethylene antagonized the effect of sugar responses; on the other hand, the equation between glucose and ethylene relies on the glucose concentration and the developmental stage of the plant (Cho et al. 2010). Bui et al. (2012) provided evidence for the effects of ethylene on apospory and regeneration of *Ceratopteris richardii*. Unlike apogamy, the molecular basis of apospory is still being untangling.

## 13.5 Somatic Embryogenesis in Ferns

Somatic embryogenesis (SE) is one of the morphogenetic processes through which the somatic cells of plants develop into somatic embryos. Somatic embryos look like zygotic embryos but lack syngamy. In plant biotechnology, SE is one of the more important and versatile methods used for the large-scale clonal propagation, genetic improvement, and experimental material for morphogenetic studies (Yang and Zhang 2010; Mahdavi-Darvari et al. 2015). The induction of somatic embryogenesis is not a simple process, and it involves multifactorial influenced sequential events (Jimenez 2005). This model is thought ideal for inducing embryogenic responses under in vitro conditions (Gaj 2004).

The maturation of somatic embryos and their translation into plants is influenced by media and plant growth regulators supplementation (Filonova et al. 2000; Jimenez 2005). In addition to PGR and media, osmotic shock, pH, wounding, and temporary immersion in a liquid media also triggered the SE induction (Kamada et al. 1989; Smith and Krikorian 1990; Kamada et al. 1993; Akula et al. 2000; Zavattieri et al. 2010).

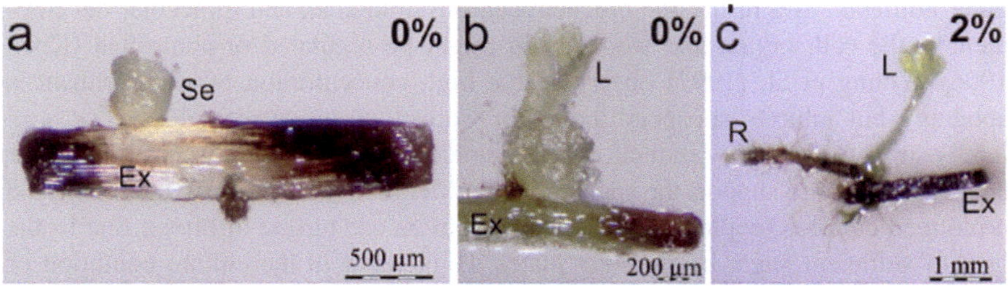


**Fig. 13.4** Somatic embryogenesis in *Cyathea delgadii* (source: Mikuła et al. (2015a, b). (**a**) Somatic embryo growth stopped at the early embryonic leaf stage when grown in darkness; (**b**) somatic embryo which reached the late embryonic leaf stage under photoperiod conditions; (**c**) somatic embryo-derived sporophyte with the first embryonic leaf and root developing in the presence of light and sucrose. (Ex – explants; L – leaf; R – root; Se – somatic embryo). Source: © Mikuła et al. (2015). The figure is reproduced under the terms of creative common attribution 4.0 International License. (Mikuła A, Pożoga M, Grzyb M. et al. (2015) An unique system of somatic embryogenesis in the tree fern *Cyathea delgadii* Sternb.: the importance of explant type, and physical and chemical factors. Plant Cell Tiss Organ Cult 123, 467–478. https://doi.org/10.1007/s11240-015-0850-z)

Steward et al. (1958) reported the first somatic embryogenesis induction from flowering plants, since a number of reports are available on these plants. Limited observations are reported on SE induction from non-seed plants, viz., *Lycopodiella inundata* (L.) Holub (Atmane et al. 2000), *Huperzia selago* (L.) Bernh. Ex Schrank & Mart., *Lycopodium selago* L. (Szypuła et al. 2005), *Ceratopteris richardii* Brongn. (Johnson and Renzaglia 2008, 2009), and *Cyathea delgadii* (Mikuła et al. 2015a, b). Mikuła et al. (2015a, b) confirmed the potential of *C. delgadii* stipe epidermal cells to initiate somatic embryo development (Fig. 13.4). They added that the induction somatic embryo is much easier than in spermatophytes. Mikuła et al. (2015a, b) raised the somatic embryo without any exogenous hormonal supplementation. They pointed out a specific type of explants, i.e., a piece of very young etiolated sporophyte, and maintaining the initial culture in constant darkness supported to regain their totipotency and attained the capability to convert the somatic cells to embryogenic cells. Mikuła et al. (2015a, b) have done an attempt to describe the three different morphogenetic stages of somatic embryo in ferns: (1) linear stage, from the first cell division until the formation of several-celled pro-embryos; (2) early embryonic leaf stage, until the emergence of the first leaf and; and (3) late embryonic leaf stage, until the emergence of the second leaf primordium.

Also the salt requirement for the induction of somatic embryo in ferns is lesser than in spermatophytes. Atmane et al. (2000) achieved the induction of nodular callus and somatic embryos of *Lycopodiella inundata* on half strength mineral salt content supplemented with PGRs. Szypuła et al. (2005) and Mikuła et al. (2015a, b) obtained nodular callus and somatic embryos of *Huperzia selago* on half strength MS hormone-free medium. The induction of somatic embryo may be due to the influence of non-hormonal factors, viz., osmotic shock, drought, wounding, macrosalts, heavy metal ions, and heat or cold shock (Smith and Krikorian 1989; Choi et al. 1998; Patnaik et al. 2005; You et al. 2007; Mikuła et al. 2011).

## 13.6 Polyembryony

Polyembryony (Fig. 13.5) means the development of two or more sporophytes upon a single gametophyte, and it plays an important role in the practical breeding. Among the pteridophytes, however, the phenomenon is less frequent (Klekowski 1970, 1972). The available literature says that two to five embryos were regularly developed on each gametophyte of *Pteridium aquilinum* (L.) Kuhn cultured in Moore's medium containing dimethyl sulfoxide (DMSO) solvent (Sheffield 1984).

Baskaran et al. (2016) showed polyembryony development from a single gametophyte. They observed Di, Tri, Tetra, Hexa, and Octa numbers of polyembryony from 3-month-old spore-derived gametophyte (single) of *P. tripartita* cultured on half strength MS basal media. Baskaran et al. (2016) noted the Di to Octa numbers of polyembryony or juvenile sporophytes in the midrib of gametophytes of *P. tripartita*. Baskaran et al. (2016) studied the influence of BA concentration on the development of Di, Tri, Tetra, Hexa, and Octa numbers of polyembryony or juvenile sporophytes from the gametophytes cultured on half strength MS medium. The gametophytes of *P. tripartita* cultured on 13.2 μM of BA and above the concentration showed the positive signal for the induction of Di, Tri, Tetra, Hexa, and Octa numbers of polyembryony in *P. tripartita* (Baskaran et al., 2016). The multiple numbers of juvenile sporophytes on *D. mollis* and *P. longifolia* gametophyte were reported by Austin (1923). The gametophyte of *D. mollis* and *P. longifolia* was split into two parts, and each portion is able to act separately to do their physiological function. Subsequently each portion of the gametophyte can produce young plants (Austin 1923). Cousens (1979) reported the polyembryony induction in the prolonged cultures of *Blechnum spicant* gametophytes. Recently

**Fig. 13.5** Spontaneous polyembryony *in Dryopteris affinis* ssp. *affinis*

Baskaran et al. (2016) also noted the polyembryony induction in the prolonged cultures of *P. tripartita*. Mottier (1925) recorded the maximum number of sporophytes (11 sporophytes per gametophyte) on the gametophytes of *Matteuccia nodulosa* Fernald, *D. mollis* Maxon, *Osmunda claytoniana* L., and *P. longifolia* L. Sakamaki and Ino (2007) focused on the role of organic matter on the polyembryony induction, and they noticed Di and Tri polyembryony induction from a gametophyte of *Thelypteris palustris* Schott on the knop culture medium. They noted the competition among the polyembryonic sporophytes to absorb the nutrients from the culture medium as well as the light energy for photosynthesis (Sakamaki and Ino 2007).

## 13.7 Vegetative Propagation of Ferns

Ferns are propagated through vegetative methods. The vegetative propagation of fern is by means of stolons (ground layering) and rhizomes (division). Vegetative bud formation on ferns is one of the instances of vegetative reproduction. In proliferous ferns, vegetative buds or gemmae formation takes place on the rachis. These vegetative buds fall on to the ground and start to grow as a new separate plant, and some of the buds start to grow while being attached to the mother plant rachis. Occasionally, some of the ferns are following vegetative bud mediated reproduction, e.g., walking ferns – *Adiantum caudatum*, *Bolbitis appendiculata*, *Bolbitis semicordata*, *Woodwardia radicans*, and *Asplenium zankaranum* (Fig. 13.6). The rhizomes of ferns grow from several inches (*Pteris multiaurita* and *Leptochilus decurrens*) to several feet (*Pteridium*) under the soil surface. The creeping rhizomes of the ferns grow horizontally forward, branching frequently and decaying behind; the adventitious buds of rhizome lead as a separate plant. Rhizome growth may be rapid, as in *Dennstaedtia punctilobula* (Hayscented Fern) and *Adiantum caudatum* (Walking fern) which may become a nuisance in a small garden area. *Osmunda cinnamomea* (cinnamon fern) have very slow rhizome growth and tend to form clumps fronds, forming a good foliage accent in a wildflower garden. Hammen (1990) suggested a propagating method for ferns using rhizome divisions that "rhizome divisions should possess a forward growing tip/sporophytes/leaves." Anita and Srivastava (2007) successfully multiplied the Fishtail holly fern (*Cyrtomium caryotideum*) by rhizome divisions. The matured rhizomes are employed as explants, and they are split into rhizomes with leaves and without leaves. Anitha and Srivastava (2007) observed 90% mortality in the rhizomes without leaves, whereas rhizome with leaves has shown 90–90.2% new sporophytes proliferations. Anita and Srivastava (2007) results were supplemented and validated the Hammen (1990) suggestion. Anita (2012) recorded the survival percentage 70.0–76.7 after 2 years of field establishment. According to Hughes (1966) and Evans et al. (1976), growth of a plant species is influenced by various environmental factors of which light is one of the most important factors. Physical parameters of environment such as light and temperature determine the success and failure of a species in a particular locality. Anita and Srivastava (2007) reported that

**Fig. 13.6** Vegetative buds in ferns. (**a**) *Asplenium zankaranum*; (**b**) *Woodwardia radicans*; (**c**) *Bolbitis semicordata*

*C. caryotideum* grows well in 50% shade. Based on the results of *C. caryotideum*, Anita and Srivastava (2007) suggested that multiplication of ferns using rhizomes is a cost-effective technique, which is easy to conduct, economic to adopt, and convenient to transport also. Johnson et al. (2015) optimized the vegetative propagation protocol for the economically and ecologically important ferns *Elaphoglossum stigmatolepis* (Fée) T. Moore and *Lepisorus nudus* (Hook.) Ching. The effect of auxins on sporophyte and rhizoids formation on rhizomes of *Elaphoglossum stigmatolepis* (Fée) T. Moore and *Lepisorus nudus* (Hook.) Ching is also evaluated. Johnson et al. (2015) employed 5–7 cm in length healthy disease-free and wound-free rhizome divisions with adventitious sprouts for the vegetative propagation. Johnson et al. (2015) employed various concentrations (285.5–1142 μM for IAA; 246–984 μM for IBA; 268.5–1074 μM NAA) of auxin and dipped for 10 min, and distilled water-dipped rhizome served as control. They employed tree fern fibers, coconut husk fibers, soil, and farm yard manure (1:1:1:1) as potted mixture. Among all the treatment, IAA at 856.5 μM found highly effective in inducing adventitious sporophyte formation in the studied two ferns. Similarly, Shibila and Johnson (2015) produced vegetative propagation protocol for the rare and endangered fern *Pteris multiaurita* J. Agardh using rhizome segments as explants. The isolated rhizomes with adventitious sprouts were drenched individually with various concentrations of IBA, IAA, NAA, and BA. Maximum percentage (98 ± 4.47) of sporophyte proliferations were observed in 14.27 μM of IAA supplemented rhizomes of *Pteris multiaurita* (Fig. 13.7). Shibila et al. (2018) produced an alternative vegetative propagation protocol for the medicinally important fern *Pteris otaria* Beddome using rhizome segments as explants. Maximum percentage (79 ± 2.23) of sporophyte proliferations were observed in 14.27 μM of IAA supplemented rhizomes of *Pteris otaria.* Next to that, 76 ± 4.18% of sporophyte proliferation was observed in 11 μM of BA augmented rhizomes. The observed results suggest that the rhizome division-mediated vegetative propagation method may be employed as an alternative protocol to multiply the economically and medicinally important ferns especially the ferns which possess creeping rhizome.

## 13.8 Conclusion

The present review summarizes various morphogenetic events in ferns, such as apogamy, apospory, somatic embryogenesis, polyembryony, and vegetative propagules. The available literature on the different morphogenetic events says about the importance of culture conditions and the influential role on culture conditions, either chemical or physical on these developmental processes. Therefore, each morphogenetic episode requires its own protocol when cultured in vitro. In vitro culture of ferns is deemed as an excellent tool for deepening on the mechanisms involved behind these pathways.

**Fig. 13.7** Vegetative propagation of *Pteris* sp. (**a**) Creeping rhizome of *Pteris* sp.; (**b**) rhizome division cultured on soil media; (**c** and **d**) sporophytes proliferations from rhizome divisions; (**e** and **f**) sporophytes

## References

Abeles FB, Morgan PW, SaltveitJr ME (1992) Ethylene in plant biology, 2nd edn. Academic Press, Inc., San Diego, CA, pp 1–13

Akula A, Becker D, Bateson M (2000) High-yielding repetitive somatic embryogenesis and plant recovery in a selected tea clone TRI-20250 by temporary immersion. Plant Cell Rep 19:1140–1145

Aldea F, Banciu C, Brezeanu A et al (2016) *In vitro* micropropagation of fern species (Pteridophyta) of biotechnological interest, for *ex situ* conservation Muzeul Olteniei Craiova. Oltenia Studii şi comunicări Ştiinţele Naturii Tom 32:27–35

Amaki W, Higuchi H (1992) A possible propagation system of *Nephrolepis*, *Asplenium*, *Pteris*, *Adiantum* and *Rumohra* (Arachniodes) through tissue culture. Acta Hortic 300:237–244

Amanda LG (2016) A current perspective on apomixis in ferns. Syst Evol 54(6):656–665. https://doi.org/10.1111/jse.12228

Anita T (2012) Impact of different treatments in *Gentiana kurroo* and *Cyrtomium caryotideum* rhizome macro-proliferation. Nat J Chembiosis 3(1):11–13

Anita T, Srivastava RK (2007) Vegetative propagation of less known fern - *Cyrtomium caryotideum*- a new approach. Indian For 133(2):261–262

Archana GN, Pradeesh S, Nikhila GS et al (2013) In vitro propagation of a rare medicinal fern of Western Ghats – *Diplazium esculentum* (Reytz.). Ind J Exp Biol 51:919–923

Archana GN, Pradeesh S, Nikhila GS et al (2014) Conservation of vegetable fern (*Diplazium esculentum*) of Western Ghats through *in vitro* spore culture. J Aquat Biol Fish 2(2):43–50

Ashraf MF, Aziz MA, Kadir MA et al (2013) *In vitro* Tuberization of *Chlorophytum Borivilianum* Sant& Fern (Safed Musli) as influenced by sucrose, CCC and culture systems. Plant Cell Physiol 54(8):1356–1364. https://doi.org/10.1093/pcp/pct083

Atallah NM, Vitek O, Gaiti F et al (2018) Sex determination in *Ceratopteris richardii* is accompanied by transcriptome changes that drive epigenetic reprogramming of the Young gametophyte. G3 (Bethesda) 8(7):2205–2214. https://doi.org/10.1534/g3.118.200292

Atmane N, Blervacq AS, Michaux-Ferriere N et al (2000) Histological analysis of indirect somatic embryogenesis in the Marsh clubmoss *Lycopodiella inundata* (L.) Holub (Pteridophytes). Plant Sci 156(2):159–167. https://doi.org/10.1016/s0168-9452(00)00244-2

Austin E (1923) Polyembryony developed under experimental conditions in certain polypodiaceous ferns. Bull Torrey Bot Club 50(3):95–108

Aya K, Kobayashi M, Tanaka J et al (2015) *De novo* transcriptome assembly of a Fern, *Lygodium japonicum*, and a web resource database, Ljtrans DB. Plant Cell Physiol 56(1): e5. https://doi.org/10.1093/pcp/pcu184

Baskaran X, Jeyachandran R, Melghias G (2014) *In vitro* spore germination and gametophytic growth development of a critically endangered fern *Pteris tripartita* Sw. Afr J Biotechnol 13 (23):2350–2358

Baskaran XR, Geo Vigila AV, Rajan K et al (2015) Apogamous sporophyte development through spore reproduction of a South Asia's critically endangered fern: *Pteris tripartita* Sw. Asian Pac J Reprod 4(2):135–139

Baskaran XR et al (2016) *In vitro* polyembryony induction in a critically endangered fern, *Pteris tripartite* Sw. Asian Pac J Reprod 5(4):345–350

Bhambie S, Gupta B (1994) Induced apogamy and apospory in *Adiantum capillus-veneris* L. J Ind Bot Soc 73:25–28

Bharati SK, Manabendra DC, Behari MP (2013) In vitro propagation in Pteridophytes: a review. Int J Res Ayurveda Pharm 4:297–303

Biradar MS, Angadi SG, Mokashi AN et al (2011) *In vitro* performance of Safed musli (*Chlorophytum borivilianum* Sant and Fern.) explants Karnataka. J Agric Sci 24(3):369–370

Bolouri-Moghaddam MR, Roy KL, Xiang L et al (2010) Sugar signaling and anti-oxidant network connections in plant cells. FEBS J 277(9):2022–2037. https://doi.org/10.1111/j.1742-4658.2010.07633.x

Braithwaite AF (1964) A new type of apogamy in ferns. New Phytol 63:293–305

Bui LT, Hurst A, Irish EE et al (2012) The effects of sugars and ethylene on apospory and regeneration in *Ceratopteris richardii*. Am J Plant Sci 3:953–961. https://doi.org/10.4236/ajps.2012.37113

Bui LT, Pandzic D, Youngstrom CE et al (2017) A fern AINTEGUMENTA gene mirrors BABY BOOM in promoting apogamy in *Ceratopteris richardii*. Plant J 90(1):122–132

Cheema HK (1997) Ferns as an excellent experimental system for morphogenesis: an overview. Indian Fern J 14:1–9

Chen X, Chen Z, Huang W et al (2019) Proteomic analysis of gametophytic sex expression in the fern *Ceratopteris thalictroides*. PLoS One 14(8):e0221470. https://doi.org/10.1371/journal.pone.0221470

Cho YH, Sheen J, Yoo SD (2010) Low glucose un-couples Hexokinase1-dependent sugar signaling from stress and defense hormone abscisic acid and C2H4 responses in Arabidopsis. Plant Physiol 152(3):1180–1182. https://doi.org/10.1104/pp.109.148957

Choi YE, Yang DC, Choi KT (1998) Induction of somatic embryos by macrosalt stress from mature zygotic embryos of Panax ginseng. Plant Cell Tissue Org Cult 52:177–181

Cordle AR, Irish EE, Cheng C-L (2007) Apogamy induction in *Ceratopteris richardii*. Int J Plant Sci 168:361–369. https://doi.org/10.1086/511049

Cordle AR, Bui LT, Irish EE et al (2011) Laboratory-induced Apogamy and Apospory in *Ceratopteris richardii*. In: Fernández H, Kumar A, Revilla MA (eds) Working with ferns. Springer, New York, pp 25–36. https://doi.org/10.1007/978-1-4419-7162-3_3

Cordle AR, Irish EE, Cheng CL (2012) Gene expression associated with apogamy commitment in *Ceratopteris richardii*. Sex Plant Reprod 25:293–304

Cousens MI (1979) Gametophytic ontogeny, sex expression and genetic load as measures of population divergence in *Blechnum spicant*. Am J Bot 66:116–132

Cyaria G, Anirudra G (2019) *Ex-situ* conservation of three Himalayan ferns through hydroponic culture under greenhouse conditions. Pleione 13(2):216–225. https://doi.org/10.26679/Pleione13.2.2019.216-225

Der JP, Barker MS, Wickett NJ et al (2011) *De novo* characterization of the gametophyte transcriptome in bracken fern, *Pteridium aquilinum*. BMC Genomics 12:99. https://doi.org/10.1186/1471-2164-12-99

Dolinsek AJ, Camloh M (1997) Gametophytic and sporophytic regeneration from scales of the fern *Platycerium bifurcatum* (Cav.) Chr *in vitro*. Ann Bot 80:23–28

Dolinsek AJ, Camloh M, Bohanec B et al (2002) Apospory in leaf culture of staghorn fern (*Platycerium bifurcatum*). Plant Cell Rep 20:791–796. https://doi.org/10.1007/s00299-001-403-2

Druery CT (1884) Observations on a singular development in the lady fern. J Linn Soc 21:354–360

Dyer RJ, Savolainen V, Schneider H (2012) Apomixis and reticulation in the *Asplenium monanthes* fern complex. Ann Bot 110:1515–1529. https://doi.org/10.1093/aob/mcs202

Ekrt L, Koutecky P (2016) Between sexual and apomictic: unexpectedly variable sporogenesis and production of viable polyhaploids in the pentaploid fern of the *Dryopteris affinis* agg. (Dryopteridaceae). Ann Bot 117:97–106

Elmore HW, Whittier DP (1973) The role of ethylene in the induction of apogamous buds in *Pteridium* gametophytes. Planta 111:85–90. https://doi.org/10.1007/BF00386738

Evans GC, Bainbridge R, Rackham D (eds) (1976) Light as an ecological factor. Blackwell's, Oxford, UK

Fernández H, Revilla MA (2003) *In vitro* culture of ornamental ferns. Plant Cell Tissue Organ Cult 73:1–13

Fernández H, Bertrand AM, Sánchez-Tamés R (1993) *In vitro* regeneration of *Asplenium nidus* L. from gametophytic and sporophytic tissue. Sci Hortic 56:71–77

Fernández H, Bertrand AM, Sánchez-Tamás R (1996) Influence of tissue culture conditions on apogamy in *Dryopteris affinis sp. affinis*. Plant Cell Tissue Organ 45(1):93–97

Fernández H, Bertrand AM, Sánchez-Tamés R (1997) Plantlet regeneration in *Asplenium nidus* L., and *Pteris ensiformis* L., by homogenization of BA treated rhizomes. Sci Hortic 68:243–247

Filonova LH, Bozhkov PV, von Arnold S (2000) Developmental pathway of somatic embryogenesis in *Picea abies* as revealed by time-lapse tracking. J Exp Bot 51:249–264

Fonseka DLCK (2020) Propagation of leather leaf Fern (*Rumohra adiantiformis*) from rhizomes by *in vitro* techniques. Int J Agric Biol Sci 113:113–119

Gaj M (2004) Factors influencing somatic embryogenesis induction and plant regeneration with particular reference to *Arabidopsis thaliana* (L.) Heynh. Plant Growth Reg 43:27–47

Gayathiri M, RamyaRoselin I, Sujatha S et al (2018) *In vitro* life cycle, micromorphological studies and reproductive biology of an anti-diabetic Fern. Int J Curr Microbiol App Sci 7(7):3590–3603

George EF (1994) Plant propagation by tissue culture. Part 1: the technology. Exegetics, Edington, UK, pp 231–272

Goller K, Rybcynski JJ (2007) Gametophytes and sporophyte of tree ferns in vitro culture. Acta Soc Bot Pol 76(3):193–199

Grossmann J, Fernandez H, Chaubey PM et al (2017) Proteogenomic analysis greatly expands the identification of proteins related to reproduction in the Apogamous Fern *Dryopteris affinis* ssp. *affinis*. Front Plant Sci 8:336. https://doi.org/10.3389/fpls.2017.00336-

Grusz AL, Pryer KM (2015) Development of microsatellite markers for the apomictic triploid fern *Myriopteris lindheimeri* (Pteridaceae). Appl Plant Sci. https://doi.org/10.3732/apps.1500061

Haddad S, Bayerly R (2014) In vitro propagation of ferns (*Asplenium nidus*) via spores culture. Jordan J Agric Sci 10:144–150

Hammen LS (1990) Fern propagation. Rhodes Island Wild Plant Society News Letter, 7

Hara K, Yagi M, Koizumi N et al (2000) Screening of wound-responsive genes identifies an immediate-early expressed gene encoding a highly charged protein in mechanically wounded tobacco plants. Plant Cell Physiol 41:684–691

Higuchi H, Amaki W, Suzuki S (1987) In vitro propagation of Nephrolepis cordifolia Prsel. Sci Hortic 37:105–113

Hirsch AM (1975) The effect of sucrose on the differentiation of excised Fern leaf tissue into either gametophytes or sporophytes. Plant Physiol 56(3):390–393. https://doi.org/10.1104/pp.56.3.390

Hirsch AM (1976) The development of aposporous gametophytes and regenerated sporophytes from epidermal cells of excised fern leaves: an anatomical study. Am J Bot 63:263–271

Hughes AP (1966) The importance of light compared with other factors affecting plant growth. In: Bainbridge R (ed) Light as an ecological factor. Blackwell's, Oxford, pp 121–146

Jimenez VM (2005) Involvement of plant hormones and plant growth regulators on in vitro somatic embryogenesis. Plant Growth Reg 47:91–110

Johnson M (2003) In vitro studies on some rare and endangered ferns of Western Ghats, South India. PhD., Thesis Submitted to Manonmaniam Sundaranar University, Tirunelveli, Tamil Nadu, India. https://shodhganga.inflibnet.ac.in/handle/10603/61782

Johnson M, Manickam VS (2006) *In vitro* studies on normal and abnormal life cycle of *Metathelypteris flaccida* (Bl.) Ching. Ethiop J Sci Technol 4(1):37–44

Johnson M, Manickam VS (2011) Ex situ conservation of two threatened ferns of the Western Ghats through *in vitro* spore culture. J Threat Taxa 3(7):1919–1928

Johnson GP, Renzaglia KS (2008) Embryology of *Ceratopteris richardii* (Pteridaceae, tribe Ceratopterideae), with emphasis on placental development. J Plant Res 121:581–592

Johnson GP, Renzaglia KS (2009) Evaluating the diversity of pteridophyte embryology in the light of recent phylogenetic analyses leads to new inferences on character evolution. Plant Syst Evol 283:149–164

Johnson M, Shibila T (2018) In vitro spore culture and ontogeny of *Elaphoglossum stigmatolepsis* (fee.). Moore Phytomorphol 68(1&2):35–38

Johnson M, Revathy I, Shibila T et al (2015) Effect of auxins on vegetative propagation of *Elaphoglossum stigmatolepis* (Fee) T. Moore and *Lepisorus nudus* (Hook.) Ching. Indian Fern J 32(1):16–28

Kamada H, Kobayashi K, Kiyosue T et al (1989) Stress induced somatic embryogenesis in carrot and its application to synthetic seed production. In Vitro Cell Dev Biol 25:1163–1166

Kamada H, Ishikawa K, Saga H et al (1993) Induction of somatic embryogenesis in carrot by osmotic stress. Plant Tissue Cult Lett 10:38–44

Khan S, Raziq M, Kayani HM (2008) *In vitro* propagation of bird's nest fern (*Asplenium nidus)* from spores. Pak J Bot 40(1):91–97

Klekowski EJJ (1970) Populational and genetic studies of a homosporous fern-*Osmunda regalis.* Am J Bot 57:1122–1138

Klekowski EJ (1972) Evidence against genetic self-incompatibility in the homosporous fern *Pteridium aquilinum.* Evolution 26:66–73

Koch KE (1996) Carbohydrate-modulated gene expression in plants. Annu Rev Plant Mol Biol 47:509–540

Kwa SH, Wee YC, Loh CS (1988) Production of aposporous gametophytes from *Drymoglossum piloselloides* (L.) Presl frond strips. Plant Cell Rep 7:142–143

Kwa SH, Wee YC, Loh CS (1990) Production of first and second generation aposporous gametophytes from *Pyrrosia piloselloides* (L.) Price frond strips cultured in vitro. Plant Cell Rep 8:530–533

Kwa SH, Wee YC, Loh CS (1991) Production of aposporous gametophytes and calli from *Pteris vittata* L. pinnae strips cultured *in vitro*. Plant Cell Rep 10:392–393

Kwa SH, Wee YC, Kumar PP (1995) Role of ethylene in the production of sporophytes from *Platycerium coronarium* (Koenig) Desv. Frond and rhizome pieces cultured *in vitro.* J Plant Growth Regul 14(4):183–189. https://doi.org/10.1007/BF00204910

Liao KY, Wu YH (2011) *In vitro* propagation of *Platycerium bifurcatum* (Cav.) C. Chr. via green globular body initiation. Bot Stud 52:455–463

Liu HM, Dyer RJ, Guo ZY et al (2012) The evolutionary dynamics of apomixis in ferns: a case study from polystichoid ferns. J Bot:1–11. https://doi.org/10.1155/2012/510478

Macklis RM, Beresford B (1991) Radiation Hormesis. J Nucl Med 32(2):350–359

Mahdavi-Darvari F, Mohd Noor N, Ismanizan I (2015) Epigenetic regulation and gene markers as signals of early somatic embryogenesis. Plant Cell Tissue Organ Cult 120:407–422

Manickam VS, Irudayaraj V, Johnson M (2003a) Incidence of Apospory in *Pteris confusa* T.G. Walker. Phytomorphol 53(1):73–77

Manickam VS, Vallinayagam S, Johnson M (2003b) Micro propagation and conservation of rare and endangered ferns of the Southern Western Ghats through in vitro culture. In: Chandra S, Srivastava M (eds) Pteridology of new millennium. Kluwer Academic Publishers, Dordrecht

Manton I (1950) Problems of cytology and evolution in the Pteridophyta. Cambridge University Press, Cambridge

Martin KP, Sini S, Zhang CL et al (2006) Efficient induction of apospory and apogamy *in vitro* in silver fern (*Pityrogramma calomelanos* L.). Plant Cell Rep 25:1300–1307

Materi DM, Dionne LA, Cumming BG (1995) In vitro polyploidization of mature ostrich fern sporophytes through rejuvenation. Plant Cell Tissue Organ Cult 42:27–31

Mazumder PB, Mazumder B, Choudhury DM et al (2011) *In vitro* propagation of *Drynaria quercifolia* (L.) J Sm a medicinal Fern. Assam Univ J Sci Technol Biol Environ Sci 7(I):79–83

Mehra PN, Sulklyan DS (1969) *In vitro* studies on apogamy, apospory and controlled differentiation of rhizome segments of the fern, *Ampelopteris prolifera* (Retz.) Copel. Bot J Linn Soc 62:431–443

Menendez V, Villacorta NF, Revilla MA et al (2006) Exogenous and endogenous growth regulators on apogamy in *Dryopteris affinis* (Lowe) Fraser-Jenkins sp. *affinis*. Plant Cell Rep 25:85–91

Menéndez V, Abul Y, Bohanec B et al (2011) The effect of exogenous and endogenous phytohormones on the *in vitro* development of gametophyte and sporophyte in *Asplenium nidus* L. Acta Physiol Plant 33:2493–2500

Mikuła A, Tomiczak K, Rybczyński JJ (2011) Cryopreservation enhances embryogenic capacity of *Gentiana cruciata* (L.) suspension culture and maintains (epi)genetic uniformity of regenerants. Plant Cell Rep 30:565–574

Mikuła A, Pożoga M, Grzyb M et al (2015a) An unique system of somatic embryogenesis in the tree fern Cyathea delgadii Sternb.: the importance of explant type, and physical and chemical factors. Plant Cell Tissue Organ Cult 123:467–478. https://doi.org/10.1007/s11240-015-0850-z

Mikuła A, Pozoga M, Tomiczak K et al (2015b) Somatic embryogenesis in ferns: a new experimental system. Plant Cell Rep 34:783–794

Miller PM, Sweet HC, Miller JH (1970) Growth regulation by ethylene in Fern gametophytes. I. Effects on protonemal and rhizoidal growth and interaction with auxin. Am J Botany 57(2):212–217. https://doi.org/10.2307/2440515

Mottier DM (1925) Polyembryony in certain Polypodiaceae and Osmundaceae. Bot Gaz 80:331–336

Ootsuki R, Shinohara W, Suzuki T et al (2011) Genetic variation in the apogamous fern *Cyrtomium fortune* (Dryopteridaceae). Acta Phytotaxon Geobot 62:1–14

Park K, Jang BK, Lee HM et al (2021) Effective propagation of *Selaginella tamariscina* through optimized medium composition. Agron 11:578. https://doi.org/10.3390/agronomy11030578

Patnaik D, Mahalakshmi A, Khurana P (2005) Effect of water stress and heavy metals on induction of somatic embryogenesis in wheat leaf base cultures. Ind J Exp Biol 43:740–745

Peredo EL, Mendez-Couz M, Revilla M et al (2013) Mating system in *Blechnum spicant* and *Dryopteris affinis* ssp. *affinis* correlates with genetic variability. Am Fern J 103:27–39

Raghavan V (1989) Developmental biology of fern gametophytes. Cambridge University Press, Cambridge, pp 259–295

Rivera A, Conde P, Cañal MJ et al (2018) Biotechnology and Apogamy in *Dryopteris affinis* ssp. *affinis*: the influence of tissue homogenization, auxins, Cytokinins, gibberellic acid and polyamines. In: Fernández H (ed) Current advances in Fern research. Springer, Cham, Switzerland. https://doi.org/10.1007/978-3-319-75103-0_5

Rybczyński J, Tomiczak K, Grzyb M et al (2018) Morphogenic events in ferns: single and multicellular explants *in vitro*. In: Fernández H (ed) Current advances in Fern research. Springer, Cham, Switzerland. https://doi.org/10.1007/978-3-319-75103-0_5

Sajeev S, SavioMelo J, Hegde S (2018) Gamma radiation-induced *in vitro* hormetic apogamy in the fern *Pityrogramma calomelanos* (L.) link. BioSystems. https://doi.org/10.1016/j.biosystems.2018.08.004

Sakamaki Y, Ino Y (2007) Gametophyte contribution to sporophyte growth on the basis of carbon gain in the fern *Thelypteris palustris*: effect of gametophyte organic matter production on sporophytes. J Plant Res 120:301–308

Salmi ML, Bushart TJ, Roux SJ (2010) Cellular, molecular and genetic changes during the development of *Ceratopteris richardii* gametophytes. In: Fernández H, Kumar A, Revilla MA (eds) Working with ferns: issues and applications. Springer, London, UK, pp 11–24

Sara SC, Manickam VS, Antonisamy R (1998) Regeneration in kinetin - treated gametophytes of *Nephrolepis multiflora* (Roxb.). Jarret Morton Curr Sci 75(5):503–508

Schneller J, Krattinger K (2010) Genetic composition of Swiss and Austrian members of the apogamous *Dryopteris affinis* complex (Dryopteridaceae, Polypodiopsida) based on ISSR markers. Plant Syst Evol 286:1–6

Seo S, Sano H, Ohashi Y (1997) Jasmonic acid in wound signal transduction pathways. Physiol Plant 101:740–745

Sheffield E (1984) Effects of dimethyl sulphoxide on the gametophytic and sporophytic phases of *Pteridium aquilinum* (L.) Kuhn. Ann Bot 54:531–536

Sheffield E, Bell PR (1981) Cessation of vascular activity correlated with aposporous development in *Pteridium aquilinum* (L.) Kuhn. New Phytol 88:533–538

Shibila T, Johnson M (2015) Effect of plant growth regulators on vegetative propagation of *Pteris multiaurita* J. Agardh. Indian Fern J 32:31–35

Shibila T, Johnson M (2016) Ex situ conservation of an endangered Fern *Elaphoglossum nilgiricum* (Krajina ex sledge and Bonner) T. Moore ex Alston & Bonner using *in vitro* spore culture. Austin J Biotechnol Bioeng 3(2):1064

Shibila T, Johnson M, Adaikalaraj G et al (2018) Effect of plant growth regulators on vegetative propagation of Pteris otaria Beddome. In: Pteridology today: challenges and opportunities proceedings of the symposium held at BSI, WRC, Pune, pp. 327–332

Shukla SP, Khare PB (2012) In vitro shoot regeneration via caulogenesis in fern *Pteris vittata* L. J Environ Biol 33:683–687

Shukla PS, Khare PB (2014) *In vitro* conservation of some threatened and economically important ferns belonging to the Indian subcontinent. J Bot https://doi.org/10.1155/2014/949028

Shukla AK, Upadhyay SK, Mishra M et al (2016) Expression of an insecticidal fern protein in cotton protects against whitefly. Nat Biotechnol 34(10):1046–1051. https://doi.org/10.1038/nbt.3665

Smith DL, Krikorian AD (1989) Release of somatic embryogenic potential from excised zygotic embryos of carrot and maintenance of proembryonic cultures in hormone-free medium. Am J Bot 76:1832–1843

Smith DL, Krikorian AD (1990) Somatic proembryo production from excised, wounded zygotic carrot embryo on hormone-free medium: evaluation of the effect of pH, ethylene and activated charcoal. Plant Cell Rep 9:34–37

Steward F, Mapes MO, Mears K (1958) Growth and organized development of cultured cells. II. Organization in cultures grown from freely suspended cells. Am J Bot 45:705–708

Supriya D, Choudhury MD, Mazumder PB (2013) In vitro propagation of *Cyathea gigantea* (wall ex. hook). – a tree fern. Int J Recent Sci Res 4(3):221–224

Szypuła W, Pietrosiuk A, Suchocki P et al (2005) Somatic embryogenesis and in vitro culture of *Huperzia selago* shoots as a potential source of huperzine A. Plant Sci 168:1443–1452

Teng WL, Teng MC (1997) In vitro regeneration patterns of leaf cell suspension culture. Plant Cell Rep 16:820–824

Valledor L, Menendez V, Revilla MA et al (2014) Proteomic approaches to sexual development mediated by antheridiogen in the fern *Blechnum spicant* L. Proteomics 14:2061–2071. https://doi.org/10.1002/pmic.201300166

Vallinayagam S, Manickam VS, Seeni S (2002) Multiplication of *Diplazium cognatum* through *in vitro* culture. Phytomorphology 52:23–27

Vargas IB, Droste A (2014) *In vitro* propagation of *Cyathea atrovirens* (Cyatheaceae): spore storage and sterilization conditions. Rev Biol Trop 62(1):299–308

Von Aderkas P (1986) Enhancement of apospory in liquid culture of *Matteuccia struthiopteris*. Ann Bot 57:505–510

Walker TG (1985) Some aspects of agamospory in ferns—the Braithwaite system. Proc R Soc B Biol Sci 86:59–68

Wang X, Li H, Wang M et al (2010) Regulatory networks of somatic embryogenesis in plant. Chin J Biotechnol 26:141–146

Ward JA, Ponnala L, Weber CA (2012) Strategies for transcriptome analysis in non-model plants. Am J Bot 99(2):267–276

Whittier DP (1964) The influence of cultural conditions on induced apogamy in *Pteridium* gametophytes. Am J Bot 51:730–736

Whittier DP, Steeves TA (1960) The induction of apogamy in the bracken fern. Can J Bot 38:925–931

Winarta B, Da Silva JAT (2012) Sterilization procedure for in vitro culture of leather leaf fern (*Rumohra adiantiformis*). Int J Plant Dev Biol 6(1):46–50

Wyder S, Rivera A, Valdés AE et al (2020) Differential gene expression profiling of one- and two-dimensional apogamous gametophytes of the fern *Dryopteris affinis* ssp. affinis. Plant Physiol Biochem 148:302–311. https://doi.org/10.1016/j.plaphy.2020.01.021

Yang X, Zhang X (2010) Regulation of somatic embryogenesis in higher plants. Crit Rev Plant Sci 29:36–57

You XL, Han JY, Choi YE (2007) Plant regeneration via direct somatic embryogenesis in *Panax japonicus*. Plant Biotechnol Rep 1:5–9

Young BD, Weber T, Hass T et al (1997) Generating auto tetraploid sporophytes and their use in Ana-lyzing mutations affecting gametophyte development in the Fern *Ceratopteris*. Genetics 147(2):809–814

Zavattieri MA, Frederico AM, Lima M et al (2010) Induction of somatic embryogenesis as an example of stress-related plant reactions. Electron J Biotech:13. https://doi.org/10.2225/vol13-issue1-fulltext-4

Zenkteler E (2006) Micropropagation of *Matteuccia struthiopteris* (L.) Tod through meristem proliferation from rhizomes. Biodivers Res Conserv 1-2:167–1A73

**Johnson Marimuthu**, F.I.F.S., completed his Doctorate in Botany (Plant Tissue Culture) from Manonmaniam Sundaranar University, Tirunelveli District, Tamil Nadu, and is working as Assistant Professor of Botany and Director, Centre for Plant Biotechnology, St. Xavier's College (autonomous), Palayamkottai, Tamil Nadu. He has 21 years of research experience and 18 years of teaching experience. He has published 223 papers in reputed journals, 27 papers in proceedings, and 37 papers as book chapters. He is a recipient of the prestigious SERB Young Scientist—2013, Prof. S. S. Bir Gold Medal—2017, and Certificate of Merit 2014 by Indian Fern Society, India.

**Helena Fernández** is Associate Professor of Plant Physiology at the University of Oviedo, Spain. She received her PhD in Biology in 1993 from Oviedo University, followed by an EU-funded post-graduate research sojourn at the INRA in Orléans, France, training in HPLC-MS techniques. She got the Ramón y Cajal National Excellence Research Program, dealing with reproduction in ferns. She is guest researcher at the Institute of Plant Biology, University of Zurich, working on apomixes, through omics techniques. She is a member of the dean's team, in the Faculty of Biology at the University of Oviedo. She publishes in relevant journals, having edited two books with Springer.

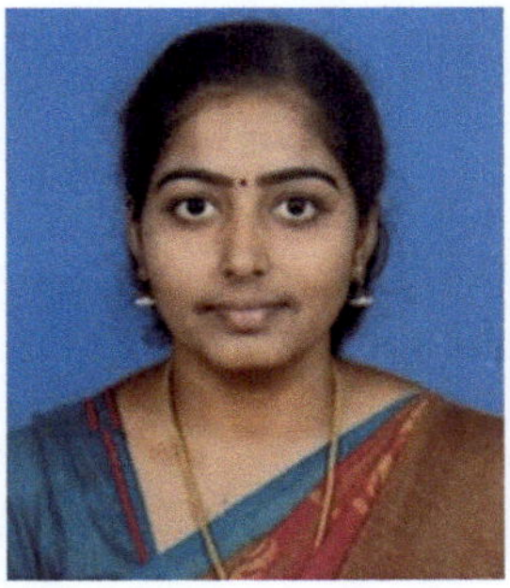

**Shibila Thangaiah** obtained her PhD in Botany from Manonmaniam Sundaranar University, Tirunelveli District, Tamil Nadu, and is working as an Assistant Professor at Sri Paramakalyani Centre for Excellence in Environmental Sciences, Manonmaniam Sundaranar University, Alwarkurichi, Tamil Nadu. She has published more than two book chapters, 20 proceedings, and 25 research papers in international journals. She has attended 18 national and international conferences and seminars as well as received best paper presentation award in a national conference. She has 7 years of research experience in plant tissue culture and optimized the large-scale propagation protocol for four ferns.

# Antagonistic Action of Yucasine and DMSO on Apogamy in the Fern *Dryopteris affinis* ssp. *affinis*

14

Eugenio Sánchez, Alejandro Menéndez, Alejandro Rivera, María Jesús Cañal, and Helena Fernández

**Abstract**

Apogamy is a peculiar case of apomixis, very frequent in ferns, in which an asexual embryo evolves from somatic cells of the gametophyte generation. This work reports about the effect of the yucasine, an inhibitor of the auxin biosynthesis, on vegetative and reproductive development of the gametophyte of *Dryopteris affinis* ssp. *affinis*, an obligate apogamous species. To cope with it, homogenized gametophytes were cultured in vitro in liquid Murashige and Skoog medium, supplemented with sucrose 2% (w/v), and tested up to three concentrations of yucasine (0.47, 4.7, and 23.6 μM). This compound was dissolved using dimethyl sulfoxide (DMSO), and therefore a control with the maximum dose (0.5%) used of this agent was included in the experiment. While yucasine has demonstrated an inhibitory power of gametophyte development as expected, surprisingly, DMSO exhibited an opposite effect, increasing either vegetative development as apogamy. The results inspired us to think that both substances could have been counteracted, throwing new evidence about the role of auxins and the compound DMSO on gametophyte development and apogamy by first time. New research lines are waiting to be started in the coming future.

**Keywords**

Apogamy · Apomixis · Asexual reproduction · Auxin · In vitro culture · Gametophyte

E. Sánchez · A. Menéndez · A. Rivera · M. J. Cañal
Department of BOS, Area of Plant Physiology, Oviedo, Spain
e-mail: eugenio.sanchez01@goumh.umh.es; mjcanal@uniovi.es

H. Fernández (✉)
Department of Organisms and Systems Biology, Area of Plant Physiology, University of Oviedo, Oviedo, Spain
e-mail: fernandezelena@uniovi.es





J. Marimuthu et al. (eds.), *Ferns*, https://doi.org/10.1007/978-981-16-6170-9_14

## Abbreviations

| | |
|---|---|
| DMSO | Dimethyl sulfoxide |
| L/A | Length/width rate |
| MS | Murashige and Skoog basal medium (1962) |
| YUC | 5-(4-clorofenil)-1, 2-dihidro-1, 2, 4-triazol-3-tiocetona or yucasine |

## 14.1 Introduction

Angiosperms form seeds asexually to reproduce, avoiding meiosis and fertilization, thanks to a process called apomixis, whose study attracts many researchers belonging to either public institutions or private companies. A better understanding of the mechanisms that control plant reproduction will not only provide vital information on key aspects of plant development but will also provide tools for the reproductive system manipulation. In agriculture, apomixis has the potential to transform plant breeding, allowing new varieties to retain valuable traits through asexual reproduction (Grossniklaus et al. 1998; Koltunow and Grossniklaus 2003; Ozias-Akins 2006). Certainly, achieving the production of seeds without sexual union is deemed a great challenge; however, the clues operating behind remain elusive.

In ferns, apomixis combines two processes: apogamy (spore formation from gametophyte somatic cells without gamete fusion) and diplospory (production of non-reduced spores) (Ekrt and Koutecký 2015). The life cycle of apomictic ferns begins with the formation of non-reduced chromosomal diplospores that germinate, resulting in gametophytes capable of generating new sporophytes from somatic cells located near the apical zone or meristematic area. The taxonomic distribution of apomixis is uneven, with estimates of 0.1% in angiosperms, and up to 10% in ferns, and with little or no evidence of its existence in gymnosperms, mosses, or livers (Lovis 1978; Asker and Jerling 1992; Dyer et al. 2012). The highest frequency of apomixis in ferns is probably due to repeated and independent evolution in the largest families of Leptosporangiate: Dryopteridaceae, Polypodiaceae, Pteridaceae, and Aspleniaceae (Grusz 2016).

Apogamy is often mandatory due to non-functional formation of archegonia (female sex organs), which prevents sexual reproduction (White 1979). The absence of sexual reproduction in most apogamous ferns makes reproduction less complex than in angiosperms, where apomixis and sexuality often coexist (Liu et al. 2012; Kandemir and Saygili 2015). Recently, many reviews have focused on several aspects of apomixis, including cytological, genetic, and molecular (Fei et al. 2019), and mutagenesis has been shown to induce sex plants to produce results similar to apomictic traits. The later further supports the hypothesis that apomictic and sexual growth might share common elements (Tucker et al. 2003).

Our fern species, *Dryopteris affinis* ssp. *affinis*, is a diploid monilophyte with an obligatory apomictic life cycle, originated through the cross between *Dryopteris oreades* and some sexual ancestor of the species *D. wallichiana* or *D. caucasian* (Fraser-Jenkins 1980), and it is widely distributed in the Mediterranean region, Macaronesia, and western Eurosiberia. When gametophyte acquires the cordiform or shell appearance, it is observed near the apical slit, a group of small, dark brown cells, which will eventually generate an embryo (Menéndez et al. 2006). Being an exclusively apomictic species, it is of particular interest for the study of apogamy and gametophyte development processes.

The activation of cell division and the consequent processes of organogenic differentiation, as well as the organogenesis in the gametophyte of *D. affinis*, are influenced by phytohormones, including auxins (Menéndez et al. 2006; Grossmann et al. 2017; Wyder et al. 2020), which can act among other processes, through the establishment of local gradients (Casanova-Sáez and Voß 2019). The auxin IAA is recognized as a master bioregulator that is involved in a large number of aspects related to plant development, whose endogenous distribution is controlled by biosynthesis, catabolism, and polar transport (Woodward and Bartel 2005). Hormone biosynthesis and transport inhibitors (HBTIs) are useful tools for studying the effect of phytohormones. Yucasine is an inhibitor of the tryptophan-dependent indole-3-pyruvic acid (IPyA) biosynthesis route of IAA in *Arabidopsis* (Mashiguchi et al. 2011; Won et al. 2011). In this pathway, tryptophan is initially converted by Trp aminotransferases (AT) to produce IPyA, which is then oxidized to IAA by YUCCA proteins, a class of flavin monooxygenases present in plants (Dai et al. 2013). YUCCA genes have also been shown to be involved in the biosynthesis of IAA in monocotyledons (Yamamoto et al. 2007; Fujino et al. 2008; Gallavotti et al. 2008). Genetic studies have shown that YUCCA functions as the limiting enzyme for the speed of the IPyA pathway, indicating that YUCCA genes play a crucial role in IAA-regulated development processes (Cheng et al. 2007; Stepanova et al. 2008; Tao et al. 2008; Yamada et al. 2009; Nishimura et al. 2014). Recently, we have identified a gene counterpart that encodes monooxygenase containing *A. thaliana* YUCCA8-like flavin, in the gametophyte of *D. affinis*, which was upregulated in two-dimensional gametophytes tenfold respect to the one-dimensional (Wyder et al. 2020).

The fern gametophyte is a very easy organism to grow and multiply in vitro, being able to provide enough samples to perform molecular analysis (Fernández and Revilla 2003; Grossmann et al. 2017; Wyder et al. 2020). The objective of this work is to explore the effect of yucasine on the vegetative and reproductive development of the gametophyte of the apogamic fern *Dryopteris affinis* ssp. *affinis*.

## 14.2 Material and Methods

### 14.2.1 Collection, Cleaning, and Culture of Spores and Gametophytes

Spores of *D. affinis* ssp. *affinis* were collected from mature fronds obtained in the Valley of Turón (Asturias, Spain), coordinates 43°12′ N and 5°43′ W. Removed fronds were placed between sheets of paper, in a dry environment. After that, spores and sporangia were sieved to separate spores from the rest of plant material, and spores were kept in vials and stored at 4 °C until using. Spores (10 mg) were soaked in water for 2 h, disinfected for 10 min with a solution of NaClO (0.5% w/v) containing Tween 20 (0.1% w/v), and rinsed three times with sterile distilled water. Spores were centrifuged at 700 g for 3 min between rinses. Prior in vitro culture of spores, density was adjusted to around 4000 spores per flask, by using an optical microscope (Nikon Eclipse E-600) and a Fuchs-Rosenthal (Brand) chamber.

Spores were cultured in 100 mL flasks containing 20 mL Murashige and Skoog medium (MS) supplemented with 2% (w/v) sucrose and 0.7% (w/v) agar, and, unless otherwise noted, the pH was adjusted to 5.7 with 1 or 0.1 N NaOH (Murashige and Skoog 1962). The cultures were maintained at 25 °C under cool white fluorescent light (70 $\mu mol/m^2/s^1$) with a 16:8 h light: dark photoperiod. After spore germination, gametophyte development took place, and they were subcultured into the same medium composition, monthly.

### 14.2.2 Homogenized Gametophyte Cultures

Gametophytes (60 mg) were placed into Falcon tubes with 2 ml of liquid MS medium to crush the plant tissue using an Ultra Turrax T 25 Basic homogenizer for 15 s under aseptic conditions at 9500 rpm. Then 2 ml aliquots of sterile Milli-Q water were used to remove the pieces of gametophytes attached to the walls of the tube. The homogenized tissue samples were cultured in 250 ml Erlenmeyer flasks, containing 50 mL of liquid MS medium with 2% (w/v) sucrose, and supplemented with one of the following concentrations of yucasine (0.47, 4.7, and 23.6 μM). To this end, a stock of yucasine was prepared, using dimethyl sulfoxide (DMSO) as a solvent, so another control treatment with this substance was included. Cultures were placed on an orbital shaker and kept at the same conditions cited above for spore cultures. Finally, the flasks were left in the growing chamber for 60 days (Fig. 14.1).

### 14.2.3 Microscopical Examination

Regenerated and fresh gametophytes, derived from the abovementioned treatments, were observed under an optical microscope (Nikon Eclipse E-600), after 60 days of culture initiation. Data about vegetative development and apogamy were scored. The primer refers to the shape of gametophyte (filamentous, spatula and heart-shaped)

**Fig. 14.1** Evolution of homogenized cultures derived from gametophytes of *Dryopteris affinis ssp. affinis*, during 60 days

and the length/width ratio and the latter to apogamy (percentage of apogamous gametophytes and embryo size) (Fig. 14.2).

One hundred individuals from each treatment were taken randomly to determine the frequency of morphotype and the length/width ratio. Apogamy data were determined from 50 heart-shaped gametophytes, at which the embryo is usually well defined, although in some cases, spatulas were also considered as well, with those treatments accelerating the apogamic process.

### 14.2.4 Statistical Analysis

The Chi-square ($\chi^2$) test was applied to non-parametric data, such as morphotype and number of apogamous embryos. The parametric data, i.e., rate length/width and the embryo size, were analyzed by ANOVA, using the Levene and Bartlett test for homogeneity of variances and Shapiro-Wilk for normality. The tests post hoc Tukey HSD and Duncan were used in case there were differences among treatments, in the ANOVA test. Analyses were carried out by RStudio (Team 2016) and set to a significance level of $\alpha = 0.05$.

## 14.3 Results

### 14.3.1 Effect of Yucasine on Vegetative and Apogamous Development

After 60 days of cultivation under controlled conditions, of homogenized gametophyte, data about shape, length/width rate of gametophyte, apogamy frequency, and embryo area were recorded (Fig. 14.3).

(a) **Morphotype**. This parameter refers to the state of development of the gametophyte, i.e., whether they were one-dimensional filaments or two-dimensional, i.e., spatula or heart-shaped, and data are shown in the Fig. 14.3a. Almost all of the filaments present in the cultures were grouped as can be seen in the Fig. 14.2a, which prevented their quantification and measurement. For this reason, we decided to count these groupings as individuals. In addition, the

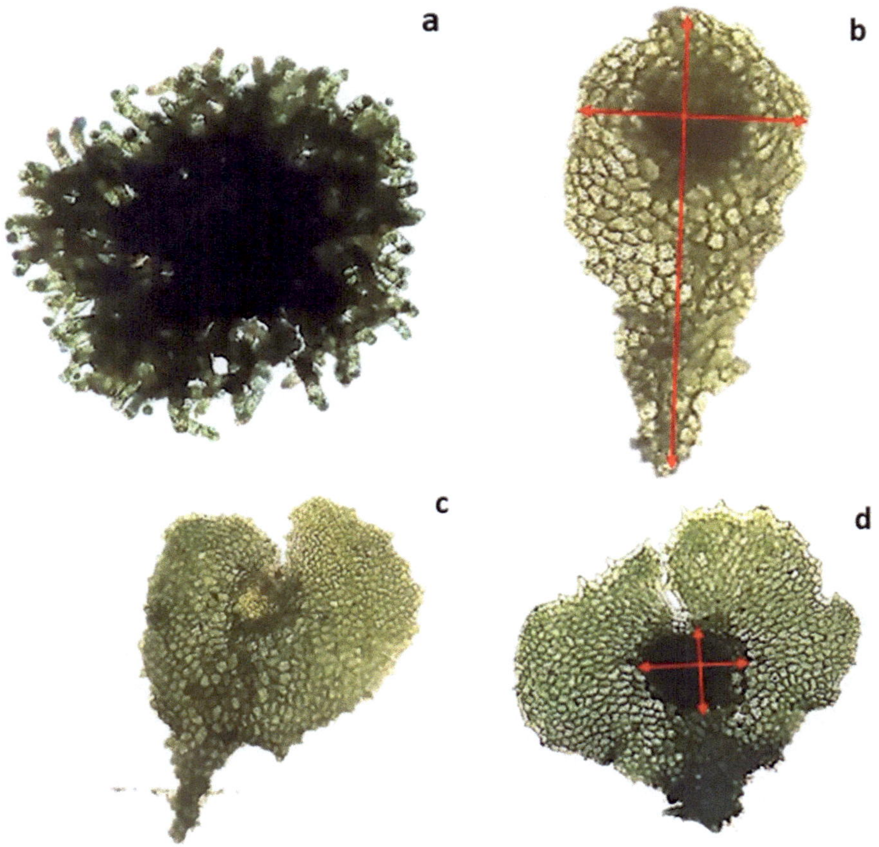


**Fig. 14.2** Morphotypes of regenerated gametophytes derived from homogenized cultures of gametophytes of *Dryopteris affinis* ssp. *affinis*: (**a**) cluster of filaments, (**b**) spatula, (**c**) cordate with a spot embryo, (**d**) cordate with a button embryo. Length and width measures in two-dimensional gametophytes (**c**) and in the apogamous embryo (**d**) are represented by arrows

presence of these filamentous aggregates was virtually negligible and proved not significant because they represented a similar amount in all treatments, always at a value much lower than other morphotypes. With regard to the presence of spatula or heart-shaped gametophytes, statistical data obtained between all treatments showed significant differences ($\chi^2 = 33{,}087$, $p$-value $<0.001$). Comparing each treatment with the control, antagonistic effects were observed by DMSO and yucasine. This solvent agent favored the heart shape ($\chi^2 = 8627$, $p$-value $= 0.003$), while with the treatment with yucasine 0.47 μM gametophytes remained spatulated ($\chi^2 = 5115$, $p$-value $= 0.024$). At the highest dose of yucasine, no significant differences were observed with respect to the control.

(b) **Length/width rate**. Taken together, significant differences were observed for the comparison of all treatment levels ($F = 19.95$, $p$-value $<0.001$) (Fig. 14.3b).

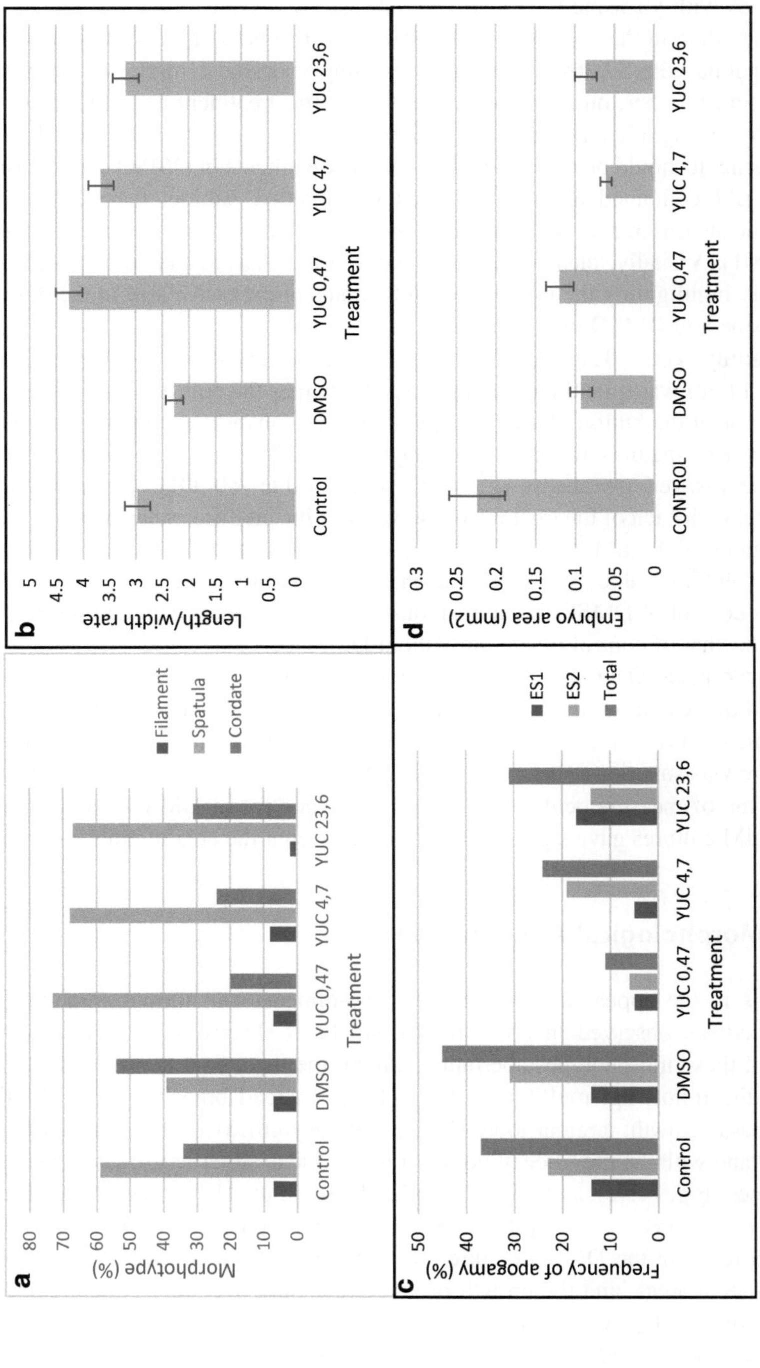


**Fig. 14.3** Effect of yucasine and DMSO in homogenized cultures derived from gametophytes of Dryopteris affinis ssp. affinis on (**a**) morphotype; (**b**) length/width ratio of gametophyte; (**c**) apogamy rate, and (**d**) embryo area. Data after 60 days of culture.

The effect of applying the two lower concentrations of the yucasine inhibitor was also evident in the length/width rate against the control culture. In the cultures with yucasine 0.47 μM, L/A rate was observed 1.49 times higher, on average than in the control ($p$-value $<0.001$). The predominant morphotype is the spatula with 89.36% presence. By adding yucasine 1 mg/L, we get a rate 0.76 times higher, on average, than in the control treatment ($p$-value $<0.001$). Finally, no significant differences in control were found with the highest dose of yucasine. It should be noted that although the treatments of DMSO and yucasine 23.6 μM contained a similar amount of dimethyl sulfide, there is a large significant difference between them in terms of the long/wide rate ($p$-value $<0.001$). Visually, clear differences were also observed between treatments, mainly highlighting the presence of individuals of excessive size in the dose of yucasine and DMSO treatments.

(c) **Apogamy rate**. The presence or absence of embryos in gametophytes (Fig. 14.3c) was quantified, taking note of whether the embryo was in the spot phase or in the button phase. This parameter was inversely proportional to the presence of spatulas in each culture. Significant differences were found between all the assayed treatments ($\chi^2 = 33{,}146$, $p$-value $<0.001$). Comparison of control with each of the treatments showed significant differences with yucasine treatments 0.47 μM ($\chi^2 = 18{,}982$, $p$-value $<0.001$) and yucasine 4.7 μM ($\chi^2 = 3927$, $p$-value $= 0.0475$), causing an inhibition of apogamy. In addition, in the control of DMSO, an apogamous response was observed that practically doubles that of control treatment without DMSO.

(d) **Embryo area**. Data about the embryo area are shown in the Fig. 14.3d. With regard to the control, the treatment of yucasine 23.6 μM is the only one that had significant differences ($p$-value $= 0.0217$). The area of embryos in the control culture was superior to the other treatments, and they were more asymmetrical, in terms of measurements, than the rest. The control of DMSO and yucasine 23.6 μM cultures gave significant differences, being the area of embryo smaller

### 14.3.2 Morphological Abnormalities

Alterations in the appearance of a typical two-dimensional normal gametophyte were sometimes observed in both spatula and heart shapes (Fig. 14.4), and we considered these morphotypes aberrants. Among the most common modifications, there are the following: multilobed (Fig. 14.4a), chained branching (Fig. 14.4b), aberrant basal growth, circular lobes (Fig. 14.4c), asymmetric bilobed (Fig. 14.4d), spiralized and with the presence of hooks or horns due to superior growth in the area of trichomes (Fig. 14.4e), and rhizoids randomly distributed by gametophyte. These aberrations appeared at a low frequency but were representative in two of the treatments used: in the DMSO culture (with the appearance of meristematic cells, in non-habitual areas, and the growth of amorphous and oversized structures), and the cultivation with yucasine 4.7 μM (with the appearance of a large number of filamentous horn-shaped protrusions). Regarding apogamy, treatments of yucasine

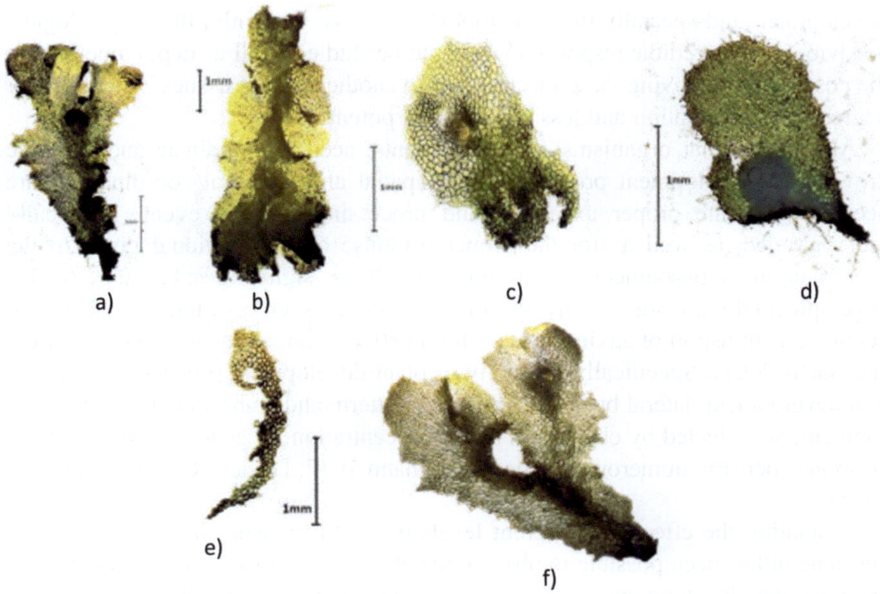


**Fig. 14.4** Abnormalities observed in the regenerated gametophytes derived from homogenized cultures of gametophytes of *Dryopteris affinis* ssp. *affinis*. Data after 60 days of culture

23.6 μM and DMSO had noticeable irregularities in gametophytes with several embryos in each (polyembryony). The treatment with the highest polyembryony was that of DMSO (Fig. 14.4f).

## 14.4 Discussion

This work has evaluated the effects of the auxin synthesis inhibitor, yucasine, on vegetative growth and apogamy in the gametophyte of *Dryopteris affinis* ssp. *affinis*. The results obtained have shed some light on apomixis, a peculiar and frequently occurring process in ferns. Once again, in vitro culture has proven to be useful in dealing with complex biological processes in plants, such as asexual reproduction. Crushing the homogenized tissue from gametophyte breaks the connection between cells and, consequently, the organization of the prothallus, which is committed to maintaining and preserving the set of interrelationships between how many structural and functional elements comprise it. Despite the stress that mechanical fragmentation exerted on the tissue can entail, some cells are able to resume their ability to divide and differentiate, generating a new individual and demonstrating that they contain the genetic information necessary for it. In this case, the gametophyte of *D. affinis* ssp. *affinis* shows a regenerative potential comparable to other species that have been cultivated in our laboratory (Fernández and Revilla 2003; Somer et al. 2010). This demonstrates, once again, the great plasticity that characterizes plant development and specially the gametophyte of ferns. Certainly, this physiological behavior is an incredible response that should be studied in all its depth, opening up the possibility of moving these mechanisms to another type of tissues, with a greater degree of differentiation and less regeneration potential.

All multicellular organisms, including plants, need to coordinate and integrate growth and development processes. Both spatial and temporal coordinations are necessary for the proper execution and processing of each event, genetically preprogramed, as well as for the proper response of an individual given to the ever-changing environmental conditions. In plants, signaling substances (called developmental regulators or phytohormones) mediate such coordinations. The production and transport of auxins have profound effects on plant development (Berleth and Sachs 2001). Specifically, a variety of plant development processes, including root development, lateral bud onset, vascular pattern, and embryonic polarity establishment, are affected by changes in auxin concentration. In general, different plant hormones perform numerous functions (Thimann 1977; Davies 2010; Pacifici et al. 2015).

Evaluating the effects of different levels of treatment with the auxin inhibitor, yucasine, it has been possible to observe the effects that this type of substance may have on the development of fern gametophyte, which is novel. As far as our knowledge, no previous work has been done in ferns and even less in the gametophyte. Our results showed a different morphology between the different treatment levels, where the control differed with the treatment of yucasine 0.47 μM significantly and to a greater extent with that of DMSO alone. Specifically, the treatment of the lowest level of yucasine turned out to be the one with the least heart-shaped gametophytes, and on the contrary, that of DMSO was the treatment with the greatest number of them, being the only one where more heart shapes were found than spatula. The parameters of the frequency of morphotypes and length/width rate are closely related, since in the presence of spatulas the ratio will be slightly higher than 1 because they are estimated to be longer than wider structures, while in heart-shaped gametophytes, it will be slightly lower than 1. The control showed significant differences in this rate with yucasine 0.47 μM and 4.7 μM, and DMSO treatment. In fact, in the presence of DMSO increases the development of gametophytes, even to the point of obtaining individuals with enormous proportions.

Both parameters reflect how yucasine at low level could slow the gametophyte development, affecting their vegetative growth. The same goes for apogamy, where the lowest concentration treatment of yucasine was the one that obtained the fewest embryos and with significant differences with the control. The embryo area was the least significant parameter, and yucasine 23.6 μM was the only treatment showing significant difference with the control. This could be due to the large difference in the number of embryos present in each treatment, as we have seen with the frequency of apogamy, and also the fact that certain treatments do not have an adequate sample size to make a rigorous comparison. In addition, it might be said that due to the way the embryos are measured two-dimensionally (where elliptical shapes are assumed to calculate their area), the size of the buttons is underestimated as it is in certainly a three-dimensional structure.

Auxins are phytohormones that have been used in in vitro culture in numerous studies for the development of embryogenesis and organogenesis in all types of plants and in turn have been associated with organogenesis in ferns (Menéndez et al. 2006; Mikuła et al. 2015; Grossmann et al. 2017). Taking into account the inhibitory effect of yucasine on auxin biosynthesis and the data obtained, yucasine could be considered to have affected gametophytes development. Although the effect has clearly been greater on the concentration of 0.47 μL, we were not able to determine at which concentrations it is most effective, since on the other hand the presence of DMSO could be affecting the results.

In numerous plant studies, DMSO has been widely used as an efficient solvent for water-insoluble compounds (Thao et al. 2003: Limera et al. 2016). But studies such as Zhao and Gu (1984) are also known to report that parthenogenesis in non-pollinated maize was induced by combinations of DMSO, colchicine, and maleic hydrazide (MH), in which 19 homozygous diploids were obtained. Recently in a study of Chen et al. (2018), it was possible to increase apomixis with DMSO in cassava, but still the effects that DMSO may have on apogamy in plants are somewhat unknown as far as apomixis is concerned. In our case, it seems that the presence of DSMO has influenced gametophytes, favoring the increase of both vegetative and apogamous development. In addition, it was the treatment of DMSO that showed a greater number of irregularities related to apogamy, inducing sometimes the formation of several embryos in the same gametophyte.

Looking at the results of all the parameters studied, it is possible that the inhibitory effect of yucasine is affecting the vegetative and reproductive development of gametophytes. When yucasine is present in low concentrations, it is clear that growth becomes delayed. However, due to the results obtained with the treatment where DMSO was used as a solvent for yucasine, and knowing that it could have flattering properties for the development of apogamy, it could have been counteracting the effect of yucasine as its concentrations increased.

## 14.5 Conclusion

Yucasine has demonstrated an inhibitory power in the development of gametophytes. Otherwise, DMSO has affected gametophytes with an opposite effect, increasing their vegetative and reproductive development. It would be possible that both substances could have been offset. The results bring new knowledge about the role of these compounds on vegetative and reproductive development of this apogamous fern.

## References

Asker SE, Jerling L (1992) Apomixis in plants. CRC Press, Boca Ratón. ISBN 978-0-8493-4545-6

Berleth T, Sachs T (2001) Plant morphogenesis: long-distance coordination and local patterning. Curr Opin Plant Biol 4(1):57–62

Casanova-Sáez R, Voß U (2019) Auxin metabolism controls developmental decisions in land plants. Trends Plant Sci 24(8):741–754

Chen X, Lai HG, Sun Q, Liu JP, Chen SB, Zhu WL (2018) Induction of apomixis by dimethyl sulfoxide (DMSO) and genetic identification of apomictic plants in cassava. Breed Sci 68(2): 227–232. https://doi.org/10.1270/jsbbs.17089

Cheng Y, Dai X, Zhao Y (2007) Auxin synthesized by the YUCCA flavin monooxygenases is essential for embryogenesis and leaf formation in *Arabidopsis*. Plant Cell 19(8):2430–2439

Dai X, Mashiguchi K, Chen Q, Kasahara H, Kamiya Y, Ojha S, DuBois J, Ballou D, Zhao Y (2013) The biochemical mechanism of auxin biosynthesis by an *Arabidopsis* YUCCA flavin-containing monooxygenase. J Biol Chem 288(3):1448–1457

Davies PJ (ed) (2010) Plant hormones: biosynthesis, signal transduction, action. Springer, Dordrecht, p 802. ISBN 978-1-4020-2684-3

Dyer RJ, Savolainen V, Schneider H (2012) Apomixis and reticulate evolution in the Asplenium monanthes fern complex. Ann Bot 110(8):1515–1529

Ekrt L, Koutecký P (2015) Between sexual and apomictic: unexpectedly variable sporogenesis and production of viable polyhaploids in the pentaploid fern of the *Dryopteris affinis* agg. (Dryopteridaceae). Ann Bot 117(1):97–106

Fei X, Shi J, Liu Y, Niu J, Wei A (2019) The steps from sexual reproduction to apomixis. Planta 249(6):1715–1730

Fernández H and Revilla MA. (2003). *In vitro* culture of ornamental ferns. Plant Cell Tissue Organ Cult 73, 1–13

Fraser-Jenkins CR (1980) *Dryopteris affinis*: a new treatment for a complex species in the European Pteridophyte Flora, vol 10. Botanischer Garten und Botanisches Museum, Berlin

Fujino K, Matsuda Y, Ozawa K, Nishimura T, Koshiba T, Fraaije MW, Sekiguchi H (2008) NARROW LEAF 7 controls LEAF shape mediated by auxin in rice. Mol Genet Genomics 279(5):499–507

Gallavotti A, Barazesh S, Malcomber S, Hall D, Jackson D, Schmidt RJ, McSteen P (2008) Sparse inflorescence1 encodes a monocot-specific YUCCA-like gene required for vegetative and reproductive development in maize. PNAS 105(39):15196–15201

Grossmann J, Fernández H, Chaubey PM, Valdés AE, Gagliardini V, Cañal MJ, Russo G, Grossniklaus U (2017) Proteogenomic analysis greatly expands the identification of proteins related to reproduction in the apogamous fern *Dryopteris affinis* ssp *affinis*. Front Plant Sci. 8: 336

Grossniklaus U, Koltunow AM, Van Lookeren M (1998) A bright future for apomixis. Trends Plant Sci 3:415–416

Grusz AL (2016) A current perspective on apomixis in ferns. J Syst Evol 54:656–665

Kandemir N, Saygili I (2015) Apomixis: new horizons in plant breeding. Turk J Agric For 39:549–556

Koltunow AM, Grossniklaus U (2003) Apomixis: a developmental perspective. Ann Rev Plant Biol 54(1):547–574

Limera C, Wang K, Xu L, Wang Y, Zhu X, Feng H, Sha Y, Gong Y, Liu L (2016) Induction of autotetraploidy using colchicine and its identification in radish (*Raphanus sativus* L.). J Hortic Sci Biotechnol 91:63–70

Liu HM, Dyer RJ, Guo ZY, Meng Z, Li JH, Schneider H (2012) The evolutionary dynamics of apomixis in ferns: a case study from Polystichoid ferns. J Bot:510478. https://doi.org/10.1155/2012/510478

Lovis JD (1978) Evolutionary patterns and processes in ferns. Adv Bot Res 4(January):229–415

Mashiguchi K, Tanaka K, Sakai T, Sugawara S, Kawaide H, Natsume M, Hanada A, Yaeno T, Shirasu K, Yao H, McSteen P, Zhao Y, Hayashi K, Kamiya Y, Kasahara H (2011) The main auxin biosynthesis pathway in Arabidopsis. PNAS 108(45):18512–18517

Menéndez V, Villacorta NF, Revilla MA, Gotor V, Bernard P, Fernández H (2006) Exogenous and endogenous growth regulators on apogamy in *Dryopteris affinis* (Lowe) Fraser-Jenkins sp. affinis. Plant Cell Rep 25(2):85–91

Mikuła A, Pożoga M, Tomiczak K, Rybczyński JJ (2015) Somatic embryogenesis in ferns: a new experimental system. Plant Cell Rep 34:783–794

Murashige T, Skoog F (1962) A revised medium for rapid growth and bioassays with tobacco tissue cultures. Plant Physiol. 15:473–497

Nishimura T, Hayash KI, Suzuki H, Gyohda A, Takaoka C, Sakaguchi Y, Matsumoto S, Kasahara H, Sakai T, Kato JI, Kamiya Y, Koshiba T (2014) Yucasin is a potent inhibitor of YUCCA, a key enzyme in auxin biosynthesis. Plant J 77(3):352–366

Ozias-Akins P (2006) Apomixis: developmental characteristics and genetics. Crit Rev Plant Sci 25: 199–214

Pacifici E, Polverari L, Sabatini S (2015) Plant hormone cross-talk: the pivot of root growth. J Exp Bot 66:1113–1121

Somer M, Arbesú R, Menéndez V, Revilla MA, Fernández H (2010) Sporophyte induction studies in ferns *in vitro*. Euphytica 171(2):203–210

Stepanova AN, Robertson-Hoyt J, Yun J, Benavente LM, Xie DY, Doležal K, Schlereth A, Jürgens G, Alonso JM (2008) TAA1-mediated auxin biosynthesis is essential for hormone crosstalk and plant development. Cell 133(1):177–191

Tao Y, Ferrer JJ, Ljung K, Pojer F, Hong F, Long JA, Li L, Moreno JE, Bowman ME, Ivans LJ, Cheng Y, Lim J, Zhao Y, Ballaré CL, Sandberg G, Noel JP, Chory J (2008) Rapid synthesis of auxin via a new tryptophan-dependent pathway is required for shade avoidance in plants. Cell 133(1):164–176

Team Rs (2016) RStudio: integrated development for R. RStudio, Boston

Thao NTP, Ureshino K, Miyajima I, Ozaki Y, Okubo H (2003) Induction of tetraploids in ornamental *Alocasia* through colchicine and oryzalin treatments. Plant Cell Tissue Organ Cult 72:19–25. https://doi.org/10.1023/A:1021292928295

Thimann KV (1977) Hormone action in the whole life of plants. The University of Massachusetts Press, Amherst, p 464. ISBN 978-0870232244

Tucker MR, Araujo ACG, Paech NA, Hecht V, Schmidt EDL, Rossell JB, De Vries SC, Koltunow AMG (2003) Sexual and apomictic reproduction in *Hieracium* subgenus Pilosella are closely interrelated developmental pathways. Plant Cell 15:152437

White RA (1979) Experimental investigations of fern sporophyte development. BioSci 21(6): 271–275

Won C, Shen X, Mashiguchi K, Zheng Z, Dai X, Cheng Y, Kasahara H, Kamiya Y, Chory J, Zhao Y (2011) Conversion of tryptophan to indole-3-acetic acid by TRYPTOPHAN AMINOTRANSFERASES OF ARABIDOPSIS and YUCCAs in Arabidopsis. PNAS 108(45):18518–18523

Woodward AW, Bartel B (2005) Auxin: regulation, action, and interaction. Ann Bot 95(5):707–735

Wyder S, Rivera A, Valdés AE, Cañal MJ, Gagliardini V, Fernández H, Grossniklaus U (2020) Differential gene expression profiling of one- and two-dimensional apogamous gametophytes of the fern *Dryopteris affinis* ssp. *affinis*. Plant Physiol Biochem 148(January):302–311

Yamada M, Greenham K, Prigge MJ, Jensen PJ, Estelle M (2009) The TRANSPORT INHIBITOR RESPONSE2 gene is required for auxin synthesis and diverse aspects of plant development. Plant Physiol 151(1):168–179

Yamamoto Y, Kamiya N, Morinaka Y, Matsuoka M, Sazuka T (2007) Auxin biosynthesis by the YUCCA genes in Rice. Plant Physiol 143(3):1362–1371

Zhao Z, Gu MG (1984) Production of diploid pure lines of maize through parthenogenesis induced by chemicals. Acta Genet Sin 11:39–46

**Eugenio Sánchez** has a master's degree in Biotechnology Applied to Conservation & Sustainable Management of Plant Resources. At present, he is an environmental auditor. He is expertise in conservation and management of plant species. He usually drives environmental impact studies.

**Alejandro Menéndez** has a degree in Biology at the University of Oviedo. He is expertise in the field of in vitro culture. He is training in plant ecophysiology of Canary Islands, through the study of the different bioclimatic floors, and also he has expended some time in the Sierra Madre Oriental of the United Mexican States, where he had the opportunity to work with a diverse group of exploration of Mexican forest botany, in addition to collaborating in different forestry works and the dissemination of the Los Mármoles National Park together with important American documentary film production chains.

**Alejandro Rivera** has a master's degree in Biotechnology Applied to Conservation & Sustainable Management of Plant Resources and a PhD from Department of Organisms and Systems Biology, at the Oviedo University. He has published in SCR and authored book chapters from Springer.

**Ma Jesús Cañal** is Professor at the University of Oviedo, Coordinator of master's degree "Biotechnology Applied to Conserva tion & Sustainable Management of Plant Resources," and Director of the Department of Organisms and Systems Biology, at the University of Oviedo. Lead of EPIPHYSAGE scientific group, involved in Biology of Systems. Professor Cañal is expert in Epigenomics (epigenetics, transcriptomics, and proteomics of plant development), aging reinvigoration, cellular reprograming, and response to antibiotic stress. In addition, cloning, regardless of age, of species of agroforestry interest and industrial plant production in temporary immersion systems. She counts a great amount of SCR publications, awards, and other academic honors.

**Helena Fernández** is Associate Professor of Plant Physiology at the University of Oviedo, Spain. She received her PhD in Biology in 1993 from Oviedo University, followed by an EU-funded post-graduate research sojourn at the INRA in Orléans, France, training in HPLC-MS techniques. She got the Ramón y Cajal National Excellence Research Program, dealing with reproduction in ferns. She is guest researcher at the Institute of Plant Biology, University of Zurich, working on apomixes, through omics techniques. She is a member of the dean's team, in the Faculty of Biology at the University of Oviedo. She publishes in relevant journals, having edited two books with Springer.

# The Effect of Phytohormones and Inhibitors on Apogamous Gametophytes of *Dryopteris affinis* ssp. *affinis* 15

S. Granados, A. Rivera, María Jesús Cañal, and Helena Fernández

**Abstract**

The fern *Dryopteris affinis* ssp. *affinis* exhibits a peculiar case of apomixis, in which an embryo develops from somatic cells of the prothallium (apogamy). The effect on vegetative and apogamous gametophyte development of 14 phytohormones and inhibitors of their biosynthesis or transport (HBTIs) was analyzed in homogenized gametophytes cultured in liquid Murashige and Skoog (MS) medium. The list includes balances of the auxins: indol-butyric (IBA) or naphthalene acetic (NAA), with the cytokinin 6-benzylaminopurine (BA); gibberellic acid ($GA_3$), spermidine (S); the inhibitor of auxin polar transport N,1-naphthylthalamic acid (NPA); the inhibitor of GAs biosynthesis, flurprimidol (F); and the inhibitor of spermidine biosynthesis, cyclohexylamine (CHA). The results point out a phytohormonal effect on gametophyte development and the apogamy event. The auxin IBA exhibited a wide-ranging impact on morphogenesis, including length/width ratio, apogamy, embryo size, and rhizoid localization. Indeed, intervention of auxin transport is suggested as wider gametophytes and bigger and conical embryos were observed, when adding the inhibitor NPA to the medium. Gametophyte elongation is promoted by $GA_3$, being this result reinforced by the use of flurprimidol; and gametophyte growth in length or width, and embryo size resulted also affected by spermidine. In some cultures, the homogenization and/or the addition of growth regulators induced callus, which then proliferated in the presence of 2,4-dichlorophenoxyacetic acid. The findings

S. Granados · A. Rivera · M. J. Cañal
Area of Plant Physiology. Dpt. B.O.S., Faculty of Biology, University of Oviedo, Oviedo, Spain
e-mail: mjcanal@uniovi.es

H. Fernández (✉)
Department of Organisms and Systems Biology, Area of Plant Physiology, University of Oviedo, Oviedo, Spain
e-mail: fernandezelena@uniovi.es



J. Marimuthu et al. (eds.), *Ferns*, https://doi.org/10.1007/978-981-16-6170-9_15

might contribute to widen our understanding about the role of phytohormones on vegetative gametophyte development and the asexual embryogenesis presents in apogamy; and make too available a protocol to induce callus useful in further molecular analyses.

**Keywords**

Apomixis · Apogamy · *Dryopteris affinis* ssp. *affinis* · Gametophyte · Fern · Morphogenesis

## Abbreviations

| | |
|---|---|
| BA | 6-Benzylaminopurine |
| CHA | Cyclohexylamine |
| 2,4-D | 2,4-Dichlorophenoxyacetic acid |
| F | Flurprimidol |
| $GA_3$ | Gibberellic acid |
| IBA | Indole-3-butyric acid |
| NAA | Naphthalene acetic acid |
| NPA | N-1-naphthylphthalamic acid |
| MS | Murashige and Skoog (1962) basal medium |
| S | Spermidine |

## 15.1 Introduction

Plant research mostly focuses into angiosperms models. Historically, ferns had been forgotten, being regarded basically by their ornamental, conservational, or therapeutical applications (Fernández and Revilla 2003). A handful of species have been selected to study basic plant developmental processes, such as photomorphogenesis (Wada 2007), germination (Salmi et al. 2005, 2007; Suo et al. 2015), cell polarity (Salmi et al. 2010), cell wall composition (Eeckhout et al. 2014), or reproduction (Fernández and Revilla 2003; Menéndez et al. 2006a, b, c, 2009; Cordle et al. 2007, 2010; Kazmierczak 2010; López and Renzaglia 2014; Valledor et al. 2014; de Vries et al. 2016; Grossmann et al. 2017).

Asexual reproduction by clonal seeds would allow to maintain agriculturally important phenotypes, including that of highly productive F1 hybrids, which has been regarded as an ambitious goal for plant culturing (Grossniklaus et al. 1998). In ferns, it is suggested apomixis to have evolved recurrently and individualistically, being a very frequent event, and bigger than in Angiosperms (Grusz 2016). In ferns, apomixis comprises apogamy, in which an asexual embryo is formed from somatic cells of the gametophyte, and a restitution meiotic event to produce unreduced (diplo) spores (Ekrt and Koutecký 2016). Indeed, the absence of sexual reproduction, in most of apogamous ferns, makes reproduction less complex than in Angiosperms, where often coexists apomixis and sexuality (Liu et al. 2012; Kandemir and Saygili 2015).

The gametophyte of ferns is a small free-living organism, being handy for in vitro culture and sample collection (Fernández and Revilla 2003). Indeed, we have reported it is suitable to deal with studies on reproduction (asexual and sexual) (Rivera et al. 2018a). In particular, apogamy represents an attractive process to deepen on asexual reproduction, which otherwise involved a process of somatic embryogenesis. Apogamy can be induced in vitro, playing with environmental factors such as carbohydrates (Whittier and Steeves 1960, 1962; Whittier 1964a; Ekrt and Koutecký 2016), osmotic conditions (Elmore and Whittier 1975, von Aderkas (1984), and growth regulators if supplied with sucrose (Whittier 1964b; Kato 1967; Elmore and Whittier 1975; von Aderkas 1984). Studies on both apogamy and somatic embryogenesis in ferns are scarce (von Aderkas 1984; Makowski et al. 2015; Rybczynski et al. 2018).

Homogenized gametophytes represent a good choice to deal with reproduction, taking advantage of its great regeneration capacity (Finnie and van Staden 1987; Fernández et al. 1993; Menéndez et al. 2006a, 2010; Somer et al. 2010; Rivera et al. 2018b). In ferns, organogenesis and embryogenesis are also influenced by phytohormones, and for years we have explored the morphogenic capacity exhibited by spore, gametophyte, and sporophyte (Fernández and Revilla 2003; Menéndez et al. 2009; Menéndez et al. 2010; Rivera et al. 2018a, b). Indeed, their biosynthesis and transport are subject to enzymatic reactions, which can be blocked by using inhibitors, such as 2,3,5-triiodobenzoic acid or TIBA, flurprimidol, and cyclohexylamine, interfering on auxin transport, ent-kaurene oxidation, or spermidine biosynthesis, respectively (Klíma et al. 2016; Grzyb et al. 2018; Rivera et al. 2018b).

*Dryopteris affinis* ssp. *affinis* is an obligate apomictic diploid fern, derived from the crossing of *D. oreades* and some sexual ancestor of the species *D. wallichiana* or *D. caucasica*, widely distributed in the regions of the Mediterranean, Macaronesia, and West of Eurosiberia (Salvo 1990). The gametophyte forms male sexual organs or antheridia but not archegonia. When cultured in vitro, the gametophyte draws the typical heart shape, near the apical notch, as a group of small and dark brown isodiametric cells progress, forming an embryo (Menéndez et al. 2006c).

Rivera et al. (2018b) reported the effect of some phytohormones and HBTIs on vegetative and reproductive development in the gametophyte of *D. affinis* ssp. *affinis*, in cultures starting either from spore or from homogenized gametophyte. The results revealed a greater plasticity exhibited by the gametophytes regenerated from homogenized gametophytes than from spore. Now, the research goal is to insist in studying the effect of auxin/cytokinin combinations, as well as the gibberellic acid $GA_3$, and the polyamine spermidine, by comparison with their counterpart HBTIs: N-1-naphthylthalamic for transport; and flurprimidol and cyclohexylamine for biosynthesis, respectively, on both vegetative development and asexual reproduction (apogamy) in homogenized gametophytes of *D. affinis* ssp. *affinis* cultured in vitro.

## 15.2 Materials and Methods

### 15.2.1 Collection, Cleaning, and Culture of Spores and Gametophytes

Spores of *D. affinis* ssp. *affinis* were collected from sporophytes growing in the Valley of Turón (Asturias, Spain), with coordinates 43°12′ N and 5°43′ W. Fronds were placed between sheets of paper, at room temperature. After that, spores and sporangia were filtered to remove the rest of sporangia tissues, and spores were stored in glass vials and kept at 4 °C. To guarantee aseptic cultures, spores (10 mg) were soaked in water for 2 h, cleaned for 10 min with a solution of NaClO (0.5% w/v) with some drops of Tween 20 (0.1% w/v), and finally washed three times with sterile distilled water. Given the small size of spores, we changed the solutions by centrifugation at 700 g. Spore density in the cultures was adjusted to around 4000 spores per flask, by using an optical microscope (Nikon Eclipse E-600) and a Fuchs-Rosenthal (Brand) chamber.

Spores were cultured in 100 mL flasks containing 20 mL Murashige and Skoog medium (Murashige and Skoog 1962), or MS, with 2% (w/v) sucrose and 0.7% (w/v) agar, and the pH was adjusted to 5.7 with 1 or 0.1 N NaOH. The cultures were maintained at 25 °C under cool white fluorescent light (70 μmol/m$^2$/s) with a 16:8 h light: dark photoperiod. After spore germination, gametophyte development occurs, being transferred to the same fresh medium, monthly.

### 15.2.2 Homogenized Gametophyte Cultures

Gametophyte tissue (0.2 g) was broken with a sterile blender for 15 s under aseptic conditions at 9500 rpm. The homogenized samples were cultured in 250 mL Erlenmeyer flasks, containing 50 mL of liquid MS medium with 2% (w/v) sucrose and without (control) or with the assayed phytohormones and HBTIs. The auxins and cytokinins were added into the medium in combination as follows (IBA 2.5 μM + BA 2.2 μM; IBA 5 μM + BA 0.5 μM; NAA 0.5 μM + BA 4.4 μM; NAA 2.7 μM + BA 2.2 μM); the rest of the compounds were added alone, and two concentrations were tested ($GA_3$ 0.3 and 3 μM; S 0.7 and 7 μM; NPA 0.3 and 3.5 μM; F 0.3 and 3 μM; CHA 1 and 10 μM). Cultures were put on an orbital shaker and kept at the aforementioned conditions for spore cultures. Experiments were repeated at least twice.

### 15.2.3 Callus Culture

Cellular aggregates obtained from homogenized gametophytes cultured in MS medium without or with IBA 5 μM + BA 2.2 μM, or $GA_3$ 0.3 μM, were transferred to MS medium with the auxin 2,4D 2.3 μM plus cytokinin (BA 2.2 μM or kinetin

2.3 μM). In other experiment, gametophytes were wounded with a scalpel and placed on MS media supplied with 2,4D 2.3 μM + BA 2 0.2 μM.

### 15.2.4 Microscopical Examination

Gametophytes developed in the abovementioned treatments were observed under an optical microscope (Nikon Eclipse E-600), after 50 days of culture. Data about vegetative development and apogamy were scored. The primer includes the morphotype or gametophyte shape (filamentous, spatula, and heart or cordate) and the length/width ratio; the latter refers to the percentage of apogamous gametophytes and embryo size. One hundred gametophytes were picked up randomly to determine the frequency of morphotype and the length/width ratio. Apogamy data were recorded in 50 heart-shaped gametophytes, at which an embryo is usually well defined, although in the case of those treatments accelerating the apogamic process, spatulas were also considered.

### 15.2.5 Statistical Analysis

The Chi-square ($\chi^2$) test was applied to non-parametric data, such as morphotype and number of apogamous embryos. The parametric data, i.e., rate length/width and the embryo size, were analyzed by ANOVA, using the Levene and Bartlett test for homogeneity of variances and Shapiro-Wilk for normality. The tests post hoc Tukey HSD and Duncan were used in case there were differences among treatments, in the ANOVA test. Analyses were carried out by RStudio (Team 2016) and set to a significance level of $\alpha = 0.05$.

### 15.2.6 Histological Examination

Callus tissue was fixed using 4% paraformaldehyde-PBS (v/v) (phosphate buffer, pH = 5.8), at 4 °C, and transferred, after 48 h, to a solution of 0.1% paraformaldehyde-PBS (v/v), to keep them until use. Callus were dehydrated in a series of tert-butyl alcohol solutions, embedded in *Paraplast* (Fisher Scientific Co.), as described by Jensen (1962) and sectioned into 10 μm slides using a microtome (JUNG, mod. 1130). Sections were placed in a microscope slide, covered with albumin until dried at 40 °C. Then, they were stained with Nile red, to bind lipids inside the callus, for 10 min, in a dark chamber. Gametophyte samples were stained with Toluidine blue O for 1 h (Jensen 1962).

## 15.3 Results

### 15.3.1 Effects of Assayed Treatments on Vegetative Development

Vegetative development of regenerated gametophytes was annotated in terms of their shape or morphotype, i.e., one-dimensional (filamentous) versus two-dimensional (spatulate and heart), and the length/width ratio (Fig. 15.1).

(a) *Morphotype*. The transition from one to two-dimensional growth of gametophytes was favoured either by the addition of phytohormones such as the NAA/BA combinations ($\chi^2 = 91$, $p$-value $<0.001$), the gibberellin $GA_3$ or its inhibitor flurprimidol, in this case at the lowest dose ($\chi^2 = 10$, $p$-value $=$ 0.001, and $\chi^2 = 5$, $p$-value $= 0.021$), respectively, and also by adding the polyamine spermidine at the highest concentration ($\chi^2 = 35$, $p$-value $<0.001$). The inhibitor CHA reduced the two-dimensional transition at the lowest doses ($\chi^2 = 12$, $p$-value $<0.001$) and induction at the highest ($\chi^2 = 35$, $p$-value $<0.001$).

(b) *Length/width ratio*. This relation indicates in what direction (apical-basal or lateral) the gametophyte expands as it is growing. Gametophyte elongation, following the apical-basal polarity, was promoted by IBA ($p$-value $<0.001$) and $GA_3$ ($p$-value $<0.001$), being most of them spatula-shaped (Fig. 15.2a). In those treatments with the auxin inhibitor transport NPA, gametophytes were longer than wider at the lowest dose, and wider than longer at the highest, drawing a circle in the last case ($p$-value $<0.001$) (Fig. 15.2b). The inhibitor of the biosynthesis of GAs, flurprimidol, at 0.3 μM, induced the regeneration of gametophytes wider than longer, decreasing the proportion of spatulas to 18%

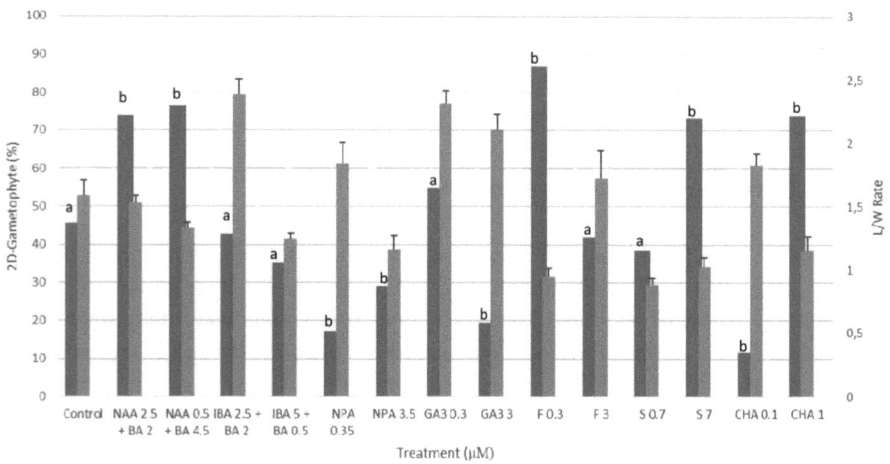


**Fig. 15.1** Effect of phytohormones and inhibitors of their biosynthesis or transport, on two-dimensional transition (black bars) and length/width rate (grey bars), in regenerated gametophytes of *Dryopteris affinis* ssp. *affinis*. Data after 50 days. Different letters mean significant differences with respect to the control

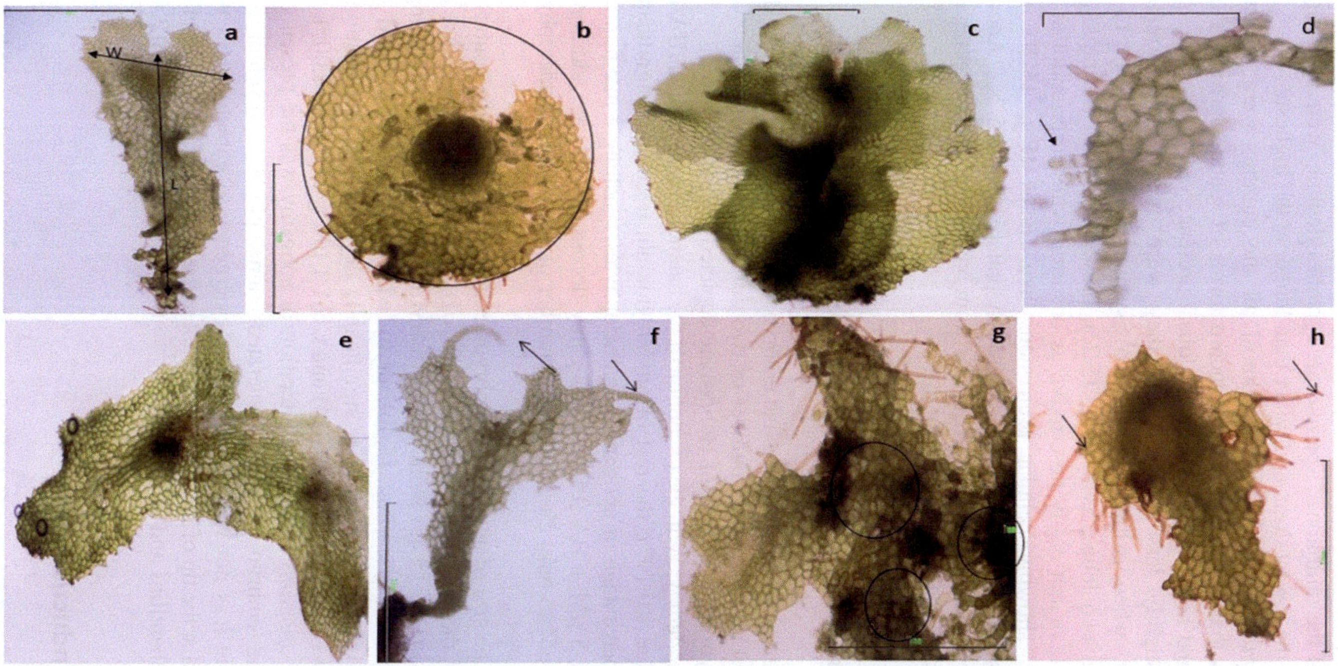


**Fig. 15.2** Morphological features in regenerated gametophytes of *Dryopteris affinis* ssp. *affinis*, cultured in MS liquid medium with the following phytohormones or inhibitors of their biosynthesis or transport: (**a**) elongation with $GA_3$ 0.3 μM (w = width; l = length); (**b**) circular shape with NPA 3.5 μM; (**c**) growing in width with flurprimidol 0.3 μM; (**d**) antheridia with $GA_3$ 3 μM; (**e**) asymmetrical lobes in a free hormone/inhibitors medium; (**f**) thorn-like protrusions (arrows); (**g**) undefined growth pattern (circles) with flurprimidol 3 μM; (**h**) rhizoids placed at the apical region (arrows) in presence of IBA 2.5 μM + 2.2 μM. Bar = 1 mm

(Fig. 15.2c). IBA favored the antheridia formation in filamentous-shaped gametophytes (Fig. 15.2d). As occurred with NPA, the gametophytes growing in presence of the inhibitor of spermidine, cyclohexylamine, were also longer than wider at the lowest dose, and wider than longer at the highest.

(c) *Aberrations in the morphotype*. In almost all the cultures were observed gametophytes that deviate from the standard forms of spatula (greater length than width) and heart (two symmetrical lobes, presence of notch and length/width rate slightly less than 1), and they were annotated as irregular morphologies, when exhibiting: (1) two asymmetrical lobes (Fig. 15.2e, (2) more than two lobes, as in (Fig. 15.2c), presenting up to 8 wings, and (3) amorphous or not easily descriptive as occurred with flurprimidol at the highest dose, showing cellular protrusions like spikes (Fig. 15.2f), and gametophytes following an undefined growth pattern (Fig. 15.2g). Another irregularity was the presence of rhizoids far from basal areas of the regenerated gametophytes (Fig. 15.2h).

### 15.3.2 Effect of Assayed Treatments on Apogamy

Regenerated gametophytes formed asexual embryos from vegetative cells, and data about frequency of apogamy, i.e., number of two-dimensional gametophytes with apogamous embryos, and the area of embryo are depicted in Fig. 15.3.

1. *Frequency of apogamy*. With the exception of the balanced combination IBA/BA ($\chi^2 = 2$, $p$-value <0.197), the rest of auxin/cytokinin treatments significantly increased the percentage of apogamous gametophytes, reaching a maximum of 50% with IBA 5 μM + BA 0.5 μM. Apogamy decreased with the addition of flurprimidol 3 μM to only 2% of the total number of regenerated gametophytes forming embryos.
2. *Embryo size*. The size of apogamous embryos enlarged respect to the control by the addition of the balanced combination IBA/BA to the culture medium, ($p$-value = 0,003) (Fig. 15.4a), with NPA 3.5 μM ($p$-value <0.001), adopting a conic aspect in this case (Fig. 15.4b), or by adding spermidine 0.7 μM ($p$-value = 0,001). In contrast, the size significantly decreased by the addition of flurprimidol ($p$-values 0.023 and 0.034) and CHA 1 μM ($p$-value = 0.007).
3. *Aberrations in apogamy*. A notable alteration of apogamy was the presence of more than one embryo in the regenerated gametophytes (polyembryony) (Fig. 15.4c), but it is not very frequent. The emergence of gametophytes with embryos placed at the tip of lobes was recurrent at the lowest concentration of the inhibitor of gibberellins and flurprimidol ($\chi^2 = 6$; $p$-value = 0.014) (Fig. 15.4d).

### 15.3.3 Callus Induction

1. *Homogenized gametophytes*. Cellular aggregation was observed in a MS hormone or inhibitor-free medium and also in MS medium with the following

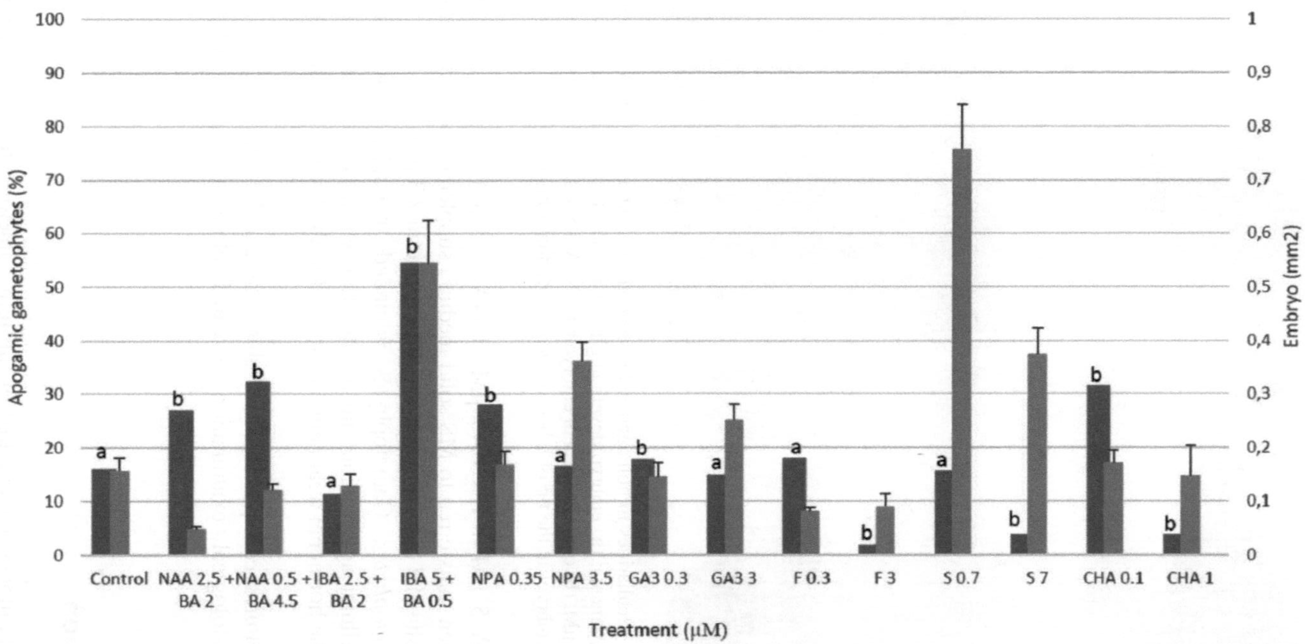


**Fig. 15.3** Effect of phytohormones and inhibitors of their biosynthesis or transport, on apogamy frequency (black bars) and embryo size (grey bars), in regenerated gametophytes of *Dryopteris affinis* ssp. *affinis*. Data after 50 days. Different letters mean significant differences with respect to the control ($\chi^2 = 8$, $p$-value = 0.004), and also with spermidine at the highest concentration ($\chi^2 = 7.08$, $p$-value = 0.01). In opposite, apogamy increased with CHA 0.1 μM, to 30% ($\chi^2 = 11$, $p$-value <0.001), dropping with the highest concentration, to only 4% of gametophytes forming apogamous embryo ($\chi^2 = 5$, $p$-value = 0.033).

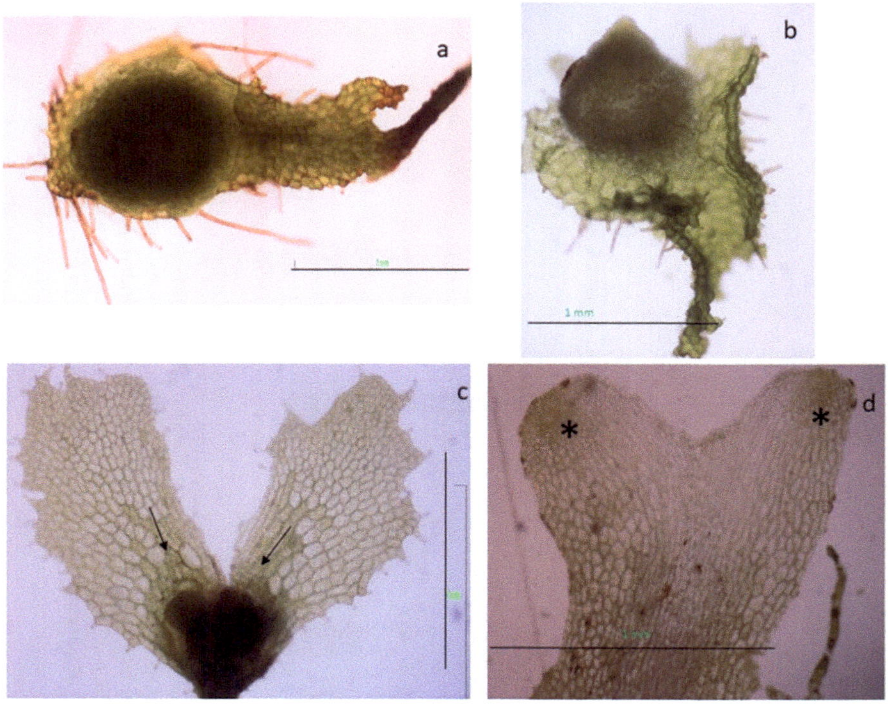


**Fig. 15.4** Apogamy peculiarities in homogenized gametophytes, cultured in liquid MS medium with the following treatments: (**a**) embryo developing from a spatulate-shaped gametophyte with IBA 2.5 μM + BA2.2 μM; (**b**) conical aspect of embryo with NPA 3.5 μM; (**c** and **d**) polyembryony and displacement of embryos at the tip of lobes, observed with flurprimidol 0.3 μM. Bar = 1 mm

treatments: IBA 5 μM + BA 0.45 μM or $GA_3$ 0.3 μM (Fig. 15.5a), which proliferated when transferred to MS solid medium supplemented with the auxin 2,4-D 2.3 μM plus the cytokinin BA 2.2 μM or kinetin 2.3 μM.

2. *Hand-cut gametophytes*. Gametophytes wounded manually with a scalpel showed also cellular aggregation and proliferation when cultured on solid MS medium with the hormonal treatments mentioned in the preceding paragraph.

In both types of cultures, *calli* presented a friable texture and a yellowish green color (Fig. 15.5a). Histological examination revealed abundant bodies of lipids (Fig. 15.5b).

## 15.4 Discussion

In this work, the homogenates cultures of the fern gametophyte are shown as an interesting tool to deepen on apogamy, a peculiar case of apomixis, otherwise which is very frequent in ferns. The homogenized gametophyte of *D. affinis* ssp. *affinis*,

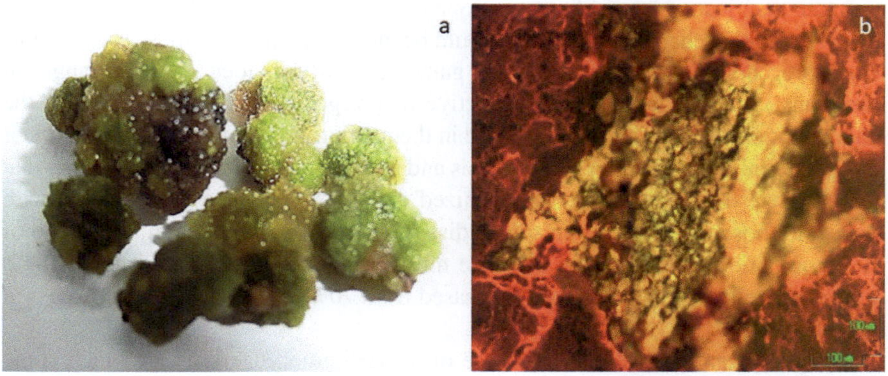


**Fig. 15.5** Callus derived from regenerated gametophytes of *Dryopteris affinis* ssp. *affinis*, showing (**a**) compact aspect and yellowish green color; (**b**) lipidic bodies as spots of green color, stained with Nile red

exhibits a great plasticity, being able to vary its vegetative and reproductive behavior by means of the addition of phytohormones or HBTIs, to the culture medium. Moreover, the high regeneration capacity of the fern gametophyte represents an excellent supply of plant material to carry out further analyses for different purposes coping with plant development.

The mechanical disruption caused by the prothallus breaks somehow the connection between cells and the interrelationships between many structural and functional elements. Despite the stress involved in the mechanical fragmentation exerted, some cells retake the capacity of division and differentiation, generating a new individual and demonstrating, once again, the plasticity that characterizes the plant organisms and how the fern gametophyte has evolved and retained the ability for reprograming cell identity to facilitate tissue repair and developmental plasticity (Shin and Seo 2018). In line with it, the gametophyte of *D. affinis* ssp. *affinis* exhibits a great regenerative potential comparable to other species that have been previously cultivated in our laboratory (Fernández and Revilla 2003; Menéndez et al. 2006c; Somer et al. 2010; Rivera et al. 2018b). In general terms, basal land plants display high regenerative capacities (Ikeuchi et al. 2016).

The mature gametophyte exhibits certain degree of organization (apical-basal polarity, apical meristem, rhizoids, trichomes, antheridia, lateral wings, dorsoventral symmetry, and asexual embryo). However, once fragmented, each little piece of tissue resets somehow its differentiation pattern and rebuilds a new individual, thanks to the remaining toti/pluripotency of some cells that rest alive after being broken.

In previous work with cultures of gametophytes derived from spores, the addition of compounds like those used here, to the medium, did not prove to cause major variations or changes, in terms of gametophyte morphology and the apogamy process (Rivera et al. 2018b). The reason could be found in the explant itself; both spore and the entire gametophyte represent a more differentiated and organized structure than just a piece of few gametophyte cells, which, after the disorganization caused by the mechanical disruption, could be more influenced by these compounds. Therefore, homogenized cultures of gametophytes provided a wide range of variations, affecting both the reproductive and vegetative growth, caused by the fragmentation of tissue and the presence in the medium of plant growth regulators or compound inhibitors of their biosynthesis and transport. However, we must state that gametophyte development in homogenized cultures experiences some delay with respect to those originated from spores disseminated on solid medium, at the end of the same period, as each piece of tissue must face a long way until regenerating a new gametophyte. Also, apogamy decreased from 70% to 30%, as we will discuss in the following.

Apogamy can be regarded as a case of asexual embryogenesis, in which some vegetative cells in the gametophyte form an embryo, maintaining the same ploidy level gametophyte and sporophyte. In this work, a balanced ratio IBA/BA increased either the formation of apogamic embryos or their size. In general, auxins have shown an important role in the somatic embryogenesis (Johri 2008), and, in particular, the inducer effect of IBA has been demonstrated (Jha et al. 2007; Johri 2008). Moreover, differences among the auxins IBA and NAA were evident, so that while IBA seems to have a major role on embryogenesis, NAA resulted more effective to favour the transition from one- to two-dimensional growth, which could also mean to speed up the formation of apogamous embryos. A positive effect of the same concentrations of NAA on the number of apogamous sporophytes after 3 months of culture has been reported (Menéndez et al. 2006c). The variations found in our research among NAA and IBA, compounds belonging to the same family of chemicals, may be due to different factors such as genetic mechanisms. It was reported in rib1 mutants of *Arabidopsis*, which are resistant to the induction of adventive roots by IBA but not NAA, and not finding variations between them in the wild genotype (Ludwig-Müller 2000). On the other hand, they might experience differences in the absorption by cells, being favored the IBA intake. In this sense, it has been found that NAA is not a good substrate for influx auxin transports (AUX1) (Klíma et al. 2016).

Rhizoids are present in the gametophyte of ferns and clubmosses, and it is thought that they may have a function similar to the root hairs from the roots of angiosperms, favoring the intake of nutrients and also serving as anchor of the gametophyte to substrate, although they are not formed neither in pollen grain nor in embryo sac (Jones and Dolan 2012). Rhizoids and root hairs seem to share genetic mechanisms, in particular, transcription factors of the family of the ROOT HAIR DEFECTIVE (RHD) and ROOT HAIR DEFECTIVE SIX-LIKE (RSL) (Vijayakumar et al. 2016), having been identified as members of these families, in the gametophytes of this species (Wyder et al. 2020). Auxins control the formation of roots and hair roots in angiosperms (Ludwig-Müller 2000) and the formation of rhizoids on algae, ferns, and liverworts (Hickok and Kirilux 1984; Klämbt et al. 1992; Jones and Dolan 2012; Atallah et al. 2018), regulating the expression of these transcription factors.

Observing the anomalous position of rhizoids in the regenerated gametophytes cultured in presence of IBA + BA, we can deduce, that effectively, auxins can mediate the development of these structures.

The inhibitor of auxin transport, NPA, is classified as a phytotropin, having a controversial mode of action on polar auxin transport (Teale and Palme 2018). In our work, neither the two-dimensional transition nor the percentage of apogamous gametophytes did not seem to affect, but the L/W ratio of gametophytes, and the embryo size and shape, when added at 3.5 μM, promotes wider gametophytes and bigger and conical embryos.

The effect of the chemical compounds added to the culture medium, on embryo size, was underestimated, as the area of the apogamous center was considered. The embryos grown in presence of NPA displayed a prominent size, and the measures might have been deviated, as occurred also with highest dose of IBA and the lowest of spermidine. The auxin polar transport has a profound effect on plant development as stem and root growth, the initiation of lateral buds, vascular pattern, and embryogenic polarity are influenced by auxins (Su et al. 2011). In studies with stem corn (Scanlon 2003), the use of NPA altered the transport of the auxin IAA at the root, causing decrease in elongation and a globular form in *Picea abies*. The same authors, working with stalks of corn grown with NPA, reported an elongation of the apical meristem of the stem, as well as a delay in the initiation of the leaves. The circular form observed in regenerated gametophytes of *D. affinis* ssp. *affinis* (and the lack of multilobes), as the bigger size and elongated form of embryos, could reflect an alteration in the transport of auxins and their involvement on gametophyte elongation and on embryogenesis.

$GA_3$ has an important role on both vegetative and reproductive plant development, including the fern gametophyte (Menéndez et al. 2006a; Valledor et al. 2014; Rivera et al. 2018b). In this work, the effect on gametophyte elongation caused by $GA_3$ was reinforced by the addition to the culture medium of the biosynthesis inhibitor flurprimidol, which causes the opposite effect, stimulating growth in width. Flurprimidol, at the lowest concentration, caused irregularities in the gametophyte morphology, such as the presence of several lobes, −which has been previously reported (Menéndez et al. 2006a), in the gametophyte of *Blechnum spicant*-, and also were observed amorphous gametophytes, growing without a well-defined pattern, at the highest concentration of this inhibitor. The combined action of the stress caused by the mechanical tissue disruption and the inhibitor could lead to these anomalies, which were absent with $GA_3$ treatments. Thus, gibberellins are involved in the vegetative development of the gametophyte in *D. affinis* ssp. *affinis*.

The importance of the GAs on apogamy was revealed also by using flurprimidol, whose addition to the medium decreased the size of the embryo. Moreover, the highest concentration of flurprimidol inhibited apogamy, in agreement with that reported in our lab (Menéndez et al. 2006c). However, the role of GAs on apogamy is not clear since the exogenous addition of $GA_3$ seems to have not affected the process, not discarding that another gibberellin might have a more decisive function. It has been stated that $GA_4$ could have a more active role on apogamy, scoring higher endogenous levels of this compound when the apogamous center starts developing (Menéndez et al. 2006c).

The polyamines are involved in vegetative growth, promoting division and differentiation (Hickok and Kirilux 1984). In our work, it was noted that the addition of spermidine stimulates two-dimensional transition and gametophytes wider than longer, being these results unmodified by CHA addition, remaining not clear the role of these compounds.

Polyamines have been considered essential for the embryogenic development in *Arabidopsis* and somatic embryogenesis (De-la-Peña et al. 2008; Dutra et al. 2013). In our species, spermidine influenced embryo development, and its inhibitor, CHA, on the induction of apogamy. All these results agree with previous studies reporting the presence of some proteins connected to the action of the main phytohormones (Grossmann et al. 2017).

Homogenized gametophyte cells were able to divide and differentiate a new gametophyte and to form disorganized cell masses, like callus, which were observed also in cultures derived from gametophytes wounded with scalpel. This response was induced spontaneously, probably caused by culturing under a stressful situation, also when pieces of gametophyte tissue grew in presence of phytohormones like $GA_3$ 0.3 μM or the balance IBA 5 μM + BA 5 μM. Although the auxin IBA favored the formation of these cellular aggregates, it was ineffective to keep them proliferating, and it was only possible when cultured on solid MS medium with the auxin 2,4-D. Rybczynski et al. (2018) found this compound also suitable in cultured gametophytes of *Adiantum capillus-veneris*. Initially, the callus was compact and green, and after being subcultured for several times, it becomes brown and softer, showing symptoms of toxicity. The presence of lipid bodies in the histological sections of callus represents a line of research about the profile of fatty acids that comprise them, due to the interest that exists today to replace more polluting energy (Jouzani et al. 2018).

In brief, the experimental system used in this work along with the addition of certain phytohormones and inhibitors of their synthesis or transport provided a better understanding of these chemicals on both vegetative development and apogamy, in the gametophyte of *Dryopteris affinis* ssp. *affinis*, resulting to be more efficient than previous studies, starting from spore or the entire gametophyte. A balance auxin/cytokinin favorable to IBA increases elongation, the frequency of apogamy, and the size of the embryos. $GA_3$ shows an important role in vegetative growth, promoting elongation, evidenced by the inhibitor of GAs biosynthesis, flurprimidol. Although the application of flurprimidol, inhibited apogamy, is not clear, the possible involvement of GAs since their addition did not give any positive or negative result. The role of the polyamine spermidine on both vegetative development and apogamy remains more controversial. Now, a valid protocol is available to induce and proliferate callus of *D. affinis* ssp. *affinis*, which can lead to a biological system to deepen on apogamy and to conduct further molecular analyses on this topic.

## References

Atallah NM, Vitek O, Gaiti F et al (2018) Sex determination in *Ceratopteris richardii* is accompanied by transcriptome changes that drive epigenetic reprogramming of the young gametophyte. Genes Genomes Genet 8:2205–2214

Cordle AR, Irish EE, Cheng CL (2007) Apogamy induction in *Ceratopteris richardii*. Int J Plant Sci 168:361–369

Cordle AR, Bui LT, Irish EE, Cheng CL (2010) Laboratory-induced Apogamy and Apospory in *Ceratopteris richardii*. In: Fernández H, Kumar A, Revilla MA (eds) Working with Ferns. Issues and applications. Springer, New York, pp 25–36. ISBN 978-1-4419-7161-6

de Vries J, Fischer AM, Roettger M, Rommel S, Schluepmann H, Bräutigam A, Carlsbecker A, Gould SB (2016) Cytokinin-induced promotion of root meristem size in the fern Azolla supports a shoot-like origin of euphyllophyte roots. New Phytologist 209(2):705–720. https://doi.org/10.1111/nph.13630

De-la-Peña C, Galaz-Avalos RM, Loyola-Vargas VM (2008) Possible role of light and polyamines in the onset of somatic embryogenesis of *Coffea canephora*. Mol Biotechnol 39:215–224

Dutra NT, Silveira V, de Azevedo IG, Gomes-Neto LR, Façanha AR, Steiner N, Guerra MP, Floh EIS, Santa-Catarina C (2013) Polyamines affect the cellular growth and structure of pro-embryogenic masses in *Araucaria angustifolia* embryogenic cultures through the modulation of proton pump activities and endogenous levels of polyamines. Physiol Plant 148:121–132

Eeckhout S, Leroux O, Willats WGT, Popper ZA, Viane RLL (2014) Comparative glycan profiling of *Ceratopteris richardii* "C-Fern" gametophytes and sporophytes links cell-wall composition to functional specialization. Ann Bot 114:1295–1307

Ekrt L, Koutecký P (2016) Between sexual and apomictic: unexpectedly variable sporogenesis and production of viable polyhaploids in the pentaploid fern of the *Dryopteris affinis* agg. (Dryopteridaceae). Ann Bot 117:97–106

Elmore HW, Whittier DP (1975) The involvement of ethylene and sucrose in the inductive and developmental phases of apogamous bud formation in *Pteridium* gametophytes. Can J Bot 53: 375–381

Fernández H, Revilla MA (2003) *In vitro* culture of ornamental ferns. Plant Cell Tissue Organ Cult 73:1–13

Fernández H, Bertrand A, Sánchez-Tamés R (1993) *In vitro* regeneration of *Asplenium nidus* L. from gametophytic and sporophytic tissue. Sci Hortic 56:71–77

Finnie JF, van Staden J (1987) Multiplication of the tree fern *Cyathea dregei*. HortSci 22:665

Grossmann J, Fernández H, Chaubey PM, Valdés AE, Gagliardini V, Cañal MJ, Russo G, Grossniklauss U (2017) Proteogenomic analysis greatly expands the identification of proteins related to reproduction in the apogamous fern *Dryopteris affinis* ssp *affinis*. Front Plant Sci 8: 336

Grossniklaus U, Koltunow AM, Van Lookeren M (1998) A bright future for apomixis. Trends Plant Sci 3:415–416

Grusz AL (2016) A current perspective on apomixis in ferns. J Syst Evol 54:656–665

Grzyb M, Kalandyk A, Mikuła A (2018) Effect of TIBA, fluridone and salicylic acid on somatic embryogenesis and endogenous hormone and sugar contents in the tree fern *Cyathea delgadii* Sternb. Acta Physiol Plant 40:1

Hickok A, Kirilux RM (1984) Effects of auxins on gametophyte development and sexual differentiation in the fern *Ceratopteris thalictroides* (L.) Brongn. Bot Gaz 145:37–42

Ikeuchi M, Ogawa Y, Iwase A, Sugimoto K (2016) Plant regeneration: cellular origins and molecular mechanisms. Development 143:1442–1451

Jensen WA (1962) Botanical histohemistry: principles and practice. W.H. Freeman and Co., San Francisco, CA. ISBN 10:9383285257

Jha TB, Mukherjee P, Datta MM (2007) Somatic embryogenesis in *Jatropha curcas* Linn., an important biofuel plant. Plant Biotechnol Rep 1:135–140

Johri M (2008) Hormonal regulation in green plant lineage families. Physiol Mol Biol Plants 14:23–38

Jones VAS, Dolan L (2012) The evolution of root hairs and rhizoids. Ann Bot 110:205–212

Jouzani GS, Sharafi R, Soheilivand S (2018) Fueling the future; plant genetic engineering for sustainable biodiesel production. Biofuel Res J 19:829–845

Kandemir N, Saygili İ (2015) Apomixis: new horizons in plant breeding. Turk J Agric For 39:549–556

Kato Y (1967) Physiological and morphogenetic studies of fern gametophytes and sporophytes in aseptic culture – VIII. Behaviour of *Pteris vittata* gametophytes on sugar and complex media in red light. Planta 77:127–134

Kazmierczak A (2010) Gibberellic acid and ethylene control male sex determination and development of *Anemia phyllititdis* gametophytes. In: Fernández H, Kumar A, Revilla A (eds) Working with Ferns. Issues and applications. Springer, New York, pp 49–66. ISBN 978-1-4419-7161-6

Klämbt D, Knauth B, Dittmann I (1992) Auxin dependent growth of rhizoids of *Chara globularis*. Physiol Plant 85:537–540

Klíma P, Laňková M, Zažímalová E (2016) Inhibitors of plant hormone transport. Protoplasma 253: 1391–1404

Liu H-M, Dyer RJ, Guo Z-Y et al (2012) The evolutionary dynamics of apomixis in ferns: a case study from *Polystichoid* ferns. J Bot 2012:1–11

López RA, Renzaglia KS (2014) Multiflagellated sperm cells of *Ceratopteris richardii* are bathed in arabinogalactan proteins throughout development. Am J Bot 101:2052–2061

Ludwig-Müller J (2000) Indole-3-butyric acid in plant growth and development. Plant Growth Regul 32:219–230

Makowski D, Rybczyński JJ, Mikula A, Klimaszewska K (2015) A simple way to overcome the recalcitrance of the water fern *Ceratopteris thalictroides* (L.) Brongn. to cryopreservation. Acta Soc Bot Pol 84:385–388

Menéndez V, Revilla MA, Bernard P, Gotor V, Fernández H (2006a) Gibberellins and antheridiogen on sex in *Blechnum spicant* L. Plant Cell Rep 25:1104

Menéndez V, Revilla MA, Fernández H (2006b) Growth and gender in the gametophyte of *Blechnum spicant* L. Plant Cell Tissue Organ Cult 86:47–53

Menéndez V, Villacorta NF, Revilla MA et al (2006c) Exogenous and endogenous growth regulators on apogamy in *Dryopteris affinis* (Lowe) Fraser-Jenkins sp. affinis. Plant Cell Rep 25:85–91

Menéndez V, Revilla MA, Fal MA, Fernández H (2009) The effect of cytokinins on growth and sexual organ development in the gametophyte of *Blechnum spicant* L. Plant Cell Tissue Organ Cult 96:245–250

Menéndez V, Arbesú R, Somer M, Revilla MA, Fernández H (2010) From spore to sporophyte: how to proceed in vitro. In: Fernández H, Kumar A, Revilla MA (eds) Working with Ferns: Issues and applications. Springer, New York, pp 97–110. ISBN 978-1-4419-7161-6

Murashige T, Skoog F (1962) A revised medium for rapid growth and bioassays with tobacco tissue cultures. Plant Physiol 15:473–497

Rivera A, Cañal MJ, Grossniklaus U, Fernández H (2018a) The gametophyte of fern: born to reproduce. In: Fernández H (ed) Recent advances in fern research. Springer, New York, pp 3–20. ISBN 978-3-319-75102-3

Rivera A, Conde P, Cañal MJ, Fernández H (2018b) Biotechnology and apogamy in *Dryopteris affinis* ssp. *affinis*: the influence of tissue homogenization, auxins, cytokinins, gibberellic acid and polyamines. In: Fernández H (ed) Recent advances in Fern research. Springer, New York, pp 139–152. ISBN 978-3-319-75102-3

Rybczynski J, Tomiczak K, Grzyb M, Mikula A (2018) Morphogenic events in ferns: single and multicellular explants in Vitro. In: Fernández H (ed) Recent advances in Fern research. Springer, New York, pp 99–120. ISBN 978-3-319-75102-3

Salmi ML, Bushart TJ, Stout SC, Roux SJ (2005) Profile and analysis of gene expression changes during early development in germinating spores of *Ceratopteris richardii*. Genet Genomics Mol Evol 138:1734–1745

Salmi ML, Morris KE, Roux SJ, Porterfield DM (2007) Nitric oxide and cGMP signaling in calcium-dependent development of cell polarity in *Ceratopteris richardii*. Plant Physiol 144: 94–104

Salmi ML, Bushart TJ, Roux S (2010) Cellular, molecular, and genetic changes during the development of *Ceratopteris richardii* gametophytes. In: Fernández H, Kumar A, Revilla MA (eds) Working with ferns. Issues and applications. Springer, New York, pp 11–24

Salvo E (1990) Guía de helechos de la Península Ibérica y Baleares. Ediciones Pirámide S.A, Madrid. ISBN 84-368-0548-8

Scanlon MJ (2003) The polar auxin transport inhibitor N -1- naphthylphthalamic acid disrupts leaf initiation , KNOX protein regulation, and formation of leaf margins in maize. Plant Physiol 133: 597–605

Shin J, Seo PJ (2018) Varying auxin levels induce distinct pluripotent states in callus cells. Front Plant Sci 9:1653

Somer M, Arbesú R, Menéndez V, Revilla MA, Fernández H (2010) Sporophyte induction studies in ferns *in vitro*. Euphytica 171:203–210

Su Y, Liu Y, Zhang X (2011) Auxin- cytokinin interaction regulates meristem development. Mol Plant 4:616–625

Suo J, Zhao Q, Zhang Z, Chen S, Cao J, Liu G, Wei X, Wang T, Yang C, Dai S (2015) Cytological and proteomic analyses of *Osmunda cinnamomea* germinating spores reveal characteristics of fern spore germination and rhizoid tip growth. Mol Cell Proteomics 14:2510–2534

Teale W, Palme K (2018) Naphthylphthalamic acid and the mechanism of polar auxin transport. J Exp Bot 69:303–312

Team Rs (2016) RStudio: integrated development for R. RStudio, Boston

Valledor L, Menéndez V, Cañal MJ, Revilla MA, Fernández H (2014) Proteomic approaches to sexual development mediated by antheridiogen in the fern *Blechnum spicant* L. Proteomics 14: 2061–2071

Vijayakumar P, Datta S, Dolan L (2016) ROOT HAIR DEFECTIVE SIX-LIKE4 ( RSL4) promotes root hair elongation by transcriptionally regulating the expression of genes required for cell growth. New Phytol 4:944–953

von Aderkas P (1984) Promotion of apogamy in *Matteuccia struthiopteris*, the ostrich fern. Am Fern J 74:1

Wada M (2007) The fern as a model system to study photomorphogenesis. J Plant Res 120:3–16

Whittier DP (1964a) The effect of sucrose on apogamy in *Cyrtomium falcatum* Presll. Amer Fern J 54:20–25

Whittier DP (1964b) The influence of cultural conditions on the induction of apogamy in *Pteridium* gametophytes. Am J Bot 51:730–736

Whittier DP, Steeves TA (1960) The induction of apogamy in the bracken fern. Can J Bot 38:925–930

Whittier DP, Steeves TA (1962) Further studies on induced apogamy in ferns. Can J Bot 40:1525–1531

Wyder S, Rivera A, Valdés AE, Gagliardini V, Cañal MJ, Fernández H, Grossniklaus U (2020) Plant physiology and biochemistry differential gene expression profiling of one- and two-dimensional apogamous gametophytes of the fern *Dryopteris affinis* ssp. *affinis*. Plant Physiol Biochem 148:302–311

**S Granados** has a Master's degree in Biotechnology Applied to Conservation & Sustainable Management of Plant Resources. She is well trained in in vitro culture techniques and explored the effect of plant growth regulators and inhibitors on regeneration of homogenized cultures of apogamous gametophytes. She showed excellent qualities to work in science.

**A Rivera** has a Master's degree in Biotechnology Applied to Conservation & Sustainable Management of Plant Resources and a PhD from Department of Organisms and Systems Biology, at the Oviedo University. He has published in SCR and authored book chapters from Springer.

**MJ Cañal** is Professor at the University of Oviedo, Coordinator of Master's degree "Biotechnology Applied to Conservation & Sustainable Management of Plant Resources," and Director of the Department of Organisms and Systems Biology, at the University of Oviedo. Lead of EPIPHYSAGE scientific group, involved in Biology of Systems. Professor Cañal is expert in Epigenomics (epigenetics, transcriptomics, and proteomics of plant development), aging reinvigoration, cellular reprograming, and response to antibiotic stress. In addition, cloning, regardless of age, of species of agroforestry interest and industrial plant production in temporary immersion systems. She counts a great amount of SCR publications, awards, and other academic honors.

**Helena Fernández** is Associate Professor of Plant Physiology at the University of Oviedo, Spain. She received her PhD in Biology in 1993 from Oviedo University, followed by an EU-funded postgraduate research sojourn at the INRA in Orléans, France, training in HPLC-MS techniques. She got the Ramón y Cajal National Excellence Research Program, dealing with reproduction in ferns. She is guest researcher at the Institute of Plant Biology, University of Zurich, working on apomixes, through omics techniques. She is a member of the dean's team, in the Faculty of Biology at the University of Oviedo. She publishes in relevant journals, having edited two books with Springer.

# Part III

# Medicinal Uses

# Ferns: A Potential Source of Medicine and Future Prospects

16

Sonia Abraham and Toji Thomas

**Abstract**

Ferns are good resource of medication for a variety of infirmities. Even though they possess immense medicinal potential, ferns are less used as medicine compared to angiosperms. Medicinal properties of some ferns are mentioned in various ancient literatures by Theophrastus, Sushruta, Charaka, Dioscorides, etc. Information regarding few ferns used as drugs is available in pharmacopoeias of different countries. Ethnic communities all over the world use ferns for various ailments such as dysentery, malaria, stomach ache, urinary disorders, burns, etc. Ayurvedic, Homeopathic and Unani medicines utilise ferns for various medicinal preparations. Recently, many phytochemical and pharmacological studies of ferns are carried out by various workers, and information with respect to the bioactive components of important medicinal ferns is also available. The chapter delineates the traditional ethnomedicinal uses of ferns along with potential medicinal properties like antimicrobial, anti-inflammatory, antidiabetic, anticancerous, etc. In addition to this, the chapter reviews various chemical compounds isolated and characterised from ferns and analyses future prospects of fern research.

**Keywords**

Anticancerous · Ethnomedicinal · Pharmacological · Phytochemical

S. Abraham
Department of Botany, St. Dominic's College Kanjirappally, Kanjirappallay, Kerala, India

T. Thomas (✉)
Department of Botany, St. Thomas College Palai, Pala, Kerala, India





J. Marimuthu et al. (eds.), *Ferns*, https://doi.org/10.1007/978-981-16-6170-9_16

## 16.1 Introduction

Ferns are less regarded as useful plants compared to angiosperms except a few species (Good 1933). Nonetheless, medicinal properties of some ferns such as *Adiantum caudatum* L., *A. capillus-veneris* L., *Actiniopteris radiata* (Sw.) Link, *Pteridium aquilinum* (L.) Kuhn, *Dryopteris filix-mas* (L.) Schott, etc. are mentioned in ancient literatures which date back from Theophrastus, Sushruta, Charaka, Dioscorides, etc. The Pharmacopoeias of USSR, Germany and India narrate the medicinal properties of many ferns (Puri 1970). *Adiantum capillus-veneris* L. is listed as an official drug in Austrian, Belgian, French, Portuguese, Russian, Serbian, Spanish, Swedish and Swiss Pharmacopoeias (May 1978). Recently, there are many literatures (Kaushik and Dhiman 1995; Vasudeva 1999) that delineate the economic importance of ferns and their allies. Ethnic communities all over the world use ferns for various ailments such as dysentery, malaria, stomach ache, urinary disorders, burns, etc. Ayurvedic, Homeopathic and Unani medicines utilise ferns for various medicinal preparations. Apart from the economic utility, many studies are there regarding the phytochemical constituents, bioactive components and pharmacological properties (Singh et al. 2008a; Ho et al. 2011; Pan et al. 2011; Chai et al. 2012; Xu et al. 2012; Aulakh et al. 2019) carried out by various workers. This chapter reviews the potential medicinal uses of ferns along with their traditional ethnomedicinal uses and antimicrobial, anti-inflammatory, anticancerous and antidiabetic activities in various ferns. Apart from this, the chapter narrates the various chemical compounds identified and characterised from ferns and also analyse its future prospects.

## 16.2 Ethnomedicinal Uses of Ferns

Traditional and ethnobotanical uses of ferns in India and also from many parts of the world are reported by many workers (Nwosu 2002; Sharma 2002; Kholia and Punetha 2005; Benjamin and Manickam 2007; Srivastava 2007; Upreti et al. 2009; Rai and Lalramnghinglova 2010; Benniamin 2011; Ho et al. 2011; Prathibha et al. 2011; Revathi et al. 2013; Sathiyaraj et al. 2015). Ethnic communities use ferns for treating various diseases such as malaria, jaundice, cough, asthma, diarrhoea, dysentery, fever, leprosy, stomach disorders, inflammations, snake bite, ulcers, etc.

### 16.2.1 Anthelmintic

Ethnic communities in different countries use large number of ferns to destroy parasitic worms in different formulations. Few species, viz. *Actiniopteris radiata* (Sw.) Link, *Asplenium adiantum-nigrum* L., *Drynaria quercifolia* (L.) J. Sm. and *Oleandra musifolia* (Blume) C. Presl, were utilised as entire plant for anthelmintic medication, whereas rhizomes of *Acrostichum aureum* L., *Blechnum orientale* L., *Pityrogramma calomelanos* (L.) Link, *Pteridium aquilinum* (L.) Kuhn and *Tectaria wightii* (C.B. Clarke) Ching are utilised for anthelmintic medication (Nwosu 2002; Benjamin and Manickam 2007). Other ferns used as vermifuge or anthelmintic medication include *Dicranopteris linearis* (Burm. F.) Underw., *Cyrtomium falcatum* (L.f.) C. Presl and *Cibotium barometz* (L.) J. Sm. (Benniamin 2011).

### 16.2.2 Asthma and Bronchitis

Fronds of *Dicranopteris linearis* (Burm. F.) Underw. (Benniamin 2011) and *Pityrogramma calomelanos* (L.) Link and entire plant of *Tectaria coadunata* (J. Sm.) C. Chr. are used in the treatment of asthma. *T. coadunata* (J. Sm.) C. Chr. is also used in the treatment of bronchitis (Benjamin and Manickam 2007). Decoction prepared from the leaves of *Marsilea minuta* L. along with ginger is used for bronchitis. Infusion of the young fronds of *Adiantum incisum* Forssk. is employed to treat bronchial diseases (Nwosu 2002).

### 16.2.3 Bone Fracture

Branches of *Equisetum ramosissimum* Desf. (Benjamin and Manickam 2007) and fronds of *Lygodium flexuosum* (L.) Sw. (Nwosu 2002) are made into a paste, and it is locally applied as a remedy of bone rupture and disarticulation of bones.

### 16.2.4 Burns and Boils

Crushed juice obtained from the fronds of *Parahemionitis arifolia* (Burm.) Moore and fronds and rhizomes of *Stenochlaena palustris* (Burm. f.) Bedd. are used to treat burns (Benjamin and Manickam 2007). Rhizome paste of *Leucostegia immersa* C. Presl., *Ophioglossum gramineum* Willd. and *Pteris quadriaurita* Retz. and fresh fronds of *Blechnum orientale* L., *Hypolepis glandulifera* Brownsey & Chinnock and *Pityrogramma calomelanos* (L.) Link were applied as a poultice to treat boils by the tribal communities in different parts of the world (Benjamin and Manickam 2007). The use of *Blechnum orientale* L. as poultice in boils is also reported by Benniamin (2011).

### 16.2.5 Cough and Cold

*Marsilea minuta* L. plant is used to cure cough. Fronds of *Adiantum poiretii* Wikstr. and *Pityrogramma calomelanos* (L.) Link and, in addition to this, rhizomes of *Drynaria quercifolia* (L.) J. Sm., *Lygodium flexuosum* (L.) Sw., *Nephrolepis cordifolia* (L.) C. Presl, *Phlebodium aureum* (L.) J. Sm., *Pleopeltis macrocarpa* (Willd.) Kaulf. and *Asplenium nidus* L. are used as a treatment of cough (Benjamin and Manickam 2007; Benniamin 2011). The rhizome paste of *Helminthostachys zeylanica* (L.) Hook. is utilised in the treatment of whooping cough. The leaves of *Asplenium nidus* L. are smoked to get relief from cold. Frond decoction of *Pleopeltis macrocarpa* (Willd.) Kaulf., *Pityrogramma calomelanos* (L.) Link and *Pyrrosia lanceolata* (L.) Farw. is also used for treating cold in South Africa (Benjamin and Manickam 2007). *Adiantum aethiopicum* L., *A. capillus-veneris* L. and *A. caudatum* L. are used as expectorant (Nwosu 2002; Benniamin 2011).

### 16.2.6 Chest Diseases

Rhizome of *Asplenium nidus* L. is useful in treating chest diseases (Benjamin and Manickam 2007). In Malaya, leaf and rhizome decoction of *Adiantum philippense* L. is used to treat chest complaints (Benniamin 2011). Rhizome of *Nephrolepis cordifolia* (L.) C. Presl and fronds of *Pityrogramma calomelanos* (L.) Link are used in chest congestion (Benjamin and Manickam 2007). Rhizome decoction of *Ophioglossum vulgatum* L. is taken internally for heart diseases (Nwosu 2002).

### 16.2.7 Constipation

*Leucostegia immersa* C. Presl. rhizome is useful for treating constipation (Benjamin and Manickam 2007). Decoction of *Adiantum caudatum* L. is also taken internally as a remedy for constipation (Nwosu 2002).

### 16.2.8 Cuts and Wounds

Wound healing property is exhibited by fronds of many ferns, viz. *Adiantum caudatum* L., *Nephrolepis cordifolia* (L.) C. Presl, *Microsorum punctatum* (L.) Copel., *Ophioglossum gramineum* Willd., *Ophioglossum reticulatum* L., *Pteris cretica* L. and rhizome of *Pteridium aquilinum* (L.) Kuhn. These plants are used to make ointment, which is used for wound healing (May 1978; Benjamin and Manickam 2007).

### 16.2.9 Diabetes

Fronds of *A. poiretii* Wikstr. (Benjamin and Manickam 2007) and *A. caudatum* L. (Benniamin 2011) are utilised to rectify diabetes.

### 16.2.10 Diarrhoea

For the treatment of diarrhoea, *Actiniopteris radiata* (Sw.) Link and *Marsilea minuta* L. plants are used. *Drynaria quercifolia* (L.) J. Sm. and *Pteridium aquilinum* (L.)

Kuhn rhizomes are used to prevent diarrhoea (Benjamin and Manickam 2007). Extract of fresh rhizome of *Tectaria coadunata* (J. Sm.) C. Chr. and oily spores of *Psilotum nudum* (L.) P. Beauv. are given to prevent diarrhoea in children (Benjamin and Manickam 2007). Internal administration of the decoction prepared from the rhizome of *Cheilanthes albomarginata* C. B. Clarke is also effective against diarrhoea (Nwosu 2002).

### 16.2.11 Dysentery

Entire plants of *Actiniopteris radiata* (Sw.) Link and *Botrychium lanuginosum* Wall. ex Hook & Grev. and rhizome of *Helminthostachys zeylanica* (L.) Hook. (Benjamin and Manickam 2007) are useful in dysentery. Apart from this, frond extract of *Angiopteris evecta* (G. Forst.) Hoffm. (Benjamin and Manickam 2007; Benniamin 2011) and *Lygodium microphyllum* (Cav.) R. Br. is also used to treat dysentery (Benjamin and Manickam 2007). Rhizome juice of *Dryopteris cochleata* (D. Don) C. Chr. is used to treat amoebic dysentery (Benjamin and Manickam 2007).

### 16.2.12 Fever

Plants of *Actiniopteris radiata* (Sw.) Link, *Adiantum caudatum* L., *Lygodium flexuosum* (L.) Sw. and *Marsilea minuta* L. are used for treating fever (Benjamin and Manickam 2007). In the case of *Adiantum poiretii* Wikstr., *Pityrogramma calomelanos* (L.) Link and *Stenochlaena palustris* (Burm. f.) Bedd. fronds are the useful part for the treatment of fever, whereas *Asplenium nidus* L., *Phlebodium aureum* (L.) J. Sm. and *Trigonospora caudipinna* (Ching) Sledge rhizome juices are used to treat fever (Benjamin and Manickam 2007). *Actiniopteris radiata* (Sw.) Link and *Adiantum capillus-veneris* L. are also used as febrifuge (Benjamin and Manickam 2007; Benniamin 2011).

### 16.2.13 Gonorrhoea

Aqueous extract of rhizome of *Lygodium flexuosum* (L.) Sw. and infusion of the roots of *Tectaria macrodonta* (Fee) C. Chr. are used for the treatment of gonorrhoea (Nwosu 2002; Benjamin and Manickam 2007).

### 16.2.14 Gynecological Problems

Crushed juice of *Parahemionitis arifolia* (Burm.) Moore (Benjamin and Manickam 2007) and *Cheilanthes farinosa* (Forssk.) Kaulf. (Benniamin 2011) are used to treat menstrual disorders.

### 16.2.15 Headache and Migraine

Tribals also smoked dried plants of *Hymenophyllum javanicum* Spreng. along with garlic and onions to cure headache (Benjamin and Manickam 2007). *Drynaria quercifolia* (L.) J. Sm. plants are effective remedy intended for migraine (Benjamin and Manickam 2007).

### 16.2.16 Jaundice

Decoction prepared from fresh fronds of *Lygodium flexuosum* (L.) Sw. and *Nephrolepis cordifolia* (L.) C. Presl is given as a drink to treat jaundice (Benjamin and Manickam 2007). *Osmunda regalis* L. plants are also useful in jaundice (Nwosu 2002).

### 16.2.17 Leprosy

Spores of *Angiopteris evecta* (G. Forst.) Hoffm. (Benniamin 2011) and small quantity of powdered rhizome of *Dryopteris cochleata* (D. Don) C. Chr. are effectively used in the treatment of leprosy. *Marsilea minuta* L. is also useful in treating leprosy (Benjamin and Manickam 2007).

### 16.2.18 Malaria

Infusion obtained from the rhizome of *Helminthostachys zeylanica* (L.) Hook. (Benjamin and Manickam 2007), young fronds of *Adiantum incisum Forssk.* and roots of *Osmunda regalis* L. (Nwosu 2002) are used as a therapeutic for malaria.

### 16.2.19 Respiratory Diseases

Rhizome decoction of *Ophioglossum vulgatum* L. is useful in treating pulmonary diseases (Nwosu 2002).

### 16.2.20 Rheumatism

*Christella parasitica* (L.) H. Lev. and *Equisetum ramosissimum* Desf. plants are useful in the treatment of rheumatism, whereas various preparations of rhizome of *Dryopteris cochleata* (D. Don) C. Chr., *Lygodium flexuosum* (L.) Sw. and *Nephrolepis cordifolia* (L.) C. Presl and fronds of *Osmunda hugeliana* Presl and *Vittaria elongata* Sw. are used for rheumatic treatment (Benjamin and Manickam

2007; Benniamin 2011). Tonic obtained from roots of *Dryopteris wallichiana* (Spreng.) Hyl. was taken internally for the treatment of rheumatism (Nwosu 2002).

### 16.2.21 Scabies

Decoction of *Adiantum caudatum* L. (Nwosu 2002) is made into a paste and applied externally to heal scabies.

### 16.2.22 Skin Diseases

A large number of plants are employed in treating skin diseases, viz. *Marsilea minuta* L., spores of *Angiopteris evecta* (G. Forst.) Hoffm., *Adiantum poiretii* Wikstr. (Benjamin and Manickam 2007), etc. Fronds of *Adiantum caudatum* L., *Ceratopteris thalictroides* (L.) Brongn., *Lygodium microphyllum* (Cav.) R. Br. and *Stenochlaena palustris* (Burm. f.) Bedd. are also applied as poultice in skin diseases (Benjamin and Manickam 2007; Benniamin 2011).

### 16.2.23 Snakebite

Whole plant extract of *Dryopteris cochleata* (D. Don) C. Chr., along with rhizome of *Alsophila gigantea* Wall. ex Hook., *Adiantum philippense* L. and *Oleandra musifolia* (Blume) C. Presl, is used against snakebite (Benjamin and Manickam 2007). Powdered rhizome of *Adiantum philippense* L. is used as an antidote towards dog bite (Benniamin 2011).

### 16.2.24 Sore Throat

Plant decoction of *Pleopeltis macrocarpa* (Willd.) Kaulf. and *Pyrrosia lanceolata* (L.) Farw. is employed to cure sore throat (Benjamin and Manickam 2007).

### 16.2.25 Sprain

*Pyrrosia heterophylla* (L.) M. G. Price and *Lygodium flexuosum* (L.) Sw. (Benjamin and Manickam 2007) are used in the management of sprains.

### 16.2.26 Stomach Disorders

Powdered stem of *Equisetum ramosissimum* Desf. is used as enema in children when they have stomach disorders. Rhizome of *Pteridium aquilinum* (L.) Kuhn is employed to treat intestinal inflammations and other chronic disorders of the spleen and viscera. Tender parts of *Tectaria coadunata* (J. Sm.) C. Chr. are used after cooking to reduce stomach problems. Root of *Cheilanthes farinosa* (Forssk.) Kaulf. is used to rectify stomach ache (Benjamin and Manickam 2007; Benniamin 2011).

### 16.2.27 Swelling and Inflammation

*Adiantum capillus-veneris* L. (Benniamin 2011), *Pyrrosia heterophylla* (L.) M. G. Price and *Ophioglossum reticulatum* L. are useful for treating inflammations (Benjamin and Manickam 2007). Other plants which have anti-inflammatory effects include the rhizome of *Drynaria quercifolia* (L.) J. Sm. and *Pteridium aquilinum* (L.) Kuhn and the fronds of *Helminthostachys zeylanica* (L.) Hook. Fronds of *Lygodium microphyllum* (Cav.) R. Br. are also employed to cure swellings (Benjamin and Manickam 2007).

### 16.2.28 Tonic for General Health Problems

*Cheilanthes tenuifolia* (Burm. f.) Sw. and *Cibotium barometz* (L.) J. Sm. (Benniamin 2011) plants are used as a general tonic for health.

### 16.2.29 Urinary Problems

Rhizome of *Acrostichum aureum* L. is used to cure bladder complaints. Fresh fronds of *Blechnum orientale* L. are also utilised for urinary bladder complaints in different countries (Benjamin and Manickam 2007). Decoction of the fronds of *Asplenium adiantum-nigrum* L. (Nwosu 2002) is useful in treating urinary trouble.

### 16.2.30 Ulcer

Various preparations of rhizome of *Acrostichum aureum* L., *Drynaria quercifolia* (L.) J. Sm., *Lygodium flexuosum* (L.) Sw. and *Stenochlaena palustris* (Burm. f.) Bedd. are used in the treatment of ulcers (Benjamin and Manickam 2007).

### 16.2.31 Other Ailments

*Thelypteris parasitica* (L.) Tardieu is useful in treating gout (Benniamin 2011). Ferns used to treat various other ailments include *Dryopteris cochleata* (D. Don) C. Chr. for epilepsy; *Lygodium flexuosum* (L.) Sw. used to treat piles; *Pityrogramma calomelanos* (L.) Link used for renal disorders; and *Asplenium nidus* L., *Acrostichum aureum* L., *Isoetes coromandelina* L. f. and *Diplazium esculentum*

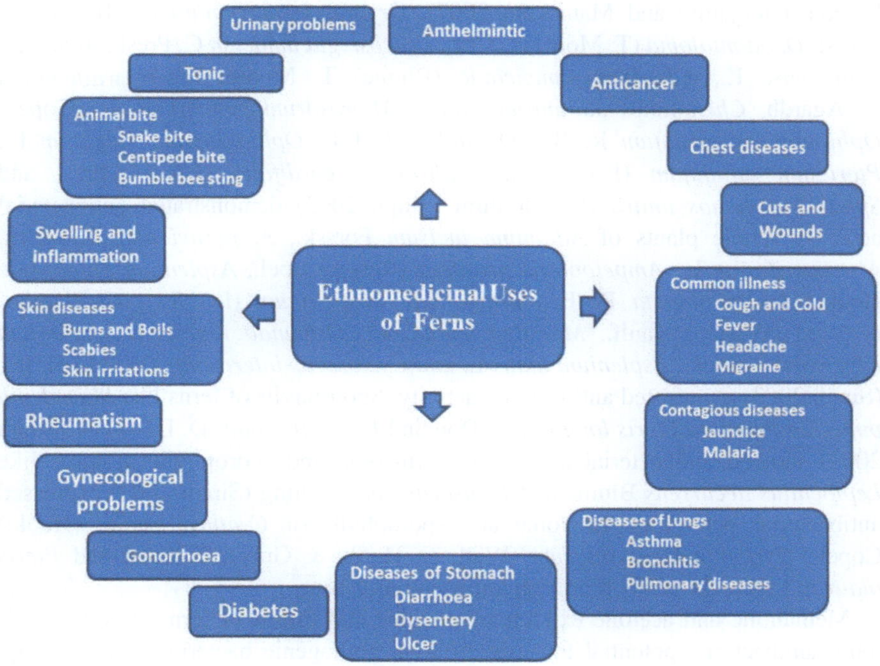


**Fig. 16.1** Ethnomedicinal uses of ferns

(Retz.) Sw. used for elephantiasis, syphilis, liver diseases and toothache, respectively (Nwosu 2002; Benjamin and Manickam 2007). The ethnobotanical uses of ferns are summarised in Fig. 16.1.

## 16.3 Bioactivities of Ferns

Even though ferns have high medicinal potential, the biomedical activities of different ferns are least studied except a few genera. Many of the medicinally useful ferns have antimicrobial, antidiabetic, anticancerous, antioxidant and anti-inflammatory properties. Some of these reported activities are summarised here.

### 16.3.1 Antibacterial Activities

Several ferns are reported to have antibacterial activities. Reports on antibacterial properties of ferns were given in the works of Banerjee and Sen (1980) and Singh (2003). Fronds show antibacterial property in *Pteris cretica* L., *Davallia bullata* Wall. ex Hook., *Dryopteris wallichiana* (Spreng.) Hyl., *Microlepia villosa* (D. Don) Ching, *Lygodium flexuosum* (L.) Sw. and *Microsorum alternifolium* (Willd.) Copel. (Singh 2003). Rhizomes of plants like *Hypodematium crenatum* (Forssk.) Kuhn &

Decken (Benjamin and Manickam 2007), *Dryopteris marginata* (C. B. Clarke) Christ, *D. odontoloma* (T. Moore) C.Chr., *Leucostegia immersa* C. Presl., *Adiantum philippense* L., *Athyrium puncticaule* (Blume) T. Moore, *Pteris wallichiana* C. Agardh, *Cheilanthes dalhousiae* Hook., *Microsorum punctatum* (L.) Copel., *Ophioglossum costatum* R. Br., *O. nudicaule* L.f., *Ophioglossum vulgatum* L., *Pteridium aquilinum* (L.) Kuhn, *Nephrolepis cordifolia* (L.) C. Presl and *Sphaerostephanos unitus* (L.) Holttum (Singh 2003) demonstrated antibacterial activity. Whole plants of *Adiantum incisum* Forssk., *A. peruvianum* Klotzsch, *A. trapeziforme* L., *Ampelopteris prolifera* (Retz.) Copel., *Asplenium dalhousiae* Hook., *Azolla pinnata* R. Br., *Botrychium virginianum* (L.) Sw., *Cheilanthes farinosa* (Forssk.) Kaulf., *Marsilea minuta* L., *Osmunda regalis* L., *Adiantum edgeworthii* Hook., *Asplenium nidus* L. and *Cyclosorus interruptus* (Willd.) H. Ito. (Singh 2003) manifested antibacterial activity. Sporophylls of ferns like *Microlepia puberula* Alderw., *Pteris longipes* D. Don and *P. squarrosum* (D. Don) Fee (Singh 2003) showed antibacterial action. Both rhizome and sporophylls of ferns like *Leptochilus decurrens* Blume and *Lepisorus nudus* Ching (Singh 2003) expressed antibacterial behaviour. Rhizome and sporophylls of *Cyathea crinita* (Hook.) Copel., *Cyrtomium caryotideum* (Wall. ex Hook. & Grev.) C. Presl and *Pteris biaurita* L. (Singh 2003) produced antibacterial vigour.

Methanolic and acetone extracts of rhizome and fronds of ferns showed maximum antibacterial potential towards different pathogenic bacteria. Butanol, ethyl acetate and aqueous portions of the methanolic decoction of *Blechnum orientale* L. showed significant activity towards gram-positive bacteria while they were not effective towards gram-negative bacteria (Lai et al. 2010). Preliminary screening experiments using the methanolic extract of the fronds of *Pteris quadriaurita* Retz. showed antibacterial activity towards *Pseudomonas aeruginosa* (Thomas 2011). Thomas (2012) demonstrated antibacterial activity of methanolic extracts of *Acrostichum aureum* L. towards the multidrug-resistant gram-negative bacteria, *Pseudomonas aeruginosa.* Singh et al. (2008b) verified the antimicrobial agents against gram-positive and gram-negative bacteria in the methanolic decoction of *Adiantum venustum* D. Don, *A. peruvianum* Klotzsch, *A. capillus-veneris* L. and *A. caudatum* L.; they concluded that the perceived antibacterial activity could be correlated with the amount of total phenolics. Decoctions obtained from the aerial parts and roots of *Adiantum capillus-veneris* L. are effective towards various multidrug-resistant bacteria (Ishaq et al. 2014). *Tectaria gemmifera* (Fee) Alston is a medicinally useful fern with potential antibacterial property. Acetone extracts of both rhizome and fronds of the plant showed antimicrobial activity towards pathogenic bacteria (Neel et al. 2017). Methanolic extracts of *Adiantum caudatum* L. and *A. latifolium* Lam. showed antibacterial activities towards *Staphylococcus aureus*, *E. coli* and *Klebsiella pneumoniae* (Johnson et al. 2017). Antibacterial action of the aqueous and alcoholic decoction of the fronds of 12 species of ferns was evaluated against *Escherichia coli*, *Agrobacterium tumefaciens*, *Salmonella arizonae*, *S. typhi* and *Staphylococcus aureus* (Parihar et al. 2010).

Even though there are numerous reports regarding antibacterial properties of ferns, limited reports are there with respect to the characterisation of the active principles. Singh et al. (2008a) reported the antimicrobial potential of 70% methanolic decoction of *Pteris vittata* L. towards gastrointestinal pathogens. The active principle behind this property may be established as flavonoid and rutin, which exhibited potential activity towards *B. cereus*, *P. aeruginosa* and *K. pneumoniae*. Antibacterial potential acylated flavonol glycosides, viz. stenopalustrosides A-D, were reported in *Stenochlaena palustris* (Burm. f.) Bedd. (Liu et al. 1999). Methanolic decoction of *Pteris biaurita* L. fronds demonstrated significant action towards *Bacillus pumilus* (Dalli et al. 2007). Proanthocyanidins and other flavonoids are the key components isolated from the fronds and rhizome of *Phymatopteris triloba* (Houtt.) Pic. Serm. provided the antibacterial potential (Chai et al. 2013). The essential oils from *Lygodium microphyllum* (Cav.) R. Br. showed inhibitory effects on *Staphylococcus aureus*, *Escherichia coli* and *Pyricularia oryzae* (Wang et al. 2014).

### 16.3.2 Antifungal Activities

Crude ethanolic extract and phenols extracted from the immature fronds, rhizome and the gametophyte of *Adiantum capillus-veneris* L. and *A. lunulatum* Burm. f. demonstrated significant bioactivity towards the fungal strains such as *Aspergillus niger* and *Rhizopus stolonifer* (Guha et al. 2005). Antifungal activity of four species of *Adiantum* L., viz. *Adiantum caudatum* L., *A. capillus-veneris* L., *A. peruvianum* Klotzsch and *A. venustum* L., was tested by Singh et al. (2008b). Dalli et al. (2007) characterised the antifungal components from a medicinal fern *Pteris biaurita* L. and reported strong antimicrobial potential due to the presence of active components like eicosenes and heptadecanes. *Stenochlaena palustris* (Burm. f.) Bedd. exhibited significant antifungal activity towards *Aspergillus niger*, a food-borne pathogen, and this is the reason why the plant can be utilised as a natural source of food preservatives (Sumathy et al. 2010). Even though there are limited reports concerning antifungal activity of ferns, few reports are available for the fronds of *Cheilanthes dalhousieae* Hook.; whole plant extract of *Adiantum peruvianum* Klotzsch and *Salvinia molesta* D. Mitch. (Singh 2003) demonstrated antifungal activity.

### 16.3.3 Antiviral Activities

Whole plant extracts of *Adiantum venustum* D. Don, *Ampelopteris prolifera* (Retz.) Copel., *Dryopteris filix-mas* (L.) Schott, *Equisetum arvense* L. and *Ophioglossum costatum* R. Br. (Singh 2003) reported to have antiviral potential. In addition to this, fronds of *Helminthostachys zeylanica* (L.) Hook., *Pteris vittata* L. and *Adiantum capillus-veneris* L. (Benjamin and Manickam 2007) also exhibited antiviral property. Essential oil extracted from *Osmunda regalis* L. showed significant antiviral activity towards Coxsackievirus B4 (CV-B4) (Bouazzi et al. 2018). Uddin et al. (2013) isolated a phthalic acid ester from the aerial parts of *Acrostichum aureum*

L. which showed potential activity against viruses which cause dengue fever, chikungunya and influenza.

### 16.3.4 Anticancer Activities

Whole plants of *Adiantum incisum* Forssk., *A. venustum* D. Don, *Botrychium lunaria* (L.) Sw., *Dicranopteris linearis* (Burm. F.) Underw., *Dryopteris filix-mas* (L.) Schott, *Ophioglossum vulgatum* L. and *Osmunda regalis* L. (Singh 2003) reported to have anticancerous property. Fronds of *Adiantum capillus-veneris* L. and rhizome of *Adiantum philippense* L. (Singh 2003) were also known for their anticancerous activity. The diterpenoids obtained from the ethanolic extract of *Pteris semipinnata* L. possess anticancer activity (Zhang et al. 1999). The anticancerous property of *Adiantum venustum* D. Don may be because of the occurrence of flavonoids and terpenoids detected in it (Viral et al. 2011). *Blechnum orientale* L. could be used for developing new therapeutic drugs to treat colon cancer as its butanol fraction of the methanolic extract obtained from fronds showed anticancerous properties (Lai et al. 2010). Both ethyl acetate and dichloromethane extracts of *Equisetum ramosissimum* Desf. indicated protective activity against human melanoma (Li et al. 2016). Two pterosin compounds isolated from *Pteris cretica* L. exhibited cytotoxic activity towards human tumour cell line, HCT-116 (Lu et al. 2019).

### 16.3.5 Anti-Inflammatory Activities

Su et al. (2016) isolated prenylated flavonoids from the whole plant extract of *H. zeylanica* (L.) Hook, which could provide anti-inflammatory activity. Fronds of *Alsophila gigantea* Wall. ex Hook. and rhizome of *Pteridium aquilinum* (L.) Kuhn (Singh 2003) also reported to have anti-inflammatory activity. *Adiantum venustum* D. Don displayed anti-inflammatory activity (Hussain et al. 2008). Methanolic extract of *Adiantum latifolium* Lam. possessed an inhibitory effect on inflammatory response (Nonato et al. 2011). Pharmacological studies on the anti-inflammatory activity of the medicinal fern *Blechnum occidentale* L., using the methanolic extract of the fronds, strongly support its use as a medication for the cure of inflammatory and pulmonary diseases in folk medicine (Nonato et al. 2009). Compounds isolated from *Pteris ensiformis* Burm. render anti-inflammatory potential as reported by Shi et al. (2017). Baskaran et al. (2019) evaluated anti-inflammatory property of the ethanolic extract of fronds of *Pteris tripartita* Sw.

Along with two other triterpenoids, Haider et al. (2013) isolated two triterpenoids, which possessed significant anti-inflammatory activity observed in ethyl alcohol decoction obtained from the fronds of *Adiantum capillus-veneris* L. Anti-inflammatory property was reported for the hexane fraction of total alcoholic extract (Ibraheim et al. 2011) and ethyl acetate fraction of the methanolic decoction (Haider et al. 2011) of *A. capillus-veneris* L. The butanol and ethyl acetate fractions of the ethanolic extract obtained from fronds of *Dryopteris filix-mas* (L.) Schott confirmed anti-inflammatory activity. The biologically active component responsible for this property is identified as quercetin-3-O-α-L-rhamnopyranoside, a flavonoid (Erhirhie et al. 2019).

### 16.3.6 Antidiabetic Activities

Hypoglycaemic and anti-diabetic properties of *Hemionitis arifolia* (Burm.) Moore is evaluated by Nair et al. (2006). Ibraheim et al. (2011) stated the antidiabetic action of the total alcoholic extract of *Adiantum capillus-veneris* L. However, Ranjan et al. (2014) reported that in addition to methanol extract, aqueous extract also showed significant antibacterial activity. Pterosin-type compounds isolated from *Ceratopteris thalictroides* (L.) Brongn., *Hypolepis punctata* (Thunb.) Mett. and *Pteridium revolutum* (Blume) Nakai exhibited the same extent of antioxidant potential as the known pterosins (Chen et al. 2015). The antidiabetic potential of Pterosin A isolated from the whole plant extract of *Hypolepis punctata* (Thunb.) Mett. was already reported by Hsu et al. (2013).

### 16.3.7 Antioxidant Activities

Several species of ferns are reported to have antioxidant activities. Chen et al. (2007) identified the antioxidant potential of phenolic compounds from *Pteris ensiformis* Burm. and observed that the major phenolic antioxidants were the derivatives of caffeic acid. The antioxidant potential of caffeic acid was already proved by Gulcin (2006). Methanolic extracts of sterile stems of *Equisetum arvense* L. possessed strong antioxidant potential. Different assessment methods highlighted its protective activity towards oxidative agents, free radicals and lipid peroxidation (Mimica et al. 2008). Methanolic decoction of *Equisetum ramosissimum* Desf. also exhibited significant antioxidant activity, and hence it could be used as a natural source of antioxidants (Paulsamy et al. 2013). Methanolic extract of *Dryopteris filix-mas* (L.) Schott fronds manifested antioxidant strength (Ali et al. 2012). Aqueous extract of the whole plant of *Pteris multifida* Poir. showed strong antioxidant potential. High phenolic content in the extract was responsible for its antioxidant property (Lan et al. 2011). Antioxidant activity of the aerial parts of *Pteris tripartita* Sw. was ascertained by DPPH assay, which showed that the acetone extract of the plant possessed free radical scavenging activity and these actions are mainly because of the occurrence of phenolic compounds and flavonoids (Baskaran and Jeyachandran 2010). There is a significant correlation among the antioxidant capacity and amount of phenolic compounds present in the samples (Cai et al. 2004; Chang et al. 2007; Soare et al. 2012). Chowdhary et al. (2010) assessed the antioxidant action of *Cheilanthes anceps* Sw. and established that the flavonoid quercetin exhibited prominent antioxidant property. The antioxidant potential of the fronds and rhizome of *Phymatopteris triloba* (Houtt.) Pic. Serm. increased with an augmentation in total flavonoids and hydroxycinnamic acid content (Chai et al. 2013). Fronds of *Stenochlaena palustris*

(Burm. f.) Bedd. contained several phenolic constituents having antioxidant activities, and therefore, they demonstrated the potential of the plant to be employed as a natural supply of antioxidants (Chai et al. 2012). *Pteridium aquilinum* (L.) Kuhn is a widely distributed fern, and it contained compounds which possessed carcinogenic properties (Vetter 2009). Xu et al. (2009) purified a water-soluble polysaccharide from it which exhibited significant antioxidant activity.

Ethyl acetate partition of methanolic decoction of the fronds of *Blechnum orientale* L. showed strong antioxidant activity equivalent to Trolox C, whereas the methanolic crude extract and its butanol and water fractions also possessed significant antioxidant activity (Lai et al. 2010). *Athyrium filix-femina* (L.) Roth, *Dryopteris affinis* (Lowe) Fraser-Jenk. and *D. filix-mas* (L.) Schott might be used as a resource of natural antioxidants as the methanolic extracts from the fronds of these ferns showed antioxidant activity and the antioxidant action accelerated with an increment in the strength of phenolic compounds (Soare et al. 2012).

## 16.4 Phytochemical Constituents of Ferns

Phytochemical constituents and active components of ferns were studied by several workers. Soeder (1985), Ho et al. (2011) and Cao et al. (2017) reviewed the phytochemical constituents of ferns. Aulakh et al. (2019) made a review on the phytochemicals present in medicinally important pteridophytes. The most widely distributed bioactive components are flavonoids and phenolic compounds (Baskaran et al. 2018). Salatino and Prado (1998) isolated and ascertained that flavonoid glycosides from the fronds of four genera, viz. *Pteris* L., *Pellea* Andre, *Doryopteris* J. Sm. and *Cheilanthes* Sw., belonged to Pteridaceae. They concluded that Cheilanthoideae and Pteridoideae members could be distinguished by means of the flavonoid forms. Vogler et al. (2012) evaluated the distribution form of caffeic acid derivatives and other flavonoids in *Dryopteris filix-mas* (L.) Schott and *D. dilatata* (Hoffm.) A. Gray. Lu et al. (1999) isolated some bioactive phenolic compounds from *Pteris multifida* Poir. The traditional medicinal and edible fern, *Stenochlaena palustris* (Burm. f.) Bedd., reported to have phenolic constituents, flavonoids, hydroxycinnamic acid and anthocyanins. However, the quantity and effectiveness of the phenolic compounds varied with the type of leaf used for making extract (Chai et al. 2012).

Flavonoid glycosides and di-E-caffeoyl-meso-tartaric acid are the main constituents identified in *Equisetum arvense* L. (Mimica et al. 2008). Glucosides, phenolic compounds, steroids and terpenoids were reported in *Cibotium barometz* (L.) J. Sm. (Xu et al. 2012). The antimicrobial activity of *Pteris vittata* L. is partly due to the occurrence of the flavonoid rutin (Singh et al. 2008a). Baskaran and Jeyachandran (2010) reported the occurrence of bioactive compounds α-caryophyllene and octadecanoic acid in *Pteris tripartita* Sw. Phytochemical and spectral analysis culminated in the isolation and documentation of new flavanone glycosides obtained from the aerial parts of *Macrothelypteris torresiana* (Gaud.) Ching. (Fu et al. 2009). Haider et al. (2013) characterised two novel triterpenoids isolated from the ethanolic extract of *Adiantum capillus-veneris* L. fronds. Phytochemical analyses of the frond extract of *Dryopteris filix-mas* (L.) Schott was done by Uwumarongie et al. (2016) and observed that it contains glycosides, flavonoids, steroids and tannins.

### 16.4.1 Chemical Compounds Isolated and Characterised from Ferns

A large number of phytochemical compounds were isolated and characterised from ferns. It is cumbersome to represent and describe all of them; therefore, some important phytochemicals characterised from ferns are described here, and these compounds are listed in PubChem database.

Zhao et al. (2011) identified a novel flavan-4-ol glycoside, abacopterin K, and another novel dihydrochalcone glycoside, named as abacopterin L, which were identified in the rhizomes of *Abacopteris penangiana* (Hook.) Ching. *Adiantum lunulatum* Burm.f. provided filicenol B, adiantone (Reddy et al. 2001). Rhizomes of *Drynaria fortunei* (Kunze ex Mett.) J. Sm. yielded single monomer of flavan-3-ol, 4α-carboxymethyl-(+)-catechin methyl ester (Liang et al. 2011). Palustrine and palustridiene alkaloids were characterised from *Equisetum palustre* L. (Cramer et al. 2015). Rhizomes of the fern *Helminthostachys zeylanica* (L.) Hook. yielded flavonoids like ugonins E–L (Huang et al. 2003). Flavonoids called ugonins M, N and T were also characterised from *H. zeylanica* (L.) Hook. (Huang et al. 2009). Underground parts of *Lygodium japonicum* (Thunb.) Sw. provided a novel compound 1,4-naphthoquinone (Chen et al. 2010). The underground parts of *Lygodium japonicum* (Thunb.) Sw. also provided a novel ecdysteroside, called as lygodiumsteroside A (Zhu et al. 2009a). The underground parts of *Lygodium japonicum* (Thunb.) Sw. yielded a novel compound ecdysteroside, identified as 2beta,3beta,14alpha,20R,22R-pentahydroxy-24R-methyl-5beta-cholest-7-en-6-one-3-O-beta-D-glucopyranoside (Zhu et al. 2009b). Quercetin, isoquercetin and β-sitosterol were characterised from *Lygodium microphyllum* (Cav.) R. Br. (Kuncoro et al. 2017). Extraction mediated through ethanol from whole plant parts of *Ophioglossum pedunculosum* Desv. yielded seven novel homoflavonoid glucosides named as pedunculosumosides A–G (Wang et al. 2011a).

Ophioglonin, ophioglonol prenyl ether, ophioglonol 4′-O-beta-D-glucopyranoside, ophioglonol and ophioglonin 7-O-beta-D-glucopyranoside were characterised from *Ophioglossum petiolatum* Hook., all of them were homoflavonoids (Lin et al. 2005). The fern *Pityrogramma calomelanos* (L.) Link produced three structurally varying novel flavonoids with structural complexity identified and named as calomelanols A–C (Asai et al. 1991). Two novel illudane-type sesquiterpene glycosides, viz. isoptaquiloside and caudatoside, were identified from *Pteridium aquilinum* var. *caudatum* (L.) Christ along with a known compound ptaquiloside (Castillo et al. 1997). The plant also yielded protoilludane-type sesquiterpene glucoside named as pteridanoside and pteridanone (Castillo et al. 1999). *Pteridium aquilinum* var. *latiusculum* (Desv.) Underw. ex A. Heller yielded compounds like palmitylpterosin A and B with carcinogenicity (Yoshihira et al.

1978). Pedunculosumosides A and C provided medium level of action in preventing HBsAg secretion. Fletcher et al. (2010) isolated a new norsesquiterpene glucoside called as ptesculentoside from *Pteridium esculentum* (G.Forst.) Cockayne, along with carcinogen ptaquiloside and caudatoside. Arial portion of *Pteris cretica* L. yielded creticoside A, a category of pterosin (Lu et al. 2019). From the plant *Pteris ensiformis* Burm., henrin A and 2′-hydroxy-4′-methoxychalcone were isolated and characterised by various spectroscopic methods (Shi et al. 2017), and the plant also indicated the occurrence of cyclolaudenol (Chen et al. 2008). Three novel C 14 pterosin sesquiterpenoids are identified as multifidoside A–C from *Pteris multifida* Poir., of which multifidoside A and B exhibited cytotoxicity activity (Ge et al. 2008). The plant also produced A novel C (14) pterosin sesquiterpenoid, called as (2R)-pterosin P, as well as a novel compound called as dehydropterosin B. The dehydropterosin B exhibited cytotoxicity towards PANC-1 (human pancreatic cancer) and NCI-H446 (human small cell lung cancer) cell lines (Ouyang et al. 2010).

*Pteris semipinnata* L. (Pteridaceae) ethanol extract yielded six scientifically novel ent-15-oxokauran-19-oic acid derivatives, called as pterisolic acids A–F (Wang et al. 2011b). Stem and fronds of *Pteris semipinnata* L. also provided three illudalane sesquiterpenoids, called as (2R)-norpterosin B, (2R)-12-O-β-D-glucopyranosylnorpterosin B as well as semipterosin A (Zhan et al. 2010). Stem and fronds of *Pteris semipinnata* L. provided three already revealed sesquiterpenoids having single 1-indanone skeleton, viz. (2R)-pterosin B, norpterosin C and (2S, 3S)-pterosin C (Zhan et al. 2009). Flavonoids like kaempferol and quercetin were isolated from *Pteris vittata* L. (Lin et al. 2016). The chemical details are provided in Table 16.1.

## 16.5 Future Prospects of Fern Research

Ferns have immense potential to be exploited as a natural source of drugs. Medicinal potential of ferns is less studied, as compared to angiosperms. Even though there are voluminous reports regarding the traditional uses of ferns, few are studied in detail and active chemical compounds isolated and characterised. Tribal people from different parts of the world utilise aerial parts and rhizomes of ferns for treating various diseases. Majority of these ferns are not yet evaluated for their bioactivities, and few attempts have been made to isolate active components present in it. Further studies will improve the information regarding various phytochemical components and their corresponding bioactivities, which will or will not authenticate their traditional uses. Future researches also aid in the development of new drugs with fewer side effects and increase efficiency for curing various ailments. Besides this, the characterisation of the active principles will be a natural alternative to certain commercially available alternatives. In addition, the newly identified active compounds from ferns can be used as a suitable candidate for computer-aided drug development.

**Table 16.1** Few important chemical compounds isolated, characterised from ferns

| Compound name | Molecular formula and molecular weight | Structure | Reference |
|---|---|---|---|
| *Abacopteris penangiana* (Hook.) Ching | | | |
| Abacopterin K | $C_{30}H_{36}O_{14}$ 620.6 g/mol | 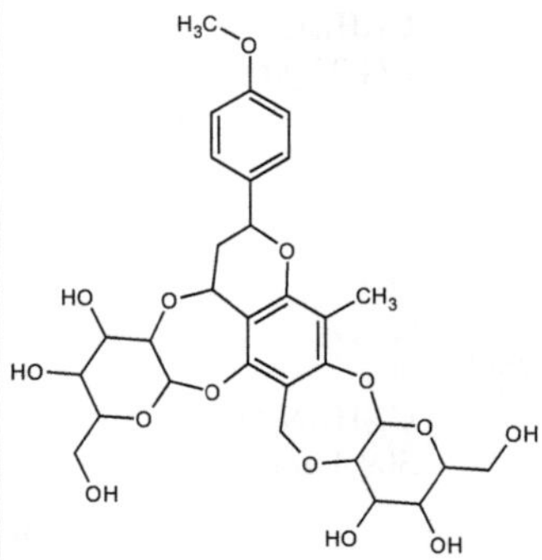 | Zhao et al. (2011) |
| Abacopterin L | $C_{26}H_{32}O_{11}$ 520.5 g/mol | 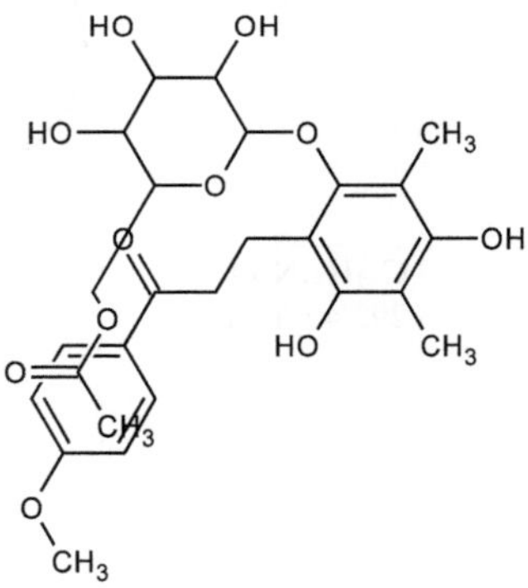 | Zhao et al. (2011) |
| *Adiantum lunulatum* Burm.f. | | | |
| Filicenol-B | $C_{30}H_{50}O$ 426.7 g/mol | 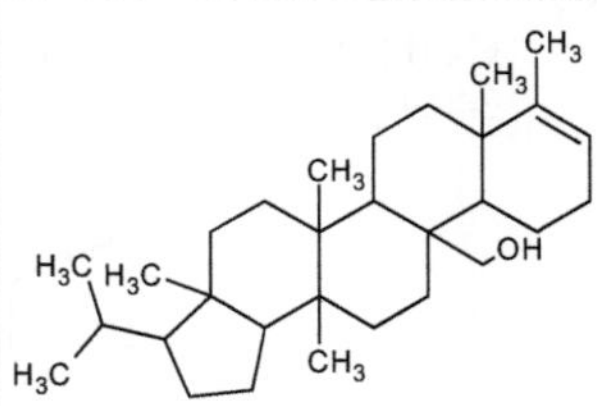 | Reddy et al. (2001) |
| Adiantone | $C_{29}H_{48}O$ 412.7 g/mol | 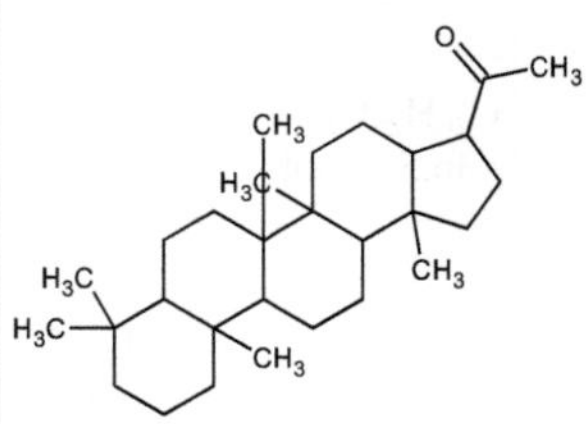 | Reddy et al. (2001) |

(continued)

**Table 16.1** (continued)

| Compound name | Molecular formula and molecular weight | Structure | Reference |
|---|---|---|---|
| *Drynaria fortunei* (Kunze ex Mett.) J. Sm. | | | |
| Flavan-3-Ols | $C_{15}H_{14}O_5$ 274.27 g/mol | 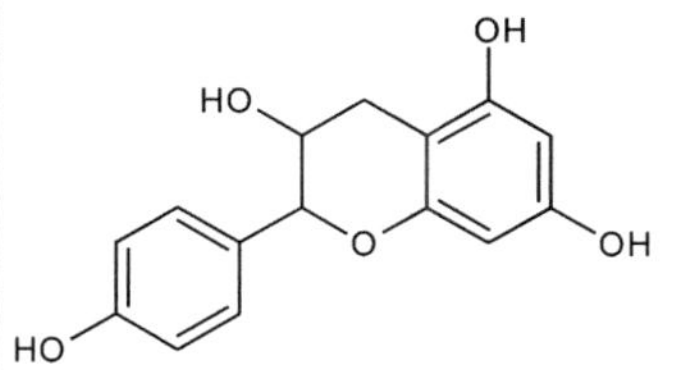 | Liang et al. (2011) |
| *Equisetum palustre* L. | | | |
| Palustrine | $C_{17}H_{31}N_3O_2$ 309.4 g/mol | 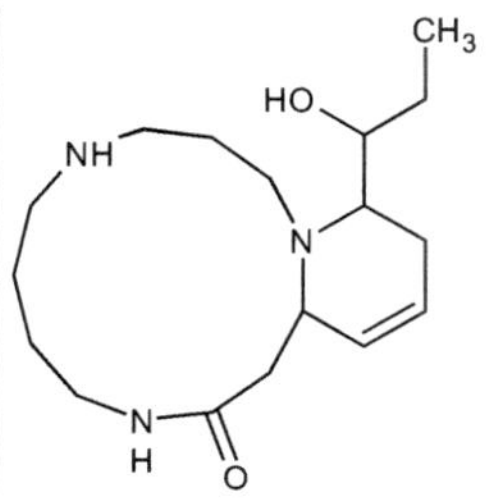 | Cramer et al. (2015) |
| Palustridiene | $C_{17}H_{31}N_3O$ 293.4 g/mol | 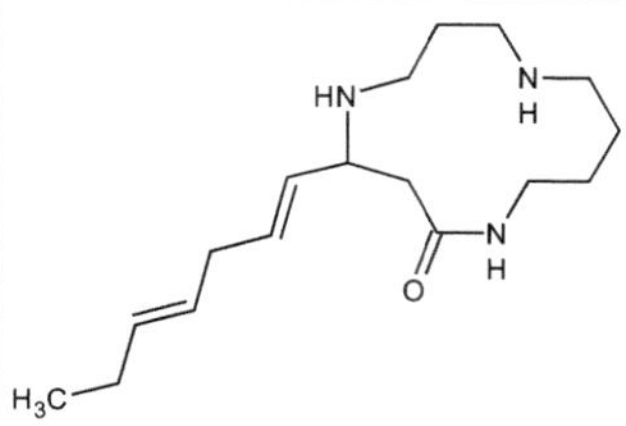 | Cramer et al. (2015) |
| *Helminthostachys zeylanica* (L.) Hook. | | | |
| Ugonin E | $C_{20}H_{20}O_5$ 340.4 g/mol | 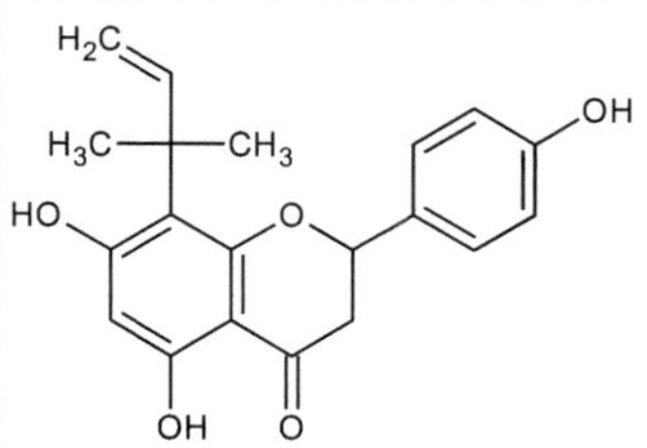 | Huang et al. (2003) |
| Ugonin L | $C_{26}H_{28}O_6$ 436.5 g/mol | 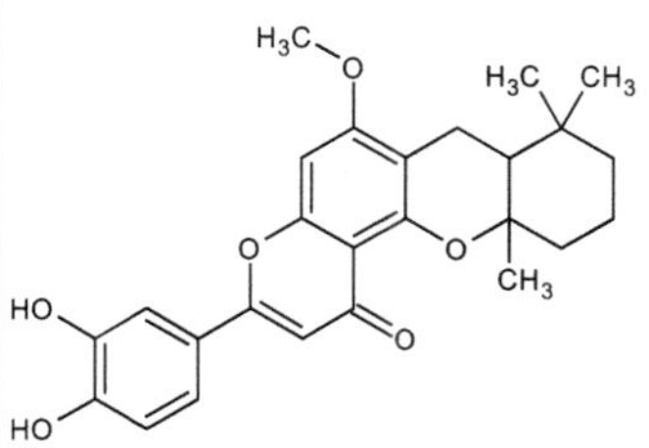 | Huang et al. (2003) |

(continued)

**Table 16.1** (continued)

| Compound name | Molecular formula and molecular weight | Structure | Reference |
|---|---|---|---|
| Ugonin M | $C_{25}H_{24}O_7$ 436.5 g/mol | 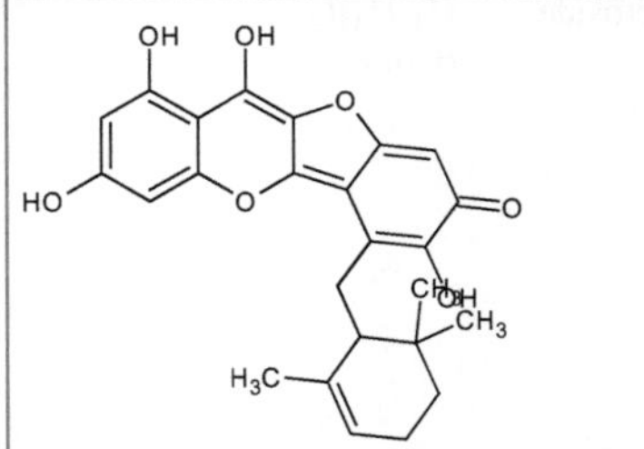 | Huang et al. (2009) |
| Ugonin N | $C_{25}H_{26}O_7$ 438.5 g/mol | 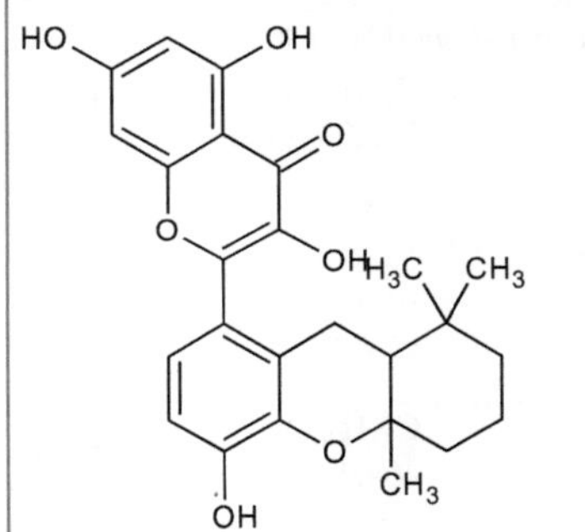 | Huang et al. (2009) |
| Ugonin T | $C_{20}H_{18}O_6$ 354.4 g/mol | 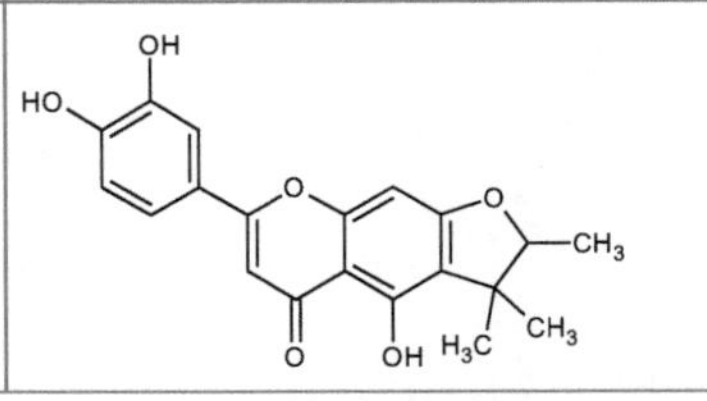 | Huang et al. (2009) |
| *Lygodium japonicum* (Thunb.) Sw. | | | |
| 1,4-naphthoquinone | $C_{10}H_6O_2$ 158.15 g/mol | 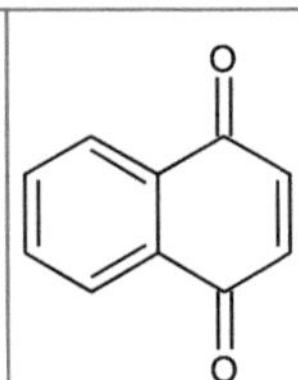 | Chen et al. (2010) |
| Ecdysteroside | $C_{39}H_{64}O_{17}$ 804.9 g/mol | 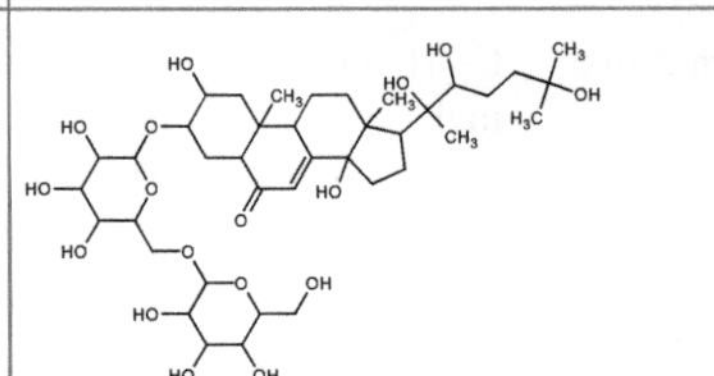 | Zhu et al. (2009b) |

(continued)

**Table 16.1** (continued)

| Compound name | Molecular formula and molecular weight | Structure | Reference |
|---|---|---|---|
| Lygodiumsteroside A | $C_{35}H_{54}O_{12}$ 666.8 g/mol | 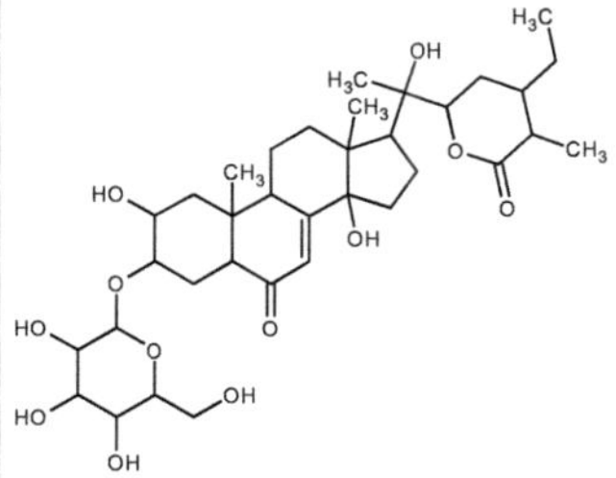 | Zhu et al. (2009a) |
| *Lygodium microphyllum* (Cav.) R.Br. | | | |
| Quercetin | $C_{15}H_{10}O_{7}$ 302.23 g/mol | 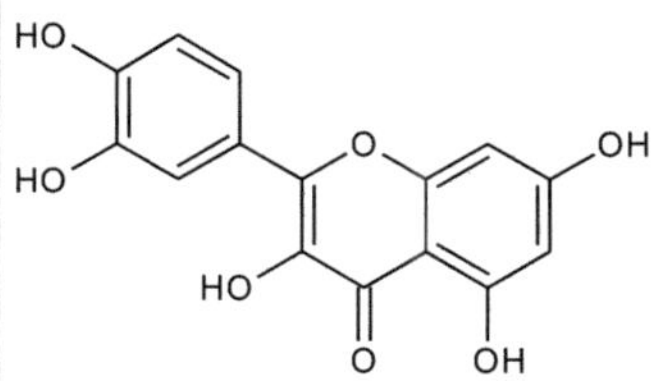 | Kuncoro et al. (2017) |
| Isoquercetin | $C_{21}H_{20}O_{12}$ 464.4 g/mol | 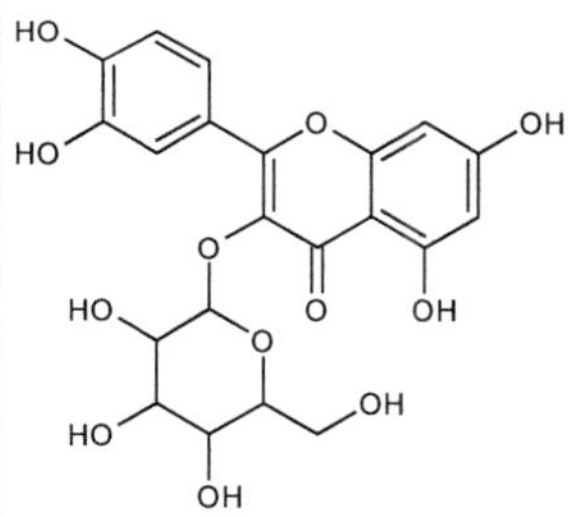 | Kuncoro et al. (2017) |
| β-Sitosterol | $C_{29}H_{50}O$ 414.7 g/mol | 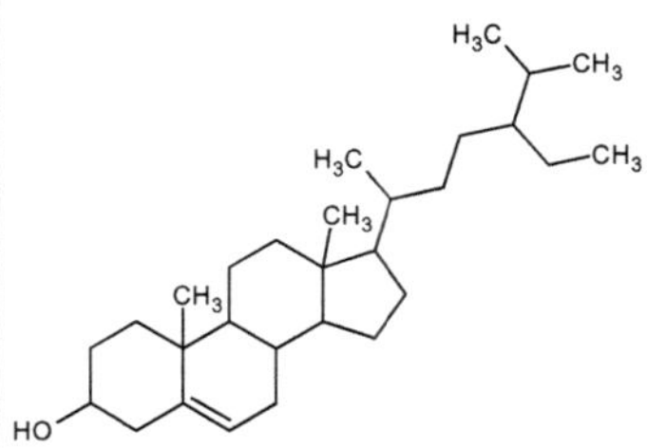 | Kuncoro et al. (2017) |
| Stigmast-4-en-3-one | $C_{29}H_{48}O$ 412.7 g/mol | 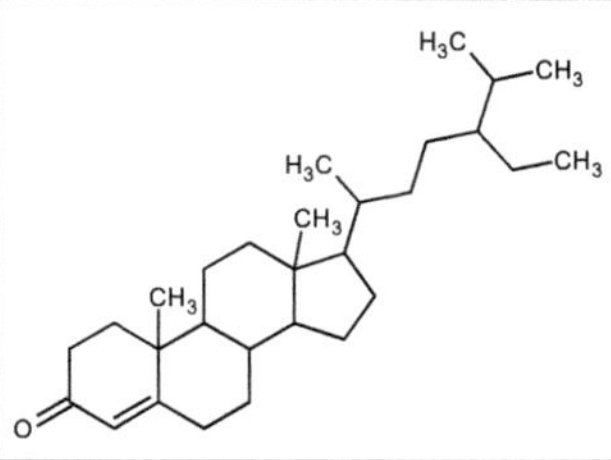 | Kuncoro et al. (2017) |

(continued)

**Table 16.1** (continued)

| Compound name | Molecular formula and molecular weight | Structure | Reference |
|---|---|---|---|
| *Ophioglossum pedunculosum* Desv. | | | |
| Pedunculosumoside A | $C_{39}H_{50}O_{22}$ 870.8 g/mol | 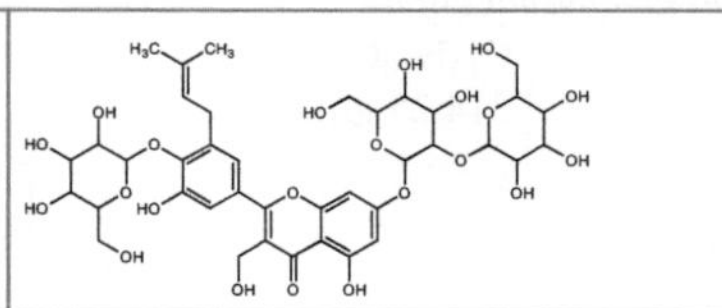 | Wang et al. (2011a) |
| Pedunculosumoside B | $C_{33}H_{40}O_{17}$ 708.7 g/mol | 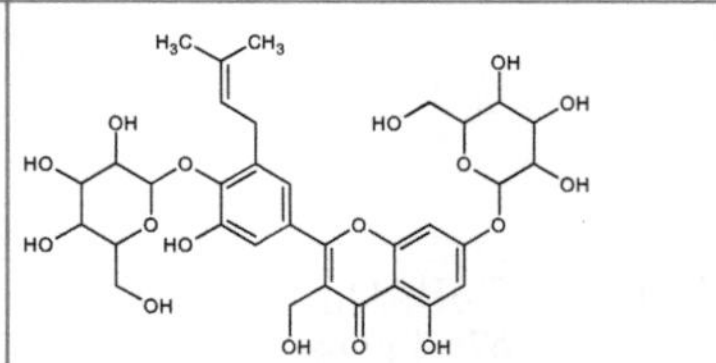 | Wang et al. (2011a) |
| Pedunculosumoside C | $C_{33}H_{40}O_{17}$ 708.7 g/mol | 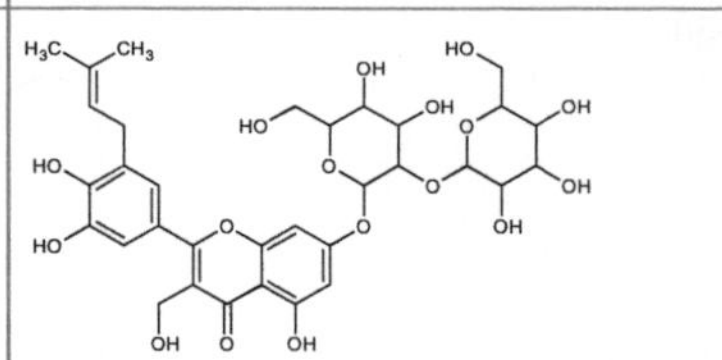 | Wang et al. (2011a) |
| Pedunculosumoside D | $C_{27}H_{30}O_{12}$ 546.5 g/mol | 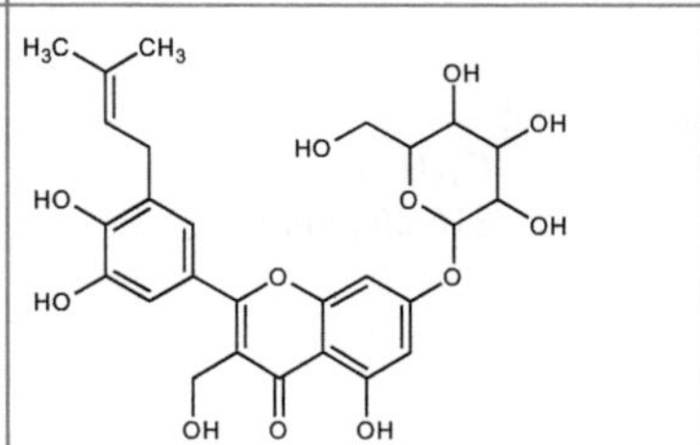 | Wang et al. (2011a) |
| Pedunculosumoside E | $C_{34}H_{42}O_{22}$ 802.7 g/mol | 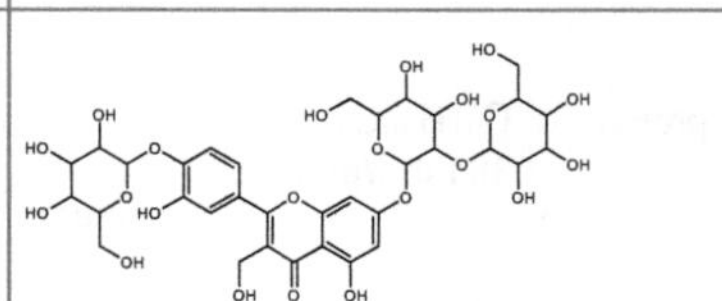 | Wang et al. (2011a) |
| Pedunculosumoside G | $C_{28}H_{32}O_{17}$ 640.5 g/mol | 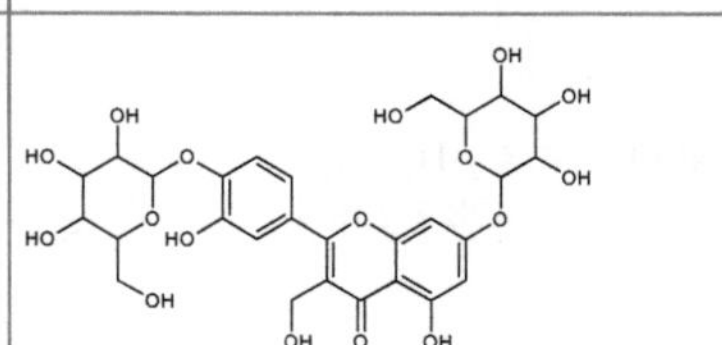 | Wang et al. (2011a) |
| Pedunculosumoside G | $C_{22}H_{22}O_{12}$ 478.4 g/mol | 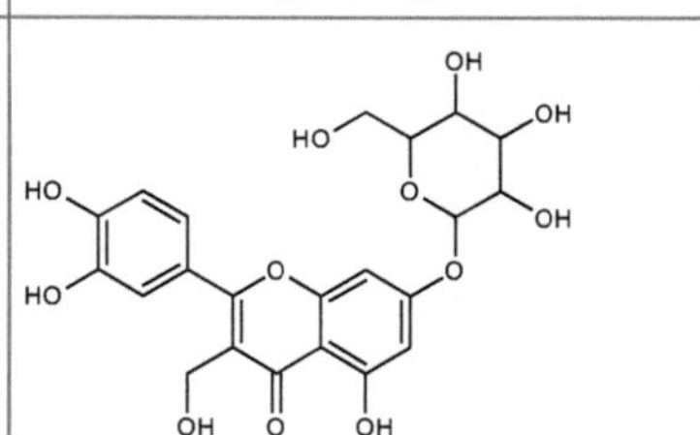 | Wang et al. (2011a) |

(continued)

**Table 16.1** (continued)

| Compound name | Molecular formula and molecular weight | Structure | Reference |
|---|---|---|---|
| *Ophioglossum petiolatum* Hook. | | | |
| Ophioglonin | $C_{16}H_{10}O_7$ 314.25 g/mol | 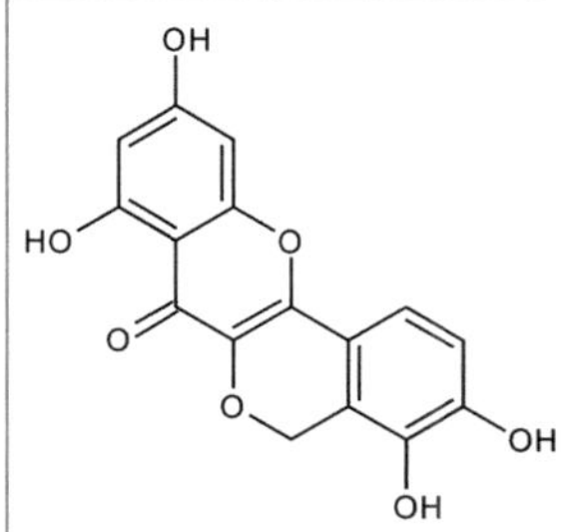 | Lin et al. (2005) |
| Ophioglonin 7-O-beta-D-glucopyranoside | $C_{22}H_{20}O_{12}$ 476.4 g/mol | 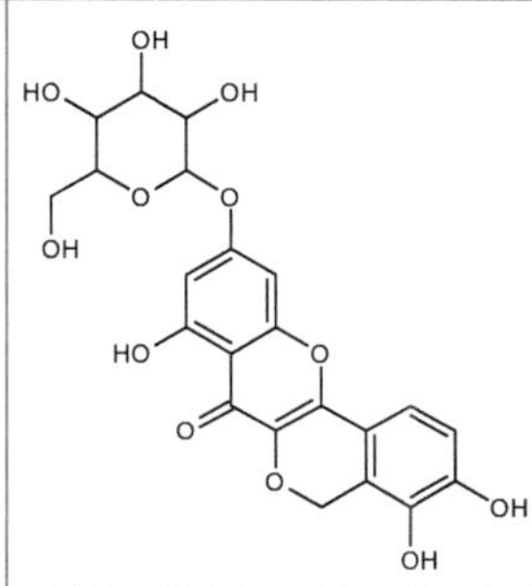 | Lin et al. (2005) |
| Ophioglonol | $C_{16}H_{12}O_7$ 316.26 g/mol | 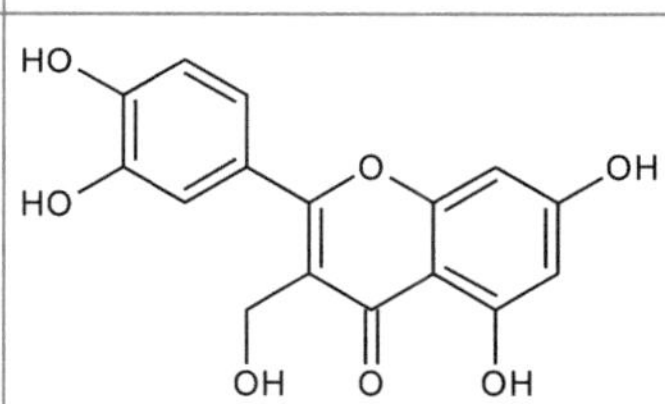 | Lin et al. (2005) |
| Ophioglonol prenyl ether | $C_{21}H_{20}O_7$ 384.4 g/mol | 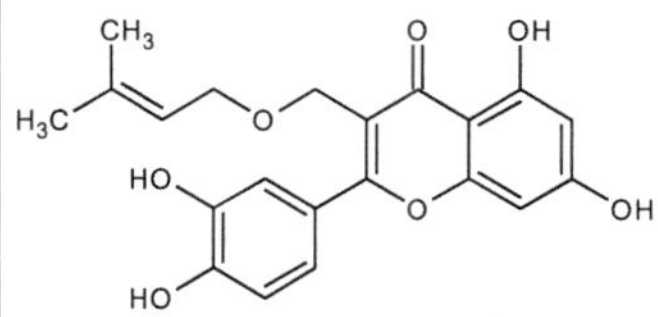 | Lin et al. (2005) |
| Ophioglonol 4′-O-beta-D-glucopyranoside | $C_{22}H_{22}O_{12}$ 478.4 g/mol | 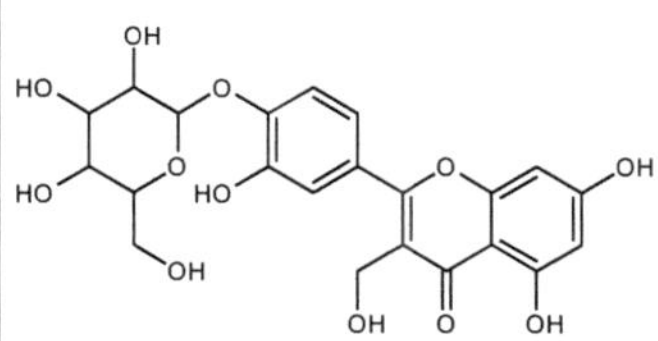 | Lin et al. (2005) |

(continued)

**Table 16.1** (continued)

| Compound name | Molecular formula and molecular weight | Structure | Reference |
|---|---|---|---|
| *Pityrogramma calomelanos* (L.) Link | | | |
| Calomelanol A | $C_{25}H_{22}O_6$ 418.4 g/mol | 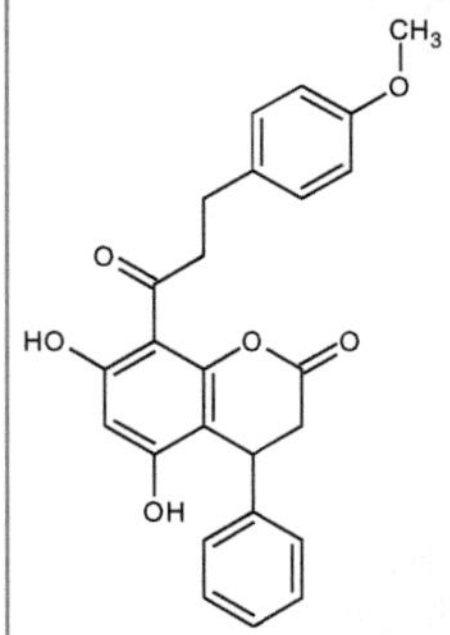 | Asai et al. (1991) |
| Calomelanol B | $C_{24}H_{20}O_6$ 404.4 g/mol | 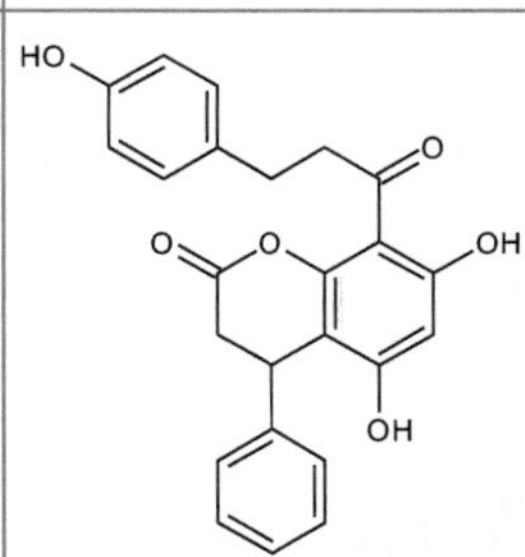 | Asai et al. (1991) |
| Calomelanol C | $C_{24}H_{20}O_6$ 404.4 g/mol | 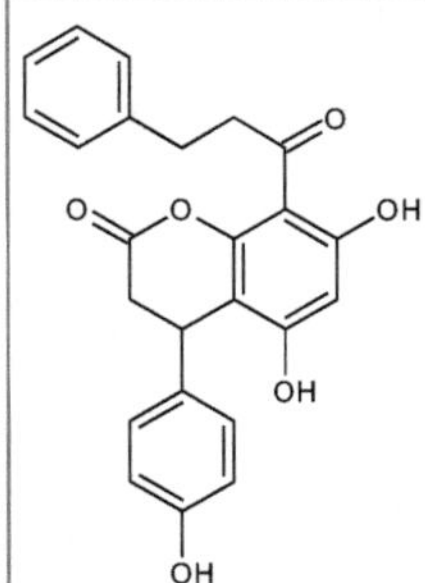 | Asai et al. (1991) |
| *Pteridium aquilinum* var. *caudatum* (L.) Christ | | | |
| Ptaquiloside | $C_{20}H_{30}O_8$ 398.4 g/mol | 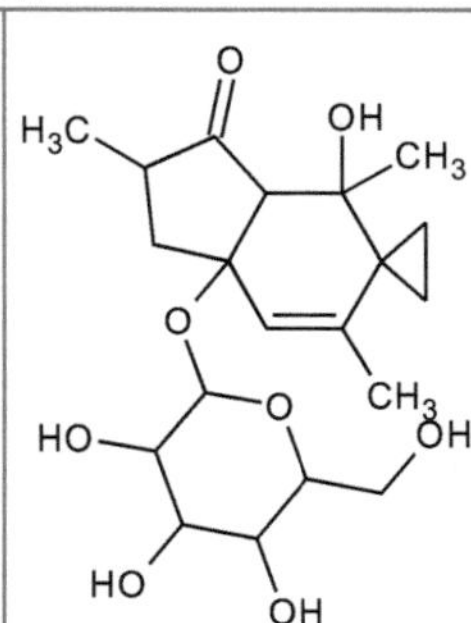 | Castillo et al. (1997) |

(continued)

**Table 16.1** (continued)

| Compound name | Molecular formula and molecular weight | Structure | Reference |
|---|---|---|---|
| Isoptaquiloside | $C_{20}H_{30}O_8$ 398.4 g/mol | 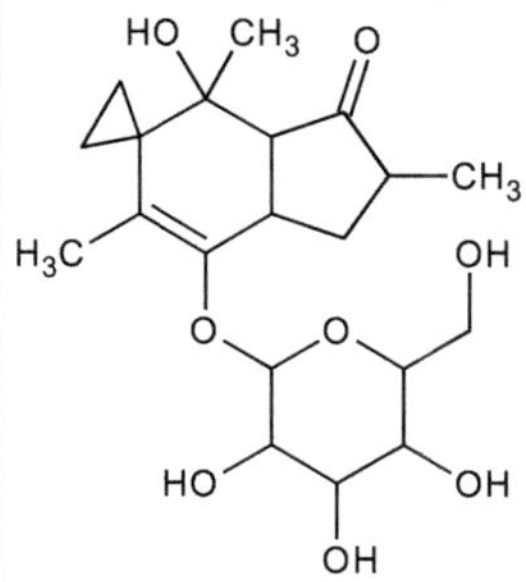 | Castillo et al. (1997) |
| Caudatoside | $C_{21}H_{32}O_9$ 428.5 g/mol | 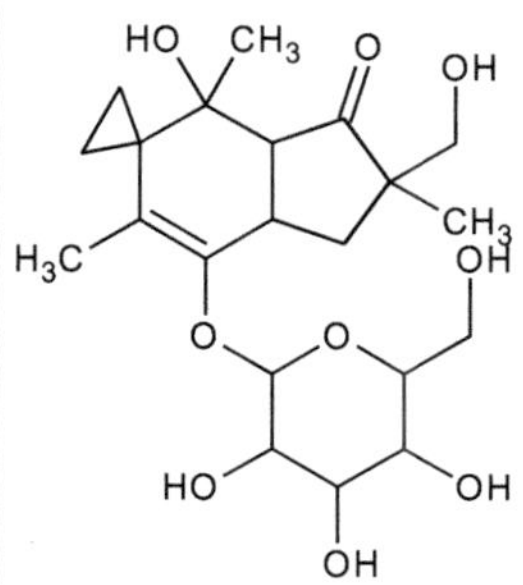 | Castillo et al. (1997) |
| Pteridanoside | $C_{21}H_{32}O_8$ 412.5 g/mol | 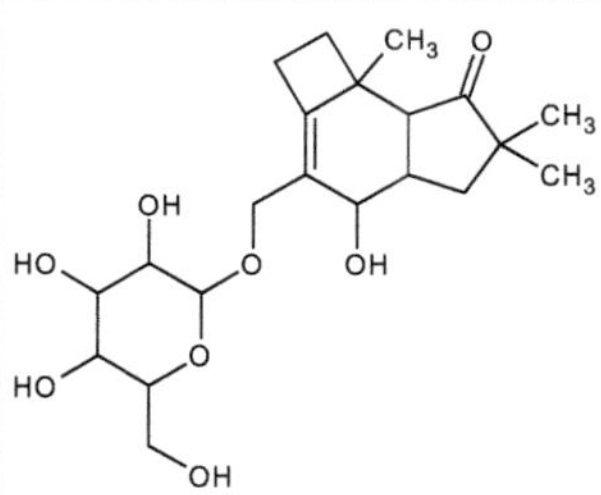 | Castillo et al. (1999) |
| Pteridanone | $C_{15}H_{22}O_3$ 250.33 g/mol | 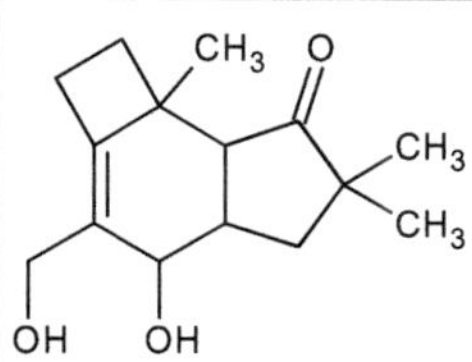 | Castillo et al. (1999) |

(continued)

**Table 16.1** (continued)

| Compound name | Molecular formula and molecular weight | Structure | Reference |
|---|---|---|---|
| *Pteridium aquilinum* var. *latiusculum* (Desv.) Underw. ex A. Heller | | | |
| Palmitylpterosin A | $C_{31}H_{50}O_4$ 486.7 g/mol | 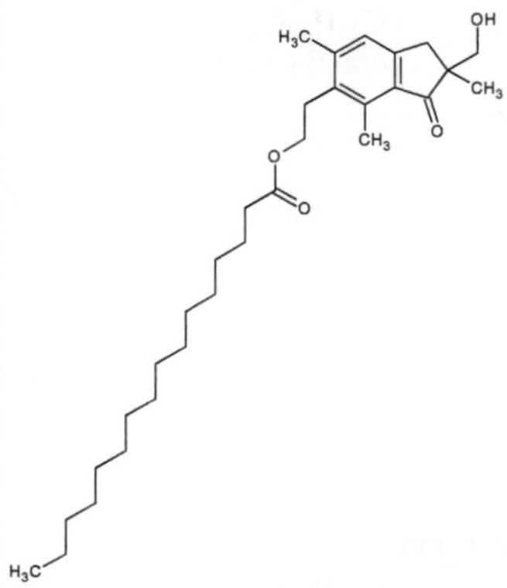  | Yoshihira et al. (1978) |
| Palmitylpterosin B | $C_{30}H_{48}O_3$ 456.7 g/mol | 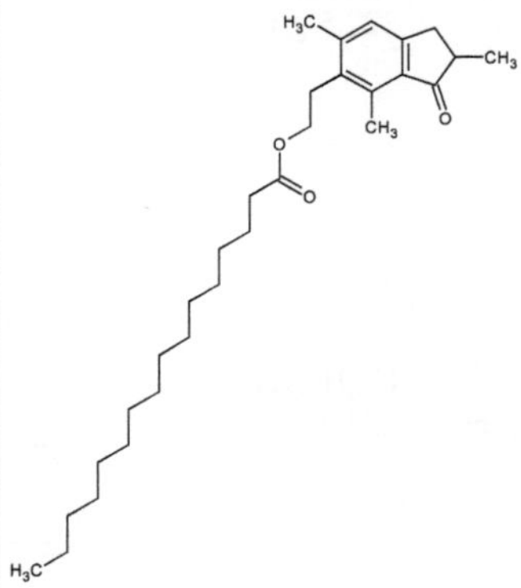  | Yoshihira et al. (1978) |
| *Pteridium esculentum* (G.Forst.) Cockayne | | | |
| Ptesculentoside | $C_{20}H_{30}O_9$ 414.4 g/mol | 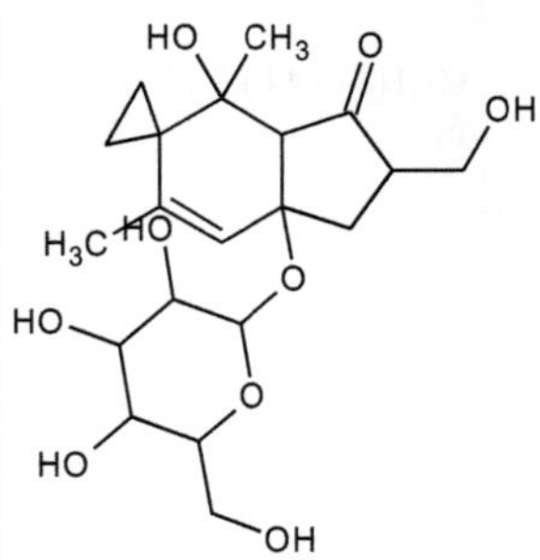  | Fletcher et al. (2010) |

(continued)

**Table 16.1** (continued)

| Compound name | Molecular formula and molecular weight | Structure | Reference |
|---|---|---|---|
| *Pteris cretica* L. | | | |
| Creticoside A | $C_{26}H_{42}O_7$ 466.6 g/mol | 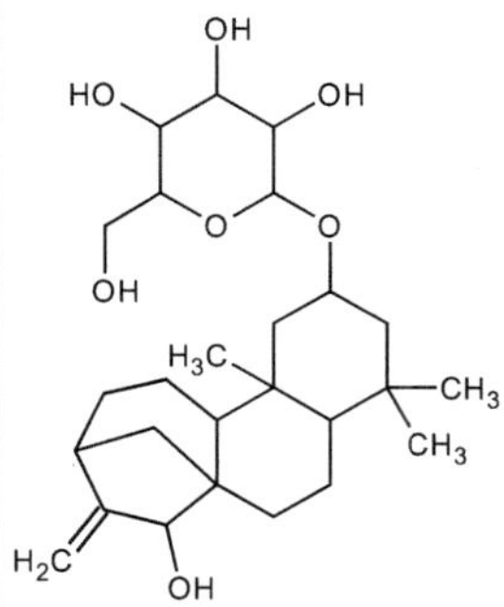  | Lu et al. (2019) |
| *Pteris ensiformis* Burm. | | | |
| Henrin A | $C_{20}H_{34}O_3$ 322.5 g/mol | 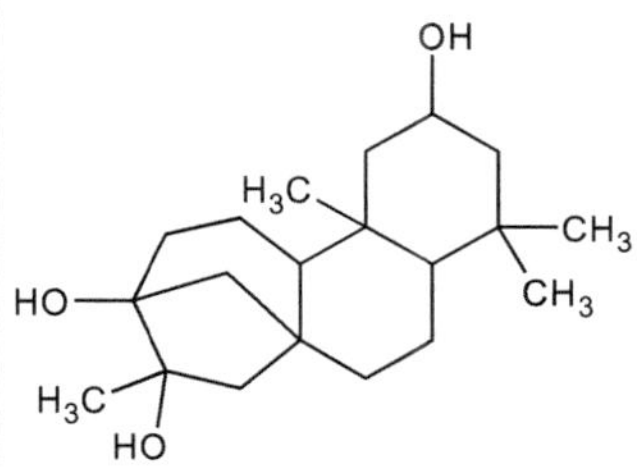  | Shi et al. (2017) |
| 2′-Hydroxy-4′-methoxychalcone | $C_{16}H_{14}O_3$ 254.28 g/mol | 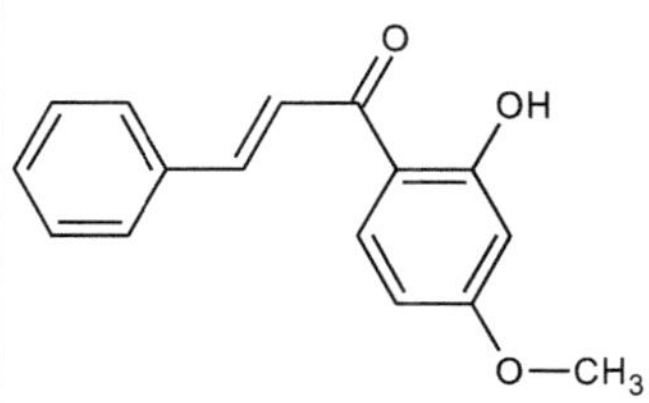  | Shi et al. (2017) |
| Cyclolaudenol | $C_{31}H_{52}O$ 440.7 g/mol | 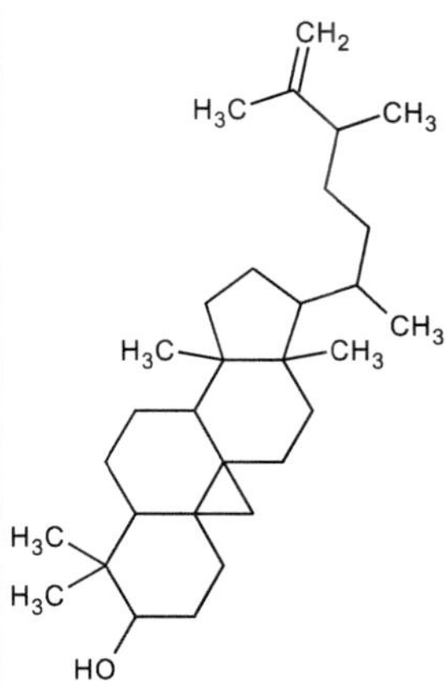  | Chen et al. (2008) |

(continued)

**Table 16.1** (continued)

| Compound name | Molecular formula and molecular weight | Structure | Reference |
|---|---|---|---|
| *Pteris multifida* Poir. | | | |
| Multifidoside A | $C_{29}H_{36}O_{15}$<br>624.6 g/mol | 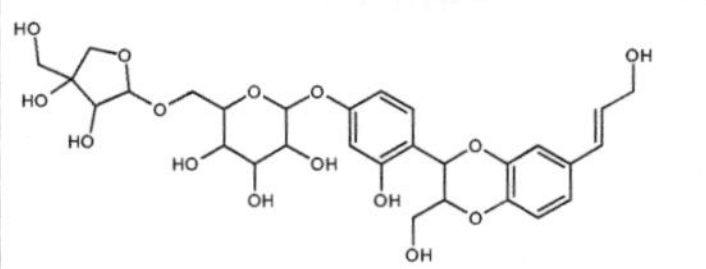 | Ge et al. (2008) |
| Multifidoside B<br>or<br>(R)-pterosin B | $C_{14}H_{1}8O_{2}$<br>218.29 g/mol | 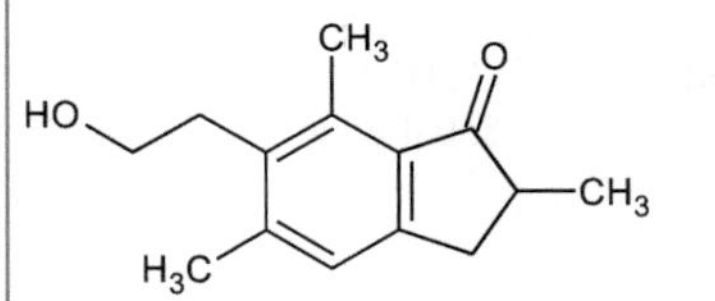 | Ge et al. (2008) |
| Multifidoside C | $C_{29}H_{34}O_{9}$<br>526.6 g/mol | 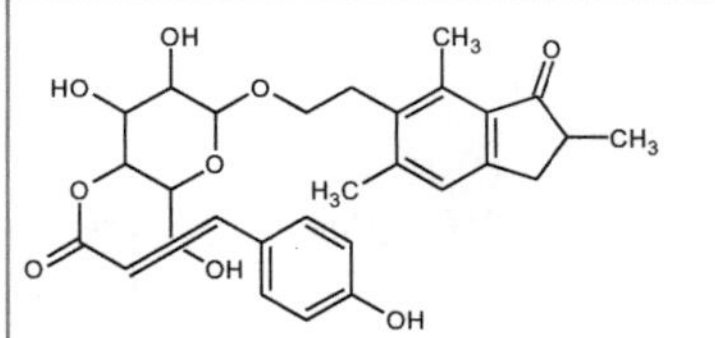 | Ge et al. (2008) |
| (2R)-Pterosin P | $C_{14}H_{18}O_{3}$<br>234.29 g/mol | 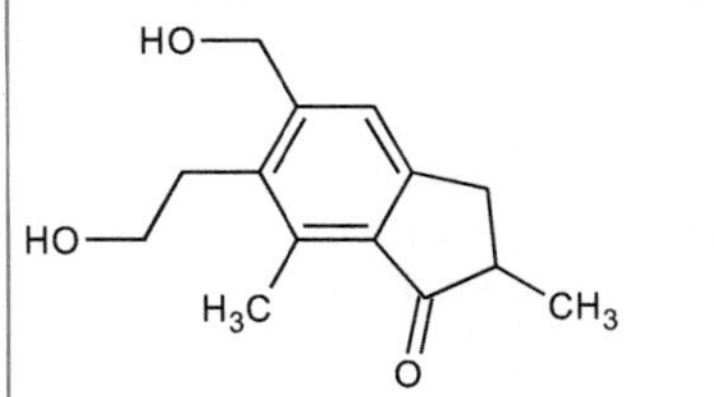 | Ouyang et al. (2010) |
| Dehydropterosin B | $C_{16}H_{18}O_{3}$<br>258.31 g/mol | 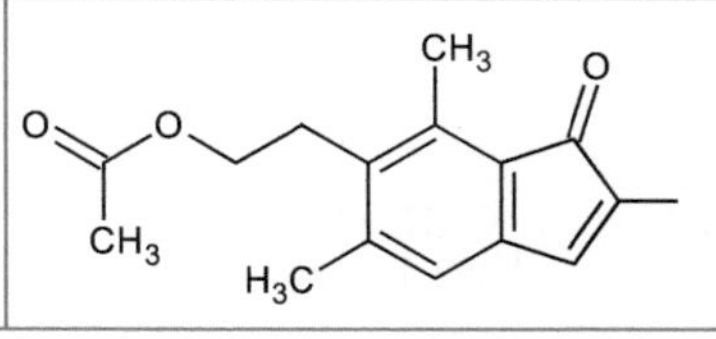 | Ouyang et al. (2010) |
| *Pteris semipinnata* L. | | | |
| Pterisolic acid A | $C_{20}H_{26}O_{5}$<br>346.4 g/mol | 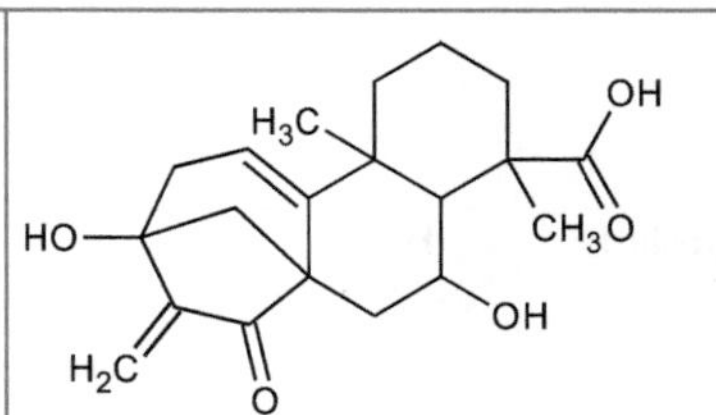 | Wang et al. (2011b) |
| Pterisolic acid B | $C_{20}H_{26}O_{4}$<br>330.4 g/mol | 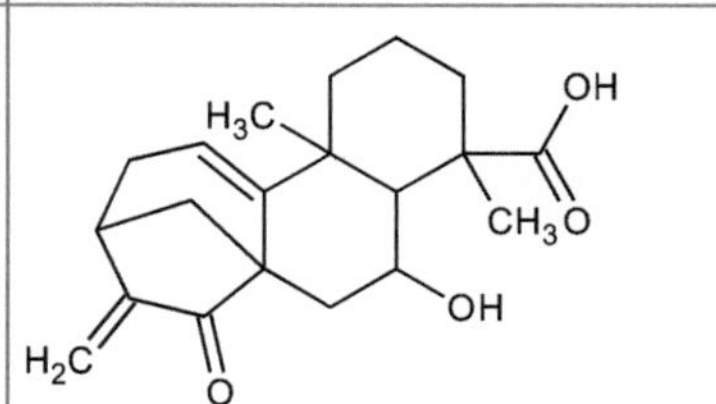 | Wang et al. (2011b) |

(continued)

**Table 16.1** (continued)

| Compound name | Molecular formula and molecular weight | Structure | Reference |
|---|---|---|---|
| Pterisolic acid C | $C_{20}H_{26}O_4$ 330.4 g/mol | 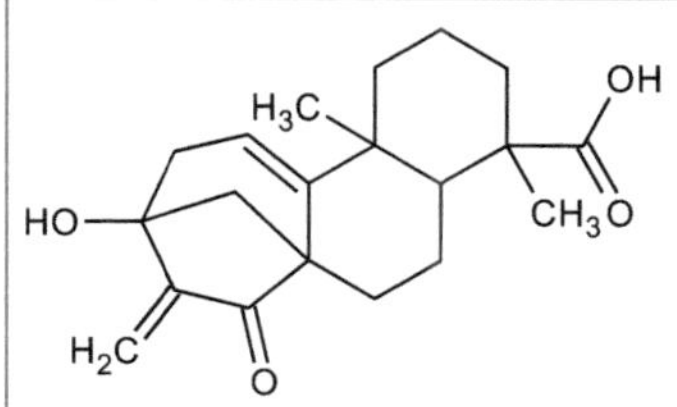 | Wang et al. (2011b) |
| Pterisolic acid D | $C_{20}H_{30}O_5$ 350.4 g/mol | 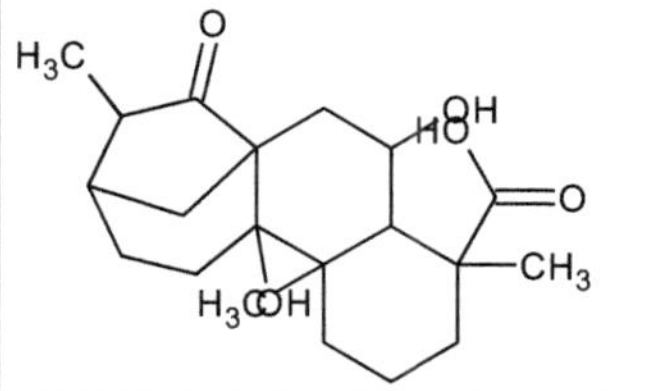 | Wang et al. (2011b) |
| Pterisolic acid E | $C_{20}H_{30}O_5$ 350.4 g/mol | 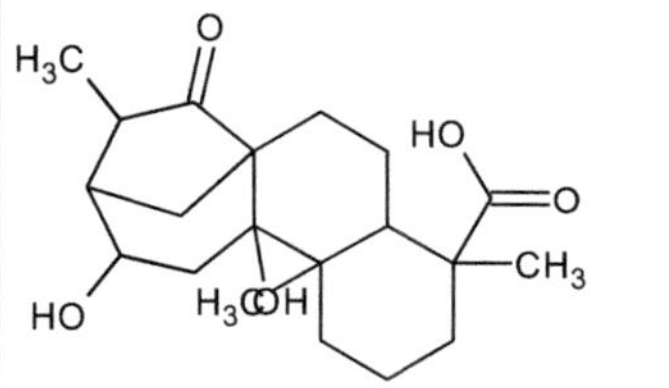 | Wang et al. (2011b) |
| Pterisolic acid F | $C_{20}H_{30}O_6$ 366.4 g/mol | 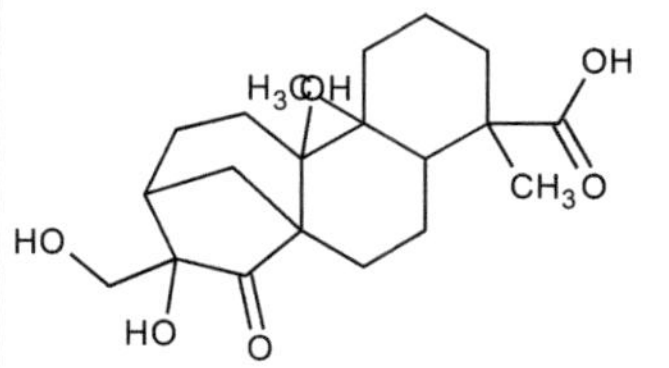 | Wang et al. (2011b) |
| Semipterosin A | $C_{14}H_{20}O_4$ 252.31 g/mol | 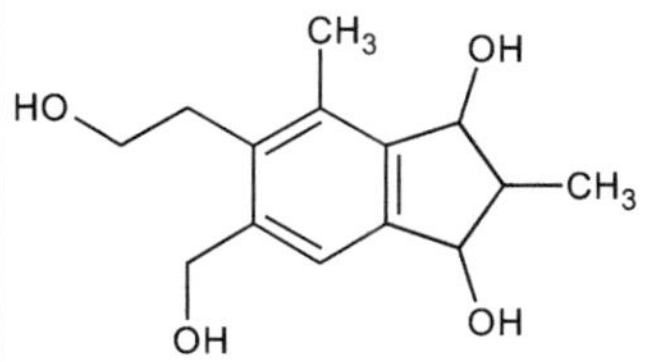 | Zhan et al. (2010) |
| (2R)-Norpterosin B | $C_{13}H_{16}O_2$ 204.26 g/mol | 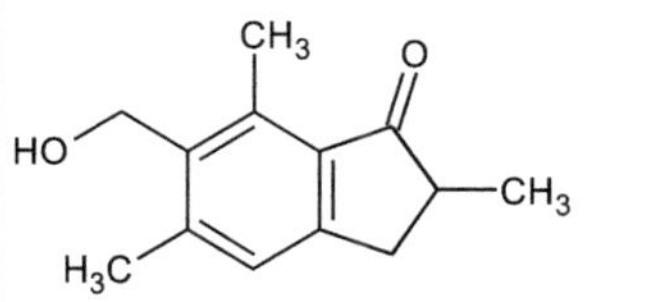 | Zhan et al. (2010) |

(continued)

**Table 16.1** (continued)

| Compound name | Molecular formula and molecular weight | Structure | Reference |
|---|---|---|---|
| *Pteris semipinnata* L. | | | |
| (2R)-Pterosin B | $C_{14}H_{18}O_2$ 218.29 g/mol | 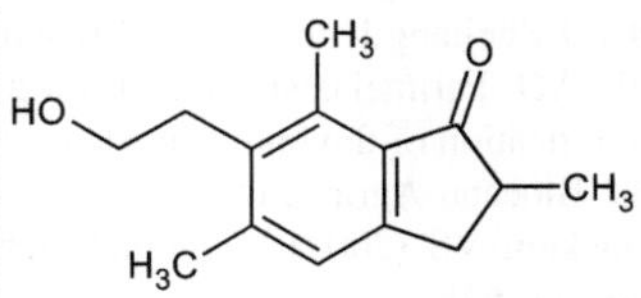 | Zhan et al. (2009) |
| (2S, 3S)-Pterosin C | $C_{14}H_{18}O_3$ 234.29 g/mol | 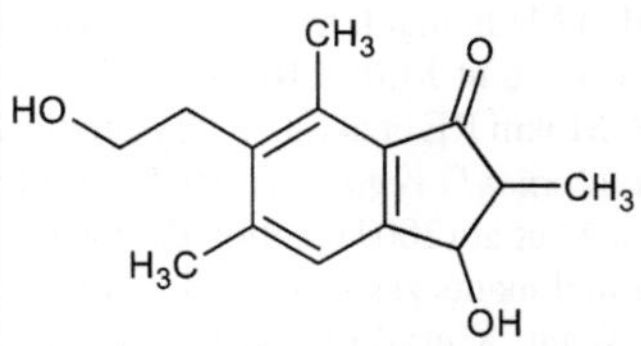 | Zhan et al. (2009) |
| Norpterosin C | $C_{13}H_{16}O_3$ 220.26 g/mol | 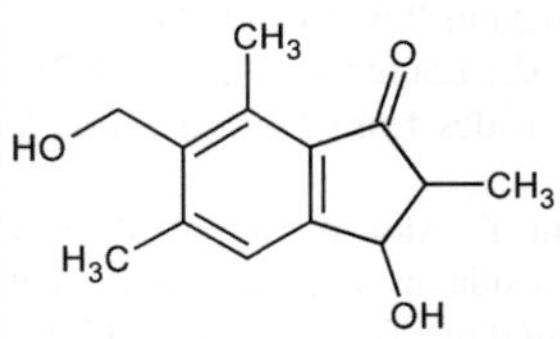 | Zhan et al. (2009) |
| *Pteris vittata* L. | | | |
| Kaempferol | $C_{15}H_{10}O_6$ 286.24 g/mol | 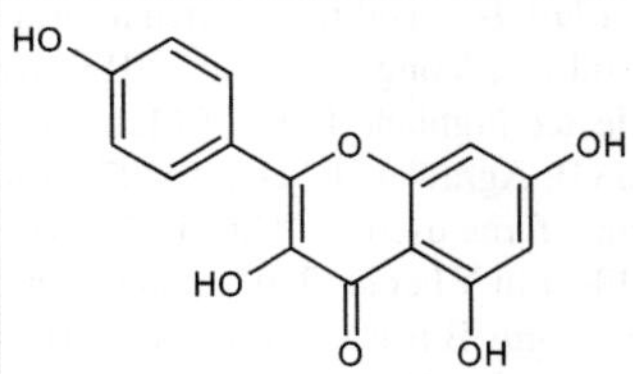 | Lin et al. (2016) |
| Quercetin | $C_{15}H_{10}O_7$ 302.23 g/mol | 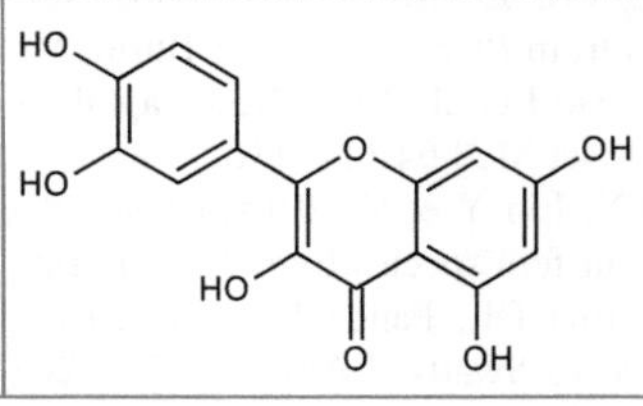 | Lin et al. (2016) |

## References

Ali MS, Mostafa K, Raihan MO et al (2012) Antioxidant and cytotoxic activities of methanolic extract of *Dryopteris filix-mas* (L.) Schott leaves. Int J Drug Dev Res 4:223–229

Asai F, Iinuma M, Tanaka T et al (1991) Complex flavonoids in farinose exudate from *Pityrogramma calomelanos*. Phytochemistry 30:3091–3093

Aulakh MK, Kaur N, Saggoo MIS (2019) Bioactive phytoconstituents of pteridophytes - a review. Indian Fern J 36:37–79

Banerjee RD, Sen SP (1980) Antibiotic activity of pteridophytes. Econ Bot 34:284–298

Baskaran X, Jeyachandran R (2010) Evaluation of antioxidant and phytochemical analysis of *Pteris tripartita* Sw. A critically endangered fern from South India. J Fairy Lake Bot Gard 9:28–34

Baskaran X, Vigila AG, Zhang S et al (2018) A review of the use of Pteridophytes for treating human ailments. J Zhejiang Univ Sci B (Biomed Biotechnol) 19:85–119

Baskaran X, Vigila AG, Parimelazhagan T et al (2019) Antioxidant, anti-inflammatory activities and HPLC quantification of flavonoids in *Pteris tripartita* Sw a critically endangered medicinal fern from India. Biocatal Agric Biotechnol 21:101304

Benjamin A, Manickam VS (2007) Medicinal pteridophytes from the Western Ghats. Indian J Tradit Knowl 6:611–618

Benniamin A (2011) Medicinal ferns of north eastern India with special reference to the Arunachal Pradesh. Indian J Tradit Knowl 10:516–522

Bouazzi S, Jmii H, Mokni RE et al (2018) Cytotoxic and antiviral activities of the essential oils from Tunisian Fern, *Osmunda regalis*. S Afr J Bot 118:52–57

Cai Y, Luo Q, Sun M et al (2004) Antioxidant activity and phenolic compounds of 112 traditional Chinese medicinal plants associated with anti-cancer. Life Sci 74:2157–2184

Cao H, Chai TT, Wang X et al (2017) Phytochemicals from fern species: potential for medicine applications. Phytochem Rev 16:379–440

Castillo UF, Wilkins AL, Lauren DR et al (1997) Isoptaquiloside and caudatoside, illudane-type sesquiterpene glucosides from *Pteridium aquilinum* var. *caudatum*. Phytochemistry 44:901–906

Castillo UF, Sakagami Y, Alonso-Amelot M et al (1999) Pteridanoside, the first protoilludane sesquiterpene glucoside as a toxic component of the neotropical bracken fern *Pteridium aquilinum* var. *caudatum*. Tetrahedron 55:12295–12300

Chai TT, Panirchellvum E, Ong HC et al (2012) Phenolic contents and antioxidant properties of *Stenochlaena palustris*, an edible medicinal fern. Bot Stud 53:439–446

Chai TT, Elamparuthi S, Yong AL et al (2013) Antibacterial, anti-glucosidase, and antioxidant activities of selected highland ferns of Malaysia. Bot Stud 54:55

Chang HC, Huang GJ, Agrawal DC et al (2007) Antioxidant activities and polyphenol contents of six folk medicinal ferns used as "Gusuibu". Bot Stud 48:397–406

Chen YH, Chang FR, Lin YJ et al (2007) Identification of phenolic antioxidants from Sword Brake fern (*Pteris ensiformis* Burm.). Food Chem 105:48–56

Chen YH, Chang FR, Lu MC et al (2008) New benzoyl glucosides and cytotoxic Pterosin Sesquiterpenes from *Pteris ensiformis* Burm. Molecules 13:255–266

Chen L, Zhang G, He J et al (2010) New naphthoquinone from the root of *Lygodium japonicum* (Thunb.) Sw. J Nat Med 64:114–116

Chen CY, Chiu FY, Lin Y et al (2015) Chemical constituents analysis and antidiabetic activity validation of four fern species from Taiwan. Int J Mol Sci 16:2497–2516

Chowdhary S, Verma DL, Pande R et al (2010) Antioxidative properties of flavonoids from *Cheilanthes anceps* Swartz. J Am Sci 6:203–207

Cramer L, Ernst L, Lubienski M et al (2015) Structural and quantitative analysis of *equisetum* alkaloids. Phytochemistry 116:269–282

Dalli AK, Saha G, Chakraborty U (2007) Characterization of antimicrobial compounds from a common fern *Pteris biaurita*. Indian J Exp Biol 45:285–290

Erhirhie EO, Emeghebo CN, Ilodigwe EE, et al. (2019) *Dryopteris filix-mas* (L.) Schott ethanolic leaf extract and fractions exhibited profound anti-inflammatory activity. Avicenna J Phytomed 9: 396–409

Fletcher MT, Hayes PY, Somerville MJ et al (2010) Ptesculentoside, a novel norsesquiterpene glucoside from the Australian bracken fern *Pteridium esculentum*. Tetrahedron Lett 51:1997–1999

Fu W, Fang W, Ruan J (2009) Two new flavanone glycosides from *Macrothelypteris torresiana* (Gaud.) Ching. Chin Chem Lett 20:579–581

Ge X, Ye G, Li P et al (2008) Cytotoxic diterpenoids and sesquiterpenoids from *Pteris multifida*. J Nat Prod 71:227–231

Good R (1933) Plants and human economics. Cambridge University Press, Cambridge

Guha P, Mukhopadhyay R, Gupta K (2005) Antifungal activity of the extracts and extracted phenols from gametophytes and sporophytes of two species of *Adiantum*. Taiwania 50:272–283

Gulcin I (2006) Antioxidant activity of caffeic acid (3,4-dihydroxycinnamic acid). Toxicology 217: 213–220

Haider S, Nazreen S, Alam MM et al (2011) Anti-inflammatory and anti-nociceptive activities of ethanolic extract and its various fractions from *Adiantum capillus-veneris* Linn. J Ethnopharmacol 138:741–747

Haider S, Kharbanda C, Alam MS et al (2013) Anti-inflammatory and anti-nociceptive activities of two new triterpenoids from *Adiantum capillus-veneris* Linn. Nat Prod Res 27:2304–2310

Ho R, Teai T, Bianchini J et al (2011) Ferns: from traditional uses to pharmaceutical development, chemical identification of active principles. In: Fernandez H, Kumar A, Revilla MA (eds) Working with ferns issues and applications. Springer, Berlin, pp 321–346

Hsu FL, Huang CF, Chen YW et al (2013) Antidiabetic effects of pterosin a, a small-molecular-weight natural product, on diabetic mouse models. Diabetes 62:628–638

Huang YL, Yeh PY, Shen CC et al (2003) Antioxidant flavonoids from the rhizomes of *Helminthostachys zeylanica*. Phytochemistry 64:1277–1283

Huang YC, Hwang TL, Chang CS, Yang YL, Shen CN, Liao WY, Chen SC, Liaw CC (2009) Anti-inflammatory flavonoids from the rhizomes of *Helminthostachys zeylanica*. J Nat Prod 72: 1273–1278

Hussain MM, Muthuprasanna P, Srinivasarao T et al (2008) Analgesic and anti-inflammatory activity of *Adiantum venustum*. Res Rev Biosci 2:102–104

Ibraheim ZZ, Ahmed AS, Gouda YG (2011) Phytochemical and biological studies of *Adiantum capillus-veneris* L. Saudi Pharm J 7:1–10

Ishaq MS, Hussain MM, Afridi MS et al (2014) In vitro phytochemical, antibacterial, and antifungal activities of leaf, stem, and root extracts of *Adiantum capillus-veneris*. Sci World J 2014:269793

Johnson MJ, Mahesh M, Janakiraman N (2017) Antibacterial potentials of *Adiantum* species against the UTI pathogens. J Microbiol Exp 4:00104

Kaushik P, Dhiman K (1995) Common medicinal pteridophytes. Indian Fern J 12:139–145

Kholia BS, Punetha NN (2005) Useful pteriophytes of kumaon central Himalaya, India. Indian Fern J 22:1–6

Kuncoro H, Farabi K, Rijai L (2017) Steroids and isoquercetin from *Lygodium microphyllum*. J App Pharm Sci 7:136–141

Lai HY, Lim YY, Kim KH. (2010) *Blechnum orientale* Linn.—a fern with potential as antioxidant, anti-cancer and antibacterial agent. BMC Complement Altern Med 10: 1–8

Lan KP, Shen YP, Lee SH et al (2011) Antioxidant and free radical-scavenging activities of *Pteris multifida* Poiret aqueous extract. J Food Qual 34:252–258

Li PH, Chiu YP, Shih CC et al (2016) Biofunctional activities of *Equisetum ramosissimum* extract: protective effects against oxidation, melanoma, and melanogenesis. Oxidative Med Cell Longev 2016:2853543

Liang YH, Ye M, Yang WZ et al (2011) Flavan-3-ols from the rhizomes of *Drynaria fortunei*. Phytochemistry 72:1876–1882

Lin YL, Shen CC, Huang YJ et al (2005) Homoflavonoids from *Ophioglossum petiolatum*. J Nat Prod 68:381–384

Lin L, Xb H, Zc L (2016) Isolation and identification of flavonoids components from *Pteris vittata* L. Springerplus 5:1649

Liu H, Orjala J, Sticher O et al (1999) Acylated flavonol glycosides from leaves of *Stenochlaena palustris*. J Nat Prod 62:70–75

Lu H, Hu J, Zhang LX et al (1999) Bioactive constituents from *Pteris multifida*. Planta Med 65: 586–587
Lu J, Peng C, Cheng S et al (2019) Four new pterosins from *Pteris cretica* and their cytotoxic activities. Molecules 24:2767
May LW (1978) The economic uses and associated folklore of ferns and fern allies. Bot Rev 44: 491–528
Mimica DN, Natasa S, Jelena C et al (2008) Phenolic compounds in field horsetail (*Equisetum arvense* L.) as natural antioxidants. Molecules 13:1455–1464
Nair AS, Shylesh BS, Gopakumar B et al (2006) Anti-diabetes and hypoglycaemic properties of *Hemionitis arifolia* (Burm.) Moore in rats. J Ethnopharmacol 106:192–197
Neel RS, Jadhao AS, Bhuktar AS (2017) Antibacterial activity of rhizomes and leaf of *Tectaria gemmifera* (Fee.) Alston. Int J Bot Stud 2:89–92
Nonato FR, Barros TAA, Lucchese AM et al (2009) Antiinflammatory and antinociceptive activities of *Blechnum occidentale* L. extract. J Ethnopharmacol 125:102–107
Nonato FR, Nogueira TM, Barros TA et al (2011) Antinociceptive and antiinflammatory activities of *Adiantum latifolium* Lam.: evidence for a role of IL-1β inhibition. J Ethnopharmacol 136: 518–524
Nwosu MO (2002) Ethnobotanical studies on some Pteridophytes of Southern Nigeria. Econ Bot 56:255–259
Ouyang DW, Ni X, Xu HY et al (2010) Pterosins from *Pteris multifida*. Planta Med 76:1896–1900
Pan C, Chen YG, Ma XY et al (2011) Phytochemical constituents and pharmacological activities of plants from the genus *Adiantum*: a review. Trop J Pharm Res 10:681–692
Parihar P, Parihar L, Bohra A (2010) In vitro antibacterial activity of fronds (leaves) of some important pteridophytes. J Microbiol Antimicrob 2:19–22
Paulsamy S, Moorthy D, Nandakumar K et al (2013) Evaluation of in vitro antioxidant potential of methanolic extracts of the ferns, *Actiniopteris radiata* (Sw.) link. And *Equisetum ramosissimum* Desf. Int J Res Dev Pharm Life Sci 2:451–455
Prathibha K, Arman Mahmoudi OH, Govindapyari Y et al (2011) Some ethnomedically important Pteridophytes of India, Uttarakhand India. Int J Med Aromat Plants 1:18–22
Puri HS (1970) Indian pteridophytes used in folk remedies. Am Fern J 60:137–143
Rai PK, Lalramnghinglova H (2010) Lesser known ethnomedicinal plants of Mizoram, north East India: an indo Burma hotspot region. J Med Plants Res 4:1301–1307
Ranjan V, Vats M, Gupta N et al (2014) Antidiabetic potential of whole plant of *Adiantum capillus-veneris* L. in streptozotocin induced diabetic rats. Int J Pharm Clin Res 6:341–347
Reddy VL, Ravikanth V, Rao TP et al (2001) A new triterpenoid from the fern *Adiantum lunulatum* and evaluation of antibacterial activity. Phytochemistry 56:173–175
Revathi R, Muthuraja R, Thomas B et al (2013) Ethno medicinal fern and fern-allies used by tribe *Malayalis* of Kolli Hills, Eastern Ghats, Tamil Nadu. Pteridol Res 2:1–10
Salatino MLF, Prado J (1998) Flavonoid glycosides of Pteridaceae from Brazil. Biochem Syst Ecol 26:761–769
Sathiyaraj G, Muthukumar T, Ravindran KC (2015) Ethnomedicinal importance of fern and fern allies traditionally used by tribal people of Palani Hills (Kodaikanal), Western Ghats, South India. J Med Herbs Ethnomed 1:4–9
Sharma NK (2002) Ethnomedicinal studies of ferns and fern allies of Hadoti plateau, South Eastern Rajasthan. Zoos Print J 17:732–734
Shi YS, Zhang Y, Hu WZ et al (2017) Dihydrochalcones and Diterpenoids from *Pteris ensiformis* and their bioactivities. Molecules 22:1413
Singh HB (2003) Economically viable Pteridophytes of India. In: Chandra S, Srivastava M (eds) Pteridology in the new millennium. Kluwer Academic Publishers, Dordrecht, pp 421–446
Singh M, Govindarajan R, Rawat AKS et al (2008a) Antimicrobial flavonoid rutin from *Pteris vittata* L. against pathogenic gastrointestinal microflora. Am Fern J 98:98–103
Singh M, Singh N, Khare PB et al (2008b) Antimicrobial activity of some important *Adiantum* species used traditionally in indigenous systems of medicine. J Ethnopharmacol 115:327–329

Soare LC, Ferdes M, Stefanov S et al (2012) Antioxidant activity, polyphenols content and antimicrobial activity of several native pteridophytes of Romania. Not Bot Hortic Agrobot Cluj-Napoca 40:53–57

Soeder RW (1985) Fern constituents: including occurrence, chemotaxonomy and physiological activity. Bot Rev 51:442–536

Srivastava K (2007) Ethnobotanical studies of some important ferns. Ethnobot Leafl 11:164–172

Su LH, Li YP, Li HM et al (2016) Anti-inflammatory prenylated flavonoids from *Helminthostachys zeylanica*. Chem Pharm Bull 64:497–501

Sumathy V, Lachumy SJ, Zuraini Z et al (2010) Effects of *Stenochlaena palustris* leaf extract on growth and morphogenesis of food borne pathogen, *Aspergillus niger*. Malays J Nutr 16:439–446

Thomas T (2011) Preliminary antibacterial evaluation of fronds of *Pteris quadriaurita* Retz. Towards bacteria involved in dermatological diseases. J Appl Pharm Sci 1:214–216

Thomas T (2012) *In vitro* evaluation of antibacterial activity of *Acrostichum aureum* L. Indian J Nat Prod Resour 3: 135–138

Uddin SJ, Bettadapura J, Guillon P et al (2013) In-vitro antiviral activity of a novel phthalic acid ester derivative isolated from the Bangladeshi mangrove fern *Acrostichum aureum*. J Antivir Antiretrovir 5:139–144

Upreti K, Jalal JS, Tewari LM et al (2009) Ethnomedicinal uses of Pteridophytes of Kumaun Himalaya, Uttarakhand, India. J Am Sci 5:167–170

Uwumarongie HO, Enike MA, Bafor EE (2016) Pharmacognostic evaluation and gastrointestinal activity of *Dryopteris filix-mas* (L.) Schott (Dryopteridaceae). Ew J Herbal Chem Pharmacol Res 2:19–25

Vasudeva SM (1999) Economic importance of Pteridophytes. Indian Fern J 16:130–152

Vetter J (2009) A biological hazard of our age: bracken fern [*Pteridium aquilinum* (L.) Kuhn]-a review. Acta Vet Hung 57:183–196

Viral D, Shivanand P, Jivani NP (2011) Anticancer evaluation of *Adiantum venustum* Don. J Young Pharm 3:48–54

Vogler G, Donath O, Saukel J et al (2012) Polar phenolic compounds in *Dryopteris filix-mas* und *Dryopteris dilatata*. Verh Zool-Bot Ges Osterreich 148(149):279–289

Wang CX, Zhang PH, Luo JG et al (2011a) Homoflavonoid glucosides from *Ophioglossum pedunculosum* and their anti-HBV activity. J Nat Prod 74:683–689

Wang F, Li YJ, Ren FC et al (2011b) Pterisolic acids A-F, new ent-kaurane diterpenoids from the fern *Pteris semipinnata*. Chem Pharm Bull (Tokyo) 59:484–487

Wang RL, Zheng ZH, Lu J et al (2014) Allelopathic potential of invasive climbing fern *Lygodium microphyllum* against native plants and antibacterial activity of essential oils. Allelopath J 33: 97–106

Xu W, Zhang F, Luo Y et al (2009) Antioxidant activity of a water-soluble polysaccharide purified from *Pteridium aquilinum*. Carbohydr Res 344:217–222

Xu JX, Wang YL, Wang JJ et al (2012) Chemical constituents of *Cibotium barometz* and their bioactivities: a review. Nat Prod Res Dev 24:134–140

Yoshihira K, Fukuoka M, Kuroyannagi M et al (1978) Chemical and toxicological studies on bracken Fern, *Pteridium aquilinum* var. *latiusculum*. I. Introduction, extraction and fractionation of constituents and toxicological studies including carcinogenicity tests. Chem Pharm Bull 26: 2346–2364

Zhan ZJ, Zhang FY, Li CP et al (2009) A novel ent-kaurane diterpenoid from Pteris semipinnata. J Chem Res 2009:149–150

Zhan ZJ, Ying YM, Zhang FY et al (2010) Three new Illudalane Sequiterpenoids from *Pteris semipinnata*. Helvetica 93:550–554

Zhang X, Li JH, He CW (1999) Study on the diterpenoid constituents and anticancer action of *Pteris semipinnata*. Chin Pharm J 34:512–514

Zhao Z, Ruan J, Jin J et al (2011) Two new flavonoids from the rhizomes of *Abacopteris penangiana*. Helv Chim Acta 94:446–452

Zhu L, Zhang G, Chen L et al (2009a) A new ecdysteroside from *Lygodium japonicum* (Thunb.) Sw. J Nat Med 63:215–219

Zhu L, Zhang G, Wang S et al (2009b) A new compound from *Lygodium japonicum* (Thunb.) Sw. Nat Prod Res 23:1284–1288

**Sonia Abraham** is a native of Kerala, India. She has a Master's degree in Botany from Mahatma Gandhi University, Kottayam (Kerala, India), and PhD from Dept. of Botany, Bharathiar University, Coimbatore (Tamil Nadu, India). She is currently working as a guest lecturer in Botany at St. Dominic's College, Kanjirapally (Kerala, India). She has published eight research papers in national and international journals and three book chapters and co-authored a book. She has experience in the taxonomy and ecology of ferns and lycophytes.

**Toji Thomas** works as Associate Professor in Botany, at St. Thomas College Palai, affiliated to Mahatma Gandhi University, Kottayam, Kerala, India. He was born in Pala, Kerala (India). He did his graduation and post-graduation in Botany from Mahatma Gandhi University, Kottayam. He did his PhD from School of Biosciences, Mahatma Gandhi University, Kottayam. As a part of his PhD programme, he characterised a novel antibacterial terpenoid (1E, 4E)-1,5-bis (4-hydroxy-3,5-dipropyl phenyl) penta-1,4- dien-3-one from *Drynaria quercifolia* (L.) J. Sm. He has 19 years of teaching experience in UG and PG levels. He supervises students for their doctoral degree in various research areas related to botany like plant biochemistry, bionanotechnology, microbiology, plant tissue culture, hairy root production and secondary metabolites and secondary metabolite isolation from soil fungi. He is the author of several books, and he published more than 35 research papers, including papers regarding bioactivities of ferns.

# Pteridophytes Used by Peoples of Indian Himalayan Region and Northern India: An Overview

**17**

B. S. Kholia and Acharya Balkrishna

**Abstract**

Pteridophytes are being used by human beings from time immemorial, but compared to flowering plants, their uses and economic potential are still neglected; therefore, they need to be highlighted and popularised. Out of ca. 1200 taxa of Indian pteridophytes, nearly 800 occur in the Himalayan region. Traditionally different ethnic groups of the Himalayas are using various species of pteridophytes as vegetable, food, fodder and folk medicine and in making yeast cakes as a starter and flavouring in country-made beer and wine, in construction of hill houses and cattle sheds, etc. Besides its traditional uses, in modern societies, pteridophytes are also used by the Himalayan peoples for decoration, in making flower bouquet and as stuffing material; further they are also introduced in horticulture, as stock in orchid culture, in mulching and in other nursery practices. However, recent development and modernisation of societies completely changed the social status and way of life of the Himalayan people, and this traditional knowledge is being diminished generation by generation; therefore it needs to be documented correctly. Keeping this in view, the socioeconomically useful pteridophytes of the Indian Himalayan regions are highlighted here.

**Keywords**

Ethnobotany · Himalayan pteridophytes · Food · Fodder · Folk medicine

---

B. S. Kholia (✉)
Botanical Survey of India, Dehradun, Uttarakhand, India

A. Balkrishna
Patanjali Herbal Research Department, Patanjali Research Foundation Trust, Haridwar, Uttarakhand, India
e-mail: pyp@divyayoga.com



J. Marimuthu et al. (eds.), *Ferns*, https://doi.org/10.1007/978-981-16-6170-9_17

## 17.1 Introduction

Throughout the world, pteridophytes are associated with the folklores of different ethnic groups and are being used as food and medicine and in other socioeconomic activities from time immemorial. However, compared to flowering plants, they are of little use, and their economic value and potential are neglected. Ancient Indian Ayurvedic literature Sushruta Samhita and Charak Samhita (ca ± 900–600 BC.) documented medicinal ferns under the name Mayursikha, Mayurpankha, Hansraj and Hanspadi. Based on 'Doctrine of Signature', these are assumed to be *Actiniopteris radiata* (Sw.) Link, various species of *Adiantum* L. and *Marsilea* L. Qualities of ferns were also mentioned by Theophrastus ca. 300 BC., and similarly various species of brake ferns (*Pteris* L.), male ferns (*Dryopteris* Adans.) and others were also known to Greek physician, pharmacologist and botanist Dioscorides. In the medieval time, it was thought the ferns have several magical powers and religious values. By this belief, *Botrychium* species are still used to remove the evil spirits by Buddhist monks in Sikkim and other part of the Himalayas during several religious practices (Kholia 2014).

Pteridophytes are used as food, fodder, vegetables, medicine, fibres, fats, oils, fragrance, dyes and flavours and in other socioeconomic and cultural values throughout the world. Besides its traditional uses, in modern societies, they are also cultivated as horticulture plant or pot plants in gardens and introduced in nursery trade (Chakraverty et al. 2003; Olsen 2007; Kholia 2010a, 2010b, 2017). Recently, ferns and lycophytes are used in soil reclamation as hyperaccumulator (Ma et al. 2001; Francesconi et al. 2002; Wang et al. 2002; Zhao et al. 2002; Meharg 2003; Tu and Ma 2003, 2004; Srivastava et al. 2005, 2005). Their pharmacological studies (Alam et al. 2000; Zangara 2003; Matsuda et al. 2002; Maridass and Ghanthikumar 2008), antimicrobial characters (Zhang et al. 2002; Parihar and Bohra 2003; Mandal and Mondal 2011), antibiotic properties (Nath et al. 2016), antifungal properties (Parihar and Bohra 2002), antioxidant and immunomodulatory properties (Gayathri et al. 2005; Wu et al. 2005), toxicity (Ugochukwu 2019; Lin et al. 2000), nutritional (Leterme et al. 2009; Chettri et al. 2018; Gupta et al. 2020) and antioxidant potential (Chang et al. 2007; Dvorakova et al. 2021), chemical studies for active agents (Tsuzuki et al. 2001; Toyota et al. 2002; Tong et al. 2003; Lin et al. 2005; Gupta et al. 2000; Hirasawa et al. 2006; Aulakh et al. 2019), cancer research (Siman et al. 2000), etc. are being explored and highlighted throughout the globe.

Pteridophytes are also part of the society in India from the ancient period and documented in Ayurvedic literature, and their economic and medicinal properties are also listed in almost all encyclopaedias of Indian useful and medicinal plants (Dymock 1890; Caius 1935; Kirtikar and Basu 1935; Nadkarni 1954; Chopra et al. 1956, 1958, 1969; Nayar 1959; Wealth of India 1948–1969; Jain 1991; Jain and DeFilipps 1991; Ambast 2000). Based on these encyclopaedias and other published literature, Singh et al. (2002) compiled 173 species of pteridophytes used in India for various purposes like food and flavour, dye and tans, biofertilisers, oils, fibre, biogas production, traditional medicines, etc. In recent years, many researches (Kirn and Kapahi 2001; Singh et al. 2001, 2002; Sharma 2002; Gogoi and Das 2002; Kumar and Ramesh 2003; Kholia and Punetha 2005a; Pandey and Rout 2006; Rao et al. 2007; Fraser-Jenkins 2008; Upreti et al. 2009; Kholia 2010b, 2014; Singh and Singh 2012; Singh and Upadhyay 2012; Mir et al. 2015; Shah 2015; Singh 2018, etc.) have also added many species of useful pteridophytes from different parts of India.

The Indian Himalayan Region (IHR) falls in the Himalayan biodiversity hotspot and Indo-Burma biodiversity hotspot (Myers et al. 2000), ranging from Kashmir to Arunachal Pradesh and north-eastern states of India including the Himalayan Kingdom of Nepal and Bhutan. Though in some literature the south of Brahmaputra River which includes Meghalaya and other NE states of Arakan Yoma is recognised as a distinct phytogeographic zone (North East India), however, in the present study it is treated under IHR owing to its similar topography and floristic composition. The region is with rich, diverse and luxurious flora and fauna along with its unique ethnicity. The Himalayan region is also inhabited by more than 100 aboriginal tribes (Wikipedia) of various ethnic groups. This bulk and diverse population of the Indian Himalayan Region is privileged by wealth of traditional inherited knowledge. However, modernisation of the society badly affected the modern generations, and this inherit knowledge are gradually diminishing. Nevertheless, several ethnobotanists, plant taxonomists and naturalists started documenting this knowledge. During ethnobotanical studies, flowering plants are given much emphasis, but the ferns and fern allies are often neglected or reported very seldom and often vaguely. Most of the species reported as medicine or other uses are erroneously identified and with incomplete, unclear and ambiguous information because most of the ethnobotanists were not professional pteridophyte taxonomists. Furthermore, very few of them prepared voucher specimens of useful pteridophytes and deposited them in recognised herbaria for further studies or future consultation, and as such many of them are wrongly identified and published. On further reviewing some of these published ethnobotanical literatures on ferns and fern allies, sometimes without vouchers, it is very difficult to guess the exact identity of the useful pteridophyte.

As compared to higher plants, the ethnobotanical study of Indian Himalayan pteridophytes is limited and scattered in few local or regional reports. The food value of trunks of tree ferns (*Cyathea* J.E. Smith) and rhizome of gigantic ferns (*Angiopteris* Hoffm.) are reported in almost all literature and encyclopaedias, and these are the major food and ingredients of country-made beer in Sikkim too (Anonymous a 2009; and Kholia 2010a, 2010b, 2014). Useful pteridophytes of Sikkim are documented by Kholia (2010a, 2010b, 2014) and Uttarakhand Himalayas by Kholia and Punetha (2005a), Upreti et al. (2009), etc. Similarly, ethnobotany of pteridophytes of Nepal was published by Uprety et al. (2012), Luitel and Pathak (2013), Parajuli (2013), Kunwar et al. (2013), etc. Matsushima et al. (2008, 2012) reported the uses of pteridophytes from Bhutan; Benniamin (2011) from Arunachal Pradesh; Gogoi and Das (2002) from Garo Hills; Sen and Ghosh (2011) and Nath et al. (2016) from Assam; Das and Dutta Choudhury (2012) from Tripura; and Singh et al. (2001) and Yumkham and Singh (2011) from Manipur regions of NE India and Eastern Himalayas. Present paper is based on nearly 34-year study and research experience of senior author on fern taxonomy and ethnobotany of the Himalayan and northern Indian ferns and lifelong experience of junior author as Ayurvedic doctor. The ethnobotany literature on pteridophytes from this region is also consulted and analysed during this study. In different communities, ferns and fern allies are used as following purposes.

## 17.2 Material and Method

Present article is based on the available literature trawl as well as data collected during pteridophyte research carried out by authors throughout the Himalayan regions during the last three decades. During field survey, authors collected ethnobotanical information from elderly villagers and some tribal and herbal practitioners. Local markets where ferns are sold as vegetables and folk medicine are also surveyed and explored for the collection of ethnobotanical data along with available vernacular names and methods of use. The specimens gathered are housed in the Herbarium Botanical Survey of India Gangtok Sikkim (BHSC), Herbarium Botanical Survey of India Dehradun (BSD) and Herbarium Patanjali Research Foundation Herbarium Haridwar (PRFH).

## 17.3 Results

The socioeconomic and ethnobotanical uses of pteridophytes by peoples belonging to different ethnic groups of the Indian Himalayan regions are described below.

### 17.3.1 Vegetable

The Indian Himalayan Region harbours a variety of wild edibles including ferns, and recipe of fern vegetable was also discussed variously (Kholia and Punetha 2005a, Fraser-Jenkins 2008, etc.). These edible/vegetable ferns are known under various traditional names by different peoples of the Himalayas and Eastern India. In Jammu and Kashmir and Himachal Pradesh, croziers of edible fern are called *HIMLI*, *KHANDHOR*, *KASROR*, *KANDAI*, *KAKAIE*, *LINGRA*, *LINGHORA*, etc.; however, in Uttarakhand, they are called *LIUNO LIUNDO*, *KUTHURKA* and *LINGRO*. Similarly, in Nepal and Sikkim, edible ferns are called *NIURO* or *NINGRO* or *NIHURO*. Some other local names of edible ferns are *TYRKHANG* by the Khasi people; *DHEKIA* by Bengalis of Bengal Assam and NE India; *MAKHUNSEK* and *SUSANAT* and *PALOI* by tribes of Garo Hills; *LI-CHANGKHARANG* in Manipuri; *ANPHAN* by Lotha tribe of Nagaland; etc. Similarly, different tribes of Nagaland, Mizoram, Arunachal Pradesh, Tripura and Meghalaya have different local or traditional names for edible ferns. During his studies, present authors also studied the folk nomenclature of pteridophytes by the peoples of Uttarakhand, Sikkim, Darjeeling and Nepal and found that their nomenclature is perfectly logical. In Uttarakhand, edible ferns are called *LIYUNO*. These names are based on the structure of fern crozier which is with stooping head. In Nepalese and Kumauni languages, the term *LIYUNI* (in Uttarakhand) or *NIYURI* (in Nepal) are used for bending head or body forwards and downwards (generally at the time of salutation or greeting). In Kumaun region of Uttarakhand, loose and open crozier without curled head is called *KUTHURKA*. This term originated from the Kumaoni word *KUTHURK* or *LUTHURAK* used for tender things which cannot stand erect; often it is also used for a very weak and ill person who is unable to stand for a longer time. There are also various further classified local names for edible ferns in Nepali language, as the ferns growing in valleys are named as *AULE NIURO* (Aule = warmer valley or low hills), and those growing in higher elevations in montane forests are named as *LEKH NIURO* (Lekh = cooler places or up hills). The other criteria used for naming the ferns in local Nepalese languages are colour, taste, shape, smoothness, availability of moths, etc. The details on local nomenclature of ferns in Sikkim are described by Kholia (2010a). Likewise, on the availability in the fourth and fifth months (Srawan and Bhadrapad) of the Hindu calendar, the edible ferns are named as *SAWNE NIURO* and *BHADORE NIURO*. On the basis of colour or shades of the croziers, black, white and red coloured ferns are called as *KALO NIURO* (Kalo = black), *SETO NIURO* (Seto = white) and *RATO NIURO* (Rato = red), respectively. The croziers with thick stick-like appearance are called DANTHE NIURO (Dantha = stick); however, croziers with smooth and shining oily appearance are called *CHIPLEY NIURO* (Chippley = smooth and slippery) and *GHIU NIURO* (Ghiu = ghee or melted butter).

Throughout the Himalayas and Northern India, all the available species of *Diplazium* Sw. (ca. 35 spp.) are edible as vegetable; however, *Diplazium esculentum* (Retz.) Sw. and *Diplazium maximum* (D. Don) C.Chr. are most commonly used and sold in local markets. The species of *Diplazium* used as vegetable in Sikkim, Darjeeling and Nepal have different local names; for example *Diplazium esculentum* (Retz.) Sw. is named there as *PANI NIURO, MACHI NIURO, CHIPLO NIURO, SAWNE NIURO*, etc. based on applying local logic. As it is frequent near water (= loc. Pani) sources, it is named *PANI NIURO*, similarly for its fish (loc. Machi) like taste named as *MACHI NIURO*. Due to its smooth texture (Chiple = smooth) and frequent availability in the fifth month of Hindu calendar Shrawan, this is called *CHIPLO NIURO* and *SAWNE NIURO* respectively. On the other hand, Tharu tribal of Nepal Terai call it *KOCHIYA* (Uprety et al. 2012), but *Kochiya* is called for vegetable species of *Colocasia* by Kumauni peoples. Similarly, *Diplazium kawakamii* Hayata is called *JIRE NINGRO* due to the resemblance of its sori with the cumin (=Jira) seeds. All the species of edible ferns with smooth shining stipe are called *CHIPLEY NINGRO* (chipley = smooth), species growing in lower altitudes or warmer valleys are called *AULEY NINGRO* (auley = warm), and species growing in higher elevations are called *LEKH NINGRO* (lekh = upper hill). The other edible species are *Diplazium doederleinii* (Luerss.) Makino (*LEKH CHIPLEY NINGRO*), *Diplazium dilatatum* Blume and *Diplazium laxifrons* Rosenst. (*LEKH CHIPLEY NINGRO* and *BHADORE NINGRO*), *Diplazium forrestii* (Ching) Fras. -Jenk. (*AULEY CHIPLEY NINGRO*), *Diplazium maximum* (D. Don) C.Chr. (*SWANEY NINGRO*), *Diplazium javanicum* (Blume) Makino (*SANO CHIPLEY NINGRO*),

*Diplazium himalayense* (Ching) Panigrahi (*DNATHEY NINGRO*), *Diplazium heterophlebium* (Mett. *ex* Baker) Diels (*LEKH CHIPLEY NINGRO*), *Diplazium spectabile* (Wall. *ex* Mett.) Ching (*SANO AULEY NINGRO*), *Diplazium succulentum* (C.B. Clarke) C.Chr. (*LEKH CHIPLEY NIGRO*), *Diplazium sikkimense* (C.B.Clarke) C.Chr. (*SWANE NINGRO*), *Diplazium stoliczkae* (*LEKH KALO NINGRO*), etc. The thick straight 2– 4 feet long croziers of *Diplazium himalayense* (Ching) Panigrahi resemble with the walking stick hence locally known as *DANTHEY NINGRO* (dantha = stick). In Kumaun region of Uttarakhand, loose and open crozier without curled head is called *KUTHURKA*. Generally, *Botrychium* Sw. *Botrychium lanuginosum* Wall. *ex* Hook. & Grev., *B. multifidum* (S.G. Gmel.) Rupr. and *B. ternatum* (Thunb.) Sw.], *Athyrium* spp., *Dryopteris cochleata* and *Helminthostachys zeylanica* (L.) Hook. are called *KUTHURKA*; on the contrary, sometimes similar-appearing loose crozier of *Diplazium esculentum* (Retz.) Sw. is vaguely also named *KUTHURKA* in Kumaun region of the Himalayas. On the other hand, in Nepal *Diplazium esculentum* is variously named as *PANI NIURO*, *MACHI NIURO*, *CHIPPLEY NIURO*, *SAUNE NIOURO* and *KOCHIYA* (Kandel 2020); however, *Botrychium* Sw. is called *DEGAANI*, *Helminthostachys zeylanica* (L.) Hook. is called *MUJUR KHUTTE* (mujur = peacock, khutte = feet; the fronds resemble with paw of peacock), and *Athyrium* spp. is called *LIKURO* and *NIHURO*.

Besides, the different species of *Diplazium* Sw., the species of *Tectaria* (*KALI NIURO*), *Botrychium* Sw., *Athyrium* Roth., *Deparia* Hook. & Grev., *Cornopteris* Nakai, *Ceratopteris* Brongniart, *Marsilea* L., *Pteris* L. etc. are also used as vegetable in the Indian Himalayas, Nepal, Bhutan and North-Eastern India. Further, there are reports on edibility of *Ampelopteris prolifera* (Retz.) C.F.Reed, *Thelypteris (Christella) arida* (D.Don) C.V.Morton, and *Nephrolepis biserrata* (Sw.) Schott from Arunachal Pradesh (Benniamin 2011). Though *Pteridium* Gled. ex Scop. etc. is avoided from eating in west Himalayas and other parts, however, in Sikkim, Darjeeling and Nepal it is edible. *Ophioglossum reticulatum* L. is used in foot hills of Kumaun and Nepal as vegetable and salad and locally called as *JIBRE SAG* (Jibro = tongue, sag = vegetable) due to the resemblance with tongue. In Sikkim and Darjeeling, this is considered as one of the tastiest fern vegetables where it becomes very rare due to overexploitation. Similarly, *Helminthostachys zeylanica* (L.) Hook. is also eaten raw or cooked in many parts of India (Ambast 2000) and Nepal (Kandel 2020). The tubers of *Nephrolepis auriculata* (L.) Trimen are also eaten by cowboys to quench the thirst in western Himalayas (Kholia and Punetha 2005a; Shah 2015) where they are locally called *RASHMULA* (Ras = juice or sap, Mula = root). In Sikkim these tubers are also eaten (Kholia 2010b, 2014) and called *PANI AMLA* (Pani = water, Amla = Emblic or Indian gooseberry) due to resemblance with *Phyllanthus emblica* L. In Kashmir, young rhizomes and crozier of *Pteridium revolutum* (Blume) Nakai are used as vegetable and in preparation of pickle after boiling and roasting (Khan et al. 2009); similarly in Sikkim they are also used and locally called *TITE NINGRO* (Tite = bitter). In foot hills of Uttarakhand, Darjeeling, North Bengal and Assam Himalayas, fresh leaves of the clover leaf fern *Marsilea minuta* L. and *Ceratopteris pteridoides* (Hook.) Hieron. are also used as green leafy vegetable and commonly called *SUSNI SAG* and *SAG*, respectively, by Bengali community. The *Marsilea* L. is used in Bengal as sedative and treatment of duodenal and gastric ulcers, eye diseases and insomnia (Nirupama Goswami, per comm.). Due to its sedative nature, it is locally called *SUSHUNI* in Bengal (Shusni = lying down); on the other hand, all leafy vegetables in Bengal and other Hindi-speaking belt are called *SAG*. In Tanakpur and Banbasa area of Kumaun bordering west Nepal, *Ceratopteris thalictroides* (L.) Brongn. is called Pani Palak (Pani = water, Palak = spinach) due to its marshy habitat and its resemblance to spinach and is desirably eaten.

The other ferns used as vegetable in Sikkim and Nepal are *Dryopteris cochleata* (D. Don) C.Chr. (*GHIU NINGRO* or *CHIPHLEY NINGRO*), *Deparia boryana* (Willd.) M.Kato (*KALO NINGRO, KALO GHIU NINGRO*), *Cornopteris decurrentialata* (Hook.) Nakai (*RATO NINGRO*), *Pleocnemia submembranacea* (Hayata) Tagawa & K.Iwats. (*RATO NINGRO*), *Microlepia platyphylla* (D. Don) J. Sm. (*SETO CHIPHLEY NINGRO*), *Microlepia rhomboidea* (Wall. *ex* Kunze) Prantl (*TITE NINGRO*), *Microlepia speluncae* (L.) T. Moore (*TITE NINGRO*), *Tectaria coadunata* (J.Sm.) C.Chr. (*AULE KALO NINGRO* or *KALO CHETTY, KALO NIHURO*), *Tectaria ingens* (Atk. *ex* C.B.Clarke) Holttum. (*RATO LEKH NINGRO, KUHETTA KATTAKWA NOK* by Limboo tribe of Sikkim), *Pteridium revolutum* (Blume) Nakai (*TITE NINGRO*), *Pteris* spp. (*THARE NINGRO*), *Thelypteris* spp. (*SANO CHIPHLEY NINGRO* or *SANO DANTHEY NINGRO*), *Stenochlaena palustris* (Burm f.) Bedd., etc. Some of these species are also eaten in other states of Eastern Himalayas and NE India under different local names. Though species of *Pteris* L., *Microlepia* C.Presl, *Thelypteris*, *Dryopteris* Adans., *Pteridium revolutum* (Blume) Nakai, etc. are thought to be low-grade vegetable and rarely consumed, sometimes these are eaten mistakenly due to its similar appearance with croziers of edible species. These species are often also used as adulteration of fern vegetables by local sellers. From Bhutan, nearly a dozen species of ferns are reported as edible by Giri (2004) and Matsushima et al. (2008, 2012); among these species, besides common edible *Cyathea* spp. (*DENG*), *Diplazium* spp. (loc. Name *NAGKEY*), *Pteris excelsa* Gaudich., *Pteris wallichiana* J. Agardh (loc. Name *NIMIN DAWAY*), *Thelypteris prolifera* (Retz.) C.F. Reed, *Athyrium* spp., (loc.name *GHASA DAWAY*), *Botrychium* spp., *Microlepia* spp., *Cornopteris* spp., etc. are common edible ferns.

### 17.3.2 Food and Beverages

Large bulbous rhizomes or corm of different species of giant fern *Angiopteris* Hoffm. and the pith of tree ferns *Cyathea* J.E. Smith are used as food by tribal peoples of Eastern Himalayas, Bhutan and Nepal (Fig. 17.1). The rhizomes and trunk pith are washed, boiled, chopped, sun dried and grind with cereals to make flour which is used in making bread. Chopped pieces are also fermented alone or mixed with millet and cereals which are used in making country-made beer and wine. According to Lepcha tribe of Sikkim, they use only three species of tree ferns,

**Fig. 17.1** Edible ferns: (**a–c**) selling of vegetable ferns. (**a**) *Diplazium esculentum* on board and *Ophioglossum vulgatum* on ground; (**b**) *Diplazium doederleinii* and *Diplazium laxifrons*; (**c**) *Ophioglossum vulgatum*; (**d**) tubers of *Nephrolepis cordata*; (**e**) sample of data collection during ethnobotanical study by present author; (**f**) rhizome of *Angiopteris* sp.

viz. *Cyathea andersonii* (J. Scott ex Bedd.) Copel., *Cyathea khasyana* (T. Moore ex Kuhn) Domin and *Cyathea spinulosa* Wall. ex Hook., in making country-made beer or wine. On the other hand, they told *Cyathea sollyana* (Griff.) Fraser-Jenk. and

*Cyathea gigantea* (Wall. ex Hook.) Holttum are harmful and cause nausea, vomiting and stomach problems. Besides its magical and religious values, the tender shoots of moonworts (*Botrychium* spp.) are also eaten raw as salad or used in making delicious vegetable or pickle (Leach 2003; Kholia 2014; Kandel 2020).

Tender leaves of *Thelypteris procera* (D.Don) Fraser-Jenk. (Nepali name, *PEERE SOTAR*; Limboo, *KHESUNGLOOPMA*) are used as starter for making local yeast cakes (loc.name *MARCHA*), used for alcoholic fermentation and used as flavouring for country-made beer and wine in Sikkim. The fronds of *Thelypteris procera* (D.Don) Fraser-Jenk. are also used to cover the boiled millets, cereal, fruits, tubers, bulbs or other starchy material after mixing the country-made yeast cakes (*MARCHA*) during the alcoholic fermentation before keeping them in vats. These fronds are often also used as lid for fermentation pots or vats. Young croziers of *Cyathea* spp. (Nepali name, *RUKH UNIYON* or *CHATEY UNION*) along with *Angiopteris* spp., *Diplazium maximum* and *Diplazium himalayensis* and other vegetable ferns are also used for making pickles. After removing the scales, cleaning, drying and chopping in large pieces, the croziers are slightly roasted in mustard oil with different spices (garlic, chillies, turmeric, salt, funnel, fenugreek, etc.), kept in airtight glass jars after pouring some mustard oil and put in sun for 2–3 weeks.

### 17.3.3 Fodder

The cattle are selective while eating fern fodder; tender leaves of almost all the edible or vegetable ferns are agreeably consumed by livestock throughout the year; however, tender leaves of some other species are also taken unpleasantly during the non-availability of green fodder (Fig. 17.2). Besides vegetable ferns, leaves of different species of *Angiopteris* Hoffm., *Cyathea* J. E. Smith, *Coniogramme* Fee., *Arthromeris* (T. Moore) J. Sm., *Araiostegia* Copeland, *Lepisorus* (J. Smith) Ching, *Selaginella* Pal. Beauv, *Stenochlaena palustris* (Burm.f.) Bedd., *Dryopteris cochleata* (D.Don) C.Chr., *Deparia japonica* (Thunb.) M.Kato, *Deparia petersenii* (Kunze) M. Kato, *Microlepia platyphylla* (D. Don) J.Sm, etc. are also eaten or grazed by animals and hence can be recognised as fodder ferns. Similarly, some soft species and tender leaves of thelypteroid, pteroid and dryopteroid ferns are eaten or grazed by some animals, but the mature leaves of these species are not eaten by livestock. While passing from old Rajpur-Mussoorie Road, Dehradun, senior author also found *Selaginella subdiaphana* (Wall. *ex* Hook. & Grev.) Spring grazed by goats and collected by locals as goat fodder along with tender *Aleuritopteris bicolor* (Roxb.) Fraser-Jenk. Many ferns mixed accidently with palatable fodder at the time of fodder collection which are left behind by livestock are also can be considered as non-fodder ferns. These non-fodder fern species left behind are *Bracken* [*Pteridium revolutum* (Blume) Nakai] and species of *Aleuritopteris* Fee, *Adiantum* L., *Onychium* Kaulf., *Dryopteris* Adans., *Pteris* L., *Polystichum* Roth., *Thelypteris* Schmidel, *Athyrium* Roth., *Lygodium* Sw., *Dennstaedtia* Bernhardi, *Dicranopteris* Bernhardi, *Gleichenia* J.E. Smith, *Cyathea* J. E. Smith, *Peranema* D. Don, *Woodwardia* J. E. Smith, etc. Some species of ferns like *Pteridium revolutum*

**Fig. 17.2** (**a**) Collection of ferns for fodder and cattle bed in Kumaun region of Uttarakhand; (**b**, **c**) selling of yeast cakes (with a major ingredient *Thelypteris procera*) at Ravongla market of South Sikkim; (**d**) mixture of millets, chopped dried rhizome of *Angiopteris* sp., powdered starter (yeast cake Fig **c**) with fern leaves used for flavouring and covered before keeping for fermentation in vat

(Blume) Nakai, *Dryopteris juxtaposita* Christ, *Pteris cretica* L., *Polystichum squarrosum* (D. Don) Fee, *Thelypteris* (*Pseudocyclosorus*) *cana* (J.Sm.) Ching, *Athyrium setiferum* C.Chr., *Onychium tenuifrons* Ching, *Lygodium japonicum* (Thunb.) Sw., *Thelypteris* (*Pseudocyclosorus*) *tylodes* (Kunze) Ching, *Thelypteris* (*Christella*) *arida* (D. Don) C.V.Morton, *Dryopteris cochleata* (D.Don) C.Chr., *Dennstaedtia appendiculata* (Wall. *ex* Hook.) J.Sm., *Onychium cryptogrammoides* Christ, *Onychium lucidum* (D.Don) Spreng., etc. are tested and found to be toxic and harmful to livestock, and few cause serious diseases like enzootic bovine haematuria and cancer due to certain fern toxins (Somvanshi and Kataria 2002; Somvanshi et al. 2006; Pathania et al. 2012; Hidano et al. 2017).

The starchy rhizomes of *Angiopteris crassipes* Wall. *ex* C. Presl, *Angiopteris evecta* (G.Forst.) Hoffm., *Angiopteris helferiana* C.Presl and *Deparia boryana* (Willd.) M.Kato and pith of tree ferns *Cyathea andersonii* (J.Scott ex Bedd.) Copel. and *Cyathea brunoniana* (Wall.) Fraser-Jenk. (Bak.) Copel. are also used to feed the cattle and pigs in Sikkim after removing the outer scales and root mass, slicing, pounding and boiling for hours in large pans which is served after pouring salt; sometimes flour of millets or cereals are also added to make it tastier.

### 17.3.4 Cattle Bed

Generally, cattle bed is used throughout the year, but during rainy and winter season, thick cattle bed is required to protect livestock from moisture and cold (Fig. 17.2). As a matter of fact, almost all the species of available ferns both edible and non-edible are used for bedding of cattle and pigs. Besides edible and fodder ferns, tender leaves of other ferns are occasionally also taken as fodder, and the remaining part left behind serves as cattle bed. Generally, the commonly available fern species with large fronds, having thick coriaceous texture and cuticle, which needs least effort for making a head load are used as cattle bed. The cattle bed is locally called *SOTAR* in Uttarakhand, Nepal and Sikkim. Some common ferns used as cattle bedding in Himalayas are *Diplazium* spp., *Arthromeris wallichiana* (Spreng.) Ching, *Cibotium barometz* (L.) J.Sm., *Cyathea brunoniana* (Wall.) Kholia & Fraser-Jenk., *Cyathea khasyana* (T. Moore *ex* Kuhn) Domin, *Cyathea henryi* (Baker) Copel., *Cyathea gigantea* (Wall. ex Hook.) Holttum, *Cyathea sollyana* (Griff.) Fraser-Jenk., *Deparia boryana* (Willd.) M.Kato, *Dicranopteris lanigera* (D. Don) Fraser.-Jenk, *Dicranopteris linearis* (Burm.f.) Underw., *Dicranopteris splendida* (Hand.-Mazz.) Ching, *Dipteris wallichii* (R.Br.) T. Moore, *Drynaria quercifolia* (L.) J. Sm., *Dryopsis apiciflora* (Wall. ex Mett.) Holttum & P.J. Edwards, *Dryopteris carolihopei* Fraser-Jenk., *Dryopteris redactopinnata* S.K. Basu & Panigrahi, *Dryopteris sparsa* (D.Don) Kuntze, *Dryopteris wallichiana* (Spreng.) Hyl., *Dennstaedtia appendiculata* (Wall. *ex* Hook.) J.Sm., *Gleichenia gigantea* Wall. *ex* Hook. & Bauer, *Histiopteris incisa* (Thunb.) J. Sm., *Hypolepis polypodioides* (Blume) Hook., *Nephrolepis cordifolia* (L.) C. Presl, *Onychium cryptogrammoides* Christ, *Pteris spinescens* C. Presl, *Osmunda claytoniana* L., *Peranema cyatheoides* D. Don, *Plagiogyria pycnophylla* (Kunze) Mett., *Polystichum discretum* (D.Don) J. Sm., *Polystichum longipaleatum* Christ, *Polystichum squarrosum* (D.Don) Fee, *Pteridium revolutum* (Blume) Nakai, *Pteris aspericaulis* Wall. *ex* J. Agardh, *Pteris biaurita* L., *Pteris puberula* Ching, *Pteris terminalis* Wall. ex J. Agardh, *Pteris wallichiana* J. Agardh, *Thelypteris procera* (D. Don) Fras.-Jenk., *Thelypteris* (*Pseudocyclosorus*) *esquirolii* (Christ) Ching, *Thelypteris* (*Pseudocyclosorus*) *cana* (J.Sm.) Ching, *Thelypteris* (*Pronephrium*) *lakhimpurensis* (Rosenst.) K. Iwats., *Thelypteris* (*Pronephrium*) *nudata* (Roxb.) C.V. Morton, etc.

The other ferns which are also used as cattle bed are *Arachniodes henryi* (Christ) Ching, *Arachniodes carvifolia* (Kunze) Ching, *Arachniodes cornucervi* (D. Don) Fraser-Jenk., *Athyrium atkinsonii* Bedd., *Athyrium foliolosum* T. Moore *ex* R. Sim,

*Athyrium flabellulatum* (C.B.Clarke) Tardieu, *Athyrium wallichianum* Ching, *Blechnum orientale* L., *Coniogramme affinis* Hieron., *Coniogramme fraxinea* (D. Don) Fée *ex* Diels, *Coniogramme intermedia* Hieron., *Coniogramme procera* (D. Don) Fras.-Jenk., *Coniogramme serrulata* (Blume) Fée, *Deparia allantodioides* (Bedd.) M.Kato, *Deparia boryana* (Willd.) M.Kato, *Deparia japonica* (Thunb.) M.Kato, *Deparia petersenii* (Kunze) M. Kato, *Drynaria coronans* (Wall. *ex* Mett.) J. Sm. ex T. Moore, *Drynaria propinqua* (Wall. ex Mett.) J. Sm. and *Drynaria quercifolia* (L.) J. Sm., *Dryopteris chrysocoma*, *Dryopteris cochleata* (D. Don) C. Chr., *Dryopteris panda*, *Dryopteris splendida* (Hand.-Mazz.) Ching, *Dryopteris xanthomelas* (Christ) C.Chr., *Dryopteris zayuensis* Ching & S.K.Wu, *Hypodematium polypodioides* (Blume) Hook., *Microlepia* spp., *Monachosorum henryi* Christ, *Nothoperanema hendersonii* (Bedd.) Ching, *Nothoperanema squamiseta* (Hook.) Ching, *Osmunda japonica* Thunb., *Peranema aspidioides* (Blume) Mett., *Polystichum lentum* (D. Don) T. Moore, *Polystichum piceopaleaceum* Tagawa, *Polystichum discretum* (D.Don) J.Sm., *Pteridium revolutum* (Blume) Nakai, *Pteris biaurita* L., *Pteris pellucens* J.Agardh, *Pteris normalis* D. Don, *Pteris puberula* Ching, *Pteris spinescens* C.Presl, *Pteris vittata* L., *Stenochlaena palustris* (Burm.f.) Bedd., *Tectaria coadunata* ((J.Sm.) C.Chr., *Tectaria ingens* (Atk. *ex* C.B. Clarke) Holttum, *Thelypteris* (*Christella*) *arida* (D. Don) C.V. Morton, *Thelypteris*(*Christella*) *crinipes* (Hook.) K. Iwats., *Thelypteris* (*Christella*) *dentata* (Forssk.) E.P.St. John, *Thelypteris* (*Christella*) *papilio* (C. Hope) K. Iwats., *Thelypteris* (*Christella*) *procera* (D. Don) Fras.-Jenk., *Thelypteris* (*Glaphyropteridopsis*) *erubescens* (Wall. *ex* Hook.) Ching, *Thelypteris* (*Oreopteris*) *elwesii* Baker, *Thelypteris* (*Pronephrium*) *penangiana* (Hook.) C.F. Reed, *Thelypteris* (*Pseudocyclosorus*) *cana* (J.Sm.) Ching, *Thelypteris* (*Pseudocyclosorus*) *esquirolii* (Christ) Ching, *Thelypteris* (*Pseudocyclosorus*) *ornatipes* (Holttum & J.W. Grimes) Fras.-Jenk., *Thelypteris* (*Pseudocyclosorus*) *tylodes* (Kunze) Ching, *Thelypteris* (*Pseudophegopteris*)*levingei* (C.B.Clarke) Ching, *Thelypteris* (*Pseudophegopteris*) *levingei* (C.B.Clarke) Ching, *Thelypteris* (*Macrothelypteris*) *ornata* (J.Sm.) Ching, *Thelypteris* (*Macrothelypteris*) *torresiana* (Gaudich.) Alston, *Thelypteris* (*Cyclogramma*) *auriculata* (J.Sm.) K. Iwats., *Woodwardia unigemmata* (Makino) Nakai etc. Along with the dunghill, these ferns are ultimately converted into organic manure or used in mulching. In Bhutan and Arunachal Pradesh, green leaves of ferns are harvested, dried and stored along with naturally dried fern leaves for the use in winters.

### 17.3.5 Folk Medicines

After Ayurvedic literature, perhaps Dymock (1890) was the first worker who documented six species of pteridophytes [*Actiniopteris dichotoma* Beddome = *Actiniopteris radiata* (Sw.) Link; *Adiantum venustum* D. Don.; *Asplenium parasiticum* Willd. = *Thelypteris parasitica* (L.) Fosberg; *Asplenium falcatum* Willd. =*Athyrium falcatum* Bedd.; *Polypodium quercifolium* Sprng. =*Drynaria quercifolia* (L.) J. Sm. and *Polypodium vulgare* L.] with medicinal and pharmacological properties from India as well as from Indian Himalayas. Though the abovementioned *Polypodium vulgare* L. is a European fern and is not present in India, it is considered a species of *Polypodium* sensu *lato* from Malabar or South India. In a monumental work, Caius (1935) reported 37 species of Indian pteridophytes belonging to 24 genera with medicinal properties, and all these are also present in the Himalayas. All these medicinal pteridophytes were also listed in the subsequent compilations (*Indian Medicinal Plants*, *Glossary of Indian Medicinal Plants*, *Chopra's Indigenous Drugs of India*, *Bhavaprakasa Nighantu*, *Medicinal Plants of India*, etc.) and research papers along with some additions. Based on literature trawl, Singh (1999, 2002) enumerated nearly 160 species of Indian pteridophytes used in medicine and other purpose; however, Ambast (2000) listed only 80 species while compiling his 'Useful Plants of India'. Thereafter, in recent two decades, suddenly there is a boom in publication on ethnobotany of pteridophytes from different parts of India including the Himalayas (Gupta et al. 2000; Samant et al. 2001; Singh et al. 2001; Gogoi and Das 2002; Sharma 2002; Singh 2002; Kumar and Ramesh 2003; Dixit and Singh 2004; Kholia and Punetha 2005a; Benniamin and Manickkam 2007; Srivastava 2007; Rao et al. 2007; Mannan et al. 2008; Kholia 2010b; Benniamin 2011; Kumari et al. 2011; Sen and Ghosh 2011; Yumkham and Singh 2011; Das and Dutta Choudhury 2012; Singh and Singh 2012; Singh and Upadhyay 2012; Kholia 2013; Wani et al. 2016; Chander et al. 2017; Hidano et al. 2017; Singh 2018; Balkrishna et al. 2019; Rajagopal 2019, etc.), which added many species in the list of the Indian and Himalayan medicinal pteridophytes, and at present, nearly 225 species of medicinal pteridophytes are known from India (Goswami et al. 2016).

From Jammu and Kashmir, Kirn and Kapahi (2001) reported 11 medicinal pteridophytes; on the other hand, only 8 species were reported by Gairola et al. (2014) from the state including Ladakh. Further, Mir et al. (2015) reported 9 medicinal pteridophytes from only Shopian district of Jammu and Kashmir, and Wani et al. (2016) reported 38 species from Kashmir valley. Very few species of medicinal pteridophytes were known from Himanchal Pradesh, and among them, Chander et al. (2017) reported 12 species from Hamirpur, 2 species (*Adiantum venustum* D. Don and *Equisetum arvense* L.) from Manali Wildlife Sanctuary (Rana and Samant 2011) and 1 species from Joginder Nagar, Mandi (Kumar 2014). About 40 species of medicinal pteridophytes are also known from Uttarakhand (Kholia and Punetha 2005a; Upreti et al. 2009), 21 species from Sikkim (Kholia 2014) and 51 from Arunachal Pradesh (Benniamin 2011). On the other hand, in a detailed study, Choudhury et al. (2009) reported 108 species of pteridophytes used against various diseases from southern Assam.

There are many reports on medicinal pteridophytes from different parts of Nepal. Uprety et al. (2012) reported *Tectaria coadunata* (J.Sm.) C.Chr. and *Tectaria zeilanica* (Houtt.) Sledge as medicinal, but *Tectaria zeilanica* is not present in Nepal; therefore, the second medicinal species from Nepal may be similar to *Tectaria fuscipes* (Bedd.) C.Chr., *Tectaria heterocarpa* (Bedd.) C.V. Morton or juvenile *T. polymorpha* (Wall. *ex* Hook.) Copel. Similarly, four species of medicinal pteridophytes [*Adiantum venustum* D. Don, *Lycopodium japonicum* Thunb.,

*Lygodium japonicum* (Thunb.) Sw. and *Nephrolepis cordifolia* (L.) C. Presl] were reported from Dhorpatan Hunting Reserve of Nepal by Luitel and Pathak (2013) and three species [*Cheilanthes dalhousieae* Hook. = *Aleuritopteris bicolor* (Roxb.) Fras.-Jenk., *Equisetum diffusum* D. Don and *Nephrolepis auriculata* (L.) Trimen] from Ilam district of east Nepal bordering Sikkim by Parajuli (2013). Kunwar et al. (2013) reported *Adiantum capillus-veneris* L. as antidote of scorpion sting in Nepal. Nearly a dozen species of medicinal pteridophytes are recorded from Bhutan in limited and scattered publications (Wangchuk et al. 2007, 2016, 2017; Dorji and Thomas 2017; Chetri 2019), where *Dryopteris cochleata* (D. Don) C.Chr., *Drynaria propinqua* (Wall. ex Mett.) J. Sm., *Drynaria quercifolia* (L.) J. Sm., *Equisetum arvense* L., *Lepisorus contortus* (Christ) Ching, *Pteris biaurita* L., *Selaginella involvens* and *Lycopodium* sp. are used against various human ailments.

During the present study, it is observed that the information given in the literature pertaining to edible/vegetable ferns seems to be almost correct and verified by senior author in field; on the other hand, the information on medicinal pteridophytes looks obscure. Almost all the genera and species hitherto reported as medicinal in previous publications (encyclopaedias, glossaries, research reports) were also documented in subsequent literature of Indian medicinal plants along with the addition of few more allied species. Therefore, it seems that many recent ethnobotanical publications are the repetitions and compilations of previous works instead of genuine research. For example, in all recent publications, *Drynaria quercifolia* (L.) J. Sm. is reported useful against typhoid which is ensued from Caius (1935). Furthermore, many of these publications seems to be ambiguous where same species listed effective against one disease in one literature and for different disease in others (Parihar and Parihar 2006, Mannan et al. 2008, Upreti et al. 2009, Goswami et al. 2016 etc.). For example, a single fern *Adiantum capillus-veneris* L. is reported with many medicinal properties, as astringent, expectorant, emmenagogue, febrifuge, anticancerous, anti-diabetic, purgative, catarrhal, etc. It is also mentioned as used in the treatment of menstrual problem, fever, cough, cold, bronchial diseases, mouth blisters, infertility, women's disease, dandruff, tumours of the spleen, liver diseases, etc. The powdered rhizome of this fern is also mixed with camphor and palm kernel oil and used as eye ointment and against lice and other external parasites (Goswami et al. 2016).

Fairly a large number of pteridophytes (ca.30% species of Indian pteridophytes) are recorded as medicinal in ethnobotanical literature, but very few are in pharmaceutical or medicine industry. Therefore, a series of questions arises on the ethnobotanical studies of pteridophytes in India. Whether the standard methodologies used in these researches and the data generated from local informer are correct or not? Was the local informer or healer genuine or not? If genuine, he/she had provided the correct information or not? Was the data obtained from informer is rechecked or verified by researcher? Do these potential medicinal ferns correctly identified and documented following the standards of the field and herbarium techniques? etc. Sometimes, the local informer may not be a genuine person and for the sake of publicity he claims himself as herbal healer and give fake information. Again the genuine local healer may not give the correct information during the survey, because there is a myth among Indian societies that medicinal property of plant is ravaged if disclosed to the public, so they transfer this secret knowledge to their next generation only, but during this transformation of knowledge to the next generation, there may be chances of knowledge loss which depends on the quality, learning capability and interest of knowledge adopters. Therefore, the ethnobotanical studies depend upon the knowledge and honesty of informer or local healer along with the skills of ethnobotanist for eliciting the data during the survey. It also depends on genuine research or researchers that had published fake information for carrier enhancement. Such fake publications for carrier enhancement in predatory journals are harmful for both sciences and society.

Generally, it is assumed that the plants used by any human community is definitely known or identified by a popular local name, but in almost all the recent ethnobotanical studies on Indian pteridophytes, the local names of plants are not recorded. Further, many ethnobotanists do not maintain the proper record of investigation or filed interview-like datasheets including name and address of the healer or informer, field pictures, proper voucher specimen, etc. for future verification. If some vouchers are maintained, they are not properly identified by genuine pteridophyte taxonomist before publication. For example, in a recent publication from Lakhimpur, Assam (Singh 2018), six coloured images of medicinal ferns are published, but among these five (83%) are misidentified at genera level. This blunder is even not corrected at the time of reviewing and editorial level by the journal authorities. This pathetic identification problem in Indian pteridophyte taxonomy was already highlighted from time to time (Kholia 2009; Kholia and Fraser Jenkins 2011).

Initially species of *Actiniopteris* Link, *Adiantum* L., *Asplenium* L., *Athyrium* Roth., *Blechnum* L., *Botrychium* Sw., *Cheilanthes* Sw., *Cibotium* Kaulf., *Cystopteris* Bernh., *Drynaria* (Bory) J. Sm., *Dryopteris* Adans., *Gleichenia* Sm., *Helminthostachys* Kaulf., *Lygodium* Sw., *Ophioglossum* L., *Osmunda* L., *Pellaea* Link., *Pleopeltis* Humb. & Bonpl. ex Willd., *Pteris* L., *Stenochlaena* J. Sm., etc. were recorded as medicinal pteridophytes from India (Caius 1935); later, species of few more genera like *Acrostichum* L., *Ceratopteris* Brongn., *Cyathea* J.Sm., *Cyrtomium* C. Presl, *Davallia* Sm., *Dicranopteris* Bernh., *Drymoglossum* C. Presl, *Equisetum* L., *Hemionitis* L., *Hypodematium* Kunze., *Hypolepis* Bernh., *Lemmaphyllum* C. Presl., *Lycopodiella* Holub, *Lycopodium* L., *Marsilea* L., *Nephrolepis* Schott., *Odontosoria* Fee, *Oleandra* Cav., *Onychium* Kaulf., *Polypodium* s.l., *Psilotum* Sw., *Selaginella* P. Beauv., *Tectaria* Cav., *Thelypteris* Schmidel., *Trichomanes* L., *Woodwardia* Sm., etc. are added with available English, Hindi, Sanskrit and regional names (Nadkarni 1954; Chopra et al. 1956, 1969; Nayar 1959; Ambast 2000).

The species of maidenhair fern (*Adiantum* L.) and its medicinal properties are known to human beings from the time of early Greek philosophers. In India, reports on the use of different species of *Adiantum* L. against various ailments are also documented in early Ayurvedic literature like Charak and Sushruta Samhita where it is named as *HANSRAJ* and *HANSPADI*. The other regional or local names of *Adiantum* L. mentioned in various ethnobotanical literature are *MUBARAK*, *KANBADHWA*, *HANSPADI*, *KHAS*, *DUMTI*, *MYLEKONDI*, *GAJKARNA*,

*ADHSARITA KI JAHRI*, *GUNKARI KANGHAI*, *KALIJHANT*, *NAYALAD*, etc. The species of *Marsilea* L. and *Actiniopteris radiata* (Sw.) Link are called *MAYURSIKHA* (Mayur = peacock, sikha = head crest) in Sanskrit or Hindi language, but sometimes *Adiantum incisum* Forssk. is also called *MAYURSIKHA* due to similar whorl of proliferating fronds on sub-terminal rachis bud.

Medicinal properties of nine species of Indian *Adiantum* L. [*A. aethiopicum* L. (=*A. poiretii* Wikstr.), *A. edgeworthii* Hook., *A. incisum* Forssk., *A. pedatum* L., *A. cuneatum* Langsd. & Fisch. (=*A. raddianum* C. Presl), *A. capillus-veneris* L., *A. flabellulatum* (C.B.Clarke) Tardieu, *A. lunulatum* Burm.f. (=*A. philippense* L.), *A. soboliferum* Wall. ex Hook.] distributed throughout India were highlighted by Caius (1935). Thereafter, in almost all the Indian ethnobotanical literature along with these species, medicinal uses of different other species of *Adiantum* L. are also documented, with additional uses. The uses of *Adiantum* spp. listed are as astringent, demulcent, expectorant, diuretic, purgative, aphrodisiac, emollient, anthelmintic, styptic, febrifuge, anti-cancer, emmenagogue etc. and they are used in treatment of cough, chest diseases, glandular swelling, dysentery, diseases of blood, ulcers, epilepsy, leprosy, skin diseases, kidney diseases, inflammations, styptic, fever, tumours, boils, headache, piles, wounds, hair fall, scorpion stinging, malaria, bacterial, viral and fungal infections, etc. (Ambast 2000). Kholia (2010a, 2010b, 2014) reported the use of species *Adiantum* L. from Sikkim (local name *SIMSERY UNIYON*) in pneumonia, wounds and cuts. Mir and John (2014) mentioned that *Adiantum capillus-veneris* L. is used for increasing libido in the Pulwama Tehsil of Jammu and Kashmir; however, it is used in scorpion sting by the peoples of far western Nepal bordering to India (Kunwar et al. 2013). On the other hand, several other uses of *Adiantum capillus-veneris* are documented by Goswami et al. (2016). They also reported the uses of *Adiantum lunulatum* Burm.f. (=*Adiantum philippense* L.), *Adiantum aethiopicum* L., *Adiantum caudatum* L., *Adiantum flabellulatum* L., *Adiantum incisum* L., *Adiantum soboliferum* L., *Adiantum pedatum* L. and *Adiantum venustum* D. Don along with few exotic maidenhair ferns.

*Lycopodium japonicum* Thunb., one of the common fern allies of the Indian Himalayas, is recorded as medicinal against many human ailments (Mannan et al. 2008; Upreti et al. 2009; Dubal and Kale 2014). In Nepal and Sikkim, it is locally called *NAG BELI*, and spores are used in skin diseases and itching. The resurrecting club moss *Selaginella bryopteris* (L.) Baker is often locally called *MRIT-SANJEEVANI* or *SANJEEVANI BOOTI* and considered as highly useful in unconsciousness and other weaknesses (Kholia and Punetha 2005a). According to the Hindu epic, Ramayana a medicinal plant named as *SANJEEVANI* is bought by Lord Hanuman from the Himalayas for restoring the life of Lord Lakshmana during the Ram Ravan war at Sri Lanka. Therefore, since long back along with other herbs, *Selaginella bryopteris* (L.) Baker is one of the candidate plants of mythological *SANJEEVANI* (Ganeshaia et al. 2009; Balasubramanian 2010). On the other hand, Fraser-Jenkins (2013) advocated that *Selaginella bryopteris* is a common tropical plant distributed in South India, central India and the foothills of the Himalayas; therefore, at the time of emergency, in spite of bringing medicine from the nearest place, Lord Hanuman brought it from cooler Himalayas which is rather far away.

Therefore, he postulated that a rare club moss endemic to Central Himalaya and adjacent Tibet, *Selaginella pulvinata* (Hook. & Grev.) Maxim., might be the correct candidate for *SANJEEVANI* instead of *S. bryopteris* (L.) Baker. Both *Selaginella bryopteris* (L.) Baker and *Selaginella pulvinata* (Hook. & Grev.) Maxim. are resurrection plants and also listed in Indian and Chinese medicinal systems (Kholia and Punetha 2005a, 2005b; Kholia 2008, 2009; Sah 2008; Balasubramanian 2010; Antony and Thomas 2011; Aulakh et al. 2019; Anonymous b, in Tumour Research 2010; Xu et al. 2003). Besides this mythological belief, *Selaginella bryopteris* (L.) Baker is also used for gonorrhoea, venereal diseases, urinary tract diseases, stomach ache, beriberi and dysentery; is used in curing wounds, jaundice, colitis, indigestion and constipation to regain vigour and vitality; is used in treatment of infertility; and is used as a hallucination agent with tobacco (Sah et al. 2005; Kholia and Punetha 2005a; Chandrakant et al. 2019). *Selaginella involvens* Spring, *Selaginella wallichii* Spring and *Selaginella willdenowii* (Desv.) Bak. are other Himalayan species of *Selaginella* which are also said to have property in prolonging life, and their decoction is used after parturition. These club mosses are recorded against lowering the fever and reducing body pain and for treatment of mental diseases, rickets, blood vomiting and spermatorrhoea (Chunekar and Pandey 1969; Ambast 2000).

Based on the interview with three local herbal healers (Shayam Lal Bista, Lila Ram Dhakal and Nar Bahadur Subba), *Equisetum diffusum* D. Don and *Equisetum ramosissimum* Desf. locally known as *SALLI BISALLI* in Sikkim are used in the treatment of epilepsy, cattle haematuria and dog bite (Kholia 2010a, 2014); however in Jammu and Kashmir, it is used for the treatment of jaundice (Khan et al. 2009). The species of *Equisetum* L. from the Arunachal Pradesh are reported effective in the treatment of urinary problems, haemostatic, acidity, dyspepsia, bone fracture, bleeding, gonorrhoea, joint pain, fungal infection, kidney disorders and hydrophobia (Balkrishna et al. 2019). In almost all the parts of the Himalayas, species of *Equisetum* L. are used in the treatment of bone diseases. Adi and other tribes of Arunachal use it in the treatment of dislocation and bone fractures (Balkrishna et al. 2019), and similar uses of *Equisetum* L. are also reported from Bhutan (Chetri 2019). In Nepal and Kumaun, it is noted that after over eating the rotten meat of dead animal, the dogs eat the *Equisetum* plant locally called Kukkure Jhar (Kukur = dog, Jhar = grass); perhaps it acts to stimulate the vomiting and also treat the indigestion and food poisoning. In Sikkim the juice of *Equisetum* L. is also used as an appetiser (Kholia 2014). Several other medicinal uses of *Equisetum* L. are also compiled by Gairola et al. (2014), Mannan et al. (2008), Parihar and Parihar (2006) and Upreti et al. (2009). Some bioactive compounds of *Equisetum* L. with anti-inflammatory, antimicrobial and antioxidant properties are also documented recently (Aulakh et al. 2019). Similarly, some chemical constituents of *Equisetum* L. are also listed by Wani et al. (2016), etc. According to herbal practitioners of Sikkim, the roots (= underground creeping rhizome) of *Equisetum diffusum* D. Don are mixed with the roots of *Acorus calamus* L., *Clematis buchananiana* DC and *Phytolacca* sp., and after grinding or pounding, a glass of juice is taken for 15 days for the treatment of epilepsy. Similarly, in dog bite, whole plant of *Equisetum diffusum* D. Don is pounded with the fruits of *Datura* spp., rhizomes of *Smilax zeylanica* L., rhizome of *Pteris* sp. (loc. *THARE UNIYON*) and rhizome of *Curcuma zedoaria* (Christm.) Roscoe, and half spoon juice is given to children and two to three spoon to adults once a day, and the residue is also applied on dog bitten wound (information from Lila Ram Dhakal herbal healer of Sikkim).

In almost all literature, species of *Botrychium* Sw. are considered a good vulnerary with anti-dysentery properties (Ambast 2000). They are also applied as antibiotics in cut, wounds, ruptures, bruises and sores, and its juice is used in treating breast cancer (Singh 2002; Ambast 2000). Boiled rootstock of *Botrychium* spp. is used as expectorant and in treating pneumonia and catarrh. The paste from its roots is also used as facial mask (Gaur and Bhatt 1994). *Botrychium virginianum* (L.) Sw. is reported effective against snakebites in the Himalayas by Srivastava (2007), but authors did not come to know about its use against snakebite during his more than 30 years studies on pteridophytes of the Himalayas. *Helminthostachys zeylanica* (L.) Hook. locally known as *KAMRAJ* or *DHIMRAJ* is used in impotency, blisters of tongue and waist pain and is used as brain tonic (Singh et al. 1989); it is also used in sciatica (Mannan et al. 2008), whooping cough, curing impotency, dysentery, cataract, improving memory power, jaundice, malaria, sciatica and early stages of phthisis (Ambast 2000; Kholia and Punetha 2005a; Begum et al. 2016).

*Actiniopteris radiata* (Sw.) Link is another oldest known medicinal fern of India believed to be named in Ayurveda as Mayurshikha or Mayurpankha. Its anthelmintic and astringent properties are listed in British Indian literature (Watt 1889–1892; Dymock 1890; Caius 1935; Kirtikar and Basu 1935) and also followed in almost all subsequent post-independence publications (Ambast 2000) along with trails in livestock (Kalidas et al. 2009). In the Himalayas, *Actiniopteris* is distributed in the foot hills of Uttarakhand and Nepal. Besides its anthelmintic properties, it is listed as antibacterial, antifungal and antimicrobial and effective to stop bleeding, diarrhoea and dysentery. Furthermore, it is also used for the treatment of bronchitis, gynaecological disorders, tuberculosis, infertility, prolonged malarial fever, diabetes, stress, allergy, haemoptysis, etc. (Ambast 2000; Upreti et al. 2009; Aulakh et al. 2019, Begum et al. 2016).

Rhizomes of *Tectaria polymorpha* (Wall. ex Hook.) Copel. are also reported as anthelmintic from British time (Caius 1935), and besides being anthelmintic, its additional medicinal properties against eczema, scabies and fever are also mentioned by Balkrishna et al. (2019). *Tectaria coadunata* (J.Sm.) C.Chr. is another medicinal species widely used throughout the Himalayas. It is used against diarrhoea in Darjeeling hills; diabetes, piles and pneumonia in Sikkim (Kholia 2010a, 2014); and blood dysentery and stomach ache in Nepal and Uttarakhand (Upreti et al. 2009; Uprety et al. 2012). In some Indian ethnobotanical literature, this species is erroneously named as African *Tectaria macrodonta* (Fee) C. Chr. wood fern or male fern *Dryopteris* Adans. and is one of the common fern genera of the Indian Himalayas with nearly 55 species. There are reports on its anthelmintic properties and effectiveness against all types of parasitic worms (Goswami et al. 2016). Besides vermifuge, species of *Dryopteris* are antibacterial and also useful in the treatment of eczema, acne, rheumatism, diarrhoea and dysentery (Gaur and Bhatt 1994; Singh 1999, 2002; Rajagopal 2019). The other Himalayan ferns reported with anthelmintic properties are *Athyrium* spp., *Blechnum orientale* L., *Cibotium barometz* (L.) J.Sm.,

*Cyrtomium anomophyllum* (Zenker) Fraser-Jenk., *Cyrtomium caryotideum* (Wall. ex Hook. & Grev.) C.Presl, *Cyrtomium fortunei* J.Sm., *Cyrtomium macrophyllum* (Makino) Tagawa, *Dicranopteris lanigera* (D.Don.) Fraser-Jenk., *Dicranopteris linearis* (Burm.f.) Underw., *Pteris multifida* Poir., *Adiantum flabellulatum* (C.B. Clarke) Tardieu, *Actiniopteris radiata* (Sw.) Link., *Asplenium trichomanes* L., *Asplenium adiantum-nigrum* L., *Athyrium pectinatum* (Wall. ex Mett.) T. Moore, *Asplenium ruta-muraria* L., *Blechnum orientale* L., *Athyrium falcatum* Bedd., *Pityrogramma calomelanos* (L.) Link, *Pteridium revolutum* (Blume) Nakai, etc. (Singh 1999, 2002; Goswami et al. 2016; Rajagopal 2019).

In the Indian Himalayas, the climbing fern *Lygodium* Sw. (*LAHRE UNIYON* in Nepalese) is represented by about seven species. In almost all the ethnobotanical literature from India, it is documented as an expectorant along with its uses as diuretic and in dysentery, cuts and wounds, rheumatism, sprains, scabies, eczema, gonorrhoea, piles, etc. (Singh 1999, 2002; Ambast 2000). The species of *Aleuritopteris* Fee (silver ferns) are also used throughout the Himalayas as antiseptic in cuts and wounds (Kholia 2010a). In Sikkim they are called RANI UNIYON (Rani = Queen, Uniyon = fern) due to the resemblance of the crown of queen; however, in Kumaun, silver ferns are locally known as *DUMI SINKO*. The stipes of silver ferns are traditionally used as antibiotic sticks in the freshly pierced nose and ear before wearing ornaments; perhaps these sticks protect them from infections and are useful in hardening of newly pricked pores. Besides quenching of thirst, tubers of Boston fern *Nephrolepis cordifolia* (L.) C. Presl (*PANI AWOLA*) are also used in Sikkim as appetiser after fever or illness; juice from its leaves is applied on wounds and cuts for blood coagulation; the decoction is given in coughs for reducing the chest congestion; and it is often believed that it gives strength to the kidneys and also gives relief in hernia too. Paste of leaves of *Thelypteris procera* (D.Don) Fras.-Jenk. (Nepali Name- *PEERE SOTAR*. Limboo- *KHESUNGLOOPMA*) is used to eradicate bed bugs and lice in Sikkim.

The spleenwort genus *Asplenium* L. is represented in the Himalayas by about 50 species, but very few are reported as medicinal. Caius (1935) reported that the Himalayan *Asplenium trichomanes* L., *Asplenium ruta-muraria* L. and *Asplenium adiantum-nigrum* L. have expectorant and are useful in pulmonary diseases, and *Asplenium falcatum* L. is useful in treatment of prolonged malaria fever. The other uses of Himalayan spleenwort are reported as diuretic, anthelmintic, laxative, astringent, antibacterial, depurative and sedative and in treatment of jaundice, skin disease, spleen diseases, urinary disorders, malaria, dysentery, rickets, etc. (Singh 2002, Ambast 2000, Balkrishna et al. 2019).

The juice of tender leaves of *Pteris* spp. (*THARE UNIYON*) is used as antibiotic against pneumonia and to treat the dysentery, piles and fissure (Anonymous a 2009). Besides being used as an antibiotic and medicine for pneumonia, *Pteris biaurita* L. is used in Sikkim against and cuts and wounds (Kholia 2010a). *Pteris aspericaulis* Wall. ex J. Agardh, *Pteris ensiformis* Burm. f., *Pteris longipes* D.Don, *Pteris multifida* Poir., *Pteris spinescens* C.Presl. and *P. vittata* L. is used in dysentery, high blood pressure, diarrhoea, fever, sores and wounds, body pain and piles, in treatment of glandular swelling of the neck and as vermifuge (Wealth of India; Benniamin 2011; Kholia 2014; Balkrishna et al. 2019). In Sikkim, paste of rhizome of *Drynaria quercifolia* (L.) J. Sm., *Drynaria propinqua* (Wall. ex Mett.) J. Sm. and *Drynaria coronans* (Wall. *ex* Mett.) J. Sm. ex T. Moore (loc. Name *KAMRE LARA*) is applied in the waist to relieve the pain after child birth (Kholia *vide* Lila Ram Dhakal). Uses of *Drynaria quercifolia* (L.) J. Sm. as tonic and astringent to the bowels and in the treatment of phthisis, typhoid, dyspepsia, cough, diarrhoea, gastritis, tuberculosis, throat infections and fever were reported by Caius (1935). *Drynaria quercifolia* (L.) J. Sm. is also reported as antibacterial, antipyretic, anthelmintic, anti-inflammatory, analgesic, antidermatophytic and laxative (Sen and Bhattacharya 2007; Das and Dutta Choudhury 2012; Aulakh et al. 2019). In addition, the paste of its rhizome is also used as plaster for the treatment of fractured and dislocated bones (Shil and Dutta Choudhury 2009). Other Pteridophytes reported as medicinal from the Himalayas by different workers (Chopra et al. 1956; Nayar 1959; Singh 2002; Dutta Choudhury and Choudhury 2002; Shil and Dutta Choudhury 2009; Dutta Choudhury et al. 2009; Kholia 2010a, 2010b; Das and Dutta Choudhury 2012; Goswami et al. 2016; Ambast 2000; Wani et al. 2016) against miscellaneous ailments are as follow: *Aleuritopteris albomarginata* (C. B. Clarke) Fraser-Jenk. is used for the treatment of tuberculosis; *Angiopteris* spp. used for flavouring and intoxication of countrymade beer and wine; species of *Asplenium nidus* complex are used against sores, ulcers and other skin diseases; *Blechnum orientale* L. in urinary disorders, bladder complaints, impotence, boils, worms, typhoid, stomach ache and ulcers; *Cibotium barometz* (L.) J.Sm. in wounds, lumbago, as general tonic; *Cyathea brunoniana* (Wall. ex Hook.) C.B. Clarke & Baker used as blood coagulant; *Dicranopteris linearis* (Burm.f.) Underw. for the treatment of asthma; *Dipteris wallichii* (R. Br.) T. Moore in jaundice; *Drymoglossum piloselloides* (L.) C.Presl used as anti-inflammatory to reduce the pain, swelling and sprains; *Microsorum punctatum* (L.) Copel. in snake bite; *Odontosoria chinensis* (L.) J.Sm for treating enteritis and other gastric problems; *Oeosporangium tenuifolium* (Burm.f.) Fraser-Jenk. & Pariyar taken as general tonic; *Osmunda japonica* Thunb. used in rickets, rheumatism and intestinal problems; *Thelypteris* (*Ampelopteris*) *prolifera* (Retz.) C.F. Reed used in constipation; *Thelypteris* (*Pronephrium*) *nudata* (Roxb.) C.V.Morton in headache and pyorrhoea, etc. The paste of *Ophioglossum reticulatum* L. (= O. vulgatum L.) is locally applied in headache and cut and wounds and the paste of *Stenochlaena palustris* (Burm.f.) Bedd. is used as a cooling agent in burn and ulcers.

In recent times, certain active compounds with antioxidant, anti-inflammatory, anti-cancer, antidiabetic, antiviral, antimicrobial, antibacterial and anti-Alzheimer properties are discovered in many species of ferns which are also present in the Indian Himalayan Region (Kale 2015; Nath et al. 2016). From the ethnobotanical literature and uses of ferns as folk medicines, it is clear that ferns have many medicinal properties, but this medicinal potential needs to have proper scientific and pharmaceutical research towards the exploration of drugs for human welfare. We are still hopeful that besides traditional therapeutic uses of ferns and their allies, in the future, through more sophisticated scientific research, we will be able to find such compounds or drugs from ferns which are useful to humans, and drugs made of pteridophyte will occupy its respective place in the medicine market and trade (Fig. 17.3).

**Fig. 17.3** Selling of pteridophytes in markets: (**a**) A herbal practitioner is selling *Equisetum diffusum* in Sikkim; (**b**) a herbal practitioner is selling *Drynaria rhizome* in Sikkim; (**c**, **d**) selling of *Selaginella bryopteris* at Nehru place, New Delhi

### 17.3.6 Horticulture

The beauty of fern plant has fascinated people and society from time immemorial (Fig. 17.4). Ferns are popular indoor and outdoor plants in pots, gardens and nurseries due to their aesthetic value. Cultivation and trade of orchids, ferns, succulents and other ornamental plants are the main occupation of many SE Asian

**Fig. 17.4** (**a**) Cultivation of tree ferns as avenue plant in gardens and park; (**b**, **c**) use of tree fern trunks for cultivation of orchids, staghorn fern and other ornamental; (**d**–**i**) cultivation or domestication of wild pteridophytes as ornamentals – (**d**) *Asplenium phyllitidis*, (**e**) *Selaginella monospora*, (**f**) *Bolbitis heteroclita*, (**g**) *Tectaria coadunata*, (**h**) *Asplenium crinicaule*, (**i**) *Polystichum acutidens*

countries; however, despite being diverse and rich native Pteridophytes, in India the interest on pteridophyte cultivation is still at gestation stage. Few decades back in Indian gardens, only limited exotic species, cultivars and selection were under cultivation (Chakraverty et al. 2003). These commonly cultivated ferns are Bird nest ferns [Asplenium nidus L. and aggregate], Mexican Maidenhair ferns [few Mexican species of *Adiantum* L.], Hares foot fern [*Davallia fejeensis* Hook. and *Davallia mariesii* T. Moore ex Baker], Japanese Holly fern [*Cyrtomium falcatum* (L.) C.Presl], Mexican Silver fern [*Pityrogramma calomelanos* (L.)Link], East Indian polypody [*Phymatosorus scolopendria* (Burm.f.) Pic.Serm.], Leathery Shield Fern [*Rumohra adiantiformis* (G. Forst.) Ching], Staghorn fern [*Platycerium alcicorne* Desv. and *P. bifurcatum* (Cav.) C.Chr.], Chinese lace fern [*Selaginella braunii* Baker], Electric Fern or Peacock fern [*Selaginella willdenovii* (Desv. ex Poir.) Baker], Tender spike moss [*Selaginella vogelii* Spring], and different species of Boston and their cultivars [*Nephrolepis exaltata* (L.) Schott and its cultivars (cv. 'Cotton Candy', cv. 'Hillii', cv. 'Bostoniensis', cv. 'Fluffy Ruffle'), *Nephrolepis cordifolia* (L.) C.Presl and its cultivars (cv. 'Duffii', cv. 'Compacta' cv. 'Plumosa'), etc.]. However, in recent decades, besides these limited exotics and selections, many native species of ferns and lycophytes are also introduced in Indian gardens or nurseries. The horticultural potential and the possibilities of trade of Himalayan pteridophytes was highlighted (Kholia 2018) with the details on domestication of nearly 150 species of wild pteridophytes (Kholia 2017).

During his more than three decade's experience on Himalayan pteridophytes, authors found that along with the abovementioned exotics, only native fern *Nephrolepis cordifolia* (L.) C.Presl is under cultivation in nurseries in the western Himalayas. However, some wild and native ferns from different parts of the Himalayas and India are under cultivation in botanical gardens and ferneries of Panjab University Chandigarh, Punjabi University Patiala and National Botanical Research Institute Lucknow along with few exotics and cultivars. Further in the last decade, the Ministry of Environment, Forest and Climate Change launched Assistant to Botanic Garden Programme, and with this financial support, ferneries are established at Govt. Post Graduate College Pithoragarh, Uttarakhand, IHBT Palampur, and Botanical Survey of India, Northern Regional Centre, Dehradun. Fairly a large number of native ferns are domesticated, multiplied and researched for cultivation and nursery trade in these centres (Punetha et al. 2007; Kumari et al. 2010). The common ferns under cultivation in these ferneries are *Nephrolepis cordifolia* (L.) C. Presl, *Selaginella adunca* A.Braun ex Hieron., *Selaginella aitchisonii* Hieron., *Selaginella braunii* Baker, *Psilotum nudum* (L.) P.Beauv., *Lygodium japonicum* (Thunb.) Sw., *Hypolepis polypodioides* (Blume) Hook., *Microlepia setosa* (Sm.) Alston, *Microlepia speluncae* (L.) T. Moore, *Adiantum concinnum* Humb. & Bonpl. ex Willd., *Onychium lucidum* (D. Don) Spreng, *Pteris cretica* L., *Pteris ensiformis* Burm.f., *Pteris multifida* Poir., *Pteris vittata* L., *Asplenium dalhousieae* Hook., *Thelypteris arida* (D. Don) C.V.Morton, *Thelypteris dentata* (Forssk.) E.P.St.John, *Thelypteris papilio* (C.Hope) K.Iwats., *Diplazium esculentum* (Retz.) Sw., *Diplazium maximum* (D. Don) C.Chr., *Cyrtomium caryotideum* (Wall. ex Hook. & Grev.) C.Presl, *Dryopteris caroli-hopei* Fraser-

Jenk., *Dryopteris cochleata* (D. Don) C.Chr., *Polystichum discretum* (D.Don) J.Sm., *Polystichum lentum* (D.Don) T.Moore, *Polystichum obliquum* (D.Don) T.Moore, *Polystichum squarrosum* (D.Don) Fée, *Tectaria coadunata* (J.Sm.) C.Chr., *Adiantum capillus-veneris* L., *Adiantum concinnum* Humb. & Bonpl. ex Willd., *Adiantum raddianum* C. Presl (often confused with *Adiantum venustum* D.Don), *Adiantum tenerum* Sw., *Microsorum punctatum* (L.) Copel., *Leptochilus pteropus* (Blume) Fraser-Jenk., etc.

As compared to western Himalayas, a good fern craze is observed in the eastern Himalayan states. In this part, many nurseries are cultivating orchids, rhododendrons, primulas, succulents, palms, cobra lilies and other wild ornamentals including ferns. Many nurseries (owned by Ganesh Mani Pradhan, Udai C. Pradhan, Holumba nursery, Pawan Pradhan nursery, etc. at Kalimpong and K.C. Pradhan nursery Ranipool, Hidden Forest Retreat, Gangtok) and gardens like Himalayan Botanic Garden at Rumtek, Saramsa Garden at Ranipool, Bhanu Park and Millennium Park at Gangtok, etc. have fine collection and stock of ferns. Few of these nurseries are also registered plant traders who export orchids and other ornamentals as per government policies. Along with the exotic bird's-nest ferns, staghorn ferns, Mexican maidenhair ferns, Boston hare's foot fern, spleenworts, *Pteris ensiformis* cv. 'Victoriae', *Pteris cretica* cv. 'Albolineata', etc., these nurseries are also cultivating tree ferns like *Cyathea spinulosa* Wall. ex Hook., *Cyathea gigantea* (Wall. ex Hook.) Holttum, *Cyathea brunoniana* (C.B. Clarke) C.B.Clarke & Baker, *Cyathea sollyana* (Griff.) Fraser-Jenk, *Brainea insignis*, *Cibotium barometz*, *Dryopteris wallichiana* (Spreng.) Hyl., *Diplazium esculentum* (Retz.) Sw., *Diplazium himalayensis* (Ching) Panigrahi, *Diplazium laxifrons* Rosenst., *Diplazium maximum* (D. Don) C.Chr., *Angiopteris evecta* (G.Forst) Hoffm., *Angiopteris helferiana* C.Presl, *Microlepia speluncae* (L.) T.Moore, *Thelypteris arida* (D.Don) C.V.Morton, *Thelypteris dentata* (Forssk.) E.P.St.John, *Thelypteris papilio* (C.Hope) K.Iwats., etc. as ground ferns or landscape species. They have also cultivated many native mesophytes, lithophytes and epiphytes as pot plants like fox tail ferns [*Huperzia squarrosa* (G.Forst.) Trevis., *Huperzia pulcherrima* (Wall. ex Hook. & Grev.) Pic.Serm., *Huperzia hamiltonii* (Spreng.) Trevis]; heart fern [*Mickelopteris cordata* (Roxb. ex Hook. & Grev.) Fraser-Jenk. = *Parahemionitis cordata* (Roxb. ex Hook. & Grev.) Fraser-Jenk.]; club mosses [*Selaginella bisulcata* Spring, *Selaginella braunii* Baker, *Selaginella helferi* Warb., *Selaginella helferi* Warb., *Selaginella inaequalifolia* (Hook. & Grev.) Spring, *Selaginella involvens* (Sw.) Spring, *Selaginella monospora* Spring, *Selaginella picta* A.Braun ex Baker, *Selaginella vogelii* Spring, *Selaginella wallichii* (Hook. & Grev.) Spring]; bamboo leaf fern [*Conigramme fraxinea* (D.Don) Fée ex Diels, *Coniogramme intermedia* Hieron., *Coniogramme procera* Fée, *Coniogramme pubescens* Hieron., *Coniogramme serrulata* (Blume) Fée]; lady fern [*Athyrium clarkei* Bedd., *Athyrium distans* (D.Don) T.Moore, *Athyrium drepanopterum* (Kunze) A.Br. *ex* Milde, *Athyrium falcatum* Bedd.]; wood ferns [*Dryopteris cochleata* (D.Don) C.Chr., *Dryopteris hirtipes* (Kunze) C.Chr., *Dryopteris fructuosa* (Christ) C.Chr., *Dryopteris gamblei* (C.Hope) C.Chr., *Dryopteris juxtaposita* Christ, *Dryopteris scottii* (Bedd.) Ching, *Dryopteris sparsa* (D.Don) Kuntze] Holly ferns [*Polystichum discretum* (D.Don) J.Sm., *Polystichum lentum* (D.Don) T.Moore, *Polystichum obliquum* (D.Don) T.Moore, *Polystichum piceopaleaceum* Tagawa, *Polystichum semifertile* (C.B.Clarke) Ching]; shield ferns [*Cyrtomium hookerianum* (C.Presl) C.Chr., *Cyrtomium falcatum* (L.f.) C.Presl, *Cyrtomium caryotideum* (Wall. ex Hook. & Grev.) C.Presl, *Cyrtomium fortunei* J.Sm.]; *Osmunda japonica* Thunb. (Japanese Royal fern); lacy ground fern (*Dennstaedtia appendiculata* (Wall. *ex* Hook.) J.Sm.); downy ground fern (*Hypolepis polypodioides* (Blume) Hook.); *Platycerium wallichii* Hook. (staghorn fern); *Antrophyum callifolium* Blume (dagger fern); *Pityrogramma calomelanos* (L.) Link (silver fern); *Onychium siliculosum* (Desv.) C.Chr. (golden carrot fern); *Microlepia hallbergii* (d'Almeida) C.Chr. (Hallberg's ground fern); *Microlepia hookeriana* (Wall. ex Hook.) C.Presl (hookers ground ferns); *Microlepia rhomboidea* (Hook.) C.Presl ex Prantl (rhomboid lacy fern); *Microlepia speluncae* (L.) T. Moore (lacy ground ferns); *Pteris cretica* L. (Cretan brake); *Pteris ensiformis* Burm.f. (slender brake fern); *Pteris vittata* L. (Chinese brake fern); *Pteris spinescens* C. Presl; *Pteris semipinnata* L.; *Pteris aspericaulis* Wall. *ex* J. Agardh.; *Drynaria coronans* (Wall. *ex* Mett.) J. Sm. ex T. Moore; *Doryopteris concolor* (Langsd. & Fisch.) Kuhn.; *Doryopteris ludens* (Wall. ex Hook.) J.Sm.; and davalliod ferns (hare foot ferns) and different species of *Bolbitis* spp., *Elaphoglossum* spp., etc. Many of these pot ferns can be also grown as hanging basket plants. But these nurserymen know very few scientific or English names of the cultivated species. Under maidenhair ferns, they have mixed native *Adiantum capillus-veneris* L. with *Adiantum concinnum* Humb. and *Adiantum venustum* D. Don with *Adiantum raddianum* C. Presl. Similarly they are also confused with exotic bird's-nest fern *Asplenium nidus* L. with native cultivated species *Asplenium nidoides* Fraser-Jenk., *Asplenium phyllitidis* D.Don and *Asplenium simonsianum* Hook.

The regional centres of Botanical Survey of India located at Shillong, Gangtok and Itanagar have also good cultivation of tree ferns [*Cyathea brunoniana* (C.B. Clarke) C.B. Clarke & Baker, *Cyathea gigantea* (Wall. ex Hook.) Holttum, *Cyathea henryi* Copel, *Cyathea khasyana* (T. Moore ex Kuhn) Domin, *Cyathea sollyana* (Griff.) Fraser-Jenk, *Cyathea spinulosa* Wall. ex Hook, *Cibotium barometz* (L.) J.Sm., *Brainea insignis* (Hook.) J. Sm.], giant fern (*Angiopteris* spp.) and staghorn ferns [*Platycerium alcicorne* Desv., *Platycerium bifurcatum* (Cav.) C. Chr., *Platycerium wallichii* Hook.] along with few exotics and naturally growing native species in their premises. One of the authors (BSK) also domesticated nearly 150 native species of pteridophytes in Botanical Survey of India experimental garden at Gangtok, Sikkim (Kholia 2017), in hanging baskets and as avenue plants, landscape plants, pot plants, hedge plants and wall plants in Sikkim. He also recommended for their introduction in different nurseries and trade (Kholia 2018). In Sikkim, tree ferns are cultivated as avenue trees in parks, lawns, gardens, along roadside, in front of government offices and private houses. *Nephrolepis cordifolia* (L.) C. Presl is cultivated along the foot paths as hedge plants; it is also cultivated as pot plant too. Generally, the tree ferns, large-sized ferns (*Diplazium* spp., *Microlepia* spp., *Angiopteris* spp., *Thelypteris* spp.), nest ferns (*Athyrium wallichianum* Ching, *Dryopteris atrata* (Kunze) Ching, *Dryopteris scottii* (Bedd.) Ching, *Dryopteris gamblei* (C. Hope) C.Chr., *Dryopteris wallichiana* (Spreng.) Hyl., *Dryopteris zayuensis* Ching & S.K. Wu, *Dryopsis apiciflora* (Wall. ex Mett.) Holttum & P.J. Edwards, *Dryopsis clarkei* (Baker) Holttum & P. J. Edwards, *Dryopsis nidus* (Baker) Holttum & P. J. Edwards, *Matteuccia struthiopteris* (L.) Tod., *Polystichum squarrosum* (D. Don) Fee), etc. are suitable for avenue and landscape ferns. Almost all the medium- to small-sized ground ferns can be easily cultivated as pot plants. The epiphytes are best suited for hanging baskets, but they can be easily cultivated in pots too if potting material and pots are more porous. The ferns successfully grown in hanging basket and pot plant are *Antrophyum callifolium* Blume (dagger fern), *Vittaria amboinensis* Fée (ribbon fern), *Vittaria doniana* Hieron. (ribbon fern), *Vittaria sikkimensis* Kuhn (Sikkim shoestring fern), bird's-nest fern complex [*Asplenium nidus* L., *Asplenium phyllitidis* D. Don, *Asplenium nidoides* Fraser-Jenk, *Asplenium simonsianum* Hook.], *Asplenium normale* D. Don (rainforest spleenwort), *Asplenium pellucidum* Lam. (transparent spleenwort), *Platycerium wallichii* Hook. (staghorn fern), *Nephrolepis cordifolia* (L.) C.Presl (Boston), polypods (polypody), davalliod ferns (hare foot ferns), *Asplenium trichomanes* L. (common or maidenhair spleenwort), tongue ferns [*Elaphoglossum stelligerum* (Wall. *ex* Baker) T. Moore *ex* Salomon, *Elaphoglossum marginatum* (Wall. *ex* Fee) T. Moore *ex* Salomon] and oak ferns [*Drynaria coronans* (Wall. *ex* Mett.) J. Sm. ex T. Moore, *Drynaria propinqua* (Wall. ex Mett.) J. Sm., *Drynaria quercifolia* (L.) J. Sm.].

The trunk of tree ferns is used as substrate for the cultivation of orchids, stag-horn ferns and other epiphytic ornamental plants. These epiphytes are either directly tightened on the poles made from tree fern trunk and on sliced trunk chips or cultivated in pots made from trunk pieces of tree ferns after removing the pith. Chopping of fully grown tree ferns for this purpose is one of the major causes for the decline in natural population of tree ferns. Therefore, for conservation of these majestic and conservation priority plants, which are also included in CITES and different categories of IUCN, an appeal has already been made by the present author along with suggesting the alternative substrate (Kholia 2010b, 2013; Kholia and Joshi 2010) for cultivation of these ornamentals.

### 17.3.7 Other Miscellaneous Uses

Socioeconomically several species of pteridophytes are used in different parts of the Himalayas during the construction of traditional hill houses. In Uttarakhand, the traditional houses are constructed from stone and slates; during the roofing and flooring of these houses the fronds of *Dicranopteris lanigera* (D. Don) Fras.-Jenk., *Dicranopteris linearis* (Burm.f.) Underw., *Pteridium revolutum* (Blume) Nakai and *Hypolepis polypodioides* (Blume) Hook are used as stuffing material. At the time of roofing, wooden planks are laid on the sleepers, and a thick layer of these ferns is placed on it followed by a thick mud layer to fix the slates at the top. In multi-storied mud floor houses, these ferns are also used at the time of flooring; thick fern bed is placed on wooden plank followed by thick mud before pargeting the topmost layer.

These ferns give good insulation and prevent temperature fluctuations. In eastern Himalayan states, these hardy ferns along with *Dicranopteris taiwanensis* Ching & P.S. Chiu, *Diplazium giganteum*, *Dipteris wallichii* (R.Br.) T. Moore, *Diplopterygium giganteum* (Wall. ex Hook.) Nakai, *Histiopteris incisa* (Thunb.) J. Sm. and *Peranema cyatheoides* D. Don and leaves of tree ferns (*Cyathea* spp. and *Cibotium barometz* (L.) J.Sm.) are also used as roofing material of cattle sheds as well as making walls with bamboo and mud. The tree fern trunks are also used as poles and sleepers in the construction of cattle sheds and fencing. In many countries, the climbing fern *Lygodium* spp., forked ferns *Dicranopteris* spp. and *Gleichenia* spp. are used for weaving and making baskets, hats, fish traps, bangles and rings, etc. (Christensen 1997), but there is no such report from India; however, these utilities of ferns may be explored in India.

Certain species of ferns are also thought to be insect repellent and anti-helminthic and therefore used for the controlling of pests during the storage of grains and seeds. The seeds are generally stored in large wooden boxes (*BHAKARS*) or bamboo baskets. The dried fern leaves were kept at the base of the basket before pouring the seeds and also covered at the top before putting the lid. The ferns used for this purpose are *Thelypteris* spp., *Pteridium revolutum* (Blume) Nakai, *Dicranopteris lanigera* (D. Don) Fras.-Jenk, *Gleichenia gigantea* Wall. *ex* Hook. & Bauer, *Pteris* spp., *Thelypteris* spp., *Hypolepis polypodioides*, large species of *Pteris*, etc. In earlier times generally, these species were used as shakeable bed, albeit still used at camping sites, cattle pen, etc. either as shakeable bed or for making indigenous mattress by filling the leaves of these ferns inside jute bags. Besides bedding material, they also keep away the bedbugs, lice and other external parasite that are plentiful in these sites. These species along with *Lycopodium japonicum* Thunb. (Nepali name *NAG-BELI*) are also used for the bedding of domesticated fowl.

In Kumaun, species of *Thelypteris*, *Dicranopteris*, *Pteridium* and *Pteris* are used in paddy cultivation. In paddy cultivation, nursery beds are prepared, and young plants prepared in these beds are transplanted in rice fields. Before sowing the seeds in nursery beds, the seeds are sprouted. For sprouting, soaked seeds are kept in bamboo or *Arundinaria* (*RINGAL*) baskets for a week in a dark place. For sprouting the semidried or dried fern leaves are kept at the bottom and along the inner walls of basket before putting soaked seeds. After putting soaked seeds in basket, fern leaves are again kept at top before putting the lid. It is believed that fern leaves enhance the sprouting and protect the sprouting seeds and its juvenile parts from fungi, bacteria and insects.

Throughout the Himalayas, fronds of *Nephrolepis cordifolia* (L.) C. Presl, *Polystichum squarrosum* (D. Don) Fee, *Polystichum discretum* (D. Don) J.Sm. and *Dryopteris* spp. along with fronds of easily available hard and coriaceous ferns are used as decorative material during festivals and ceremonies and in flower bouquet. Curled up plants of *Selaginella bryopteris* (L.) Baker are often sold in Indian markets; this resurrection plant turns green after dipping in water and is placed as decorative in drawing rooms. Silver ferns are also used in making greeting cards.

**Acknowledgements** For the facilities and encouragements, the senior author is thankful to various directors of Botanical Survey of India (BSI), viz. Dr. M. Sanjappa, Dr. P. Singh and Dr. A. A. Mao, and those in charge of different regional centres of BSI where he worked during this study. The second author (AB) is thankful to Yog Guru Baba Ramdev and his research scientist Dr. Bhaskar Joshi.

## References

Alam MS, Chopra N, Ali M et al (2000) Nonnethyl pentacyclic and lanostane-typetriterpenes from *Adiantum venustum*. Phytochemistry 54(2):215–220

Ambast SP (2000) The useful plants of India. In: Kashyap K, Chand R (eds) NISCOM. CSIR, New Delhi, pp 1–918

Anonymous a (2009) Traditional Herbal Healers of Sikkim. In: State medicinal Plants Board, Forest Env. and Wildlife Management Department. Government of Sikkim, Gangtok

Anonymous b (2010) Studies in the Chemical Constituents of *Selaginella pulvinata* (Hook et Grev) Maxim. Tumour Research

Antony R, Thomas R (2011) A mini review on medicinal properties of the resurrecting plant *Selaginella bryopteris* (Sanjeevani). Int J Pharm Life Sci 2(7):933–939

Aulakh MK, Kaur N, Saggoo MIS (2019) Bioactive Phytoconstituents of Pteridophytes - a review. Indian Fern J J36:37–79

Balasubramanian D (2010) In search of the Sanjeevani plant of Ramayana internet paper via dbala@lvpeiorg (original not seen)

Balkrishna A, Joshi B, Srivastava A et al (2019) Some economic aspects of ferns and fern-allies of Seijosa forest area of Pakke-Kessang District, Arunachal Pradesh. Int J Adv Res Bot 5 (3):14–20. https://doi.org/10.20431/2455-4316.0503003

Begum SN, Ravi Kumar K, Ved DK (2016) Chapter-6: Ethnobotanicals of Western Ghats. In: Pullaiah T, Krishnamurthy KV, Bahadur B (eds) Ethnobotany of India, Western Ghats and West Coast of Peninsular India, vol 2. Apple Academic Press, Canada & USA, pp 107–203

Benniamin A (2011) Medicinal ferns of Nroth Eastern Indian with special reference to Arunachal Pradesh. Indian J Trad Knowl 10(3):516–522

Benniamin A, Manickkam VS (2007) Medicinal Pteridophytes from Western Ghats. Indian J Trad Knowl 6(4):611–618

Caius JF (1935) The medicinal and poisonous ferns of India. J Bombay Nat Hist Soc 38(2):341–361

Chakraverty RK, Pandey DS, Mukhopadhyay DP (2003) Directory of plants in the botanic gardens of India. Flora India 4:1–555

Chander H, Choudhary N, Sharma N (2017) Taxonomic and ethnobotanical notes on some ferns and Fern allies of Hamirpur (H.P.) North-Western Himalaya. J Biol Chem Chron 3(1):28–40

Chandrakant J, Shety AN, Mety S et al (2019) In vivo assessment of antifertility potential of pteridophytic plants: *Selaginella bryopteris* (Sw) L. and *Actiniopteris radiata* (L) Baker in Swiss albino mice. Asian J Pharm Pharmacol 5(1):152–158

Chang HC, Huang GJ, Agrawal DC et al (2007) Antioxidant activities and polyphenol contents of six folk medicinal ferns used as "Gusuibu". Bot Stud 48:397–406

Chetri BK (2019) Ethnobotanical study of South Eastern foothills of Bhutan. Asian Plant Res J 2 (1):1–20

Chettri S, Manivannan S, Muddarsu VR (2018) Nutrient and elemental composition of wild edible ferns of the Himalaya. Am Fern J 108(3):95–106

Chopra RN, Chopra IC, Handa KL et al (1958) Chopra's indigenous drugs of India. U. N. Dhur and Sons, Calcutta

Chopra RN, Chopra IC, Verma BS (1969) Supplement to glossary of Indian medicinal plants. New Delhi, CSIR

Chopra RN, Nayar SL, Chopra IC (1956) Glossary of Indian medicinal plants. Council of Scientific and Industrial Research, New Delhi

Choudhury MD, Das B, Mazumdar PB (2009) Fern flora and fern-allies of southern Assam: ethno-Medicobotanical studies and certain conservation aspects. Scichem Publishing House, Udaipur, pp 1–184

Christensen R (1997) Use of ferns in two indigenous communities in Sarawak, Malaysia. In: Jones RJ (ed) Holttum Memorial Volume. Royal Botanic Gardens Kew, London, pp 177–192

Chunekar KC, Pandey GS (1969) Bhavparkash Nighantu (in Hindi). Varanasi, The Chowkhamba Vidya Bhawan

Das S, Dutta Choudhury M (2012) Ethnomedicinal uses of pteridophytes of West Tripura District, India. In: Talukdar AD, Choudhury S (eds) Researches in medicinal and aromatic plants, by Dutta Choudhury M, Sharma GD. Swastik Publication, Delhi

Dixit RD, Singh S (2004) Medicinal pteridophytes-an overview. In: Trivedi PC (ed) Medicinal plants utilization and conservation. Aavishkar Publication, Jaipur, pp 268–297

Dorji R, Thomas MB (2017) Pteridophytes of Bhutan. Bhutan Biodiversity Portal Accessed 15 Sep 2019

Dubal K, Kale M (2014) GC-MS profile of *Lygodium flexuosum* L. (Lygodiaceae): a medicinal fern from North Western Ghats. Indian Fern J 31:112–117

Dutta Choudhury M, Choudhury S (2002) Ethnomedico-botanical aspects of Reang tribe of Assam, part- II, New ethnomedicinal plants. In: Bhattacharya MK (ed) Biodiversity of Assam and its conservation. Today & Tomorrow's Printers & Publishers, New Delhi

Dutta Choudhury M, Das B, Mazumder PB (2009) Fern Flora and Fern allies of southern Assam-ethno-medico botanical studies and certain conservation aspects. Scichem Publishing House, Udaipur, India

Dvorakova M, Pumprova K, Antonínová Ž et al (2021) Nutritional and antioxidant potential of fiddleheads from European ferns. Foods 10(460):1–16. https://doi.org/10.3390/foods10_2021_460

Dymock W (1890) Pharmacographia Indica, A history of the principal drugs of vegetable origin met within British India, vol 1. Kegan Paul, London

Francesconi K, Visoottiviseth P, Sridokchan W et al (2002) Arsenic species in an arsenic hyperaccumulating fern, *Pityrogramma calomelanos:* a potential phytoremediator of arsenic-contaminated soils. Sci Tot Environ 284:27–35

Fraser-Jenkins CR (2008) Taxonomic revision of three hundred Indian subcontinental Pteridophytes with a revised census list, a new picture of fern-taxonomy and nomenclature in the Indian subcontinent. Bishen Singh Mahendra Pal Singh, Dehra Dun, pp 1–679

Fraser-Jenkins CR (2013) Sanjeevani cannot be *Selaginella bryopteris*. Indian Fern J J29:183–195

Gairola S, Sharma J, Bedi YS (2014) A cross-cultural analysis of Jammu, Kashmir and Ladakh (India) medicinal plant use. J Ethnopharmacol 155:925–986

Ganeshaia KN, Vasudeva R, Uma Shanker R (2009) In search of Sanjeevani. Curr Sci 97 (4):484–489

Gaur RD, Bhatt BP (1994) Folk utilisation of some Pteridophytes of Deoprayag area in Garhwal Himalaya: India. Econ Bot 48(2):146–151

Gayathri V, Asha VV, Subamonium A (2005) Preliminary studies on the immunomodulatory and antioxidant properties of *Selaginella* specie. Indian J Phannacol 37(6):381–385

Giri S (2004) The vital link Monpas and their forests. The Center of Bhutanese Studies, Thimphu Bhutan

Gogoi R, Das MK (2002) Ethnobotanical studies of some ferns used by the Garo tribal of Meghalaya. Adv Pl Sci 15:401–405

Goswami HK, Sen K, Mukhopadhyay R (2016) Pteridophytes: evolutionary boon as medicinal plants. Plant Genet Res Char Util 14(4):328–355

Gupta SM, Ballabh B, Yadav PK et al (2020) Nutrients analysis of *Diplazium esculentum*: underutilized wild wetland Pteridophytes ensure food and nutritional security. Acta Scientific Nutr Health 4(11):46–49

Gupta RS, Kumar P, Sharma A et al (2000) Hypocholesterolemic activity of *Marsilea minuta* in gerbils. Fitoterapia 71(2):113–117

Hidano A, Sharma B, Rinzin K et al (2017) Revisiting an old disease? Risk factors for bovine enzootic haematuria in the Kingdom of Bhutan. Prev Vet Med 1(140):10–18

Hirasawa Y, Kobayashi J, Morita H (2006) Lycoperine a: a novel C27N3-type pentacyclic alkaloid from *Lycopodium hamiltonii,* inhibiting acetylcholinesterase. Organ Lett Jan 8(1):123–126

Jain SK (1991) Dictionary of Indian folk medicine and Ethanobotany. Deep Publications, New Delhi

Jain SK, DeFilipps RA (1991) Medicinal plants of India, vol I &II. Reference Publications, USA

Kale M (2015) Analysis of bioactive components of *Dicranopteris linearis* (Burm.F.) underwood, whole plant ethanol extract by GC-MS. Bulletin of Pure and Applied Sciences Botany 34 (b):9–12

Kalidas C, Muthukumar K, Mohan VR et al (2009) Ethnoveterinary medicinal uses of plants from Agasthiamalai biosphere reserve (KMTR), Tirunelveli District, Tamil Nadu, India. My Forest 45(1):7–14

Kandel DR (2020) *Nepal Ka NiuroSagharu* (in Nepali) (=vegetable Fern of Nepal). National Herbarium and botanical laboratory, Godavari, Lalitpur, Govt. of Nepal, p 1–16

Khan M, Kumar S, Hamal IA, Koul S (2009) Wild edible plants of Sewa catchment area in northwest Himalaya. J Plant Dev Sci 1, 2:1–7

Kholia BS (2008) *Selaginella bryopteris* (L) Bak. An endemic fern-ally of India under threat. Indian Fern J 25:(1–2):73-78

Kholia BS (2009) A note on mistaken identity of 'Sanjeevani Booti'. Indian Fern J 26:199–200

Kholia BS (2010a) Ferns and Fern-allies of Sikkim: a pictorial handbook Part-1, Gangtok: Sikkim state biodiversity board and botanical survey of India

Kholia BS (2010b) Sikkim Himalayan Pteridophytes - traditional uses, economic potentials, threats and conservation. In: International symposium on ferns and fern -allies: diversity, Bioperspectation and conservation on 10–12 November 2010. Institute of Himalayan Bioresources and Technology (CSIR), Palampur HP, p 71

Kholia BS (2013) Sikkim Himalayan tree ferns: diversity, distribution, uses, economic potentials and conservation. In: Singh RA (ed) Geology, biodiversity & natural resources of Himalaya. Gyanodaya Publication, Nainital, pp 274–287

Kholia BS (2014) Traditional uses of pteridophytes in Sikkim Himalaya. In: Kumar S (ed) Ethanobotanical studies in India. Deep Publications, Delhi, pp 291–302

Kholia BS (2017) Pteridophytic wealth in botanical survey of India, experimental garden Gangtok-Sikkim; with a note on domestication of Himalayan ferns. In: Singh P, Das SS Indian botanic gardens role in conservation Botanical Survey of India, Dehradun 261–280

Kholia BS (2018) Domestication of Himalayan Pteridophytes: A plea for Horticulture and ex-situ Conservation. Abstracts: International Conference On Climate Change, Biodiversity and Sustainable Agriculture (ICCBSA-2018) Jorhat, Assam; 13–16 December 2018, pp: 135–136

Kholia BS, Fraser Jenkins CR (2011) Misidentification makes scientific publications worthless - save our taxonomy and taxonomists. Curr Sci 100(4):458–461

Kholia BS, Joshi R (2010) International year of biodiversity - adopting innovative measures for conservation. NeBIO 1(2):55–61

Kholia BS, Punetha N (2005a) Useful pteridophytes of Kumaon (central Himalaya, India). Indian Fern J 22:1–6

Kholia BS, Punetha N (2005b) Notes on distribution and conservation of the rare *Selaginella pulvinata* (Hook.& Grev.) Maxim, India. Indian J Forestry 28(4):389–391

Kirn HS, Kapahi BK (2001) Ethnobotanical notes on some ferns and fern-allies of Jammu and Kashmir State, India. Indian Fern J 18:35–38

Kirtikar KR, Basu BD (1935) Indian medicinal plants, vol 1–4, 2nd edn. Lalit Mohan Basu, Allahabad

Kumar N (2014) Common medicinal plants available at Joginder Nagar, District Mandi (H.P.), India. Res Pharm 4(4):16–24

Kumar M, Ramesh M (2003) Medicinal pteridophytes of Kerala, South India. Indian Fern J 20:1–28

Kumari P, Otaghvari AM, Govindpyari H et al (2011) Some ethnomedicinally important Pteridophytes of India. Int J Med Arom Plants 1(1):18–22

Kumari A, Prakash O, Kumar A, Uniyal SK et al (2010) Ferns and Fern-allies of IHBT Campus Palampur, Himanchal Pradesh. Indian Fern J 27(1–2):53–64

Kunwar RM, Mahat L, Acharya RP et al (2013) Medicinal plants, traditional medicine, markets and management in far-West Nepal. J Ethnobiol Ethnomed 9:1–10. http://www.ethnobiomed.com/content/9/1/24. Accessed 24 Feb 2021

Leach H (2003) Fern consumption in Aotearoa and its oceanic precedents. J Polyn Soc 112 (2):141–156

Leterme P, Londono AM, Munoz JE et al (2009) Nutritional value of aquatic ferns (*Azolla filiculoides* Lam. and *Salviniamolesta* Mitchell) in pigs. Anim Feed Sci Technol 149:135–148

Lin LC, Kuo YC, Chou CJ (2000) Cytotoxic bioflavonoids from *Selaginella deliculatra*. J Nat Prod 63(5):627–630

Lin YL, Shen CC, Huang YJ et al (2005) Homonaflavonoids from *Ophioglossum petiolatum*. J Nat Prod 68(3):381–384

Luitel DR, Pathak M (2013) Documentation of medicinal and aromatic plants of Dhorpatan hunting reserve, Western Nepal. Plant Res J Dept Pl Res Nepal 35:36–43

Ma LQ, Komar KM, Tu C et al (2001) A fern that hyperaccumulates arsenic: a hardy, versatile, fast-growing plant helps to remove arsenic from contaminated soils. Nature 409:579–579

Mandal A, Mondal AK (2011) Studies on antimicrobial activities of some selected ferns and lycophytes in eastern India with special emphasis on ethno-medicinal uses. African J Plant Sci 5(7):412–420. http://www.academicjournals.org/ajps

Mannan MM, Maridass M, Victor B (2008) A review on the potential uses of ferns. Ethnobotanical Leaflets 12:281–285

Maridass M, Ghanthikumar S (2008) Antibacterial activity of leaves of *Blechnum orientale* L. Pharmacologyoline 3:58–60

Matsuda H, Yamazaki M, Namo S et al (2002) Anti-androgenic and hair growth promoting activities of Lygodii spora (spore of *Lygodium japonicum*) I. active constituents inhibiting testosterone 5alphareductase. Biol Pharm Bull 25(5):622–626

Matsushima K, Minami M, Nemoto K (2012) Use and conservation of edible wild plants in Bhutan. J Faculty Agr Shinshu Univ 48(1–2):75–81

Matsushima K, Minami M, Nemoto K, et al (2008) Edible wild plants of Bhutan and their associated traditional knowledge. Council for RNR research in Bhutan, Thimphu: Ministry of Agriculture, Royal Government of Bhutan p 1–130

Meharg AA (2003) Variation in arsenic accumulation - hyperaccumulation in ferns and their allies. New Phytol 157:25–31

Mir GM, John SA (2014) Ethno-medicinal study of Pulwama tehsil (Jammu and Kashmir). J Med Plants Stud 2(4):5–8

Mir SA, Mishra AK, Pala SA et al (2015) Ferns and fern allies of district Shopian, Kashmir Valley, India. Biodiversitas 16(1):27–43

Myers N, Mittermeier RA, Mittermeier CG et al (2000) Biodiversity hotspots for conservation priorities. Nature 403(6772):853–858

Nadkarni KM (1954) Indian Materia Medica, vol I. Popular Book Depot, Bombay

Nath K, Bhattacharya MK, Kar S (2016) Antibacterial activity of some ethno-botanically important ferns of southern Assam, India. Taiwania 61(3):260–268

Nayar BK (1959) Medicinal ferns of India. Bull Nat Bot Gardn (Lucknow) India 29:1–36

Olsen S (2007) Encyclopedia of garden ferns. Timber Press, Portland, OR

Pandey AK, Rout SD (2006) Ethnobotanical uses of plants by tribals of Similipal biosphere reserve, Orissa. Ethnobotany 18:102–106

Parajuli RR (2013) Indigenous knowledge on medicinal plants: Maipokhari, Maimajhuwa and Mabu VDCs of Ilam District, Eastern Nepal. Int Plant Res J Dept Pl Res Nepal 35:50–58

Parihar P, Bohra A (2002) Antifungal efficacy of various Pteridophytic plant parts: a study in vitro. Adv Plant Sci 15(1):35–38

Parihar P, Bohra A (2003) Effect of some Pteridophytic plant extracts on human pathogenic bacteria *salmonella typhi*. Indian Fern J 20:39–41

Parihar P, Parihar L (2006) Some pteridophytes of medicinal importance from Rajasthan. Nat Prod Radiance 5:297–301

Pathania S, Kumar P, Singh S et al (2012) Detection of ptaquiloside and quercetin in certain Indian ferns. Curr Sci 102(12):1683–1691

Punetha N, Punetha R, Bhakuni K (2007) Wealth of ferns and fern-allies in the botanical garden at Pithoragarh (Central Himalayas). Indian Fern J 24:54–59

Rajagopal PK (2019) Potentials and prospects of ferns and lycophytes of Western Ghats (presidential address delivered at the international conference on pteridophytes, held at Dr. Bhupendra Nath Dutta Smriti Mahavidyalaya, Hatgobindapur, Purba-Barddhaman, West Bengal). Indian Fern J 36:8–17

Rana MS, Samant SS (2011) Diversity, indigenous uses and conservation status of medicinal plants in Manali wildlife sanctuary, north western Himalayas. Ind J Trad Knowl 10(3):439–459

Rao KT, Reddy KN, Pattanaik C et al (2007) Ethnomedicinal importance of Pteridophytes used by Chenchus of Nallamalais, Andhra Pradesh, India. Ethnobotanical Leaflets 11:6–10

Sah P (2008) Does the magical herb "Sanjevani Booti" really exist in nature? J Am Sci 3:65–67

Sah NK, Singh SMP, Sahdev S et al (2005) Indian herb 'Sanjeevani' (*Selaginella bryopteris*) can promote growth and protect against heat shock and apoptotic activities of ultra violet and oxidative stress. J Biosci 30:499–505

Samant SS, Dhar U, Palni LMS (2001) Himalayan medicinal plants: potential and prospects. Gyanodaya Prakashan, Nainital (India)

Sen A, Bhattacharya MK (2007) Further contribution to the fern flora of Barak Valley, Assam, N. E. India. Indian Fern J 24:156–160

Sen A, Ghosh PD (2011) A note on the ethnobotanical studies of some Pteridophytes in Assam. Indian J Tradit Know 10(2):292–295

Shah R (2015) Edible plants of north west Himalaya (Uttarakhand). Uttarakhand Biodiversity Board and Bishen Singh Mahendra Pal Singh, Dehradun

Sharma NK (2002) Ethnomedicinal studies on ferns and fern-allies of Hadoti plateau, Southern Rajasthan. Zoos Prin J 17(3):732–734

Shil S, Dutta Choudhury M (2009) Ethnomedicinal importance of Pteridophytes used by Reang tribe of Tripura, North East India. Ethnobotanical Leaflets 13:634–643

Siman SE, Povery AC, Ward TH, Margison G, Sheffield E (2000) Fern spore extracts can damage DNA. Br J Cancer 83(1):69–73

Singh HB (1999) Potential medicinal Pteridophytes of India and their chemical constituents. J Econ Tax Bot 23(1):63–78

Singh HB (2002) Economically valuable Pteridophytes of India. In: Chandra S, Srivastava M (eds) Pteridology in New Millennium NBRI Golden Jubilee. Kluwer Academic Publisher, Dordrecht, pp 421–426

Singh AV (2018) Ethnobotany of ferns of Lakhimpur, Assam, India. Int Res J Biol Sci 7(7):27–34

Singh BH, Prasad P, Rai LK (2002) Folk medicinal plants in the Sikkim Himalayas of India. Asian Folklore Studies 61:295–310

Singh KK, Sah S, Maheshwari JK (1989) Uses of Pteridophytes by some tribals of Uttar Pradesh. Indian Fern J 6:63–67

Singh S, Singh R (2012) Ethnomedicinal use of Pteridophytes in reproductive health of tribal women of Pachmari Biosphere Reserve, Madhya Pradesh, India. Int J Pharm Sci Res 3:4780–4770

Singh LS, Singh KS, Singh EJ (2001) Ethnobotanical uses of some pteridophytic species in Manipur. Indian Fern J 18(1–2):14–17

Singh BP, Upadhyay R (2012) Ethno-botanical importance of Pteridophytes used by the tribe of Pachmarhi, Central India. J Med Plants Res 6(1):14–18

Somvanshi R, Kataria M (2002) Characterisation of long-term feeding effects of *Dryopteris juxtaposita* fern in hill cattle. Indian Fern J 19:40–54

Somvanshi R, Lauren DR, Smith BL et al (2006) Estimation of the fern toxin, ptaquiloside, in certain Indian ferns other than bracken. Curr Sci 91:1547–1552

Srivastava K (2007) Ethnobotanical studies of some important ferns. Ethnobotanical Leaflets 11: 164–172

Srivastava M, Ma LQ, Cotruvo JA (2005) Uptake and distribution of selenium in different fern species. Int J Phytoremed 7(1):33–42

Srivastava M, Ma LQ, Singh N et al (2005) Antioxidant responses of hyper-accumulator and sensitive fern species to arsenic. J Exp Bot 56(415):1335–1342

Tong XT, Tan CH, Ma XQ et al (2003) Miyoshianines A and B, two new lycopodium alkaloids from *Huperzia miyoshiana*. Planta Med 69(6):576–579

Toyota M, Oiso Y, Asakawa Y (2002) New glycosides from the Japanese fern *Hymenophyllum barbatum*. Chem Pharmaceut Bull (Tokyo) 50(4):508–514

Tsuzuki K, Ohashi A, Arai Y et al (2001) Triterpenoids from *Adiantum caudatum*. Phytochemistry 58(2):363–367

Tu S, Ma LQ (2003) Interactive effects of pH, arsenic & phosphorus on uptake of As and P and growth of the arsenic hyperaccumulator *Pteris vittata* L. under hydroponic conditions. Environ Exp Bot 50:243–251

Tu S, Ma LQ (2004) Comparison of arsenic uptake and distribution in arsenic hyperaccumulator *Pteris vittata* and non-hyperaccumulator *Nephrolepis exaltata*. J Plant Nutr 27:1227–1242

Ugochukwu ICI (2019) Bracken fern toxicity and its associated clinicopathological effects in humans and animals: a review. Comp Clin Pathol 28:593–597. https://doi.org/10.1007/s00580-018-2636-2

Upreti K, Jalal JS, Tiwari LM et al (2009) Ethnomedicinal uses of Pteridophytes of Kumaun Himalaya, Uttarakhand, India. J Am Sci 5(4):167–170

Uprety Y, Poudel RC, Shrestha KK et al (2012) Diversity of use and local knowledge of wild edible plant resources in Nepal. J Ethnobiol Ethnomed 8:1–15. http://www.ethnobiomed.com/content/8/1/16

Wang J, Zhao FJ, Meharg AA et al (2002) Mechanisms of arsenic Hyperaccumulation in *Pteris vittata*. Uptake kinetics, interactions with phosphate, and arsenic speciation. Plant Physiol 130: 1552–1561

Wangchuk P, Namgay K, Gayleg K et al (2016) Medicinal plants of Dagala region in Bhutan: their diversity, distribution, uses and economic potential. J Ethnobiol Ethnomed 12:1–28. https://doi.org/10.1186/s13002-016-0098-7

Wangchuk P, Wangchuk D, Hansen JA (2007) Bhutanese traditional medicine (gso-Ba-rig-pa): an integrated part of the formal health care services. South East Asian J Trop Med Publ Health 38 (1):161–167

Wangchuk P, Yeshi K, JamphelK (2017) Pharmacological, ethnopharmacological, and botanical evaluation of subtropical medicinal plants of lower Kheng region in Bhutan. Integrative Med Res 6:372–387

Wani MH, Shah MY, Naqshi AR (2016) Medicinal ferns of Kashmir, India. Int J Bioassays 5:4677–4685

Watt G (1889–1892) Dictionary of economic products of India, parts I–IV. Superintendent of Government Printing, India, Calcutta

Wealth of India (1948–1969) Parts I-VIII. Council of Scientific and Industrial Research, India, New Delhi

Wikipedia (2020) List of scheduled tribes in India. https://en.wikipedia.org/wiki. Accessed 1 Sep 2020

Wu MJ, Weng CY, Wang L et al (2005) Immunomodulatory mechanism of the aqueous extract of sword brake fern (*Pteris ensiformis* Burm.). J Etlmopharmacol 98(1–2):73–81

Xu KP, Xu S, Deng YH et al (2003) A new Bioflavone from *Selaginella pulvinata* Maxim. Chinese J Organic Chem 2:395–396

Yumkham SD, Singh PK (2011) Less known fern and fern-allies of with ethnobotanic uses. Indian J Tradit Know 10(2):287–291

Zangara A (2003) The psychopharmacology of huperzine a: an alkaloid with cognitive enhancing and neuroprotective properties of interest in the treatment of Alzheimer's disease. Pharmacol Biochem Behav 75(3):675–686

Zhang Z, El Sohly HN, Jacob MR, Pasco DS, Walker LA, Clark AM (2002) Natural products inhibiting Candida albicans secreted aspartic proteases from *Lycopodium cernuum*. J Nat Prod 65(7):979–985

Zhao FJ, Dunham SJ, McGrath SP (2002) Arsenic hyper accumulation by different fern species. New Phytol 156:27–31

**B. S. Kholia** is a trained pteridologists working in Botanical Survey of India, a primer institute of taxonomy and plant research. He surveyed all the phytogeographical regions of India including Indian Islands, collected specimens and gathered information on Indian pteridophytes. He has experience of 35 years of research in the field of pteridology and published 5 authored books and more than 100 research papers on taxonomy, ethnobotany and conservation of pteridophytes.

**Acharya Balkrishna** is a saint of the Indian spiritual tradition and a scholar of Ayurveda, Sanskrit language and ancient Indian philosophy, literature, Upaniṣads, yoga, etc. He is the founder Secretary/MD/Chairman of Patanjali Trust, Patanjali Research, Patanjali Bio Research Institute, Divya Pharmacy, Patanjali Ayurved Ltd., etc. leading organisations in India working for revival of indigenous knowledge for complete well-being of human. He is also vice chancellor of Patanjali University. Acharya Balkrishna is an Aayurvedic scholar, herbal practitioner and yoga teacher; under his able guidance *Aayurvedic* treatments and research are being carried out at the Patanjali Aayurvedic Hospitals, Patanjali Yogpeeth and Patanjali University. He has also scored 29 patents on Yoga & Ayurveda and published 41 research papers and many books on Yoga, *Ayurveda* and medicinal plants.

# Ethnobotanical Uses of Ferns and Lycophytes of Kerala

# 18

Raju Antony and S. Suresh

**Abstract**

The present study focusses on the ethnobotanical uses of ferns and lycophytes that are traditionally used by the tribal and local people of Kerala. Based on the field study, it is found that a total of 41species of ferns and lycophytes are used for various purposes. This is 14.6% of the pteridophyte species recorded so far from Kerala. Rhizomes are the major officinal parts used. Out of this 22 species are used as medicinal, 19 as edibles and 12 have miscellaneous uses. Even though there are 34 tribal groups in Kerala, only 14 tribal groups are found to depend on ferns and lycophytes. *Kani* tribe among them used the highest number of species followed by Malapandaram and Muthuvan. The quality and quantity of information obtained in this short study reveal the scope of conducting more ethnopteridological investigations in Kerala State.

**Keywords**

Ethnobotany · Ferns · Lycophytes · Kerala

## 18.1 Introduction

Since time immemorial, man has been using plants as a source of food, medicine and many other necessities of life. The term ethnobotany is precisely used to cover all studies which describe the use of plants by human beings. Ethnobotanical investigations conducted all over the world have significantly helped the humanity for developing life-saving drugs (Schultes 1993; Mendelsohn and Balick 1995), understanding utilization patterns of plant resources (Phillips et al. 1994) and

R. Antony (✉) · S. Suresh
Jawaharlal Nehru Tropical Botanic Garden and Research Institute (JNTBGRI), Thiruvananthapuram, Kerala, India





J. Marimuthu et al. (eds.), *Ferns*, https://doi.org/10.1007/978-981-16-6170-9_18

searching out promising new economic plants and land races (Arora 1997). Thus, ethnobotanical surveys have gained remarkable significance all the time.

Ferns and lycophytes (fern allies), together called pteridophytes, are non-flowering, vascular and spore-bearing plants. They are originated in Silurian period and were a dominant group during the Carboniferous period. They are known to human beings for more than 2000 years for their medicinal and other values. The discipline that studies the relationship between the uses of ferns and humans is termed ethnopteridology. Presently there are more than 12,000 species of pteridophytes reported in the world (Chapman 2009), of which many are used by different people as medicine, food, ornamental, etc. Keller and Prance (2015) commented that ethnobotany of pteridophytes is not a new topic and there are enough records about the use of pteridophytes by local cultures in ancient literature. Georgius Everhardus Rumphius (1627–1702) provided ethnobotanical information about uses of pteridophytes by the Ambonese people and natives of other islands of Indonesia in the seventeenth century (Rumphius 2011). Van Rheede (1678–1703) recorded the ethnomedicinal uses of pteridophytes by the traditional healers of Malabar. The studies came out on ethnopteridology of human groups, and individual species of pteridophytes from Ecuador, Peru and Bolivia (Boom 1985; Navarrete et al. 2006), Honduras (Hernandez and Sutherland 2007), Argentina (Keller et al. 2011), Malaysia (Christensen 1997), Africa and Nigeria (Nwosu 2002), China (Chang et al. 2010; Liu et al. 2012), Brazil (Prudente 2007), etc. show that pteridophytes have wider use among people than it was believed earlier. A global survey of pteridophytes used by folks for medicinal purposes revealed that most of them are used to cure common diseases because of their purgative and antibacterial properties (Ho et al. 2010).

As for India, more than 1200 species of pteridophytes (Subhash Chandra et al. 2008; Fraser-Jenkins 2012) have been reported so far. Many among them are being used as food and medicine. A study of the ethnobotanical uses of pteridophytes occurring in Jammu and Kashmir listed 17 species, of which 11 were medicinal, 4 were thatching material for roofs and 3 were foods (Kirn and Kapahi 2001). Joshi (1997) studied ethnobotanical uses of 44 species of pteridophytes in the state of Uttar Pradesh. Pandey and Pangtey (1987) listed seven species of pteridophytes consumed by the people of Western Himalaya.

*Hortus Malabaricus* by Van Rheede (1678–1703) could be the earliest record which refers ethnomedicinal uses of pteridophytes of Kerala State. He mentioned 18 pteridophytes used by the traditional healers of Malabar region (Table 18.1). Mathew et al. (1996, 1999), Kumar et al. (1999) and Manilal et al. (2000) have reported the uses of pteridophytes in the traditional system of treatments of tribals in Kerala. Binu (1999) reported six pteridophytes used by the tribes in Pathanamthitta District.

**Table 18.1** The list of pteridophytes described in *Hortus Malabaricus*

| Roman script | Malayalam script | Scientific name | Family |
|---|---|---|---|
| Patitisjivi-maravara | Pattichevi | *Parahemionitis cordata* | Pteridaceae |
| Panna-Kelango-Maravara Welli-panna-kelengu Maravara | Pannakizhengu | *Drynaria quercifolia* | Polypodiaceae |
| Welli-Panna-Kelangu-Marvara | Velipanna | *Phymatosorus mebranifolius* | Polypodiaceae |
| Tama-Pouel-Paatsja-Maravara | Tamapalpacha | *Huperzia phlegmaria* | Lycopodiaceae |
| Para-panna | Parapanna | *Diplazium esculentum* | Athyriaceae |
| Kal-Panna-Maravara | Kalpanna | *Cheilanthes tenuifolia* | Pteridaceae |
| Kari-Welli-Panna Paravara | Karivelipanna | *Christella parasitica* | Thelypteridaceae |
| Nella-Panna-Maravara | Nellapanna | *Asplenium decrescens* | Aspleniaceae |
| Pannamara-Maravara | Panna | *Bolbitis terminans* | Lomariopsidaceae |
| Maretta-Mala-Maravara | Marathemala | *Pyrrosia heterophylla* | Polypodiaceae |
| Valli-Varakody-Maravara | Vallivathakody | *Leptochilus bahupunctika* | Polypodiaceae |
| Arana-panna | Aranapanna | *Nephrolepis brownii* | Nephrolepidaceae |
| Valli-Panna | Vallipanna | *Lygodium flexuosum* | Lygodiaceae |
| Tsjeru-valli-panna | Vallipanna | *Lygodium microphyllum* | Lygodiaceae |
| Panna-Valli | Pannavalli | *Stenochlaena palustris* | Blechnaceae |
| Avenka | Avenka | *Adiantum philippense* | Pteridaceae |
| Bellan-patsja | | *Lycopodiella cernua* | Lycopodiaceae |
| Tiri-panna | Tiripanna | *Pyrrosia lanceolata* | Polypodiaceae |

## 18.2 Study Area and Methods

Kerala is the southernmost state of India. It lies along the extreme south-western coast of India excluding the peninsular tip of Kanyakumari District of Tamil Nadu. Lying between northern latitudes 8°18′ and 12°48′ and eastern longitudes 74°52′ and 77°22′, the state covers an area of 38,855 sq. km. The greatest length is 550 km, and breadth is 121 km in near about the middle from where it diminishes gradually towards the north and the south. It is bounded on the north and the north-east by the State of Karnataka, the east and the south by Tamil Nadu State and the west by Arabian Sea. It enjoys a wet and maritime tropical climate influenced by seasonal heavy rains of the southwest summer monsoon and northeast winter monsoon. The humid tropical climate and luxuriant forests support over 250 species of pteridophytes (Manickam and Irudayaraj 1992; Nair et al. 1988, 1992a, b; 1994; Madhusoodanan 2015). There are 34 tribal groups in Kerala (Fig. 18.1).

The information presented here are gathered during the plant explorations throughout the forests of Kerala to collect pteridophytes for establishing an ex situ

**Fig. 18.1** Images tribes of Kerala. (**a**) Cholanaikkan; (**b**) Kani; (**c**) Malappandarum; (**d**) Mannan; (**e**) Malayan; (**f**) Muthuvan females; (**g**) Muthuvan male; (**h**) Ulladan

**Table 18.2** The list of tribes from which the information was collected

| Name of tribes | General information |
|---|---|
| *Adiyan* | A matrilineal tribal group seen in the states of Kerala and Karnataka. In Kerala, they live in the districts of Kannur and Wayanad districts. They were treated as bonded slave labourers by the landlords earlier |
| *Cholanaikkan* | One of the last remaining hunter-gatherer tribes, living in the forests of Malappuram District of Kerala State. They speak the Cholanaikkan language, which belongs to the Dravidian family |
| *Irular* | A Dravidian ethnic group inhabiting the area of the Nilgiri Mountains, in the states of Tamil Nadu and Kerala. Traditionally, their main occupation is snake and rat catching and honey collection |
| *Kani* | A tribe traditionally with nomadic communities that have settled in the forests of the Agasthyamalai Hills of the Western Ghats' part of Kerala State |
| *Kattunaikan* | One of the earliest known inhabitants of the Western Ghats. They are engaged in the collection of forest produce, mainly wild honey and wax. They are the tribal group who had been the true inhabitants of forest in Wayanad District in Kerala State |
| *Malappandarum* | A tribal group settled along the Pamba river, Achankovil river in Kollam and Pathanamthitta districts. Nomadic gathering and hunting in the forest have traditionally provided the basis of their living |
| *Malavedar* | One of the tribal groups in Kerala. Many of them live in the Ernakulam, Kollam, Kottayam, Idukki, Pathanamthitta and Thiruvananthapuram districts |
| *Malayan* | A tribal community mostly found in Edamalayar of Kuttampuzha panchayat in Idukki District. They are adapted to occupations such as bamboo cutting, fishing, manual labour and collection of forest produce |
| *Mannan* | One of the tribal groups who live in Idukki District. They have a hereditary king who leads them. They are said to be the warrior tribe of Pandyan king |
| *Muthuvan* | They live on the border hill forests of Kerala and Tamil Nadu. In Kerala, they are found in the Adimali and Devikulam forest regions of Idukki District. They are very independent and reluctant to interact with the outside world and cultivate ragi and lemon grass for their living |
| *Paliya* | Tribes living in the montane rain forests of the Southern Western Ghats of Tamil Nadu and Kerala states. They are traditional nomadic hunter-gatherers, honey hunters and foragers |
| *Paniya* | They are mainly found in Kerala and Karnataka states. In Kerala they are mainly inhabit in the villages of Wayanad, Kozhikode, Kannur and Malappuram districts of Kerala. They speak Paniya language, closely related to Malayalam |
| *Pathinaikkan* | They are the tribes residing near the streams in Kerala part of Nilgiri biosphere reserve. They are mainly hunter-gatherers. Both men and women are engaged in fishing and collection of forest produces |
| *Ulladan* | They are the tribes seen in Idukki, Kottayam, Pathanamthitta and Quilon districts of Kerala State. Most of them are agricultural workers and forest gatherers. They are hunters and gatherers of root vegetables |

conservatory at Jawaharlal Nehru Tropical Botanic Garden and Research Institute (JNTBGRI). Table 18.2 provides the brief details of the tribal groups from where the information were obtained. Apart from the tribal people, data from the local people in villages of Kerala were also gathered. Information were collected from persons accompanied for the plant collection team as guides. Plants were identified, and the nomenclature was updated consulting The Plant List (2013: version 1.1) and International Plant Names Index (IPNI). Herbarium specimens of species were prepared following Forman and Bridson (1991). The voucher specimens were deposited in Jawaharlal Nehru Tropical Botanic Garden and Research Institute Herbarium (TBGT). The data are presented in the alphabetical order of scientific names of species. Latin name, family, common name, Malayalam name, brief description of the species and uses are provided for medicinal ferns.

## 18.3 Results

***Acrostichum aureum* L.,** Pteridaceae (Fig. 18.2a).

Common name, golden leather fern; Mal. name, *Kandalpanna.*

A large terrestrial fern commonly found near brackish water and mangroves. They have tufted, simply pinnate and leathery leaves and erect rhizomes. The lower surface of the terminal part of leaves is fully covered with sori.

**Uses**: A paste made from the rhizome is applied to heal wounds by local people of Ernakulam District.

***Actiniopteris radiata*** (Sw.) Link, Pteridaceae (Fig. 18.2b).

Common name, peacock fern; Mal. name, *Mayurishikha.*

A rare species found growing on dry rocks in exposed areas. Herbaceous with erect rhizomes. Leaves are semi-circular or wedge shaped with pale green colour. Sporangia are borne within the marginal grooves of leaves.

**Uses**: The *Irular* tribes in Palakkad District use leaf juice of this species against dysentery in children. The juice is mixed with breast milk and given internally for 3 days.

***Adiantum caudatum*** L., Pteridaceae (Fig. 18.2c).

Common name, walking fern; Mal. name, *Sancharapannal.*

This is a common species found on earth banks, road cuttings and rock crevices. They are herbs having erect rhizomes with tufted stipes. Leaves are simply pinnate and pale green and are hairy on both surfaces. The basal pairs of leaves are slightly reduced. Sori are linear.

**Uses:** The decoction made from the rhizome is used for the treatment of chest complains by *Kani* tribes in Thiruvananthapuram District. The juice made from the rhizome is used to treat fever by *Kani* tribes in Thiruvananthapuram District.

***Adiantum philippense*** L., Pteridaceae. (Fig. 18.2d).

Common name, maidenhair fern; Mal. name, *Avenka.*

A common fern found growing in moist shady localities on earth banks from all forests. Small herbs with erect rhizomes. Leaves are simply pinnate carrying dark brown sori arranged linearly.

**Uses**: The decoction made from the roots and rhizomes is used against fever by *Malappandaram* tribes in Kollam District.

***Angiopteris helferiana*** C. Presl, Marattiaceae (Fig. 18.2e).

Common name. Giant fern; Mal. name, *Valiyachurali.*

**Fig. 18.2** Ethnomedicinal ferns and lycophytes. (**a**) *Acrostichum aureum*; (**b**) *Actiniopteris radiata*; (**c**) *Adiantum caudatum*; (**d**) *Adiantum philippense*; (**e**) *Angiopteris helferiana*; (**f**) *Asplenium decrescens*; (**g**) *Asplenium nidus*

A common species seen in well-shaded humid areas of forest near water sources. They are large terrestrial herbs with erect and fleshy rhizomes. Leaves are bipinnate with swollen petioles carrying stipules at the base. Sori are marginal.

**Uses**: *Adiyan* tribes of Wayanad District collect fresh crosiers from the plant and ground to a paste on granite and applied on tumours.

***Asplenium decrescens*** Kunze, Aspleniaceae (Fig. 18.2f).

Common name. Spleenwort; Mal. name, *Nellapanna.*

Small terrestrial herbs with long creeping rhizomes. Leaves are simply pinnate with dark brown and polished petioles. The lowermost leaves are slightly reduced. Sori are linear and arranged on veins.

**Uses:** The decoction made from rhizome is given to treat dysentery by *Irular* tribes in Palakkad District.

***Asplenium nidus*** L., Aspleniaceae (Fig. 18.2g).

Common name, bird's-nest fern; Mal. name, *Pakshikoodupannal.*

A rare species found in shola and evergreen forests. They are epiphytic herbs with erect rhizomes. Leaves are simple tufted and nest like. Sori are linear and arranged in rows.

**Uses**: The decoction made from rhizome is used to treat fever by *Malappandaram* tribes in Kollam District.

***Cheilanthes tenuifolia*** (Burm. f.) Sw., Pteridaceae (Fig. 18.3a).

Common name, rock fern; Mal. name, *Kalpanna.*

This common species is seen on semi-arid earth banks. They are small terrestrial herbs with short creeping rhizomes. Leaves are dark green, quadripinnate below and tripinnate above with tufted and dark brown petioles. Sori are marginal on ultimate lobes.

**Uses**: The decoction made from the whole plant is used as hair wash by *Kani* tribes in Thiruvananthapuram District.

***Cyathea gigantea*** (Wall. ex Hook.) Holttum, Cyatheaceae (Fig. 18.3b).

Common name. Tree fern; Mal. name, *Marapannal.*

A rare species found growing in well-shaded areas of the banks of rivulets and streams. They are large terrestrial tree ferns having trunk-like erect rhizomes with scales densely covering the apex of the trunk. Leaves are bipinnate and dark green. Sori are median on the veins and are arranged in inverted 'V' shape.

**Uses**: Fresh leaves are made into a paste and applied externally for reducing the inflammation caused due to snake bite by *Irular* tribes in Palakkad District.

***Diplazium esculentum*** (Retz.) Sw., Athyriaceae (Fig. 18.3c).

Common name, vegetable fern; Mal. name, *Parapanna.*

They are medium terrestrial herbs with erect rhizomes. Leaves are bipinnate with tufted and polished petioles. Basal leaflets are progressively reduced. Sori are linear and dark brown.

**Uses**: The decoction made from fresh rhizome is given internally to children against stomach pain by *Malappandarum* tribes in Kollam District.

***Drynaria quercifolia*** (L.) J. Sm., Polypodiaceae (Fig. 18.3d).

Common name, oak leaf fern; Mal. name, *Annanpothu.*

**Fig. 18.3** Ethnomedicinal ferns and lycophytes. (**a**) *Cheilanthes tenuifolia*; (**b**) *Cyathea gigantea*; (**c**) *Diplazium esculentum*; (**d**) *Drynaria quercifolia*; (**e**) *Helminthostachys zeylanica*

They are large epiphytic herbs with long creeping and fleshy rhizomes ornamented with golden shining scales. Leaves are seasonal, dimorphic and simple. Sterile leaves are nest like and collect humus. Fertile leaves are pinnatifid with numerous sori.

**Uses**: The decoction made from rhizome is taken internally against fever by the *Kani* tribes in Thiruvananthapuram District. The local people in many districts of Kerala use this plant against jaundice.

***Helminthostachys zeylanica*** (L.) Hook., Ophioglossaceae (Fig. 18.3e).

Common name. Flowering fern; Mal. name, *Pazhutharakkali.*

A rare species seen in semi-shaded localities in the plains and forests of low altitude. They are small herbs with creeping, thick and fleshy rhizomes which resembles centipedes. Leaves are palmately pinnate and ternately divided. Sporangia carry green spores and are arranged on the spikes.

**Uses**: The *Kattunaikan* tribe in Malappuram District use the rhizomes for the treatment of various hepatic disorders. Besides, they use the paste made from rhizome to heal wounds. The *Kani* tribes in Thiruvananthapuram District use this plant for the treatment of centipede bite.

***Lygodium flexuosum*** (L.) Sw., Lygodiaceae (Fig. 18.4a).

Common name, climbing fern; Mal. name, *Vallipanna.*

They are terrestrial climbing herbs with long creeping rhizomes and climbing rachis. Young leaves are palmate, but mature leaves are tripinnate. Sporangia are arranged on a fingerlike spikes present along the margins of leaflets.

**Uses**: The *Malappandaram* tribes in Kollam District use the plant to cure burns. The paste made from the entire plant is mixed with coconut oil and applied on burns. The roots and rhizomes of the plant are made in to a paste and taken internally thrice a day against leprosy by the *Pathinaikkan* tribes in Malappuram District.

***Lygodium microphyllum*** (Cav.) R. Br., Lygodiaceae (Fig. 18.4b).

Common name, climbing fern; Mal. name, *Vallipanna.*

They are terrestrial climbing herbs with long creeping rhizomes and twining rachis. Leaves are bipinnate with zigzag costa. Sporangia are arranged on fingerlike spikes.

**Uses**: The decoction made from leaves is given to treat dysentery by *Kani* tribes in Thiruvananthapuram District.

***Parahemionitis cordata*** (Roxb. ex Hook. & Grev.) Fraser-Jenk. (*Hemionitis arifolia*) Pteridaceae (Fig. 18.4c).

Common name, heart fern; Mal. name, *Pattichevi.*

A common species seen in semi-exposed localities of plains to high altitude areas. They are small herbs with short creeping rhizomes. Leaves are heart shaped, simple and dimorphic with dark brown petioles. Sori are dark brown and fully covered on underside of leaves.

**Uses**: Fresh leaves are made into a paste with water and consumed in empty stomach in the morning half an hour before meals for controlling sugar level in blood by *Kani* tribals of Thiruvananthapuram District. The *Ulladan* tribes in Pathanamthitta District used to take the decoction made from entire plant to cure fever.

**Fig. 18.4** Ethnomedicinal ferns and lycophytes. (**a**) *Lygodium flexuosum*; (**b**) *Lygodium microphyllum*; (**c**) *Parahemionitis cordata*; (**d**) *Phymatosorus membranifolius*; (**e**) *Pityrogramma calomelanos*; (**f**) *Pyrrosia heterophylla*

***Phymatosorus membranifolius*** (R.Br.) Tindale, Polypodiaceae (Fig. 18.4d).

Mal. name, *Velipanna.*

A common species found growing on trees in open places in hilly regions. They are medium-sized herbs with long, creeping, thick and fleshy rhizomes. The pinnatifid leaves are arranged in two rows dorsally on the rhizomes. Sori are round, deeply sunken and are projected on the upper side.

Uses: The juice made from the rhizome kills worms. It is used by *Ulladan* tribes in Pathanamthitta District.

***Pityrogramma calomelanos*** (L.) Link, Pteridaceae (Fig. 18.4e).

Common name, silver fern; Mal. name, *Vellipanna.*

A common species found growing in both semi-shaded and fully exposed areas in all habitats. Small terrestrial or lithophytic herbs with erect rhizomes. Leaves are bipinnate with tufted black purple coloured petioles. The lower surface of leaves is covered with white farina. Sori are fully covered.

**Uses**: The *Malayan* tribes in Ernakulum District administer the juice made from whole plant for smooth expulsion of placenta after delivery.

***Pyrrosia heterophylla*** (L.) M.G. Price, Polypodiaceae (Fig. 18.4f).

Common name, dragon's scale fern; Mal. name, *Seethathali.*

A common epiphytic climbing herb seen from seal level to 800 m alt. They have long creeping rhizomes. Leaves are simple and dimorphic. Sterile leaves are ovate and leathery, and fertile leaves are linear. Sori are acrostichoid.

**Uses**: *Kani* tribes in Thiruvananthapuram District use the fronds for treating jaundice. The fronds are made into a paste and mix with cow milk. This mixture is given internally thrice a day. The local people in different parts of Kerala also use this plant for treating jaundice.

***Selaginella delicatula*** (Desv. ex Poir.) Alston., Selaginellaceae (Fig. 18.5a).

Common name, small clubmoss; Mal. name, *Seevothi.*

A common terrestrial herb with sub-erect and fleshy stem. Leaves are heterophyllous. Sporophylls are unimorphic.

**Uses**: The paste made from the entire plant is applied on ulcers by the *Malayan* tribes in Ernakulam District.

***Selaginella involvens*** (Sw.) Spring, Selaginellace (Fig. 18.5b).

Common name, small clubmoss; Mal. name, *Garudapacha.*

A rare species found growing on shaded stream banks and moist rocks in semi evergreen and evergreen forests in high altitudes. They are small herbs with erect stems. Leaves are heteromorphic. Sporophylls are unimorphic.

**Uses**: The leaves ground to a paste is mixed with water and given internally for snake bites by the *Kani* tribes in Thiruvananthapuram District. They also use the dried leaves for treating leucorrhoea in woman.

***Selaginella wightii*** Hieron., Selaginellaceae (Fig. 18.5c).

Common name, small clubmoss; Mal. name, *Cheriya garudapacha.*

Rare xerophytic lithophytes seen on the crevices of dry rocks along road sides in high altitudes. They are small herbs with creeping stems. Leaves are narrow, uniform, dense and spirally arranged. Strobili are borne on ultimate branches. Sporophylls are unimorphic.

**Fig. 18.5** Ethnomedicinal ferns and lycophytes. (**a**) *Selaginella delicatula*; (**b**) *Selaginella involvens*; (**c**) *Selaginella wightii*; (**d**) *Trigonospora caudipinna*

**Uses**: The *Muthuvan* tribes in Idukki District use this plant against urinary infections. The decoction made from the entire plant and cumin seeds are taken internally twice a day for 1 week.

***Trigonopsora caudipinna*** (Ching) Sledge, Thelypteridaceae (Fig. 18.5d).

A rare species seen in moist shady places along banks of streams above 700 m alt. They are medium-sized terrestrial herbs with erect rhizomes. Leaves are tufted, bipinnatifid and dark green carrying dark brown sori.

**Uses**: The juice prepared from rhizome is given against fever by *Irular* tribes in Palakkad District.

### 18.3.1 Miscellaneous Uses

Besides the above medicinal uses, pteridophytes are found widely being used as food throughout Kerala State by tribes and local people. Data of such species are summarised in Table 18.3. The plant parts are used after cooking. The *Muthuvan* tribes in Idukki District use the leaves of *Cyathea crinita* Copel., for thatching huts. The leaves of *Cyathea nilgirensis* Holttum and *C. spinulosa* Wall. ex Hook. are also used to thatch huts by *Kani* and *Muthuvan* tribes in Kerala. The trunks of above two species are also used as a medium for growing orchids by local people. In northern districts of Kerala, *Lygodium flexuosum* (L.) Sw. is considered as a sacred plant and is commonly used in different rituals. It is locally known as *Poluvally* (*Polu* means prosperity, *vally* means vine). This species is used in *Nira*, a festival associated with harvest. The new spikes of rice and ten medicinal plants are tied together with the vine of this species and hung inside the house. This is considered to be a good omen and believed to bring prosperity in the coming years. *Selaginella delicatula* (Desv. ex Poir.) Alston is also treated as a holy plant and is known as *Seevothy* meaning *Sribhagavathy*, the Goddess of prosperity. So, this species is used for floral arrangements in the foreyards of houses during post *Onam* days, an important festival celebrated in Kerala State. The fronds of *Dicranopteris linearis* (Burm. f.) Underw. are widely used by the *Kani* tribes for decorating the premises during all their festivals. The dried sterile leaves of *Drynaria quercifolia* (L.) J. Sm. painted with silver and golden colours are used for dried flower arrangements and other decorative purposes by local people in Kerala. Children take impression of leaves of 'silver fern' (*Pityrogramma calomelanos* (L.) Link) on their hands as leaves having white farina on back surface while playing. The leaves of *Nephrolepis brownii* (Desv.) Hovenkamp & Miyam. and *Adiantum latifolium* Lam. are widely used in flower arrangements. *Adiantum raddianum* C. Presl and *Nephrolepis cordifolia* (L.) C. Presl are used for indoor decorations.

**Table 18.3** The list of edible ferns used by tribal and local peoples in Kerala

| Sl. no. | Botanical name | Parts used | Used by |
|---|---|---|---|
| 1 | *Acrostichum aureum* L. | Tender leaves | Local people |
| 2 | *Ampelopteris prolifera* (Retz.) Copel. | Fronds | Ulladan tribes |
| 3 | *Angiopteris helferiana* C. Presl | Tender stem and leaves | Adiyan and Cholanaikkan tribes |
| 4 | *Blechnum orientale* L. | Rhizome | Malayan tribes |
| 5 | *Botrychium lanuginosum* wall. Ex hook. & Grev. | Tender portions | Muthuvan tribes |
| 6 | *Ceratopteris thalictroides* (L.) Brongn. | Young fronds | Local people |
| 7 | *Christella parasitica* (L.) H. lev. | Young fronds | Muthuvan tribes |
| 8 | *Diplazium esculentum* (Retz.) Sw. | Young fronds | Paniya tribes and local people. |
| 9 | *Dryopteris cochleata* (D. Don) C. Chr. | Tender portions | Cholanaikkan tribes |
| 10 | *Dryopteris sparsa* (Blume) Kunze | Fronds | Malavedar tribes |
| 11 | *Helminthostachys zeylanica* (L.) Hook. | Tender portions | Kani tribes |
| 12 | *Lygodium flexuosum* (L.) Sw. | Tender portions | Paliya tribes |
| 13 | *Marsilea minuta* L. | Tender leaves | Local peoples |
| 14 | *Nephrolepis cordifolia* (L.) C. Presl | Tubers | Mannan tribes |
| 15 | *Ophioglossum reticulatum* L. | Fronds | Mannan tribes |
| 16 | *Phymatosorus membranifolius* (R. Br.) S.G. Lu | Tender portions | Muthuvan tribes |
| 17 | *Stenochlaena palustris* (Burm. f.) Bedd. | Young shoots | Local peoples |
| 18 | *Tectaria coadunata* (J. Sm.) C. Chr. | Tender portions | Malappandaram tribes |
| 19 | *Trigonospora caudipinna* (Ching) sledge | Tender portions | Mannan tribes |

## 18.4 Discussion

Though ethnobotanical investigations of pteridophytes were conducted in different parts of the world (reviewed by Keller and Prance 2015), such studies received little attention in India though the country harbours 1200 species of pteridophytes. As for the State of Kerala also, though she harbours 250 species, their use by tribal and local people has never become a topic of serious investigation. Ethnobotanical investigations generally found to avoid pteridophytes, may be because of the lack of taxonomic experts to identify the plant species.

This study reports 41species of pteridophytes used by tribes and local people in Kerala State. This comes to 14.6% of the pteridophyte species reported so far from Kerala. Among these, 22 species are used for medicinal purpose, 19 as food, 2 for rituals, 5 for decorative purposes and 3 for hut making and 2 for miscellaneous purposes. Among the medicinal pteridophytes, four species are used in the treatment of fever; two each against snake bite, jaundice, dysentery and healing wounds; one each against high blood sugar, stomach pain, hepatic disorders, centipede bite, burns, leprosy, ulcer, leucorrhoea, worms and urinary infection; one employed for the smooth expulsion of the placenta; and one as hair wash. The officinal parts used are rhizomes, leaves and whole plants. Rhizomes are the major officinal parts used (12 uses) followed by whole plant in 11 and leaves in 8 uses. Out of the 22 species used for medicinal purposes, only 3 species are used through media. Leaf juice of *Actiniopteris radiata* is mixed with breast milk for treating dysentery in children, the entire plant of *Lygodium flexuosum* is mixed with coconut oil to treat burns, and fresh leaves of *Parahemionitis cordata* are mixed with water to control blood sugar. All the plant species used are those which are readily available in the premises of inhabited areas or settlements. People collect fresh plant parts from the natural population and use. Nobody is cultivating these plants in their premise, and they are not available in the market. So, they directly depend on natural populations whenever required.

Among the 41 species recorded in the present study, only 4 species were subjected to pharmacological validation. Gayathri et al. (2005) have found that *Selaginella involvens* possesses remarkable thymus growth stimulatory activity and anti-lipid peroxidation property. Suja et al. (2004) have established that *Helminthostachys zeylanica* possesses hepatoprotective activity. Ajikumaran et al. (2006) found that the fresh leaves of *Parahemionitis cordata* have anti-diabetes and hypoglycaemic properties. Anuja et al. (2014) observed that the fertile fronds of *Drynaria quercifolia* have anti-edematous and analgesic properties.

Among the 34 tribal groups in Kerala, information from only 14 tribal could be gathered during our explorations. *Kani* tribe among them used the highest number of species (nine species) followed by Malapandaram and Muthuvan tribes (six species each). Eight species recorded multiple uses.

This is not an exclusive study on ethnopteridology of Kerala State. So, an attempt spending more time in the field covering different forest areas incorporating all tribal groups would certainly yield more data, both quantitative and qualitative. Such an attempt would help to identify more pharmacologically potential species. On the other side, being a plant group which need specialised habitats to survive, exploitation of pteridophytes from natural populations should be done with great caution. For example, among the five species frequently used for the various Ayurvedic preparations in Kerala, *Parahemionitis cordata* (Roxb. ex Hook. & Grev.) Fraser-Jenk. and *Selaginella involvens* (Sw.) Spring are now becoming rare plants due to overexploitation. So, when potential plants are identified, their sustainable utilisation is a challenge because (1) these plant groups have complex reproductive mechanisms and (2) habitats of pteridophytes are shrinking due to deforestation and climate change. Thus, appropriate ex situ conservation and mass multiplication mechanisms need to be developed for the continuous supply of raw materials of the potential species.

**Acknowledgement** We are thankful to the Director of Jawaharlal Nehru Tropical Botanic Garden and Research Institute (JNTBGRI) for the facilities provided for the study. We express our sincere gratitude to Dr. Vinod Kumar, T. G., Scientist, Ethnomedicine Division, JNTBGRI, for sharing the images of tribes.

## References

AjiKumaran NS, Shylesh BS, Gopakumar B et al (2006) Anti-diabetes and hypoglycaemic properties of *Hemionitis arifolia* (Burm.) Moore in rats. J Ethnopharmacol 106:192–197

Anuja GI, Latha PG, Shine VJ et al (2014) Antioedematous and analgesic properties of fertile fronds of *Drynaria quercifolia*. ISRN Inflammation 4:302089

Arora RK (1997) Ethnobotany and its role in the conservation and use of plant genetic resources in India. Ethnobotany 9:6–15

Binu S (1999) Ethnobotany of Pathanamthitta District, Kerala, India. Doctoral Thesis Submitted to the University of Kerala

Boom BM (1985) Ethnopteridology of the Chacobo Indians in Amazonian Bolivia. Am Fern J 75: 19–21

Chandra S, Fraser-Jenkins CR, Kumari A et al (2008) A summary of the status of threatened pteridophytes of India. Taiwania 53(2):170–209

Chang H-C, Gupta SK, Tsay H-S (2010) Studies on folk medicinal ferns: an example of "Gu-Sui-Bu". In: Fernandez H, Kumar A, Angeles R (eds) Working with ferns. Springer, New York

Chapman AD (2009) Numbers of living species in Australia and the world, 2nd edn. Australian Biodiversity Information Services, Australia

Christensen H (1997) Uses of ferns in two indigenous communities in Sarawak, Malaysia. In: Johns RJ (ed) Holttum Memorial Volume. Royal Botanic Gardens, Kew, pp 177–192

Forman L, Bridson L (1991) The herbarium hand book. Royal Botanic Gardens, Kew

Fraser-Jenkins CR (2012) Rare and threatened Pteridophytes of Asia 2. Endangered species of India- the higher IUCN categories. Bull Nat Mus Nat Sci B 38(4):153–181

Gayathri V, Asha VV, Subramoniam A (2005) Preliminary studies on the immunomodulatory and antioxidant properties of *Selaginella* species. Indian J Pharmacol 37:381–385

Hernandez CRK, Sutherland CHN (2007) Etnobotánica de los Helechos de Honduras. Ceiba 48(1–2):1–10

Ho R, Teai T, Bianchini J-P, Lafont R, Raharivelomanana P (2010) Ferns: from traditional uses to pharmaceutical development, chemical identification of active principles. In: Fernandez H, Kumar A, Angeles R (eds) Working with ferns. Springer, New York

Joshi P (1997) Ethnobotany of Pteridophytes of hilly districts of Uttar Pradesh, India. Indian Fern J 14:14–18

Keller HA, Meza-Torres EI, Prance GT (2011) Ethnopteridology of the Guaranis of Misiones Province. Argentina Am Fern J 101(3):193–204

Keller HA, Prance GT (2015) Ethnobotany of ferns and lycophytes. Fern Gaz 20(1):1–13

Kirn HS, Kapahi BK (2001) Ethnobotanical notes on some ferns and fern-allies of Jammu and Kashmir State, India. Indian Fern J 18:35–38

Kumar KK, Sasidharan N, Swarupanandan K (1999) Ethnobotanical studies on the hill tribes in the shoal forests of high ranges Kerala. South India J Econ Tax Bot 23(2):451–466

Liu Y, Wujisguleng W, Long C (2012) Food uses of ferns in China: a review. Acta Soc Bot Poloniae 81:263–270

Madhusoodanan PV (2015) Handbook on ferns and fern allies of Kerala. Malabar Botanic Garden and Institute for Plant Sciences

Manickam VS, Irudayaraj V (1992) Pteridophyte flora of the Western Ghats South India. B.I. Publications, New Delhi

Manilal KS, Remesh M, Kumar M (2000) Ethnobotanical studies of wild leafy vegetables used by two tribal groups of Attappadi Palakkad Dt. Kerala. Paper presented in National seminar on

Plant Biodiversity Systematics Conservation and Ethnobotany, November 9–11. Siliguri, West Bengal

Mathew PJ, Mathew D, Unnithan CM et al (1999) Ethnomedical information on some pteridophytes of the Western Ghats. Biodiversity, Conservation and Utilization of Spices, Medicinal & Aromatic Plants:265–268

Mathew PJ, Pushpangadan P, Unnithan CM (1996) Four candidate plant species for evaluation of anticancer property from Adian tribals of Kerala, India. In: Jain SK (ed) Ethnobiology in human welfare. Deep Publications, New Delhi, pp 91–92

Mendelsohn R, Balick MJ (1995) The value of undiscovered pharmaceuticals in tropical forests. Econ Bot 49:223–228

Nair NC, Ghosh SR, Bhargavan P (1988) Fern-allies and ferns of Kerala-part I. J Econ Tax Bot 12(1):171–209

Nair NC, Ghosh SR, Bhargavan P (1992a) Fern allies and ferns of Kerala-part II. J Econ Tax Bot 16(2):251–282

Nair NC, Ghosh SR, Bhargavan P (1992b) Fern allies and ferns of Kerala-part III. J Econ Tax Bot 16(3):501–549

Nair NC, Ghosh SR, Bhargavan P (1994) Fern allies and ferns of Kerala-part IV. J Econ Tax Bot 18(2):449–476

Navarrete H, Leon B, Gonzales J et al (2006) Helechos. In: Moraes RM, Ollgaard B, Kvist LP, Borchsenius F, Balslev H (eds) Botanica Economica de los Andes Centrales. Universidad Mayor de San Andres, La Paz, pp 385–411

Nwosu MO (2002) Ethnobotanical studies on some Pteridophytes of southern Nigeria. Econ Bot 56:255–259

Pandey PC, Pangtey YPS (1987) Studies on ethnobotany –I. On some less known edible and economic ferns of Kumaun region of Western Himalaya. J Econ Tax Bot 11:81–85

Phillips O, Gentry AH, Reynel C et al (1994) Quantitative ethnobotany and Amazonian conservation. Conserv Biol 8(I):225–248

Prudente LT (2007) Arquitetura Mbyá-Guaraní na Mata Atlântica do Rio Grandedo Sul. Estudo do caso do Tekoá Nhüu Porã. Tese (Mestrado) em engenharia. Universidades Federal do Rio Grande do Sul. Escola de Engenharia, Porto Alegre, p 164

Rumphius GE. (2011). The Ambonese herbal vol 5: 121–165. English translation by E.M. Beekman of the Ambonese herbal first published in 1741, Yale University Press

Schultes RE (1993) The virgin field in psychoactive plant research. Ethnobotany 5:5–61

Suja SR, Latha PG, Pushpangadan P, Rajasekharan S (2004) Evaluation of hepatoprotective effects of *Helminthostachys zeylanica* (L.) hook. Against carbon tetrachloride-induced liver damage in Wistar rats. J Ethnopharmacol 92(1):61–66

**Raju Antony** is working as a Senior Technical Officer in Jawaharlal Nehru Tropical Botanic Garden and Research Institute (JNTBGRI). He did his PhD on 'The studies of the Pteridophyte Flora of Agasthyamala, Western Ghats', from Kannur University. He joined the Division of Garden Management in 1992 and has 28 years of research experience in the field of conservation and systematic studies of ferns and lycophytes of the Western Ghats. He has developed a fernery in JNTBGRI, which is one of the largest ferneries in India. He is a Life Member of the Indian Fern Society and a member of the International Association of Pteridologists, USA. He is an accomplished pteridologist who listed in the Directory of Annual Review of Pteridological Research (ARPR) of the International Association of Pteridologists, USA. He has 35 research publications to his credit.

**S. Suresh** had his Master's degree in Botany from the University of Mumbai with specialization in Systematic Palynology. He joined the Division of Conservation Biology at Jawaharlal Nehru Tropical Botanic Garden and Research Institute (JNTBGRI) in 1992 and trained in the areas of seed dispersal ecology, pollination ecology, plant population demography and analysis of pollen morphology in relation to plant animal interaction. He is currently working as a technical officer and has got more than 15 years of research experience in the above fields. He is pursuing his doctoral programme in Kannur University in Kerala on Demography of *Cullenia exarillata*. He has authored/co-authored 30 research articles and 20 oral/poster presentations. Currently he is conducting research on plant-animal interaction in the tropical forest ecosystems of the Western Ghats giving special emphasis on pollination and ecology and seed dispersal.

# Phytochemistry of Indian Pteridophytes: A Review

19

Johnson Marimuthu, N. Janakiraman, J. Chandra Saleride, A. Sivaraman, B. Shivananthini, and K. Paulraj

**Abstract**

In the present review, an attempt is made to provide an update on the phytochemistry of Indian pteridophytes, covering the recent findings concerning the phytochemical composition of crude extracts and their histochemical, spectroscopic and chromatographic profiles. Primary and secondary metabolites of Indian pteridophytes qualitative and quantitative profiles are taken into consideration. A report on the preliminary phytochemical analysis of 170 species, primary and secondary metabolites quantitative profiles of 115 species, histochemical profiles of 61 species, chromatographic profile (amino acids and sugar) of 43 species, TLC profile of 14 species, HPLC and HPTLC profiles of 23 species and GC-MS profiles of 32 species is included for the present review. The results confirmed the existence of the secondary metabolites, viz. phenolics, tannins, flavonoids, steroids, saponins, triterpenoids, glycosides and alkaloids, in the Indian pteridophytes. The available literature on phytochemistry confirmed that Indian pteridophytes are a pool of therapeutic agents. The outcome of the preliminary phytochemical studies on the qualitative and quantitative profiles of Indian pteridophytes revealed the chemical constituents and therapeutic values and

J. Marimuthu (✉) · B. Shivananthini
Department of Botany, Centre for Plant Biotechnology, St. Xavier's College (Autonomous), Palayamkottai, India

N. Janakiraman
Department of Botany, The Madura College (Autonomous), Madurai, Tamil Nadu, India

J. Chandra Saleride
Department of Botany, Women's Christian College, Nagercoil, Tamil Nadu, India

A. Sivaraman
Maria Research Foundations, Tirunelveli, Tamil Nadu, India

K. Paulraj
Department of Botany, Nesamony Memorial Christian College, Marthandam, Tamil Nadu, India





J. Marimuthu et al. (eds.), *Ferns*, https://doi.org/10.1007/978-981-16-6170-9_19

provided the chemical marker for the studied pteridophytes. These profiles will be used as phytochemical markers for the identification of the species in the pharmaceutical industries and to find a solution for the taxonomical disputes. Further studies on the isolation and characterization of active principles responsible for the bioactivity are needed.

**Keywords**

Chromatography · Histochemistry · Metabolites · Spectroscopy · Ferns · Fern allies

## 19.1 Introduction

Pteridophytes (ferns and fern allies) are the primitive vascular plants that appeared on the earth in the mid-Paleozoic Era. They are the pioneer colonizers on earth and one of the ubiquitous vegetation that dominated the land in the Carboniferous period. Pteridophytes with their specialized water and food- conducting tissues heralded an era of great colonization in terrestrial ecosystems. They successfully established themselves as land plants and dominated most of the forests on earth's surface. It was a very rapid rate of evolution. It possesses simple organization and is unique in being characterized by cryptogamic mode of reproduction. They are very conspicuous and gorgeous elements of biodiversity which occurs in various kinds of habitats ranging from sea level to mountain top and tropical to subpolar regions (Dudani and Ramachandra 2010). The decline of pteridophytes began with the evolution and dominance of the flowering plants in the late Cretaceous. Since then, pteridophytes act as a bridge between the non-vascular cryptogams and the phanerogams. It continues to occupy several niches on the land, in marshes and even in water bodies (Dudani et al. 2011).

Pteridophytes represent a broad spectrum of biological types from small filmy to arborescent tree ferns and from submerged aquatics to epiphytes and xerophytes (Kumar 1998). They are very conspicuous and elegant elements of our present- day flora. They had a very glorious past being the dominant vegetation on the earth about 280–230 million years ago (Bir 1987; Khare 1996). They are the beautiful elements of biodiversity and occur in a wide range of habitats from sea level to mountain top and tropical to subpolar regions. Pteridophytes attracted the attention of botanists because of their characteristic life cycle in which the sporophytic and gametophytic generations not only are independent to each other but also are usually autotrophic and morphologically distinct. They are the wonderful group of plants with fascinating foliage architecture and have drawn special attention of the plant lovers, researchers and horticulturists. In the world flora, about 13,600 species of extant pteridophytes are recorded (Moran 2008). Among which 1300 species into 70 families and 191 genera occur in different biogeographical regions of India (Chandra et al. 2008) with the main centres being the Himalayas, the Western Ghats and the Eastern Ghats (Chandra 2000; Dixit 2000). Unfortunately, ferns are considered as a neglected group of plants in biodiversity as far as their economic value is concerned. However, with the introduction of ethnobotany by Hershberger (1896) for the study of relationship which exists between peoples of primitive societies and their plant environment, numerous attempts were made to study the relationships of pteridophytes with human beings for their medicinal value. Theophrastus (327–287 BC) and Dioscorides (50 AD) cited the medicinal values of pteridophytes. They have been successfully used in the homoeopathic, ayurvedic, unani and tribal systems of medicines (Das 2003; Rout et al. 2009). Ferns had an important role in folklore medicine. Singh (1999) described 160 useful ferns in India based on the pharmacological, phytochemical and ethnobotanical studies. Apart from the medicinal properties, ferns have great aesthetic value for their graceful, delicate beauty and are very good for interior decoration and green houses (Chandra and Kaur 1974).

## 19.2 Phytochemistry

Plants synthesize a wide variety of chemical compounds which can be sorted by their chemical class, biosynthetic origin and functional groups. The medicinal value of plants lies in chemical substances or group of compounds that produce a definite physiological action in the human body (Edeoga et al. 2005). The beneficial effects of plants are usually due to the secondary metabolites present in it. Medicinal plants contain active ingredients that can be used for therapeutic purposes and are precursors of chemotherapeutical semisynthesis (WHO 1979). It also provides temporary relief to symptomatic problems and health- promoting characteristics and has curative properties. Medicinal plants are under chemical scrutinization due to the advent of modern scientific methods. This leads to the isolation of active compounds. After their isolation and characterization, these active principles in pure form or well-characterized extracts became useful in several countries. Pteridophytes play an important role in the ecological niches of forest ecosystem as an integral part of biogeochemical cycling of minerals. They also help as ecological indicators for the understanding of a particular habitat. Pteridophytes synthesize a wide array of primary and secondary metabolites enriched with various bioactivities to defend and acclimatize various ecological conditions that could potentially be used as an alternative medicine for the treatment of many diseases. However, the available literature on the phytochemicals and their pharmacological applications in pteridophytes are limited. The periodical update on the occurrence, chemotaxonomy and physiological activity of fern secondary metabolites was published by Soeder (1985) and Santos et al. (2010). Irudayaraj and Johnson (2012) summarized the phytochemical and pharmacological studies on spike mosses. Cao et al. (2017) summarized the bioactive phytochemicals and pharmacology of fern species. Baskaran et al. (2018) listed the medicinal uses of ferns. The phytochemical composition and their pharmacological activity of *Athyrium* were summarized by Salehi et al. (2019). In the present review, an attempt is made to provide an update on the phytochemistry of Indian pteridophytes, covering the recent findings concerning the

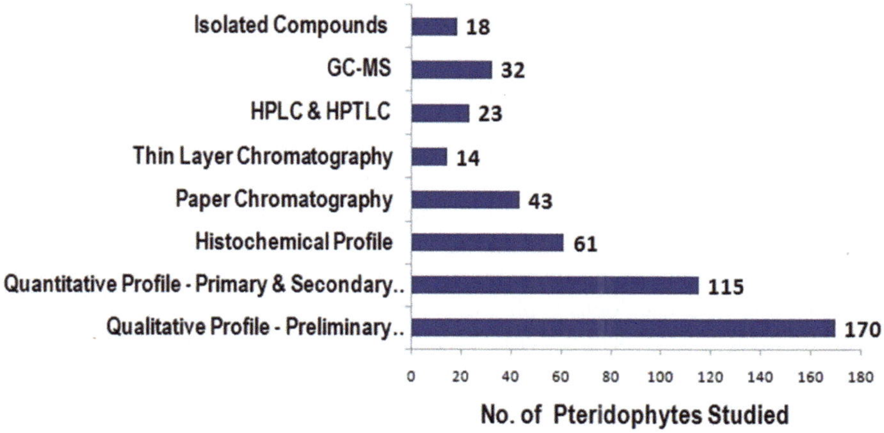


**Fig. 19.1** Phytochemistry of Indian pteridophytes

phytochemical composition of crude extracts and their histochemical, spectroscopic and chromatographic profile. Qualitative and quantitative profiles with their primary and secondary metabolites are also included (Fig. 19.1).

## 19.3 Phytochemistry of Pteridophytes

Plant synthesizes a wide variety of chemical compounds, which can be sorted by their chemical class, biosynthetic origin and functional groups. Many of the phytocompounds provide protection against insect attacks and plant diseases. They also exhibit a number of protective functions for humans. Plants normally produce these secondary metabolites not only to adapt to their environment but also to defend themselves against biotic or abiotic stress, such as high light intensity, extremely high or low temperature, high salinity, drought and natural enemies (Bennett and Wallsgrove 2006). These metabolites are polyphenols, flavonoids, terpenoids, steroids, quinones, alkaloids, polysaccharides and so on (Swain 1977). Ferns are reported to have many useful phytochemicals (secondary metabolites) such as flavonoids, steroids, alkaloids, phenols, triterpenoid compounds, varieties of amino acids and fatty acids (Zeng-fu et al. 2008). In recent years, phytochemical investigations on pteridophytes, especially the ferns, have been progressing steadily and assumed extraordinary importance. The progress in this field led to the development of sophisticated techniques in chemical analysis which can detect even traces of chemical compounds or their precursors in spite being unstable. Quantitative and qualitative estimations of these compounds can be used for finding out which particular fern taxon can be exploited effectively. With increasing awareness or locating additional raw materials for the plant -based industry with particular reference to the sources of proteins, oils, starch, tannins, etc., the knowledge of the distribution pattern of chemical ingredients can play an effective role in exploiting the plant resources. Knowledge on these phytoconstituents present in plants are used for the discovery of therapeutic agents. Phytochemical studies on pteridophytes attracted the attention of plant scientists due to the ethnomedicinal values. Ferns are expected to have many useful secondary metabolites than other plants.

## 19.4 Histochemistry of Indian Pteridophytes

Histology is the microscopic study of the structure of biological cells and tissues. Histochemistry is devoted to study the identification and distribution of chemical compounds within and between biological cells, using stains, indicators and light and electron microscopy (Wick 2012). Histochemical methods are employed for qualitative and quantitative analysis of all cellular components including proteins, carbohydrates, lipids, nucleic acids, lignins, tannin, phenolics substances, flavonoids, triterpenoids, etc. (Gahan 1984; Kiernan 1989). Irudayaraj et al. (2014) confirmed the presence of lipids, polyphenols, lignin and tannins in *Lycopodiella cernua*, *Selaginella involvens*, *S. inaequalifolia*, *S. tenera*, *Angiopteris evecta*, *Marattia fraxinea*, *Lygodium microphyllum*, *Pteris argyraea*, *P. confusa*, *Dryopteris concolor*, *Cheilanthes viridis*, *Pellaea boivini*, *Hemionitis arifolia*, *Pityrogramma calomelanos*, *Adiantum raddianum*, *Pteridium aquilinum*, *Histiopteris incisa*, *Hypolepis glandulifera*, *Microlepia speluncae*, *Odontosoria chinensis*, *Lindsaea ensifolia*, *Araiostegia hymenophylloides*, *Nephrolepis multiflora*, *Trichomanes obscurum*, *Dicranopteris linearis* and *Cyathea nilgirensis*. Johnson et al. (2014a, b) revealed the histochemical profile, viz. lipids, polyphenols, lignins and tannins, on *Pseudophegopteris pyrrhorhachis*, *Macrothelypteris torresiana*, *Trigonospora ciliate*, *Cyclosorus interruptus*, *Amphineuron terminans*, *Sphaerostephanos arbuscula*, *Christella parasitica*, *C. dentata*, *Asplenium cheilosorum*, *Diplazium muricatum*, *D. travancoricum*, *D. brachylobum*, *Tectaria paradoxa*, *Arachniodes tripinnata*, *A. ambailis*, *A. aristata*, *Dryopteris sparsa*, *Bolbitis appendiculata*, *Blechnum orientale*, *Leptochilus decurrens* and *Pyrrosia porosa*.

Sivaraman and Johnson (2014) localized the polyphenols, lignin and tannins on *S. involvens* (Sto.) Spring, *S. intermedia* (Bl.) Spring, *S. inaequalifolia* (Hook. and Grev.) Spring and *S. tenera* (Hook. and Grev.). SaiLakshmi et al. (2015) revealed the phytochemical constituents of selected *Asplenium* species using histochemical analysis. SaiLakshmi et al. (2015) revealed the existence of lignin, tannin, cutin, suberin and polyphenols in the stipes of *Asplenium aethiopicum*, *A. affine*, *A. decrescens* and *A. viviparum* using histochemical analysis. Vidyarani et al. (2015) revealed the histochemical profiles, viz. lignin, cutin, suberin and polyphenols, on *Adiantum hispidulum*, *A. raddianum*, *A. peruvianum*, *A. pubescens*, *A. tripinnatum* and *A. pacotii*. They recorded varied degree of occurrence of lignin, cutin, suberin and polyphenols in all the studied species. They noted the presence of tannin in all the studied species except *A. raddianum*. Histochemical analysis results revealed the variations in the occurrence of various metabolites, viz. phenolics, tannins and lipids, with varied quantities in different tissues of the rachis. High concentration of lignin, cutin and suberin presence was observed in the epidermis, hypodermis and xylem of studied pteridophytes. Lignin provides structural rigidity to the plant body and plays a vital defensive role against pathogens and insects.

## 19.5 Indian Fern Phytochemistry: Preliminary Phytochemical Analysis

As far as Indian pteridophytes are concerned, phytochemical studies have been started by Mehra and Mittal (1960–1961). They screened the Indian species of *Dryopteris* and *Polystichum* for 'filicin' which is used as vermifuge and reported that *Polystichum* lacked 'filicin' completely while *Dryopteris* species gave positive results. Phytochemical analysis on rare, endangered and medicinally important spleenworts, *Asplenium* and *Psilotum*, was investigated by Lal (1979) and Rohtagi et al. (1984). Phytochemical analysis of the edible ferns, *Ampelopteris prolifera* and *Diplazium esculentum*, shows its nutritional value (Shankar and Khare 1985; Singh et al. 1989). Kaur et al. (1986) made a comparative investigation of amino acids and free proline of some Rajasthan ferns.

During the period of 1991–2000, a total of 25 papers were published on the qualitative phytochemical profile. The phytochemical profiles of 61 Indian pteridophytic species, viz. 19 *Thelypteris* sp. of Western Ghats, 3 *Pteris* sp., 3 *Cyathea* sp., 2 *Diplazium* sp., 2 *Adiantum* sp., 3 *Polystichum* sp., 3 *Sphaerostephanos* sp., 3 *Ophioglossum* sp., *Christella* sp., *Tinospora* sp., *Marsilea minuta*, *Diplazium brahylobhum*, *D. polypodioides*, *Pteris vittata*, *Actiniopteris radiata*, *Asplenium pumilum*, *Tectaria macrodonta*, *Cheilanthes farinosa*, *Hypodesmatium crenatum* and *Cyclosorus dentatus*, *Histiopteris incisa*, *Hypolepis glandulifera*, *Microlepia speluncae* and *Pteridium aquilinum*, *Polystichum harpophyllum*, *P. piceopaleaceum* and *P. kunthianum*, were revealed (Chark and Dhir 1991; De Britto et al. 1991; Ramachandran et al. 1991; Rathore and Sharma 1991; Antonisamy et al. 1992; De Britto et al. 1992; Patric Raja et al. 1992; Sharma and Sharma 1992; De Britto et al. 1993; Gopalakrishnan et al. 1993; Jesudass et al. 1993; Henry Joseph et al. 1993; De Britto et al. 1994, a, b, c; Harsh and Sharma 1994; Kumar 1995; Vyas et al. 1995; Yadav 1995 and Irudayaraj 1996). The authors revealed the qualitative and quantitative profile of pigment composition, sugars, starch, amino acids and secondary metabolites of Indian pteridophytes. During this period a good number of ferns have been subjected to preliminary phytochemical screening (Irudayaraj and Raja 1998).

Paulraj (2007) studied the chemical composition of epidermal glands of *S. unitus*, *S. arbuscula* and *S. subtruncatus*. Korwar et al. (2010) revealed the phytochemical profile of *Drynaria quercifolia* Linn rhizome. Talukdar et al. (2010) disclosed the qualitative profile of *Cyathea brunoniana* and *Cyathea gigantea*. *Selaginella* is a rich source of steroids, biflavonoids, alkaloids, secolignans, neolignans and caffeoyl derivatives. Other compounds such as alkaloidal glycosides, phenylpropanones and lignans were also reported (Sa et al. 2012). Phytoprofile of *Blechnum orientale*, *Ceratopteris thalictroides*, *Drymoglossum heterophyllum*, *Dicranopteris linearis*,

*Drynaria quercifolia*, *Hemionitis arifolia*, *Pityrogramma calomelanos*, *Lindsaea ensifolia*, *Nephrolepis multiflora* and *Pteris confusa* (Mithraja et al. 2012), *Adiantum caudatum*, *A. latifolium*, *A. lunulatum*, *Christella dentata* and *C. parasitica* (Mithraja et al. 2012a), *Acrostichum aureum* (Thomas 2012), *A. capillus-veneris* (Ahmed et al. 2012; Kumar and Nagarajan 2012), *Azolla microphylla* (Abraham and Aeri 2012), *Cyathea gigantea* (Kiran et al. 2012), *H. arifolia* (Bindu et al. 2012) and *Sphenomeris chinensis* (Aceret 2012) are revealed. The authors observed the varied degree of chemical diversity in the studied extracts. The existence of the metabolites is based on the solvents and self-defence mechanism of the ferns. Shakoor et al. (2013) divulged the qualitative phytoprofile of pteridophytes from Shopian district (Jammu and Kashmir), viz. *Adiantum venustum*, *Asplenium adiantum-nigrum*, *A. trichomanes*, *A. pseudofontanum*, *A. septentrionale*, *Athyrium atkinsonii*, *A. attenuatum*, *A. mackinnonii*, *A. wallichianum*, *Azolla cristata*, *Coniogramme affinis*, *Cystopteris fragilis*, *Deparia acuta*, *D. allantodioides*, *Diplazium maximum*, *D. sibiricum*, *Dryopteris blanfordii*, *D. barbigera*, *D. nigropaleacea*, *D. ramosa*, *D. xanthomelas*, *Equisetum arvense*, *Gymnocarpium dryopteris*, *Onychium cryptogrammoides*, *Osmunda claytoniana*, *Phegopteris connectilis*, *Polystichum bakerianum*, *P. discretum*, *P. prescottianum*, *P. yunnanense*, *Pseudophegopteris levingei*, *Pteridium aquilinum*, *Pteris cretica* and *Salvinia natans* aqueous, methanolic, ethanolic and acetone extracts. Phytochemical analysis confirmed the presence of carbohydrates, proteins and free amino acids in 34 species, flavonoids in 27 species, phenolic compounds and tannins in 26 species, glycosides in 24 species, terpenoids in 23 species, saponins in 22 species, volatile oils in 18 species, alkaloids in 15 species, phlobatannins in 12 species and resins in 3 species. Awadhesh Kumar et al. (2014) exposed the phytoprofile for *Adiantum* and *Pteris*. Mismawati et al. (2015) reported the phytochemical constituents of *Angiopteris evecta* leaves methanolic extracts. Ahmed et al. (2015) reported the phytochemical profile of *Drynaria quercifolia*. Kunnathupara et al. (2016) determined the quantitative phytochemical profile of *Azolla microphylla* ethanolic extract.

A good number of preliminary phytochemical analyses were carried out on Indian ferns and fern allies. The available literature on the phytochemical analysis of Indian pteridophytes explains that the phytochemical composition of 170 species belongs to 56 genera and 27 families (Fig. 19.2). The phytochemical analysis on primary and secondary metabolites confirmed the presence of carbohydrates, sugars, reducing sugars, amino acids, proteins, lipids, steroids, saponins, terpenoids, triterpenoids, alkaloids, phenolic compounds, tannins, flavonoids, catechins, glycosides, cardiac glycosides and anthraquinone with varied degree.

### 19.5.1 Carbohydrates

The presence of carbohydrates is reported in *Isoetes* (Rathore and Sharma 1990), *Dryoathyrium boryanum*, *Athyrium puncticaule*, *A. solenopteris*, *A. lanceum*, *Deparia petersenii*, *Diplazium muricatum*, *D. sylvaticum*, *D. dilatatum*, *D. travancoricum*, *D. polypodioides*, *D. esculentum* and *D. brachylobum*, *Pteris*

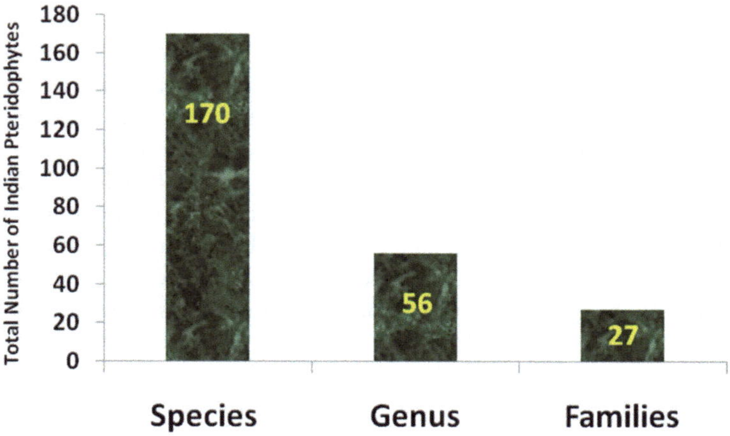


**Fig. 19.2** Preliminary phytochemical analysis at family, genus and species level (1960–2021)

*vittata*, *P. multiaurita*, *P. pellucida*, *P. cretica*, *P. scabripes*, *P. biaurita*, *P. kleiniana*, *P. aspericaulis*, *P. mertensoides*, *P. otaria*, *P. argyraea*, *P. linearis*, *P. confusa*, *Acrostichum aureum*, *Tectaria paradoxa*, *T. wightii*, *T. coadunata*, *Lastreopsis tenera*, *Polystichum harpophyllum*, *P. piceopaleaceum*, *P. moluscense*, *P. squarrosum*, *Arachniodes tripinnata*, *A. aristata*, *A. amabilis*, *Dryopteris hirtipes*, *D. madrasensis*, *D. cochleata*, *D. sparsa*, *D. approximate*, *Histiopteris incisa*, *Hypolepis glandulifera*, *Odontosoria chinensis*, *Nephrolepis auriculata*, *N. multiflora*, *Trichomanes obscurum*, *Dicranopteris linearis* var. *sebastiana*, *D. linearis* var. *tenuis*, *D. linearis* var. *brevis*, *Blechnum orientale*, *B. occidentale*, *Cyathea crinita*, *C. nilgirensis*, *Osmunda regalis*, *Pteridium aquilinum*, *Leptochilus decurrens* and *L. decurrens* f. *lanceolatus* (Antonisamy 1997; Joseph 1997; Jesudass 1997; Patric Raja 1997), *Azolla pinnata*, *Marsilea minuta* and *Salvinia molesta* (Mithraja et al. 2011), *Tectaria zeylanica* (Sukumaran et al. 2012), *Selaginella bryopteris* (Rupa and Bhavani 2014), *Pteris vittata* (Maneesha et al. 2015), *Dryopteris cochleata* (Das et al. 2015), *Cyathea nilgirensis* (Sahaya Mary and Mahesh 2015), *Salvinia molesta* (Gaya et al. 2016), *Diplazium esculentum* (Dash et al. 2017), *Tectaria cicutaria* (Preeti and Namdeo 2018), *Azolla pinnata* (Thiripurasundari and Padmini 2018), *Azolla pinnata* (Farook et al. 2019), *Asplenium indicum*, *Lepisorus nudus* and *Microsorum membranaceum* (Jadhav et al. 2019), *Ceratopteris thalictroides* (Smitha and Vadivel 2019) and *Tectaria wightii* (Vineesh et al. 2021).

## 19.5.2 Sugar

The presence of sugar and reducing sugar in the *Dryoathyrium boryanum*, *Athyrium puncticaule*, *A. solenopteris*, *A. lanceum*, *Deparia petersenii*, *Diplazium muricatum*, *D. sylvaticum*, *D. dilatatum*, *D. travancoricum*, *D. polypodioides*, *D. esculentum* and

*D. brachylobum* (Antonisamy 1997), *Pteris vittata*, *P. multiaurita*, *P. pellucida*, *P. cretica*, *P. scabripes*, *P. biaurita*, *P. kleiniana*, *P. aspericaulis*, *P. mertensoides*, *P. otaria*, *P. argyraea*, *P. linearis*, *P. confusa*, *Acrostichum aureum* (Jesudass 1997), *Tectaria paradoxa*, *T. wightii*, *T. coadunata*, *Lastreopsis tenera*, *Polystichum harpophyllum*, *P. piceopaleaceum*, *P. moluscense*, *P. squarrosum*, *Arachniodes tripinnata*, *A. aristata*, *A. amabilis*, *Dryopteris hirtipes*, *D. madrasensis*, *D. cochleata*, *D. sparsa*, *D. approximate* (Joseph 1997), *Histiopteris incisa*, *Hypolepis glandulifera*, *Odontosoria chinensis*, *Nephrolepis auriculata*, *N. multiflora*, *Trichomanes obscurum*, *Dicranopteris linearis* var. *sebastiana*, *D. linearis* var. *tenuis*, *D. linearis* var. *brevis*, *Blechnum orientale*, *B. occidentale*, *Cyathea crinita*, *C. nilgirensis*, *Osmunda regalis*, *Pteridium aquilinum*, *Leptochilus decurrens* and *L. decurrens* f. *lanceolatus* (Patric Raja 1997), *Elaphoglossum beddomei* (Maridass and Raju 2010), *Selaginella inaequalifolia* (Irudayaraj et al. (2010), *S. involvens* (Irene Pearl et al. (2011), *S. tenera* (Suganya et al. 2011), *Adiantum lunulatum*, *A. capillus-veneris*, *Pteris otaria*, *P. aspericaulis*, *P. kleiniana*, *P. confusa*, *P. multiaurita*, *P. vittata*, *Asplenium polyodon* and *Hypodematium crenatum* (De Britto et al. 2012a), *Pteris argyraea*, *P. confusa*, *P. vittata*, *P. biaurita* and *P. multiaurita* (Herinsheeba et al. 2013), *Actiniopteris radiata* (Manonmani and Sara 2013), *Marsilea quadrifolia* (Sivagurunathan and Xavier Innocent 2014), *Cheilanthes farinosa*, *C. anceps*, *C. tenuifolia* and *C. albomarginata* (Pradnya et al. 2015), *Pneumatopteris callosa* (Daryono and Rhomawati 2020) and *Ceratopteris thalictroides* (Smitha and Vadivel 2019) is reported.

### 19.5.3 Amino Acids

Goswami and Khandelwal (1976) and Khandelwal and Goswami (1976) studied the occurrence of amino acids in different species of Indian *Ophioglossum* while Khare and Shankar (1987) studied the amino acids and proteins in *Psilotum nudum* collected from Pachmarhi. The amino acid profiles of Indian pteridophytes are reported by various localities and researchers, viz. Rajasthan ferns (Kaur et al. 1986). Amino acids existence in the *Dryoathyrium boryanum*, *Athyrium puncticaule*, *A. solenopteris*, *A. lanceum*, *Deparia petersenii*, *Diplazium muricatum*, *D. sylvaticum*, *D. dilatatum*, *D. travancoricum*, *D. polypodioides*, *D. esculentum* and *D. brachylobum* (Antonisamy 1997), *Pteris vittata*, *P. multiaurita*, *P. pellucida*, *P. cretica*, *P. scabripes*, *P. biaurita*, *P. kleiniana*, *P. aspericaulis*, *P. mertensoides*, *P. otaria*, *P. argyraea*, *P. linearis*, *P. confusa*, *Acrostichum aureum* (Jesudass 1997), *Tectaria paradoxa*, *T. wightii*, *T. coadunata*, *Lastreopsis tenera*, *Polystichum harpophyllum*, *P. piceopaleaceum*, *P. moluscense*, *P. squarrosum*, *Arachniodes tripinnata*, *A. aristata*, *A. amabilis*, *Dryopteris hirtipes*, *D. madrasensis*, *D. cochleata*, *D. sparsa*, *D. approximate* (Joseph 1997), *Histiopteris incisa*, *Hypolepis glandulifera*, *Odontosoria chinensis*, *Nephrolepis auriculata*, *N. multiflora*, *Trichomanes obscurum*, *Dicranopteris linearis* var. *sebastiana*, *D. linearis* var. *tenuis*, *D. linearis* var. *brevis*, *Blechnum orientale*, *B. occidentale*,

*Cyathea crinita*, *C. nilgirensis*, *Osmunda regalis*, *Pteridium aquilinum*, *Leptochilus decurrens* and *L. decurrens* f. *lanceolatus* (Patric Raja 1997), *Selaginella* sp. (Paramanik et al. 2002), *Bolbitis appendiculata*, *B. virens*, *Osmunda regalis*, *Ceratopteris thalictroides* and *Drynaria quercifolia* (Kale and Dongarae 2007), *Azolla pinnata*, *Marsilea minuta* and *Salvinia molesta* (Mithraja et al. 2011), *Selaginella involvens* (Irene Pearl et al. 2011), *Diplazium muricatum*, *Diplazium travancoricum* and *Diplazium brachylobum* (Sivaraman et al. 2011), *Pteris argyraea*, *Pteris confusa*, *Pteris vittata*, *Pteris biaurita* and *Pteris multiaurita* (Herinsheeba et al. 2013), *Actiniopteris radiate* (Manonmani and Sara 2013), *Marsilea quadrifolia* (Sivagurunathan and Xavier Innocent 2014), *Selaginella bryopteris* (Rupa and Bhavani 2014), *Pteris vittata* (Maneesha et al. 2015), *Cheilanthes farinosa*, *Cheilanthes anceps*, *Cheilanthes tenuifolia* and *Cheilanthes albomarginata* (Pradnya et al. 2015), *Diplazium esculentum* (Dash et al. 2017) and *Tectaria wightii* (Vineesh et al. 2021) is validated.

### 19.5.4 Proteins

Proteins are found in *Psilotum nudum* (Khare and Shankar 1987), *Isoetes* (Rathore and Sharma 1990), *Dryoathyrium boryanum*, *Athyrium puncticaule*, *A. solenopteris*, *A. lanceum*, *Deparia petersenii*, *Diplazium muricatum*, *D. sylvaticum*, *D. dilatatum*, *D. travancoricum*, *D. polypodioides*, *D. esculentum* and *D. brachylobum* (Antonisamy 1997), *Pteris vittata*, *P. multiaurita*, *P. pellucida*, *P. cretica*, *P. scabripes*, *P. biaurita*, *P. kleiniana*, *P. aspericaulis*, *P. mertensoides*, *P. otaria*, *P. argyraea*, *P. linearis*, *P. confusa*, *Acrostichum aureum* (Jesudass 1997), *Tectaria paradoxa*, *T. wightii*, *T. coadunata*, *Lastreopsis tenera*, *Polystichum harpophyllum*, *P. piceopaleaceum*, *P. moluscense*, *P. squarrosum*, *Arachniodes tripinnata*, *A. aristata*, *A. amabilis*, *Dryopteris hirtipes*, *D. madrasensis*, *D. cochleata*, *D. sparsa*, *D. approximate* (Joseph 1997), *Histiopteris incisa*, *Hypolepis glandulifera*, *Odontosoria chinensis*, *Nephrolepis auriculata*, *N. multiflora*, *Trichomanes obscurum*, *Dicranopteris linearis* var. *sebastiana*, *D. linearis* var. *tenuis*, *D. linearis* var. *brevis*, *Blechnum orientale*, *B. occidentale*, *Cyathea crinita*, *C. nilgirensis*, *Osmunda regalis*, *Pteridium aquilinum*, *Leptochilus decurrens* and *L. decurrens* f. *lanceolatus* (Patric Raja 1997), *Bolbitis appendiculata*, *B. virens*, *Osmunda regalis*, *Ceratopteris thalictroides* and *Drynaria quercifolia* (Kale and Dongarae 2007), *Azolla pinnata*, *Marsilea minuta* and *Salvinia molesta* (Mithraja et al. 2011), *Tectaria zeylanica* (Sukumaran et al. 2012), *Marsilea quadrifolia* (Sivagurunathan and Xavier Innocent 2014), *Diplazium esculentum* (Dash et al. 2017), *Adiantum lunulatum* (Jenat and Suresh 2018), *A. aureum* (Arockia Badhsheeba and Vadivel 2020) and *Tectaria wightii* (Vineesh et al. 2021).

### 19.5.5 Lipids

The available literature confirmed the occurrence of lipids in *Dryoathyrium boryanum*, *Athyrium puncticaule*, *A. solenopteris*, *A. lanceum*, *Deparia petersenii*, *Diplazium muricatum*, *D. sylvaticum*, *D. dilatatum*, *D. travancoricum*, *D. polypodioides*, *D. esculentum* and *D. brachylobum* (Antonisamy 1997), *Pteris vittata*, *P. multiaurita*, *P. pellucida*, *P. cretica*, *P. scabripes*, *P. biaurita*, *P. kleiniana*, *P. aspericaulis*, *P. mertensoides*, *P. otaria*, *P. argyraea*, *P. linearis*, *P. confusa*, *Acrostichum aureum* (Jesudass 1997), *Tectaria paradoxa*, *T. wightii*, *T. coadunata*, *Lastreopsis tenera*, *Polystichum harpophyllum*, *P. piceopaleaceum*, *P. moluscense*, *P. squarrosum*, *Arachniodes tripinnata*, *A. aristata*, *A. amabilis*, *Dryopteris hirtipes*, *D. madrasensis*, *D. cochleata*, *D. sparsa*, *D. approximate* (Joseph 1997), *Histiopteris incisa*, *Hypolepis glandulifera*, *Odontosoria chinensis*, *Nephrolepis auriculata*, *N. multiflora*, *Trichomanes obscurum*, *Dicranopteris linearis* var. *sebastiana*, *D. linearis* var. *tenuis*, *D. linearis* var. *brevis*, *Blechnum orientale*, *B. occidentale*, *Cyathea crinita*, *C. nilgirensis*, *Osmunda regalis*, *Pteridium aquilinum*, *Leptochilus decurrens* and *L. decurrens* f. *lanceolatus* (Patric Raja 1997), *Diplazium muricatum*, *D. travancoricum* and *D. brachylobum* (Sivaraman et al. 2011). Vineesh et al. (2021) confirmed the existence of oils and fats in *Tectaria wightii*.

### 19.5.6 Steroid

The occurrence of steroids is noted in *Dryoathyrium boryanum*, *Athyrium puncticaule*, *A. solenopteris*, *A. lanceum*, *Deparia petersenii*, *Diplazium muricatum*, *D. sylvaticum*, *D. dilatatum*, *D. travancoricum*, *D. polypodioides*, *D. esculentum* and *D. brachylobum* (Antonisamy 1997), *Pteris vittata*, *P. multiaurita*, *P. pellucida*, *P. cretica*, *P. scabripes*, *P. biaurita*, *P. kleiniana*, *P. aspericaulis*, *P. mertensoides*, *P. otaria*, *P. argyraea*, *P. linearis*, *P. confusa*, *Acrostichum aureum* (Jesudass 1997), *Tectaria paradoxa*, *T. wightii*, *T. coadunata*, *Lastreopsis tenera*, *Polystichum harpophyllum*, *P. piceopaleaceum*, *P. moluscense*, *P. squarrosum*, *Arachniodes tripinnata*, *A. aristata*, *A. amabilis*, *Dryopteris hirtipes*, *D. madrasensis*, *D. cochleata*, *D. sparsa*, *D. approximate* (Joseph 1997), *Histiopteris incisa*, *Hypolepis glandulifera*, *Odontosoria chinensis*, *Nephrolepis auriculata*, *Trichomanes obscurum*, *Dicranopteris linearis* var. *sebastiana*, *Dicranopteris linearis* var. *tenuis*, *Dicranopteris linearis* var. *brevis*, *Blechnum orientale*, *Blechnum occidentale*, *Nephrolepis multiflora*, *Cyathea crinita*, *Cyathea nilgirensis*, *Osmunda regalis*, *Pteridium aquilinum Leptochilus decurrens* and *L. decurrens* f. *lanceolatus* (Patric Raja 1997), *Selaginella lepidophylla* (Chikmawati and Miftahudin 2008), *S. inaequalifolia* (Irudayaraj et al. 2010), *Azolla pinnata*, *Marsilea minuta* and *Salvinia molesta* (Mithraja et al. 2011), *Selaginella involvens* (Irene Pearl et al. 2011), *Diplazium muricatum*, *D. travancoricum* and *D. brachylobum* (Sivaraman et al. 2011), *Selaginella tenera* (Suganya et al. 2011), *Adiantum lunulatum*, *A. capillus-veneris*, *Pteris otaria*, *P. aspericaulis*, *P. kleiniana*,

*P. confusa*, *P. multiaurita*, *P. vittata*, *Asplenium polyodon* and *Hypodematium crenatum* (De Britto et al. 2012a), *Tectaria zeylanica* (Sukumaran et al. 2012), *Pteris argyraea*, *P. confusa*, *P. vittata*, *P. biaurita* and *P. multiaurita* (Herinsheeba et al. 2013), *Actiniopteris radiata* (Manonmani and Sara 2013), *A. radiata*, *Acrostichum aureum*, *Dryopteris cochleata*, *Drynaria quercifolia*, *Hemionitis arifolia* and *Pityrogramma calomelanos* (Kalpana Devi et al. 2014a, b), *Selaginella bryopteris* (Rupa and Bhavani 2014), *Salvinia auriculata* (Suvarnalatha et al. 2015), *Cyathea nilgirensis*, *C. gigantea* and *C. crinita* (Janakiraman and Johnson 2015a, b), *Cheilanthes farinosa*, *C. anceps*, *C. tenuifolia* and *C. albomarginata* (Pradnya et al. 2015), *Bolbitis virens*, *B. appendiculata* and *B. presliana* (Manisha 2015), *Salvinia molesta* (Nithya et al. 2015), *Dryopteris cochleata* (Das et al. 2015), *Drynaria quercifolia* (Padhy and Dash 2015), *Dicranopteris linearis* (Kalpana Devi et al. 2016a, b), *Adiantum capillus-veneris* (Ranjan and Vats 2016), *Adiantum latifolium*, *Angiopteris evecta* and *Marattia fraxinea* (Chandra et al. 2017), *Azolla pinnata* (Thiripurasundari and Padmini 2018), *Asplenium indicum*, *Lepisorus nudus* and *Microsorum membranaceum* (Jadhav et al. 2019), *Ceratopteris thalictroides* (Smitha and Vadivel 2019), *Asplenium aureum* (Arockia Badhsheeba and Vadivel 2020) and *Sphaerostephanos unitus* (Johnson et al. 2020a, b, c).

### 19.5.7 Saponins

The available literature confirmed the presence of saponins in *Dryoathyrium boryanum*, *Athyrium puncticaule*, *A. solenopteris*, *A. lanceum*, *Deparia petersenii*, *Diplazium muricatum*, *D. sylvaticum*, *D. dilatatum*, *D. travancoricum*, *D. polypodioides*, *D. esculentum* and *D. brachylobum* (Antonisamy 1997), *Histiopteris incisa*, *Hypolepis glandulifera*, *Odontosoria chinensis*, *Nephrolepis auriculata*, *Trichomanes obscurum*, *Dicranopteris linearis* var. *sebastiana*, *D. linearis* var. *tenuis*, *D. linearis* var. *brevis*, *Blechnum orientale*, *B. occidentale*, *Nephrolepis multiflora*, *Cyathea crinita*, *C. nilgirensis*, *Osmunda regalis*, *Pteridium aquilinum*, *Leptochilus decurrens* and *L. decurrens* f. *lanceolatus* (Patric Raja 1997), *Selaginella lepidophylla* (Chikmawati and Miftahudin 2008), *Elaphoglossum beddomei* (Maridass and Raju 2010), *Azolla pinnata*, *Marsilea minuta* and *Salvinia molesta* (Mithraja et al. 2011), *Selaginella involvens* (Irene Pearl et al. 2011), *Diplazium muricatum*, *D. travancoricum* and *D. brachylobum* (Sivaraman et al. 2011), *Christella parasitica* (Paulraj et al. 2011), *Tectaria zeylanica* (Sukumaran et al. 2012), *Pteris argyraea*, *P. confusa*, *P. vittata*, *P. biaurita* and *P. multiaurita* (Herinsheeba et al. 2013), *Actiniopteris radiata* (Manonmani and Sara 2013), *Actiniopteris radiata*, *Acrostichum aureum*, *Dryopteris cochleata*, *Drynaria quercifolia*, *Hemionitis arifolia* and *Pityrogramma calomelanos* (Kalpana Devi et al. 2014a, b), *Marsilea quadrifolia* (Sivagurunathan and Xavier Innocent 2014), *Nephrolepis cordifolia* (Rukmini and Suvarnalatha 2014), *Salvinia auriculata* (Suvarnalatha et al. 2015), *Cyathea nilgirensis*, *C. gigantea* and *C. crinita* (Janakiraman and Johnson 2015a, b), *Cheilanthes farinosa*, *C. anceps*, *C. tenuifolia* and *C. albomarginata* (Pradnya et al. 2015), *Lygodium flexuosum*

(Singh 2017), *Salvinia molesta* (Nithya et al. 2015), *Drynaria quercifolia* (Padhy and Dash 2015), *Dicranopteris linearis* (Kalpana Devi et al. 2016a, b), *Salvinia molesta* (Gaya et al. 2016), *Adiantum capillus-veneris* (Ranjan and Vats 2016), *Stenochlaena palustris* (Arullappan et al. 2017), *Adiantum latifolium*, *Angiopteris evecta* and *Marattia fraxinea* (Chandra et al. 2017), *Tectaria cicutaria* (Preeti and Namdeo 2018), *Adiantum lunulatum* (Jenat and Suresh 2018), *Azolla pinnata* (Thiripurasundari and Padmini 2018; Farook et al. 2019), *Asplenium indicum*, *Lepisorus nudus* and *Microsorum membranaceum* (Jadhav et al. 2019), *Ceratopteris thalictroides* (Smitha and Vadivel 2019), *A. aureum* (Arockia Badhsheeba and Vadivel 2020), *Adiantum philippense* (Adnan et al. 2020), *Sphaerostephanos unitus* (Johnson et al. 2020b, b, c) and *Tectaria wightii* (Vineesh et al. 2021).

### 19.5.8 Terpenoids and Triterpenoids

*Tectaria paradoxa*, *T. wightii*, *T. coadunata*, *Lastreopsis tenera*, *Polystichum harpophyllum*, *P. piceopaleaceum*, *P. moluscense*, *P. squarrosum*, *Arachniodes tripinnata*, *A. aristata*, *A. amabilis*, *Dryopteris hirtipes*, *D. madrasensis*, *D. cochleata*, *D. sparsa*, *D. approximate* (Joseph 1997), *Leptochilus decurrens* f. *lanceolatus* and *L. decurrens* (Patricraja 1997), *Selaginella lepidophylla* (Chikmawati and Miftahudin 2008), *S. inaequalifolia* (Irudayaraj et al. 2010), *Elaphoglossum beddomei* (Maridass and Raju 2010), *S. involvens* (Irene Pearl et al. 2011), *Asplenium affine*, *A. decrescens* and *A. zenkeranum* (Irudayaraj and Johnson 2011), *Diplazium muricatum*, *D. travancoricum* and *D. brachylobum* (Sivaraman et al. 2011), *S. tenera* (Suganya et al. 2011), *Christella parasitica* (Paulraj et al. 2011), *Pteris argyraea*, *P. confusa*, *P. vittata*, *P. biaurita* and *P. multiaurita* (Herinsheeba et al. 2013), *Actiniopteris radiata* (Manonmani and Sara 2013), *A. radiata*, *Acrostichum aureum*, *Dryopteris cochleata*, *Drynaria quercifolia*, *Hemionitis arifolia* and *Pityrogramma calomelanos* (Kalpana Devi et al. 2014a, b), *Pyrrosia lanceolata* (Ruby and Sara 2014), *Selaginella bryopteris* (Rupa and Bhavani 2014), *Cheilanthes farinosa*, *C. anceps*, *C. tenuifolia* and *C. albomarginata* (Pradnya et al. 2015), *Dryopteris cochleata* (Das et al. 2015), *Pteris vittata* (Maneesha et al. 2015), *Salvinia auriculata* (Suvarnalatha et al. 2015), *S. molesta* (Nithya et al. 2015), *Adiantum capillus-veneris* (Ranjan and Vats 2016), *Dicranopteris linearis* (Kalpana Devi et al. 2016a, b), *Salvinia molesta* (Gaya et al. 2016), *A. latifolium*, *Angiopteris evecta* and *Marattia fraxinea* (Chandra et al. 2017), *Diplazium esculentum* (Dash et al. 2017), *Stenochlaena palustris* (Arullappan et al. 2017), *Adiantum lunulatum* (Jenat and Suresh 2018), *Tectaria cicutaria* (Preeti and Namdeo 2018), *Azolla pinnata* (Farook et al. 2019), *Asplenium indicum*, *Lepisorus nudus* and *Microsorum membranaceum* (Jadhav et al. 2019), *Ceratopteris thalictroides* (Smitha and Vadivel 2019), *Pneumatopteris callosa* (Daryono and Rhomawati 2020), *A. aureum* (Arockia Badhsheeba and Vadivel 2020) and *Sphaerostephanos unitus* (Johnson et al. 2020b, b, c) showed the existence of terpenoids and triterpenoids in the ferns and fern allies of India.

### 19.5.9 Alkaloids

The existing literature authenticated the existence of alkaloids in the crude extracts of *Dryoathyrium boryanum*, *Athyrium puncticaule*, *A. solenopteris*, *A. lanceum*, *Deparia petersenii*, *Diplazium muricatum*, *D. sylvaticum*, *D. dilatatum*, *D. travancoricum*, *D. polypodioides*, *D. esculentum* and *D. brachylobum* (Antonisamy 1997), *Tectaria paradoxa*, *T. wightii*, *T. coadunata*, *Lastreopsis tenera*, *Polystichum harpophyllum*, *P. piceopaleaceum*, *P. moluscense*, *P. squarrosum*, *Arachniodes tripinnata*, *A. aristata*, *A. amabilis*, *Dryopteris hirtipes*, *D. madrasensis*, *D. cochleata*, *D. sparsa*, *D. approximate* (Joseph 1997), *Selaginella lepidophylla* (Chikmawati and Miftahudin 2008), *Elaphoglossum beddomei* (Maridass and Raju 2010), *Azolla pinnata*, *Marsilea minuta* and *Salvinia molesta* (Mithraja et al. 2011), *Asplenium affine*, *A. decrescens* and *A. zenkeranum* (Irudayaraj and Johnson 2011), *Christella parasitica* (Paulraj et al. 2011), *Adiantum lunulatum*, *A. capillus-veneris*, *Pteris otaria*, *P. aspericaulis*, *P. kleiniana*, *P. confusa*, *P. multiaurita*, *P. vittata*, *Asplenium polyodon* and *Hypodematium crenatum* (De Britto et al. 2012a), *Pteris argyraea*, *P. confusa*, *P. vittata*, *P. biaurita* and *P. multiaurita* (Herinsheeba et al. (2013), *Actiniopteris radiata* (Manonmani and Sara 2013), *A. radiata*, *Acrostichum aureum*, *Dryopteris cochleata*, *Drynaria quercifolia*, *Hemionitis arifolia* and *Pityrogramma calomelanos* (Kalpana Devi et al. 2014a, b), *Marsilea quadrifolia* (Sivagurunathan and Xavier Innocent 2014), *Selaginella bryopteris* (Rupa and Bhavani 2014), *Nephrolepis cordifolia* (Rukmini and Suvarnalatha 2014), *Salvinia auriculata* (Suvarnalatha et al. 2015), *Pteris vittata* (Maneesha et al. 2015), *Cyathea nilgirensis*, *C. gigantea* and *C. crinita* (Janakiraman and Johnson 2015a, b), *Cheilanthes farinosa*, *C. anceps*, *C. tenuifolia* and *C. albomarginata* (Pradnya et al. 2015), *Bolbitis virens*, *B. appendiculata* and *B. presliana* (Manisha 2015) *Cyathea gigantea* (Manisha 2015), *Selaginella bryopteris* (Jyothi et al. 2015), *Dryopteris cochleata* (Das et al. 2015), *Actiniopteris radiata* (Mathad and Modi 2015), *Salvinia molesta* (Nithya et al. 2015), *Lygodium flexuosum* (Singh 2017), *Cyathea nilgirensis* (Mary and Mahesh 2015), *Dicranopteris linearis* (Kalpana Devi et al. 2016a, b), *Adiantum latifolium*, *Angiopteris evecta* and *Marattia fraxinea* (Chandra et al. 2017), *Salvinia molesta* (Gaya et al. 2016), *Stenochlaena palustris* (Arullappan et al. 2017), *Pneumatopteris callosa* (Daryono and Rhomawati 2020), *Asplenium indicum*, *Lepisorus nudus* and *Microsorum membranaceum* (Jadhav et al. 2019), *Ceratopteris thalictroides* (Smitha and Vadivel 2019), *Azolla pinnata* (Farook et al. 2019), *Adiantum philippense* (Adnan et al. 2020) and *Tectaria wightii* (Vineesh et al. 2021).

### 19.5.10 Phenolics Compounds

The phenolics are found in *Selaginella* (Bohra et al. 1979), *Isoetes* (Rathore and Sharma 1990), *Dryoathyrium boryanum*, *Athyrium puncticaule*, *A. solenopteris*, *A. lanceum*, *Deparia petersenii*, *Diplazium muricatum*, *D. sylvaticum*, *D. dilatatum*, *D. travancoricum*, *D. polypodioides*, *D. esculentum* and

*D. brachylobum* (Antonisamy 1997), *Pteris vittata*, *P. multiaurita*, *P. pellucida*, *P. cretica*, *P. scabripes*, *P. biaurita*, *P. kleiniana*, *P. aspericaulis*, *P. mertensoides*, *P. otaria*, *P. argyraea*, *P. linearis*, *P. confusa*, *Acrostichum aureum* (Jesudass 1997), *Tectaria paradoxa*, *T. wightii*, *T. coadunata*, *Lastreopsis tenera*, *Polystichum harpophyllum*, *P. piceopaleaceum*, *P. moluscense*, *P. squarrosum*, *Arachniodes tripinnata*, *A. aristata*, *A. amabilis*, *Dryopteris hirtipes*, *D. madrasensis*, *D. cochleata*, *D. sparsa*, *D. approximate* (Joseph 1997), *Histiopteris incisa*, *Hypolepis glandulifera*, *Odontosoria chinensis*, *Nephrolepis auriculata*, *Trichomanes obscurum*, *Dicranopteris linearis* var. *sebastiana*, *D. linearis* var. *tenuis*, *D. linearis* var. *brevis*, *Blechnum orientale*, *B. occidentale*, *Nephrolepis multiflora*, *Cyathea crinita*, *Cyathea nilgirensis*, *Osmunda regalis*, *Pteridium aquilinum*, *Leptochilus decurrens* and *L. decurrens* f. *lanceolatus* (Patric Raja 1997), *S. lepidophylla* (Chikmawati and Miftahudin 2008), *S. inaequalifolia* (Irudayaraj et al. 2010), *Azolla pinnata*, *Marsilea minuta* and *Salvinia molesta* (Mithraja et al. 2011), *S. tenera* (Suganya et al. 2011), *Adiantum lunulatum*, *A. capillus-veneris*, *Pteris otaria*, *P. aspericaulis*, *P. kleiniana*, *P. confusa*, *P. multiaurita*, *P. vittata*, *Asplenium polyodon* and *Hypodematium crenatum* (De Britto et al. 2012a), *Pteris argyraea*, *P. confusa*, *P. vittata*, *P. biaurita* and *P. multiaurita* (Herinsheeba et al. 2013), *Actiniopteris radiata* (Manonmani and Sara 2013), *A. radiata*, *Acrostichum aureum*, *Dryopteris cochleata*, *Drynaria quercifolia*, *Hemionitis arifolia* and *Pityrogramma calomelanos* (Kalpana Devi et al. 2014a, b), *Marsilea quadrifolia* (Sivagurunathan and Xavier Innocent 2014), *Pyrrosia lanceolata* (Ruby and Sara 2014), *Nephrolepis cordifolia* (Rukmini and Suvarnalatha 2014), *Salvinia auriculata* (Suvarnalatha et al. 2015), *Pteris vittata* (Maneesha et al. 2015), *Cyathea nilgirensis*, *C. gigantea* and *C. crinita* (Janakiraman and Johnson 2015a, b), *Cheilanthes farinosa*, *C. anceps*, *C. tenuifolia* and *C. albomarginata* (Pradnya et al. 2015), *Bolbitis virens*, *B. appendiculata* and *B. presliana* (Manisha 2015), *Selaginella bryopteris* (Jyothi et al. 2015), *Dryopteris cochleata* (Das et al. 2015), *Actiniopteris radiata* (Mathad and Modi 2015), *Salvinia molesta* (Nithya et al. 2015), *Drynaria quercifolia* (Padhy and Dash 2015), *Adiantum capillus-veneris* (Ranjan and Vats 2016), *Dicranopteris linearis* (Kalpana Devi et al. 2016a, b), *Salvinia molesta* (Gaya et al. 2016), *Adiantum latifolium*, *Angiopteris evecta* and *Marattia fraxinea* (Chandra et al. 2017), *Stenochlaena palustris* (Arullappan et al. 2017), *Adiantum lunulatum* (Jenat and Suresh 2018), *Azolla pinnata* (Thiripurasundari and Padmini 2018), *Tectaria cicutaria* (Preeti and Namdeo 2018), *Asplenium indicum*, *Lepisorus nudus* and *Microsorum membranaceum* (Jadhav et al. 2019), *Ceratopteris thalictroides* (Smitha and Vadivel 2019), *Azolla pinnata* (Farook et al. 2019), *Acrostichum aureum* (Arockia Badhsheeba and Vadivel 2020), *Adiantum philippense* (Adnan et al. 2020) and *Tectaria wightii* (Vineesh et al. 2021).

### 19.5.11 Tannins

The tannins showed their presence in the crude extracts of *Pteris vittata*, *P. multiaurita*, *P. pellucida*, *P. cretica*, *P. scabripes*, *P. biaurita*, *P. kleiniana*, *P. aspericaulis*, *P. mertensoides*, *P. otaria*, *P. argyraea*, *P. linearis*, *P. confusa*, *Acrostichum aureum* (Jesudass 1997), *Tectaria paradoxa*, *T. wightii*, *T. coadunata*, *Lastreopsis tenera*, *Polystichum harpophyllum*, *P. piceopaleaceum*, *P. moluscense*, *P. squarrosum*, *Arachniodes tripinnata*, *A. aristata*, *A. amabilis*, *Dryopteris hirtipes*, *D. madrasensis*, *D. cochleata*, *D. sparsa*, *D. approximate* (Joseph 1997), *Histiopteris incisa*, *Hypolepis glandulifera*, *Odontosoria chinensis*, *Nephrolepis auriculata*, *Trichomanes obscurum*, *Dicranopteris linearis* var. *sebastiana*, *D. linearis* var. *tenuis*, *D. linearis* var. *brevis*, *Blechnum orientale*, *B. occidentale*, *Nephrolepis multiflora*, *Cyathea crinita*, *C. nilgirensis*, *Osmunda regalis*, *Pteridium aquilinum*, *Leptochilus decurrens* and *L. decurrens* f. *lanceolatus* (Patric Raja 1997), *Selaginella lepidophylla* (Chikmawati and Miftahudin 2008), *S. inaequalifolia* (Irudayaraj et al. 2010), *Elaphoglossum beddomei* (Maridass and Raju 2010), *Azolla pinnata*, *Marsilea minuta* and *Salvinia molesta* (Mithraja et al. 2011), *Diplazium muricatum*, *D. travancoricum* and *D. brachylobum* (Sivaraman et al. 2011), *Christella parasitica* (Paulraj et al. 2011), *Selaginella tenera* (Suganya et al. 2011), *Adiantum lunulatum*, *A. capillus-veneris*, *Pteris otaria*, *P. aspericaulis*, *P. kleiniana*, *P. confusa*, *P. multiaurita*, *P. vittata*, *Asplenium polyodon* and *Hypodematium crenatum* (De Britto et al. 2012a), *Tectaria zeylanica* (Sukumaran et al. 2012), *Pteris argyraea*, *P. confusa*, *P. vittata*, *P. biaurita* and *P. multiaurita* (Herinsheeba et al. 2013), *Actiniopteris radiata* (Manonmani and Sara 2013), *A. radiata*, *Acrostichum aureum*, *Dryopteris cochleata*, *Drynaria quercifolia*, *Hemionitis arifolia* and *Pityrogramma calomelanos* (Kalpana Devi et al. 2014a, b), *Marsilea quadrifolia* (Sivagurunathan and Xavier Innocent 2014), *Selaginella bryopteris* (Rupa and Bhavani 2014), *Nephrolepis cordifolia* (Rukmini and Suvarnalatha 2014), *Salvinia auriculata* (Suvarnalatha et al. 2015), *Pteris vittata* (Maneesha et al. 2015), *Cyathea nilgirensis*, *C. gigantea* and *C. crinita* (Janakiraman and Johnson 2015a, b), *Cheilanthes farinosa*, *C. anceps*, *C. tenuifolia* and *C. albomarginata* (Pradnya et al. 2015), *Bolbitis virens*, *B. appendiculata* and *B. presliana* (Manisha 2015), *Selaginella* bryopteris (Jyothi et al. 2015), *Actiniopteris radiata* (Mathad and Modi 2015), *Salvinia molesta* (Nithya et al. 2015), *Drynaria quercifolia* (Padhy and Dash 2015), *Adiantum capillus-veneris* (Ranjan and Vats 2016), *Diplazium esculentum* (Dash et al. 2017), *Adiantum latifolium*, *Angiopteris evecta* and *Marattia fraxinea* (Chandra et al. 2017), *Stenochlaena palustris* (Arullappan et al. 2017), *Adiantum lunulatum* (Jenat and Suresh 2018), *Tectaria cicutaria* (Preeti and Namdeo 2018), *Asplenium indicum*, *Lepisorus nudus* and *Microsorum membranaceum* (Jadhav et al. 2019), *Ceratopteris thalictroides* (Smitha and Vadivel 2019), *Pneumatopteris callosa* (Daryono and Rhomawati 2020), *Sphaerostephanos unitus* (Johnson et al. 2020b, b, c) and *Tectaria wightii* (Vineesh et al. 2021).

### 19.5.12 Flavonoids

Flavonoids existence was observed in the crude extracts of *Dryoathyrium boryanum*, *Athyrium puncticaule*, *A. solenopteris*, *A. lanceum*, *Deparia petersenii*, *Diplazium muricatum*, *D. sylvaticum*, *D. dilatatum*, *D. travancoricum*, *D. polypodioides*, *D. esculentum* and *D. brachylobum* (Antonisamy 1997), *Pteris vittata*, *P. multiaurita*, *P. pellucida*, *P. cretica*, *P. scabripes*, *P. biaurita*, *P. kleiniana*, *P. aspericaulis*, *P. mertensoides*, *P. otaria*, *P. argyraea*, *P. linearis*, *P. confusa*, *Acrostichum aureum* (Jesudass 1997), *Tectaria paradoxa*, *T. wightii*, *T. coadunata*, *Lastreopsis tenera*, *Polystichum harpophyllum*, *P. piceopaleaceum*, *P. moluscense*, *P. squarrosum*, *Arachniodes tripinnata*, *A. aristata*, *A. amabilis*, *Dryopteris hirtipes*, *D. madrasensis*, *D. cochleata*, *D. sparsa*, *D. approximate* (Joseph 1997), *Histiopteris incisa*, *Hypolepis glandulifera*, *Odontosoria chinensis*, *Nephrolepis auriculata*, *N. multiflora*, *Trichomanes obscurum*, *Dicranopteris linearis* var. *sebastiana*, *D. linearis* var. *tenuis*, *D. linearis* var. *brevis*, *Blechnum orientale*, *B. occidentale*, *C. crinita*, *C. nilgirensis*, *Osmunda regalis*, *Pteridium aquilinum*, *Leptochilus decurrens* and *L. decurrens* f. *lanceolatus* (Patric Raja 1997), *Selaginella lepidophylla* (Chikmawati and Miftahudin 2008), *Azolla pinnata*, *Marsilea minuta* and *Salvinia molesta* (Mithraja et al. 2011), *Diplazium muricatum*, *D. travancoricum* and *D. brachylobum* (Sivaraman et al. 2011), *Asplenium affine*, *A. decrescens* and *A. zenkeranum* (Irudayaraj and Johnson 2011), *Christella parasitica* (Paulraj et al. 2011), *Adiantum lunulatum*, *A. capillus-veneris*, *Pteris otaria*, *P. aspericaulis*, *P. kleiniana*, *P. confusa*, *P. multiaurita*, *P. vittata*, *Asplenium polyodon* and *Hypodematium crenatum* (De Britto et al. 2012a), *Pteris argyraea*, *P. confusa*, *P. vittata*, *P. biaurita* and *P. multiaurita* (Herinsheeba et al. 2013), *Actiniopteris radiata* (Manonmani and Sara 2013), *A. radiata*, *Acrostichum aureum*, *Dryopteris cochleata*, *Drynaria quercifolia*, *Hemionitis arifolia* and *Pityrogramma calomelanos* (Kalpana Devi et al. 2014a, b), *Marsilea quadrifolia* (Sivagurunathan and Xavier Innocent 2014), *Selaginella bryopteris* (Rupa and Bhavani 2014), *Nephrolepis cordifolia* (Rukmini and Suvarnalatha 2014), *Salvinia auriculata* (Suvarnalatha et al. 2015), *Cyathea nilgirensis*, *C. gigantea* and *C. crinita* (Janakiraman and Johnson 2015a, b), *Cheilanthes farinosa*, *C. anceps*, *C. tenuifolia* and *C. albomarginata* (Pradnya et al. 2015), *Selaginella bryopteris* (Jyothi et al. 2015), *Actiniopteris radiata* (Mathad and Modi 2015), *Dryopteris cochleata* (Das et al. 2015), *Salvinia molesta* (Nithya et al. 2015), *Drynaria quercifolia* (Padhy and Dash 2015), *Cyathea nilgirensis* (Sahaya Mary and Mahesh 2015), *Dicranopteris linearis* (Kalpana Devi et al. 2016a, b), *Salvinia molesta* (Gaya et al. 2016), *Adiantum capillus-veneris* (Ranjan and Vats (2016), *Diplazium esculentum* (Dash et al. 2017), *Adiantum latifolium*, *Angiopteris evecta* and *Marattia fraxinea* (Chandra et al. 2017), *Stenochlaena palustris* (Arullappan et al. 2017), *Azolla pinnata* (Thiripurasundari and Padmini 2018), *Pneumatopteris callosa* (Daryono and Rhomawati 2020), *Asplenium indicum*, *Lepisorus nudus* and *Microsorum membranaceum* (Jadhav et al. 2019), *Ceratopteris thalictroides* (Smitha and Vadivel 2019), *Acrostichum aureum* (Arockia Badhsheeba and Vadivel

2020), *Adiantum philippense* (Adnan et al. 2020), *Sphaerostephanos unitus* (Johnson et al. 2020b, b, c) and *Tectaria wightii* (Vineesh et al. 2021).

### 19.5.13 Catechins

The presence of catechins was documented in the crude extracts of *Dryoathyrium boryanum*, *Athyrium puncticaule*, *A. solenopteris*, *A. lanceum*, *Deparia petersenii*, *Diplazium muricatum*, *D. sylvaticum*, *D. dilatatum*, *D. travancoricum*, *D. polypodioides*, *D. esculentum* and *D. brachylobum* (Antonisamy 1997), *Pteris vittata*, *P. multiaurita*, *P. pellucida*, *P. cretica*, *P. scabripes*, *P. biaurita*, *P. kleiniana*, *P. aspericaulis*, *P. mertensoides*, *P. otaria*, *P. argyraea*, *P. linearis*, *P. confusa*, *Acrostichum aureum* (Jesudass 1997), *Tectaria paradoxa*, *T. wightii*, *T. coadunata*, *Lastreopsis tenera*, *Polystichum harpophyllum*, *P. piceopaleaceum*, *P. moluscense*, *P. squarrosum*, *Arachniodes tripinnata*, *A. aristata*, *A. amabilis*, *Dryopteris hirtipes*, *D. madrasensis*, *D. cochleata*, *D. sparsa*, *D. approximate* (Joseph 1997), *Selaginella inaequalifolia* (Irudayaraj et al. 2010), *Selaginella involvens* (Irene Pearl et al. 2011), *Actiniopteris radiata* (Manonmani and Sara 2013) and *Pyrrosia lanceolata* (Ruby and Sara 2014).

### 19.5.14 Glycosides and Cardiac Glycosides

The available literature confirms the occurrence of glycosides in the crude extracts of *Diplazium muricatum*, *D. travancoricum* and *D. brachylobum* (Sivaraman et al. 2011), *Pyrrosia lanceolata* (Ruby and Sara 2014), *Selaginella bryopteris* (Rupa and Bhavani 2014), *Bolbitis virens*, *B. appendiculata* and *B. presliana* (Manisha 2015), *Cyathea gigantea* (Manisha 2015), *Dryopteris cochleata* (Das et al. 2015), *Dicranopteris linearis* (Kalpana Devi et al. 2016a, b), *Stenochlaena palustris* (Arullappan et al. 2017), *Adiantum lunulatum* (Jenat and Suresh 2018), *Asplenium indicum*, *Lepisorus nudus* and *Microsorum membranaceum* (Jadhav et al. 2019), *Ceratopteris thalictroides* (Smitha and Vadivel 2019), *Acrostichum aureum* (Arockia Badhsheeba and Vadivel 2020) and *Adiantum philippense* (Adnan et al. 2020). The cardiac glycosides existence is confirmed in *Actiniopteris radiata*, *Acrostichum aureum*, *Dryopteris cochleata*, *Drynaria quercifolia*, *Hemionitis arifolia* and *Pityrogramma calomelanos* (Kalpana Devi et al. 2014a, b), *Cyathea nilgirensis*, *C. gigantea* and *C. crinita* (Janakiraman and Johnson 2015a, b), *Dicranopteris linearis* (Kalpana Devi et al. 2016a, b), *Adiantum latifolium*, *Angiopteris evecta* and *Marattia fraxinea* (Chandra et al. 2017) and *Sphaerostephanos unitus* (Johnson et al. 2020b, b, c).

### 19.5.15 Anthraquinone

The anthraquinone presence was reported in *Lastreopsis tenera*, *Polystichum piceopaleaceum*, *Arachniodes tripinnata*, *Dryopteris madrasensis*, *D. approximata*, *D. cochleata*, *D. sparsa* (Joseph 1997) and *Ceratopteris thalictroides* (Smitha and Vadivel 2019).

### 19.5.16 Coumarin

*Azolla pinnata*, *Marsilea minuta* and *Salvinia molesta* (Mithraja et al. 2011), *Tectaria zeylanica* (Sukumaran et al. 2012), *Actiniopteris radiata*, *Acrostichum aureum*, *Dryopteris cochleata*, *Drynaria quercifolia*, *Hemionitis arifolia* and *Pityrogramma calomelanos* (Kalpana Devi et al. 2014a, b), *Dicranopteris linearis* (Kalpana Devi et al. 2016a, b), *Ceratopteris thalictroides* (Smitha and Vadivel 2019) and *Azolla pinnata* (Farook et al. 2019) crude extracts showed the presence of coumarin.

### 19.5.17 Betacyanin

The presence of betacyanin is reported in *Actiniopteris radiata*, *Acrostichum aureum*, *Dryopteris cochleata*, *Drynaria quercifolia*, *Hemionitis arifolia* and *Pityrogramma calomelanos* (Kalpana Devi et al. 2014a, b) and *Dicranopteris linearis* (Kalpana Devi et al. 2016a, b).

### 19.5.18 Quinone

The existing literature confirms the occurrence of quinine in *Azolla pinnata*, *Marsilea minuta and Salvinia molesta* (Mithraja et al. 2011), *Actiniopteris radiata*, *Acrostichum aureum*, *Dryopteris cochleata*, *Drynaria quercifolia*, *Hemionitis arifolia* and *Pityrogramma calomelanos* (Kalpana Devi et al. 2014a, b), *Dicranopteris linearis* (Kalpana Devi et al. 2016a, b), *Salvinia molesta* (Nithya et al. 2015), *Asplenium indicum*, *Lepisorus nudus* and *Microsorum membranaceum* (Jadhav et al. 2019) and *Ceratopteris thalictroides* (Smitha and Vadivel 2019) crude extracts.

Phytochemical analysis has been done on a large number of Indian fern and fern allies. But the majority of the reports pertain to quantitative estimation of primary metabolites which are universal in occurrence and are highly variable with the environment. Figure 19.3 summarized the preliminary phytochemical studies on Indian pteridophytes.

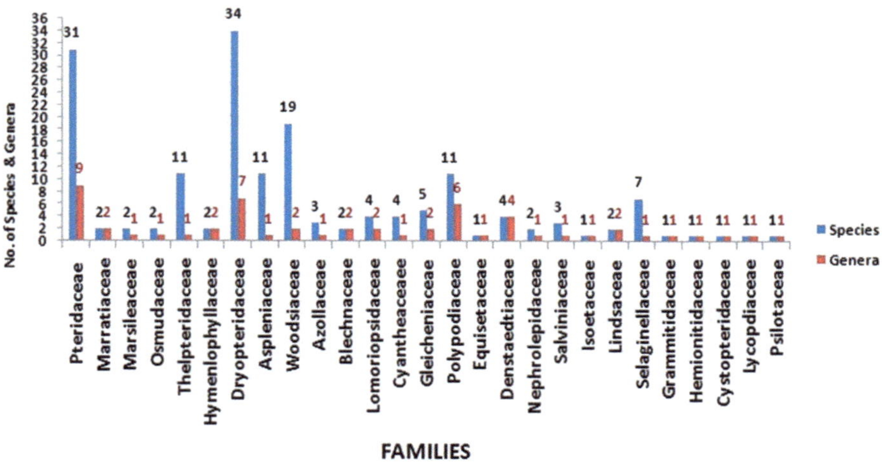


**Fig. 19.3** Phytochemical studies on Indian pteridophytes: f amily, genus and species level

## 19.6 Chromatographic Studies on Pteridophytes

Plant extracts are evaluated by different biological methods to find out the pharmacological activity. Qualitative and quantitative chemical examination is designed to detect and isolate the active ingredients (AOAC 2005). Due to the advances in analytical instrumentation, the quantitative determination of phytochemicals is made very easy. Recent developments in the isolation, purification and structure elucidation of naturally occurring metabolites have made it possible to establish appropriate strategies for the process of standardization. The modern pharmacognosist adapt various analytical tools for the quality assessment of the crude drugs, which includes fluorescence, spectroscopic (UV-Vis, FT-IR, MS, NMR), chromatographic (paper, TLC, HPLC, HPTLC and GC-MS) and electrophoretic analysis. The results from these sophisticated techniques provide a chemical fingerprint as to the nature of chemicals or impurities present in the plant extract (Bilia et al. 2002; Rozylo et al. 2002; WHO 2002). The chromatographic techniques are accepted as a strategy for identification and evaluation of the quality of plant medicines (Farnsworth et al. 1985). In chemotaxonomy, they are used to distinguish the medicinal sources from its adulterants and standardization of plant products and are used as taxonomic tool to classify the medicinal plants. Information on these chemical constituents not only aid in discovering new therapeutic drugs, but such information can also help in disclosing new sources of economic materials which are precursors for the synthesis of complex chemical substances (Farnsworth 1996). In the recent years, advancement in chromatographic and spectral fingerprints plays an important role in the quality control of complex herbal medicines. Chemical fingerprints obtained using chromatographic methods are used not only for the purpose of quality control of herbal medicines but also for authentication and identification of herbal plant.

Amino acids and sugar profiles of *Tectaria paradoxa*, *T. wightii*, *T. coadunata*, *Lastreopsis tenera*, *Polystichum harpophyllum*, *P. piceopaleaceum*, *P. moluscense*, *P. squarrosum*, *Arachniodes tripinnata*, *A. aristata*, *A. amabilis*, *Dryopteris hirtipes*, *D. madrasensis*, *D. cochleata*, *D. sparsa*, *D. approximata*, *Pteris vittata*, *P. multiaurita*, *P. pellucida*, *P. cretica*, *P. scabripes*, *P. biaurita*, *P. kleiniana*, *P. aspericaulis*, *P. mertensoides*, *P. otaria*, *P. argyraea*, *P. linearis*, *P. confusa*, *Acrostichum aureum*, *Dryoathyrium boryanum*, *Athyrium puncticaule*, *Athyrium solenopteris*, *A. lanceum*, *Deparia petersenii*, *Diplazium muricatum*, *D. sylvaticum*, *D. dilatatum*, *D. travancoricum*, *D. polypodioides*, *D. esculentum* and *D. brachylobum* have been studied by ascending and descending types of paper chromatographic methods (Antonisamy 1997; Joseph 1997; Jesudass 1997). They calculated the paired affinity indices of sugar and amino acids among the studied ferns by the method described by Ellison et al. (1962). These profiles are employed as marker to differentiate the species, and these profiles solved the taxonomical conflict between and among the ferns.

TLC and HPTLC are efficient tools for the phytochemical evaluation and widely accepted technique for its high accuracy, precision and reproducibility of results. It has many advantages because of high sample throughput at low operating cost, easy sample preparation, short analysis time and analytical assurance (Liebler et al. 1996). TLC has the special ability to assay many samples at the same time on a single plate. TLC enables trustworthy separation and analysis of compounds from a wide range of classes in many types of biological samples (Sherma and Fried 2005). TLC enables reliable separation and analysis of compounds from a wide variety of classes in many types of biological samples (Sherma and Fried 2005). Fingerprint analysis approach using HPTLC has become the most potent tool for quality control of herbal medicines because of its simplicity and reliability (Bobby et al. 2012). It can serve as a tool for identification, authentication and quality control of herbal drugs (Prema et al. 2015). HPTLC is utilized for linking the chemical constituents of the plant with high proficiency which in turn provides unique profiling of that particular plant (Bobby et al. 2012). It is more efficient and faster, and the results are more reliable and reproducible. HPTLC provides the record of separation in the form of a chromatogram with fractions represented as peaks with defined parameters including absorbance (intensity), $R_f$, height and area. Furthermore, the feature of a pictorial fluorescence image of HPTLC coupled with a digital scanning profile is more attractive to herbal analysts for constructing a herbal chromatographic fingerprint. These provide adequate information and parameters for comprehensive identification, assessment and comparison of major active constituent fingerprints in the samples, and it serves as a basis for their use in medicinal preparations. By using chromatographic fingerprints, the authentication of herbal medicines can be accurately conducted even if the amount or concentration of the chemical constituents is not exactly the same for different samples of drug.

Saito et al. (1989) reported the distribution of ptaquiloside and ptaquiloside-like compounds in Pteridaceae by chemical assay (TLC) and observed the widespread occurrence in a variety of ferns, viz. *Cheilanthes myriophylla*, *Cibotium barometz*, *Dennstaedtia scabra*, *Histiopteris incisa*, *Pityrogramma calomelanos*, *Pteris cretica*, *P. nipponica*, *P. oshimensis*, *P. tremula* and *P. wallichiana*. Sharma and Sharma (1992) identified various flavonoids in eight different ferns from Rajasthan by paper chromatography. Krishna and Dawra (1994) reported ptaquiloside in *Pteris quadriaurita* and *Onychium contiguum* for the first time in India by TLC method. Dalli et al. (2007) studied the methanolic extract of *Pteris biaurita* using TLC bioassay, and the active fraction exposed the inhibition zone with $R_f$ value 0.5–0.65. Irudayaraj and Johnson (2011) revealed the interspecific relationship among *Asplenium affine*, *A. decrescens* and *A. zenkeranum*. Pathania et al. (2012) analysed the flavonoid quercetin in various ferns growing in northern India using TLC. The existence of quercetin in *Christella arida*, *Deparia japonica*, *Dryopteris cochleata*, *D. juxtaposita*, *Hypodematium crenatum*, *Polystichum squarrosum* and *Pteridium revolutum* is confirmed using TLC. Chikmawati et al. (2012) divulged the alkaloids, flavonoids and steroids profile of *Selaginella* species. Mandal and Mondal (2012) analysed the free amino acids present in the leaf glands of pteridophytes using TLC. DL-methionine is the common free amino acid of *Pteris vittata*, *Drynaria quercifolia*, *Ampelopteris prolifera* and *Dryopteris filix-mas*. L-tyrosine monohydrochloride is common in *D. filix-mas* and *Selaginella indica*. L-arginine monohydrochloride is also common in *D. quercifolia*, *Ceratopteris thalictroides* and *Marsilea quadrifolia*. Glycine is the only amino acid found in *Helminthostachys zeylanica*. Pauline Vincent et al. (2012) determined the phytochemical profile for *A. terminans*, *C. parasitica* and *C. interruptus* using TLC. Johnson et al. (2013) revealed the phenolic and steroid profile of two varieties of *Blechnum orientale* L., viz. var. *grandis* and var. *orientale*. Selvaraj et al. (2013) isolated bioactive polyphenolic compounds from methanolic extract of *Azolla microphylla*. Ranjan and Vats (2016) confirmed the presence of diverse type of phytochemicals in *Adiantum capillus-veneris* using TLC. Singh (2017) revealed the TLC profile of *Lygodium flexuosum* ethanolic extract.

Sharad et al. (2008) and Srivastava et al. (2008) disclosed the phytochemical composition of *Lycopodium clavatum* using HPTLC and observed the presence of ferulic acid. Pauline Vincent et al. (2012) determined the phytochemical profile for *A. terminans*, *C. parasitica* and *C. interruptus* using HPTLC. The phenolics and flavonoids HPTLC profile of *Pteris vittata* was determined (Paul and Banerjee 2013). Selvaraj et al. (2013) isolated and quantified the rutin and quercetin from methanolic extract of *Azolla microphylla*. Shweta et al. (2013) determined the interspecific variation among four *Adiantum* species using HPTLC profile. Sivagurunathan and Xavier Innocent (2014) revealed the presence of 11 compounds from *Marsilea quadrifolia*. HPTLC fingerprinting of rhizome extracts of *Dryopteris cochleata* showed several peaks with different $R_f$ values (Das et al. 2015). Janakiraman and Johnson (2016a, b) revealed the HPTLC phenolics, flavonoids and tannins profile of *Cyathea nilgirensis*, *C. gigantea* and *C. crinita*. Johnson et al. (2020b, b, c) validated and revealed the phenolic, flavonoids, alkaloids and tannins profile of *Asplenium aethiopicum* (Burm. f.) Becherer methanolic extracts using HPTLC. The observed profile will help us to identify the crude drugs and improve the therapeutic potentials of *A. aethiopicum*. In addition, the TLC and HPTLC profiles will be used as phytomarker to distinguish the drug from its adulterants.

These profiles play a vital role in chemosystematic to find a solution for the taxonomical problems.

HPLC is a highly sensitive method for detection, identification and quantification of any chemical in a particular sample using UV and visible absorbance (Hanachi and Golkaho 2009). HPLC offers advantages in identifying the isolated compounds (Zhang et al. 2010) and in the quantitative determination (Douat et al. 2011). HPLC is an important qualitative and quantitative technique generally used for quality control testing and estimation of pharmaceutical and biological samples (Guillaume and Veuthey 2012; Snyder et al. 2009; Ahuja and Rasmussen 2007). De Britto et al. (1994) performed the HPLC studies on the species of Thelypteridaceae and produced the phytomarker for the studied ferns.

HPLC analysis of *Pteris biaurita* following elution from TLC plate revealed a single peak with the retention time of 8.1 min (Dalli et al. 2007). Paulraj et al. (2011) studied the presence of various kinds of terpenoids, alkaloids, tannins, saponins and flavonoids on the epidermal glands extract of the glandular morphotype *Christella parasitica* using HPLC. Mostafa and Ibrahim (2012) analysed α-tocopherol in *Azolla caroliniana* using HPLC which showed great quantitative variations, where as ascorbic acid and β-carotene showed marked changes in both number and area of the characterized peaks subjected to UV-B. Chikmawati et al. (2012) determined the highest amentoflavone (6.87 ppm) content in *Selaginella subalpina* using HPLC analysis. Pathania et al. (2012) quantified the carcinogen and ptaquiloside in various ferns growing in certain enzootic areas of Himachal Pradesh and Uttarakhand. The presence of ptaquiloside in *Dryopteris cochleata*, *Hypodematium crenatum*, *Pseudocyclosorus canus* and *Pteris cretica* was identified and quantified. Pauline Vincent et al. (2012) determined the phytochemical profile for *A. terminans*, *C. parasitica* and *C. interruptus* using HPLC. Selvaraj et al. (2013) isolated bioactive polyphenolic compounds from methanolic extract of *Azolla microphylla* using HPLC. Kamboj and Kalia (2014) studied the detailed pharmacognostical parameters for the histological and physicochemical standardization of *Drynaria quercifolia* using HPTLC and HPLC methods. Jaishee and Chakraborty (2015) identified the phenolics, viz. catechin, caffeic, ferulic, salicylic and vanillic acid occurrence in *P. vittata* and *D. linearis* by HPLC analysis.

Knowledge of the chemical constituents of plants is used for the discovery of therapeutic agents, and this information may be helpful in finding new sources of economically important phytocompounds for the synthesis of complex chemical substances (Milne 1993). Mass spectrometry, coupled with chromatographic separations such as gas chromatography (GC-MS), is normally used for direct analysis of components existing in traditional medicines and medicinal plants. GC-MS is a powerful technique having many applications including high sensitivity and specificity. The combination of a principle separation technique (GC) with the best identification technique (MS) made GC-MS ideal for qualitative and quantitative analysis for volatile and semi-volatile compounds (Karthishwaran et al. 2012). GC-MS is proved to be an effective tool for analysing medicinal plants for the presence of non-polar components, volatile essential oil, fatty acids, lipids, etc. (Jie and Choi 1991) and alkaloids (Bertz et al. 1997). Kurumatani et al. (2001) isolated the gibberellins $A_{73}$ methyl ester, the most abundant antheridiogen, and the methyl esters of $GA_9$ and several monohydroxy $GA_{73}$ derivatives from the schizaeaceous ferns *Lygodium microphyllum* and *Lygodium reticulatum*. Ramesh et al. (2001) isolated friedelin, epifriedelinol, β-amyrin, β-sitosterol, β-sitosterol 3-β-D-glucopyranoside and naringin from the dried rhizome of *Drynaria quercifolia*. Juliani et al. (2004) detected the presence of sesquiterpenes (75%) with lower amounts of monoterpenes. Niko et al. (2006) identified 25 compounds in *Equisetum arvense* stem essential oil. Hashemi et al. (2007) studied the volatile components of *Artemisia aucheri* using GC-MS which includes 1,8-cineole, chrysanthenone, α-pinene and mesitylene. Choudhary et al. (2008) isolated two glycosides, viz. 6′-O-(3,4-dihydroxy benzoyl)-β-d-glucopyranosyl ester and 4-O-β-d-glucopyranoside-3-hydroxy methyl benzoate, along with five known compounds, viz. methyl benzoate, hypogallic acid, caffeic acid, paeoniflorin and pikuroside from a fresh water fern *Salvinia molesta* showing potent antioxidant radical scavenging activity.

Kumar et al. (2011) reported more than 13 individual compounds from *Polypodium decumanum* ethanolic extracts, and the main compound identified was long -chain fatty acids along with the flavonoids. Dubal et al. (2013) identified the presence of octadec-9 -enoic acid (oleic acid), n- hexadecanoic acid (palmitic acid), octadecanoic acid (stearic acid), di-n-octyl phthalate, hexadecanoic acid methyl ester, hexadecanoic acid and ethyl ester from the methanolic extracts of *Tectaria coadunata* rhizome. Babu (2013) confirmed the metabolites namely Pentadecanoic acid, 14-methyl-, methyl ester, Hexadecanoic acid, methyl ester, 9, 12-Octadecadienoic acid, methyl ester, (E, E)- 4) 9, 12-Octadecadienoic acid, methyl ester, 9-Octadecenoic acid (Z)-, methyl ester, Octadecanoic acid, methyl ester, Squalene and 4, 8,12,16-Octadecatetraen-1-ol,4,9,13,17-tetramethyl presence in *Adiantum latifolium*, 3,7,11,15-Tetramethyl-2-hexadecen-1-ol, 4 Cyclopropylcarbonyloxydodecane, Phytol, 1,2-Benzenedicarboxylic acid, mono (2-ethylhexyl) ester and Squalene existence in *Acrostichum aureum*, 6-Octen-1-ol, 3,7-dimethyl-,propanoate (Synonym-Citronellyl propionate), Phytol and Squalene presence in C. interruptus and 1,8-Octanediol, Naphthalene, 1-butyl-1,2,3,4-tetrahydro-4-pentyl, 1-Indanone, 3,3,4,5,7-Pentamethyl, Phytol and Squalene existence in *Histiopteris incisa* using GC-MS analysis. Kumar et al. (2014) identified 5 -7A-isoprpenyl-4, 5-dimethyloctahydro-1 h-inden-4yl)-3-methyl-2-penta (24.49%), n-hexadecanoic acid (18.29%), gamma-sitosterol (10.61%) and cis-vaccenic acid (9.25%) from the methanolic extracts of *Adiantum capillus-veneris*. Rukmini and Suvarnalatha (2014) observed the existence of neophytadiene and 2,6,10-trimethyl,14,ethylene,2 hexadecen-1-ol and 3,7,11,15 tetramethyl RR, hexadecanoic acid, palmitic acid, 9 -octadecenoic acid (Z)-, oleic acid, octadecanoic acid, stearic acid, stigmast-4-en-3-one and 4-stigmasten-3-one from *Nephrolepis cordifolia*.

Santhosh et al. (2014) identified the presence of 5- 7A-isoprpenyl-4, 5-dimethyloctahydro-1 h-inden-4yl)-3-methyl-2-penta, n-hexadecanoic acid, gamma-sitosterol, cis-vaccenic acid (9.25%), 5-7A-isopropenyl-4,5-dimethyl-octahydro-inden-4-yl)-3-methyl-pent-2-EL, tetradecanoic acid and phenanthrene in the methanolic extracts of *Adiantum capillus-veneris*. Sivagurunathan and Xavier

(2014) reported the existence of hexadecanoic acid, ethyl ester, phytol, 9,12-octadecadienoic acid (Z,Z), 1,2-benzenedicarboxylic acid, diisooctyl ester, 3,7,11,15-tetramethyl-2-hexadecen-1-ol, 2(3H)-furanone, dihydro-3-hydroxy-4,4-dimethyl and octadecanoic acid, ethyl ester in the ethanolic extract of *Marsilea quadrifolia*. Prasanna and Chitra (2014) studied the phytochemical constituents and chemical composition of *Drynaria quercifolia* rhizome methanolic and petroleum ether extracts.

Karikalan and Rajangam (2014) reported the occurrence of 4H-pyran-4-one,2,3-dihydro-3,5-dihydroxy-6-methyl, n-hexadecanoic acid, 9,12,15-octadecatrienoic acid, methyl ester (Z,Z,Z), 2-furancarboxaldehyde, 5-(hydroxymethyl), 4H-pyran-4-one and 2, 3-dihydro-3, 5-di hydroxy-6-methyl from *Marsilea quadrifolia*. Rukmini et al. (2015) validated the presence of mesityl oxide, 2- pentanone, benzene, neophytadiene, phytol, oleic acid, stearic acid and stigmast-4-en-3-one in the acetone extracts of *Hemionitis arifolia*.

GC-MS analysis of *Salvinia molesta* confirmed the presence of bioactive components apiol, hexadecanoic acid, pentadecanoic acid and octadecatriene (Nithya et al. 2015). Manonmani and Catharin Sara (2015) revealed the existence of hexadecanoic acid, ethyl ester (20.40%), 9,12-octadecadienoic acid, methyl ester, (E,E), (E)-9-octadecenoic acid ethyl ester, docosanoic acid, ethyl ester and heptadecanoic acid heptadecyl ester in *A. radiata*. Kanchan et al. (2015) reported the occurrence of germacrene D; 1, 3- cyclohexanedione, 2- methyl −2- (3-o xobutyl); and neoisolongifolene, 8, 9- dehydro 7 from *Tectaria coadunata* ethanolic extract. Pradnya et al. (2015) reported the presence of 3, 7, 11, 15 -tetramethyl-2-hexadecen-1-ol, n- hexadecanoic acid, ethyl ester 9-octadecenoic acid, 1, 2-benzenedi carboxylic acid, discotyl ester, n-tetracontane and diploptene in the *Cheilanthes farinosa* ethanolic extracts. Kunnathupara et al. (2016) determined the presence of 2-butanone, 3-methoxy-3-methyl, 2,2-dimethyl propionic acid, cyclopentyl ester, butane, 1-bromo-2-methyl, 2-hexen-1-ol,2-ethyl, 5-hydroxy-2,2-dimethyl hexan-3-one, phthalic acid, ethyl pentyl ester and pentanoic acid, 2-methyl from *Azolla microphylla*. Janakiraman and Johnson (2016a, b) revealed the presence of methyloctadecyl dichlorosilane and 2-methylbutane-1,4-diol in *C. nilgirensis*, 3-(1-ethoxyethoxy) in *C. gigantea* and 2-hydroxy-5-methylbenzaldehyde in *C. crinita*. GC-MS analysis of *Acrostichum aureum* revealed five phytosterols, viz. campesterol, stigmasterol, γ-sitosterol, cycloartanol, methylene and cycloartanol (Anitta et al. 2016). Quercetin, 7, 3, 4 -trimethyl ether, phytol, feren-8-ene and hexadecanoic acid are reported from the methanolic extract of *Dicranopteris linearis* (Kalpana Devi et al. 2016a, b). Nayak and Padhy (2017) revealed the presence of gliricidin7-O-hexoside, quercetin-7-O-rutinoside, kaempferol-3-O-rutinoside and myricetin-3-O-rhamnoside from *A. caroliniana* and the major compound 3-O-methyl -d-glucose with peak area 91.89% (Jarial et al. 2018b, b). GC-MS analysis of methanolic extract of *Adiantum capillus-veneris* identified major bioactive phytochemical compounds, viz. dodecanoic acid, nonadecane, tetradecanoic acid, 3,7,11,15-tetramethyl-2-hexadecen-1-ol, acetic acid, 3,7,11,15-tetramethyl-hexadecyl ester, docosane, 1,2-benzenedicarboxylic acid, butyl octyl ester, n-hexadecanoic acid, 9-octadecenoic acid, di-n-octyl phthalate and tetracontane

(Chhaya et al. 2020). Vineesh et al. (2021) validated the presence of octadec-9-enoic acid (oleic acid), n-hexadecanoic acid (palmitic acid), octadecanoic acid (stearic acid), di-n-octyl phthalate, hexadecanoic acid methyl ester and hexadecanoic acid ethyl ester from *Tectaria wightii*.

## 19.7 Spectroscopic Studies on Indian Pteridophytes

Spectroscopic (UV-Vis and FT-IR) methods together or separately can be used as a predictable method for detection of biomolecular composition (Socaciu et al. 2005; Schultz and Baranska 2007; Ibrahim and Osman 2009). FT-IR is one of the most widely used methods to identify the functional groups and pave a way to elucidate the compounds structures. It has been used as a requisite method to identify medicines in pharmacopoeia of many countries (Liu et al. 2006). FT-IR is a rapid, non-destructive technique with minimum sample preparation necessary (Smith 2011). It allows the qualitative determination of organic compounds as the appearance of the bands in the infrared spectrum. Sant et al. (2013) reported FT-IR analysis of *Adiantum philippense* frond to identify the biogroups that bound distinctively on the gold and silver surface. The major peaks observed in FT-IR of the extract were 3369, 2360, 1585, 1384, 1076 and 514 $cm^{-1}$. Peak at 3369 $cm^{-1}$ is attributed to OH stretch in phenols. Gowtham (2013) analysed the UV-Vis spectra of different extracts of *Asplenium aethiopicum* and identified the presence of metabolites and the functional groups. The results of FT-IR analyses confirmed the presence of alkanes, esters, primary amines alkyl halides, carboxylic acids and secondary amines in *Asplenium aethiopicum*.

Janakiraman and Johnson (2014) performed UV-Vis spectroscopic analysis for different extracts of *Cyathea nilgirensis*, *C. gigantea* and *C. crinita*. They used macroscopical fingerprint characters not only to identify the chemical constituents but also to distinguish the morphologically similar species of *Cyathea*. Janakiraman and Johnson (2015a, b) revealed the functional groups present in *C. nilgirensis*, *C. gigantea* and *C. crinita* using FT-IR. The analytical evaluation of the FT-IR spectra exposed the significant differences in band position and absorbance intensities. The comparative FT-IR spectra showed that there is an apparent change in relative intensity of the bands. Prasanna and Anuradha (2016) analysed the FT-IR profile of methanolic extracts of *Drynaria quercifolia* rhizome and confirmed the presence of amines, alkanes, denatured amines, alkynes, carboxylic acids and alkenes. John Peter Paul (2018) revealed the FTIR profile of *Blechnum orientale* methanolic extract and confirmed the presence of functional groups such as aldehydes, monosubst benzenes, 1, 2, 4 trisubstituted benzenes, organophosphorus, siloxanes, pyridine- n- oxides, aliphatic nitro, amino acids, hydrochlorides, amino acid esters, aliphatic and aromatic primary amines. In addition, spectroscopic methods are employed to quantify the secondary metabolites of ferns (Johnson et al. 2020b, b, c) and to confirm the plant extracts mediated nanoparticle synthesis (Johnson et al. 2017a, b, 2020b, b, c).

The quantity of chlorophylls, carotenoids and phenolics is estimated in drought resistance ferns and fern allies (*Selaginella*) from Rajasthan (Bohra et al. 1979; Vyas et al. 1989). Yadav (1990) estimated the proline of five *Ophioglossum* species growing in Rajasthan. Rathore and Sharma (1990) determined the soluble proteins, total phenols and carbohydrates on three species of *Isoetes* collected from Rajasthan. Phytochemical composition of Rajasthan ferns and fern allies was determined by Kaur et al. (1986); Vyas and Sharma (1988); Sharma (1989); and Rathore and Sharma (1990). Kale and Upadhye (2003) estimated the chlorophylle contents and inorganic constituents of dimorphic ferns namely, *Bolbitis appendiculata*, *Gymnopteris contaminoides*, *Ceratopteris thalictroides* and *Drynaria quercifolia* grown in different habitats of Western Ghats. Kale and Dongare (2007) analysed total nitrogen, crude proteins and nitrate reductase from *Bolbitis appendiculata*, *Bolbitis virens*, *Osmunda regalis*, *Ceratopteris thalictroides* and *Drynaria quercifolia*.

*Tectaria paradoxa*, *T. wightii*, *T. coadunata*, *Lastreopsis tenera*, *Polystichum harpophyllum*, *P. piceopaleaceum*, *P. moluscense*, *P. squarrosum*, *Arachniodes tripinnata*, *A. aristata*, *A. amabilis*, *Dryopteris hirtipes*, *D. madrasensis*, *D. cochleata*, *D. sparsa*, *D. approximata* chlorophyll, carotenoids, anthocyanin, flavonoids, sugar, reducing sugar, starch, amino acids, phenols, lipids and proline contents are determined (Joseph 1997). In line with the pigments, primary metabolites and secondary metabolites of *Pteris vittata*, *P. multiaurita*, *P. pellucida*, *P. cretica*, *P. scabripes*, *P. biaurita*, *P. kleiniana*, *P. aspericaulis*, *P. mertensoides*, *P. otaria*, *P. argyraea*, *P. linearis*, *P. confusa* and *Acrostichum aureum* are estimated (Jesudass 1997). Patric Raja (1997) validated the quantitative presence of chlorophyll, carotenoids, anthocyanin, flavonoids, sugar, reducing sugar, starch, amino acids, phenols, lipids and proline in *Histiopteris incisa*, *Hypolepis glandulifera*, *Odontosoria chinensis*, *Nephrolepis auriculata*, *N. multiflora*, *Trichomanes obscurum*, *Dicranopteris linearis* var. *sebastiana*, *D. linearis* var. *tenuis*, *D. linearis* var. *brevis*, *Blechnum orientale*, *B. occidentale*, *Cyathea crinita*, *C. nilgirensis*, *Osmunda regalis*, *Pteridium aquilinum*, *Leptochilus decurrens* and *L. decurrens* f. *lanceolatus*. Antonisamy (1997) determined the total contents of pigments and the primary and secondary metabolites in *Dryoathyrium boryanum*, *Athyrium puncticaule*, *A. solenopteris*, *A. lanceum*, *Deparia petersenii*, *Diplazium muricatum*, *D. sylvaticum*, *D. dilatatum*, *D. travancoricum*, *D. polypodioides*, *D. esculentum* and *D. brachylobum*.

Numerous reports were recorded in the quantification of primary and secondary metabolites of Indian ferns, viz. *Christella* spp. (Paulraj et al. 2011), *Selaginella intermedia*, *S. tenera*, *S. involvens*, *S. inaequalifolia* (Sivaraman et al. 2013), *Asplenium aethiopicum* (Johnson et al. 2014a, b), *Dryopteris cochleata* (Kathirvel et al. 2014), *Pteris vittata* (Kaur et al. 2014), *Marsilea quadrifolia* (Sivagurunathan and Xavier Innocent 2014), *Actiniopteris radiata*, *Acrostichum aureum*, *Dryopteris cochleata*, *Drynaria quercifolia*, *Hemionitis arifolia* and *Pityrogramma calomelanos* (Kalpanadevi et al. 2014), *Cheilanthes farinosa*, *C. anceps*, *C. tenuifolia* and *C. albomarginata* (Ghorpade et al. 2015), *Actiniopteris radiata* (Pratima et al. 2015), *Salvinia auriculata* (Suvarnalatha Devi et al. 2015),

*Dicranopteris linearis* and *Blechnum orientale* (Kalpanadevi et al. 2016), *Salvinia molesta* (Nithya et al. 2015), *Cyathea nilgirensis*, *C. gigantea* and *C. crinita* (Janakiraman and Johnson 2015), *Cheilanthes tenuifolia* and *Asplenium nidus* (Jarial et al. 2018), *Azolla pinnata* (Thiripurasundari and Padmini 2018), *Tectaria cicutaria* (Preeti and Namdeo 2018), *Asplenium indicum*, *Lepisorus nudus*, *Microsorum membranaceum* (Damaynati et al. 2019), *Tectaria coadunata* (Shrestha et al. 2019), *Sphaerostephanos unitus* (Johnson et al. 2020b, b, c), *Dicranopteris linearis* (Zakaria et al. 2020), *Tectaria paradoxa* (Manivannan et al. 2020), *Bolbitis appendiculata* (Manivannan et al. 2021), *Selaginella repanda* (Adnan et al. 2021), *A. caudatum*, *A. hispidulum*, *A. raddianum*, *Anemia wightiana*, *Arachniodes aristata*, *Asplenium aethiopicum*, *A. auritum*, *A. inaequilaterale*, *A. indicum*, *A. laciniatum*, *A. nidus*, *A. nitidum*, *A. polyodon* var. *bipinnatum*, *Athyrium solenopteris*, *Botrychium daucifolium*, *Cyathea nilgirensis*, *Dicranopteris linearis* var. *sebastiana*, *Diplazium cognatum*, *D. esculentum*, *Doryopteris concolor*, *Dryopteris cochleata*, *D. sparsa*, *Hymenophyllum javanicum*, *Hypolepis glandulifera*, *Lastreopsis tenera*, *Lepisorus nudus*, *Leptochilus decurrens*, *Lycopodiella cernua*, *Nephrolepis cordifolia*, *Odontosoria tenuifolia*, *Parahemionitis cordata*, *Phanerophlebia caryotidea* var. *micropteris*, *Pityrogramma austroamericana*, *Pteris linearis*, *P. confusa*, *P. vittata*, *Pyrrosia lanceolata*, *Pyrrosia porosa* var. *porosa*, *Selaginella wightii*, *Thelypteris caudipinna*, *T. meeboldii*, *T. ochthodes* and *T. papilio* (Johnson et al. 2021).

Patric Raja et al. (1992), De Britto et al. (1994), Antonisamy (1997), Joseph (1997) and Jesudass (1997) revealed the relationship between photosynthetic pigments and habitat of the fern. They observed high amount of chlorophylls in the shade- loving ferns followed by partially exposed ferns. Contrary to the chlorophyll content, carotenoid, anthocyanin and flavonoid contents are more in many of the species exposed to sunlight than shaded species. The reduction in the amount of total chlorophylls in the sunlight -exposed ferns may be due to stress that prevailed in the habitat. Antonisamy (1997) observed the flavonoids and anthocyanins in high altitude ferns. Similar findings were reported by Rathore and Sharma (1991) and Patric Raja et al. (1992). High content of flavonoids is a natural protector from the UV rays. Anthocyanin and flavonoid contents are found to be higher in the ferns of high altitude than in the ferns of low altitude (Patric Raja 1997). Variation is not only observed in the phytochemical content based on the habitat, but also the vegetative fronds and fertile ones showed the variation (Vyas and Sharma 1988; Rathore and Sharma 1990; Malik and Bhardwaja 1992; Patric Raja et al. 1992; De Britto et al. 1994; Antonisamy 1997; Joseph 1997; Jesudass 1997). The vegetative frond of ferns possesses more amounts of carbohydrates and sugars than fertile fronds. Varma (1992) observed fewer amounts of total soluble sugars in sporophylls of *Asplenium* than vegetative leaves. Ramachandran et al. (1991) and Antonisamy (1997) observed the variation of starch content in the rhizome and fronds of ferns. They observed more amount of starch in rhizome than aerial portions. In contrary, De Britto et al. (1994) observed more sugar contents in the fronds than the rhizomes, and reverse situation was observed with reference to starch. Beri and Bir (1995) observed the variations in the reserve food materials among the ferns growing in partial sun / shade or growing in open sun and deep forest shade.

The large quantity of soluble sugars is providing cytoplasmic osmoticum during conditions of intracellular potential and adverse conditions, viz. desiccation (Yadav 1992). Antonisamy (1997) and Jesudass (1997) noted that the ferns growing on sunlight- exposed area possess more amounts of total free amino acids than shade -loving fern. Rathore and Sharma (1990), Antonisamy et al. (1992), Joseph et al. (1993) and De Britto et al. (1994) correlated the concentrations of amino acids, proline, phenolics and lipids with reference to the habitats of the species. Wang De qun (1988) observed the existence of different types of flavonoids in ferns and alkaloids in microphyllous pteridophytes (mostly fern allies). In addition, the presence of phenols, steroids, triterpenoids and tannins is also observed in ferns (Antonisamy et al. 1992; Joseph et al. 1993; De Britto et al. 1994; Johnson et al. 2020b, b, c). Vyas et al. (1989) observed that soft and herbaceous lamina possesses more total phenolics than coriaceous lamina. Jesudass (1997) stated that the variations in the chemical constituents are due to nature of the soil, age of the plant and seasonal and climatic changes.

## 19.8 Bioactive Principles Isolated from Indian Pteridophytes

Patric Raja (1997) identified and isolated the following flavonoids, viz. flavonol-3-0-glycosides, afzelin and quercitrin from *Dicranopteris linearis* var. *brevis* (Fig. 19.4). Patric Raja (1997) confirmed the presence of quercitrin and isoquercitrin from *Dicranopteris linearis* var. *tenuis* (Fig. 19.4). *Dicranopteris linearis* var. *sebastiana* showed the presence of astragalin, isoquercitrin, rutin and kaempferol-3-O-(4-O-pcoumaroyl-3-O-a-L-rhamnopyranosyl)-a-L-rhamnopyranosyl (1→6)-13-glycopyranoside (Patric raja 1997; Fig. 19.4). Antonisamy (1997) isolated and identified the luteolin glucoside, kaempferol-3-O-glucoside, kaempferol-3-O-diglucoside, quercetin-3-O-glucoside, quercetin-3-O-diglucoside and quercetin-3-O-diglucoside3′-O-xyloside existence from *Athyrium solenopteris*. Luteolin7-O-β-D-glucoside and kaempferol −3-O-β-D-glucoside are isolated and identified from *Pteris confusa* (Jesudass 1997). The flavonoids orientin, isoorientin, vitexin, isovitexin and rutin are isolated from the fronds of *Tectaria paradoxa* (Joseph 1997; Fig. 19.4). De Britto et al. (1995) determined the aglycone chirality in dihydroflavonol −3-O-α-L-rhamnosides from Thelypteridaceae members. Narasimhaiah et al. (2012) isolated two new glycosidic compounds, viz. 2-(3,4-O-diglucos cinnamoyl)-4-hydroxyl furan and 1-heptaloyl, 8-hexyl, 3-(O-diglucos), 10-methyl, 9,10-dihydro naphthalene from ethyl acetate extract of *Actiniopteris radiata*.

Pteridophytes produce secondary metabolites not simply to adapt to their environment but also to resist themselves against several environmental stresses and also for the process of co-evolution with various interacting organisms. Pteridophytes have survived from Paleozoic times and come across various environmental and climatic conditions. Due to the adaptability, pteridophytes are expected to synthesize

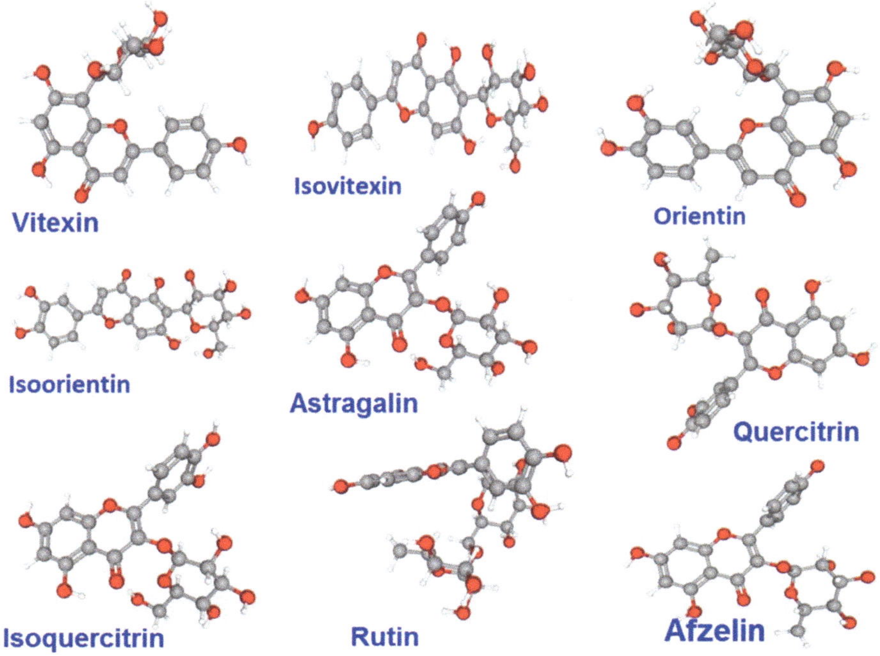


**Fig. 19.4** Isolated chemical compound structure of Indian pteridophytes. Source: National Center for Biotechnology Information (2021i). PubChem Compound Summary for CID 5280441, Vitexin. Retrieved July 6, 2021 from https://pubchem.ncbi.nlm.nih.gov/compound/Vitexin. National Center for Biotechnology Information (2021e). PubChem Compound Summary for CID 162350, Isovitexin. Retrieved July 6, 2021 from https://pubchem.ncbi.nlm.nih.gov/compound/Isovitexin. National Center for Biotechnology Information (2021f). PubChem Compound Summary for CID 5281675, Orientin. Retrieved July 6, 2021 from https://pubchem.ncbi.nlm.nih.gov/compound/Orientin. National Center for Biotechnology Information (2021c). PubChem Compound Summary for CID 114776, Isoorientin. Retrieved July 6, 2021 from https://pubchem.ncbi.nlm.nih.gov/compound/Isoorientin. National Center for Biotechnology Information (2021b). PubChem Compound Summary for CID 5282102, Astragalin. Retrieved July 6, 2021 from https://pubchem.ncbi.nlm.nih.gov/compound/Astragalin. National Center for Biotechnology Information (2021a). PubChem Compound Summary for CID 5316673, Afzelin. Retrieved July 6, 2021 from https://pubchem.ncbi.nlm.nih.gov/compound/Afzelin. National Center for Biotechnology Information (2021h). PubChem Compound Summary for CID 5280805, Rutin. Retrieved July 6, 2021 from https://pubchem.ncbi.nlm.nih.gov/compound/Rutin. National Center for Biotechnology Information (2021d). PubChem Compound Summary for CID 5280804, Isoquercitrin. Retrieved July 6, 2021 from https://pubchem.ncbi.nlm.nih.gov/compound/Isoquercitrin. National Center for Biotechnology Information (2021g). PubChem Compound Summary for CID 5280459, Quercitrin. Retrieved July 6, 2021 from https://pubchem.ncbi.nlm.nih.gov/compound/Quercitrin

and possess many useful secondary metabolites than other land plants. The available literature confirms the existence of many useful phytochemicals (secondary metabolites) such as flavonoids, steroids, alkaloids, phenolics, triterpenoids, amino acids and fatty acids from ferns and fern allies (De Britto et al. 2012a). Due to the existence of these secondary metabolites, the ferns and fern allies harbours various therapeutical potentials, viz. antioxidant, antibacterial, anti-diabetic, antitumor, antifungal, antiseptic, anti-inflammatory activities, etc. (Irudayaraj et al. 2010; Paulraj et al. 2011; De Britto et al. 2012a, b; Sivaraman et al. 2013; Johnson et al. 2014a, b; Janakiraman and Johnson 2015a, b, 2016a, b; Johnson et al. 2017a, b, 2020b, b, c, 2021). The medicinal properties of various plant -derived metabolites are depicted in the Figs. 19.5 and 19.6.

## 19.9 Medicinal Value of Pteridophytes

The medicinal values of certain ferns in India have been studied (Caius 1935; Nair 1959). Rhizome and roots of *Cheilanthes tenuifolia* are used by the tribal people (Dixit 1959). The young shoots of *Lygodium flexuosum*, a common climbing fern species, is used as vegetable, whereas the rhizome of the plant is boiled with mustard oil and locally applied in scabies, ulcers, rheumatism, sprains, eczema and cuts (Dixit 1959; Dixit and Vohra 1984). *Blechnum orientale* is used as a poultice for boils in Malaya; the rhizome is used as an anthelmintic in China, bladder complaints in India and aromatic in the Philippines (Dixit and Vohra 1984). The rhizome of *Dicranopteris linearis* is used as anthelmintic in Assam while the fronds are used for asthma in Madagascar (Manickam and Irudayaraj 1992). The rhizome of *Pteridium revolutum* (syn. *P. aquilinum*) is astringent, is anthelmintic and is useful in diarrhoea and inflammation of gastric and intestinal mucous membranes. The decoction of rhizome and fronds *Pteridium revolutum* (syn. *P. aquilinum*) is used to control chronic disorders of viscera and spleen. The tender fronds of *P. revolutum* are consumed as vegetables and also employed in soups preparation (Manickam and Irudayaraj 1992).

*Drynaria quercifolia* rhizome is bitter and used as antibacterial and anti-inflammatory and also for treating constipation, ulcers and other inflammations. The decoction of plant is used in typhoid fever, and fronds are useful in treating swellings (Dixit and Vohra 1984; Warrier et al. 1996). Singh et al. (2001) reported that 14 common pteridophytic species were used by the local people of Manipur. Kirn and Kapathi (2001) observed the ethnomedical uses of 19 pteridophytes in Jammu and Kashmir. Pteridophytes possess various economic values including food, fodder, biofertilizers and insect repellents (Ghosh et al. 2004). Khare and Kumar (2007) studied the ethnobotanical usage of five pteridophytes, viz. *Helminthostachys zeylanica*, *Adiantum philippense*, *Diplazium esculentum*, *Lygodium flexuosum* and *Ophioglossum reticulatum* used by the Tharu tribe of Dudhwa National Park, Lakhimpur, Kheri (Uttar Pradesh). Medicinal values of 61 different ferns and fern allies of the Western Ghats have been prepared by Benjamin and Manickam (2007). Karthik et al. (2011) noticed the medicinal values of 30 pteridophytes from Tamil Nadu. They were used to treat various ailments, viz. wound healing, body sickness, diarrhoea, skin problems, body pain, knee problem, etc. The leaf and root decoction of *Adiantum lunulatum* and *Adiantum philippense* is very effective in the treatment of chest complaints (Rout et al. 2009). The recent ethnobotanical and pharmacological studies confirmed and validated the versatile

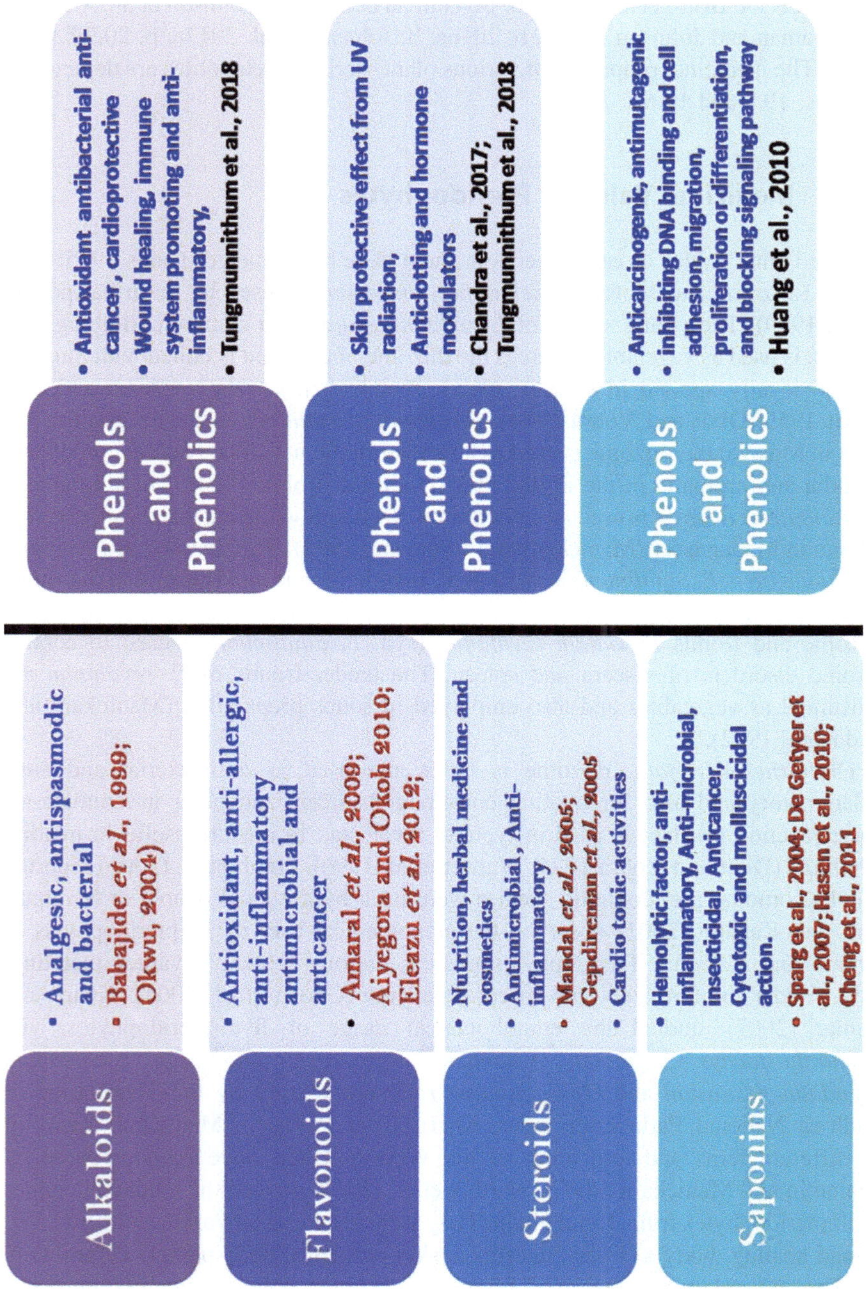


**Fig. 19.5** Medicinal properties of alkaloids, flavonoids, steroids, saponins, phenols and phenolics

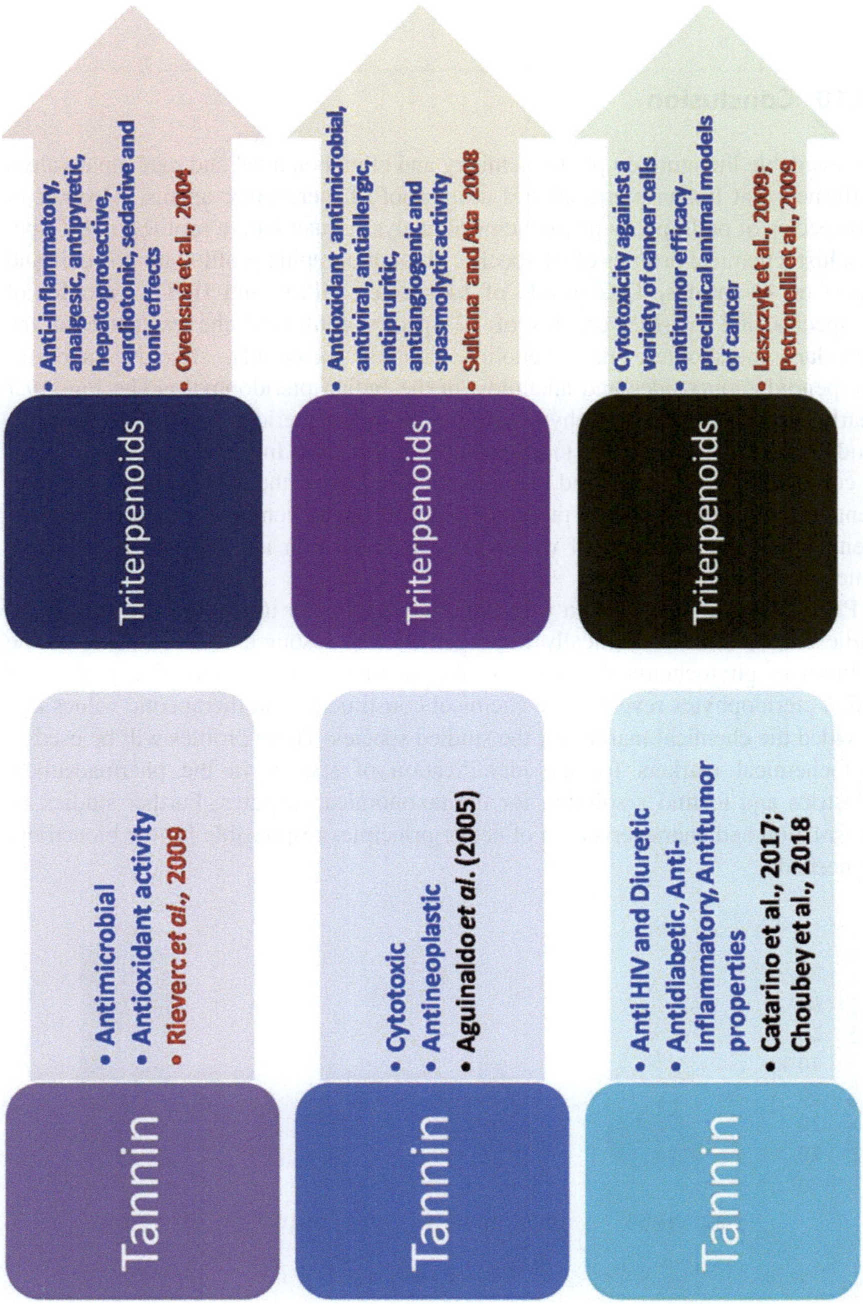


**Fig. 19.6** Medicinal properties of tannin and triterpenoids

medicinal application of ferns. The observations and experimental data confirmed the medicinal properties of ferns.

## 19.10 Conclusion

The available literature on phytochemistry and ethnobotanical and medicinal values confirmed that Indian pteridophytes are a pool of therapeutic agents. A report of 170 species of preliminary phytochemical analysis, quantitative profiles of 115 species, histochemical profiles of 61 species, chromatographic profile (amino acids and sugar) of 43 species, TLC profile of 14 species, HPLC and HPTLC profiles of 23 species and GC-MS profiles of 32 species confirmed the existence of the secondary metabolites, viz. phenolics, tannins, flavonoids, steroids, saponins, triterpenoids, glycosides and alkaloids, in the Indian pteridophytes. The Fig. 19.7 clearly explains the trend of phytochemistry of Indian pteridophytes. Even though a good number of preliminary studies are carried out from Indian pteridophytes, only 21 compounds are isolated and identified so far. Due to the advancement in instrumentation techniques, Indian pteridologists and phytochemists now focussed their attention on quantification of metabolites and isolation and separation of active principles from pteridophytes.

Phytochemical analysis is a very important parameter to identify the new therapeutic source and economically important natural resource. The outcome of the preliminary phytochemical studies on the qualitative and quantitative profiles of Indian pteridophytes revealed the chemical constituents and therapeutic values and provided the chemical marker for the studied species. These profiles will be used as phytochemical markers for the identification of species in the pharmaceutical industries and to find a solution for the taxonomical disputes. Further studies on the isolation and characterization of active principles responsible for the bioactivity are needed.

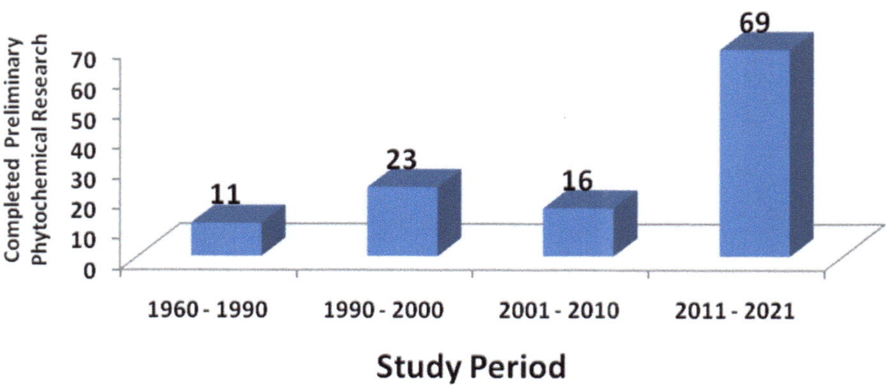


**Fig. 19.7** Trend of Indian pteridophytes phytochemistry: preliminary. (Based on the published papers)

## References


Abraham G, Aeri V (2012) A preliminary examination of the phytochemical profile of *Azolla microphylla* with respect to seasons. Asian Pac J Trop Biomed 2(3):S1392–S1395

Aceret RS (2012) Determination of the secondary metabolites and antimicrobial capability of *Sphenomeris chinensis* (L.) Maxon. Graduate School J 1(1):1–13

Adnan M, Patel M, Deshpande S et al (2020) Effect of *Adiantum philippense* extract on biofilm formation, adhesion with its antibacterial activities against foodborne pathogens, and characterization of bioactive metabolites: An *in vitro*-in silico approach. Front Microbiol 11:823. https://doi.org/10.3389/fmicb.2020.00823

Adnan M, Siddiqui AJ, Hamadou WS et al (2021) Phytochemistry, bioactivities, pharmacokinetics and toxicity prediction of *Selaginella repanda* with its anticancer potential against human lung, breast and colorectal carcinoma cell lines. Molecules 26:768. https://doi.org/10.3390/molecules26030768

Ahmed MN, Gowan M, Azam MNK et al (2015) Clinical appraisals and phytochemical potential of ethnomedicinal pteridophyte: *Drynaria quercifolia* (l.) J. Smith (Polypodiaceae). Pharmacol Online News Lett 1:4–17

Ahmed A, Jahan N, Wadud A et al (2012) Physicochemical and biological properties of *Adiantum capillus-veneris* Linn: an important drug of Unani system of medicine. Int J Curr Res Rev 4 (21):70–75

Ahuja S, Rasmussen H (2007) HPLC method development for pharmaceuticals. In: Analytical chemistry, Separation science and technology, 1st edn. Academic Press, Cambridge, MA

Anitta T, Prashob PN, Kumar C (2016) A profiling of anti-tumour potential of sterols in the mangrove Fern *Acrostichum aureum*. Int J Pharmacogn Phytochem Res 8(11):1828–1832

Antonisamy R (1997) Studies on some Athyrioid ferns of the Western Ghats of South India. Ph. D Thesis, Manonmaniam Sundaranar University, Tirunelveli, Tamil Nadu, India

Antonisamy R, Manickam VS, Gopalakrishnan S (1992) Phytochemical studies on two species of *Diplazium* viz. *Diplazium polypodioides* BI and *Diplazium brachylobum* (sledge) Manickam and Irudayaraj. Indian Fern J 9:269–273

AOAC (ed) (2005) Official methods of analysis of AOAC international, 18th edn. AOAC International, Gaithersburg, MD

Arockia Badhsheeba M, Vadivel V (2020) Physicochemical and phytochemical contents of the leaves of *Acrostichum aureum* L. J Global Biosci 9(4):7003–7018

Arullappan S, Sawai S, Chee LA et al (2017) Phytochemical screening and evaluation of cytotoxic effect and antioxidant activity of fractions isolated from *Stenochlaenapalustri*(Burm.f.) Bedd. Leaves. Indian Journal of Pharmaceutical Education and Research 51(4s Suppl):s735–s740

Awadhesh Kumar S, Bhawana P, Deepti C (2014) Phytochemical analysis of Adiantum and Pteris ferns & its role as antioxidant. Indian J Sci Res 4(1):31–38

Babu C (2013) Screening for Biopesticidal activity in locally available ferns to control lepidopteron pests in Kanniyakumari District. PhD thesis, Manonmaniam Sundaranar University, Tirunelveli, India

Baskaran XR, Geo Vigila AV, Zhang SZ (2018) A review of the use of pteridophytes for treating human ailments. J Zhejiang Univ-Sci B (Biomed & Biotechnol) 19(2):85–119

Benjamin A, Manickam VS (2007) Medicinal pteridophytes from the Western Ghats. Indian J Trad Knowl 6(4):611–618

Bennett RN, Wallsgrove RM (2006) Secondary metabolites in plant defense mechanisms. New Phytol 127(4):617–633

Beri A, Bir SS (1995) Chemical analysis of fern spores from North West Himalaya. Indian Fern J 12:188–194

Bertz JM, Gay ML, Mossoba MM et al (1997) Chiral gas chromatographic determination of ephedrine-type alkaloids in dietary supplements containing Ma Huang. J AOAC Int 80 (2):303–315

Bilia AR, Bergonzi MC, Lazari D, Vincieri FF (2002) Characterization of commercial kava-kava herbal drug and herbal drug preparations by means of nuclear magnetic resonance spectroscopy. J Agric Food Chem 50(18):5016–5025

Bindu NH, Devi PS, Rukmini K, Chgarya MAS (2012) Phytochemical screening and antibacterial activity of *Hemionitis arifolia* (Burm.) Moore. Indian J Nat Prod Resour 3(1):9–13

Bir SS (1987) Pteridophytic flora of India: rare and endangered elements and their conservation. Indian Fern J 4(1&2):95–101

Bobby MDN, Wesely EG, Johnson M (2012) High performance thin layer chromatography profile studies on the alkaloids of *Albizia lebbeck*. Asian Pac J Trop Biomed 2(1):S1–S6

Bohra DR, Soni SR, Sharma BB (1979) Ferns of Rajasthan: behaviour of chlorophyll and carotenoids in drought resistance. Experientia 35(3):332–333

Britto AJD, Manickam VS, Gopalakrishnan S et al (1995) Determination of Aglycone chirality in Dihydroflavonol 3-0-a-L-Rhamnosides by H-NMR spectroscopy. Chem Pharm Bull 42 (2):338–339

Caius JF (1935) The medicinal and poisonous ferns of India. J Bombay Nat Hist Soc 38:341–361

Cao H, Chai TT, Wang X et al (2017) Phytochemicals from fern species: potential for medicine applications. Phytochem Rev 16(3):379–440. https://doi.org/10.1007/s11101-016-9488-7

Chandra S (2000) The ferns of India. International Book Distributors, Dehradun

Chandra S, Fraser-Jenkins CR, Kumari A et al (2008) A summary of the status of threatened pteridophytes of India. Taiwania 53(2):170–209

Chandra P, Kaur S (1974) Ornamental ferns of India. Indian J Hortic 18:14–19

Chandra S, Paul Raj K, Johnson M (2017) Phytochemical profiles of *Adiantum latifolium* LAM. *Angiopteris evecta* (FORST) HOFFM and *Marattia fraxinea* SM. EJBPS 4(12):571–575

Chark KS, Dhir KK (1991) Electrophoretic analysis of soluble proteins in *Pteris vittata* L. during rhizoidal differentiation. Proc Natl Acad Sci India 61:15–120

Chhaya S, Neha C, Sushil Kumar U, Raj S (2020) Phytochemistry and ethnopharmacological study of *Adiantum capillus-veneris* l. (maiden hair fern). Plant Archiv 20(2):3391–3398

Chikmawati T, Miftahudin (2008) Biodiversity and potential many *Selaginella* as anti-oxidants and anticancer, research reports. Institute for Research and Community Service, Bogor Agricultural Institute, Bogor

Chikmawati T, Setyawan AD, Miftahudin (2012) Phytochemical composition of *Selaginella* spp. from Java Island Indonesia. Makara J Sci 16(2):129–133

Choudhary MI, Naheed N, Abbaskhan A et al (2008) Phenolic and other constituents of fresh water fern *Salviniamolesta*. Phytochemistry 69(4):1018–1023. https://doi.org/10.1016/j.phytochem.2007.10.028

Dalli AK, Saha G, Chakraborty U (2007) Characterization of antimicrobial compounds from a common fern, Pteris biaurita. Indian J Exp Biol 45:285–290

Damaynati J, Manda G, Vanita K (2019) Phytochemicals studies on three epiphytic ferns from Mahabaleshwar and Panchgani hills. RJLBPCS 5(3):680–689

Daryono RNH, Rhomawati M (2020) An examination of medicinal potential of *Pneumatopteris callosa*: phytochemical screening, antibacterial, and antioxidant activity. IOP Conf Ser Earth Environ Sci 456:012068

Das S (2003) Usefulness of pteridophytes in India with special reference to medicine and conservation. J Econ Taxon Bot 27:7–16

Das S, Lariya S, Mistry S (2015) Chromatographic finger print analysis of *Dryopteris cochleata* by HPTLC technique. Res J Pharm Tech 8(10):1394–1398. https://doi.org/10.5958/0974-360X.2015.00250.4

Dash KG, Siti KJK, Ahmad FS (2017) Physico-chemical analysis of the aerial parts of *Diplazium esculentum*(Retz.) SW. (family: Athyriaceae). IJPPR 9(6):772–774

De Britto AJ, Gracelin DHS, Rathna Kumar PBJ (2012a) Phytochemical studies on five medicinal ferns collected from southern Western Ghats, Tamil Nadu. Asian Pac J Trop Biomed 2(2):S536–S538

De Britto AJ, Gracelin DHS, Rathna Kumar PBJ (2012b) Antibacterial screening of a few medicinal ferns against antibiotic resistant phyto pathogen. IJPSR 3(3):868–873

De Britto AJ, Manickam VS, Gopalakrishnan S (1992) Preliminary phytochemical analysis of *Sphaerostephanos* species of Western Ghats in South India. Indian Fern J 9:144–146

De Britto AJ, Manickam VS, Gopalakrishnan S (1993) Phytochemistry of *Christella* and *Trigonospora* of Western Ghats, South India. Indian Fern J 10:214–283

De Britto AJ, Manickam VS, Gopalakrishnan S (1994) Chemotaxonomical studies on Thelypteroid ferns of the Western Ghats of South India. J Econ Tax Bot 18:3

De Britto AJ, Manickam VS, Gopalakrishnan S (1994a) Preliminary phytochemical analysis of *Sphaerostephanos* species of Western Ghats in South India. Indian Fern J 11:116–123

De Britto AJ, Manickam VS, Gopalakrishnan S (1994b) Phytochemistry on members of Thelypteridaceae of Western Ghats of South India. Indian Fern J 11:124–125

De Britto AJ, Manickam VS, Gopalakrishnan S (1994c) Preliminary phytochemical screening of Thelypteroid ferns of the Western Ghats of South India. Acta Bot Ind 22:276–278

De Britto AJ, Ramchandran E, Rajendran S, Manickam VS (1991) Phytochemical studies on some ferns from Kothayar Hills, South India. Indian Fern J 8:5–8

De Qun W (1988) A tentative investigation of medicinal ferns in China. Proc ISSP: 313–317

Devi PS, Rukmini K, Himabindu SVSSSLN, Savithramma N (2015) Antibacterial activity and phytochemical screening of *Salvinia auriculata* Aubl. From Tirumala Hills, Tirupati. Int J Pharm Sci Rev Res 30(1):35–38

Dixit RD (1959) Medicinal ferns of India. Bull Natl Bot Gard Lucknow 29:1–36

Dixit RD (2000) Conspectus of pteridophytic diversity in India. Indian Fern J 17:77–91

Dixit RD, Vohra JN (1984) A dictionary of the Pteridophytes of India. Department of Environment, Government of India, Howrah, Botanical Survey of India Publication, Howrah, pp 1–177

Douat J, Vachon P, Beaudry F (2011) Characterization of *in vitro* metabolism of capsazepine, a vanilloid transient receptor potential channel antagonist, by liquid chromatography quadrupole ion trap mass spectrometry. Biomed Chromatogr 25(4):479–492

Dubal KN, Ghorpade PN, Kale MV (2013) Studies on bioactive compounds of *Tectaria coadunate* (wall. Ex hook. & Grev.) C. Chr Asian J Pharm Clin Res 6(2):186–187

Dudani SN, Chandran MDS, Mahesh MK et al (2011) Diversity of Pteridophytes of Western Ghats. http://wgbis.ces.iisc.ernet.in/biodiversity/sahyadri_ news/newsletter/issue33/index.htm

Dudani S, Ramachandra TV (2010) Pteridophytes of Western Ghats. First Indian Biodiversity Congress 156

Edeoga HO, Okwu DE, Mbaebie BO (2005) Phytochemical constituents of some Nigerian medicinal plants. Afr J Biotechnol 4(7):685–688

Ellison WL, Alston RE, Turner BL (1962) A systematic study of the genus Bahia. Am J Bot 49: 599–604

Farnsworth NR (1996) Biological and phytochemical screening of plants. J Pharmacol Sci 55 (3):225–276

Farnsworth NR, Akerele O, Bingel AS et al (1985) Medicinal plants in therapy. Bull World Health Organ 63(6):965–981

Farook MA, Muthu Mohamed HS, Santhosh Kumar G et al (2019) Phytochemical screening, antibacterial and antioxidant activity of *Azolla pinnata*. IJRAR 6(2):240–247

Gahan PB (1984) Plant histochemistry and cytochemistry: an introduction. Academic Press, Orlando

Gaya KS, Ramesh Babu MG, Lizzy M (2016) A preliminary phytochemical study of *Salvinia molesta*, north Paravur, Ernakulam, Kerala. Int J Glob Sci 5(1):3437–3441

Ghorpade PN, Thakar SB, Dongare MM et al (2015) Phytochemical analysis of four *Cheilanthes* species from northern Western Ghats of India. RJLBPCS 1(2):92

Ghosh SR, Ghosh B, Biswas A et al (2004) The pteridophytes flora of Eastern India. Flora of India. BSI Ser 4(1):1–591

Gopalakrishnan S, Rama V, Angelin S, Manickam VS (1993) Phytochemical studies on tree ferns of Western Ghats. Indian Fern J 10:206–213

Goswami HK, Khandelwal S (1976) Free amino acids in Ophioglossum leaves at the time of spike initiation. Israel J Bot 25:211–213

Gowtham J (2013) Evaluation of chemical constituents and biopotential of *Asplenium aethiopicum* (Burm.F.) Becherer. B.Sc Dissertation, St. Xavier's College (Autonomous), Palayamkottai

Guillaume APG, Veuthey JL (2012) Comparison of ultra-high performance supercritical fluid chromatography and ultra-high performance liquid chromatography for the analysis of pharmaceutical compounds. J Chromatogr 1266:158–167

Hanachi P, Golkaho SH (2009) Using HPLC to determination the composition and antioxidant activity of *Berberis vulgaris*. Eur J Sci Res 29(1):47–54

Harsh R, Sharma BD (1994) Pteridophytes of Rajasthan: Phytochemistry of *Azolla pinnata* Brown. Indian Fern J J11:150–152

Hashemi P, Abolghasemi MM, Fakhari AR, Ebrahimi SN, Ahmadi S (2007) Hydrodistillation - solvent microextraction and GC-MS identification of volatile components of *Artemisia aucheri*. Chromatographia 66(3–4):283–286

Herinsheeba GD, Britto AJD, Kumar P (2013) Qualitative and quantitative analysis of phytochemicals in five Pteris species. Int J Pharm Pharm Sci 5:105–107

Hershberger JW (1896) The purpose of ethnobotany. Bot Gaz 31:146–154

Ibrahim M, Osman O (2009) Spectroscopic analyses of cellulose: Fourier transform infrared and molecular modelling study. J Comput Theor Nanosci 6(5):1054–1058

Irene Pearl J, Syed Ismail T, Irudayaraj V et al (2011) Pharmacognostical studies on anti-cancer spike moss *Selaginella involv*ens (sw.), Spring. IJDFR 2(6):195–211

Irudayaraj V (1996) Studies on intra-specific variation in south Indian ferns VI: preliminary phytochemical analysis of epidermal glands in *Christella parasitica* (L.) Lev. Flora Fauna 2 (2):1–4

Irudayaraj V, Janaky M, Johnson M et al (2010) Preliminary phytochemical and antimicrobial studies on a spike-moss *Selaginella inaequalifolia* (hook. & Grev.) Spring. Asian Pac J Trop Med 3(12):957–960

Irudayaraj V, Johnson M (2011) Pharmacognostical studies on three *Asplenium* species. J Phytology 3(10):1–9

Irudayaraj V, Johnson M (2012) Morpho- anatomy, phytochemsitry and pharmacology of anticancer spike – mosses: a review. In: Gupta VK (ed) Natural products research reviews, vol 2. Daya Publishing House, New Delhi, pp 213–237. 9788170358664

Irudayaraj V, Johnson M (2014) Morpho- anatomy, phytochemsitry and pharmacology of anticancer spike –mosses: a review. In: Gupta VK (ed) Natural products research reviews, vol 2. Daya Publishing House, New Delhi, pp 213–237. 9788170358664

Irudayaraj V, Johnson M, Kala A et al (2014) In: Gupta VK, Kaul A (eds) Histochemical studies on selected medicinally important primitive ferns of Western Ghats, South India, Utilization and management of medicinal plants, vol 2. Daya Publishing House, New Delhi, pp 115–152

Irudayaraj V, Raja DP (1998) Phytochemical studies on Indian ferns. Indian Fern J 15:149–164

Jadhav D, Ghatage M, Karande V (2019) Phytochemicals studies on three epiphytic ferns from Mahabaleshwar and Panchgani hills. RJLBPCS 5(3):680–689

Jaishee N, Chakraborty U (2015) Comparative assessment of phytochemicals and HPLC analyses of phenolics present in *Dicranopteris linearis* (N. Burm.) Underw and *Pteris vittata* L. Int J Pharm Pharm Res 5(1):1–7

Janakiraman N, Johnson M (2014) UV-vis spectroscopic profile as taxonomic criteria to distinguish the tree ferns (*Cyathea*). IJREB 2(6):203–212

Janakiraman N, Johnson M (2015a) Functional groups of tree ferns (Cyathea) using FT-IR: chemotaxonomic implications. Romanian J Biophys 25(2):131–141

Janakiraman N, Johnson M (2015b) In vitro antioxidant properties of natural products isolated from selected species of Cyathea. J Clin Nephrol Res 2(2):1027

Janakiraman N, Johnson M (2016a) GC–MS analysis of ethanolic extracts of *Cyathea nilgirensis*, *C. gigantea*, and *C. crinite*. Egypt Pharm J 15:43–47

Janakiraman N, Johnson M (2016b) HPTLC fingerprint profile (Phenolics) of selected Cyathea species from Western Ghats, South India. Chinese J Biol 7:6420371. https://doi.org/10.1155/2016/6420371

Jarial R, Shard A, Thakur S et al (2018b) Characterization of flavonoids from fern *Cheilanthes tenuifolia* and evaluation of antioxidant, antimicrobial and anticancer activities. J King Saud Univ Sci 30:425–432

Jarial R, Sveta T, Sakinah M et al (2018a) Potent anticancer, antioxidant and antibacterial activities of isolated flavonoids from *Asplenium nidus*. J King Saud Univ Sci 30:185–192

Jenat PJ, Suresh SN (2018) Phytochemical analysis and antimicrobial activity of fern *Adiantum lunulatum* Burm. F. WJPR 7(3):640–648

Jesudass LL (1997) Micromorphological ecological and phytochemical studies of the family Pteridaceae of the Western Ghats, South India. Ph. D Thesis, Manonmaniam Sundaranar University, Tirunelveli, Tamil Nadu, India

Jesudass L, Manickam VS, Gopalakrishnan S (1993) Phytochemistry of three taxa of *Pteris* L. Indian Fern J 10:1–5

Jie MSF, Choi CYC (1991) Characterization of picolinyl and methyl ester derivatives of isomeric thia fatty acids. J Int Fed Clin Chem 3:122

John Peter Paul J (2018) Screening of phytochemicals of *Blechnum orientale* L. collected from Kothiyar, Kanyakumari District, Tamil Nadu. India IAJPS 05(03):1671–1674

Johnson M, Amutha S, Shibila T et al (2017b) Green synthesis of silver nanoparticles using Cyathea nilgirensis Holttum and their cytotoxic and phytotoxic potentials. Part Sci Technol 36:5. https://doi.org/10.1080/02726351.2016.1278292

Johnson M, Anwardeen I, Periasamy A et al (2021) Evaluation of phenolic constituents and toxicity of lycophytes and ferns of Shervarayan Hills aqueous extracts. Chemistry Africa 4:513–523. https://doi.org/10.1007/s42250-021-00240-0

Johnson M, Gowtham J, Janakiraman N et al (2020b) Phytochemical profile of *Asplenium aethiopicum* (Burm. F.)Becherer using HPTLC. Separations 7(1):8

Johnson M, Gowtham J, Sivaraman A et al (2014a) Antioxidant, Larvicidal, and cytotoxic studies on *Asplenium aethiopicum* (Burm. F.) Becherer. Int Sch Res 12:1–6. https://doi.org/10.1155/2014/876170

Johnson M, Irudayaraj V, Linson D et al (2014b) In: Gupta VK, Kaul A (eds) Histochemical studies on selected medicinally important advanced ferns of Western Ghats, South India, Utilization and Management of Medicinal Plants, vol 2. Daya Publishing House, New Delhi, pp 153–187

Johnson M, Ramakrishnan P, Perumal S et al (2017a) Anti-inflammatory activity of selected pteridophytes from Western Ghats. Int J Complementary Alternative Med 9(4):00307

Johnson M, Revathi I, Janakiraman N, Renisheya Joy Jeba Malar T, Narayani M (2013) Phenolic and steroid profile of two *Blechnum* varieties. IAJPS 3(3):3336

Johnson M, Shibila T, Amutha S et al (2020c) Evaluation of cytotoxic effect of silver nanoparticles (AgNPs) synthesized from Phlebodium aureum (L.) J. Smith extracts. Anticancer Agents Med Chem. https://doi.org/10.2174/1871520621999201230233408

Johnson M, Xavier Madona C, Almeida RS et al (2020a) In vitro toxicity, antioxidant, anti-inflammatory, and antidiabetic potential of *Sphaerostephanos unitus* (L.) Holttum. Antibiotics 9(6):333. https://doi.org/10.3390/antibiotics9060333

Joseph LH (1997) Studies on some ferns of Dryopteridacea of the Western Ghats, South India. Ph. D Thesis, Manonmaniam Sundaranar University, Tirunelveli, Tamil Nadu, India

Joseph LH, Manickam VS, Gopalakrishnan S (1993) Phytochemistry of members of *Polystichum* from the Western Ghats. Indian Fern J 10:152–156

Juliani HR, Zygadlo JA, Scrivanti R et al (2004) The essential oil of *Anemia tomentosa* (Savigny) Sw. Var. *anthriscifolia* (Schrad.) Mickel. Flavour Fragr J 19:541–543

Jyothi C, Pratima M, Sundar M (2015) Phytochemical and antidepressant activities of *Selaginella bryopteris* (l.) baker on albino mice. Int J Appl Biol Pharm 6(4):14–19

Kale MV, Upadhye AB (2003) Study of chlorophylls in the dimorphic leaves of some homosporous ferns. Geobios 20:67–70

Kale MV, Dongare MM (2007) Comparative analysis of nitrogen metabolism in the dimorphic ferns of properties – a new approach for characterization of their bioactive compounds and establishment of structure-antioxidant activity relationships. Bioorg Med Chem 7(5):1876–1883

Kalpana Devi R, Rajesh NV, Vasantha S et al (2014a) Anthelmintic efficacy of selected ferns in sheeps (Ovis aries Linn.). Int J Ethnobiol Ethnomed 1(1):1–14

Kalpana Devi R, Vasantha S, Panneerselvam A et al (2016a) Phytochemical analysis, *in vitro* antioxidant potential and gas chromatography-mass spectrometry studies of *Dicranopteris linearis.* Asian J Pharm Clin Re*s* 9(8):220–225

Kalpana Devi R, Vasantha S, Panneerselvam A et al (2016b) Phytochemical constituents and *in vitro* trematocidal activity of *Blechnum orientale* Linn. against *Gastrothylax crumenifer.* Ann Phytomed 5(1):127–134

Kalpana Devi R, Vasantha S, Rajesh NK et al (2014b) Qualitative and quantitative phytochemical analysis in four Pteridophytes. Int J Pharm Sci Rev Res 27(2):408–412

Kamboj P, Kalia AN (2014) Morphoanatomical and physico-chemical standardization of *Drynaria quercifolia* (L.) J. Smith. World J Pharm Pharm Sci 3(4):696–711

Kanchan D, Ghorpade MD, Kale M (2015) Studies on bioactive compounds of *Tectaria coadunata* (wall. Ex hook. & Grev.) C. Chr. Asian J Pharm Clin Res 6(2):186–187

Karthik V, Raju K, Ayyanar M, Gowrishankar K, Sekar T (2011) Ethnomedicinal uses of Pteridophytes in Kolli Hills, Eastern Ghats of Tamil Nadu. India J Nat Prod Plant Resour 1 (2):50–55

Karikalan G, Rajangam U (2014) Antimicrobial activity of *Marsilea quadrifolia* (L.) against some selected pathogenic microorganisms. Br Microbiol Res J 4(9):2231–0886

Karthishwaran K, Muthukkumarasamy S, Sankaran M (2012) GCMS analysis of methanolic extract of aerial parts of *Pergularia daemia.* JoLS 1(1):50–55

Kathirvel A, Rai AK, Maurya G et al (2014) *Dryopteriscochleata* rhizome: a nutritional source of essential elements, phytochemicals, antioxidants and antimicrobials. Int J Pharm Pharm Sci 6 (2):179–188

Kaur A, Yadav BL, Bhardwaja TN (1986) A comparative investigation of amino acids and free proline of some Rajasthan ferns. Bionature 6:42–44

Kaur P, Kaur V, Kumar M et al (2014) Suppression of SOS response in E. coli PQ 37, antioxidant potential and antiproliferative action of methanolic extract of *Pterisvittata* L. on human MCF-7 breast cancer cells. Food Chem Toxicol 74:326–333. https://doi.org/10.1016/j.fct.2014.10.002

Khandelwal S, Goswami HK (1976) Amino acids differentiation in *Ophioglossum* L. by paper chromatography. Curr Sci 45:62–63

Khare PB (1996) Ferns and fern allies - their significance and fantasies. Appl Bot 26:52

Khare PK, Kumar S (2007) Studies on some pteridophytes used by the Tharu tribe of Dudhwa National Park, Lakhimpur Kheri, Uttar Pradesh, India. Indian Fern J 24:137–147

Khare PK, Shankar R (1987) On some phytochemical investigations of *Psilotum nudum* L. Proc Nat Acad Sci India 57(1):78–82

Kiernan JH (ed) (1989) Histological and histochemical methods. Theory and practice, 2nd edn. Berlin, Springer

Kiran PM, Raju AV, Rao BG (2012) Investigation of hepatoprotective activity of *Cyathea gigantea* (wall. Ex. hook.) leaves against paracetamol-induced hepatotoxicity in rats. Asian Pac J Trop Biomed 2(5):352–356

Kirn HS, Kapathi BK (2001) Ethnobotanical notes on some ferns and fern allies of Jammu and Kashmir State, India. Indian Fern J 18:35–38

Korwar PG, Beknal A, Patil B et al (2010) A study on phytochemical investigation of Drynaria quercifolia Linn rhizome. IJPSR 1(12):148–158

Krishna L, Dawra RK (1994) Bracken fern induced carcinoma in Guinea pigs. Indian J Vet Pathol 18:21–26

Kumar M (1995) Phytochemical investigations of cultivated and wide populations of some selected species of *Marsilea.* National Symposium on Researches in Pteridology, Jai Narain Vyas University, Jodhpur, India, p 117

Kumar MSM (1998) Studies on the fern flora of Kerala with special reference to sylvan valley, Munnar. KFRI Res Report 145:1–46

Kumar N, Kshirsagar MD, Saini V (2011) GC-MS analysis of ethanolic extract of *Polypodium decumanum*. IRJP 2(9):155–156

Kumar SS, Nagarajan N (2012) Screening of preliminary phytochemical constituents and antimicrobial activity of *Adiantum capillus-veneris*. J Res Antimicrobials 1(1):56–61

Kumar S, Samydurai P, Ramakrishnan R et al (2014) Gas chromatography and mass spectrometry analysis of bioactive constituents of *Adiantum capillus-veneris* L. Int J Pharm Sci 6 (4):0975–1491

Kunnathupara BS, Sowmya S, Gopalakrishnan VK et al (2016) Quantitative phytochemical analysis, in vitro antioxidant potential and gas chromatography-mass spectrometry studies in ethanolic extract of *Azolla microphylla*. Asian J Pharm Clin Sci 9(2):318–323

Kurumatani M, Yagi K, Murata T et al (2001) Isolation and identification of antheridiogens in the ferns, *Lygodium microphyllum* and *Lygodium reticulatum*. Biosci Biotechnol Biochem 65 (10):2311–2314

Lal SD (1979) Phenolic constituents of the therapeutic fern *Asplenium trichomanes*. Sci Cult 45: 452–453

Liebler DC, Burr JA, Philips L, Ham AJL (1996) Gas chromatography-mass spectrometry analysis of vitamin E and its oxidation products. Anal Biochem 236(1):27–34

Liu H, Sun S, Lv G et al (2006) Study on Angelica and its different extracts by Fourier transform infrared spectroscopy and two-dimensional correlation IR spectroscopy. Spectrochim Acta A Mol Biomol Spectrosc 64(2):321–326

Malik SK, Bhardwaja TN (1992) Biochemical systematics of three species of genus *Athyrium* from Mt Abu India. In: Bhardwaja TN, Gena CB (eds) Aspects of plant sciences perspective in pteridology present and future, Today and tomorrow's printers and publishers New Delhi Vol, vol 14, pp 467–472

Mandal A, Mondal AK (2012) Qualitative analysis of free amino acids of some pteridophytes with special reference to their ethnomedicinal uses in West Bengal, India. IJSN 3(4):819–823

Maneesha S, Kangna M, Chhaya S (2015) Phytochemical evaluation and antioxidant activity of different samples of *Pteris Vittata* in Doon Valley, Uttarakhand region. Int J Pure App Biosci 3 (4):296–304

Manickam VS, Irudayaraj V (1992) Pteridophyte Flora of the Western Ghats-South India. B.I Publications Pvt. Ltd, New Delhi

Manisha KV (2015) Qualitative and quantitative analysis of *Cyathea gigantea* ferns from South Western Ghats. RJLBPCS 1(4):183–187. https://doi.org/10.26479/2015.0104.02

Manivannan V, Johnson M, Almeida RS, Coutinho HDM (2021) Chemical profiling of *Tectaria paradoxa* (fee.) sledge and *Bolbitis appendiculata* (Willd.) K. Iwats using UHPLC. Biocatal Agric. Biotechnology 34:102043

Manonmani R, Catharin Sara SC (2015) GC-MS analysis of bioactive components of an important medicinal Fern *Actiniopteris radiata* (Swartz) link. World J Pharm Res 4(4):1860–1869

Manonmani R, Sara SC (2013) Preliminary phytochemical analysis of *Actiniopteris radiata* (Swartz) link. IRJP 4(6):214–216

Maridass M, Raju G (2010) Evaluation of phytochemicals and brine shrimp activity from *Elaphoglossum Beddomei* sledge. Pharmacologyonline 1:640–644

Mary S, Mahesh MK (2015) Determination of antibacterial and antifungal activities of endemic tree fern *Cyathea nilgirensis* plant extract. IJCR 7(07):18276–18279

Mathad CJ, Modi R (2015) Phytochemical and antioxidant activity of *Actiniopterisradiata* Linn.an important pteridophytic medicinal plant of Gulbarga region Pratima. IMPACT: IJRANSS 3 (2):27–34

Milne A (1993) Inhalational and local anesthetics reduce tactile and thermal responses in *Mimosa pudica*. Can J Anaesth 46(3):287–289

Mismawati A, Suwannaket CS, Mingvanish W et al (2015) Phytochemical screening and bioactivity of *Angiopteris evecta* leaves from East Kalimantan. Pure and Applied Chemistry International Conference, (PACCON2015)
Mithraja MJ, Johnson M, Mahesh M et al (2012) Chemical diversity analysis on some selected medicinally important pteridophytes of Western Ghats, India. Asian Pac J Trop Biomed 2(1): S34–S39
Mithraja MJ, Johnson M, Mahesh M et al (2012a) Inter-specific variation studies on the phytoconstituents of *Christella* and *Adiantum* using phytochemical methods. Asian Pac J Trop Biomed 2(1):S40–S45
Mithraja MJ, Johnson M, Mony M et al (2011) Phytochemical studies on *Azolla pinnata* R. Br., *Marsilea minuta* L. and *Salvinia molesta* Mitch. Asian Pac J Trop Biomed 1(1):S26–S29
Moran RC (2008) Diversity, biogeography and floristics. In: Ranker TA, Haufler CH (eds) Biology and evolution of ferns and lycophytes. Cambridge University Press, Cambridge, pp 367–394
Mostafa EM, Ibrahim MM (2012) HPLC analysis of non-enzymatic antioxidants in *Azolla caroliniana* (Pteridopsida) subjected to UV-B Egypt. Acad J Biolog Sci 3(1):19–30. https://doi.org/10.21608/EAJBSH.2012.17003
Nair BK (1959) Medicinal ferns of India. Bull Nat Bot Gard, Lucknow 29:1–36
Narasimhaiah L, Chandrasekar MJN, Alexander S (2012) Isolation of new glycosides from pteridophyte plant. Int J Pharm Pharm Sci 4(4):615–617
National Center for Biotechnology Information (2021a) PubChem compound summary for CID 5316673. Afzelin. https://pubchem.ncbi.nlm.nih.gov/compound/Afzelin. Retrieved 6 July 2021
National Center for Biotechnology Information (2021b) PubChem compound summary for CID 5282102. Astragalin. https://pubchem.ncbi.nlm.nih.gov/compound/Astragalin. Retrieved 6 July 2021
National Center for Biotechnology Information (2021c) PubChem compound summary for CID 114776. Isoorientin. https://pubchem.ncbi.nlm.nih.gov/compound/Isoorientin. Retrieved 6 July 2021
National Center for Biotechnology Information (2021d) PubChem compound summary for CID 5280804. Isoquercitrin. https://pubchem.ncbi.nlm.nih.gov/compound/Isoquercitrin. Retrieved 6 July 2021
National Center for Biotechnology Information (2021e) PubChem compound summary for CID 162350. Isovitexin. https://pubchem.ncbi.nlm.nih.gov/compound/Isovitexin. Retrieved 6 July 2021
National Center for Biotechnology Information (2021f) PubChem compound summary for CID 5281675. Orientin. https://pubchem.ncbi.nlm.nih.gov/compound/Orientin. Retrieved 6 July 2021
National Center for Biotechnology Information (2021g) PubChem compound summary for CID 5280459. Quercitrin. https://pubchem.ncbi.nlm.nih.gov/compound/Quercitrin. Retrieved 6 July 2021
National Center for Biotechnology Information (2021h) PubChem compound summary for CID 5280805. Rutin. https://pubchem.ncbi.nlm.nih.gov/compound/Rutin. Retrieved 6 July 2021
National Center for Biotechnology Information (2021i). PubChem compound summary for CID 5280441. Vitexin. https://pubchem.ncbi.nlm.nih.gov/compound/Vitexin. Retrieved 6 July 2021
Nayak N, Padhy RN (2017) GC-MS analysis of bioactive compounds and host toxicity studies of *Azolla caroliniana* symbiotic with the cyanobacterium *Anabaena azollae*. Indian J Pharm Educ 51(2S):S24-S33. https://doi: https://doi.org/10.5530/ijper.51.2s.46
Niko R, Gordana S, Radosav P (2006) Composition and antimicrobial activity of *Equisetum arvense* L. essential oil. Phytother Res 20(1):85–88
Nithya TG, Jayanthi J, Raghunathan MG (2015) Phytochemical, antibacterial and GC-MS analysis of a floating fern *Salvinia molesta* D.S. Mitchell (1972). IJPRIF 8(9):85–90
Padhy R, Dash SK (2015) Antibacterial evaluation of methanolic rhizome extract from an *in vivo* and *in vitro* grown pteridophyte, *Drynaria quercifolia* (LINN). J Smith Asian J Pharm Clin Res 8(4):130–138

Paramanik R, Mondal AK, Mondal S (2002) Comparative study of the amino acid composition of our species of *Selaginella* P. de Beauv. Phytomorphology 52:337–345

Pathania S, Kumar P, Singh S et al (2012) Detection of ptaquiloside and quercetin in certain Indian ferns. Curr Sci 102(12):1683–1691

Patric Raja D (1997) Studies on some ferns of Kothayar and Palni hills of South India. Ph. D Thesis, Manonmaniam Sundaranar University, Tirunelveli, Tamil Nadu, India

Patric Raja D, Manickam VS, Gopalakrishnan S (1992) Ecophysiological studies on the ferns of Kothayar and Palni hills. Indian Fern J 9:138–143

Paul T, Banerjee S (2013) *In vitro* evaluation of α-amylase inhibitory activity and antioxidant potential of *Pteris vittata* L. with special reference to its HPTLC profile. Int J Pharma Bio Sci 4 (2):494–503

Pauline Vincent C, Irudayaraj V, Johnson M (2012) Anti-bacterial efficacy of macroscopic, microscopic parts of sporophyte and *in vitro* cultured gametophyte of a fern *Cyclosorus interruptus* (Willd.) H. Ito (Thelypteridaceae - Pteridophyta). JOCPR 4(2):1167–1172

Paulraj K (2007) Morphology, Biochemistry and Bioactivity of epidermal glands of selected South Indian ferns. Ph.D. Thesis, Manonmaniam university, Tirunelveli, India

Paulraj K, Irudayaraj V, Johnson M (2011) Phytochemical and anti-bacterial activity of epidermal glands extract of *Christella parasitica* (L.) H. lev. Asian Pac J Trop Biomed 1(1):8–11

Pradnya N, Ghorpade SB, Thakar MM, Dongare et al (2015) Phytochemical analysis of four *cheilanthes* species from northern Western Ghats of India. RJLBPCS 1(2):92

Prasanna G, Anuradha R (2016) Ultraviolet - visible and Fourier transform-infrared spectroscopic studies on *Drynaria quercifolia* L. Rhizome Asian J Pharm Clin Res 9(3):85–88

Prasanna G, Chitra M (2014) Phytochemical screening and GC-MS analysis of *Drynaria quercifolia* rhizome. Am J Adv Drug Deliv 3(1):72–78

Preeti K, Namdeo J (2018) Phytochemical screening and *in-vitro* anticancer activity of extracts of *Tectaria cicutaria*. Int J Pharm Sci Res 9(8):3463–3468

Prema K, Chitra V, Gopal CC, Uma Maheswara R (2015) HPTLC and GC MS analysis of bark of *Stereospermum colais*. Int J Pharm Sci Res 5(8):1–11

Ramachandran E, Rajendran S, Manickam VS et al (1991) Phytochemical studies on some ferns from Kothayar hills, South India. Indian Fern J 8:5–8

Ramesh N, Viswanathan MB, Saraswathy A et al (2001) Phytochemical and antimicrobial studies on *Drynaria quercifolia*. Fitoterapia 72(8):934–936

Ranjan V, Vats M (2016) Pharmacognostical and physico-chemical standardisation of whole plant of *Adiantum capillus veneris* linn. IJPSR 37:773–782

Rathore D, Sharma BD (1990) Pteridophytes of Rajasthan phytochemistry *of Isoetes*. Indian Fern J 7:121–122

Rathore D, Sharma BD (1991) Phytochemistry of Rajasthan Pteridophytes: study of leaf pigments in relation to stress. Indian Fern J 8:9–13

Rohtagi BK, Gupta RB, Khanna RN (1984) Chemical constituents of *Asplenium indicum*. J Nat Prod 47(5):901–901

Rout SD, Panda T, Mishra N (2009) Ethnomedicinal studies on some pteridophytes of Similipal Biosphere Reserve, Orissa, India. Int J Med Med Sci 1(5):192–197

Rozylo JK, Zabinska A, Matysiak J (2002) OPLC and HPTLC methods in physicochemical studies of a new group of antimycotic compounds. JCS 40(10):581–584

Ruby GDR, Sara SC (2014) Preliminary phytochemical analysis of *Pyrrosia lanceolata* (L.) Farwell. IRJP 5(12):879–883

Rukmini K, Devi S, Chitturi K (2015) GC-MS analysis and phytochemical screening of a fern *Hemionitis arifolia* (BURM.) Moore from Tirumala hills. Int J Pharm Bio Sci 6(1):1360–1369

Rukmini K, Suvarnalatha DP (2014) GC-MS analysis and phytochemical screening of a rare pteridophyte *Nephrolepis cordifolia* (L.) Presl. From Tirumala Hills. IJPRR 3(12):13–19

Rupa, Bhavani L (2014) Preliminary phytochemical screening of desiccated fronds of *Selaginella bryopteris* (L) baker (Pittakalu). WJPR 3(9):1370–1378

Sa PGS, Nunes XP, Lima JT et al (2012) Antinociceptive effect of ethanolic extract of Selaginella convoluta in mice. BMC Complement Altern Med 12:187

SaiLakshmi G, Johnson M, Laly J et al (2015) Histochemical studies on *Asplenium* species from Wayanad, Western Ghats, India. In: proceeding of National seminar on plant genetic resources utilization and conservation, p 43

Saito K, Nagao T, Matoba M et al (1989) Chemical assay of ptaquiloside, the carcinogen of *Pteridium aquilinum* and the distribution of related compounds in the Pteridaceae. Phytochemistry 28(6):1605–1611

Salehi B, Ezzat SM, Fokou PVT et al. (2019) *Athyrium* plants - review on phytopharmacy properties. J Tradit Complement Med 9:201e205. https://doi.org/10.1016/j.jtcme.2018.09.001

Sant DG, Gujarathi TR, Harne SR et al (2013) *Adiantum philippense* L frond assisted rapid green synthesis of gold and silver nanoparticles J Nanoparticles 2013 182320. https://doi.org/10.1155/2013/182320

Santhosh SK, Samydurai P, Ramakrishnan R et al (2014) Gas chromatography and mass spectrometry analysis of bioactive constituents of *Adiantum capillus-veneris* L. Int J Pharm Pharm Sci 4 (4):60–63

Santos MG, Kelecom A, De Paiva SR (2010) Phytochemical studied in Pteridophytes growing in Brazil: a review. Am J Plant Sci Biotech 4(1):113–125

Schultz H, Baranska M (2007) Identification and quantification of valuable plant substances by IR and Raman spectroscopy. Vib Spectrosc 43(1):13–25

Selvaraj K, Chowdhury R, Bhattacharjee C (2013) Isolation and structural elucidation of flavonoids from aquatic fern *Azolla microphylla* and evaluation of free radical scavenging activity. Int J Pharm Pharm Sci 5(3):743–749

Shakoor AM, Anand KM, Zafar AR et al (2013) Preliminary phytochemical screening of some pteridophytes from district Shopian (J&K). Int J Pharm Pharm Sci 5(4):632–637

Shankar R, Khare PK (1985) Photochemical studies of *Ampelopteris prolifera* (Ratz) Copel and *Diplazium esculentum.* Swartz J Eco Tax Bot 6(2):499–502

Sharad SK, Adarsh SP, Ajay RSK (2008) Pharmacognostical and phytochemical evaluation of *Lycopodium clavatum* stem. J Sci Ind Res India:228–232

Sharma BD (1989) Ferns and fern allies of Rajasthan - experimental and phytochemical studies. Indian Fern J 6:195–203

Sharma A, Sharma BD (1992) Phytochemistry of Rajasthan ferns - a study of flavonoids. Indian Fern J 9:83–86

Sherma J, Fried B (2005) Thin layer chromatographic analysis of biological samples. A review. J Liq Chromatogr Relat Technol 28(15):2297–2314. https://doi.org/10.1080/10826070500187491

Shrestha SS, Sut S, Barbon Di Marco S (2019) Phytochemical fingerprinting and in vitro bioassays of the Ethnomedicinal Fern *Tectaria coadunata* (J. Smith) C. Christensen from Central Nepal. Molecules 24(24):4457. https://doi.org/10.3390/molecules24244457

Shweta S, Sayyada K, Harsh S et al (2013) A report on pharmacognostical evaluation of four *Adiantum* species, Pteridophyta, for their authentication and quality control. Rev Bras Farmacogn 23(2):207–216

Singh HB (1999) Potential medicinal pteridophytes of India and their chemical constituents. J Eco Tax Bot 23(1):63–78

Singh A (2017) Ethnobotanical and preliminary phytochemical screening of a climbing Fern *Lygodium flexuosum* (L).S.W. International Journal of Science And Research (IJSR) 6(7):25–27

Singh KK, Saha S, Maheswari B (1989) Ethnomedicinal uses of some ferns among the tribals of Uttar Pradesh. Indian Fern J 6:63–67

Singh L, Somarjt P, Singh E et al (2001) Ethnobotanical uses of some pteridophytic species in Manipur. Indian Fern J 18:14–17

Sivagurunathan A, Xavier BI (2014) GC-MS evaluation of bioactive compounds of *Marsilea quadrifolia* Linn. Aquatic Fern. Int J Pharm Res Scholars 3(1):2277–7873

Sivagurunathan A, Xavier Innocent B (2014) Phytochemical analysis and antimicrobial efficiency of *Marsilea quadrifolia* Linn (aquatic Fern). IJPRS 3(2):00256

Sivaraman A, Johnson M (2014) In: Gupta VK, Singh GD, Singh S, Kaul A (eds) Phytochemical and histochemical studies on different species of *Selaginella*, Medicinal plants: phytochemistry, pharmacology and therapeutics, vol 3. Daya Publishing House, New Delhi, pp 413–421

Sivaraman A, Johnson M, Janakiraman N et al (2011) Phytochemical studies on selected species of *Diplazium* from Tirunelveli hills, Western Ghats, South India. J Basic Appl biol 5 (3&4):241–247

Sivaraman A, Johnson M, Parimelazhagan T, Irudayaraj V (2013) Evaluation of antioxidant potential of ethanolic extracts of selected species of Selaginella. Ind J Nat Product and Resources 4(3):238–244

Smith B (ed) (2011) Fundamentals of Fourier transform infrared spectroscopy, 2nd edn. CRC Press, Boca Raton, FL. isbn:978-1-4200-6929-7

Smitha V, Vadivel V (2019) Phytochemical screening for active compounds in *Ceratopteris thalictroides* (L.) Brogn. J Pharmacogn Phytochem 8(3):3556–3559

Snyder LR, Kirkland JJ, Dolan JW (2009) Introduction to modern liquid chromatography. John Wiley and Sons, Inc., Hoboken, pp 1–7

Socaciu C, Ranga F, Diehl H (2005) UV-vis spectrometry applied for the quality and authenticity evaluation of edible oils from Romania. Bull USAMV-CN 61:295–300

Soeder RW (1985) Fern constituents: including occurrence, chemotaxonomy and physiological activity. Bot Rev 51:442–536

Srivastava SK, Singh AP, Rawat AKS (2008) Pharmacognostic and phytochemical evaluation of *Lycopodium clavatum* stem. J Sci Ind Res 67(03):228–232

Suganya S, Irudayaraj V, Johnson M (2011) Pharmacognostical studies on an endemic spike-moss *Selaginella tenera* (Hook. & Grev.) Spring from the Western Ghats, South India. J Chem Pharm Res 3(1):721–731

Sukumaran S, Mahesh M, Jeeva S (2012) Bioactive constituents of oak leaf fern-*Tectaria zeylanica* (Houtt.) sledge from southern Western Ghats. Asian Pac J Trop Biomed 2(1):S64–S66

Suvarnalatha DK, Rukmini SVSSSL, Himabindu NN et al (2015) Antibacterial activity and phytochemical screening of *Salvinia auriculata* Aubl. from Tirumala Hills, Tirupati. Int J Pharm Sci Rev Res 30(1):35–38

Swain T (1977) Secondary compounds as protective agents. Annu Rev Plant Physiol 28:479–501

Talukdar AD, Choudhury MD, Chakraborty M et al (2010) Phytochemical screening and TLC profiling of plant extracts of Cyathea *gigantea* (wall. Ex. hook.) Holt. and *Cyathea brunoniana* wall. Ex hook (cl. & Bak.). Assam Univ J Sci Technol 5(1):70–74

Thiripurasundari B, Padmini E (2018) Preliminary phytochemical screening and evaluation of antimicrobial and antioxidant activity of *Azolla pinnata*. Int J Recent Sci Res 9 (5F):26924–26930

Thomas T (2012) *In vitro* evaluation of antibacterial activity of *Acrostichum aureum* Linn. Indian J Nat Prod Resour 3(1):135–138

Varma S (1992) Phytochemical investigations of some *Asplenium* species. In: Bhardwaja TN, Gena CB (eds) Aspects of plant sciences perspective in pteridology, Present and future, vol 14. Today and Tomorrow's Printers & Publishers, New Delhi, pp 439–444

Vidyarani G, Johnson M, Laly J et al (2015) Histochemical studies on maiden hair ferns of Wayanad, Western Ghats, India. Proceeding of National seminar on plant genetic resources utilization and conservation, Wayanad, p 45

Vineesh TV, HabeebRahiman U, Venugopalakrishna Kurup V (2021) *Tectaria wightii* (C.B. Clarke) CHING- a fern with phytochemical and pharmacognostic potential. EJBPS 8 (1):362–366

Vyas MS, Rathore D, Sharma BD (1989) Phytochemistry of Rajasthan pteridophytes - study of phenols in relation to stress. Indian Fern J 6:244–246

Vyas MS, Rathore D, Sharma BD (1995) Study of amino acids in some common pteridophytes of Rajasthan. Jai Narain Vyas University, Jodhpur, India, National Symposium on Researches in Pteridology, p 105

Vyas MS, Sharma BD (1988) Phytochemistry of Rajasthan Pteridophytes. Indian Fern J 5:143–149

Warrier PK, Nambiar VPK, Raman Kutty C (1996) Indian medicinal plants Vol 5. Orient Longman Ltd, India

WHO (1979) The selection of essential drugs. Second report of the WHO Expert Committee, WHO Technical Report Series 641:1–44

WHO (2002) General guidelines for methodologies on research and evaluation of traditional medicine. World Health Organization, Geneva

Wick MR (2012) Histochemistry as a tool in morphological analysis: a historical review. Ann Diagn Pathol 16:71–78

Yadav BL (1992) In: Bhardwaja TN, Gena CB (eds) Phytochemical studies of Rajasthan species of *Ophioglossum*, Aspects of plant sciences perspective in pteridology present and future, vol 14. Today and Tomorrow's Printers & Publishers, New Delhi, pp 467–472

Yadav BL (1995) Sleep movement in *Marsilea* - a phytochemical investigation. National Symposium on researches in pteridology. Jai Narain Vyas University, Jodhpur, India

Zeng-fu LI, Huil H, Hang-yi Z et al (2008) Review on the extraction of flavonoids from fern. J Sanming Univ 25(2):200–203

Zhang Q, Hu J, Sheng L et al (2010) Simultaneous quantification of capsaicin and dihydrocapsaicin in rat plasma using HPLC coupled with tandem mass spectrometry. J Chromatogr B Analyt Technol Biomed Life Sci 878(24):2292–2297

**Johnson Marimuthu**, F.I.F.S., completed his Doctorate in Botany (Plant Tissue Culture) from Manonmaniam Sundaranar University, Tirunelveli District, Tamil Nadu, and is working as Assistant Professor of Botany and Director, Centre for Plant Biotechnology, St. Xavier's College (autonomous), Palayamkottai, Tamil Nadu. He has 21 years of research experience and 18 years of teaching experience. He has published 223 papers in reputed journals, 27 papers in proceedings, and 37 papers as book chapters. He is a recipient of the prestigious SERB Young Scientist—2013, Prof. S.S. Bir Gold Medal—2017, and Certificate of Merit 2014 by Indian Fern Society, India.

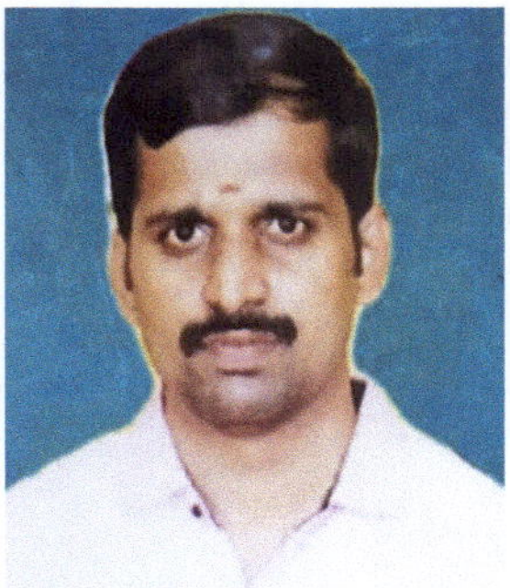

**N. Janakiraman** is working as an Assistant Professor in the Department of Botany, the Madura College (Autonomous), Madurai, from June 2019. He completed his Master's degree in St. Joseph's College (Autonomous), Tiruchirappalli. He finished his M.Phil. and PhD degrees from Manonmaniam Sundaranar University, Tirunelveli. After that, he worked as an Assistant Professor in the Department of Botany, Sri Paramakalyani College, Alwarkurichi, from January 2014 to April 2019. He published 17 research articles in Scopus indexed journals and 20 research articles in peer-reviewed national and international journals. Further, he contributed five chapters in books and deposited three sequences in GenBank.

**J. Chandra Saleride** is working as an Assistant Professor in the Department of Botany, Women's Christian College, Nagercoil, from June 2019. She completed her Master's degree in Nesamony Memorial Christian College, Marthandam. She finished her M. Phil. and PhD degrees from Manonmaniam Sundaranar University, Tirunelveli.

**A. Sivaraman** is doing research in Maria Research Foundation, Tirunelveli. He completed his Master's degree in St. Xavier's College (Autonomous), Palayamkottai. He finished his PhD degrees from Manonmaniam Sundaranar University, Tirunelveli. He published 7 research articles in peer-reviewed national and international journals. Further, he contributed two chapters in books.

**B. Shivananthini** is doing research in Centre for Plant Biotechnology (CBB) in St. Xavier's College (Autonomous), Palayamkottai. She received her education at St. Xavier's College (Autonomous) with a Master's in Botany. She also holds a Bachelor's in Botany from St. Xavier's College (Autonomous), Palayamkottai.

**K. Paulraj,** completed his Doctorate in Botany from Manonmaniam Sundaranar University, Tirunelveli District, Tamil Nadu, and is working as Associate Professor of Botany and Principal, Nesamony Memorial College, Marthandam. He has 29 years of teaching experience and 17 years of research experience. He has published 18 papers in the national and international journals.

# Biopotency of Pteridophytes: A Review 20

Johnson Marimuthu, N. Janakiraman, J. Chandra Saleride,
A. Sivaraman, B. Shivananthini, and K. Paulraj

**Abstract**

From Silurian period, pteridophytes exist in the nature and expected to harbour various useful secondary metabolites. By the presence of secondary metabolites, pteridophytes are able to survive for more than 450 million years and house various biological activities, viz. anti-bacterial, anti-cancer, anti-diabetic, anti-fungal, anti-inflammatory, anti-oxidant, hepatoprotectivity, wound healing, etc. The review intends to summarize the available biopotential of pteridophytes from 2000 to 2021. A total of 244 species are taken into account for the present review. This chapter recorded anti-oxidant potential (135), anti-bacterial and anti-fungal activities (97), cytotoxic properties (61), anti-cancer activities (39), anti-inflammatory activities (26), anti-diabetic potential (23), hepatoprotective properties (9), wound healing potential (7) and larvicidal activities (6) of pteridophytes. We made an attempt to provide an update on the biopotential of pteridophytes. This review might be useful for the pteridologist, phytochemist and pharmacist for further research.

J. Marimuthu (✉) · B. Shivananthini
Department of Botany, Centre for Plant Biotechnology, St. Xavier's College (Autonomous), Palayamkottai, India

N. Janakiraman
Department of Botany, The Madura College (Autonomous), Madurai, Tamil Nadu, India

J. Chandra Saleride
Department of Botany, Women's Christian College, Nagercoil, Tamil Nadu, India

A. Sivaraman
Maria Research Foundations, Tirunelveli, Tamil Nadu, India

K. Paulraj
Department of Botany, Nesamony Memorial Christian College, Marthandam, Tamil Nadu, India

 
J. Marimuthu et al. (eds.), *Ferns*, https://doi.org/10.1007/978-981-16-6170-9_20

**Keywords**

Anti-bacterial · Anti-cancer · Anti-diabetic · Anti-fungal · Anti-inflammatory · Anti-oxidant · Cytotoxicity · Hepatoprotective · Larvicidal · Wound healing

## 20.1 Introduction

Plants harbour various secondary metabolites as natural defence mechanism to protect them from numerous stresses and natural enemies (Bennett and Wallsgrove 2006). From Silurian period, pteridophytes are surviving in the nature and adapted with various climatic and environmental conditions (Wallace et al. 1991). Therefore, pteridophytes are expected to possess useful secondary metabolites. In addition to the primary metabolites, pteridophytes possess innumerable minerals, vitamins, alkaloids, saponins, phenols, tannins, phytosterols, triterpenes and terpenoids in a substantial amount (Patil et al. 2013). Plant-derived metabolites polyphenols (flavonoids, phenolic acids, lignans, tannins, anthocyanins, catechins, isoflavones), vitamins and pro-vitamins (ascorbic acid, α-tocopherol and β-carotene) have the potential to act as natural anti-oxidants (Yeou 2015). By the existence of secondary metabolites, the pteridophytes house biological activities such as anti-oxidant (Johnson et al. 2020), anti-microbial (Johnson et al. 2017b; Abiola et al. 2018; Farook et al. 2019; Faizal et al. 2020), anti-viral (McCutcheon et al. 1995), anti-diabetic (Johnson et al. 2020; Manivannan and Johnson 2020; Manivannan et al. 2020), hepatoprotectivity (Balne et al. 2013; Zakaria et al. 2020), anti-inflammatory (Johnson et al. 2017a, b; Zakaria et al. 2017), anti-cancer (Janakiraman and Johnson 2016) and anti-HIV (Mizushina et al. 1998). Plant-based drugs are easily available, economical and highly efficient without any side effects (Thite et al. 2013). Due to the therapeutical value, in the recent days, researchers focussed their attention towards phytochemical and pharmacological studies on pteridophytes (Soeder 1985; Santos et al. 2010: Liu et al. 2012; Johnson et al. 2014; Janakiraman and Johnson 2015, Janakiraman and Johnson 2016; Cao et al. 2017; Johnson et al. 2017a, b; Abiola et al. 2018; Baskaran et al. 2018; Halder and Chakraborty (2018) Farook et al. 2019; Salehi et al. 2019; Daryono and Rhomawati 2020; Faizal et al. 2020; Johnson et al. 2020; Manivannan and Johnson 2020, Manivannan et al., 2020; Adnan et al. 2021; Dvorakova et al. 2021; Johnson et al. 2021; Fierascu et al. 2021; Manivannan et al., 2021).

Due to the advancement in the sophisticated instrument facilities, rapid growth on the phytochemistry and pharmacological research on pteridophytes has increased dramatically. This review summarizes current knowledge regarding the biopotency of pteridophytes (Fig. 20.1).

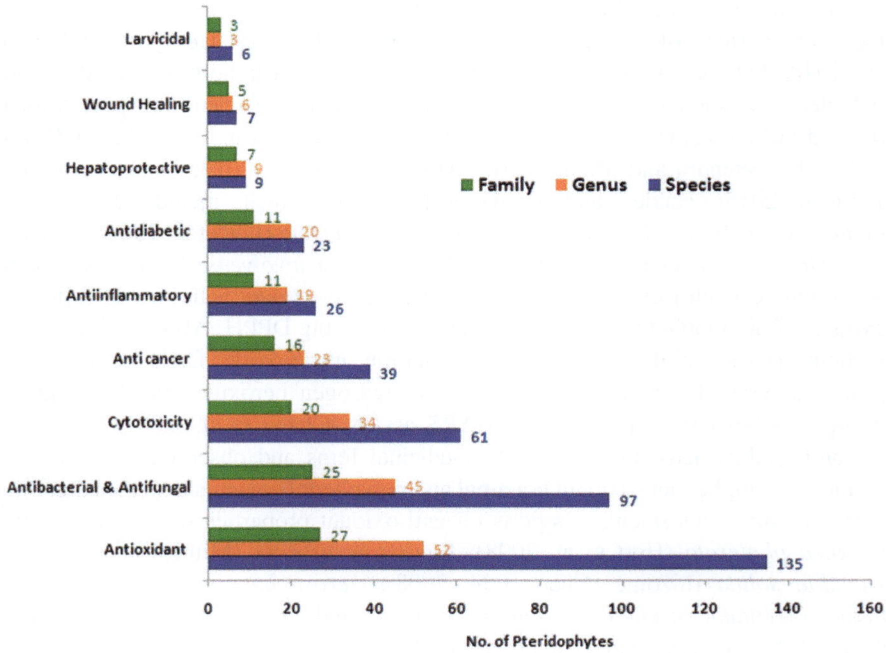


**Fig. 20.1** Summary of fern biopotency

## 20.2 Anti-oxidant Activities

Anti-oxidants are molecular substances that protect the organisms from oxidative damage by preventing the binding of various cellular target molecules. The proverb says "Too much of anything is good for nothing". It is proved practically in oxygen-mediated adverse situation (oxidative stress) in the cell, when it is present in higher concentrations. Oxidative stress is caused by an imbalance between production and accumulation of oxygen reactive species (ROS) in cells and tissues (Pizzino et al. 2017). The raise in the production of common ROS, viz. superoxide radicals, hydrogen peroxide, hydroxyl radicals and singlet oxygen, is signalling the stress condition of the tissue (Navarro-Yepes et al. 2014). The metabolic processes, namely, protein phosphorylation, activation of several transcriptional factors, apoptosis, immunity and differentiation, rely mainly on the balanced ROS production that should be maintained at a low level (Rajendran et al. 2014). The raise of ROS indicates harmful effects on important cellular structures like proteins, lipids and nucleic acids (Wu et al. 2013). Anti-oxidant detoxifies the free radicals and protects the tissues from adverse effects (Sunil 2014). Drug formulations with anti-oxidants are used for treating complex diseases, viz. Alzheimer's disease, stroke, atherosclerosis, cancer and diabetes (Devasagayam et al. 2004). Compounds responsible for anti-oxidant activities can be isolated and used for prevention and treatment of free radical-related disorders (Middleton et al. 2000). Kahl and Kappus (1993) observed the adverse effects of synthetic anti-oxidants, viz. BHT (butylated hydroxytoluene) and BHA (butylated hydroxyanisole). Due to this, the researchers are forced to find out alternative sources of natural anti-oxidants. A number of scientific reports which pointed out the natural anti-oxidants from plants (Chang et al. 2007; Lai and Lim 2011a, b; Jeetendra and Manish 2011; Mohammad et al., 2013; Janakiraman and Johnson 2015; Halder and Chakraborty 2018) which include *Selaginella tamariscina* (Miao et al. 1996); *Dryopteris crassirhizoma* (Lee et al. 2003); *Equisetum telmateia* (Helena et al. 2005); and *Selaginella involvens*, *S. delicatula* and *S. wightii* (Gayathri et al. 2005) were studied. Anti-oxidant activities of different extracts of pteridophytes were mainly determined using DPPH, ABTS, FRAP, total reducing power, inhibition of lipid peroxidation, nitric oxide, ferric ion reducing power, superoxide anion, hydroxyl radical, hydrogen peroxide, metal chelating, phosphomolybdenum, TEAC and TBARS assays. Chang et al. (2007) evaluated the anti-oxidant activities of six folk medicinal ferns and observed that aqueous extracts had higher anti-oxidant potential and polyphenol contents than the ethanolic extracts. Numerous scientific reports on anti-oxidant properties were studied, viz. *Cyathea phalerata* (Hort et al. 2008); *Equisetum arvense* (Mimica et al. 2008); *Davallia solida* rhizome (Chen et al. 2008); *Acrostichum aureum*, *Asplenium nidus*, *Blechnum orientale*, *Cibotium barametz* and *Dicranopteris linearis* (Lai et al. 2009); *Microgramma vacciniifolia* (Peres et al. 2009); *Arachniodes exilis* (Daonian et al. 2010); *Lygodium flexuosum* (Jeetendra and Manish 2011); *Blechnum orientale* (How et al. 2010); *Selaginella doederleinii* (Li et al. 2010); *Pteris multifida* (Lan et al. 2011); *P. vittata*, *P. venulosa*, *Cyathea latebrosa*, *Dicranopteris linearis*, *Cibotium barometz*, *Drynaria quercifolia*, *Blechnum orientale*, *Adiantum raddianum*, *Diplazium esculentum*, *Pityrogramma calomelanos*, *Lygodium circinnatum*, *Microsorum punctatum*, *Nephrolepis biserrata*, *Pyrrosia nummularifolia* and *Acrostichum aureum* (Lai and Lim 2011a, b); *Dicranopteris linearis* (Zakaria et al. 2011); *Cyathea latebrosa*, *Dicranopteris linearis*, *Pteris vittata*, *Cibotium barometz*, *Drynaria quercifolia*, *Blechnum orientale*, *Adiantum raddianum*, *Diplazium esculentum*, *Pityrogramma calomelanos*, *Lygodium circinnatum* and *Microsorum punctatum* (How et al. 2011); *Diplazium esculentum*, *Marsilea crenata* and *Drynaria quercifolia* (Jutarat and Jindarat 2011); *Dryopteris filix-mas*, *D. affinis* and *Athyrium filix-femina* (Soare et al. 2012); *Dryopteris filix-mas* (Sekendar et al. 2012); *Selaginella willdenowii* (Tsun and Fai 2012); *Adiantum capillus-veneris* (Rajurkar and Gaikwad, 2012); *A. philippense* (Sikder et al. 2013); *Dryopteris erythrosora* (Jianguo et al. 2013); *Pteris vittata* (Tania and Suchitra 2013); *Selaginella involvens*, *S. intermedia*, *S. inaequalifolia* and *S. tenera* (Sivaraman and Parimelazhagan 2013); *Drynaria quercifolia* (Milon et al. 2013); *Adiantum philippense* (Ali et al. 2013); *A. capillus-veneris* (Kumar et al. 2013); *Actiniopteris radiata* and *Equisetum ramosissimum* (Paulsamy et al. 2013); and *Blechnum orientale* (Antesa and Grino 2013).

Recently, the studies on anti-oxidant potential of pteridophytic extracts were increased gradually due to the presence of potential compounds. Reports include *Diplazium esculentum* (Saleha et al. 2014); *Asplenium aethiopicum* (Johnson et al.

2014); *Adiantum* and *Pteris* (Pandey et al. 2014); *Azolla pinnata* and *A. rubra* (Nawaz et al. 2014); *Diplazium esculentum*, *Marsilea crenata* and *Stenochlaena palustris* (Amoroso et al. 2014); *Pteris scabristipes*, *Aleuritopteris flava*, *Microlepia rhomboidea*, *M. hallbergii*, *Diplazium esculentum*, *Asplenium khasianum*, *Adiantum edgeworthii* and *Lindsaea odorata* (Gupta et al. 2014); *Pteris biaurita* (Jaishee and Chakraborty 2014a, b); *P. vittata* (Kaur et al. 2014); *Drynaria quercifolia* (Jinu et al. 2014; Jaishee and Chakraborty 2014a, b); *Adiantum* and *Pteris* (Pandey et al. 2014; Kathirvel et al. 2014); *Selaginella frondosa*, *Stenoloma chusanum*, *Pellaea smithii*, *Allantodia doederleinii*, *Athyrium niponicum*, *Diplazium donianum*, *Dryoathyrium boryanum*, *Woodwardia japonica*, *W. magnifica*, *Blechnum orientale*, *Brainea insignis*, *Dryopteris erythrosora*, *D. sparsa*, *D. reflexosquamata*, *D. lacera*, *Egenolfia sinensis*, *Davallia cylindrical*, *Araiostegia yunnanensis* and *Pyrrosia gralla* (Xia et al. 2014); *Cheilanthes albomarginata* (Lamichhane et al. 2014; Lamichhane et al. 2015); *Matteuccia struthiopteris*, *M. orientalis*, *Osmunda japonica* and *Pteridium aquilinum* (Dion et al. 2015); *Equisetum hyemale* (de Queiroz et al. 2015); *Pteris vittata* (Maneesha et al. 2015); *Polypodium interjectum*, *Polystichum woronowii*, *P. aculeatum*, *Asplenium scolopendrium*, *A. adiantum-nigrum*, *Dryopteris affinis*, *Athyrium filix-femina* and *Pteris cretica* (Hassan et al. 2015); *Actiniopteris radiata* (Mathad et al. 2015); *Cyathea nilgirensis*, *C. gigantea* and *C. crinita* (Janakiraman and Johnson 2015); *Marsilea quadrifolia* (Zhang et al. 2016); *Salvinia molesta* (Nithya et al. 2015); *Stenochlaena palustris* (Chai et al. 2015a, b); *Azolla microphylla* (Kunnathupara et al. 2016); *Dipteris conjugata* (Jarial et al. 2016); *Nephrolepis biserrata*, *Davallia denticulata*, *Asplenium longissimum*, *Gonioplebium percussum*, *Stenochlaena palustris*, *Vittaria elongata* and *V. ensiformis* (Khan et al. 2016); *Asplenium adiantum-nigrum* and *A. trichomanes* (Hammami et al. 2016); *Dicranopteris linearis* (Kalpana Devi et al. 2016); *Diplazium esculentum* and *D. maximum* (Wali et al. 2016); *Marsilea minuta* and *Diplazium esculentum* (Choudhury et al. 2017); *Trichomanes javanicum* and *Oleandra pistillaris* (Nofrizal and Dayar 2017); *Stenochlaena palustris* (Arullappan et al. 2017); *Drynaria quercifolia*, *Microsorum punctatum* and *Pyrrosia adnascens* (Dela Cruz et al. 2017); *Equisetum ramosissimum*, *E. arvense*, *Aleuritopteris flava*, *Pseudodrynaria coronans*, *Pyrrosia adnascens*, *P. nummularifolia*, *Microsorum punctatum*, *Drynaria fortune*, *D. quercifolia*, *Pityrogramma calomelanos*, *Adiantum lunulatum*, *A. edgeworthii*, *A. raddianum*, *Pteris biaurita*, *P. scabristipes*, *P. vittata*, *P. venulosa*, *Lindsaea odorata*, *Acrostichum aureum*, *Davallia divaricata*, *D. mariesii*, *D. solida*, *Actiniopteris radiata*, *Microlepia rhomboidea*, *M. hallbergii*, *Humata griffithiana*, *Dryopteris affinis*, *D. filix-mas*, *D. cochleata*, *D. crassirhizoma*, *Marsilea minuta*, *Dicranopteris linearis*, *Cyathea latebrosa*, *C. nilgirensis*, *C. crinita*, *C. gigantea*, *Nephrolepis biserrata*, *Cibotium barometz*, *Ampelopteris prolifera*, *Diplazium esculentum*, *D. maximum*, *Athyrium filix-femina*, *Blechnum orientale*, *Asplenium khasianum*, *Stenochlaena palustris*, *Selaginella involvens*, *S. delicatula* and *Lygodium circinnatum* (Halder and Chakraborty 2018); *Asplenium nidus* (Jarial et al. 2018a, b); *Adiantum capillus-veneris* (Abdul Qadir et al. 2018); *Azolla pinnata* (Thiripurasundari and Padmini 2018); *Dryopteris erythrosora* (Zhang et al. 2019); *A. trichomanes* and *A. scolopendrium* (Ismail et al.

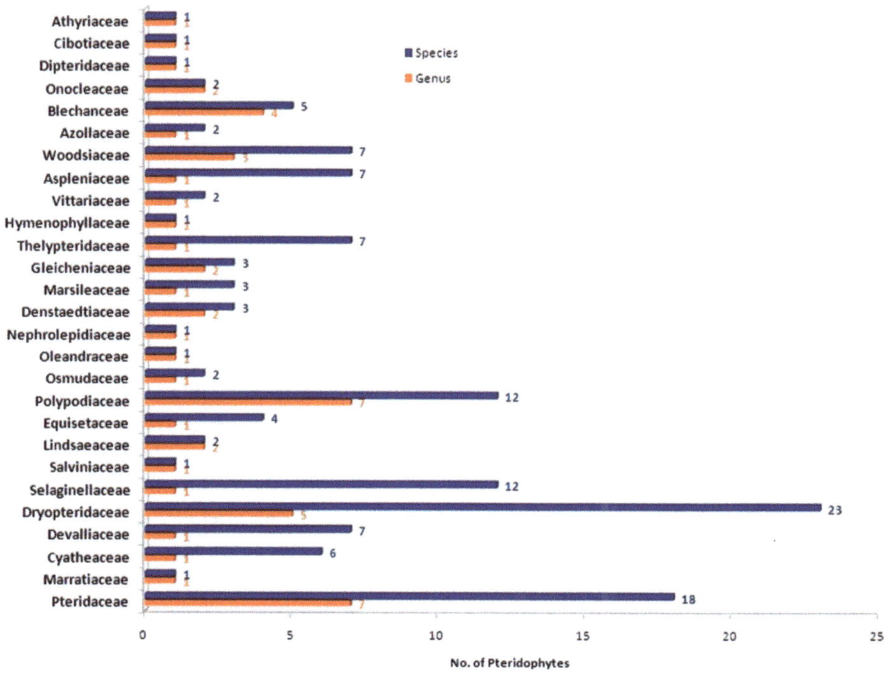


**Fig. 20.2** Anti-oxidant potential of pteridophytes

2019); *Azolla pinnata* (Farook et al. 2019); *Angiopteris helferiana* (Lamichhane et al. 2019); *Cheilanthes tenuifolia* (Mahfuz et al. 2019); *Tectaria coadunata* (Shrestha et al. 2019); *Cyathea contaminans* (Faizal et al. 2020); *Pneumatopteris callosa* (Daryono and Rhomawati 2020); *Sphaerostephanos unitus* (Johnson et al. 2020); *Asplenium scolopendrium*, *A. distentifolium*, *A. filix-femina*, *Pteridium aquilinum*, *D. aemula*, *D. affinis*, *D. borreri*, *D. cambrensis*, *D. carthusiana*, *D. caucasica*, *D. dilatata*, *D. expansa*, *D. filix-mas*, *D. oreades*, *D. remota*, *Polystichum aculeatum*, *P. setiferum*, *Matteuccia struthiopteris*, *Onoclea sensibilis*, *Osmunda regalis*, *Polypodium vulgare*, *Lastrea limbosperma*, *Phegopteris connectilis* and *Thelypteris palustris* (Dvorakova et al. 2021); and *Selaginella repanda* (Adnan et al. 2021). A detailed summary on the anti-oxidant activities of ferns and fern allies are illustrated in Fig. 20.2.

## 20.3 Anti-diabetic Activities

Diabetes mellitus is an endocrinal disorder related with exhausted insulin secretions and damaged pancreatic β-cells with altered carbohydrate, lipid and protein metabolism with increased risk of complications in various vascular diseases. People in India are more susceptible to diabetes accounting for about 40 million which will further reach to a maximum of 74 million in the year 2025 (Porter and Barrett 2005).

Diabetes becomes pandemic nowadays in India and worldwide with the highest population engulfed by this disease (Sampath Kumar et al. 2012). Diabetes mellitus Type I is characterized by chronic hyperglycemia that involves destruction of pancreatic cells. Type I diabetes affects an estimated patients of around 5 to 10% (Daneman 2006). Type II diabetes being non-genetic can be worked out in laboratory by effectively inducing diabetes in Sprague Dawley rats by a single dose of alloxan, which is a toxic glucose analogue that selectively destroys insulin-producing cells in the pancreas leading to non-insulin-dependent diabetes mellitus (Sharma and Kumar 2011). The conventional treatments of diabetes include insulin injections, oral hypoglycemic drugs, exercise, diet or combination of all of these (Rang et al. 2003). Anti-diabetic drugs that are highly preferred nowadays include sulfonylureas and biguanides. These are known to cause various adverse effects (Jain et al. 2006). In the past few decades, there is an interest in the therapeutic potential of medicinal plants. Plant extracts are known for their hypoglycemic properties (Venkataratnam et al. 2006).

Glucose-lowering effect and anti-diabetic activity are studied using glucose tolerance test in normal rats and alloxan-induced diabetic rats. Extracts of *Hemionitis arifolia* (Nair et al. 2006), *Selaginella tamariscina* (Zheng et al. 2011), *Adiantum philippense* (Paul et al. 2012), *Pteris vittata* (Tania et al. 2012; Paul and Banerjee 2013), *Actiniopteris radiata* (Chand et al. 2013), *Christella dentata* (Rabiea et al. 2013), *Selaginella bryopteris* (Singh et al. 2014), *Angiopteris evecta* (Rahmatullah et al. 2014; Samira et al. 2014), *Drynaria quercifolia* (Rajimol et al. 2014) and *Blechnum orientale*, *Davallia denticulata*, *Diplazium esculentum*, *Nephrolepis biserrata* and *Pteris vittata* (Chai et al. 2015a, b) possess strong anti-diabetic properties. Chen et al. (2015) isolated the three novel potential anti-diabetic compounds, viz. (3*R*)-Pterosin D 3-*O*-β-D-(3'-*p*-coumaroyl)-glucopyranoside, (2*R*,3*R*)-Pterosin L 3-*O*-β-D-(3'-*p*-coumaroyl)-glucopyranoside and 13-chloro-spelosin 3-*O*-β-D-glucopyranoside, from *Ceratopteris thalictroides*, *Hypolepis punctata*, *Nephrolepis multiflora* and *Pteridium revolutum*. Recent investigations on pteridophytes, viz. *Dryopteris dilatata* (Mordi Joseph et al. 2016), *Davallia formosana* (Cheng et al. 2017), *Angiopteris helferiana* (Lamichhane et al. 2019), *Sphaerostephanos unitus* (Johnson et al. 2020) and *Bolbitis appendiculata* (Manivannan et al. 2020), revealed significant anti-diabetic potentials in different extracts. The anti-diabetic properties of pteridophytes with reference to species, genus and families were displayed in Fig. 20.3.

## 20.4 Cytotoxic Activities

Cytotoxicity of plant extracts can be assessed preliminarily using brine shrimp lethality bioassay. This method is used for the determination of $LC_{50}$ values of plant extracts and to predict the cytotoxic activity (Meyer et al. 1982). *Artemia salina* L. (brine shrimp) is an invertebrate fauna of the saline aquatic and marine ecosystems. This brine shrimp in vivo lethality tests have been successively used to find out the active cytotoxic and anti-tumour agents (Krishnaraju and Tsay 2006;

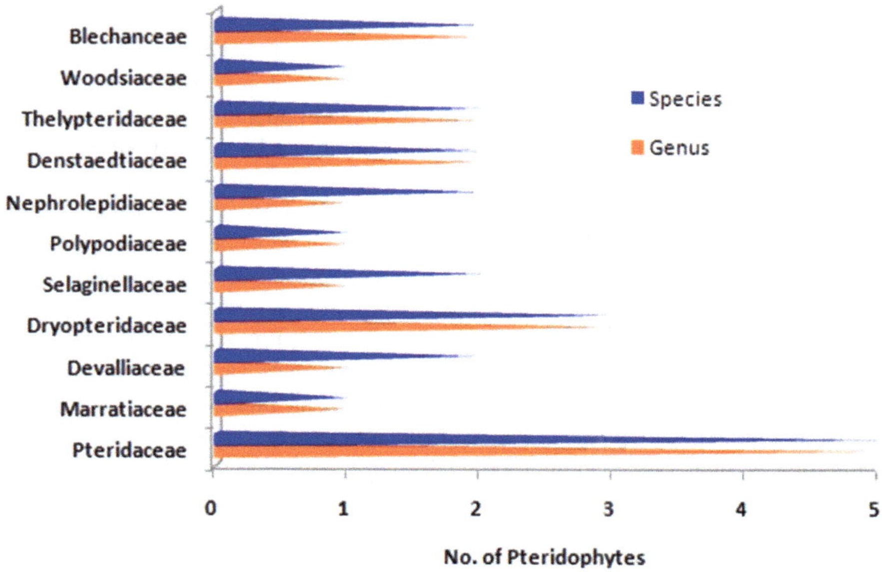


**Fig. 20.3** Anti-diabetic properties of pteridophytes

Alim et al. 2007). This bioassay was carried out to investigate the cytotoxicity of *Lygodium venustum* (Maria et al. 2013); *Adiantum philippense* (Sikder et al. 2013); *Selaginella doederleinii* (Priscilla et al. 2014); *Diplazium esculentum* (Saleha et al. 2014); *Cyathea* spp. (Janakiraman and Johnson 2016); *Nephrolepis biserrata*, *Davallia denticulata*, *Asplenium longissimum*, *Gonioplebium percussum*, *Stenochlaena palustris*, *Vittaria elongata* and *V. ensiformis* (Khan et al. 2016); *Cheilanthes tenuifolia* (Mahfuz et al. 2019); *Sphaerostephanos unitus* (Johnson et al. 2020); *Tectaria paradoxa* (Manivannan and Johnson 2020); and *Bolbitis appendiculata* (Manivannan et al. 2020). Johnson et al. (2021) investigated the toxicity of Shervarayan hill lycophytes and ferns aqueous extracts using brine shrimp and observed dose-dependent toxicity.

Cancer is one of the dreadful diseases in humans, and there is considerable scientific interest in the discovery of new anti-cancer agents from natural products. Over 50% of drugs used in clinical trials for anti-cancer activity were isolated from plants (Newman and Cragg 2007). *Pteris semipinnata* and *P. multifida* are cytotoxic and contain diterpenes (Li et al. 1998, 1999). Other species of *Pteris* possess antimutagenic, immunomodulatory and neuronal activity (Goldberg and Cooper 1975; Lee and Lin 1988; Wu et al. 2005). Phytochemicals present in plants possess anti-cancer activity (Han et al. 2008). Reports on various pteridophytes, viz. *Pteris vittata* (Simán et al. 2000); *Selaginella tamariscina* (Lee et al. 1999); *Drynaria fortunei* (Liu et al. 2001); *Christella dentata* (Somvanshi and Sharma 2005); *Selaginella involvens* (Gayathri et al. 2005); *Tectaria singaporeana*, *Blechnum orientale* and *Tacca integrifolia* (Nor Aini et al. 2008); and *B. orientale* (Lai et al. 2009), possess anti-cancer properties against various cell lines. Somjintana et al. (2005)

isolated angiopteroside from the methanolic extract of *Angiopteris evecta*. It showed significant activity for inhibition of HIV-1 reverse transcriptase and possess anti-tumour activity against lung cancer. Huanjie and Ping (2010) reported that terpenoids are able to inhibit tumour cell proliferation and induce tumour cell death by inhibiting multiple cancer-specific targets.

Amentoflavone extracted from *Selaginella tamariscina* was screened against five cancer cells, viz. HeLa, BEL-7402, MCF-7, PANC-1 and HL-60. The anti-cancer activity was determined by means of MTT assay and trypan blue cytometry. The extracts of *S. tamariscina* were effective to inhibit the proliferation tested cells and showed reliable activity against HL-60 (Jing et al. 2010). Kaempferol glycosides named palmatosides A (1), B (2) and C (3) together with three known kaempferol glycosides, viz. multiflorins A (4), B (5) and afzelin (6), were isolated from the roots of *Neocheiropteris palmatopedata*. The isolated compounds possess cancer chemopreventive potential based on their ability to inhibit tumour necrosis factor alpha (TNF-a-)induced NF-kB activity, nitric oxide production, aromatase, quinone reductase 2 (QR2) and COX-1/-2 activities (Yang et al. 2010). Numerous studies were conducted to find out the anti-cancer potential of pteridophytes, viz. *Dicranopteris linearis* against MCF-7, HeLa, HT-29, HL-60, K-562 and MDA-MB-231 cell lines (Zakaria et al. 2011); *Artocarpus heterophyllus* against HEK293, A549, HeLa and MCF-7 (Patel and Patel 2011); *Salvadora persica* (Ibrahim et al. 2011); *Selaginella ciliaris*, *Marsilea minuta* and *Thelypteris prolifera* (Sarker et al. 2011); *S. involvens* (Irene Pearl et al. 2011); *S. plana* fractions against MCF-7 cells (Sri Handayani et al. 2013); *S. doederleinii* against MDAMB231 and HepG2 (Priscilla et al. 2014); *S. labordei*, *S. tamariscina*, *S. uncinata*, *S. moellendorffii*, *S. remotifolia* and *S. pulvinata* (Li et al. 2014); *Asplenium nidus* (Tahir et al. 2014); *Athyrium filix-femina* and *Pteris cretica* (Hassan et al. 2015); *P. gralla*, *D. sparsa*, *W. japonica*, *S. chusanum*, *E. sinensis*, *D. cylindrical* and *D. erythrosora* (Xia et al. 2014); and *Blechnum orientale*, *Davallia denticulata*, *Diplazium esculentum*, *Nephrolepis biserrata* and *Pteris vittata* (Chai et al. 2015a, b). Chai et al. (2015b) studied the cytotoxic potential of six edible and medicinal ferns. Kaewsuwan et al. (2015) isolated three coumarin derivatives, viz. interruptins A, B and C, from *Cyclosorus terminans*. Interruptins A and B inhibited the growth of MCF-7 human breast and HT-29 human colon cancer cells. Recent studies showed various reports on anti-cancer properties, viz. *Acrostichum heterophyllum* (Meera and Josekumar 2016); *A. aureum* (Anitta et al. 2016); *Cyathea* spp. (Janakiraman and Johnson 2016); *Adiantum incisum* (Sayema and Dewan 2016); *A. caroliniana* (Nayak and Padhy 2017); *Stenochlaena palustris* (Arullappan et al. 2017); *A. nidus* against human hepatoma HepG2 and human carcinoma HeLa cells (Jarial et al. 2018a, b); *Cheilanthes tenuifolia* (Jarial et al. 2018a, b); *Tectaria coadunate* (Shrestha et al. 2019); *Selaginella repanda* against breast (MCF-7), colon (HCT-116) and lung (A549) cancer cells (Adnan et al. 2021); and *Polypodium vulgare* (Farràs et al. 2021) using T3 and HaCaT (non-tumour cell line) and HeLa, HepG2, MCF-7 and A549 (tumour cell line). Figures 20.4 and 20.5 summarized the cytotoxicity and in vitro anti-cancer activity of ferns and fern allies.

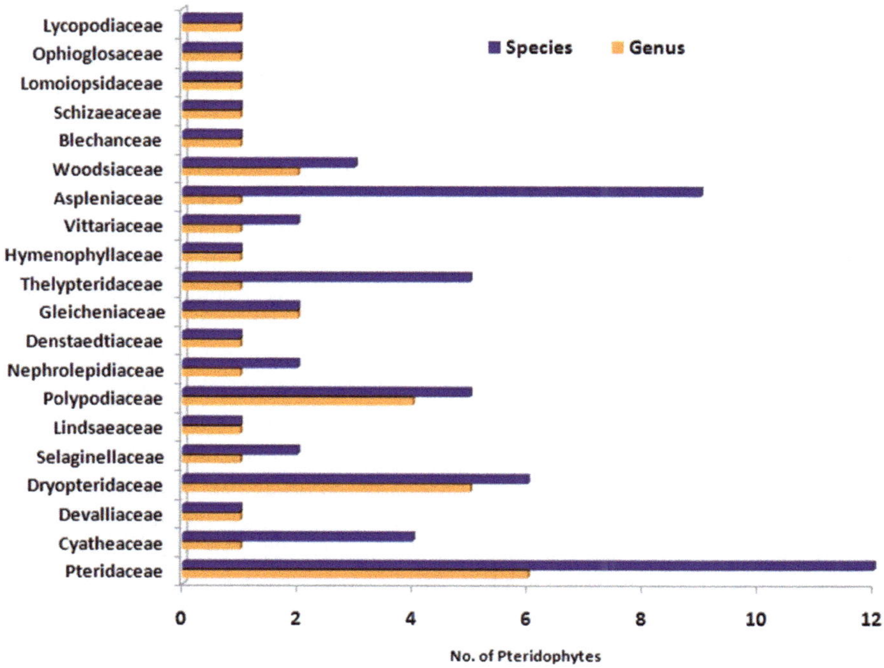


**Fig. 20.4** Cytotoxicity of ferns and fern allies

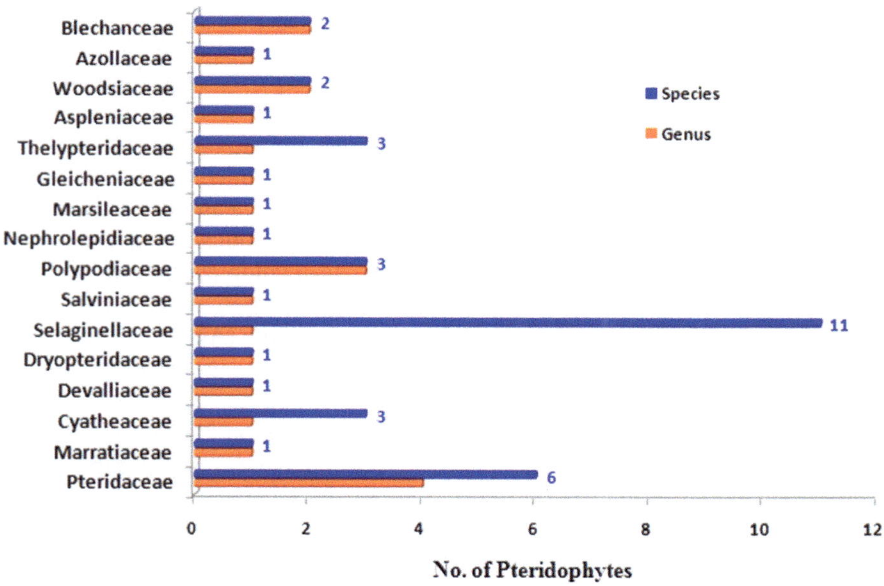


**Fig. 20.5** In vitro anti-cancer potential of pteridophytes

## 20.5 Anti-inflammatory Activities

Fabiana et al. (2009) revealed the anti-inflammatory activities of *Blechnum occidentale* extract using carrageenan-induced paw oedema and neutrophil migration method. Banani et al. (2014) evaluated the anti-inflammatory activity of aqueous and methanolic extracts of *Drynaria quercifolia* rhizome. In vitro cyclooxygenase inhibition was also carried out to investigate the pathway of anti-inflammatory activity. Anti-inflammatory activity of *D. quercifolia* could be partially due to COX-1 and COX-2 enzyme inhibition which may be attributed to the presence of flavonoids and other polyphenols in the extracts. Ali et al. (2014) studied the thrombolytic activity of methanolic extracts of *Drynaria quercifolia*. Aqueous and petroleum ether soluble fraction exhibited highest thrombolytic activity by clot lysis compared to standard streptokinase. Ismiarni et al. (2015) evaluated the anti-inflammatory activity of *Nephrolepis falcata* and *Pyrrosia lanceolata* using anti-denaturation of heat bovine serum albumin method. Ethyl acetate and methanolic extracts of *P. lanceolata* possess anti-inflammatory property at concentration of 1 to 10 μg/ml. Dion et al. (2015) evaluated the anti-inflammatory potential of *Matteuccia struthiopteris*, *Osmunda japonica*, *Matteuccia orientalis* and *Pteridium aquilinum* ethanolic extracts by the inhibition of the pro-inflammatory gene expression. The results showed a decrease of IL1-β gene expression for the five studied extracts.

Prasanna and Chitra (2015) investigated the anti-inflammatory activity of *Drynaria quercifolia* rhizome. The results showed that the maximum membrane stabilization was found to be 34.94% for methanolic extract and 41.62% for standard diclofenac at a dose of 300 mcg. Zakaria et al. (2017) investigated the anti-inflammatory potential of *Dicranopteris linearis* leaves. 100 μg/ml of methanolic extract showed a low inhibitory effect against the LOX activity. Dela Cruz et al. (2017) recorded the anti-inflammatory activity of *Drynaria quercifolia*, *Microsorum punctatum* and *Pyrrosia adnascens* rhizome and frond methanolic extracts using albumin denaturation assay. Highest activity was recorded in *M. punctatum* rhizome extracts. Johnson et al. (2017a, b) explored the anti-inflammatory property of *Cyclosorus interruptus*, *Pityrogramma calomelanos*, *Pteris vittata*, *P. confusa*, *P. biaurita*, *P. multiaurita*, *Adiantum incisum*, *A. latifolium*, *Hemionitis arifolia* and *Ceratopteris thalictroides* aqueous extracts using heat-induced haemolytic activity. The tested extracts effectively inhibited the heat-induced haemolysis. Mahfuz et al. (2019) investigated the membrane stabilizing and thrombolytic activity of *Cheilanthes tenuifolia*. The n-hexane and ethyl acetate extract demonstrated 73.79 and 69.27% of inhibition. Lamichhane et al. (2019) evaluated the anti-inflammatory activity of *Angiopteris helferiana* rhizome ethyl acetate fraction against RAW 264.7 macrophage cells for the inhibition of nitric oxide radical production after the induction by LPS. The EA fraction significantly inhibited the nitrite production in LPS-stimulated RAW 264.7 cells. The inhibition found to increase on concentration-dependent manner. Manivannan et al. (2020) revealed the anti-inflammatory potential of *Bolbitis appendiculata*. Dose-dependent anti-inflammatory activity was observed. Johnson et al. (2020) revealed the anti-inflammatory property of *Tectaria paradoxa* by adopting RBC membrane stabilization against heat-induced

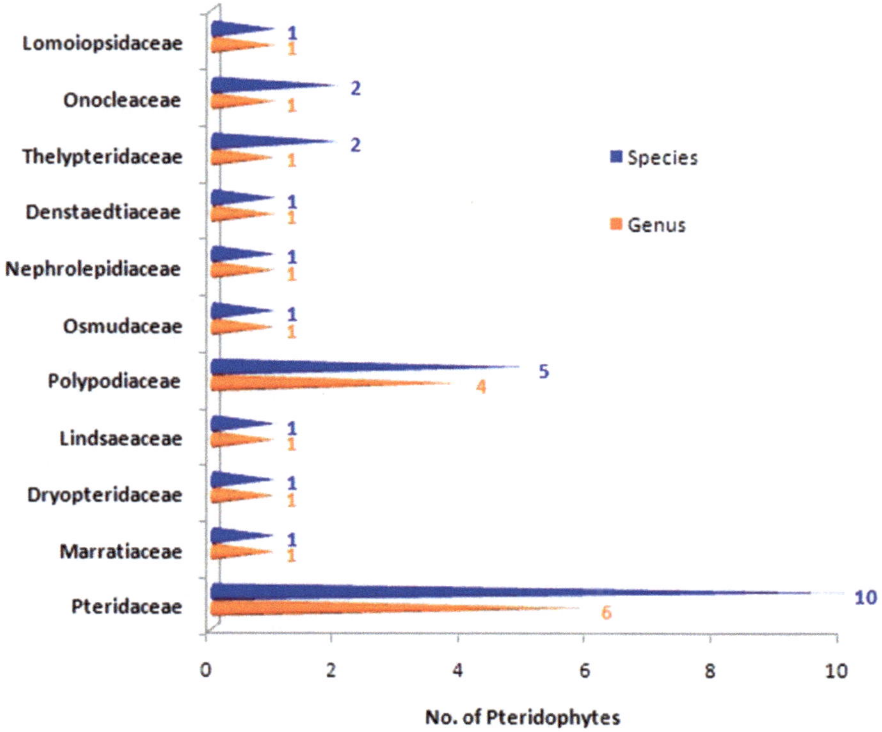


**Fig. 20.6** Summary of anti-inflammatory activity of pteridophytes

haemolysis. Johnson et al. (2020) investigated the anti-inflammatory activity of *Sphaerostephanos unitus* using membrane stabilization assays. The presence of phenolics, tannins, flavonoids, sterols and triterpenoids may be accountable for the observed anti-inflammatory properties. Figure 20.6 illustrated the anti-inflammatory properties of lycophytes and ferns.

## 20.6 Anti-bacterial and Anti-fungal Activities

A large number of microorganisms are involved in various skin infections, and plants contain phytochemicals to kill these organisms (Kar and Kumar 2010). *Gleichenia linearis* showed anti-bacterial properties in aqueous extracts (Vasudeva 1999). Methanolic extracts of *Drynaria quercifolia* showed broad concentration-dependent anti-microbial activity (Ramesh et al. 2001). *Lycopodiella cernua* was anti-virally active and had been patented as a treatment for hay fever (Zhang et al. 2002). Extracts of *D. quercifolia* are active against *Neisseria gonorrhoeae* (Shokeen et al. 2005). Amentoflavone from *S. tamariscina* inhibits several pathogenic fungi (Jung et al. 2007). Anti-microbial compounds have been characterized from *Pteris biaurita* (Dalli et al. 2007). Isocryptomerin from *Selaginella tamariscina* showed anti-bacterial potential against Gram-positive and Gram-negative bacteria (Lee et al. 2009). Ethyl acetate, butanol and aqueous extracts of *Blechnum orientale* were effective against all the tested Gram-positive bacteria (Lai et al. 2009).

Reports on various extracts of pteridophytes, viz. *Selaginella inaequalifolia* against *Staphylococcus aureus*, *Escherichia coli* and *Candida albicans* (Irudayaraj et al. 2010) and *S. involvens* and *S. inaequalifolia* against poultry pathogens (Haripriya et al. 2010), and fronds of *Pteris* (Thomas 2011); *S. involvens* (Irene Pearl et al. 2011); *Christella parasitica* (Paul Raj et al. 2011); *Cystopteris fragilis* (Soare et al. 2012); *Cyathea nilgiriensis*, *C. crinita*, *Leptochilus lanceolatus* and *Osmunda hugeliana* (Raja et al. 2012); *Cyclosorus interruptus* (Pauline et al. 2012); and *Adiantum caudatum*, *A. latifolium* and *A. lunulatum* (Johnson et al. 2017b) showed that these plants possess anti-microbial properties.

The Table 20.1 explains the anti-bacterial and anti-fungal potential of various pteridophyte extracts against the Gram-positive bacteria, Gram-negative bacteria and fungal strains. Figure 20.7 provided the family, genus and species wise anti-bacterial and anti-fungal activity of pteridophytes.

## 20.7 Larvicidal Activities

Mosquitoes transmit many dreadful diseases which cause serious health problems to human beings. The government has taken various steps using synthetic chemicals to control the mosquitoes which act as vectors for a long time. Synthetic chemicals cause pollution which results in development of resistance among mosquito species (Das and Rajagopalan 1981; Brown 1986; Thangam and Kathiresan 1990). A cautious and prolonged control of the mosquito vector can eliminate filariasis. It is not an easy task due to its natural tolerance and early development of resistance (Brown and Pal 1971). Plant extracts act as an alternative to these synthetic chemicals. Larvicidal properties of *Adiantum* ethanolic fraction showed the activity at highest concentration of 70–85% at 30,000 ppm and at 5,000 ppm. *A. craccivora* showed 70–75% activity at 30,000 ppm and 50–70% activity at 5,000 ppm between 24 and 48 h (Sood and Sharma 2010). Johnson et al. (2014) studied the larvicidal activity of different extracts of *Asplenium aethiopicum* against *Culex quinquefasciatus*. Highest larval mortality was observed in acetone extract. Janakiraman and Johnson (2017) evaluated and recorded the larvicidal potential and larval mortality of different extracts of *Cyathea*. The $LC_{50}$ values of different extracts indicated that ethanolic extracts of three *Cyathea* species are more sensitive compared to other extracts. Halimatussakdiah et al. (2018) studied the larvicidal activity of *Diplazium esculentum* methanolic extract. The results showed active larvicidal activity against *Culex* sp.

**Table 20.1** Summary of anti-bacterial and anti-fungal potential of ferns

| Name of the pteridophyte | Extract | Microorganisms tested | References |
|---|---|---|---|
| *Selaginella repanda* | Ethanol | *Staphylococcus flexneri*, *Staphylococcus aureus*, *Escherichia coli* and *Pseudomonas aeruginosa* | Adnan et al. (2021) |
| *Adiantum philippense* | Methanol | *E. coli*, *S. aureus*, *P. aeruginosa* and *S. flexneri* | Adnan et al. (2021) |
| *Pneumatopteris callosa* | Methanol | *S. aureus*, *E. coli*, *Pseudomonas* sp. and *Bacillus cereus* | Daryono and Rhomawati (2020) |
| *Cyathea contaminans* | Methanol and n-hexane | *E. coli* and *S. aureus* | Faizal et al. (2020) |
| *Azolla pinnata* | Acetone, benzene, ethanol, methanol and water | *P. aeruginosa* and *S. aureus* | Farook et al. (2019) |
| *Gleichenia pectinata* | Methanol | *E. coli*, *P. aeruginosa*, *S. aureus*, *Proteus mirabilis*, *Klebsiella pneumoniae* and *Candida albicans* | Abiola et al. (2018) |
| *Platycerium superbum* | Methanol | *S. aureus*, *P. aeruginosa*, *Bacillus subtilis*, *K. pneumoniae*, *E.*coli and *P. mirabilis* | Abiola et al. (2018) |
| *Adiantum lunulatum* | Ethanol, methanol, chloroform and water | *S. aureus*, *P. aeruginosa*, *Proteus vulgaris* and *E. coli* | Jenat and Suresh (2018) |
| *Azolla pinnata* | Ethyl acetate, chloroform, methanol, acetone and water | *S. aureus*, *E. coli*, *Salmonella* sp., *Pseudomonas* sp., *Bacillus* sp., *C. albicans*, *Aspergillus niger*, *Trichoderma viride*, *Rhizopus microsporus* and *Penicillium chrysogenum* | Thiripurasundari and Padmini (2018) |
| *Tectaria cicutaria* | Ethanol | *B. subtilis*, *S. aureus*, *E. coli* and *P. vulgaris* | Preeti and Namdeo (2018) |
| *C. parasitica*, *M. speluncae*, *P. multifida*, *M. pteropus*, *A. caudatum*, *P. vittata*, *Thelypteris appendiculata*, *Tectaria decurrens*, *Pronephrum lakhimpurense*, *Castanea dentata*, *P. cretica* | Water, methanol, acetone and petroleum ether | *B. subtilis*, *E. coli*, *P. aeruginosa*, *S. aureus* and *P. vulgaris* | Nath et al. (2018) |

(continued)

**Table 20.1** (continued)

| Name of the pteridophyte | Extract | Microorganisms tested | References |
|---|---|---|---|
| *Cheilanthes albomarginata* | Methanol | *S. aureus*, *S. citreus*, *K. pneumoniae*, *P. aeruginosa*, *Shigella flexneri*, *E. coli*, *Enterobacter* sp., *Salmonella typhimurium*, *S. paratyphi*, *S. epidermis*, *Streptococcus mutans*, *P. vulgaris* | Jarial et al. (2017) |
| *Trichomanes javanicum* | n-Hexane, ethyl acetate and n-butanol | *B. subtilis*, *Micrococcus luteus*, *S. aureus*, *S. epidermidis*, *S. mutans*, *Enterococcus faecalis*, *E. coli*, *S. typhimurium*, *P. aeruginosa* and *Vibrio cholerae* | Nofrizal and Dayar (2017) |
| *Dipteris conjugate* | Methanol | *S. typhi*, *S. paratyphi*, *P. mirabilis*, *S. typhimurium*, *E. coli*, *S. flexneri*, *P. aeruginosa*, *Enterobacter* sp., *K. pneumonia*, *S. mutans*, *S. epidermidis* and *S. aureus* | Jarial et al. (2016) |
| *Pteridium aquilinum* | n-hexane and ethanol | *P. aeruginosa*, *S. aureus*, *E. coli*, *S. typhi* and *E. aerogenes* | Awe and Amobi (2015) |
| *Polypodium interjectum*, *P. woronowii*, *P. aculeatum*, *Dryopteris affinis*, *Athyrium filix-femina*, *Asplenium scolopendrium* and *Pteris cretica* | Methanol | *E. coli* and *S. aureus* | Bahadori et al. (2015) |
| *S. molesta* | Ethanol | *P. aeruginosa*, *Aeromonas hydrophila*, *E.coli*, *B. cereus*, *B. subtilis* and *S. aureus* | Nithya et al. (2015) |
| *Equisetum hyemale* | Ethanol and methanol | *S. aureus*, *S. epidermidis*, *E. coli*, *B. subtilis*, *P. aeruginosa*, | Geisiany et al. (2014) |

(continued)

**Table 20.1** (continued)

| Name of the pteridophyte | Extract | Microorganisms tested | References |
|---|---|---|---|
| | | *C. albicans*, *T. rubrum*, *T. mentagrophytes* and *M. canis* | |
| *Salvinia auriculata* | Acetone, ethanol, methanol, ethyl acetate and benzene | *B. subtilis*, *S. aureus*, *E. coli*, *P. aeruginosa*, *P. vulgaris* and *S. typhi* | Suvarnalatha et al. (2015) |
| *Cyclosorus terminans* | n-Hexane, dichloromethane, ethyl acetate and methanol | Methicillin-sensitive *S. aureus*, methicillin-resistant *S. aureus*, *S. epidermidis*, *B. subtilis* and *E. coli* | Kaewsuwan et al. (2015) |
| *Nephrolepis cordifolia* and *Nephrolepis exaltata* | Essential oil | *S. aureus*, *S. pneumonia*, *E. faecalis*, *S. typhimurium*, *P. vulgaris*, *K. pneumonia*, *S. flexneri*, *P. aeruginosa*, *E. coli*, *M. gypseum*, *Trichophyton rubrum* and *T. mentagrophytes* | Tantawy et al. (2015) |
| *Dicranopteris linearis* | Acetone, diethyl ether, chloroform and methanol | *B. subtilis*, *S. aureus*, *E. coli*, *P. aeruginosa*, *S. typhi*, *K. pneumoniae*, *A. niger*, *A. flavus*, *Curvularia lunata*, *Macrophomina phaseolina*, *Rhizopus stolonifer* | Devi et al. (2015) |
| *Drynaria quercifolia* | Methanol | *S. aureus*, *S. epidermidis*, *S. pneumoniae*, *S. pyogenes*, *B. subtilis*, *E. coli*, *S. typhimurium*, *S. flexneri*, *S. sonnei*, *P. aeruginosa*, *P. vulgaris* and *P. mirabilis* | Padhy and Dash (2015) |
| *Azolla microphylla* | Methanol, anhydrous ethyl acetate (fraction I) and diethyl ether (fraction II) | *Xanthomonas oryzae* | Abraham et al. (2015) |
| *Adiantum capillus-veneris* | Methanol | *S. aureus*, *S. epidermidis*, *S. typhi*, *K. pneumoniae*, *S. dysenteriae*, | Ishaq et al. (2014) |

(continued)

**Table 20.1** (continued)

| Name of the pteridophyte | Extract | Microorganisms tested | References |
|---|---|---|---|
| | | *P. vulgaris*, *P. aeruginosa*, *Providencia* sp., *C. freundii* and *E. coli* | |
| *Cyathea nilgirensis* | Hexane, petroleum ether, chloroform, ethanol and methanol | *K. pneumoniae*, *S. aureus*, *E. coli*, *B. subtilis*, *C. albicans*, *F. solani*, *A. ochraceus* and *F. verticillioides* | Mary and Mahesh (2015) |
| *Marsilea minuta* | Ethanol | *E. coli* and *B. subtilis* | Nagarajan and Britto (2014) |
| *Adiantum raddianum* | Petroleum ether, acetone, methanol and water | *S. aureus* subsp. *aureus*, *E. coli*, *P. aeruginosa*, *K. pneumoniae* subsp. *pneumoniae* and *S. marcescens* | Thomas (2014) |
| *Salvinia minima*, *Thelypteris interrupta* and *Marsilea minima* | Chloroform and methanol | *S. aureus*, *S. citreus*, *E. coli*, *P. aeruginosa*, *S. typhi*, *S. paratyphi* A, *S. paratyphi* B, *Chromobacterium*, *Enterobacter*, *C. freundii*, *Klebsiella* sp., *V. cholera*, *S. sonnei*, *S. boydii*, *P. mirabilis*, *P. vulgaris*, *C. albicans*, *A. niger*, *A. flavus* and *Rhizopus* sp. | Panda et al. (2014) |
| *Drynaria quercifolia* | Methanol | *E. coli*, *B. subtilis*, *S. aureus* and *Salmonella* sp. | Prasanna et al. (2014) |
| *Adiantum capillus-veneris*, *Nephrolepis cordifolia* and *Pteris vittata* | Ethanol | *E. coli* and *B. megaterium* | Pal (2014) |
| *Marsilea quadrifolia* | Ethanol | *S. typhi*, *P. fluorescens*, *P. aeruginosa* and *E. coli* | Sethi (2014) |
| *Drynaria quercifolia* | Methanol, ethanol, chloroform andpetroleum ether | *S. aureus*, *E. coli*, *K. pneumoniae*, *P. aeruginosa*, *B. subtilis*, *S. typhi*, *V. cholerae*, *C. albicans* and *C. neoformans* | Pargavi and Sivakumar (2014) |
| *Diplazium esculentum* | | | |

(continued)

**Table 20.1** (continued)

| Name of the pteridophyte | Extract | Microorganisms tested | References |
|---|---|---|---|
| | Chloroform and methanol | *E. coli*, *S. aureus*, *V. cholerae*, *S. typhimurium*, *S. lutea*, *K. pneumoniae*, *B. subtilis* and *S. boydii* | Akter et al. (2014) |
| *Dicranopteris linearis* | Methanol | *S. aureus* | Chuah et al. (2014) |
| *Marsilea quadrifolia* | Water and methanol | Tobacco necrosis virus, *S. typhi*, *P. fluorescens*, *P. aeruginosa* and *E. coli* | Sethi (2014) |
| *Marsilea minuta* | Acetone, DMSO, ethanol, chloroform and petroleum ether | *B. subtilis*, *S. aureus*, *P. aeruginosa*, *S. pyogenes*, *E. coli*, *A. niger*, *A. flavus* and *T. rubrum* | Revathi and Sara (2014) |
| *Adiantum capillus-veneris* | Water, methanol, ethanol, ethyl acetate and hexane | *C. freundii*, *E. coli*, *Providencia* sp., *K. pneumoniae*, *P. vulgaris*, *S. typhi*, *Shigella* sp., *V. cholerae*, *P. aeruginosa*, *S. aureus*, *C. albicans*, *Pythium* sp., *A. flavus*, *A. niger* and *Trichoderma* | Ishaq et al. (2014) |
| *Marsilea quadrifolia* | Benzene, ethanol and water | *E. coli*, *K. pneumoniae*, *S. aureus*, *P. aeruginosa*, *A. hydrophila*, *A. niger*, *C. albicans* and *P. notatum* | Sivagurunathan and Innocent (2014) |
| *Angiopteris evecta* | Petroleum ether, chloroform, benzene, methanol and water | *Xanthomonas campestris* | Britto et al. (2014) |
| *Athyrium filix-femina*, *Dicranopteris linearis*, *Tectaria impressa*, *Hypolepis punctata* and *Pleopeltis macrocarpa* | Methanol, ethanol and acetone | *E. coli* and *B. megaterium* | Pal (2013) |
| *Azolla microphylla* | Methanol | Several strains of *Xanthomonas* | Gerard (2013) |
| *Lygodium flexuosum* | Petroleum ether, chloroform, methanol and water | *S. aureus*, *E. faecalis*, *E. aerogenes*, *E. coli*, *K. pneumoniae*, *P. aeruginosa* and *P. mirabilis* | Nayak et al. (2013) |
| | Ethanol | | Pal (2013) |

(continued)

**Table 20.1** (continued)

| Name of the pteridophyte | Extract | Microorganisms tested | References |
|---|---|---|---|
| *Cyclosorus interruptus*, *Gleichenia microphylla* and *Microsorum pteropus* | | *E. coli* and *B. megaterium* | |
| *Adiantum lunulatum* | Petroleum ether, acetone, methanol and water | *S. aureus* subsp. *aureus*, *E. coli*, *P. aeruginosa*, *K. pneumoniae* subsp. *pneumoniae* and *S. marcescens* | Thomas (2013) |
| *Polytrichum commune*, *Marsilea minuta*, *Dryopteris filix-mass* and *Adiantum raddianum* | Ethanol and water | *S. aureus* and *P. aeruginosa* | Sharma et al. (2013) |
| *Blechnum orientale* | Petroleum ether, chloroform, methanol and water | *S. aureus*, *B. subtilis*, *S. typhi*, *M. luteus*, *P. aeruginosa*, *E. coli*, *C. albicans*, *A. flavus* | Deepa et al. (2013) |
| *Athyrium filix-femina*, *Dryopteris affinis* and *D. filix-mas* | Ethanol and methanol | *E. coli*, *P. aeruginosa*, *S. abony*, *S. aureus* and *E. faecalis*, *Brevibacterium flavum*, *Sarcina* sp., *B. cereus*, *Saccharomyces cerevisiae* and *A. niger* | Soare et al. (2012) |
| *Pityrogramma calomelanos* | Ethanol and methanol | *E. coli*, *S. aureus*, *P. aeruginosa*, *C. albicans*, *C. krusei* and *C. tropicalis* | Souza et al. (2012) |
| *Pteris biaurita*, *Lygodium flexuosum*, *Hemionitis arifolia*, *Actinopteris radiata* and *Adiantum latifolium* | Petroleum ether, benzene, chloroform, methanol and water | *X. campestris* | Gracelin et al. (2012) |
| *Adiantum lunulatum*, *A. capillus-veneris*, *Pteris otaria*, *P. aspericaulis*, *P. kleiniana*, *P. confusa*, *P. multiaurita*, *P. vittata*, *Asplenium polyodon* and *Hypodematium crenatum* | Petroleum ether, benzene, chloroform, methanol and distilled water | *X. campestris* | Britto (2012) |
| *Adiantum capillus-veneris* | Water and ethanol | *S. aureus*, *P. mirabilis*, *K. pneumoniae*, *P. aeruginosa* and *C. albicans* | Nyarko et al. (2012) |

(continued)

**Table 20.1** (continued)

| Name of the pteridophyte | Extract | Microorganisms tested | References |
|---|---|---|---|
| *Equisetum arvense* | Ethanol | *B. subtilis*, *M. luteus*, *V. cholerae*, *E. coli*, *S. flexneri* and *S. dysenteriae* | Sinha (2012) |
| *Cyclosorus interruptus*, *Gleichenia microphylla*, *Microsorum pteropus* and *Athyrium filix-femina* | Water, methanol, ethanol and acetone | *E. coli* and *B. megaterium* | Pal (2012) |
| *Selaginella bryopteris*, *Lygodium flexuosum*, *Adiantum philippense*, *Dryopteris cochleata* and *Tectaria coadunata* | Methanol: dichloromethane | *S. aureus*, *N. gonorrhoeae*, *P. aeruginosa*, *E. coli*, *S. pyogenes* and *B. subtilis* | Malviya et al. (2012) |
| *Cyathea nilgiriensis*, *C. crinita*, *Leptochilus lanceolatus* and *Osmunda hugeliana* | Ethanol | *P. aureus*, *K. pneumoniae*, *Streptococcus* sp., *A. niger* and *Fusarium* sp. | Raja et al. (2012) |
| *Angiopteris evecta* | Chloroform, dichloromethane, acetone, ethanol and methanol | *E. coli*, *S. aureus*, *K. pneumoniae* and *S. faecalis* | Nilanthi et al. (2015) |
| *Azolla filiculoides* | Ethanol | *E. coli*, *S. aureus*, *F. oxysporum* and *T. mentagrophytes* | Angalao et al. (2012) |
| *Asplenium scolopendrium*, *Polypodium vulgare* and *Cystopteris fragilis* | Ethanol and methanol | *S. aureus*, *Streptococcus* sp., *E.coli* and *E. cloacae* | Soare et al. (2012) |
| *Selaginella involvens* | Ethanol and petroleum ether | *S. aureus* and *E. coli* | Irene Pearl et al. (2011) |
| *Pteris quadriaurita* | Petroleum ether, acetone, methanol and water | *S. aureus*, *E. coli*, *P. aeruginosa*, *M. luteus* and *Serratia marcescens* | Thomas (2011) |
| *Lygodium altum*, *Salvinia molesta*, *S. cucullata*, *Helminthostachys zeylanica* and *Dryopteris filix-mas* | Ethanol, methanol and acetone | *B. cereus*, *E. coli*, *K. pneumoniae* and *V. cholerae* | Mandal and Mondal (2011) |
| *Stenochlaena palustris* | Methanol | *S. aureus*, *B. subtilis*, *Micrococcus* sp., *E. aerogenes*, *E.coli*, *P. mirabilis*, *K. pneumoniae*, *S. typhi*, *Azospirillum lipoferum*, *Azotobacter* sp., | Zuraini et al. (2010) |

(continued)

**Table 20.1** (continued)

| Name of the pteridophyte | Extract | Microorganisms tested | References |
|---|---|---|---|
| | | *P. chrysogenum*, *Rhizopus stolonifer*, *A. niger*, *Fusarium* sp. and *S. cerevisiae* | |
| *Blechnum orientale* | Petroleum ether, chloroform, ethyl acetate, butanol and water | *B. cereus*, *M. luteus*, *S. aureus*, *S. epidermidis*, *E. coli*, *P. aeruginosa*, *K. pneumoniae* and *E. aerogenes* | How et al. (2010) |
| *Adiantum capillus-veneris*, *A. incisum*, *A. lunulatum*, *Actiniopteris radiata*, *Araiostegia pseudocystopteris*, *Athyrium pectinatum*, *Cheilanthes albomarginata*, *Cyclosorus dentatus*, *Dryopteris cochleata*, *Hypodematium crenatum*, *Marsilea minuta*, *Tectaria coadunata* and *T. macrodonta* | Methanol and water | *A. tumefaciens*, *E. coli*, *S. arizonae*, *S. typhi* and *S. aureus* | Parihar et al. (2010) |
| *Blechnum orientale* | Methanol, petroleum ether, chloroform, ethyl acetate, butanol and water | *B. cereus*, *M. luteus*, methicillin-susceptible *S. aureus*, methicillin-resistant *S. aureus*, *S. epidermidis*, *E. coli*, *P. aeruginosa*, *K. pneumoniae*, *S. choleraesuis* and *E. aerogenes* | Lai et al. (2009) |
| *Hemionitis arifolia*, *Pteridium aquilinum* and *Christella parasitica* | Chloroform | *Puccinia arachidis* and *Phaeoisariopsis personata* | Sahayaraj et al. (2009) |
| *Pteris vittata* | Water and methanol | *B. cereus*, *E. coli*, *K. pneumoniae*, *P. aeruginosa*, *S. aureus*, *S. typhimurium*, *S. pyogenes* and *S. flexneri* | Singh et al. (2008) |
| *Equisetum arvense* | Essential oil | *S. aureus*, *E. coli*, *K. pneumoniae*, *P. aeruginosa*, | Radulovic et al. (2006) |

(continued)

**Table 20.1** (continued)

| Name of the pteridophyte | Extract | Microorganisms tested | References |
|---|---|---|---|
| | | *S. enteritidis*, *A. niger* and *C. albicans* | |
| *Adiantum capillus-veneris* and *A. lunulatum* | Ethanol | *A. niger* and *Rhizopus stolonifer* | Guha et al. (2005) |
| *Drynaria quercifolia* | Methanol | *A. hydrophila*, *Chromobacterium violaceum*, *E. coli*, *K. pneumoniae*, *P. aeruginosa*, *S. typhi*, *V. cholerae*, *V. parahaemolyticus*, *B. subtilis*, *S. aureus*, *A. flavus*, *A. fumigatus*, *A. niger* and *C. albicans* | Ramesh et al. (2001) |

## 20.8 Wound Healing Properties

Wound is defined as a breaking of cellular and anatomical continuity of living tissue (Prabu et al. 2008). Wound healing is presently a clinical challenge due to inconsistencies encountered in the healing processes. Medical treatment includes administration of drugs either locally (topical) or systemically (oral or parenteral) with the objective either to shorten the time required for healing or to minimize the undesired consequences during wound repair (Myers et al. 1980). Medicinal plants are used in the treatment of various skin ailments as they are easily affordable and safe from hypersensitive reactions (Raina et al. 2008). Manjunatha et al. (2007) reported that ethanolic leave extracts of *Lycopodium serratum*-treated animals showed faster epithelization of wound. How et al. (2011) examined the wound healing activity of aqueous extract of *Blechnum orientale* on Sprague Dawley rats. The results confirmed that 25% (w/w) concentration was capable of producing significant wound healing activity. Hendy et al. (2013) studied the wound healing properties of *Acrostichum aureum* and *Acrostichum speciosum* ethanolic extract and confirmed that topical application of both extracts shown significant effect in wound healing process. Ranjan et al. (2014) studied wound healing activity of *Drynaria quercifolia* rhizome and reported that the rate of wound contraction was significantly higher in the animals treated with methanol extracts compared to the reference drug neosporin. Further, the extracts exhibited significantly decreased period of epithelisation compared to controls. Azam et al. (2015) studied the effect of *Equisetum arvense* on wound healing activity, and the results relieved pain during the 10 days period after episiotomy. Clericuzio et al. (2012) isolated the flavonoid oligoglycosides from *Ophioglossum vulgatum* which possesses wound healing properties. Two new glycosylated and acylated flavonols, viz. quercetin-3-O-

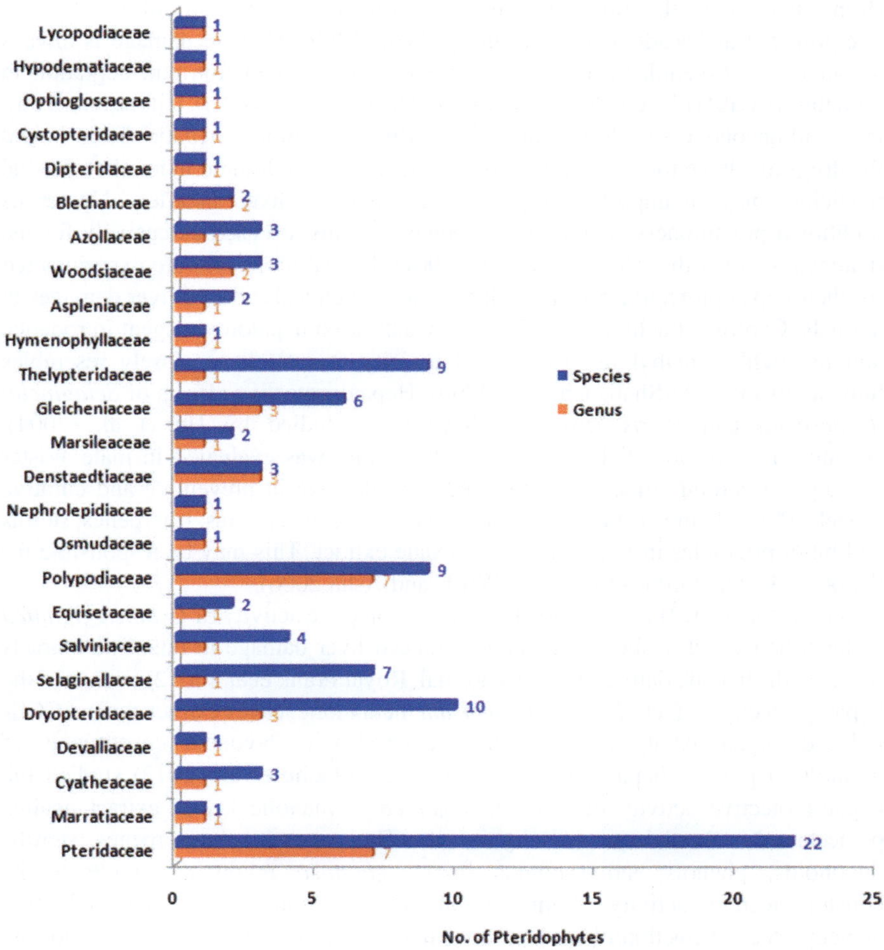


**Fig. 20.7** An illustrated account of anti-bacterial and anti-fungal potential of pteridophytes

[(6-caffeoyl)-β-glucopyranosyl (1→3) α-rhamnopyranoside]-7-O-α-rhamnopyranoside (2) and kaempferol-3-O-[(6-caffeoyl)-β-glucopyranosyl (1→3) α-rhamnopyranoside]-7-O-α-rhamnopyranoside (3), together with the known quercetin-3-O-methyl ether were isolated from the aerial parts of *Ophioglossum vulgatum*.

## 20.9 Hepatoprotective Activities

Hepatic disorders are a significant cause of mortality worldwide. The liver is a vital organ involved in maintaining metabolic functions and also detoxifies the exogenous and endogenous challenges like viral infections, chronic alcoholism, etc.

(Ramachandra et al. 2007). The natural protective mechanism of the liver is overpowered and leads to hepatic injury (Wolf 1999). Hepatic damage is always associated with cellular necrosis, increase in lipid peroxidation and depletion in glutathione (GSH) levels (Sandy and Ben 1998). Liver disorders still remain common and unconquered. Modern medicines offer alleviation of hepatic diseases, and the drugs available for the treatment of liver disorders are limited in number. Herbal medicines play an important role in the treatment of liver disorders. Numerous traditional practitioners claimed that medicinal plants are used extensively for the treatment of liver disorders (Dash et al. 2007). Herbal medicines are experimented for their hepatoprotective effects against various chemical-induced liver damages in animals. Carbon tetrachloride ($CCl_4$) is a widely used hepatotoxic agent in rodents, and its trichloromethyl radical-induced toxicity in rat's liver closely resembles human cirrhosis (Al-Shabanah et al. 2000). Hepatoprotective effects of *Selaginella doederleinii* and *Pteris ensiformis* have been studied by Hu et al. (2004). Hepatoprotective potential of *Lygodium flexuosum* was evaluated in male Wistar rats against carbon tetrachloride-induced liver damage in preventive and curative models. Phytochemical studies revealed the presence of saponins, triterpenes, sterols and bitter principles in *L. flexuosum* n-hexane extract. This may be responsible for the possible hepatoprotective effect (Wills and Asha 2006).

Kiran et al. (2012) investigated the hepatoprotective activity of *Cyathea gigantea* methanolic extract against paracetamol-induced liver damage in rats. The extracts reversed the hepatic damage towards normal. Royal Frank et al. (2012) evaluated the hepatoprotective effect of *Adiantum incisum* methanolic leaves extract against $CCl_4$-induced hepatotoxicity in rats. They reported that flavonoid constituents of *A. incisum* possess hepatoprotective properties. Madhu et al. (2012) studied the hepatoprotective activity of *Cyathea gigantea* methanolic leaves extract against paracetamol-induced hepatotoxicity in rats. The presence of triterpenes, sterols, flavonoids, phenols and saponins in *C. gigantea* might contribute to its hepatoprotective activity. Balne et al. (2013) studied the most effective hepatoprotective fraction of methanolic extract of *Marsilea minuta* in three models, viz. CCl4-, paracetamol- and ethanol-induced liver damage in rats. Pretreatment with fractions significantly reversed the changes in serum biochemical parameters and histology of the liver. Balne et al. (2013) analysed the hepatoprotective effect of methanolic extract fractions of *Marsilea minuta*. Pradeep and Ajudhia (2013) evaluated the hepatoprotective effect of *Drynaria quercifolia* fronds hydroalcoholic extract and concluded that the plant exhibited hepatoprotective property due to the presence of Dq-4-like flavonoid. Suja et al. (2014) studied the ethanolic extract of *Helminthostachys zeylanica* rhizome for antihepatotoxicity. In vitro primary rat hepatocyte culture studies with *H. zeylanica* showed significant hepatoprotection by increased cell viability, lowering of hepatic enzyme levels ALT and AST. Devika and Prasanna (2016) evaluated the in vitro hepatoprotective activity of methanolic extract of *Drynaria quercifolia* rhizome and reported the presence of hepatoprotective effect due to the presence of flavonoids and alkaloids. Zakaria et al. (2017) investigated the methanolic extract of *Dicranopteris linearis* leaves to attenuate liver intoxication induced by acetaminophen in rats. Pretreatment of rats

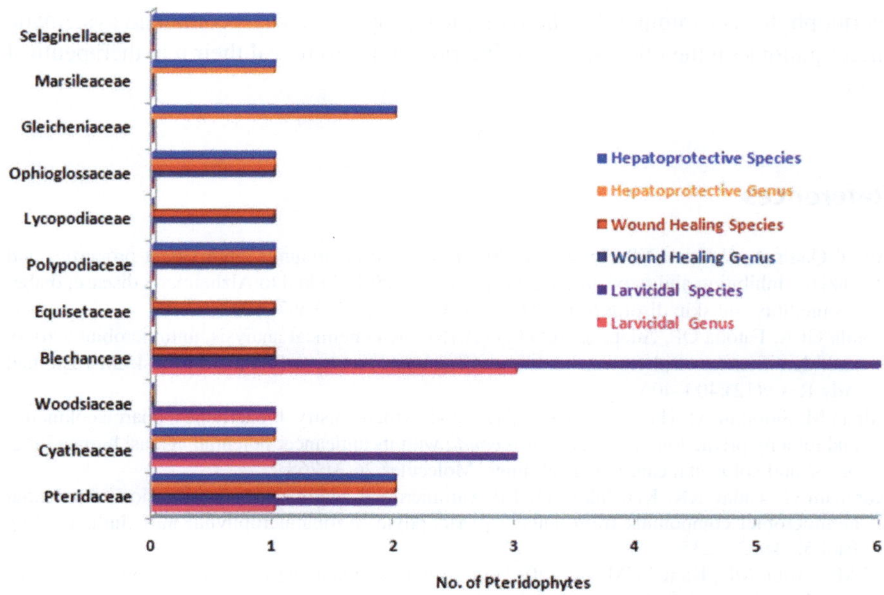


**Fig. 20.8** A graphical representation of larvicidal, wound healing and hepatoprotective properties of pteridophytes

with 10% DMSO failed to attenuate the toxic effect on the liver. This observation was supported by the significant ($p < 0.05$) increase in the level of serum liver enzymes of alanine transaminase, aspartate transaminase and alkaline phosphatase. There is a significant ($p < 0.05$) decrease in the activity of endogenous anti-oxidant enzymes of catalase and superoxide dismutase. Histopathological studies showed remarkable improvement in the liver cells architecture with increase in dose of the extract. Figure 20.8 explained the summary of larvicidal, wound healing and hepatoprotective properties of pteridophytes.

## 20.10 Conclusion

This review recorded anti-oxidant potential of 135 pteridophytes, anti-bacterial and anti-fungal activities of 97 pteridophytes, cytotoxic properties of 61 pteridophytes, anti-cancer activity of 39 pteridophytes, anti-inflammatory activity of 26 pteridophytes, anti-diabetic potential of 23 pteridophytes, hepatoprotective properties of 9 pteridophytes, wound healing potential of 7 pteridophytes and larvicidal activity of 6 pteridophytes. The demonstrated anti-bacterial, anti-cancer, anti-diabetic, anti-fungal, anti-inflammatory, anti-oxidant, hepatoprotective, larvicidal and wound healing in vitro and in vivo potencies authorized further clinical trial will be adopted in order to act as an alternative therapy. With reference to the available diversity of pteridophytes (12000 species), only very few percentage of pteridophytes is examined for the biological properties. We recommend systematic investigations on the other species of pteridophytes to reveal their real therapeutical value.

## References


Abdul Qadir A, Cakmak YS, Zengin G (2018) Phenolic compounds, antioxidant properties and enzyme inhibition ability of *Adiantum capillus-veneris* L. linked to Alzheimer's disease, diabetes mellitus and skin disorders. Curr Organic Chem 22(17):1697–1703

Abiola GFA, Fatoba OP, Adepoju AO et al (2018) Phytochemical analysis, antimicrobial activity and identification of phytoconstituents in *Gleichenia pectinata* (Willd.) C. Presl. Int J Biomed Adv Res 9(12):400–406

Adnan M, Siddiqui AJ, Hamadou WS et al (2021) Phytochemistry, bioactivities, pharmacokinetics and toxicity prediction of *Selaginella repanda* with its anticancer potential against human lung, breast and colorectal carcinoma cell lines. Molecules 26(3):768

Abraham G, Yadav RK, Kaushik G (2015) Antimicrobial activity and identification of potential antimicrobial compounds from aquatic pteridophyte, AzollamicrophyllaKaulf. Indian J Exp Biol 53(4):232–235

Ali MS, Amin MR, Kamal CM et al (2013) In vitro antioxidant, cytotoxic, thrombolytic activities and phytochemical evaluation of methanol extract of the A. philippense L. leaves. Asian Pac J Trop Biomed 3(6):464–469. https://doi.org/10.1016/S2221-1691(13)60097-0

Ali R, Hossain M, Runa JF et al (2014) Evaluation of thrombolytic potential of three medicinal plants available in Bangladesh, as a potent source of thrombolytic compounds. Avicenna J Phytomed 4(6):430–436

Alim A, Khan A, Rahman BM et al (2007) In vitro antimicrobial properties and cytotoxic activities of chromium complexes. Res J Agric Biol Sci 3(6):599–604

Al-Shabanah OA, Alam K, Nagi MN et al (2000) Protective effect of aminoguanidine, a nitric oxide synthase inhibitor, against carbon tetrachloride induced hepatotoxicity in mice. Life Sci 66 (3):265–270. https://doi.org/10.1016/s0024-3205(99)00589-5

Akter S, Monir H, Ismot A et al (2014) Investigation of *In vitro* antioxidant, antimicrobial and cytotoxic activity of *Diplazium esculentum* (Retz). Sw. IJAPBC 3(3):723–733

Amoroso VB, Lagumbay JD, Mendez RA et al (2014) Bioactives in three Philippine edible ferns. Asian Int J Life Sci 23(2):445–454

Angalao L, Doctor JGP, Banwa T (2012) Antimicrobial activities of Azollafiliculoides Lam. (Pteridophyte) and Brachytheciumbuchananii (Hook.) Jaeg (Bryophyte). Int J Sci Clin Lab 2:71–81. https://doi.org/10.7718/iamure.ijscl.v2i1.389

Anitta T, Prashob PKJ, Chandramohanakumar N (2016) A profiling of anti-tumour potential of sterols in the mangrove fern *Acrostichum aureum*. Int J Pharmacogn Phytochem Res 8 (11):1828–1832

Antesa DA, Grino YA (2013) DPPH radical scavenging activity, total phenolics, and antimicrobial screening of selected pteridophytes. Open Conf Proc J 4:259

Arullappan S, Sakunie S, Lew AC et al (2017) Phytochemical screening and evaluation of cytotoxic effect and antioxidant activity of fractions isolated from *Stenochlaena palustris* (Burm.f.) Bedd. Leaves. Indian J Pharm Educ Res 51(4S):s735–s740

Awe S, Amobi OO (2015) Antibacterial, phytochemical and proximate analysis of pteridiumaquilinum. Int J Res Pharm Biosci 2:1–7

Azam A, Bani S, Hasanpoor S et al (2015) The effect of equisetum arvense (horse tail) ointment on wound healing and pain intensity after episiotomy: a randomized placebo-controlled trial. Iran Red Crescent Med J 17(3):e25637. https://doi.org/10.5812/ircmj.25637

Bahadori MB, Mahmoodi F, Aliahmadi A et al (2015) Antibacterial evaluation and preliminary phytochemical screening of selected ferns from Iran. Res J Pharmacogn 2:53–59

Balne D, Pallerla P, Vanapatla SR et al (2013) Hepatoprotective effect of whole plant extract fractions of *Marsilea minuta* Linn. Asian J Pharm Clin Res 6(3):100–107
Banani D, Duttachoudhury M et al (2014) Antioxidant and anti-inflammatory activity of aqueous and methanolic extracts of rhizome part of *Drynaria quercifolia* (L.) J. Smith. Int J Pharm Pharm Sci 6(6):43–49
Baskaran XR, Geo Vigila AV, Zhang SZ (2018) A review of the use of pteridophytes for treating human ailments. J Zhejiang Univ-Sci B (Biomed & Biotechnol) 19(2):85–119
Bennett RN, Wallsgrove RM (2006) Secondary metabolites in plant defense mechanisms. New Phytologist 127:617–633
Britto AJD, Gracelin DHS, Benjamin PJRK (2014) Study on potential biocontrol agent—Angiopteris evecta (forst) hoff. against Xanthomonas campestris. Eur J Mol Biol Biochem 1 (5):192–195
Brown AWA (1986) Insecticide resistance in mosquitoes: a pragmatic review. J Am Mosq Control Assoc 2(2):123–140
Brown AWA, Pal R (1971) Insecticide resistance in arthropods, World Health Organization, Monger, vol 38. Switzerland, Geneva, p 491
Cao H, Chai TT, Wang X et al (2017) Phytochemicals from fern species: potential for medicine applications. Phytochem Rev 16:379–440. https://doi.org/10.1007/s11101-016-9488-7
Chai TT, Kwek MT, Ong HC et al (2015a) Water fraction of edible medicinal fern Stenochlaena palustris is a potent α-glucosidase inhibitor with concurrent antioxidant activity. Food Chem 186:26–31. https://doi.org/10.1016/j.foodchem.2014.12.099
Chai TT, Yeoh LY, Ismail NIM et al (2015b) Evaluation of glucosidase inhibitory and cytotoxic potential of five selected edible and medicinal ferns. TropJ Pharm Res 14(3):449–454
Chand Basha M, Sreenivasulu N, Pramod (2013) Anti-diabetic activity of *Actiniopteris radiata* (Linn.). Asian J Res Chem Pharm Sci 1:1–6
Chang HC, Huang GJ, Dinesh CA et al (2007) Antioxidant activities and polyphenol contents of six folk medicinal ferns used as "Gusuibu". Botanical Studies 48:397–406
Chen YH, Chang FR, Lin YJ et al (2008) Identification of antioxidants from rhizome of *Davallia solida*. Food Chem 107:684–691
Chen CY, Chiu FY, Lin Y et al (2015) Chemical constituents analysis and antidiabetic activity validation of four fern species from Taiwan. Int J Mol Sci 16(2):2497–2516. https://doi.org/10.3390/ijms16022497
Cheng L-H, Wu J-B, Jian J-Y, Shih C-C (2017) (−)-Epicatechin-3-O-β-D-allopyranoside from *Davallia formosana* prevents diabetes and dyslipidemia in streptozotocin-induced diabetic mice. PLoS One 12(3):e0173984. https://doi.org/10.1371/journal.pone.0173984
Choudhury J, Majumdar S, Roy S et al (2017) Antioxidant activity and phytochemical screening of two edible wetland Pteridophytes *Diplazium esculentum* (Retz.) Sw. and *Marsilea minuta* L.–A comparative study. WJPMR 53(9):195–203
Chuah EL, Zakaria ZA, Suhaili Z et al (2014) Antimicrobial activities of plant extracts against methicillin-susceptible and methicillin-resistant Staphylococcus aureus. J Microbiol Res 4(1):6–13
Clericuzio M, Tinello S, Burlando B, Ranzato E, Martinotti S, Cornara L, La Rocca A (2012) Flavonoid oligoglycosides from Ophioglossum vulgatum L. having wound healing properties. Planta Med 78(15):1639–1644. https://doi.org/10.1055/s-0032-1315149. Epub 2012 Aug 30
Dalli AK, Saha G, Chakraborty U (2007) Characterization of antimicrobial compounds from a common fern, *Pteris biaurita*. Indian J Exp Biol 45(3):285–290
Daneman D (2006) Type 1 diabetes. Lancet 367(9513):847–858. https://doi.org/10.1016/S0140-6736(06)68341-4
Daonian Z, Jinlan R, Yaling C et al (2010) Antioxidant and hepatoprotective activity of ethanol extract of *Arachniodes exilis* (Hance) Ching. J Ethnopharmacol 129(2):232–237
Daryono RNH, Rhomawati M (2020) An examination of medicinal potential of *Pneumatopteris callosa*: phytochemical screening, antibacterial, and antioxidant activity. IOP Conf Ser Earth Environ Sci 456:012068

Das PK, Rajagopalan PK (1981) Role of stimulated migration of mosquitoes in development and reversal of malathion resistance in *Culex pipensfatigans*. Indian J Med Res 73:139–143

Dash DK, Yeligar VC, Nayak SS et al (2007) Evaluation of hepatoprotective and antioxidant activity of *Ichnocarpus frutescens* (Linn) R.Br. on Paracetamol induced hepatotoxicity in rats. Trop J Pharm Res 6(3):755–765

Deepa J, Parashurama TR, Krishnappa M, Nataraja S (2013) Antimicrobial efficacy of Blechnum orientale L. Int J Pharm Biol Sci 4(2):475–479

de Queiroz GM, Flavio ASP, Edvanio RR et al (2015) Phytochemical characterization, antimicrobial activity, and antioxidant potential of *Equisetum hyemale* L. (Equisetaceae) extracts. J Med Food 18(7):830–834

Dela Cruz RY, Aileen May GA, Glenda ZD et al (2017) Phytochemical screening, antioxidant and anti-inflammatory activities of the three fern (Polypodiaceae) species in Bukidnon, Philippines. Bull Env Pharmacol Life Sci 6(3):28–33

Devasagayam TPA, Tilak JC, Boloor KK et al (2004) Review: free radical and antioxidants in human health: current status and future prospects. J Assoc Physicians India 52:794–804

Devika V, Prasanna G (2016) In vitro hepatoprotective activity of *Drynaria quercifilia* L. rhizome. World J Pharm Pharm Sci 5(1):512–521

Dion C, Haug C, Guan H et al (2015) Evaluation of the anti-inflammatory and antioxidative potential of four fern species from China intended for use as food supplements. Nat Prod Commun 10(4):597–603

Dvorakova M, Pumprova K, Antonínová Z et al (2021) Nutritional and antioxidant potential of fiddleheads from European ferns. Foods 10(2):460

Fabiana RN, Tais Adelita AB, Angélica ML et al (2009) Antiinflammatory and antinociceptive activities of Blechnum occidentale L. extract. J Ethnopharmacol 125(1):102–107

Faizal A, Taufik I, Anisah FR et al (2020) Antioxidant and antibacterial properties of tree fern *Cyathea contaminans*. Biodiversitas 21(5):2201–2205

Farook MA, Muthu Mohamed HS, Santhosh Kumar G (2019) Phytochemical screening, antibacterial and antioxidant activity of *Azolla pinnata*. Int J Res Anal Rev 6(2):240i–247i

Farràs A, López V, Maggi F et al (2021) *Polypodium vulgare* L.: polyphenolic profile, cytotoxicity and cytoprotective properties in different cell lines. Preprints: 2021050351 https://doi.org/10.20944/preprints202105.0351.v1

Fierascu I, Fierascu RC, Ungureanu C et al (2021) Application of polypodiopsida class in nanotechnology–potential towards development of more effective bioactive solutions. Antioxidants 10:748. https://doi.org/10.3390/antiox10050748

Gayathri V, Asha V, Subromaniam A (2005) Preliminary studies on the immunomodulatory and antioxidant properties of *Selaginella* species. Indian J Pharmacol 37:381–385

Geisiany Q, Souza-Moreira T, Salgado H et al (2014) Antimicrobial activity and toxicity in vitro and in vivo of Equisetum hyemale extracts. Rev Cienc Farm Basica Apl 35(4):559–563

Gerard A (2013) Evaluation of antimicrobial activity of methanolic extracts of Azollamicrophylla. Vegetos Int J Palnt Res 26(1):200. https://doi.org/10.5958/j.2229-4473.26.1.029

Goldberg DJ, Cooper JR (1975) Effects of thiamine antagonists on nerve conduction. I. Actions of antimetabolites and fern extract on propagated action potentials. J Neurobiol 6(5):435–452. https://doi.org/10.1002/neu.480060502

Gracelin DHS, Britto AJD, Kumar PBJR (2012) Antibacterial screening of a few medicinal ferns against antibiotic resistant phytopathogen. Int J Pharm Sci Res 3(3):868–873

Guha P, Mukhopadhyay R, Gupta K (2005) Antifungal activity of the crude extracts and extracted phenols from gametophytes and sporophytes of two species of Adiantum. Taiwania 50(4):272–283

Gupta SK, Ghosala M, Biswasb R (2014) Evaluation of *in vitro* antioxidant activity of methanolic extracts of some ferns from Mawsynram of Meghalaya, India. Int J Curr Sci 12:E87–E97

Halder K, Chakraborty S (2018) An account of antioxidant potential in pteridophytes: a biochemical perspective. Intl J Bioinform Biological Sci 6(1):15–24

Halimatussakdiah H, Amna U, Wahyuningsih P (2018) Preliminary phytochemical analysis and larvicidal activity of edible fern (*Diplazium esculentum* (Retz.)Sw.) extract against *Culex*. J Nat 18(3):141–147

Hammami S, Ali S, Ridha EM et al (2016) Essential oil constituents and antioxidant activity of *Asplenium* Ferns. J Chromatogr Sci 54(8):1341–1345

Han HJ, Tan NH, Zeng GZ et al (2008) Natural inhibitors of DNA topoisomerase I with cytotoxicities. Chem Biodivers 5(7):1364–1368. https://doi.org/10.1002/cbdv.200890124

Handayani S, Hermawan A, Meiyanto E et al (2013) Induction of Apoptosis on MCF-7 cells by *Selaginella* Fractions. J App Pharm Sci 3(4):31–34

Haripriya D, Selvan N, Jeyakumar N et al (2010) The effect of extracts of Selaginella involvens and Selaginella inaequalifolia leaves on poultry pathogens. Asian Pacific J Trop Med 3(9):678–681

Hassan V, Ali S, Fatemeh MK, Hossein D, Mir BB (2015) Cytotoxicity, antioxidant activity and phenolic content of eight fern species from North of Iran. Pharm Sci 21:18–24

Helena C, Paramas AG, Maria TA et al (2005) Characterisation of polyphenols by HPLC-PADESI/MS and antioxidant activity in *Equisetum telmateia*. Phytochem Anal 16(5):380–387

Hendy PH, Deny S, Shahbudin S et al (2013) Wound healing properties of ethanolic extract of *Acrostichum aureum* and *Acrostichum speciosum* rhizome in rats. J Trop Resour Sustain Sci 1 (2):42–48

Hort MA, Dalbó S, Brighente IMC et al (2008) Antioxidant and hepatoprotective effects of Cyathea phalerata Mart. (Cyatheaceae). Basic Clin Pharmacol Toxicol 103(1):17–24

How YL, Lim YY, Kim KH (2011) Potential dermal wound healing agent in *Blechnum orientale* Linn. BMC Complement Altern Med 11:62

How YL, Yau YL, Kah HK (2010) *Blechnum Orientale* Linn - a fern with potential as antioxidant, anticancer and antibacterial agent. BMC Complementary and Alternative Medicine 10:15. https://doi.org/10.1186/1472-6882-10-15

Hu MF, Yen MH, Liao JW et al (2004) Hepatoprotective effects of the folk medicines of "Yan-Kan-Tang No.1" and "Yan- Kan-Tang No.2" on rats. Crop Environ Bioinform 1:229–238

Huanjie Y, Ping QD (2010) Targeting apoptosis pathway with natural terpenoids: implications for treatment of breast and prostate cancer. Curr Drug Targets 11(6):733–744

Ibrahim AY, El-Gengaihi SE, Motawe HM (2011) Phytochemical and cytotoxicity investigations of *Salvadora persica* bark extracts. J Arab Soc Med Res 6(2):127–133

Irene Pearl J, Ismail TS, Irudayaraj V et al (2011) Pharmacognostical studies on anti-cancer spikemoss *Selaginella involvens* (sw.) spring. Int J Drug Dev Res 2(6):195–211

Irudayaraj V, Janaky M, Johnson M, Selvan N (2010) Preliminary phytochemical and antimicrobial studies on a spike-moss selaginella inaequalifolia (Hook. & Grev.) Spring. Asian Pacific J Trop Med 3(12):957–960

Ishaq MS, Hussain MM, Afridi MS et al (2014) In vitro phytochemical, antibacterial, and antifungal activities of leaf, stem, and root extracts of Adiantum capillus veneris. Sci World J 2:269793. https://doi.org/10.1155/2014/269793

Ismail AM, Al-Khasreji TO, Maulood BK (2019) Phytochemical and antioxidant activity of *Asplenium* Species (spleenworts) extracts from northern districts of Iraq. Eng Technol J 37 (1):248–251

Ismiarni KYA, Betha OS, Muliati F et al (2015) Antioxidant and Anti-inflammatory activity of the Indonesian ferns, *Nephrolepis falcata* and *Pyrrosia lanceolata*. Int J Pharm Pharm Sci 7 (12):162–165

Jain AK, Mehta SC, Shrivastava NM (2006) Hypoglycemic and antihyperglycemic effects of newly synthesized sulfonyloxy derivatives of Azlactone in normal and Alloxan diabetic rabbits. Indian J Pharmacol 37(6):395–396

Jaishee N, Chakraborty U (2014a) Evaluation of *in-vitro* antioxidant activities of *Pteris biaurita* L. Int J Pharm Pharm Sci 6(2):413–421

Jaishee N, Chakraborty U (2014b) Pharmacological studies and evaluation of antioxidant properties of *Drynaria quercifolia* (L.) J. Smith. Wor J Pharm Pharmac Sci 3:1205–1216

Janakiraman N, Johnson M (2015) In vitro antioxidant properties of natural products isolated from selected species of cyathea. J Clin Nephrol Res 2(2):1027

Janakiraman N, Johnson M (2016) Ethanol extracts of selected *Cyathea* species decreased cell viability and Inhibited Growth in MCF 7 cell line cultures. J Acupunct Meridian Stud 9 (3):151–155

Janakiraman N, Johnson M (2017) Larvicidal potential of Cyathea species against *Culex quinquefasciatus*. Pharmaceut Biomed Res 3(1):48–51

Jarial R, Singh L, Thakur S et al (2016) Applications of fern dipteris conjugate in anti-bacterial and anti-lipolytic purpose. In: Proceedings of the National Conference for Postgraduate Research (NCON-PGR 2016). Universiti Malaysia Pahang (UMP), Pekan, Pahang., 24–25 September 2016, pp 775–782

Jarial R, Singh L, Thakur S et al (2017) Evaluation of anti-lipolytic, anti-oxidant and anti-bacterial activities of selected ferns. J Appl Pharm Sci 7(06):150–156

Jarial R, Shard A, Thakur S et al (2018a) Characterization of flavonoids from fern *Cheilanthes tenuifolia* and evaluation of antioxidant, antimicrobial and anticancer activities. J King Saud Univ Sci 30(4):425–432

Jarial R, Sveta T, Mimi S et al (2018b) Potent anticancer, antioxidant and antibacterial activities of isolated flavonoids from *Asplenium nidus*. J King Saud Univ Sci 30(2):185–192

Jenat PJ, Suresh SN (2018) Phytochemical analysis and antimicrobial activity of fern Adiantum lunulatum Burm.f. World J Pharm Res 7(3):640–648

Jeetendra N, Manish B (2011) Correlation of antioxidant activity with phenolic content and isolation of antioxidant compound from Lygodium flexuosum (L.) SW. extracts. Int J Pharm Pharm Sci 3(2):48–52

Jianguo C, Xian X, Xuefei C et al (2013) Characterization of flavonoids from *Dryopteris erythrosora* and evaluation of their antioxidant, anticancer and acetylcholinesterase inhibition activities. Food Chem Toxicol 51:242–250

Jing Y, Zhang G, Ma E et al (2010) Amentoflavone and the extracts from *Selaginella tamariscina* and their anticancer activity. Chin Herb Med 5:226–229

Jinu U, Thayumanavan T, Singh TB et al (2014) In vitro antioxidant capacity of the fern, *Drynaria quercifolia* (L) SM, rhizome extract. Int J Pharm Pharm Sci 6(7):413–416

Johnson M, Anwardeen I, Periasamy A et al (2021) Evaluation of phenolic constituents and toxicity of lycophytes and ferns of shervarayan hills aqueous extracts. Chem Africa 4:513–523. https://doi.org/10.1007/s42250-021-00240-0

Johnson M, Gowtham J, Sivaraman A et al (2014) Antioxidant, Larvicidal, and Cytotoxic Studies on *Asplenium aethiopicum* (Burm. f.) Becherer. Int Scholarly Res Notices 876170:1–6

Johnson M, Jeeva S, Mahesh M, Janakiraman N (2017b) Antibacterial Potentials of Adiantum Species against the UTI Pathogens. J Microbiol Experiment 4(1):104

Johnson M, Madona CX, Almeida RS (2020) In Vitro toxicity, antioxidant, anti-inflammatory, and antidiabetic potential of *Sphaerostephanos unitus* (L.) Holttum. Antibiotics 9(6):333

Johnson M, Ramakrishnan P, Perumal S et al (2017a) Anti inflammatory activity of selected pteridophytes from Western Ghats. Int J Complement Alt Med 9(4):00307. https://doi.org/10.15406/ijcam.2017.09.00307

Jung HJ, Park K, Lee IS et al (2007) S-Phase accumulation of *Candida albicans* by anticandidal effect of amentoflavone isolated from *Selaginella tamariscina*. Biol Pharm Bul 130 (10):1969–1971

Jutarat T, Jindarat P (2011) Determination of flavonoid content and antioxidant activity from ferns by ultrasonic extraction. In: Proceedings of the TIChE International Conference, Hatyai, Songkhla Thailand, 10–11 November 2011, p 1–4

Kaewsuwan S, Yuenyongsawada S, Plubrukarna A et al (2015) Bioactive interruptins A and B from *Cyclosorus terminans*: antibacterial, anticancer, stem cell proliferation and ROS scavenging activities. Songklanakarin J Sci Technol 37(3):309–317

Kahl R, Kappus H (1993) Toxikologie der synthetischen Antioxidantien BHA und BHT im Vergleich mit dem natürlichen Antioxidans Vitamin E [Toxicology of the synthetic antioxidants

BHA and BHT in comparison with the natural antioxidant vitamin E]. Z Lebensm Unters Forsch 196(4):329–338. German. https://doi.org/10.1007/BF01197931

Kalpana Devi R, Vasantha S, Annamalai P et al (2016) Phytochemical analysis *in vitro* antioxidant potential and Gas chromatography-Mass spectrometry studies of *Dicranopteris linearis*. Asian J Pharm Clin Res 9(8):220–225. https://doi.org/10.22159/ajpcr.2016.v9s2.13636

Kar HK, Kumar B (2010) IAL textbook of leprosy. Jaypee Brothers Medical Publishers Private Limited, New Delhi

Kathirvel A, Rai AK, Maurya GS et al (2014) *Dryopteris cochleata* rhizome: a nutritional source of essential elements, phytochemicals, antioxidants and antimicrobials. Int J Pharm Sci 6(2):179-188

Kaur P, Kaur V, Kumar M et al (2014) Suppression of SOS response in *Escherichia coli* PQ37, antioxidant potential and antiproliferative action of methanolic extract of *Pteris vittata* L. on human MCF-7 breast cancer cells. Food Chem Toxicol 74:326–333

Khan M, Kormin F, Iwansyah AC (2016) Microwave assisted extraction; phytochemical evaluation of Malaysian palm oil trunk epiphytes ferns. Int J Pharm Pharm Sci 8(4):174–180

Kiran PM, Raju AV, Rao BG (2012) Investigation of hepatoprotective activity of Cyatheagigantea (Wall. ex. Hook.) leaves against paracetamol-induced hepatotoxicity in rats. Asian Pac J Trop Biomed 2(5):352–356. https://doi.org/10.1016/S2221-1691(12)60055-0

Krishnaraju AV, Tsay HS (2006) Biological screening of medicinal plants collected from Eastern Ghats of India using *Artemia salina*. Int J Appl Sci Eng 4(2):115–125

Kumar S, Samydurai P, Ramakrishnan R et al (2013) Polyphenols, vitamin-E estimation and in vitro antioxidant activity of *Adiantum capillus-veneris*. Int J Innov Pharm Sci Res 4(1):258–262

Kunnathupara BS, Sowmya S, Velliyur KG, Poornima K (2016) Quantitative phytochemical analysis, in vitro antioxidant potential and gas chromatography-mass spectrometry studies in ethanolic extract of *Azolla microphylla*. Asian J Pharm Clin Res 9(2):318–323

Lai H-Y, Lim Y-Y (2011a) Antioxidant Properties of Some Malaysian Ferns. 3rd International Conference on Chemical, Biological and Environmental Engineering. IPCBEE 20:8–12. http://www.ipcbee.com/vol20/2-ICBEE2011E004.pdf

Lai HY, Lim YY (2011b) Evaluation of Antioxidant Activities of the Methanolic Extracts of Selected Ferns in Malaysia. Int J Environ Sci dev 2(6):442–447

Lai HY, Limy YY, Shiau PT (2009) Antioxidative, tyrosinase inhibiting and antibacterial activities of leaf extracts from medicinal ferns. Biosci Biotechnol Biochem 73:1362–1366

Lamichhane R, Kim S, Poudel A et al (2014) Evaluation of *in vitro* and *in vivo* Biological Activities of *Cheilanthes albomarginata* Clarke. BMC Complement Altern Med 14:342

Lamichhane R, Kim SG, Poudel A et al (2015) Identification of flavonoids from *Cheilanthes albomarginata* clarke and their simultaneous determination and quantification by UPLC/DAD method. J Liq Chromatogr Relat Technol 138(19):1713–1721. https://doi.org/10.1080/10826076.2015.1091011

Lamichhane R, Pandeya PR, Lee KH et al (2019) *Angiopteris helferiana*, a fern with great potential medicinal value: Antiadipogenic, anti-inflammatory, and anti-diabetic activity. Phcog Mag 15: 423-432

Lan K-P, Shen Y-P, Lee S-H et al (2011) Antioxidant and free radical-scavenging activities of *Pteris multifida* poiret aqueous extract. J Food Qual 34:252–258. https://doi.org/10.1111/j.1745-4557.2010.00356.x

Lee J, Choi Y, Woo ER et al (2009) Antibacterial and synergistic activity of isocryptomerin isolated from *Selaginella tamariscina*. J Microbiol Biotechnol 19(2):204–207

Lee H, Lin JY (1988) Antimutagenic activity of extracts from anticancer drugs in Chinese medicine. Mutat Res 204(2):229–234. https://doi.org/10.1016/0165-1218(88)90093-6

Lee SM, Na K, An RB et al (2003) Antioxidant activity of two phloroglucinol derivatives from *Dryopteris crassirhizoma*. Biol Pharm Bull 26(9):1354–1356

Lee IS, Nishikawa A, Furukawa F et al (1999) Effects of *Selaginella tamariscina* on in vitro tumor cell growth, p53 expression, GI arrest and in vivo gastric cell proliferation. Cancer Lett 144 (1):93–99

Li JH, He CW, Liang NC et al (1999) Effects of antitumor compounds isolated from Pteris semipinnata L on DNA topoisomerases and cell cycle of HL-60 cells. Zhongguo Yao Li Xue Bao 20(6):541–545

Li J, Lei X, Chen K (2014) Comparison of cytotoxic activities of extracts from *Selaginella* species. Pharmacogn Mag 10(40):529–535

Li J, Liang N, Mo L et al (1998) Comparison of the cytotoxicity of five constituents from Pteris semipinnata L. in vitro and the analysis of their structure-activity relationships. Yao xue xue bao = Acta Pharmaceutica Sinica 33(9):641–644

Li S, Zhu R, Zhong M (2010) Effects of ultrasonic-assistant extraction parameters on total flavones yield of *Selaginella doederleinii* and its antioxidant activity. J Med Plants Res 4(17):1743–1750

Liu HC, Chen RM, Jain WC et al (2001) Cytotoxic and antioxidant effects of the water extract of the traditional Chinese herb Gusuibu (*Drynaria fortunei*) on rat osteoblasts. J Formosan Med Assoc 100:383–388

Liu Y, Wujisguleng W, Long C (2012) Food uses of ferns in China: a review. Acta Soc Bot Poloniae 81:1–10

Madhu K, Vijaya R, Ganga R (2012) Investigation of hepatoprotective activity of *Cyathea gigantea* (Wall. ex. Hook.) leaves against paracetamol-induced hepatotoxicity in rats. Asian Pac J Trop Biomed 2(5):352–356

Mahfuz A, Fayad BAS, Kanij Nahar D et al (2019) Characterization of in-vitro antioxidant, cytotoxic, thrombolytic and membrane stabilizing potential of different extracts of *Cheilanthes tenuifolia* and Stigmasterol isolation from n-hexane extract. Clin Phytoscience 5:39. https://doi.org/10.1186/s40816-019-0135-x

Malviya J, Joshi V, Singh K (2012) Antimicrobial activity of some ethno-medicinal plants used by Baiga tribes from Amarkantak, India. Adv Life Sci Technol 4:19–26

Mandal A, Mondal AK (2011) Studies on antimicrobial activities of some selected ferns and lycophytes in Eastern India with special emphasis on ethno-medicinal uses. Afr J Plant Sci 5 (7):412–420

Maneesha S, Kangna M, Chhaya S (2015) Phytochemical evaluation and antioxidant activity of different samples of *Pteris Vittata* in Doon Valley, Uttarakhand Region. Int J Pure App Biosci 3 (4):296–304

Manivannan V, Johnson M (2020) Total phenolic, tannin, triterpenoid, flavonoid and sterol contents, anti-diabetic, anti-inflammatory and cytotoxic activities of *Tectaria paradoxa* (Fee.) Sledge. Toxicol Rep 7:1465–1468. https://doi.org/10.1016/j.toxrep.2020.10.013

Manivannan V, Johnson M, Almeida R et al (2020) Phytochemical profile and bio-activity of Bolbitis appendiculata (Willd.) K. Iwats extracts. S Afr J Bot 137(12):236–241. https://doi.org/10.1016/j.sajb.2020.10.013

Manivannan V, Johnson M, Almeida RS et al (2021) Chemical profiling of *Tectaria paradoxa* (Fee.)Sledge and *Bolbitis appendiculata* (Willd.) K. Iwats using UHPLC. Biocatal Agric Biotechnol 34:102043

Manjunatha K, Vidya V, Manohara YN (2007) Wound healing activity of *Lycopodium serratum*. Indian J Pharm Sci 69(2):283–287

Maria FB, Morais-Braga T, Souza M, Karla KA et al (2013) Phenol composition, cytotoxic and anti-kinetoplastidae activities of *Lygodium venustum* SW. (Lygodiaceae). Exp Parasitol 134 (2):178–182

Mary S, Mahesh MK (2015) Anti-inflammatory activity of Cyathea nilgirensis Holttum, against Carrageenan induced Paw edema. Int J Recent Sci Res 6(8):5807–5809

Mathad P, Chandrakant J, Modi R et al (2015) Phytochemical and antioxidant activity of *Actiniopteris radiata* Linn.an important pteridophytic medicinal plant of Gulbarga region. Int J Res Appl Nat Soc Sci 3(2):27–34

McCutcheon AR, Roberts TE, Gibbons E et al (1995) Antiviral screening of British Columbian medicinal plants. J Ethnopharmacol 49:101–110

Meera G, Josekumar VS (2016) In vitro cytotoxicity screening, phytochemical profile and heavy metal analysis of different extracts of *Acrostichum heterophyllum* L. Indian J Nat Prod Res 7 (1):19–24

Meyer BN, Ferrigi NR, Putnam JE et al (1982) Brine shrimp: a convenient general bioassay for active plant constituents. Planta Med 45(5):31–34

Miao N, Tao H, Tong C et al (1996) The *Selaginella tamariscina* (Beauv.) Spring complex in the treatment of experimental diabetes and its effect on blood rheology. China J Chinese Mater Med 21(8):493–495

Middleton EJ, Kandaswami C, Theoharides TC (2000) The effects of plant flavonoids on mammalian cells: implications for inflammation, heart disease and cancer. Pharmacol Rev 52 (4):673–751

Milon CM, Avijit DSM, Abdur R, Rezwana NC (2013) Evaluation of anti-oxidant, cytotoxic and anti-microbial properties of *Drynaria quercifolia*. Int Res J Pharm 3(2):46–48

Mimica DN, Natasa S, Jelena C et al (2008) Phenolic compounds in field Horsetail (*Equisetum arvense* L.) as Natural Antioxidants. Molecules 13(7):1455–1464

Mizushina Y, Watanabe I, Ohta K et al (1998) Studies on inhibitors of mammalian DNA polymerase a and b. Biochemical Pharmacology 55:537–541

Mohammad SA, Mohammad RA, Chowdhury MIK et al (2013) In vitro antioxidant, cytotoxic, thrombolytic activities and phytochemical evaluation of methanol extract of the *A. philippense* L. leaves. Asian Pac J Trop Biomed 3(6):464–469. https://doi.org/10.1016/S2221-1691(13) 60097-0

Mordi Joseph C, Ewhre O, Lawrence E et al (2016) Aqueous leaf extract of *Dryopteris dilatata* on STZ – induced diabetic wistar rats with associated hyperlipidemic ameliorating property. IOSR J Dent Med Sci (IOSR-JDMS) 15(6):97–104

Myers KA, Marshal RD, Friedin J (1980) Principles of pathology in surgery, 1st edn. Blackwell Scientific Publications, London, pp 58–82

Nair AS, Shylesh BS, Gopakumar B et al (2006) Anti-diabetes and hypoglycaemic properties of *Hemionitis arifolia* (Burm.) Moore in rats. J Ethnopharmacol 106(2):192–197

Nagarajan G, Britto AJD (2014) Evaluation of antibacterial activity of different parts of medicinal fern—Marsilea minuta L. Life Sci Leafl 50:34–37

Nath K, Anupam DT, Mrinal KB, Deeshikha B, Shiela C, Debarati C, Mitra A, Amitabha B (2018) Antibacterial activity of certain ferns against multi drug resistant organisms. J Nat Remedies 17 (4):144–153

Navarro-Yepes J, Zavala-Flores L, Anandhan A et al (2014) Antioxidant gene therapy against neuronal cell death. Pharmacol Ther 142(2):206–230. https://doi.org/10.1016/j.pharmthera. 2013.12.007

Nayak N, Sibanarayan R, Mishra MP et al (2013) Antibacterial activity of the terrestrial fern Lygodiumflexuosum (L.) Sw. against multidrug resistant enteric- and uro-pathogenic bacteria. J Acute Dis 2(4):270–276. https://doi.org/10.1016/S2221-6189(13)60142-0

Nawaz ASN, Junaid S, Dileep N et al (2014) Antioxidant activity of *Azolla pinnata* and *Azolla rubra* – a comparative study. Sch Acad J Biosci 2(10):719–723

Nayak N, Padhy RN (2017) GC-MS analysis of bioactive compounds and host toxicity Studies of *Azolla caroliniana* symbiotic with the Cyanobacterium *Anabaena azollae*. Indian J Pharm Educ Res 51(2s):s24–s33

Newman DJ, Cragg GM (2007) Natural products as sources of new drugs over the last 25 years. J Nat Prod 70(4):461–467

Nilanthi SMLD, Wijayarathna C, Hettiarachchi GHCM (2015) Investigation of an in vitro antimicrobial activity of Angiopterisevecta (Forst.) Hoffm. RRJCHEM 4(3):47–52

Nithya TG, Jayanthi J, Raghunathan MG (2015) Phytochemical, Antibacterial and GC-MS analysis of a floating fern *Salvinia molesta* D.S.Mitchell (1972). IJPRIF 8(9):85–90

Nofrizal DPP, Dayar A (2017) Antioxidant and antibacterial constituents from two sumatran ferns, *Trichomanes javanicum* and *Oleandra pistillaris*. Nat Prod Commun 12(8):1263–1264

Nor Aini A, Merrina J, Stanslas S et al (2008) Cytotoxic potential on breast cancer cells using selected forest species found in Malaysia. Int J Cancer Res 4(3):103–109

Nyarko HD, Barku VYA, Batama J (2012) Antimicrobial examinations of Cymbopogon citratus and Adiatumcapillus veneris used in Ghanaian folkloric medicine. Int J Life Sci Pharma Res 2 (2):L 115-121

Padhy R, Dash SK (2015) Antibacterial evaluation of methanolic rhizome extract from an in vivo and in vitro grown pteridophyte, Drynariaquercifolia (LINN.) J Smith. Asian J Pharm Clin Res 8 (4):130–138

Pal SK (2013) Antimicrobial activity of ferns. IOSR J Comput Eng 12(2):01–03

Pal SK (2014) Study of antibacterial activity of some ethnomedicinal ferns of Darjeeling District, West Bengal. J Environ Sociobiol 11(2):285–290

Panda SS, Sahoo K, Rana M, Rout NC, Dhal NK (2014) Antimicrobial activities and phytochemical investigation of some native pteridophytes. Asian J Pharm Clin Res 7(1):43–45

Pandey BS, Shrivastavaa AK, Chauhan D (2014) Evaluation of *in vitro* antioxidant potential of methanolic extracts of the phytochemical analysis of *Adiantum* and *Pteris* ferns & its role as antioxidant. Indian J Sci Res 4(1):31–38

Parihar P, Leena P, Achaleshwar B (2010) In vitro antibacterial activity of fronds (leaves) of some important pteridophytes. J Microbiol Antimicrob 2(2):19–22

Pargavi B, Sivakumar T (2014) Research article phytochemical screening and antimicrobial activity of Drynariaquercifolia (L.) J. Smith. Int J Curr Res Chem Pharm Sci 1(5):68–73

Patel RM, Patel SK (2011) Cytotoxic activity of methanolic extract of *Artocarpus heterophyllus* against A549, Hela and MCF-7 cell lines. J App Pharm Sci 1(7):167–171

Patil DT, Gurav KD, Kadam AS et al (2013) Qualitative analysis of secondary metabolites from some filicales members. IJRPC 3(2):300–302

Paul T, Banerjee S (2013) In vitro evaluation of α-amylase inhibitory activity & antioxidant potential of *Pteris vittata* L. with special reference to its HPTLC profile. Int J Pharma Bio Sci 4(2):494–503

Paul T, Das B, Apte KG (2012) Evaluation of anti-hyperglycemic activity of *Adiantum Philippense* Linn, a pteridophyte in alloxan induced diabetic rats. Diabetes Metab J 3:226

Paul Raj K, Irudayaraj V, Johnson M, Patric Raja D (2011) Phytochemical and anti-bacterial activity of epidermal glands extract of Christella parasitica (L.) H. Lev. Asian Pacific J Trop Biomed 1(1):8–11

Pauline VC, Irudayaraj V, Johnson M (2012) Anti-bacterial efficacy of macroscopic, microscopic parts of sporophyte and *in vitro* cultured gametophyte of a fern Cyclosorus interruptus (Willd.) H. Ito (Thelypteridaceae - Pteridophyta). J Chem Pharmaceut Res 4(2):1167–1172

Paulsamy S, Moorthy D, Nandakumar K et al (2013) Evaluation of in vitro antioxidant potential of methanolic extracts of the Ferns *Actiniopteris radiata* (SW) Link and *Equisetum ramosissimum* Desf. Int J Res Dev Pharm Life Sci 2(3):451–455

Peres MTLP, Simionatto E, Hess SC, Bonani VFL et al (2009) Chemical and biological studies of *Microgramma vacciniifolia* (Langsd. & Fisch.) Copel (Polypodiaceae). Quimica Nova 32 (4):897–901

Pizzino G, Irrera N, Cucinotta M et al (2017) Oxidative stress: harms and benefits for human health. Oxid Med Cell Longev 2017:8416763. https://doi.org/10.1155/2017/8416763

Porter JR, Barrett TG (2005) Monogenic syndromes of abnormal glucose homeostasis: clinical review and relevance to the understanding of the pathology of insulin resistance and beta cell failure. J Med Genet 42(12):893–902

Prabu D, Nappinai M, Ponnudurai K et al (2008) Evaluation of wound-healing potential of *Pisonia grandis* R. Br: A preclinical study in Wistar rats. Int J Low Extrem Wounds 7(1):21–27

Pradeep K, Ajudhia NK (2013) Hepatoprotective effect of *Drynaria quercifolia* fronds hydroalcoholic extract and isolated constituent against CCl 4 -induced hepatocellular damage. Br J Pharm Res 3(4):563–578

Prasanna G, Chitra M, Suvitha N (2014) In vitro antimicrobial activity of Drynariaquercifolia L. rhizome. Asian J Biochem Pharm Res 4:342–9

Prasanna G, Chitra M (2015) In vitro anti-inflammatory activity of *Drynaria quercifolia* rhizome. Res J Pharmacognosy Phytochem 7(1):06–08. https://doi.org/10.5958/0975-4385.2015.00002.3

Preeti K, Namdeo J (2018) Phytochemical screening and in-vitro anticancer activity of extracts of Tectariacicutaria. Int J Pharm Sci Res 9(8):3463–3468
Priscilla J, Geethaa S, Sreeramanan S et al (2014) Brine shrimp lethality test and anti-proliferation test against human cancer-origin cell lines using ethanolic and water extracts of *Selaginella doederleinii* Hieron. J T J Biomed Pharm Res 3(4):63–69
Rabiea T, Shiblur R, Shakhawat H et al (2013) Medicinal potential of pteridophytes – an antihyperglycemic and antinociceptive activity evaluation of methanolic extract of whole plants of *Christella dentate*. Adv Nat Appl Sci 7(1):67–73
Radulović N, Stojanović G, Palić R (2006) Composition and antimicrobial activity of Equisetum arvense L. essential oil. Phytother Res 20(1):85–88. https://doi.org/10.1002/ptr.1815
Samira S, Rahmatullah MD, Nandi JK et al (2014) Preliminary antihyperglycemic and analgesic activity studies with *Angiopteris evecta* leaves in Swiss albino mice. World J Pharm Pharm Sci 3 (10):01–12
Raina R, Prawez S, Verma PK et al (2008) Medicinal plants and their role in wound healing. Vet Scan 3(1):1–7
Raja DP, Johnson M, Irudayaraj V et al (2012) Antimicrobial efficacy of selected ferns of Western Ghats, South India. Int J Curr Pharm Res 4(2):58–60
Rajendran P, Nandakumar N, Rengarajan T et al (2014) Antioxidants and human diseases. Clin Chim Acta 436:332–347. https://doi.org/10.1016/j.cca.2014.06.004
Rajimol EK, Mohammed SP, Latheef N et al (2014) Evaluation of antidiabetic and hypolipidemic potential of *Drynaria quercifolia* Linn rhizome in streptozotocin induced diabetic rats. Int J Pharm Sci Rev Res 25(1):118–124
Rajurkar NS, Gaikwad K (2012) Evaluation of phytochemicals, antioxidant activity and elemental content of *Adiantum capillus-veneris* leaves. J Chem Pharmaceut Res 4(1):365–374
Ramachandra S, Quereshi SAA, Viswanath Swamy AHM et al (2007) Hepatoprotective activity of *Calotropis procera* flowers against paracetamol-induced hepatic injury in rats. Fitoterapia 78 (7-8):451–454
Ramesh N, Viswanathana MB, Saraswathy A et al (2001) Phytochemical and antimicrobial studies on *Drynaria quercifolia*. Fitoterapia 72(8):934–936
Rang HP, Dale MM, Ritter JM et al (2003) The endocrine pancreas and the control of blood glucose. In: Hardman JG, Limbird LE (eds) *Pharmacology*, 5th edn. Churchill, Livingstone
Ranjan P, Santosh KD, Sunita P et al (2014) Studies on healing activity vis-A-vis microflora of acute induced wounds against solvent extracts of rhizome of *Drynaria quercifolia* Linn. OSR J Pharm Biol Sci (IOSR-JPBS) 9(5):38–49
Revathi M, Sara CS (2014) Preliminary phytochemical analysis and antimicrobial activity of rhizome extract of Marsileaminuta L. IJSR 3(2):46–48
Royal Frank P, Suresh V, Arunachalam G et al (2012) Evaluation of hepatoprotective effect of *Adiantum incisum* Forsk leaf extract against $CCl_4$ induced hepatotoxicity in rats. Int Res J Pharm 3(3):230–234
Sahayaraj K, Borgio JF, Raju G (2009) Antifungal activity of three fern extracts on causative. J Plant Prot Res 49:53–56
Sharma D, Vikesh KB, Sandip P et al (2013) Antimicrobial activity of selected cryptogams from solan region. Int J Biol Pharm Res 4(6):448–454
Salehi B, Ezzat MZ, Fokou PVT et al (2019) *Athyrium* plants - Review on phytopharmacy properties. J Trad Complementary Med 9:201e205. https://doi.org/10.1016/j.jtcme.2018.09.001
Sampath Kumar KP, Bhowmik D, Srivastava S et al (2012) Diabetes epidemic in India 'A comprehensive review of clinical features, management and remedies. Pharma Innov 1:17–33
Sandy O, Ben HC (1998) The effects of glutathione glycoside in acetaminophen-induced liver cell necrosis. Exp Mol Pathol 65(1):15–24
Santos MG, Kelecom A, De Paiva SR (2010) Phytochemical studied in Pteridophytes Growing in Brazil: a review. Am J Plant Sci Biotech 4(1):113–125
Sarker SQ, Mondol PC, Jahangir Md A, Sarwar PM, Firoz AM (2011) Comparative study on antitumor activity of three pteridophytes ethanol extracts. J Agric Technol 7(6):1661–1671
Sayema K, Dewan A (2016) Comparative antimicrobial and cytotoxic activity study of N-hexane, chloroform and carbon tetra-chloride extracts of *Adiantum incisum* forsk. EJPMR 3(4):100–105

Sekendar A, Kawsarul M, Obayed R, Khalilur R, Aslam H, Shah A (2012) Antioxidant and cytotoxic activities of methanolic extract of *Dryopteris filix-mas* (L.) Schott leaves. Int J Drug Dev Res 4(2):223–229

Sethi P (2014) Antibacterial activity of Marsilea quadrifolia Linn. Int Res Plant Sci 4:60–62

Sharma US, Kumar A (2011) Anti-diabetic effect of *Rubus ellipticus* fruit extracts in Alloxan induced diabetic rats. J Diabetol 2:1–6

Shokeen P, Ray K, Bala M et al (2005) Preliminary studies on activity of *Ocimum sanctum*, *Drynaria quercifolia* and *Annona squamosa* against *Neisseria gonorrhoeae*. Sexually Transm Dis 32(2):106–111

Shrestha SS, Sut S, Serena BDM et al (2019) Phytochemical fingerprinting and *in vitro* bioassays of the ethnomedicinal fern *Tectaria coadunata* (J. Smith) C. Christensen from Central Nepal. Molecules 24(24):4457

Sikder MAA, Sharmin T, Rahman AFMM et al (2013) Screenings of four medicinal plants of Bangladesh for bioactivities. Dhaka Univ J Pharm Sci 12(1):59–62

Simán SE, Povey AC, Ward TH et al (2000) Fern spore extracts can damage DNA. Br J Cancer 83 (1):69–73. https://doi.org/10.1054/bjoc.2000.1204. PMID: 10883670

Singh M, Govindarajan R, Singh RAK (2008) Antimicrobial flavonoid rutin from Pterisvittata L. against pathogenic gastrointestinal microflora. Am Fern J 98(2):98–103

Singh JK, Kumari R, Obaidullah MD et al (2014) Effect of *Selaginella bryopteris* on diabetic swiss albino mice caused by alloxan. Int J Basic Appl Sci Res 1(1):22–27

Sinha SN (2012) In vitro antibacterial activity of ethanolic extract of Equisetum arvense L. Int J Pharm Biol Res 3(1):19–21

Sivagurunathan A, Innocent BX (2014) Phytochemical analysis and antimicrobial efficiency of marsileaquadrifolialinn (Aquatic Fern). Int J Pharm Res Scholars 3(2):425–431

Sivaraman J, Parimelazhagan I (2013) Evaluation of antioxidant potential of ethanolic extracts of selected species of *Selaginella*. Indian J Nat Prod Resour 4(3):238–244

Soare LC, Mariana F, Stefan S, Zapriana D, Nicolova R, Denev P, Bejan C, Păunescu A (2012) Antioxidant activity, polyphenols content and antimicrobial activity of several native pteridophytes of Romania. Not Bot Horti Agrobo 40(1):53–57

Soeder RW (1985) Fern constituents: including occurrence, chemotaxonomy and physiological activity. Bot Rev 51:442–536

Somjintana T, Nanthita K, Narumon S et al (2005) Inhibitory activities of angiopteroside for HIV-1 reverse transcriptase and lung cancer cell-line. J Sci Res Chula Univ 30(2):187–192

Somvanshi R, Sharma VK (2005) Proliferative urocystica and adenoma in a guinea-pig. J Comp Pathol 133(4):277–280. https://doi.org/10.1016/j.jcpa.2005.04.001

Sood S, Sharma N (2010) Insect growth regulatory activity of *Adiantum Capillus-veneris* against *Plutella xylostella* and *Aphis Craccivora* in ethanol and methanol. Res J Agric Biol Sci 6 (6):785–790

Souza TM, MFB M-B, JGM C et al (2012) Enhancement of antimicrobial activity of antibiotics and antifungals by the use of natural products from Pityrogrammacalomelanos(L.) link. Arch Biol Sci Belgrade 64(1):43–48

Suja SR, Latha PG, Pushpangadan P et al (2014) Evaluation of antihepatotoxic potential of *Helminthostachys zeylanica* (Linn.) Hook. f., A Medicinal Fern against Ethanol Induced Liver Damage: *In vitro* and *In vivo* Studies. Columbia International Publishing. Am J Exp Biol 1 (1):16–30

Sunil K (2014) The importance of antioxidant and their role in pharmaceutical science - a review. Asian J Res Chem Pharmaceut Sci 1(1):27–44

Suvarnalatha DK, Rukmini SVSSSL, Himabindu NN et al (2015) Antibacterial activity and phytochemical screening of Salvinia auriculata Aubl. from Tirumala Hills, Tirupati. Int J Pharm Sci Rev Res 30(1):35–38

Tahir MM, Ibrahim N, Yaacob A (2014) Cytotoxicity and antiviral activities of Asplenium nidus. Phaleria macrocarpa Eleusine indica 1614:549–552. https://doi.org/10.1063/1.4895259

Tania P, Biswadeep D, Kishori GA et al (2012) Evaluation of anti-hyperglycemic activity of *Adiantum Philippense* Linn, a pteridophyte in alloxan induced diabetic rats. J Diabetes Metab 3:1–8

Tania P, Suchitra B (2013) In vitro evaluation of α-amylase inhibitory activity & antioxidant potential of *Pteris vittata* L. with special reference to its HPTLC profile. Int J Pharm Bio Sci 4(2):494–503

Tantawy EM, Shams M, Afifi M (2015) Chemical composition and biological evaluation of the volatile constituents from the aerial parts of Nephrolepisexaltata (L.) and Nephrolepiscordifolia (L.) C. Presl grown in Egypt. Nat Prod Res 30(10):1–5. https://doi.org/10.1080/14786419.2015.1046070

Thangam TS, Kathiresan K (1990) Synergistic effect of insecticides with plant extracts on mosquito larvae. Trop Biomed 7:135–137

Thiripurasundari B, Padmini E (2018) Preliminary phytochemical screening and evaluation of antimicrobial and antioxidant activity of *Azolla pinnata*. Int J Recent Sci Res 9 (5F):26924–26930

Thite SV, Chavan YR, Aparadh VT et al (2013) Preliminary phytochemical screening of some medicinal plants. IJPCBS 3(1):87–90

Thomas T (2011) Preliminary antibacterial evaluation of fronds of *Pteris quadriaurita* Retz. towards bacteria involved in dermatological diseases. J Appl Pharm Sci 1(8):214–216

Thomas T (2013) Antibacterial activity of Adiantum lunulatum Burm. F. towards bacteria implicated in cutaneous infections. J Biol Sci Opin 1(4):334–336. https://doi.org/10.7897/2321-6328.01411

Thomas T (2014) A study on antibacterial and phytochemical evaluation of fronds of Adiantumraddianum C. Presl. Int J Pharmacol Screen Methods 4(2):85–88

Tsun TC, Fai CW (2012) Antioxidant properties of aqueous extracts of *Selaginella willdenowii*. J Med Plants Res 6(7):1289–1296

Vasudeva SM (1999) Economic importance of Pteridophytes. Indian Fern J 16(12):130–152

Venkataratnam D, Ankola DD, Bhardwaj V et al (2006) Role of antioxidants in prophylaxis and therapy: a pharmaceutical perspective. J Control Release 113(3):189–207

Wali A, Sharma S, Walia M et al (2016) Two edible ferns of Western Himalaya: A comparative *in vitro* nutritional assessment, antioxidant capacity and quantification of lutein by UPLC/DAD. Int J Food Sci Nutr 5(3):9–18

Wallace RA, Sander GP, Ferl RJ (1991) Biology: the science of life. HarperCollins, New York, pp 547–555

Wills PJ, Asha VV (2006) Protective effect of *Lygodium flexuosum* (L.) Sw. extract against carbon tetrachloride-induced acute liver injury in rats. J Ethnopharmacol 108(3):320–326

Wolf PL (1999) Biochemical diagnosis of liver diseases. Indian J Clin Biochem 14(1):59–90. https://doi.org/10.1007/BF02869152

Wu JQ, Thomas RK, Zhang XY (2013) Free radicals, antioxidant defense systems, and schizophrenia. Progr Neuro-Psychopharmacol Biol Psych 46:200–206. https://doi.org/10.1016/j.pnpbp.2013.02.015

Wu MJ, Weng CY, Wang L et al (2005) Immunomodulatory mechanism of the aqueous extract of sword brake fern (Pteris ensiformis Burm.). J Ethnopharmacol 98(1-2):73–81. https://doi.org/10.1016/j.jep.2004.12.031. PMID: 15763366

Xia X, Cao J, Zheng Y et al (2014) Flavonoid concentrations and bioactivity of flavonoid extracts from 19 species of ferns from China. Ind Crops Prod 58:91–98

Yang JH, Kondratyuk TP, Marler LE et al (2010) Isolation and evaluation of kaempferol glycosides from the fern *Neocheiropteris palmatopedata*. Phytochem 71(5-6):641–647

Yeou NCJ (2015) Phytochemicals and antioxidative properties of edible fern, Stenochlaena palustris (Burm. f.) Bedd. Dissertation M. Sc (Pharmacy): Universiti Sains Malaysia

Zakaria ZA, Kamisan FH, Kek TL et al (2020) Hepatoprotective and antioxidant activities of Dicranopteris linearis leaf extract against paracetamol-induced liver intoxication in rats. Pharm Biol 58(1):478–489. https://doi.org/10.1080/13880209.2020.1764058

Zakaria ZA, Kamisan FH, Omar MH et al (2017) Methanol extract of *Dicranopteris linearis* L. leaves impedes acetaminophen-induced liver intoxication partly by enhancing the endogenous antioxidant system. BMC Complement Altern Med 17(1):271–284

Zakaria ZA, Mohamed AM, Jamil NS et al (2011) In vitro cytotoxic and antioxidant properties of the aqueous, chloroform and methanol extracts of *Dicranopteris linearis* leaves. Afr J Biotechnol 10(2):273–282

Zhang Z, ElSohly HN, Jacob MR et al (2002) Natural products inhibiting *Candida albicans* secreted aspartic proteases from *Lycopodium cernuum*. J Nat Prod 5(7):979–985

Zhang Y, Tian HY, Tan YF et al (2016) Isolation and identification of polyphenols from *Marsilea quadrifolia* with antioxidant properties in vitro and in vivo. Nat Prod Res 30(12):1404–1410. https://doi.org/10.1080/14786419.2015.1062377

Zhang X, Wang X, Wang M, Cao J, Xiao J, Wang Q (2019) Effects of different pretreatments on flavonoids and antioxidant activity of *Dryopteris erythrosora* leave. PLoS One 14(1):e0200174

Zheng XK, Zhang L, Wang WW et al (2011) Anti-diabetic activity and potential mechanism of total flavonoids of *Selaginella tamariscina* (Beauv.) Spring in rats induced by high fat diet and low dose STZ. J Ethnopharmacol 137(1):662–668

Zuraini Z, Sasidharan S, Roopin Kaur S et al (2010) Antimicrobial and antifungal activities of local edible fern Stenochlaena Palustris (Burm. F) Bedd. Pharmacologyonline 1:233–237

**Johnson Marimuthu**, F.I.F.S., completed his Doctorate in Botany (Plant Tissue Culture) from Manonmaniam Sundaranar University, Tirunelveli District, Tamil Nadu, and is working as Assistant Professor of Botany and Director, Centre for Plant Biotechnology, St. Xavier's College (autonomous), Palayamkottai, Tamil Nadu. He has 21 years of research experience and 18 years of teaching experience. He has published 223 papers in reputed journals, 27 papers in proceedings, and 37 papers as book chapters. He is a recipient of the prestigious SERB Young Scientist—2013, Prof. S.S. Bir Gold Medal—2017, and Certificate of Merit 2014 by Indian Fern Society, India.

**N. Janakiraman** is working as an Assistant Professor in the Department of Botany, the Madura College (Autonomous), Madurai, from June 2019. He finished his M.Phil. and PhD degrees from Manonmaniam Sundaranar University, Tirunelveli. He secured first rank in M.Phil. (gold medal) at university level. After that, he worked as an Assistant Professor in the Department of Botany, Sri Paramakalyani College, Alwarkurichi, from January 2014 to April 2019. He published 17 research articles in Scopus indexed journals and 20 research articles in peer-reviewed national and international journals. Further, he contributed five chapters in books and deposited three sequences in GenBank.

**J. Chandra Saleride** is working as an Assistant Professor in the Department of Botany, Women's Christian College, Nagercoil, from June 2019. She completed her Master's degree in Nesamony Memorial Christian College, Marthandam. She finished her M. Phil. and PhD degrees from Manonmaniam Sundaranar University, Tirunelveli.

**A. Sivaraman** is doing research in Maria Research Foundation, Tirunelveli. He completed his Master's degree in St. Xavier's College (Autonomous), Palayamkottai. He finished his PhD degrees from Manonmaniam Sundaranar University, Tirunelveli. He published seven research articles in peer-reviewed national and international journals. Further, he contributed two chapters in books.

**B. Shivananthini** is doing research in Centre for Plant Biotechnology (CBB) in St. Xavier's College (Autonomous), Palayamkottai. She received her education at St. Xavier's College (Autonomous) with a Master's in Botany. She also holds a Bachelor's in Botany from St. Xavier's College (Autonomous), Palayamkottai.

**K. Paulraj** completed his Doctorate in Botany from Manonmaniam Sundaranar University, Tirunelveli District, Tamil Nadu, and is working as Associate Professor of Botany and Principal at Nesamony Memorial Christian College, Marthandam. He has 29 years of teaching experience and 17 years of research experience. He has published 18 papers in the national and international journals.

# Omega-3 and Omega-6 Fatty Acids Distribution in Pteridophytes and its Significance in Nutraceutical, Pharmacology, and Cosmetic Industry

21

Priti Giri, Ashwani Kumar, and Prem L. Uniyal

**Abstract**

Pteridophytes are non-flowering, primitive vascular plants widely used by people all across the globe as they are easily available in large biomass in wild and often grow on relatively poor soils. The present work focuses on the omega-3 and omega-6 fatty acids (FAs) distribution in pteridophytes and its significance in *nutraceutical*, pharmaceutical, and cosmetic industry. Fatty acids (FAs) are important macromolecules that act as source of energy as well as play a vital role in human health and nutrition. Most importantly, polyunsaturated fatty acids (PUFAs) have drawn interest from lipid researchers due to their significance in human diet and well-being. The emerging depletion and contamination of traditional sources of PUFAs in human diets, the marine fisheries, have evoked an ardent engrossment in the areas of biotechnology and aquaculture to search for its alternative opportunities. The most commonly considered plant sources of PUFAs are micro- and macroalgae. However, the cultivation and handling of algae have some complication. Among the terrestrial plants, pteridophytes have some intriguing long- chain PUFAs (arachidonic and eicosapentaenoic acids). Pteridophytes possess the capability to produced arachidonic (20:4n-6, ARA) and eicosapentaenoic (20:5n-3, EPA) acids. Though pteridophytes can be a great source of fatty acids, it still posed as an ignored group of plants that require much attention due to its present and potential value in the medicinal, economic, and ecological arenas. Some pteridophytes consumed by humans have encouraged various research groups to evaluate their nutritional value, inclusive lipid compositions, of a few pteridophytes species. Based on the results of FAs

P. Giri (✉) · P. L. Uniyal
Department of Botany, University of Delhi, Delhi, India

A. Kumar
Former Head of the Department of Botany and Biotechnology, University of Rajasthan, Jaipur, Rajasthan, India





J. Marimuthu et al. (eds.), *Ferns*, https://doi.org/10.1007/978-981-16-6170-9_21

and antioxidant composition, tender fronds of ostrich fern (*M. struthiopteris*) have also endorsed as a healthy vegetable to be included in the human diet.

**Keywords**

Pteridophytes · Arachidonic acid · Polyunsaturated fatty acids · Eicosapentaenoic acid · Pharmaceutical

## List of Abbreviations

| | |
|---|---|
| ALA | α-linolenic acids |
| ARA | Arachidonic |
| DHA | Docosahexaenoic acid |
| EPA | Eicosapentaenoic |
| FAs | Fatty acids |
| GLA | γ-linolenic acid |
| LA | Linoleic acid |
| PUFAs | Polyunsaturated fatty acids |

## 21.1 Introduction

Ferns and fern allies are pteridophytes and are non- flowering, primitive vascular plants that are dispersed and reproduced through spores. Pteridophytes are classified into two groups, the Lycopodiophyta (fern allies) and Pteridophyta (true ferns) (Smith et al. 2006). The groups are known to consist almost 13,025 species in the world as updated by Groombridge and Jenkins (2002). India harbors more than 1000 species in 191 genera distributed mostly in the Himalayan region, Central India, and the Western Ghats (Dixit 2000). Pteridophytes are frequently used as ornamental plants as well as regarded as bioindicators for pollution and phytoremediators, while some have been accredited as biofertilizers. They are diversely used in various medicinal approaches such as Ayurvedic, Unani, Homeopathic and other allied branches of medicine. Since ancient times, pteridophytes have been reported as important sources of food and medicine (Lee and Shin 2010).

Fatty acids (FAs) are important macromolecules that act as source of energy as well as play a vital role in human health, nutrition, and cell signaling and in maintaining membrane fluidity. Earlier studies have also showed that the composition of FAs plays a vital role for plants to withstand different biological and physical stresses, especially cold, heat, wound healing, and salt (Nishida and Murata 1996; Yaeno et al. 2004; Guschina and Harwood 2006; Upchurch 2008; Zhang et al. 2012; Walley et al. 2013).

The FAs are important macromolecules and exist in both animals and plants. Fishes are the most common sources of FAs worldwide, among the animal species,

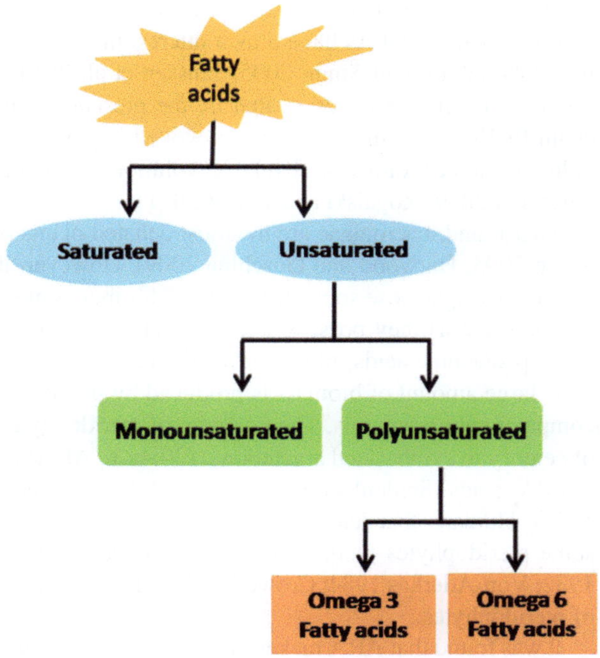


**Fig. 21.1** An overview on types of fatty acids

although there are a lots of constrains with respect to the usage of fish oil as a supply of FAs. Most problems caused by damaging pollutants such as teratogenic, mutagenic, carcinogenic, and non-carcinogenic are highlighted among them. Restrictions on the fish oil processing and marketing are commonly discussed because issues with oil stability, unpleasant odor and taste, that result in higher production costs and cause struggling in the purification of oil (Sidhu 2003; Park and Johnson 2006; Perveen et al. 2006; Li and Hu 2009; Foran et al. 2005). The production of vegetable oils provides advantages over fish oil production, since the process of extracting and purifying vegetable oils is easier and is cheaper (Dyer et al. 2008). Polyunsaturated fatty acids (PUFAs) are classified into two groups, omega-6 and omega-3 (Fig. 21.1), that are produced from linoleic acid (18:2n-6, LA) and α-linolenic acids (18:3n-3, ALA), respectively. PUFAs belonging to the omega-6 group are linoleic acid (18:2n-6), gamma-linolenic acid (18:3n-6), eicosadienoic acid (20:2n-6), dihomo-gamma-linolenic acid (20:3n-6), arachidonic acid (20:4n-6), docosadienoic acid (22:2n-6), adrenic acid (22:4n-6), osbond acid (22:5n-6), tetracosatetraenoic acid (24:4n-6), and tetracosapentaenoic acid (24:5n-6). Omega-3 group includes hexadecatrienoic acid (16:3n-3), α-linolenic acid (18:3n-3), stearidonic acid (18:4n-3), eicosatrienoic acid (20:3n-3), eicosatetraenoic acid (20: 4n-3), eicosapentaenoic acid (20:5n- 3), heneicosapentaenoic acid (21:5n), docosapentaenoic acid (22:5n-), docosahexaenoic acid 22:6n-3), tetracosapentaenoic acid (24:5n-3), and tetracosahexaenoic acid (24:6n-3) families.

Highly valued for consumption, these chemical groups are considered essential for good health and prevention of diseases and are known to be synthesized by plants but not by animals (Din et al. 2004). Owing to their significance in human nutrition and well-being, PUFAs have drawn interest in lipid research by biochemists across the globe (Ward and Singh 2005; Janssen et al. 2014). The emerging decline and pollution of marine fisheries, that is, the present extensive source of useful long-chain PUFAs in human diets, have evoked intense engrossment in developing the fields of biotechnology to develop revolutionary sources of PUFAs (Harwood and Guschina 2009; Rogalski and Carrer 2011).

Micro- and macroalgae are the most debated plants source of PUFAs (Ward and Singh 2005; Harwood and Guschina 2009). However, the cultivation and handling of aqueous algae pose some difficulties. This makes pteridophytes a desirable source for the same as they possess some desirable long -chain PUFAs (arachidonic and eicosapentaenoic acids) of any terrestrial plant.

A large amount of biomass is produced by pteridophytes, and they also grow on comparatively poor soils. The potential of pteridophytes has been known from half of century to synthesized arachidonic (20:4n-6, ARA) and eicosapentaenoic (20:5n-3, EPA) acids (Schlenk and Gellerman 1965; Haigh et al. 1996; Jamieson and Reid 1975). Humans that lead several researches to determine the nutritional value of some pteridophytes mainly of North American and Asian origin (Bushway et al. 1982; Von-Aderkas 1984; Crowe 1997; DeLong et al. 2011) consume great amounts of pteridophytes.

It was found that ARA makes up a significant proportion of total fatty acids, while EPA is rather a minor phytoconstituents in the tender fronds of the eatable pteridophytes *Pteridium aquilinum* and *Mattueccia struthiopteris* (DeLong et al. 2011). Similar findings have been identified earlier for other pteridophytes species (Schlenk and Gellerman 1965; Jamieson and Reid 1975). Tender fronds of the ostrich fern (*M. struthiopteris*) is recommended as a safe and healthy vegetable in the human diet based on their FAs and antioxidant composition (DeLong et al. 2011).

The purpose of this chapter is to compile and evaluate the currently available data on the PUFAs of pteridophytes (Tables 21.1 and 21.2). Much of this data is scattered, but the knowledge on the uses of pteridophytes by the PUFAs content helps us to assess their nutritional, pharmaceutical, and cosmetic potential and to highlight the gaps in understanding the value of pteridophytes.

## 21.2 Role of Omega-3 and Omega-6 Fatty Acids in Nutraceutical, Pharmaceutical, and Cosmetic Industry

In humans, the omega series FAs are needed under normal condition to maintain cell membranes, brain functions, and the transmission of nerve impulses. These FAs also play a vital role in atmospheric oxygen delivery to blood plasma, hemoglobin synthesis, and cell division (Fig. 21.2). FAs are also used in the treatment of rheumatoid arthritis (Connor 2000). They are called essential FAs because they are not synthesized by the human (Saravanan et al. 2010). The ALA, EPA, and docosahexaenoic acid (DHA) are the main omega-3 FAs. The precursor of EPA

**Table 21.1** Distribution of omega-6 fatty acids in pteridophytes

| Plant name Omega – 6 fatty acids | 18:2 (*n* – 6) | 18:3 (*n* – 6) | 20:2 (*n* – 6) | 20:3 (*n* – 6) | 20:4 (*n* – 6) | 22:2 (*n* – 6) | 22:4 (*n* – 6) | 22:5 (*n* – 6) | 24:4 (*n* – 6) | 24:5 (*n* – 6) | References |
|---|---|---|---|---|---|---|---|---|---|---|---|
| *Adiantum pedatum* | + | + | + | + | + | – | – | – | – | – | 1 |
| *Asplenium oblongifolium* | + | + | + | + | + | – | – | – | – | – | 1 |
| *Athyrium filix-femina* | + | + | + | + | + | – | – | – | – | – | 1 |
| *Athyrium sinense* | + | + | – | + | + | – | – | – | – | – | 1 |
| *Athyrium spinulosum* | + | + | + | + | + | – | – | | | | 1 |
| *Athyrium yokoscense* | + | + | + | + | + | – | – | – | – | – | 1 |
| *Azolla pinnata* | – | + | – | + | – | – | – | – | – | – | 7 |
| *Azolla caroliniana* | + | + | – | + | + | – | – | – | – | – | 6,7 |
| *Azolla filiculoides* | + | + | + | + | + | + | – | – | – | – | 4,8 |
| *Azolla microphylla* | + | + | + | + | + | + | – | – | – | – | 8 |
| *Cornopteris crenulatoserrulata* | + | + | + | + | + | – | – | – | – | – | 1 |
| *Cyathea dealbata* | + | + | + | + | + | – | – | – | – | – | 1 |
| *Cystopteris fragilis* | + | + | + | + | + | – | – | – | – | – | 1 |
| *Deparia pycnosora* | + | + | – | + | + | – | – | – | – | – | 1 |
| *Dryopteris expansa* | + | + | + | + | + | – | – | – | – | – | 1 |
| *Dryopteris filix-mas* | + | – | + | + | + | – | – | – | – | – | 9 |
| *Dryopteris austriaca* | + | – | + | + | + | – | – | – | – | – | 9 |
| *Dryopteris crassirhizoma* | + | + | + | + | + | – | – | – | – | – | 1 |
| *Dryopteris goeringiana* | + | + | + | + | + | – | – | – | – | – | 1 |
| *Matteuccia struthiopteris* | + | + | + | + | + | – | – | – | – | – | 1,2 |

(continued)

**Table 21.1** (continued)

| Plant name Omega – 6 fatty acids | 18: 2 (*n* – 6) | 18:3 (*n* – 6) | 20: 2 (*n* – 6) | 20:3 (*n* – 6) | 20:4 (*n* – 6) | 22: 2 (*n* – 6) | 22:4 (*n* – 6) | 22:5 (*n* – 6) | 24:4 (*n* – 6) | 24:5 (*n* – 6) | References |
|---|---|---|---|---|---|---|---|---|---|---|---|
| *Onoclea sensibilis* | + | + | + | + | + | – | – | – | – | – | 1 |
| *Osmunda claytoniana* | + | + | + | + | + | – | – | – | – | – | 1 |
| *Parathelypteris noveboracensis* | + | + | + | + | + | – | – | – | – | – | 1 |
| *Phegopteris connectilis* | + | + | + | + | + | – | – | – | – | – | 1 |
| *Phymatosorus pustulatus* | + | + | + | + | + | – | – | – | – | – | 1 |
| *Polypodium vulgare* | + | – | + | + | + | – | – | – | – | – | 9 |
| *Polystichum tripteron* | + | + | + | + | + | – | – | – | – | – | 1 |
| *Pteridium aquilinum* | + | – | + | + | + | – | – | – | – | – | 9 |
| *Pteridium esculentum* | + | + | + | + | + | – | – | – | – | – | 1 |
| *Ptisana salicina* | + | + | + | + | + | – | – | – | – | – | 1 |
| *Pyrrosia eleagnifolia* | + | + | + | + | + | – | – | – | – | – | 1 |
| *Salvinia molesta* | + | – | – | – | – | – | – | – | – | – | 5 |
| *Salvinia natans* | + | + | + | + | – | + | – | – | + | – | 3 |

((1) Eduard et al. 2019; (2) DeLong et al. 2011; (3) Rozentsvet et al. 2005; (4) Abou et al. 2011; (5) Dwiloka et al. 2015; (6) Paoletti et al. 1987; (7) Kosesakal and Yıldız 2019; (8) Bhaskaran and Kannapan 2015; (9) Jamieson and Reid 1975)
{*Presence (+); absence (–)}

**Table 21.2** Distribution of omega −3 fatty acids in pteridophytes

| Plant name Omega − 3 fatty acids | 16:3 (*n* − 3) | 18:3 (*n* − 3) | 18:4 (*n* − 3) | 20:3 (*n* − 3) | 20:4 (*n* − 3) | 20:5 (*n* − 3) | 21:5 (*n* − 3) | 22:5 (*n* − 3) | 22:6 (*n* − 3) | 24:5 (*n* − 3) | 24:6 (*n* − 3) | References |
|---|---|---|---|---|---|---|---|---|---|---|---|---|
| *Adiantum pedatum* | + | + | + | + | + | + | – | – | – | – | – | 1 |
| *Asplenium oblongifolium* | + | + | – | + | – | + | – | – | – | – | – | 1 |
| *Athyrium filix-femina* | + | + | + | + | + | + | – | – | – | – | – | 1 |
| *Athyrium sinense* | + | + | + | + | + | + | – | – | – | – | – | 1 |
| *Athyrium spinulosum* | + | + | + | + | + | + | – | – | – | – | – | 1 |
| *Athyrium yokoscense* | + | + | + | + | + | + | – | – | – | – | – | 1 |
| *Azolla pinnata* | – | + | – | – | – | – | – | – | – | – | – | 7 |
| *Azolla caroliniana* | – | + | – | – | – | – | – | – | – | – | – | 7,6 |
| *Azolla filiculoides* | – | + | – | + | – | + | – | + | + | – | – | 4,8 |
| *Azolla microphylla* | – | + | – | + | – | + | – | – | + | – | – | 8 |
| *Cornopteris crenulatoserrulata* | + | + | + | + | + | + | – | – | – | – | – | 1 |
| *Cyathea dealbata* | + | + | + | + | + | + | – | – | – | – | – | 1 |
| *Cystopteris fragilis* | + | + | + | – | – | + | – | – | – | – | – | 1 |
| *Deparia pycnosora* | + | + | + | + | + | + | – | – | – | – | – | 1 |
| *Dryopteris expansa* | + | + | + | + | + | + | – | – | – | – | – | 1 |
| *Dryopteris filix-mas* | + | + | – | – | + | + | – | – | – | – | – | 9 |
| *Dryopteris austriaca* | + | + | – | – | + | + | – | – | – | – | – | 9 |
| *Dryopteris crassirhizoma* | + | + | + | + | + | + | – | – | – | – | – | 1 |
| *Dryopteris goeringiana* | + | + | + | + | + | + | – | – | – | – | – | 1 |
| *Matteuccia struthiopteris* | + | + | + | + | + | + | – | – | – | – | – | 1,2 |
| *Onoclea sensibilis* | + | + | + | + | + | + | – | – | – | – | – | 1 |

(continued)

**Table 21.2** (continued)

| Plant name<br>Omega – 3 fatty acids | 16:3 (n – 3) | 18:3 (n – 3) | 18:4 (n – 3) | 20:3 (n – 3) | 20:4 (n – 3) | 20:5 (n – 3) | 21:5 (n – 3) | 22:5 (n – 3) | 22:6 (n – 3) | 24:5 (n – 3) | 24:6 (n – 3) | References |
|---|---|---|---|---|---|---|---|---|---|---|---|---|
| *Osmunda claytoniana* | + | + | + | + | + | + | – | – | – | – | – | 1 |
| *Parathelypteris noveboracensis* | + | + | + | + | + | + | – | – | – | – | – | 1 |
| *Phegopteris connectilis* | + | + | + | + | + | + | – | – | – | – | – | 1 |
| *Phymatosorus pustulatus* | + | + | + | + | + | + | – | – | – | – | – | 1 |
| *Polypodium vulgare* | + | + | – | – | + | – | – | – | – | – | – | 9 |
| *Polystichum tripteron* | – | + | – | + | + | + | – | – | – | – | – | 1 |
| *Pteridium aquilinum* | + | + | – | – | + | + | – | – | – | – | – | 9 |
| *Pteridium esculentum* | + | + | + | + | + | + | – | – | – | – | – | 1 |
| *Ptisana salicina* | + | + | – | + | + | + | – | – | – | – | – | 1 |
| *Pyrrosia eleagnifolia* | + | + | – | + | + | + | – | – | – | – | – | 1 |
| *Salvinia molesta* | – | + | – | – | – | – | – | – | – | – | – | 5 |
| *Salvinia natans* | + | + | + | + | + | + | – | + | + | + | – | 3 |

((1) Eduard et al. 2019; (2) DeLong et al. 2011; (3) Rozentsvet et al. 2005; (4) Abou et al. 2011; (5) Dwiloka et al. 2015; (6) Paoletti et al. 1987; (7) Kosesakal and Yıldız 2019; (8) Bhaskaran and Kannapan 2015; (9) Jamieson and Reid (1975)
{*Presence (+); absence (–)}

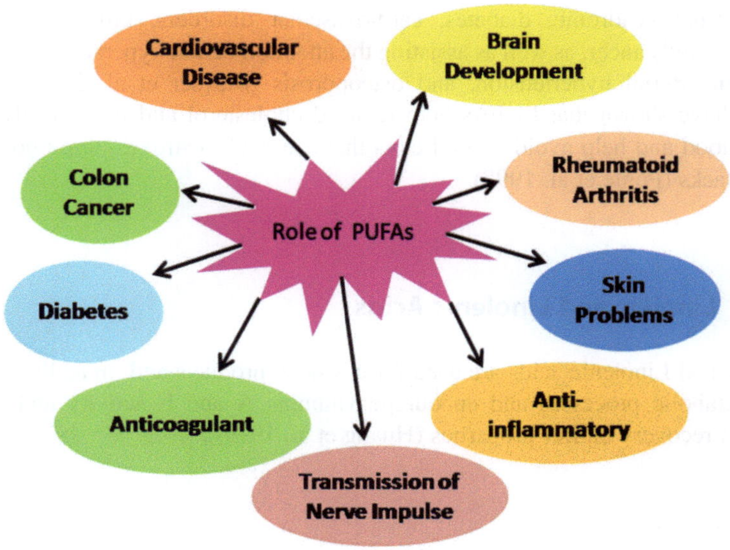


**Fig. 21.2** An overview on various roles of PUFAs

and DHA is ALA. ARA, LA, and γ-linolenic acid (18:3n-6, GLA) are the primary constituents of omega-6 FAs. Since humans do not have the enzymes to produce the essential FAs, they should be consumed through diet (Erdinest et al. 2012; Abedi and Sahari 2014; Zárate et al. 2017). The omega-3 and omega-6 PUFAs are distinguished by their beneficial effects on human health, including their role in tissue synthesis (Youdim et al. 2000). However, one of the key factors driving the global market for fatty acid supplements, which directly raises the demand for isolated omega supplements or their key plant sources, is the increased occurrence of illnesses linked to the lack of a healthy diet and the existence of sedentary lifestyles. These products are known as a lipid source to meet the energetic needs of patients needing parenteral nutrition when it is difficult, inadequate, or contraindicated for oral of enteral feeding (Leeb et al. 2006).

Due to their residual flavor, omega-3 supplements are usually consumed in the form of gelatinous tablets caused by their high oxidation instability that allows the substance to exhibit fish odor and taste. Till date linseed oil is one of the major plant sources. In addition to neutralizing stress and preventing degenerative brain disorders, these products are indicated for the prevention/treatment of cardiovascular diseases; for the reduction of triglycerides levels, total cholesterol, and arterial pressure; and also in neurological therapies, enhancing attention, memory, motivation, and motor skills (Haag 2003; Al et al. 2000). During pregnancy, they are often suggested to reducing the risk of postpartum depression and mood swings, as well as to improve health after childbirth (Lottenberg 2009).

Omega-6 supplements are commonly marketed as the common type of evening primrose oil (EPO), indicated in the prevention/treatment of problems related to premenstrual syndrome, diabetes, cardiovascular disorders, inflammation, skin problems, and cancer, as well as assisting the attention deficit hyperactivity disorder, reduction arterial hypertension, and osteoporosis (Mayser et al. 2002). Several studies have shown that PUFAs help reduced cholesterol and blood triglycerides in the blood and help avoid arterial clots that can lead to strokes, thromboses, and heart attacks (Ponte et al. 1997).

### 21.2.1 Linoleic and Linolenic Acids

Linoleic and Linolenic acids are used for cosmetic products and, in addition, affect skin metabolic processes and encourage vitamins A and E activity and stratum corneum recovery barrier properties (Huang et al. 1999).

### 21.2.2 Arachidonic Acid (ARA)

In higher plants, ARA itself has not been detected but does occur in some ferns, mosses, and algae. ARA is a precursor of the prostaglandins, thromboxanes, and leukotrienes, which display a number of pharmaceutical properties, and stimulates protein kinase redistribution in the heart cells and other eicosanoids which are inflammatory and enhance platelet aggregation (Blobe et al. 1995; Huang et al. 1997; Vangaveti et al. 2016). Lipoxins are ARA derivative products. They were first isolated in 1984 and play pivotal roles in atherosclerosis, thrombosis, inflammation, and other multicellular vascular processes (Serhan 1997).

### 21.2.3 Gamma Linolenic Acid (GLA)

GLA is an essential FA produced in the body through the enzyme delta-6 desaturase from LA. Preformed GLA is found in green leafy vegetables and nuts in trace quantities. Breast milk is the most effective source of GLA for infants. In the last four decades, GLA has become essential for its anti-inflammatory (Brzeski et al. 1991; DeLuca et al. 1999; Calder and Zurier 2001; Gillis et al. 2004; Ziboh et al. 2004) and anti-cancer activity (Sagar et al. 1992; Jiang et al. 1995; Hrelia et al. 1996; Kenny et al. 2000; Mainou-Fowler 2001; Watkins et al. 2005). It also increases the velocity of nerve conduction in diabetic patients (Coste et al. 1999; Jamal 1994), resulting in increased blood supply and decreased tingling of extremities. Various sources of GLA have commercialized due to these actions of GLA. It also reduces metastasis of cancer cells (Watkins et al. 2005).

### 21.2.4 Eicosapentaenoic Acid (EPA)

EPA is the precursor of series 3 prostaglandin (PGE 3), which provides a natural approach to lower blood cholesterol and triglycerides. PGE3 is specifically responsible for reducing the stickiness of blood platelets, leading to a simpler distribution of blood across our bodies. This natural antithrombotic anti-clotting effect of EPA has been well studied (Stefanov et al. 1997).

## 21.3 Conclusions

The application of PUFAs in the nutraceutical, pharmaceutical, and cosmetic industries has fundamental significance, in order to stimulate the production of new herbal products which offer various benefits to their consumers. Many commercial applications have evolved, and research aimed at treating cardiovascular disorders and inflammatory responses and enhancing brain functions have achieved favorable results. The usage of plants as sources of PUFAs possesses itself not as an economic option but as a possible alternative to human health. The information about the fatty acids of the pteridophytes is less documented as compared to the flowering plants. As pteridophytes are neglected groups of plants, focus is required on these plant groups that have great medicinal and economic importance. More aspects of the uses of these plants could be supported by detailed studies on the fatty acids and other phytochemicals. Domestication, nutraceutical benefit, and therapeutic ability of various pteridophytes spp. should be the subject of future research. Such characteristics would develop new possibilities for the food and pharmaceutical industries.

**Acknowledgments** The authors are thankful to University of Delhi, Delhi. Mss Priti also acknowledged the Council of Scientific and Industrial Research (CSIR) for providing fellowship.

## References

Abedi E, Sahari MA (2014) Long-chain polyunsaturated fatty acid sources and evaluation of their nutritional and functional properties. Food Sci Nutr 2:443–463. https://doi.org/10.1002/fsn3.121

Abou Y, Fiogbé ED, Beckers Y, Micha JC (2011) Approximate compositional values and tissue fatty acid profiles of Nile tilapia (*Oreochromis niloticus* L.) fed *Azolla*-diets in earthen ponds. Food Nutr Sci 2:964–973. https://doi.org/10.4236/fns.2011.29131

Al MD, Houwelingen ACV, Hornstra G (2000) Long-chain polyunsaturated fatty acids, pregnancy, and pregnancy outcome. Am J Clin Nutr 471:285S–291S. https://doi.org/10.1093/ajcn/71.1.285s

Bhaskaran SK, Kannapan P (2015) Nutritional composition of four different species of *Azolla*. Eur J Ex Biol 5(3):6–12

Blobe GC, Khan WA, Hannun YA (1995) Protein kinase C: cellular target of the second messenger arachidonic acid. Prostaglandins Leukot Essent Fatty Acids 52:129–135

Brzeski M, Madhok R, Capell HA (1991) Evening primrose oil in patients with rheumatoid arthritis and side-effects of non-steroidal anti-inflammatory drugs. Br J Rheumat 30:370–372. https://doi.org/10.1093/rheumatology/30.5.370

Bushway AA, Wilson AM, McGann DF, Bushway RJ (1982) The nutrient composition of fresh fiddlehead greens. J Food Sci 47:666–667

Calder PC, Zurier RB (2001) Polyunsaturated fatty acids and rheumatoid arthritis. Curr Opin Clin Nutr Metab Care 4(2):15–21. https://doi.org/10.1097/00075197-200103000-00006

Connor WE (2000) Importance of n-3 fatty acids in health and disease. Am J Clin 7:171S–175S. https://doi.org/10.1093/ajcn/71.1.171S

Coste T, Pierlovisi M, Leonardi J, Dufayet D, Gerbi A, Lafont H, Vague P, Raccah D (1999) Beneficial effects of gamma linolenic acid supplementation on nerve conduction velocity, Na+, K+ ATPase activity, and membrane fatty acid composition in sciatic nerve of diabetic rats. J Nutr Biochem 10:411–420. https://doi.org/10.1016/s0955-2863(99)00020-0

Crowe A (1997) A field guide to the native edible plants of New Zealand. Godwit Publishing Ltd., Auckland, New Zealand

DeLong JM, Hodges DM, Prange RP, Forney CF, Toivenon PMA, Bishop MC, Elliot ML, Jordan MA (2011) The unique fatty acid and antioxidant composition of ostrich fern (Matteuccia struthiopteris) fiddleheads. Can J Plant Sci 91:919–930. https://doi.org/10.4141/CJPS2010-042

DeLuca P, Rossetti RG, Alavian C, Karim P, Zurier RB (1999) Effects of gammalinolenic acid on interleukin-1 beta and tumor necrosis factor-alpha secretion by stimulated human peripheral blood monocytes: studies in vitro and in vivo. J Investig Med 47:246–250

Din JN, Newby DE, Flapan AD (2004) Omega-3 fatty acids and cardiovascular disease-fishing for a natural treatment. BMJ:328–330. https://doi.org/10.1136/bmj.328.7430.30

Dixit RD (2000) Conspectus of pteridophytic diversity in India. Indian Fern J 17:77–91

Dwiloka B, Setiadi A, Santoso SI, Suprijatna E, Susanti S (2015) Effects of duck feed supplemented with invasive giant salvinia (*Salvinia molesta*) on duck meat characteristics. Turk J Vet Anim Sci 39:668–675

Dyer JM, Stymne S, Green AG, Carlsson AS (2008) High-value oils from plants. Plant J 54:640–655. https://doi.org/10.1111/j.1365-313X.2008.03430.x

Eduard VN, Vasily IS, Olga VK, Mikhail VV (2019) Variability of fatty acid profiles in ferns: relation to fern taxonomy and seasonal development. Phytochemistry. https://doi.org/10.1016/j.phytochem.2019.02.015

Erdinest N, Shmueli O, Grossman Y, Ovadia H, Solomon A (2012) Anti-inflammatory effects of alpha linolenic acid on human corneal epithelial cells. Invest Ophthalmol Vis Sci 53:4396. https://doi.org/10.1167/iovs.12-9724

Foran JA, Good DH, Carpenter DO, Hamilton MC, Knuyh BA, Schwager SJ (2005) Quantitative analysis of the benefits and risks of consuming farmed and wild salmon. J Nutr 135:2639–2643. https://doi.org/10.1093/jn/135.11.2639

Gillis RC, Daley BJ, Enderson BL, Karlstad MD (2004) Inhibition of 5-lipoxygenase induces cell death in anti-inflammatory fatty acid-treated HL-60 cells. J Parenter Enter Nutr 28:308–314. https://doi.org/10.1177/0148607104028005308

Groombridge B, Jenkins MD (2002) *World atlas of biodiversity*. Prepared by the UNEP World Conservation Monitoring Centre. University of California Press, Berkeley

Guschina IA, Harwood JL (2006) Lipids and lipid metabolism in eukaryotic algae. Prog Lipid Res 45:160–186. https://doi.org/10.1016/j.plipres.2006.01.001

Haag M (2003) Essential fatty acids and the brain. Can J Psychiatr 48:195–203. https://doi.org/10.1177/070674370304800308

Haigh WG, Yoder TF, Ericson L, Pratum T, Winget RR (1996) The characterisation and cyclic production of highly unsaturated homoserine lipid in *Chlorella minutissima*. Biochim Biophys Acta 1299:183–190

Harwood JL, Guschina IA (2009) The versatility of algae and their lipid metabolism. Biochimie 91:679–684. https://doi.org/10.1016/j.biochi.2008.11.004

Hrelia S, Bordoni A, Biagi P, Rossi CA, Bernardi L, Horrobin DF, Pession A (1996) C-linolenic acid supplementation can affect cancer cell proliferation via modification of fatty acid composition. Biochem Biophys Res Commun 225:441–447

Huang FC, Ju YH, Chiang JC (1999) Gamma- linolenic acid-rich triacylglycerols derived from borage oil via lipase-catalyzed reactions. JAOCS 76:833–837

Huang XP, Pi YQ, Lokuta AJ, Greaser ML, Walker JW (1997) Arachidonic acid stimulates protein kinase C-(epsilon) redistribution in heart cells. J Cell Sci 110:1625–1634

Jamal GA (1994) The use of gamma linolenic acid in the prevention and treatment of diabetic neuropathy. Diabetic Med 11:145–149

Jamieson GR, Reid EH (1975) The fatty acid composition of fern lipids. Phytochemistry 14:2229–2232

Janssen M, Chinapaw MJM, Rauh SP, Toussaint HM, Mechelen WV, Verhagen EALM (2014) A short physical activity break from cognitive tasks increases selective attention in primary school children aged 10–11. Ment Health Phys Act 7:129–134

Jiang WG, Hiscox S, Horrobin DF, Hallett MB, Mansel RE, Puntis MC (1995) Expression of catenins in human cancer cells and its regulation by n-6 polyunsaturated fatty acids. Anticancer Res 15:2569–2573

Kenny FS, Pinder SE, Ellis IO, Gee JM, Nicholson RI, Bryce RP, Robertson JF (2000) Gamma linolenic acid with tamoxifen as primary therapy in breast cancer. Int J Cancer 85:643–648. https://doi.org/10.1002/(sici)1097-0215(20000301)85:5<643::aid-ijc8>3.0.co;2-z

Kosesakal T, Yıldız M (2019) Growth performance and biochemical profile of *Azolla pinnata* and *Azolla caroliniana* grown under greenhouse conditions. Archiv Biol Sci 71(3):475–482. https://doi.org/10.2298/ABS190131030K

Lee CH, Shin SL (2010) Functional activities of ferns for human health. In: Fernández H, Revilla MA, Kumar A (eds) Working with ferns: issues and applications. Springer, New York, pp 347–359

Leeb BF, Sautner J, Andel I, Rintelen B (2006) Intravenous application of omega-3 fatty acids in patients with active rheumatoid arthritis. The ORA-1 trial. An open pilot study. Lipids 41:29–34. https://doi.org/10.1007/11745-006-5066-x

Li D, Hu X (2009) Fish and its multiple human health effects in times of threat to sustainability and affordability: are there alternatives? Asia Pac J Clin Nutr 218:553–563

Lottenberg A (2009) Importance of the dietary fat on the prevention and control of metabolic disturbances and cardiovascular disease. Arq Bras Endocrinol Metabol 53:595–607. https://doi.org/10.1590/S0004-27302009000500012

Mainou-Fowler T (2001) Gamma-linolenic acid induces apoptosis in B-chronic lymphocytic leukaemia cells in vitro. Leuk Lymphoma 40(3–4):393–403. https://doi.org/10.3109/10428190109057939

Mayser P, Grimm H, Grimminger F (2002) n-3 fatty acids in psoriasis. Br J Nutr 87:S77–S82. https://doi.org/10.1079/BJN2001459

Nishida I, Murata N (1996) Chilling sensitivity in plants and cyanobacteria: the crucial contribution of membrane lipids. Ann Rev Plant Biol 47:541–568

Paoletti C, Bocci F, Lercker G, Capella P, Materassi R (1987) Lipid composition of *Azolla caroliniana* biomass and its seasonal variation. Phytochemistry 26:1045–1047

Park S, Johnson MA (2006) Awareness of fish advisories and mercury exposure in women of childbearing age. Nutr Rev 64:250–256. https://doi.org/10.1111/j.1753-4887.2006.tb00207.x

Perveen Z, Ando H, Ueno A, Ito Y, Yamamoto Y, Yamada Y (2006) Isolation and characterization of a novel thraustochytrid-like microorganism that efficiently produces docosahexaenoic acid. Biotechnol Lett 28:97–202. https://doi.org/10.1007/s10529-005-5335-4

Ponte E, Cafagna D, Balbi M (1997) Patologia cardiovascolare ed acidi grassi omega-3. Minerva Med 88:343–353

Rogalski M, Carrer H (2011) Engineering plastid fatty acid biosynthesis to improve food quality and biofuel production in higher plants. Plant Biotechnol J 9:554–564. https://doi.org/10.1111/j.1467-7652.2011.00621.x

Rozentsvet OA, Rezanka T, Bosenko ES, Uzhametskaya EA, Dembitsk VM (2005) Fatty acids, phospholipids, and the betaine lipid DGTS from the aquatic fern *Salvinia natans*. Chem Nat Compd 41(5):487–490

Sagar S, Das UN, Koratkar R, Ramesh G, Padma M, Kumar GS (1992) Cytotoxic action of cis-unsaturated fatty acids on human cervical carcinoma (HeLa) cells: relationship to free radicals, and lipid peroxidation and its modulation by calmodulin antagonists. Cancer Lett 63: 189–198

Saravanan P, Davidson NC, Schmidt EB, Calder PC (2010) Cardiovascular effects of marine omega-3 fatty acids. Lancet 376:540–550. https://doi.org/10.1016/S0140-6736(10)60445-X

Schlenk H, Gellerman JL (1965) Arachidonic, 5,11,14,17-eicosatetraenoic and related acids in plants – identification of unsaturated fatty acids. J Am Oil Chem Soc 42:504–511

Serhan CN (1997) Lipoxins and novel aspirin-triggered 15-epi-Lipoxins (ATL): a jungle of cell-cell interactions or a therapeutic opportunity? Prostaglandins Other Lipid Mediat 53:107–137

Sidhu KS (2003) Health benefits and potential risks related to consumption of fish or fish oil. Regul Toxicol Pharmacol 38:336–344. https://doi.org/10.1016/j.yrtph.2003.07.002

Smith AR, Pryer KM, Schuettpelz E, Korall P, Schneider H, Wolf PG (2006) A classification for extant ferns. Taxon 55:705–731

Stefanov K, Seizova K, Georgieva G, Zlatanova S, Kuleva L, Popov S (1997) Preparation of polyunsaturated fatty acid concentrates from the liver oil of dogfish (*Squalus acanthias*) from the Black Sea. Grasas Aceites 48:141–143

Upchurch RG (2008) Fatty acid unsaturation, mobilization, and regulation in the response of plants to stress. Biotechnol Lett 30:967–977

Vangaveti VN, Jansen H, Kennedy RL, Malabu UH (2016) Hydroxyoctadecadienoic acids: oxidised derivatives of linoleic acid and their role in inflammation associated with metabolic syndrome and cancer. Eur J Pharmacol 785:70–76

Von-Aderkas P (1984) Economic history of ostrich fern, Matteuccia struthiopteris, the edible fiddlehead. Econ Bot 38:14–23

Walley JW, Kliebenstein DJ, Bostock RM, Dehesh K (2013) Fatty acids and early detection of pathogens. Curr Opin Plant Biol 16:520–526

Ward OP, Singh A (2005) Omega-3/6 fatty acids: alternative sources of production. Process Biochem 40:3627–3652

Watkins G, Martin TA, Bryce R, Mansel RE, Jiang WG (2005) Gamma-linolenic acid regulates the expression and secretion of SPARC in human cancer cells. Prostaglandins Leukot Essent Fatty Acids 72:273–278. https://doi.org/10.1016/j.plefa.2004.12.004

Yaeno T, Matsuda O, Iba K (2004) Role of chloroplast trienoic fatty acids in plant disease defense responses. Plant J 40:931–941

Youdim KA, Martin A, Joseph JA (2000) Essential fatty acids and the brain: possible health implications. Int J Dev Neurosci 18:383–399. https://doi.org/10.1016/S0736-5748(00)00013-7

Zárate R, Jaber-Vazdekis N, Tejera N, Pérez JA, Rodríguez C (2017) Significance of long chain polyunsaturated fatty acids in human health. Clin Transl Med 25(6):1–19. https://doi.org/10.1186/s40169-017-0153-6

Zhang J, Liu H, Sun J, Li B, Zhu Q, Chen S, Zhang H (2012) Arabidopsis fatty acid desaturase FAD2 is required for salt tolerance during seed germination and early seedling growth. PLoS One 7(1):e30355. https://doi.org/10.1371/journal.pone.0030355

Ziboh VA, Naguwa S, Vang K, Wineinger J, Morrissey BM, Watnik M, Gershwin ME (2004) Suppression of leukotriene B4 generation by ex-vivo neutrophils isolated from asthma patients on dietary supplementation with gammalinolenic acid-containing borage oil: possible implication in asthma. Clin Dev Immunol 11:13–21. https://doi.org/10.1080/10446670410001670445

**Priti Giri** is a research scholar in Department of Botany, University of Delhi. She has completed M.Sc. and M.Phil. from Department of Botany, University of Delhi, Delhi. Her area of specialization is conservation, phytochemical, and biochemical screening of ethnomedicinally important ferns (pteridophytes), and she has published many research paper and book chapters on her subject.

**Ashwani Kumar** Alexander von Humboldt Fellow (Germany), MSc (Botany) (Gold Medalist) (1967)), PhD (1971), postdoc AvH Fellow (Germany) and JSPS fellow (Japan) Selected in the Indian Administrative Services (IAS: IPS) (1972) but he opted career in botany. 37 years of teaching & research experience. Guided 39 PhD. research students and published 320 research papers and 26 books. Elected fellow FBS, FPSI, FIMPS and FIFS. Editor in Chief IJLS. V. Puri medal (2008) CEE award (2015). President of CHEC (India), Indian Botanical Society, Society for Promotion of Plant Sciences. Professor Emeritus, Department of University of Rajasthan, Jaipur. Presently, Chair Professor at JVWU, Jaipur and Adj. Professor Modi University, Sikar.

**Prem Lal Uniyal** is presently working as a Professor in the Department of Botany, University of Delhi. He did his master's degree from HNB Garhwal University, Srinagar, Uttarakhand, and PhD in Botany from the Panjab University, Chandigarh, and has been teaching since the last 27 years. Dr Uniyal is trained in bryologist and specialized in the Biology of Archegoniates. His other areas of research are conservation biology, gametophyte development and reproductive biology of ferns, population biology of conifers, biology of Podostemaceae, and biological activity of lichens. His research group has discovered a few new taxa of mosses and *Ephedra* from India. He organized a few conservation programs with indigenous societies in Himachal Pradesh. He has published 4 books and more than 105 original research papers in peer-reviewed national and international journals; 35 review articles in proceedings, book, and scientific magazine; and 25 popular scientific articles. He has supervised 17 PhD scholars and 22 M. Phil students.

# 22 In Vitro Antiproliferative Effect of *Angiopteris evecta* (G. Forst.) Hoffm. Extracts against Cultured HT-29 Colon Cancer Cells

S. Catharin Sara and R. Gnana Deepa Ruby

**Abstract**

*Angiopteris evecta* (G. Forst.) Hoffm. belongs to the family Angiopteridaceae. It is commonly called giant or king fern. The plant is used as a traditional medicine and as diuretic, antipyretic, tonic, analgesic, antihelminthic, and antidiarrheal. The present study investigated the antiproliferative activity of *A. evecta* ethanolic extracts against malignant colon cancer cells (HT-29) and nonmalignant colon cancer cells (L929). The anticancer activity of *A. evecta* revealed that its extracts are effective against HT-29 colon cancer cells and they do not harm the L929 cells. Cytotoxicity against cell lines showed the great difference between malignant and nonmalignant cell lines, and they inhibit the cancer cells even at low concentration of 15.94 μg/ml. The inhibition concentration (IC50) value of the ethanolic extract of *A. evecta* (G. Forst.) Hoffm. treated HT-29 cancer cells at 24 h was 15.94 μg/ml and of L929 cells 81.69 μg/ml. Results showed apoptotic effect of *A. evecta* ethanolic extracts against HT-29 cancer cells in a dose-dependent manner. The apoptotic effect was confirmed by the double staining method and was observed under fluorescent spectroscopy, which revealed the apoptotic and viable cells by their fluorescence. Untreated control cells showed cells with Acridine orange stained green nucleus. In malignant cells, early and late apoptotic cells were observed in 9.5 μg/ml extract, and necrosis resulted in 21.5 μg/ml extract, and necrotic cells appeared as red stained nuclei, and thus necrosis increases as the concentration increases. DNA content and cell cycle was studied by flow cytometry. HT-29 cells treated with *A. evecta*. Showed the DNA damage or loss of DNA content in the sub G0/G1 phase, S-phase, and G2/M phase. From the results, it is very clear that maximum apoptosis took place in the G0/G1, and it is proven through the population profile index. The cell size of the treated and

S. C. Sara (✉) · R. G. D. Ruby
PG and Research Department of Botany, Holy Cross College (Autonomous), Tiruchirappalli, Tamil Nadu, India



J. Marimuthu et al. (eds.), *Ferns*, https://doi.org/10.1007/978-981-16-6170-9_22

untreated HT-29 cells differs, and moreover untreated HT-29 cells showed less population in sub G0/G1 phase, whereas in treated HT-29 cells, the population was noticed high in the G0/G1 phase than the S-phase. Hence, the results indicated that the ethanol extract of *A. evecta* plant extract has the potential antiproliferative activity with low cytotoxicity to normal cells. Our results support the ethnomedicinal observations of these plants for the management of cancer. Future study is needed about the active elements in the extracts, to be isolated and purified to investigate the synergy and additive pharmacological effect in killing cancer cells.

**Keywords**

*Angiopteris evecta* · Antiproliferation · Apoptosis · Colon cancer · Flow cytometry

## 22.1 Introduction

The plant kingdom is classified into seedless plants called cryptogams and seed bearing plants called phanerogams or spermatophyta. Cryptogams have been further divided into thallophyta, bryophyta, and pteridophyta. Pteridophytes were the first plants to extensively colonize the terrestrial environment forming forests. Fossil records indicate that they evolved about 400 million years back, i.e., in the Silurian period of late Paleozoic era and late Paleozoic era which can be regarded as "Age of Pteridophytes" (Khullar 2008).

Since most of the ferns or pteridophyte population live in a wide variety of habitats, they have a great biodiversity and, as they grow at higher altitudes, are subjected to an attack of miscellaneous hard situations including higher doses of mutagenic UV radiation, physiological drought, desiccation, and strong winds. The pteridophytic plants survive hectic environments by physiological adaptation and alter their biochemical profile inside the plant tissues, thereby producing a spectrum of secondary metabolites. Secondary metabolites are of special interest to scientists because of their unique pharmacophores and medicinal properties. Secondary metabolites like polyphenols, terpenes, and alkaloids have been reported to possess anticancer properties in many studies from these groups of plants.

Cancer is a major public health burden in both developed and developing countries. Plant-derived agents are being used for the treatment of cancer. Several anticancer agents including taxol, vinblastine, vincristine, the camptothecin derivatives topotecan and irinotecan, and etoposide derived from epipodophyllotoxin are in clinical use all over the world. A number of promising agents such as flavopiridol, roscovitine, combretastatin A-4, betulinic acid, and silvestrol are in clinical or preclinical development (Shoeb et al. 2006).

Plant-derived phytochemicals such as sulforaphane, resveratrol, geinsten, vincristine, vinblastine, curcumin, gingerol, diallyl sulfide, epigallocatechin-3-gallate, brassinin, 40-bromoflavone, brusatol, etc. have been employed for the control and restrainment of carcinogenesis (Mehta and Pezzuto 2002; Veronesi and Bonanni 2005; Saunders and Wallace 2010).

Phenolic compounds catechins and (−)-epigallocatechin-3-gallate (EGCG) possess anti-tumorigenic activity against various cancers, including melanoma. Epigallocatechin 3-o-gallate, EGCG, is the major catechin. Another catechin is GCG, gallocatechin-3-gallate. Other catechins include epicatechin gallate (ECG), epigallocatechin (EGC), and epicatechin (EC). Catechins are EGCG > ECG > EC > EGC > C. They prevent cancer and act as a protective agent against a variety of malignant disorders such as lung cancer, breast cancer, and prostate cancer. This preventive potentiality against cancer is attributed to the biologically active flavonoids called catechins (Shimizu et al. 2010).

The exploration for natural compounds from pteridophytic plants is very prosperous in antioxidant and anticancer properties, and in silico studies are getting higher attention due to their medicinal importance in controlling many interrelated chronic disorders such as cancer and cardiovascular diseases. Pteridophytes have higher number of secondary metabolites (which are potential sources of drugs) of therapeutic importance but are least studied.

*Angiopteris evecta* belongs to the family Angiopteridaceae, a large terrestrial shrub found in well-shaded, humid areas of forests near water sources. It has been used as a traditional medicine as a diuretic, antipyretic, tonic, analgesic, and antidiarrheal. Aromatic oil is extracted from large-sized fern used for perfuming coconut oil in South Sea Islands. In Central India, the rhizomes are used against scabies (Vasudeva 1999). Leaves, stem bark, stem, root, and tubers of *Angiopteris evecta* exhibited antibacterial and antifungal activity (Khan and Omoloso 2008). The preliminary phytochemical analysis of *A. evecta* resulted in the presence of flavonoids (Ruby 2018). Earlier references (Defilpps et al. (1998) and Hseu (1981)) suggest the medicinal properties and provide positive evidences regarding the present experimental plant on anticancer properties.

Owing to the ethnobotanical, phytochemical, and pharmacological importance and medicinal value of the present experimental plant *A. evecta*, the in vitro antiproliferative study was planned on this important medicinal plant against cancer cell lines such as malignant and nonmalignant cultured Ht-29 colon cancer cells.

## 22.2 Materials and Methods

### 22.2.1 Taxonomic Description of *Angiopteris evecta* (G. Forst.) Hoffm

*Angiopteris evecta* (G. Forst.) Hoffm. is commonly found in hilly areas. In India, it is highly distributed in Western Ghats, Pachmarhi and Bailadilla hills of Central India, Eastern Himalayas like Darjeeling, Sikkim, and South India, viz., Kodaikanal and Palani hills. They grow in an average elevation of 1567 meters above sea level. Taxonomic description of *Angiopteris evecta* (G. Forst.) Hoffm explains that the rachis looks massive, globose, erect, radially arranged, and 150 cm tall and 100 cm wide. Stipules measure up to 10–15 cm, densely long-scaly above, triangular, and are known as auricles. Fronds are bipinnate, up to 5–7 m long, up to 6 cm wide, and the swollen base to 15 cm, wide, without nodes. Sporangia are borne on the underside of the pinnule, very close to the margin in cluster of 5–8 opposite pairs.

### 22.2.2 Collection, Identification, and Extract Preparation

Experimental plants were collected during the month of December 2018, from Bonacaud forest, Palode, Thiruvananthapuram. *Angiopteris evecta* (G. Forst.) Hoffm. were located at 7°87′23″N 85°13′57″E, baseline of the forest. Herbarium specimens were prepared from the collected plants. The herbarium voucher specimens were identified by Dr. S. John Britto S.J., Director, The Rapinat Herbarium, Center for Molecular Systematics, St. Joseph's College, Tiruchirappalli, Tamil Nadu, India. The herbarium voucher specimens (RGDR, 002, and GDR, 001) were deposited in the Department of Botany, Holy Cross College (Autonomous), Tiruchirappalli, for future reference.

The powdered whole plant materials of *A. evecta* were extracted by Soxhlet extraction process using ethanol. According to the results, highest percentage yield value of the extraction of whole plant extract was observed in ethanol. The ethanolic extract was used for further studies.

### 22.2.3 Procurement of Nonmalignant (L929) and Malignant (HT-29) Cell Lines

Nonmalignant (L929) and malignant (HT-29) colon cancer cell lines were purchased from NCCS Pune and were maintained in Dulbecco's modified eagles media (HIMEDIA) supplemented with 10% FBS (Invitrogen) and grown to confluency at 37 °C in 5% $CO_2$ in a humidified atmosphere in a $CO_2$ incubator (NBS, EPPENDORF, GERMANY). The cells were trypsinized (500 μl of 0.025% trypsin in PBS/0.5 mM EDTA solution (Himedia)) for 2 mins and passaged to T flasks in complete aseptic conditions.

The ethanolic extracts of *A. evecta* were added to nonmalignant (L929) grown cells at a final concentration of 1, 25, 50, 75, and 100 μg/ml from a stock of 1 mg/ml and incubated for 24 h. The ethanolic extracts of *A. evecta* were added to malignant (HT-29) grown cells at a final concentration of 1.5, 5.5, 9.5, 13.5, 17.5, and 21.5 μg/ml from a stock of 1 mg/ml and incubated for 24 h. The % difference in viability was determined by standard MTT assay after 24 h of incubation. The percentage of difference in viability was determined by standard MTT assay after 24 h of incubation for nonmalignant and malignant cells.

### 22.2.4 MTT Assay

The cells were washed with 1× PBS, and then 30 μl of MTT solution were added to the culture (MTT – 5 mg/ml dissolved in PBS) (Arung et al. 2000). It was then incubated at 37 °C for 3 h. MTT was removed by washing with 1× PBS, and 200 μl of DMSO was added to the culture. Incubation was done at room temperature for 30 min until the cell got lysed and color was obtained. The solution was transferred to centrifuge tubes and centrifuged at top speed for 2 min to precipitate cell debris. Optical density was read at 540 nm using DMSO as blank in a microplate reader (ELISA SCAN, ERBA). Experiments were done in triplicates, and the results were plotted as mean (±) standard deviation.

Best fit line linear regression analysis was carried out using SPSS package. IC50 values of the two plant extracts were calculated from the best fit linear regression line plotted with percentage of cells against different tested concentrations. Higher IC50 values indicate the lesser toxicity of the plant material.

### 22.2.5 Screening of Apoptosis by Double Staining Method

DNA-binding dyes Acridine orange (AO) and ethidium bromide (EB) (Sigma, USA) were used for the morphological detection of apoptotic and necrotic cells (Zhang et al. 1998). After treatment with different concentrations (9.5 and 21.5 μg/ml) of the extracts, the cells were washed by cold PBS (phosphate buffer solution) and then stained with a mixture of AO (100 μg/ml) and ethidium bromide EB (100 μg/ml) at room temperature for 10 min. The stained cells were washed twice with 1× PBS and observed by a fluorescence microscope in blue filter of fluorescent microscope (Olympus CKX41 with Optika Pro5 camera).

Acridine orange is a vital dye and will stain both live and dead cells. Ethidium bromide will stain only cells that have lost membrane integrity. Acridine orange is taken up by both viable and nonviable cells and emits green fluorescence if intercalated into double stranded nucleic acid (DNA). Ethidium bromide is taken up only by nonviable cells and emits red fluorescence by intercalation into DNA.

### 22.2.6 Flow Cytometric Analysis of DNA Content and Cell Cycle Distribution (Martineti et al. 2010)

Monitoring a cell's ability to proliferate is critical for assessing a cell's health during toxicity studies. The most accurate method of doing this is by directly measuring DNA synthesis. The basic principle of MUSE cell cycle kit is a standard ethanol fixation and detergent permeabilization which is sufficient to gain access to the DNA during active cell cycle. The kit utilizes a premixed reagent which includes the nuclear DNA intercalating stain propidium iodide (PI) which discriminates cells at different stages of the cell cycle based on the differential DNA content in the presence of RNAse to increase the specificity of DNA staining in each phase (G0/G1, S, and G2/M).

The cell sample was transferred to a 12 × 75 mm polystyrene tube or 50 ml conical flask. The minimum recommended number of cells for fixation in a tube is 1 × 106 cells. The samples were then centrifuged at 3000 rpm for 5 min. The supernatant was removed without disturbing the pellet. After centrifugation, the cell pellet forms either a visible pellet or a white film on the bottom of the tube. Appropriate volume of phosphate buffer solution (PBS) was added to each tube (i.e., 1 ml of PBS per 1 × 106 cells), and the contents were mixed by pipetting several times or gently vortexing. Centrifuge the cells at 3000 rpm for 5 min. The supernatant was discarded without disturbing the cell pellet, leaving approximately 50 μl of PBS per 1 × 106 cells.

Resuspend the pellet in the residual PBS by repeated pipetting several times or gently vortexing. The resuspended cells were added dropwise into the tube containing 1 ml of ice-cold 70% ethanol while vortexing at medium speed. Cap and freeze the tube at −20 °C. Staining of cell cycle after overnight incubation, the samples were centrifuged at 3000 rpm for 5 min at room temperature. The supernatant was removed and 250 μl PBS was added to the pellet. Then the centrifugation was done again at the same rpm and time. The pellet was taken after discarding the supernatant; 250 μl of cell cycle reagent was added. This was incubated in the dark for 30 min (which is light sensitive). After this, it was analyzed using a flow cytometer. Gating was performed with reference to untreated control cells, and treated cells with samples were analyzed.

## 22.3 Results

Cancer therapy is still a major obstacle among medical oncologists around the world. Failure in chemotherapy of tumor cells is mainly related to occurrence of drug resistance. Cell cycle alterations and decreased apoptosis induction following chemotherapy are primarily due to molecular changes of cancer cells. Studying new compounds alone or in combination with potent chemotherapeutic drugs can open new strategic approach in cancer therapy. Among the enzyme-based assays, the MTT assay is the best known method for determining mitochondrial dehydrogenase activities in the living cells.

### 22.3.1 Determination of in Vitro Antiproliferative Effect of *Angiopteris evecta* Ethanolic Extracts on Cultured Ht-29 Cells by MTT Assay

Antiproliferative effect of plant extracts against HT-29 human colon carcinoma cell lines is able to express different feature characteristics of mature intestinal cells such as the goblet cells. The antiproliferative activity of the extracts against the human cancer cell lines was tested using the microtitration colorimetric method of MTT

**Table 22.1** Results of in vitro antiproliferative effect (apoptosis and cell viability) of *Angiopteris evecta* ethanolic extracts on cultured Ht-29 cells

| *Angiopteris evecta* concentration (μg/ml) | Percentage of apoptotic cancer cells | Percentage of cell viability |
|---|---|---|
| Control | – | 99.00 |
| 1.5 | 9.38 | 90.62 |
| 5.5 | 21.28 | 78.72 |
| 9.5 | 36.83 | 63.17 |
| 13.5 | 42.64 | 57.36 |
| 17.5 | 50.00 | 50.00 |
| 21.5 | 58.65 | 41.35 |

reduction with minor modification. MTT is used to determine cell viability in cell proliferation and cytotoxicity assays. In the current study, the antiproliferative effect of *A. evecta* ethanolic extracts on HT-29 colon cancer cell lines was studied in detail. The MTT assay is based on the conversion of yellow tetrazolium salt MTT to purple formazan crystals by metabolically active cells. The amount of formazan produced is proportional to the number of viable cells. The *A. evecta* ethanolic extract treated HT-29 colon cancer cell lines showed remarkable apoptosis or cell death. As the concentration of the extract increases, the cell viability of the cancer cells decreases. This was compared with the control; the extract treatment was not used in this control section of cancer cells. Hence there was no indication of cell death and it showed 99% cell viability. The following concentrations of the extract, 1.5, 5.5, 9.5, 13.5, 17.5, and 21.5 μg/ml, showed 9.38, 21.28, 36.83, 42.64, 50, and 58.65 percentage of apoptosis in *A. evecta* ethanolic extract treated cancer cells. As the concentration of the extract increases, the percentage of the apoptosis also increases. Regarding the cell viability of the cancer cells when treated with *A. evecta* ethanolic extract, the percentage of cell viability decreases as the concentration of the extract increases, 90.62, 78.72, 63.17, 57.36, 50, and 41.35, respectively. The $IC_{50}$ value of the *A. evecta* treated HT-29 cancer cells at 24 h was 17.5 μg/ml (Table 22.1).

### 22.3.2 Antiproliferative Effect of *Angiopteris evecta* Ethanolic Extracts on Nonmalignant Cell Line: L929 (Normal Cell Line)

Before the investigation of the anticancer activity of *Angiopteris evecta* ethanolic extract, the cytotoxicity of *Angiopteris evecta* on normal cells was evaluated using nonmalignant cell line L929, which is a fibroblast-like cell line cloned from strain L. The parent strain was derived from normal subcutaneous areolar and adipose tissue of a male C3H/A mouse. A successful anticancer drug should kill or incapacitate cancer cells without causing excessive damages to normal cells, i.e., minimum side effects. To test the toxicity level of the *A. evecta* ethanolic extracts on normal cells, the *A. evecta* ethanolic extracts were tested against a nonmalignant cell line

**Table 22.2** Results of in vitro antiproliferative effect (apoptosis and cell viability) of *Angiopteris evecta* ethanolic extracts on cultured L929 cells

| *Angiopteris evecta* concentration (μg/ml) | Percentage of apoptotic cancer cells | Percentage viability |
|---|---|---|
| Control | – | 98.79 |
| 1 | 0.48 | 98.31 |
| 25 | 19.32 | 80.68 |
| 50 | 34.45 | 65.45 |
| 75 | 50 | 50.00 |
| 100 | 61.25 | 38.75 |

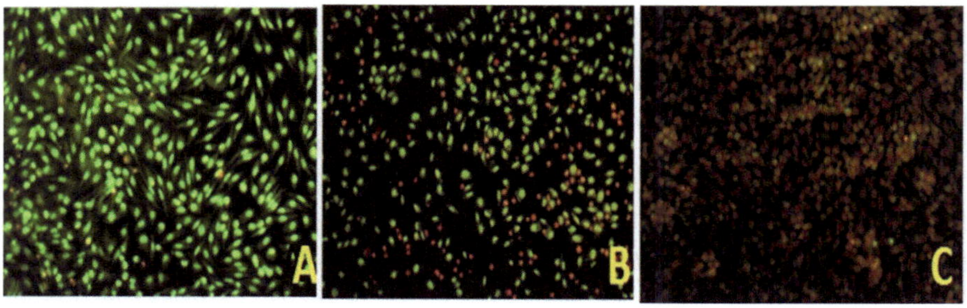


**Plate 22.1** It is shown determination of apoptosis. (**a**) Untreated control cells showing intact cells with Acridine orange stained *green* nucleus. (**b**) Cells treated with 9.5 μg/ml *Angiopteris evecta* extract showing intact cells with Acridine orange stained *green* nucleus and apoptotic cells (*red* stained nuclei). (**c**) Cells treated with 21.5 μg/ml *Angiopteris evecta* extract showing increased apoptotic cells (*red* stained nuclei)

L929, and it was evaluated by the MTT assay. The antiproliferative effect of plant extract was determined, being time- and dose-dependent.

The *A. evecta* ethanolic extract treated normal cells showed maximum cell death at the concentration of 100 μg/ml, and the percentage of cell death noticed was 61.25. As the concentration of the extract increases, the percentage of cell death of L929 also increases. 0.48, 19.32, 34.45, 50, and 61.25 percentage of L929 cells were apoptized at 1, 25, 50, 75, and 100 μg/ml concentration, respectively. The results were tabulated in Table 22.2. The $IC_{50}$ value of the ethanolic extract of *A. evecta* is 75 μg/ml. Regarding the cell viability of L929 cells, the cells were viable at low concentration.

### 22.3.3 *Angiopteris evecta* Ethanolic Extracts Induce Apoptosis in Human Colorectal Cancer (HT-29) Cells Studied by Double Staining

The result suggests the therapeutic potential of ethanolic extracts of *A. evecta* against colon cancer. (1) Live cells appear uniformly in greenshowing intact cells with Acridine orange (Plate 22.1a). (2) Early apoptotic cells stain green and intact cells with Acridine orange stained nucleus showing apoptotic cells (red stained nuclei) (Plate 22.1b). (3) Late apoptotic cells will also incorporate ethidium bromide and therefore stain orange apoptotic cells (red stained nuclei), and necrotic cells stain orange (Plate 22.1c).

### 22.3.4 Analysis of DNA Content and Cell Cycle Distribution by Flow Cytometry

Proliferating cells cycle through three major compartments such as G1, S-phase, and G2/M. The G2 and M phase includes mitosis which contains twice the DNA content before dividing to newborn G1 cells. Fraction of cells with a DNA content below G1 phase, often called "sub G1" fraction, consist of debris and fragmented cells. The degree of polyploidy of cells, meaning cells with DNA content higher than G2/M, is usually a result of failure in mitosis.

Through flow cytometric analysis, the exponential growth percentage and DNA content of untreated HT-29 cells and the HT-29 cells treated with *Angiopteris evecta* ethanolic extract of each phase such as $G_0/G_1$, S-phase, and G2/M were determined. Flow cytometry distinguishes cells in different phases of the cell cycle. Propidium iodide treated cells stain DNA quantitatively. The fluorescence intensity of the stained cells at certain wavelength correlates with the amount of DNA they contain. DNA histogram revealed that *Angiopteris evecta* ethanolic extract induced the cell death or the loss of DNA content was maximum, within the $G_0/G_1$ phase. Flow cytometric analysis of HT-29 (17.5 μg/ml) cells treated with ethanolic extract showed inhibition of cells at $G_0/G_1$ phase. There was 62% increase in cells arrested at $G_0/G_1$ phase when compared with untreated control (Table 22.3), whereas the S-phase cells decreased proportionally from 20.1 to 5.7. This finding indicates that cell cycle distribution was blocked significantly in the $G_0/G_1$ phase cells treated with *Angiopteris evecta* ethanolic extract (Tables 22.3 and 22.4).

Flow cytometric analysis also showed the DNA damage that indicates apoptosis, a form of programmed cell death. Nuclei of apoptotic cells contain less DNA in

**Table 22.3** Flow cytometric analysis of HT-29 untreated cells with ethanolic extract of *A. evecta*

| | G0/G1 | S | G2/M | Debris |
|---|---|---|---|---|
| % Gated | 61.8 | 20.1 | 11.3 | 75.6 |
| Mean | 2134.0 | 4065.5 | 5907.7 | 206.5 |
| % CV | 35.5 | 11.8 | 10.4 | 609.1 |

**Table 22.4** Flow cytometric analysis of HT-29 treated cells with ethanolic extract of *A. evecta*

| | G0/G1 | S | G2/M | Debris |
|---|---|---|---|---|
| % Gated | 62.0 | 5.7 | 2.5 | 93.2 |
| Mean | 2100.1 | 4095.7 | 5877.6 | 163.1 |
| % CV | 24.2 | 10.9 | 9.9 | 269.3 |

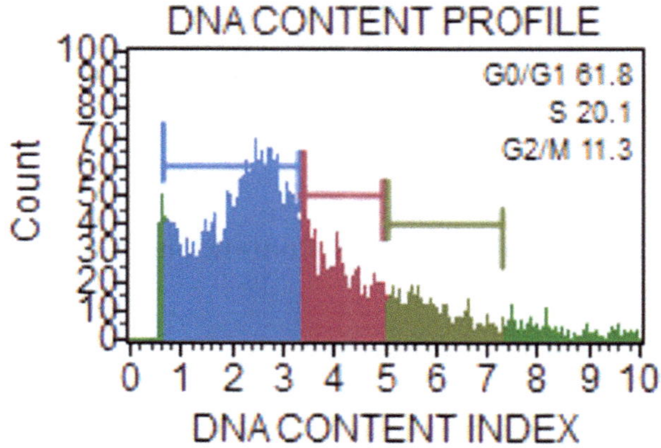


**Fig. 22.1** DNA content profile of HT-29 untreated cells with ethanolic extract of *A. evecta*

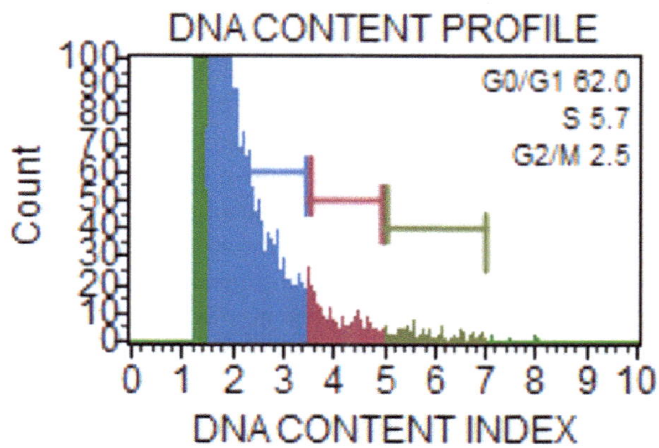


**Fig. 22.2** DNA content of HT-29 cells treated with *Angiopteris evecta* extract

*A. evecta* ethanolic extract treated HT-29 than nuclei of *A. evecta* ethanolic extract untreated HT-29 $G_0/G_1$ cells, resulting in a sub $G_0/G_1$ peak in the fluorescence histogram such as DNA content profile that was used to determine the relative amount of apoptotic cells in the *Angiopteris evecta* extract. HT-29 cells treated with the *Angiopteris evecta* ethanolic extract showed the DNA damage or loss of DNA content in the sub $G_0/G_1$ phase. In S-phase, the percentage of DNA damage is very low. G2/M phase showed very least amount of DNA loss. Hence from the result, it clearly showed that the maximum apoptosis took place in the $G_0/G_{1.}$ It also has been proven through the population profile index. The cell size of the treated and untreated HT-29 cells differs in the Figs. 22.1 and 22.2. The untreated HT-29 cells showed less population in sub $G_0/G_1$ phase (Fig. 22.3). But in the treated HT-29 cells, the population was noticed high in the $G_0/G_1$ phase than the S-phase (Fig. 22.4). Therefore, the flow cytometry clearly showed the cell distribution, percentage of cells in each phase, DNA content, and population of cells in each phase.

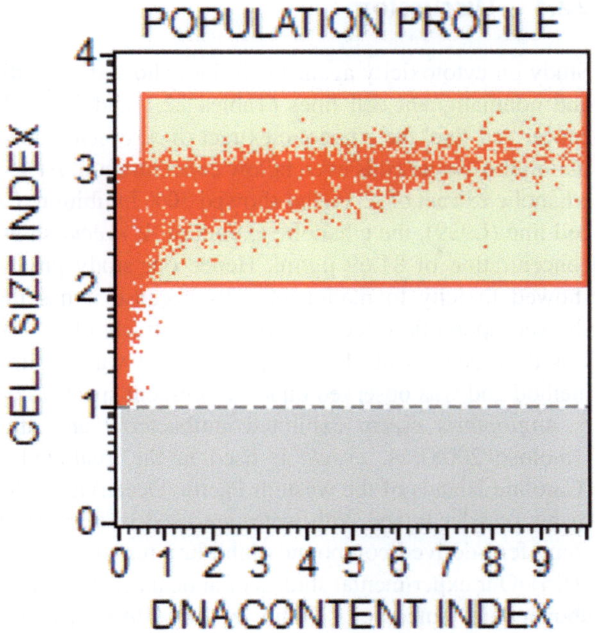


**Fig. 22.3** The population profile of HT-29 untreated cells with ethanolic extract of *A. evecta*

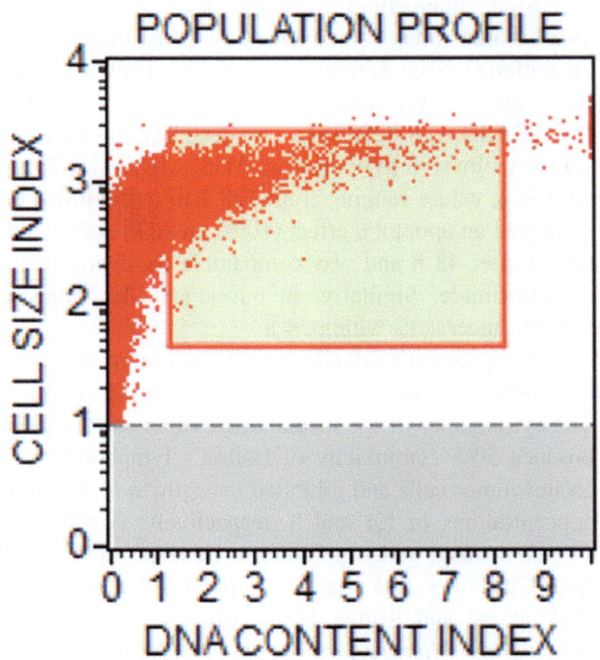


**Fig. 22.4** The population profile of HT-29 cells treated with ethanolic extract of *A. evecta*

## 22.4 Discussion

Study on cytotoxicity against cell lines showed great difference between malignant and nonmalignant cell lines (Tables 22.1 and 22.2). Regarding the HT-29 colon cancer cell line, the ethanolic extract of *A. evecta* showed higher activity, i.e., they inhibit the cancer cells even at low concentration. At 15.94 μg/ml concentration, the ethanolic extract of *A. evecta* showed 50% inhibition. Concerning the nonmalignant cell line (L929), the ethanolic extract of *A. evecta* showed 50% of inhibition at the concentration of 81.69 μg/ml. Hence our study proves that the *A. evecta* extract showed toxicity to malignant cells because it has more toxic property. Results showed apoptotic effect of ethanolic extracts of *A. evecta* on HT-29 cancer cells was dose-dependent. The apoptotic effect was confirmed by the double staining method and was observed under fluorescent spectroscopy (Plate 22.1).

*Angiopteris evecta* exhibited antibacterial and antifungal activity (Khan and Omoloso 2008). *A. evecta* is used in the treatment of headaches, in both Yap (Caroline Islands of the western Pacific Ocean) and Meghalaya. Fresh rhizome and rachis powder mixed with water are used in diarrhea. Under in vivo experiments, some fern-derived compounds inhibited tumor growth with little toxicity (Tomsik 2014). Our experimental studies of anticancer study also relate and express the same above said outcomes. Uddin et al. (1998) isolated seven known flavonoid compounds, patriscabratine, tetracosane, and five flavonoids (quercetin-3-O-β-d-glucoside, quercetin-3-O-β-d-glucosyl- (6 → 1)-α-l-rhamnoside, quercetin-3-O-α-l-rhamnoside, quercetin-3-O-α-l-rhamnosyl-7-Oβ-d-glucoside, and kaempferol), from *Acrostichum aureum*. They were tested for its cytotoxic activity. Patriscabratine was found moderately cytotoxic against AGS, MDAMB-231, and MCF-7 cells with IC50 values ranging from 69.8 to 197.3 μM. Tetracosane showed some cytotoxic activity against AGS, MDA-MB-231, HT-29, and NIH 3 T3 cells with $IC_{50}$ values ranging from 128.7 to >250 μM. Patriscabratine and tetracosane displayed an apoptotic effect (10%) on AGS cells within 24 h which was increased (20%) after 48 h and was comparable to, if not greater than the positive control, cycloheximide. Similarly, in our study, flavonoids exhibit cytotoxicity against HT-29 cancer cells within 24 h.

*Pityrogramma calomelanos* required more than 40 μg/ml of the extract to produce 50% cell death within 3 h. Our extract showed 50% of lethality at 17.5 μg/ml. The cytotoxicity of the methanolic extract of *Pityrogramma calomelanos* needs to produce 50% cytotoxicity of Dalton's lymphoma ascites tumor cells and Ehrlich ascites tumor cells and inhibited the growth of KB cells and K562 cells by 50% at concentrations of 1.1 and 8, respectively. A dihydrochalcone was isolated from *P. calomelanos* and was found to be responsible for the cytotoxic activity of this fern. Chalcones and related compounds are found to be generally toxic to cells (Sukumaran and Kuttan 1991). *Christella arida*, *Christella dentata*, *Cyclosorus interruptus*, *Microsorum punctatum*, *Nephrolepis acutifolia*, and *Pleocnemia irregularis* were tested for its cytotoxicity against K562 cell. The water extracts *of C. dentata*, *N. acutifolia*, and *P. irregularis* were strong cytotoxic agents (Chai et al. 2015).

A study was done to investigate the possible antiproliferative activity of ethanolic extract and the organic fractions from *P. commune* on murine leukemia L1210 cells. The obtained results showed that the cell viability of L1210 cells was reduced by ethanolic extract of *P. commune* in a concentration-dependent manner, and the IC50 value was about 77.22 μg/ml at 24 h posttreatment. The ethyl acetate fraction displayed higher anti-tumor effect than that of chloroform and butanol fractions with 32.29 μg/ml (IC50 value, 48 h). Microscopy studies revealed that ethanolic extract and ethyl acetate fraction treated cells showed morphological characteristics of apoptosis such as chromatin condensation and DNA aggregation. Further, Annexin V-PE/7-AAD double staining showed the leaflet of phosphatidylserine, and the decline of mitochondrial membrane potential by flow cytometry confirmed that the extracts do, in fact, induce apoptosis in L1210 cells (Xiaoxia et al. 2012). The cytotoxicity of the epiphytic fern, *Acrostichum heterophyllum* L., was investigated by George and Josekumar (2016). Preliminary scientific evaluation toward the possibility of this species as a potential bioactive agent is attempted. Phytochemical screening revealed the presence of flavonoids, alkaloids, sugars, glycosides, phenolics, and tannins as major components of ethyl acetate and methanol extracts. L929 cell line showed high $IC_{50}$ values.

## 22.5 Conclusion

The anticancer activity of experimental plants revealed that the extracts are effective against HT-29 colon cancer cells, and they do no harm to L929 cells. The inhibition concentration (IC50) value of the ethanolic extract of *A. evecta* treated HT-29 cancer cells at 24 h was 15.94 μg/ml and of L929 cells is 81.69 μg/ml. At minimum concentration, the ethanolic extract of *A. evecta* (G. Forst.) Hoffm. showed maximum effect. Hence the antiproliferative study through MTT assay, staining techniques, and DNA studies related to cell cycle through flow cytometry revealed the efficiency of *A. evecta* ethanolic extract as a powerful anticancerous agent for colon cancer cell lines HT-29. As a continuation of this study, industrial development of pharmaceutical and nutraceutical products based on secondary plant metabolites could be initiated. Hence biosynthesis of secondary metabolites especially flavonoids and terpenoids could be promoted in future. Mechanism of cancer preventing antiangiogenesis and anticancer action pathways could be studied thoroughly, and proteins related to the same could be produced through biotechnological process for plant metabolites. Antiangiogenic therapy could be explored on the experiment plants in the future, so that the cancer menace will be minimized and brought into control by plant bioactive compounds.

## References


Arung ET, Shimzu K, Kondu R (2000) Inhibitory effect of isoprenoid substituted flavanoids isolated from *Atrocarpus heterophyllus* on melanin biosynthesis. Planta Med 72:847–850

Chai TT, Yeoh LY, Ismail NIM et al (2015) Evaluation of glucosidase inhibitory and cytotoxic potential of five selected edible and medicinal ferns. Trop J Pharm Res 14(3):449–454

Defilpps RA, Maina SL, Pray LA (1998) The Palauan and yap medicinal plant studies of Masayoshi Okabe. In: The National Museum of Naturals. Smithsonian Institution, Washington, DC, pp 1941–1943

George, Meera, Josekumar VS (2016) *In vitro* cytotoxicity screening, phytochemical profile and heavy metal analysis of different extracts of *Acrostichum heterophyllum*. L. Indian J Nat Prod Resour 7(1):1–4

Hseu TH (1981) Structure of Angiopteroside (4-O-B-D-Glucopyranosyl-L-Thero-2- Hexane-5-Olide) mono-hydrate, a fern glycoside from *Angiopteris lygodiifolia*. Ros Acta Cryst B37: 2095–2098

Khan MR, Omoloso AD (2008) Antibacterial and antifungal activities of *Angiopteris evecta*. Fitoterapia 79(5):366–369

Khullar SP (2008) Diversity of microbes and cryptogams. Pteridophyta. Punjab University, Chandigarh, pp 1–62

Martineti V, Tugnarini J, Carbonell SS et al (2010) Inhibition of *in vitro* growth and arrest in the Go/G1 Phase of Hct8 line human colon cancer cells by Kaempferide Triglycoside from *Dianthus caryophyllus*. Phyto Ther Res 24(9):1302–1308

Mehta RG, Pezzuto JM (2002) Discovery of cancer preventive agents from natural products: from plants to prevention. Curr Oncol Rep 4:478–486

Ruby GD (2018) Phytochemical, Pharmacological and i*n vitro* clinical studies of *Angiopteris Evecta* (G. Forst.) Hoffm. and *Pyrrosia Lanceolata* (L.) Farwell. Ph.D Thesis, Holy Cross College, Bharathidasan University, Tiruchirappalli

Saunders FR, Wallace HM (2010) On the natural chemoprevention of cancer. Plant Physiol Biochem 48:621–626

Shimizu K, Kinouchi Shimizu N, Hakamata W et al (2010) Preventive effect of green tea catechins on experimental tumor metastasis in senescence accelerated mice. Biol Pharm Bull 33:117–121

Shoeb M, MacManus SM, Jaspars M, Trevidu J, Sarker SD (2006) Montamine, a unique dimeric indole alkaloid, from the seeds of Centaurea montana (Asteraceae), and its in vitro cytotoxic activity against the CaCo2 colon cancer cells. Tetrahedron 62(48):11172–11177

Sukumaran K, Kuttan R (1991) Screening of 11 ferns for cytotoxic and antitumor potential with special reference to *Pityrogramma calomelanos*. J Ethnopharmacol 34(1):93–96

Tomsik T (2014) Ferns and Lycopods--a potential treasury of anticancer agents but also a carcinogenic hazard. P Phyto Ther Res 28(6):798–810

Uddin MG, Mirza MM, Pasha MK (1998) The medicinal uses of Pteridophytes of Bangladesh. Bangladesh J Plant Taxon 5(2):29–41

Vasudeva SM (1999) Economic importance of Pteridophytes. Indian Fern J 16(1):2130–2152

Veronesi U, Bonanni B (2005) Chemoprevention: from research to clinical oncology. Eur J Cancer 41:1833–1841

Xiaoxia C, Xiao Y, Wang XB et al (2012) Anti-tumor and pro-apoptotic activity of ethanolic extract and its various fractions from *Polytrichum commune* L. ex Hedw in L1210 cells. J Ethnopharmacol 143(1):49–56

Zhang J, Cado D, Chen A et al (1998) Fas-mediated Apoptosis and activation-induced T-Cell Proliferation are defective in mice lacking Fadd/Mort1. Nature 392(66–73):296–300

**S. Catharin Sara** is currently working as an Assistant Professor in Botany, Holy Cross College (Autonomous), Trichy, Tamil Nadu, India. She holds a Ph.D. degree in Botany (Manonmaniam Sundaranar University) and has distinguished himself as a fern tissue culturist, fern phytochemist, and entrepreneur reaching village women folk. She has more than 16 years of teaching experience using innovative teaching methodologies and designing curriculum. She is a recipient of "Young Scientist Award" (2008) and " Outstanding Women in Science" award (2020).

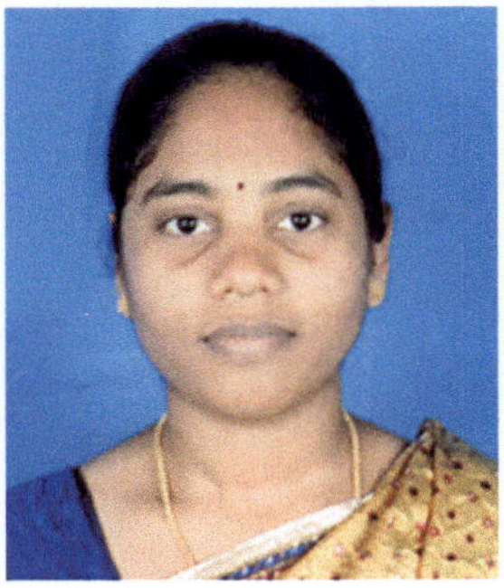

**R. Gnana Deepa Ruby** has completed her Doctoral Degree in PG and Research Department of Botany, Holy Cross College (Autonomous), Tiruchirappalli, Tamil Nadu, India. She worked as a Junior Research Fellow for UGC sanctioned project. Her area of specialization is fern phytochemistry and tissue culture. She has 2 years of teaching experience and is a recipient of "*National Children's Science Congress*" award by NCSTC, Department of Science and Technology, and Tamil Nadu Science Forum.

# *Adiantum* L.: Overview on Taxonomy, Distribution, Conservation Status, Ethnomedicinal Uses, Phytochemistry, and Pharmacognosy from India

23

Sachin Patil, Rajendra Lavate, Vinay Raole, and Kishore Rajput

**Abstract**

A perennial herb, the genus *Adiantum* (Adiantaceae) is an essential source of folk medicines, primarily used against bronchitis, cough, dysentery, dog bite, fever, leprosy, malarial fever, smallpox, and snakebite and as stomachic and diuretic. It is a rich source of triterpenoids, flavonoids, phenylpropanoids, steroids, alicyclic acids, lipids, and long-chain fatty acid compounds. The crude extract and isolated and purified compounds from the genus *Adiantum* have been verified to possess multiple pharmacological activities such as analgesic, antibacterial, anticancer, antifungal, antidiabetic, and antipyretics. Therefore, the current review aims to present comprehensive information on the taxonomy, diversity, distribution, ethnic information, and conservation status of the genus *Adiantum*, and the important findings on phytochemistry and bioactivity of the secondary metabolites present in *Adiantum* have been highlighted.

**Keywords**

Adiantaceae · Diversity · Conservation status · Phytochemistry · Pharmacognosy · Secondary metabolites

S. Patil
Department of Botany, Shivaji University, Kolhapur, Maharashtra, India

Department of Botany, Faculty of Science, The Maharaja Sayajirao University of Baroda, Vadodara, Gujarat, India

R. Lavate
Department of Botany, RajeRamrao College, Jath, Maharashtra, India

V. Raole · K. Rajput (✉)
Department of Botany, Shivaji University, Kolhapur, Maharashtra, India





J. Marimuthu et al. (eds.), *Ferns*, https://doi.org/10.1007/978-981-16-6170-9_23

## 23.1 Introduction

Knowledge appropriation is by no means a new phenomenon. Furthermore, in the broadest sense, one can argue it is a necessary aspect of the successful adaptation of past and present, as this knowledge relates to biological and cultural heritage, referred to as "collective biocultural heritage." Different groups of plants have played a pivotal role in all living organisms' lives as a source provider for basic and essential needs since time immemorial (UNESCO 2018). Even in today's era of science and technology, advanced medical science looks toward plants as a source for biomolecule that may help cure chronic or incurable diseases. Therefore, medicinal plants are the basis of remedies to cure various diseases since the earliest times, and people from all over the world have experienced the use of plant medicines. Despite the noteworthy development in synthetic drugs in the twentieth century, over 25% of permitted medications in developed countries are gained directly or indirectly from plants (Newman and Cragg 2012). Hence, information on medicinal plants may play a significant role in its potential role in drug development and people's healthcare. The use of plant-derived medicines is ubiquitous and collectively crucial in both traditional and modern medicine. About 80% of the world population in emerging countries depended on plant-based medicines for their primary healthcare needs. The medicinal pteridophytes may be an alternative source against synthetic drugs to cure various diseases with no or fewer side effects (Canter et al. 2005). Plant extracts are having numerous secondary metabolites and are known to possess antioxidant and antimicrobial properties. Due to their potential use for treating various chronic and infectious diseases in the recent decade, the search for phytochemicals with antioxidant, antimicrobial, or anti-inflammatory properties has been rising (Proestos et al. 2005).

The genus *Adiantum* is popularly known as *maidenhair fern*, an isolated, well-defined and old genus that belongs to the family Adiantaceae distributed from temperate to tropical regions of the world (Yumkham et al. 2018). It is well-known for its medicinal values and is traditionally used to cure various illnesses since long back. About 232 species are reported worldwide (Prado et al. 2007), of which 22 species and 6 subspecies are known to occur in India (Fraser-Jenkins et al. 2017). Among these, many of them are used as traditional medicines by *Vaidus*, local people, hakims, and traditional practitioners (Rastogi et al. 2018). The crude extract of the whole plant has been used in conventional medicine for the treatment of diarrhea, diabetes, malarial fever, urination problems, cough, asthma, bronchitis, hemorrhage, fractures, snakebite, wound, and burns (Samoisy and Mahomoodally 2016). The present comprehensive review throws light on the taxonomic importance of *Adiantum* species along with its diversity, phytogeographic distribution, conservation status, and pharmacological and phytochemical studies of species occurring in India.

## 23.2 Taxonomy Value

The genus *Adiantum* is described by Linnaeus (1753) and included 15 species under this name. The scientific name *Adiantum* is based on the Greek word *adiantos*, meaning *unwetted,* referring to the water-repelling properties of the surface of the fronds. The majority of *Adiantum* species are distributed in tropical, subtropical, and temperate regions of the world. Some species are reported to occur in xeric conditions, whereas no species are recorded from freezing climates. Every species shares a unique and distinctive morphological character. They are perennial, terricolous, saxicolous, or corticolous herb, having erect, short, or long creeping rhizome, unipinnate, bipinnate, tripinnate or tetrapinnate fronds, reniform, deltoid, or ovate pinnae with incised margins, sori borne along the margins in continuous or interrupted lines having various numbers per pinnae, i.e., from 5 to 15 sori per pinnae (Figs. 23.1 and 23.2). Based on morphological features, Nayar (1961) has categorized the genus into four groups, i.e., (1) *capillus-veneris* group, (2) *caudatum* group, (3) *pedatum* group, and (4) *tenerum* group.

However, many of them are either merged or excluded species, and few are added to the flora of India (Patil and Dongare 2013; Fraser-Jenkins et al. 2017). At present, based on the shape of rhizome, type of fronds, pinnae type, and sori number, the genus *Adiantum* is categorized into five complexes, viz., (1) *capillus-veneris* complex (includes the species with short-long, creeping rhizome, bi- or more pinnate, glabrous without proliferating fronds having less than ten sori per pinnae), (2) *caudatum* complex (includes the species with erect or suberect, rhizome, simple pinnate, glabrous or hairy with proliferating frond tip, pinnae deeply incised, having less than ten sori per pinnae), (3) *tenerum* complex (includes the species with erect rhizome, bi- or more pinnate, glabrous fronds without proliferating tip, pinnae incised, having more than ten sori per pinnae), (4) *pedatum* complex (rhizome erect rhizome, glabrous, two to three dichotomously branched fronds without proliferating frond tip, pinnae incised or entire, more than ten sori per pinnae), and, (5) *philippense* complex (rhizome erect, simple fronds with proliferating tip, pinnae incised but not profoundly incised) or entire, sori less than seven per pinnae). The species included in the complexes are given in Table 23.1 and Fig. 23.3.

## 23.3 Diversity and Geographic Distribution

India is one of the twelve mega biodiversity countries of the world, including two biodiversity hot spots (North-East and the Western Ghats), which harbor the majority of the Earth's species (Balasubramanian 2017). On the basis of the political map, Indian territory is divided into twenty eight state and nine union territories, while based on the geography, ecology, and plant and animal distribution, India territory is classified into ten different biogeographic areas, viz., Transhimalaya, Himalaya, North-East, Gangetic planes, Semi-Arid Desert, Deccan Peninsula, Western Ghats, Indian Coasts, and Andaman and Nicobar Islands (Singh and Chaturvedi 2017; Patil and Rajput 2017). In the present study, the authors reviewed the

**Fig. 23.1** (**a–c**) Range of variations in the rhizome (erect, suberect, and creeping, respectively); (**d–i**) range of variations in pinnae (unipinnate, fan-shaped, bipinnate, tripinnate, respectively)

distribution of the genus *Adiantum* from different biogeographic areas of India. The study revealed that maximum diversity was observed in the Himalayas, North-East and the Western Ghats. In contrast, fewer diversity species was found in the desert area (03), and other areas are having five to seven species (Map 23.1). The maximum diversity of *Adiantum* species was observed in the Western Ghats (18), Deccan Peninsula (16), and Himalayas (15), followed by Gangetic Planes (13), North-East

**Fig. 23.2** (**a–i**) Range of variation in pinnae, sori arrangement, and sori number

(13), Transhimalaya (08), and Andaman Nicobar Island (06), whereas the minimum was found in desert region (02). Most common species are *Adiantum capillus-veneris*, *A. caudatum*, *A. incisum* subsp. *incisum*, and *A. philippense*, whereas *A. venustum* and *A. wattii* restricted to Gangetic planes, North-East, and Himalaya regions. The *Adiantum capillus-veneris* and *A. philippense* were found in every biogeographic region (Map 23.1).

**Table 23.1** Genus divided into following complexes

| | |
|---|---|
| *Capillus-veneris* complex | *Adiantum capillus-veneris*, *A. poiretii*, *A. raddianum*, *A. stenochlamys*, *A. venustum*, and *A. wattii* |
| *Caudatum* complex | *Adiantum caudatum*, *A. edgeworthii*, *A. incisum*, and *A. zollingeri* |
| *Tenerum* complex | *Adiantum concinnum*, *A. formosum*, *A. latifolium*, *A. peruvianum*, *A. tenerum*, and *A. tibeticum* |
| *Pedatum* complex | *Adiantum flabellulatum*, *A. hispidulum*, *A. myriosorum*, *A. pedatum*, and *A. trapeziforme* |
| *Philippense* complex | *Adiantum philippense* and *A. soboliferum* |

## 23.4 Conservation Status

The conservation status of the *Adiantum* species is provided in Table 23.2 according to IUCN Red List category. The IUCN Red List categories (2020) define the extinction risk of species assessed. To assess the status of species, IUCN had given nine categories, viz., NE (Not Evaluated), Data Deficient (DD), Least Concerned (LC), Near Threatened (NT), Vulnerable (VU), Endangered (EN), Critically Endangered (CR), Extinct in Wild (EW), and Extinct (EX). These can be evaluated using five quantitative criteria, viz., population size reduction (past, present, or projected), geographic range size, small and declining population size, tiny population or very restricted distribution, and quantitative analysis of extinction risk. The conservation status of many species is yet to be investigated.

## 23.5 Ethnomedicinal Value

Different medicinal systems, viz., Ayurveda, Charak Samhita, Chinese Medicinal Herbs, United States Pharmacopoeia, Encyclopedia of Herbs and Herbalism, and American Folk Medicine, have provided the ethnomedicinal value of different species of *Adiantum* (Yumkham et al. 2018; Patil et al. 2020) (Table 23.3). Similarly, various institutes of the world, viz., African Traditional Medicine, Middle Eastern Traditional Medicine, Australian Traditional Medicine, and Herbal Medicine Research Centre (HMRC) in Malaysia, have documented several species of *Adiantum* that are used as a medicine (Neffati et al. 2017; HMRC 2002; Yumkham et al. 2018).

As per Indian medicinal systems (*Ayurveda*, *Charaka Samhita*, *Sushruta's Compendium*, and *Siddha*), the genus *Adiantum* L. known as hansraj/hanmspadi (Adiantaceae) has been used in various forms such as powder, decoction, paste, and other forms to cure multiple ailments, viz., cough cold, fever, pneumonia, and mucous formation (Kirtikar and Basu 2005; Singh et al. 2008; Singh et al. 2013). Its different species are also used against insect bite, snakebite, scorpion bite, and dog bite (Rai et al. 2016; Rastogi et al. 2018; Yumkham et al. 2018). It is antidysenteric,

**Fig. 23.3** (**a**) *Adiantum capillus-veneris*, (**b**) *Adiantum caudatum*, (**c**) *Adiantum concinnum*, (**d**) *Adiantum hispidulum*, (**e**) *Adiantum incisum*, (**f**) *Adiantum pedatum*, (**g**) *Adiantum philippense*, (**h**) *Adiantum poiretii*, (**i**) *Adiantum raddianum*

anti-hypoglycemic, antimicrobial, antitumor, antiulcer, antiviral, demulcent, diuretic, emmenagogue, expectorant, and tonic and also used as astringent (Reddy et al. 2001; Patil et al. 2012; Patil et al. 2013b). Though many species were used as traditional medicines, however, ethnomedicinal value of many species is yet to be investigated from India.

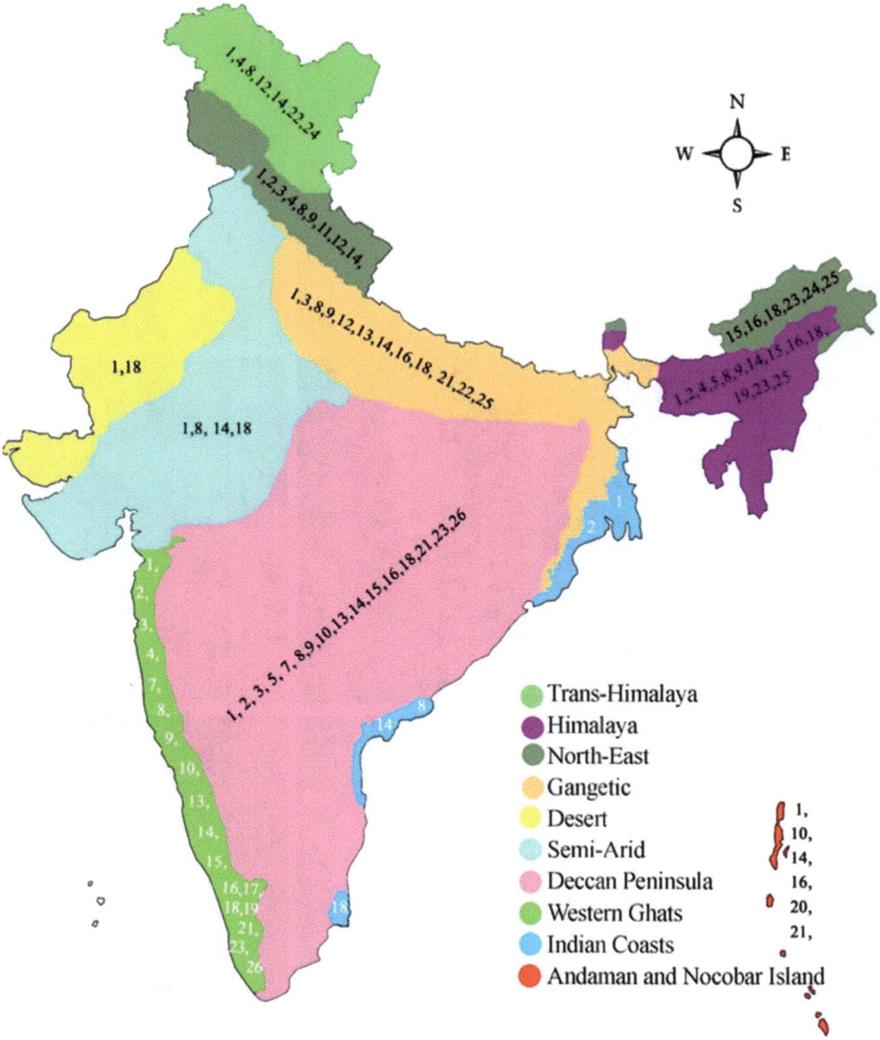


**Map 23.1** The biogeographic distribution of *Adiantum* species in India. Numbers indicate the name of species as per Table 23.2

## 23.6 Phytochemistry

According to available literature, the number of the species under the genus *Adiantum* is documented as medicinal herbs. They have been used as a traditional medicine to cure various diseases, viz., asthma, burns and scald, diabetes, diarrhea, fever, fractures, gynecology disorders, influenza, jaundice, kidney problems, pneumonia, cough, stoppage of bleeding, snakebite, and urinary tract infection, and also,

**Table 23.2** IUCN red list category of different species of *Adiantum*

| Sr. no. | Name of species | Conservation status |
|---|---|---|
| 1 | *Adiantum capillus-veneris* Linn | Least Concern |
| 2 | *A. caudatum* L. | Least Concern |
| 3 | *A. concinnum* Humb. &Bonpl. ex Willd | Data Deficient |
| 4 | *A. edgeworthii* Hook. | Data Deficient |
| 5 | *A. flabellulatum* L. | Data Deficient |
| 6 | *A. formosum* R.Br. | Data Deficient |
| 7 | *A. hispidulum* Sw. | Data Deficient |
| 8 | *A. incisum* Forssk. subsp. *incisum* | Least Concern |
| 9 | *A. incisum* subsp. *indicum* (J.Ghatak) Fraser-Jenk. | Data Deficient |
| 10 | *A. latifolium* Lam. | Data Deficient |
| 11 | *A. myriosorum* Baker | Data Deficient |
| 12 | *A. pedatum* L. subsp. *pedatum* | Data Deficient |
| 13 | *A. peruvianum* Klotzsch | Data Deficient |
| 14 | *A. philippense* L. subsp. *philippense* | Least Concerned |
| 15 | *A. philippense* L. subsp. *intermedium* S.C.Verma& Fraser-Jenk. | Endemic /Data Deficient |
| 16 | *A. philippense* L. subsp.*teestae* S.C.Verma & Fraser-Jenk. | Endemic/Data Deficient |
| 17 | *A. poiretii* Wikstr | Data Deficient |
| 18 | *A. raddianum* C.Presl. | Least Concerned |
| 19 | *A. soboliferum* Wall. ex Hook. | Data Deficient |
| 20 | *A. stenochlamys* Baker | Data Deficient |
| 21 | *A. tenerum* Sw. | Data Deficient |
| 22 | *A. tibeticum* Ching | Data Deficient |
| 23 | *A. trapeziforme* L | Data Deficient |
| 24 | *A. venustum* D.Don | Data Deficient |
| 25 | *A. wattii* Baker | Data Deficient |
| 26 | *A. zollingeri* Mett. ex Kuhn | Data Deficient |

decoction is used as blood coagulating agent, tonic, expectorant, and astringent. On the ethnic information, phytochemist isolated different triterpenes, flavonoids, phenylpropanoids, and sterols from the other species of *Adiantum* (Brahmachari et al. 2003; Pan et al. 2011; Mengane 2015). The different chemical groups, also referred to as secondary metabolites or phytoactive compounds (viz., terpenoids, flavonoids, phenylpropanoids, steroids, alicyclic acids, lipids, and long-chain compounds), have been isolated from different species of the genus (Table 23.4).

Triterpenoids and flavonoids are the dominant constituents within the genus *Adiantum* (Pan et al. 2011; Yumkham et al. 2018; Rastogi et al. 2018). The phytochemical investigations have revealed that triterpenoids belonging to the hopane, neohopane, norhopane, fernane, adiane, and filicane series form the major phytochemical groups to which the compounds isolated from different *Adiantum* species. The basic triterpenoid skeletons are commonly found in *Adiantum* species.

**Table 23.3** Ethnomedicinal value of the genus *Adiantum* L., from India

| Name of species | Plant part used | Ethnomedicinal values | References |
|---|---|---|---|
| *Adiantum capillus-veneris* Linn(CN: Venus hair fern, Hansraj, Hansapadi) | Whole plant | 1. Decoction gives relief against asthma, cough and bronchitis, fever 2. Rhizome powder is used for smallpox, boils, eczema, diabetes, menstrual problems, and snakebite 3. It has antifungal and antiviral properties 4. Dried powder is taken as herbal tea in Arunachal Pradesh, Manipur, and Nagaland | Guha et al. (2004), Ansari and Ekhlasi Kazaj (2012), Yumkham et al. (2018), Patil et al. (2020) |
| *A. caudatum* L.(CN: Hansraj, walking or tailed maidenhair fern) | Rhizome and fronds (without sori) | 1. Decoction is against cough and bronchitis, fever 2. It has microbial properties 3. Rhizome is to cure diabetes and skin diseases | Benniamin (2011), Tsuzuki et al. (2001), Patil et al. (2013b), Saha et al. (2016), Yumkham et al. (2018), Patil et al. (2020) |
| *A. concinnum* Humb. &Bonpl. ex Willd(CN: Maidenhair fern, lady fern) | Rhizome and frond | Introduced species but escaped and now found in the wild also. The extract of rhizome and fronds is used to relieve stomachache and headache | Patil et al. (2013a, 2020) |
| *A. edgeworthii* Hook. (CN: Himalayan maidenhair or Hansraj) | Rhizome and fronds (without sori) | 1. Rhizome is used as an antifungal 2. Used to cure cough and respiratory diseases 3. It has anthelminthic 4. Frond extract used in deworming | Yumkham et al. (2018) |
| *A. flabellulatum* L.(CN: Fan leaved maidenhair, Mayurshikha) | Whole plants | 1. Frond has analgesic, anthelmintic, antifungal properties 2. Used to cure cough, tumor, swelling, insect bites, and external pain | Shahriar and Kabir (2011), Rastogi et al. (2018), Yumkham et al. (2018) |
| *A. formosum* R.Br.(CN: Giant maidenhair or black stem maidenhair) | Yet to know | Ethnomedicinal value of this species is yet to be investigated | – |
| *A. hispidulum* Sw. | Whole plant | They are used in hair loss and skin diseases | Samoisy and Mahomoodally (2016), Rastogi et al. (2018), Yumkham et al. (2018) |

(continued)

**Table 23.3** (continued)

| Name of species | Plant part used | Ethnomedicinal values | References |
|---|---|---|---|
| *A. incisum* Forssk. subsp. *incisum*(CN: Mayurshikha, Hansraj, Biddapata) | Rhizome and young fronds | 1. Used in skin diseases, fever, cough, diabetes, malaria, bronchial disease, bone fracture 2. Used as diuretic, astringent, and tonic 3. Used in the treatment of headaches, snakebite and scorpion bites 4. In Manipur, the frond powder is used as herbal tea to control diabetes | Rai et al. (2016), Rastogi et al. 2018; Yumkham et al. (2018), Patil et al. (2020) |
| *A. incisum* subsp. *indicum* (J.Ghatak) Fraser-Jenk. | Yet to know | Ethnomedicinal value of this species is yet to be investigated | – |
| *A. latifolium* lam. | Fronds | Decoction stops the necrosis of tissue by snakebite | Rastogi et al. (2018) |
| *A. myriosorum* Baker | Yet to know | Ethnomedicinal value of this species is yet to be investigated | – |
| *A. pedatum* L. subsp. *pedatum*CN: Hansraj, Mayurshikha, fan leaved maidenhair fern | Whole plants | 1. Used in female disorders, chest and stomach trouble, cold, cough, chronic disorders 2. Decoction used as blood coagulant agent, tonic, expectorant, astringent | Rastogi et al. (2018) |
| *A. peruvianum* Klotzsch (CN: Silver dollar fern) | Whole plant | It has an antibacterial and antifungal properties | Yumkham et al. (2018) |
| *A. philippense* L. subsp. *philippense*(CN: Hansraj, Mayurshikha, walking maidenhair fern) | Rhizome and young fronds | 1. Rhizome powder is used in stone cases in kidneys, chest pain, bone fractures, and elephantiasis 2. Decoction is used for cough, asthma, and jaundice 3. It is also used in a dog bite, snakebite 4. Paste of fronds used for hair fall | Pallavi et al. (2011), Patil et al. (2013b), Rastogi et al. (2018), Yumkham et al. (2018), Patil et al. (2020) |
| *A. philippense* L. subsp. *intermedium* S.C.Verma & Fraser-Jenk. | Yet to know | Ethnomedicinal value of this species is yet to be investigated | – |
| *A. philippense* L. subsp. *teestae* S.C.Verma & Fraser-Jenk. | Yet to know | Ethnomedicinal value of this species is yet to be investigated | – |

(continued)

**Table 23.3** (continued)

| Name of species | Plant part used | Ethnomedicinal values | References |
|---|---|---|---|
| *A. poiretii* Wikstr | Rhizome and young plant part | 1. Aqueous infusion of rhizome is used to cure asthma, diabetes, influenza, pneumonia, kidney problems<br>2. Expectorant for cough, chest diseases, wounds of diuretic disorder | Rastogi et al. (2018), Yumkham et al. (2018), Patil et al. (2020) |
| *A. raddianum* C. Presl (CN: Hansraj, Mayurshikha, delta maidenhair fern) | Rhizome | The decoction is used as a health tonic | Yumkham et al. (2018), Patil et al. (2020) |
| *A. soboliferum* Wall. ex Hook. | Yet to know | Ethnomedicinal value of this species is yet to be investigated | – |
| *A. stenochlamys* Baker | Yet to know | Ethnomedicinal value of this species is yet to be investigate | – |
| *A. tenerum* Sw.(CN: Hansraj, Mayurshikha, Delta maidenhair fern) | Whole plant | Used as expectorant, sudorific, and demulcent<br>2. It is also used to cure gynecology disorders | Patil and Dongare (2013), Rastogi et al. (2018) |
| *A. tibeticum* Ching | Unknown | Unknown | – |
| *A. trapeziforme* L. (CN: Hansraj, Mayurshikha) | Whole plant | The plant has antibacterial property | Vasudeva (1999) |
| *A. venustum* D. Don (CN: Hansraj, Damtuli Himalayan maidenhair fern) | Rhizome and frond | 1. Decoction from the rhizome and leaves can be used to cure or treat fever, diabetes, liver problems, renal and gall bladder stone, and also scorpion bites 2. Paste made from rhizomes is used to treat cuts and wounds | Kholia and Punetha 2005, Kumari et al. 2011, Joseph and Thomas (2015) |
| *A. wattii* baker | Yet to know | Ethnomedicinal value of this species is yet to be investigated | – |
| *A. zollingeri* Mett. ex Kuhn | Yet to know | Ethnomedicinal value of this species is yet to be investigated | – |

Hopanol, neohopene, hydroxyadiantone, isoadiantone, adiantone, fernene, 7-fernene, adiantol, filicene, and ketohakonanol are some of the important triterpenoids present in *Adiantum* species (Table 23.2). Flavonoids form the second major group of compounds that have been isolated from *Adiantum* species (Table 23.2). Glycosides of kaempferol and quercetin have been abundantly

**Table 23.4** Compounds isolated from the *Adiantum* species

| Name of compounds | *Isolated* from *the species* | Source |
|---|---|---|
| **Hopane-type triterpenoids** | | |
| Hop-22(29)-ene (= Diploptene) | *A. capillus-veneris*, *A. edgeworthii* | Pan et al. (2011), Kumar et al. (2014), Rastogi et al. (2018), Yumkham et al. (2018) |
| 17b,21b-Epoxyhopane | *A. caudatum* | |
| Adininaneone | *A. incisum* | |
| Hydroxyhopane (= Hopanol) | *A. capillus-veneris*, *A. edgeworthii* | |
| Mollugogenol | *A. philippense* | |
| 6a-Acetoxy-16b,22-dihydroxy-3-ketoisohopane | *A. capillus-veneris*, *A. philippense* | |
| 17,29-Epoxyhopane | *A. capillus-veneris* | |
| Hopan-28,22-olide | *A. capillus-veneris* | |
| **Isohopane- and neohopane-type triterpenoids** | | |
| 3b-Acetoxy-6a-hydroxy-hop-15,17(21)-diene | *A. philippense* | Pan et al. (2011), Rastogi et al. (2018), Yumkham et al. (2018) |
| 3b-Acetoxy-21aH-hop-22 (29)-ene | *A. philippense* | |
| Neohop-12-ene (= Neohopene) | *A. capillus-veneris*, *A. edgeworthii*, *A. pedatum* | |
| Neohop-13(18)-ene | *A. caudatum*, *A. pedatum* | |
| Neohopa-11,13(18)-diene | *A. pedatum* | |
| **Norhopane-type triterpenoids** | | |
| Trisnorhopane | *A. capillus-veneris* | Pan et al. (2011), Kumar et al. (2014), Rastogi et al. (2018), Yumkham et al. (2018) |
| Isoglaucanone(= 17aH-Trisnorhopan-21-one) | *A. capillus-veneris*, *A. pedatum* | |
| Glaucanol A | *A. pedatum* | |
| 21-Hydroxy-30-norhopan-22-one(= 21-Hydroxyadiantone, Hydroxyadiantone) | *A. capillus-veneris*, *A. pedatum*, *A. venustum* | |
| Isoadiantone | *A. capillus-veneris*, *A. pedatum* | |
| Adiantone | *A. capillus-veneris*, *A. pedatum*, *A. philippense*, *A. venustum* | |
| 19a-Hydroxyadiantone | *A. caudatum*, *A. edgeworthii* | |
| 29-Norhopan-22-ol | *A. edgeworthii* | |
| Isoadiantol B [= (22S)-30-Norisohopan-22-ol, Isoadiantol] | *A. capillus-veneris*, *A. pedatum* | |
| 22,29x-Epoxy-30-norhopane-13-ol | *A. philippense* | |
| Adipedatol | *A. pedatum* | |
| Adipedatol Me ether | *A. pedatum* | |

(continued)

**Table 23.4** (continued)

| Name of compounds | *Isolated* from *the species* | Source |
|---|---|---|
| **Fernane-type triterpenoids** | | |
| Fern-9(11)-en-28-ol | *A. capillus-veneris*, *A. caudatum*, *A. philippense* | Pan et al. (2011), Kumar et al. (2014), Rastogi et al. (2018), Yumkham et al. (2018) |
| Fern-9(11)-en-25-oic acid | *A. edgeworthii*, *A. philippense*, *A. venustum* | |
| Fern-9(11)-en-6a-ol | *A. philippense* | |
| Fern-9(11)-en-3a-ol | *A. capillus-veneris* | |
| Fern-9(11)-en-12β-ol | *A. capillus-veneris* | |
| Fern-9(11)-en-12-one | *A. capillus-veneris* | |
| Fern-9(11)-ene (= Fernene, Davallene) | *A. capillus-veneris*, *A. edgeworthii*, *A. caudatum*, *A. pedatum*, *A. philippense* | |
| 19a-Hydroxyfern-9(11)-ene | *A. caudatum* | |
| 23-Hydroxyfernene | *A. caudatum*, *A. pedatum* | |
| Fern-7-en-3α-ol | *A. capillus-veneris* | |
| Fern-7-ene = 7-fernene | *A. capillus-veneris*, *A. edgeworthii*, *A. caudatum*, *A. pedatum* | |
| 19a-Hydroxyfern-7-ene | *A. caudatum* | |
| Ferna-7,9(11)-diene | *A. capillus-veneris*, *A. caudatum*, *A. pedatum* | |
| 19a-Hydroxyferna-7,9(11)-diene | *A. caudatum* | |
| Fern-8-ene | *A. caudatum*, *A. pedatum* | |
| 8a-Hydroxyfernan-25,7b-olide | *A. caudatum* | |
| **Adiane- and filicane-type triterpenoids** | | |
| Adian-5-en-3a-ol (=Adiantol) | *A. capillus-veneris* | Pan et al. (2011), Kumar et al. (2014), Rastogi et al. (2018), Yumkham et al. (2018) |
| Adian-5-ene (=Adianene) | *A. capillus-veneris* | |
| Adian-5(10)-en-3a-ol | *A. capillus-veneris* | |
| Filic-3-ene (= 3-Filicene, Filicene) | *A. capillus-veneris*, *A. edgeworthii*, *A. caudatum*, *A. pedatum*, *A. venustum* | |
| Filicenol B | *A. pedatum* | |
| Filicenal | *A. pedatum* | |
| Filicenoic acid | *A. pedatum* | |
| 4a-Hydroxyfilican-3-one | *A. capillus-veneris* | |
| 3a-Hydroxy-4a-methoxyfilicane | *A. caudatum* | |
| Adiantoxide = 3a,4a-Epoxyfilicane | *A. capillus-veneris* | |

reported. Several compounds belonging to the phenolic and phenylpropanoid groups have also been identified (Table 23.2). Sulfate esters of hydroxycinnamic acid sugar derivatives have also been reported to be present (Imperato 1982). A few coumarins and phytosterols have also been reported. However, phytochemical analysis of many *Adiantum* species is yet to be investigated.

## 23.7 Conclusion

Different species of *Adiantum* are medicinally essential plants and are used for long. However, several species are yet to be investigated using modern tools. It is observed that studies on ethnomedicinal values, conservation status, and phytochemistry of several species are lacking, and also species that are used in folk medicine should be scientifically validated. Details of sociocultural and phytochemical investigation are yet to be warranted in this genus. Furthermore, it has great scope to study the anatomy, micromorphology, and molecular and phylogenetic aspects of the genus *Adiantum*.

**Acknowledgments** The authors are thankful to the Science and Engineering Research Board (SERB) for financial work under NPDF program.

## References

Ansari R, Ekhlasi Kazaj K (2012). *Adiantum capillus–veneris* L.: phytochemical constituents, traditional uses and pharmacological properties: a review. J Adv Sci Res 3(4):15–20

Balasubramanian A (2017) Biodiversity profile of India. In: Technical Report submitted to Centre for Advanced Studies in Earth Science. University of Mysore, Mysore, p 11. https://doi.org/10.13140/RG.2.2.10664.57601

Benniamin A (2011) Medicinal ferns of North Eastern India with special reference to Arunachal Pradesh. Indian J Tradit Knowl 10(3):516–522

Brahmachari G, Mondel S, Chatterjee D, Brahmachari K (2003) Phytochemicals and biological activities of Adiantum species. J Sci Ind Res 62(12):1131–1137

Canter PH, Thomas H, Ernst E (2005) Bringing medicinal plants into cultivation: opportunities and challenges for biotechnology. Trends Biotechnol 23:180–185

Fraser-Jenkins CR, Gandhi KN, Kholia BS et al (2017) An annotated checklist of Indian Pteridophytes Part-1 (Lycopodiaceae to Thelypteridaceae). Bishen Singh Mahendra Pal Singh, Dehradun, p 562

Guha P, Mukhopadhyay R, Pal PK et al (2004) Antimicrobial activity of crude extracts and extracted phenols from gametophyte and sporophytic plant parts of *Adiantum capillus-veneris* L. Allelopath J 1:57–66

Herbal Medicine Research Centre (2002) Compendium of medicinal plants used in Malaysia. Institute for Medical Research, Kuala Lumpur

Imperato F (1982) kaempferol 3-sulphate in the fern adiantum capillus-veneris. Phytochemistry 21 (1):2158–2159

Joseph JM, Thomas B (2015) Chasmophytic pteridophytes in Urumbikkara hills of Idukki district, Kerala, India. Int J Res Rev 2(2):41–45

Kholia BS, Punetha NN (2005) Useful pteridophytes of Kumaon central Himalaya, India. Indian Fern J 22:1–6

Kirtikar KR, Basu BD (2005) Indian medicinal plants, Cryptogamia, vol IV, 2nd edn. Oriental Enterprises, Uttranchal, India, pp 2735–2739

Kumar S, Samydurai P, Ramakrishnan R et al (2014) Gas chromatography and mass spectrometry analysis of bioactive constituents of *Adiantum capillus-veneris* L. Int J Pharm Pharm Sci 6 (4):60–63

Kumari P, Otaghvari AM, Govindapyari H et al (2011) Some ethno-medicinally important pteridophytes of India. Int J Med Aromat Plants 1(1):18–22

Linnaeus C (1753) Species Planetarium, vol. 1–2. Imprensis Laurentii Salvii, 1200 pp.

Mengane SK (2015) Phytochemistry of some medicinal plants from Western Ghats region. J Pharm Res 1(9):1–4

Nayar BK (1961) Ferns of India 1. *Adiantum*. Bull Nat Bot Gard 5:1–38

Neffati M, Najjaa H, Mathe A (2017) Medicinal and aromatic plants of the world—Africa. Springer, Berlin

Newman DJ, Cragg GM (2012) Natural products as sources of new drugs over the 30 years from 1981 to 2010. J Nat Prod 75:311–335

Pallavi G, Gupta KL, Chate VA (2011) An ethno-Pharmaco-botanical review of Hamsapadi – Adiantum lunulatum Burm. F. (a. Philippense Linn.). IJBPA 2(6):1627–1635

Pan C, Chen YG, Ma XY et al (2011) Phytochemical constituents and pharmacological activities of plants from the genus Adiantum: a review. Trop J Pharmaceut Res 10(5):681–692

Patil SM, Dongare MM (2013) *Adiantum tenerum* Sw., (Adiantaceae-Pteridophyta): a new distributional record for India. 30:115–118

Patil SM, Rajput KS (2017) The genus Isoetes from India: an overview. Plant Sci Today 4(4):123–126

Patil SM, Mahamuni RJ, Dongare MM (2012) Diversity of fern and fern in the hills of northern Western Ghats, Maharashtra. Indian Fern J 29:158–163

Patil SM, Masal V, Dongare MM (2013a) In search of ethnomedicinal pteridophytes from the Western Ghats of Maharashtra (India). Indian Fern Jl 30:69–77

Patil SM, Patil SS, Dongare MM (2013b) The genus *Adiantum* from Maharashtra: a note on addition of two species for Maharashtra, India. Fern Gazette 19(5):159–163

Patil SM, Patil MN, Patil SS (2020) Diversity and Ethnomedicinal Pteridophytes of Kolhapur District, Maharashtra. Indian Fern J 37:224–236

Prado J, Rodrigues CN, Salatino A et al (2007) Phylogenetic relationships among Pteridaceae, including Brazilian species, inferred from rbcL sequences. Taxon 56:355–368

Proestos C, Chorianopoulos N, Nychas GJ et al (2005) RP-HPLC analysis of the phenolic compounds of plant extracts. Investigation of their antioxidant capacity and antimicrobial activity. J Agric Food Chem 53(4):1190–1195. https://doi.org/10.1021/jf040083t

Rai S, Yadav SK, Mathur K et al (2016) A review on Adiantum incisum. World J Pharm Pharm Sci 5(4):861–867

Rastogi S, Pandey MM, Rawat AKS (2018) Ethnopharmacological uses, Phytochemistry and pharmacology of genus *Adiantum*: a comprehensive review. J Ethnopharmacol 215:101–119

Reddy VLN, Ravikanth V, Rao TP et al (2001) A new triterpenoid from the fern *Adiantum lunulatum* and evaluation of antibacterial activity. Phytochemistry 56:173–175

Saha D, Ghosh SK, Das T et al (2016) Hypoglycemic and antihyperlipidemic effects of *Adiantum caudatum* in alloxan induced diabetes in rats. Asian J Pharm Clin Res 9(1):361–363

Samoisy AK, Mahomoodally F (2016) Ethnopharmacological appraisal of culturally important medicinal plants and polyherbal formulas used against communicable diseases in Rodrigues Island. J Ethnopharmacol 194:803–818. https://doi.org/10.1016/j.jep.2016.10.041

Shahriar M, Kabir S (2011) Analgesic activity of *Adiantum flabellulatum*. Dhaka Univ J Biol Sci 20 (1):91–93

Singh JS, Chaturvedi RK (2017) Diversity of ecosystem types in India: a review. Proc Indian Nat Sci Acad 83(3):569–594

Singh M, Singh N, Khare PB et al (2008) Antimicrobial activity of some important *Adiantum* species used traditionally in indigenous systems of medicine. J Ethnopharmacol 115:327–329

Singh S, Khatoon S, Singh H et al (2013) A report on pharmacognostical evaluation of four *Adiantum* species, Pteridophyta, for their authentication and quality control. Rev Bras Farmacogn Braz J Pharmacogn 23(2):207–216

Tsuzuki K, Ohashi A, Arai Y et al (2001) Terpenoids from Adiantum caudatum. Phytochemistry 58 (2):363–367

UNESCO (2018). What is Intangible Cultural Heritage? Reverted from https://ich.unesco.org/en/what-is-intangible-heritage-00003

Vasudeva SM (1999) Economic importance of pteridophytes. Indian Fern J 16:130–152

Yumkham SD, Elangbam M, Nongmaithem R et al (2018) Maiden hair ferns (*Adiantum* L., Pteridaceae-Vittarioideae) of north East India: diversity, phytochemistry and utilisation. Genet Resour Crop Evol 65(4):1269–1280. https://doi.org/10.1007/s10722-018-0612-y

**Sachin M. Patil** has 8 years of teaching and 12 years of research experience. Currently, he serves as Assistant Professor (ad hoc) in the Department of Botany, Shivaji University Kolhapur (Maharashtra). He has completed his M.Sc. (2009) and Ph.D. (2014) from Shivaji University, Kolhapur. Soon after, Dr. Sachin was awarded (2017) National Postdoctoral Fellowship by the Science and Engineering Research Board, New Delhi (SERB). He completed his postdoctoral research at the Maharaja Sayajirao University of Baroda, Vadodara. Dr. Sachin is well trained and developed expertise in the systematics of pteridophytes, algae, bryophytes, and aquatic ecology. A total of 58 articles are published in national and international journals. Due to his excellence in the field of pteridology, he has been awarded the "Certificate of Merit Award (2017)" and "Prof. S. S. Bir Gold Award (2018)" by the Indian Fern Society Chandigarh. He has also been awarded "Young Botanist Award" by Indian Botanical Society, Lucknow for his postdoctoral work in October 2019.

**Rajendra A. Lavate** Associate Professor, Department of Botany, Raje Ramrao Mahavidyalaya, Jath, has completed his M.Sc., from Shivaji University, Kolhapur, in the year 1998. He perused his M.Phil. degree in 2000 and Ph.D. in the field of bryophytes taxonomy in 2016 from Shivaji University, Kolhapur. He joined as Assistant Professor at the PDVP College, Tasgaon, in 2004. For the last 20 years, he has documented 08 new records of bryophytes and pteridophytes from Maharashtra. He has published more than 30+ papers in biodiversity and the taxonomy of bryophytes and pteridophytes. He is a life member of the Indian Fern Society and serves as a reviewer for *Plant Science Today* and *Journal of Bombay Natural History Society*.

**Vinay M. Raole** from the Department of Botany, the Maharaja Sayajirao University of Baroda, Vadodara, is serving in the department for 31 years. He has 35 years of research experience in the field of cytogenetics, grass systematics, and pteridophytes taxonomy. As a field taxonomist, he has worked on the documentation, exploration, and identification of the floristic wealth of the Gujarat state. His primary interest is in angiosperms, mainly grass systematics, as a passion and not a subject. In the last 35 years, he reported 4 new species of grasses and more than 50+ new records of Poaceae, pteridophytes, and gymnosperms from Gujarat. He has published more than 60+ papers in biodiversity and taxonomy, micromorphology, and anatomy.

**Kishore S. Rajput** Associate Professor at the Department of Botany, Maharaja Sayajirao University of Baroda, Vadodara. He has received intensive training in plant taxonomy and anatomy at the Maharaja Sayajirao University of Baroda, Vadodara. He obtained his B.Sc. degree in botany from SVS's Science College Dondaicha, University of Poona, with a first class distinction in 1991. He earned his Master's in 1993 and a Doctorate in 1998 from the Maharaja Sayajirao University of Baroda, Vadodara. He joined the Department of Botany, the Maharaja Sayajirao University of Baroda, Vadodara, in 2006. Dr. Rajput is a recipient of Prof. Y.S. Murthy Gold medal of Indian Botanical Society and Prof. S.S. Bir Gold medal of Indian Fern Society. He is an elected fellow of the Maharashtra Science Academy (Pune), Indian Botanical Society (Meerut), and Indian Fern Society (Chandigarh) and a life member of several scientific societies in India and abroad. Dr. Rajput visited different universities for academic purpose in Brazil, Costa Rica, China, and the USA. He has received research funding from DBT, DST, CSIR, UGC, GBB, and GSBTM and completed 12 major research projects. To his credit, he has 176 research articles published in journals of national and international repute.

# Phytochemical Composition and in Vitro Antioxidant and Antidiabetic Activities of *Nephrolepis auriculata* (L.) Trimen: An Unexplored Ethnomedicinal Fern

24

Jeyalatchagan Sureshkumar and Muniappan Ayyanar

**Abstract**

The present research was aimed to analyze the methanolic and aqueous extracts of *Nephrolepis auriculata* (L.) Trimen for phytochemical analysis, in vitro antioxidant activity using eight assays, and antidiabetic activity by α-glucosidase and α-amylase enzyme inhibitory assays. The qualitative phytochemical analysis showed the presence of steroids, alkaloids, phenolic, flavonoids, saponins, and tannins in methanolic and aqueous extracts. Higher amount of total phenolics (590.59 ± 1.14 mg GAE/g) and flavonoids (642.78 ± 4.83 mg QE/g) were found in methanolic extract. The methanolic extract exhibited strong antioxidant activity by DPPH ($IC_{50}$ of 24.26 ± 1.10), FRAP (29.87 ± 0.27 mg E/g), $O_2^-$ ($IC_{50}$ of 82.89 ± 5.37), metal chelating (15.24 ± 0.17 mg E/g), and nitric oxide ($IC_{50}$ of 118.00 ± 3.80) assays. However, for the phosphomolybdenum (13.77 ± 0.27 mg E/g), $OH^-$ ($IC_{50}$ of 152.44 ± 6.39), and ABTS ($IC_{50}$ of 102.00 ± 5.36) assays, aqueous extract was recorded as most active. The α-glucosidase and α-amylase inhibitory assays also showed that methanolic extract exhibited potent antidiabetic property with an $IC_{50}$ of 55.79 ± 1.01 μg/mL and 57.54 ± 1.52 μg/mL, respectively. It was revealed that aqueous and methanolic extracts of *Nephrolepis auriculata* are rich in phenolics and flavonoids which may be responsible for its antioxidant and antidiabetic activity.

**Keywords**

*Nephrolepis auriculata* · Phenolics · Flavonoids · Antioxidant · Antidiabetic

J. Sureshkumar
Department of Botany, Annai Vailankanni Arts and Science College, Bharathidasan University, Thanjavur, India

M. Ayyanar (✉)
Department of Botany, A.V.V.M. Sri Pushpam College (Autonomous), Affiliated to Bharathidasan University, Poondi, Thanjavur, India


J. Marimuthu et al. (eds.), *Ferns*, https://doi.org/10.1007/978-981-16-6170-9_24

## 24.1 Introduction

Plant-based phenolic and flavonoid compounds are playing an important role in human nutrition due to their biological properties including antioxidant, antidiabetic, and therapeutic values (Pereira et al. 2020). Most of the currently available modern drugs and their derivatives are produced based on their usage in traditional medicine. The primary benefits of using plant-derived medicines are to be relatively safer than synthetic alternatives, offering profound therapeutic benefits and also counteracting the high cost of modern medicines. Medicinal plants produce a variety of bioactive chemicals which are helpful to protect human health from various diseases and pathogenic organisms. It was well known that more than 5000 phytochemicals have been identified to date, while a large percentage still remains untapped, which must be identified for human health benefits (Abbasi et al. 2015).

The genus *Nephrolepis* (Davalliaceae, Pteridophyta) comprises more than 15 accepted species (www.theplantlist.org), and majority of them have been used in folk medicine throughout the world (Shah et al. 2014). *Nephrolepis auriculata* is one of the underexplored medicinal plants and widely used in traditional medicine practices in treating cough, wounds, stomachache, and urinary problems (Sureshkumar et al. 2018, 2021). This species is predominantly distributed in shaded and wet places such as river banks and swamps of hill tracts of Western Ghats and Eastern Ghats in India. In Kolli Hills of Eastern Ghats in Tamil Nadu, India, the plant is found in the deciduous, semievergreen, and evergreen forest regions (Sureshkumar et al. 2020). Traditional uses and phytochemical and pharmacological activities of different species of *Nephrolepis* were investigated by a very few researchers (Bassey et al. 2020; Rindita et al. 2020; Shah et al. 2015).

The studies on antioxidant and biological activity of *Nephrolepis auriculata* are scanty, which encouraged us to carry out this investigation. With this view, the present study has been planned to study the phytochemical analysis (total phenolics and flavonoids) with special attention on in vitro antioxidant and antidiabetic activities of methanolic and aqueous extracts of *Nephrolepis auriculata*.

## 24.2 Materials and Methods

### 24.2.1 Collection and Extraction of Plant Materials

The fresh leaves of *Nephrolepis auriculata* (L.) Trimen (Fig. 24.1) were collected from the Kolli Hills of Eastern Ghats (501 to 1200 m ASL; 11° 11′ to 11° 30’ N and 78° 15′ to 78° 30′ E), India. Taxonomic identification of the specimen was performed using the handbook *Ferns and Fern Allies of Kerala* (Madhusoodanan 2015) and further authenticated by the fern expert Dr. P.V. Madhusooanan at Malabar Botanical Garden and Institute for Plant Sciences, Kozhikode, Kerala. The voucher specimen (SPCH1001) of the plant was deposited in the herbarium of A.V.V.M. Sri Pushpam College, Poondi, Thanjavur, India, for future references.

**Fig. 24.1** Morphology and spore arrangement in *Nephrolepis auriculata*

The solvent extraction was done as described by Murugan and Parimelazhagan (2014) with slight modifications. Freshly collected plant materials were thoroughly washed with 90% ethanol and shade dried at 35 ± 2 °C. After complete drying, the plant materials were powdered using an electric grinder. About 25 g of powdered samples were soaked in 250 mL of methanol and aqueous solvents, successively, for 8 h, in a shaker, at 40 °C. Then the extracts were filtered through Whatman No.1 filter paper and concentrated until the solid residues were obtained. The dried extracts were stored in tightly sealed bottles at 20 °C for further studies.

### 24.2.2 Phytochemical Screening

The preliminary phytochemical screening of methanolic and aqueous extracts of *N. auriculata* leaves was done to observe the presence of different phytochemicals such as steroids, triterpenoids, alkaloids, phenolic compounds, flavonoids, saponins, and tannins using standard methods (Thangaraj 2016).

#### 24.2.2.1 Salkowski Test for Steroids

Two milliliters of $CHCl_3$ and 1 mL of concentrated sulfuric acid were added to ten drops of plant extract. Isopropyl alcohol was slowly added to the mixture until the double phase formation. A dish-brown color in the middle layer indicates the presence of steroids.

#### 24.2.2.2 Liebermann-Burchard Reaction for Triterpenoids

Approximately 50 mg of plant extract was added to 1 mL of $CHCl_3$ and mixed well. Acetic anhydride was supplemented to the mixture, and then concentrated sulfuric acid falls from the sides of tubes. Appearance of red and bluish green color indicated the presence of triterpenoids.

#### 24.2.2.3 Hager's Test for Alkaloids

Fifty milligrams of plant extract was stirred with a few mL of diluted hydrochloric acid and then filtered. To the few mL of filtrate, 2 mL of Hager's reagent was added, and prominent yellow precipitate suggests the presence of alkaloids.

#### 24.2.2.4 Lead Acetate Test for Phenolic Compounds

About 50 mg of plant extract in distilled water was added with 3 mL of 10% lead acetate solution. A bulky white precipitate indicates the presence of phenolic compounds.

#### 24.2.2.5 Alkaline Reagent Test for Flavonoids

Aqueous solution of the plant extract was added with 10% of ammonium hydroxide solution. Yellow fluorescence color in the mixture indicates the presence of flavonoids.

#### 24.2.2.6 Frothing Test for Saponins Determination

About 50 mg of plant extract was diluted with distilled water and made up to 20 mL. After continuous shaking for 15 min, formation of two layers of foam indicates the presence of saponins.

#### 24.2.2.7 Potassium Hydroxide Test for Tannins

The few mL of plant extract was added to 10 mL of freshly prepared 10% potassium hydroxide in a beaker and shaken well. The dirty precipitate formation indicates the presence of tannins.

#### 24.2.2.8 Total Phenolic Content

The total phenolic content in methanolic and aqueous extracts of *N. auriculata* leaf was estimated by following Folin-Ciocalteu reagent method (Manoharan et al. 2019) with minor modifications. Different concentrations of standard gallic acid (10, 20, 30, 40, and 50 μL) and test samples (50, 100, 150, 200, and 250 μg/mL) were prepared to form the stock solution. One milliliter of each concentration was mixed with 0.5 mL of Folin-Ciocalteu reagent and 2.5 mL of $Na_2CO_3$ (20%). After 40 min of incubation at room temperature, absorbance was recorded at 725 nm against the blank. The concentration of total phenolic content was calculated as gallic acid equivalent (mg/g) of dry extract from the calibration curve of standard solution of gallic acid.

#### 24.2.2.9 Total Flavonoid Content

The total flavonoid content was determined using aluminum chloride ($AlCl_3$) method with slight modification (Manoharan et al. 2019). Different concentrations of the standard quercetin (10, 20, 30, 40, and 50 μL) and test samples (50, 100, 150, 200, and 250 μg/mL) were prepared. Each sample of 300 μL was mixed with 150 μL of $AlCl_3$ (10%), 150 μL of $NaNO_2$, and 2 mL of 4% NaOH, and the final volume was made into 5 mL by adding distilled water. After incubation at room temperature for 30 min, absorbance was measured at 510 nm against the blank which possess water alone. Based on the quercetin calibration curve, the flavonoid content was calculated in terms of quercetin equivalents per g/mg of dry extract.

### 24.2.3 Antioxidant Activity

The DPPH (2,2-diphenyl-1-picrylhdrazyl), nitric oxide, hydroxyl, superoxide, and ABTS 2,2′-azino-bis (3-ethylbenzo-thiozoline-6-sulfonic acid) scavenging abilities of methanolic and aqueous extracts of *N. auriculata* leaf were studied according to the standard protocols (Amalraj et al. 2021a). The radical scavenging ability of *N. auriculata* leaf extract in these assays was expressed as $IC_{50}$ (inhibitory concentration). The total antioxidant activity (by phosphomolybdenum method), FRAP (ferric reducing antioxidant power), and metal chelating ability were carried out according to Amalraj et al. (2021b), and the results were expressed as ascorbic acid equivalents (AAE)/g, Fe(II) equivalents/mg, and EDTA equivalents (mg EDTA/g), respectively. In all these assays, ascorbic acid was used as a reference antioxidant.

#### 24.2.3.1 DPPH Assay

One milliliter of fresh DPPH solution (0.1 mM) was mixed with 1 mL of plant extract at different concentrations (2.5, 5, 10, 20, 40, 80, and 160 μg/mL). The reaction mixture was incubated in the dark for 30 min at 37 °C and analyzed spectrophotometrically at 517 nm against the blank. The inhibitory concentration of ($IC_{50}$) test samples and standard was determined using the percentage inhibition against concentration graph. The free radical scavenging potential of plant extract was calculated using the following formula:

$$\text{Inhibitory concentration } (\%) = [(A_0 - A_1)/A_0] \times 100$$

where $A_0$ = absorbance of the control and $A_1$ = absorbance of the sample

#### 24.2.3.2 FRAP Assay

Two milliliters of plant extract at various concentrations (2.5, 5, 10, 20, 40, 80, and 160 μg/mL, respectively) were mixed with 3 mL of FRAP reagent in the test tube. Both sample and blank were incubated at 37 °C for 10 min, and the absorbance of sample was recorded against blank at 520 nm.

#### 24.2.3.3 Phosphomolybdenum Complex Assay

Briefly, 1 mL of reagent solution (0.6 M sulfuric acid, 28 mM sodium phosphate, and 4 mM ammonium molybdate) was added to 2 mL of plant extract and incubated at 95 °C for 90 min. After cooling to room temperature, the absorbance was measured at 765 nm.

#### 24.2.3.4 Hydroxyl (OH−) Assay

One milliliter of iron-EDTA solution (0.13% ferrous ammonium sulfate and 0.26% EDTA), 0.5 mL of EDTA solution (0.018%), 1 mL of DMSO (0.85% v/v in 0.1 M phosphate buffer, pH 7.4), 0.5 mL of ascorbic acid (0.22%), and 100 μL of plant extract at different concentrations were taken in a test tube. The reaction was initiated by incubating at 80–90 °C for 15 min. One milliliter of ice-cold TCA (17.5% w/v) with 3 mL of Nash reagent was added to terminate the reaction. The absorbance was measured at 412 nm after 15 min of incubation at room temperature.

#### 24.2.3.5 ABTS$^+$ Assay

The ABTS$^+$ cation radical was generated by oxidation of 7 mM ABTS solution with 2.4 mM potassium persulfate ($K_2S_2O_8$) in the mixture which was placed in the dark for 12–24 h at room temperature before use. About 150 μL of plant extract at various concentrations (2.5, 5, 10, 20, 40, 80, and 160 μg/mL, respectively) was mixed with 1 mL of ABTS solution. The mixture was diluted in methanol, and the reaction mixture was incubated for 30 min and kept in the dark. Then the absorbance was measured at 734 nm.

#### 24.2.3.6 Superoxide ($O_2^-$) Free Radical Scavenging Assay

Three milliliters of reaction mixture containing 50 mM sodium phosphate buffer (pH 7.4), 20 μL of riboflavin, 12 mM EDTA solution, and 0.1 mg NBT were blended with 100 μL of plant extract. The reaction mixture was incubated for 15 min and the absorbance was measured at 590 nm.

#### 24.2.3.7 Metal Chelating Activity

About 50 μL of $FeCl_2$ (2 mM), 0.2 mL of ferrozine solution (5 mM), and 0.2 mL of plant extracts were mixed together and incubated at room temperature for 10 min. Then the absorbance of mixture was measured at 562 rpm.

#### 24.2.3.8 Nitric Oxide Scavenging Activity

Three milliliters of reaction mixture containing sodium nitroprusside (10 mM) and 100 μL of plant extract were incubated at 25 °C for 150 min in room temperature. After incubation, 3 mL of Griess reagent (1% sulfanilamide, 2% $H_3PO_4$, and 0.1% N-(1-naphthyl) ethylene diaminedihydrochloride) was added. The absorbance was measured at 546 nm. The phosphate-buffered saline alone was used as blank, and same reaction mixture without the sample was used as negative control.

### 24.2.4 Antidiabetic Activity

#### 24.2.4.1 α-Glucosidase Inhibitory Activity

The α-glucosidase inhibitory activity was conducted following the method described by Thangaraj (2016) with minor modifications. In brief, 3 mM *p*-nitrophenyl glucopyranoside (pNPG) and 10 μL of plant extract at various concentrations (20, 40, 60, 80, and 100 μg/mL) were mixed with 10 μL α-glucosidase. The mixture was preincubated for 30 min at 37 °C. The enzyme reaction activity was terminated by the addition of 2 mL $Na_2CO_3$. The activity of enzymes was measured at 400 nm using microplate reader. Acarbose was used as positive control. The α-glucosidase inhibitory activity was determined by the following formula:

$$\%\text{inhibition} = [(A_0 - A_1)/A_0] \times 100$$

where $A_0$ = absorbance of the control and $A_1$ = absorbance of the sample

#### 24.2.4.2 α-Amylase Inhibitory Activity

The α-amylase inhibitory activity was determined according to Thangaraj (2016) with slight modification. The reaction mixture was prepared with 20 μL of α-amylase, 50 μL of plant extract at various concentrations (20, 40, 60, 80, and 100 μg/mL), and 0.5% soluble starch. The mixture was incubated for 5 min at 37 °C, and the reaction was terminated by the addition of 2 mL of DNS (3,5-dinoitrosalicyclic acid) reagent. The reaction mixture was placed in a hot water bath for 15 min at 100 °C and cooled with 10 mL of distilled water. Finally, the absorbance was measured at 540 nm. The α-amylase inhibitory activity was determined by the following formula:

$$\%\text{inhibition} = [(A_0 - A_1)/A_0] \times 100$$

where $A_0$ = absorbance of the control and $A_1$ = absorbance of the sample

### 24.2.5 Statistical Analysis

All the test experiments were performed in triplicates. The data interpretation was done with mean and standard deviations which were calculated using Microsoft Office Professional Plus 2010 (version 14.04734.1000). The TPC, TFC, and antioxidant activity of *N. auriculata* extracts were analyzed using linear regression.

## 24.3 Results and Discussions

### 24.3.1 Ethnomedicinal Uses of N. Auriculata

The various parts of *N. auriculata* are used to cure urinary problem, stomachache, cough, and wounds by tribal people in Kolli Hills of Eastern Ghats (Sureshkumar et al. 2021). Fronds of this species are the most preferred part for preparation of medicine followed by stem. Similarly, various ethnic people use different parts of this plant to treat urinary problems, stomachache (Ganesan et al. 2004), and cough (Benniamin 2011; Thirupathi and Karthikeyan 2015) and to prevent bleeding from wound (Revathi et al. 2013).

## 24.3.2 Determination of Phytochemicals

### 24.3.2.1 Preliminary Phytochemical Analysis

The results of phytochemical screening of *N. auriculata* leaf extracts are given in Table 24.1. It was shown that phenolic compounds, flavonoids, saponins, and tannins were present in both extracts. Trace amounts of steroids and alkaloids are detected from methanolic extract and absent in aqueous extract. Triterpenoids are not recorded in both the extracts.

### 24.3.2.2 Determination of TPC and TFC

The result showed that more amount of TPC (Table 24.2) was observed in methanolic extract (590.59 ± 1.14 mg GAE/g) at the concentration of 250 mg/mL compared to aqueous extract (318.52 ± 6.67 mg GAE/g) at 250 mg/mL. The gallic acid equivalent calculated from the linear regression of the standard calibration curve is $y = 0.0113 + 0.204$ (Fig. 24.2).

The TPC in leaf extracts of *N. auriculata* was similar to other species of *Nephrolepis* as described earlier. Ethanolic leaf extract of *N. biserrata* showed the presence of TPC with 8.84 ± 0.14 mg GAE/g (Rindita et al. 2020), and methanolic extract of *N. biserrata* showed 127.28 ± 1.57 mg/g total polyphenolics (Shah et al. 2015). An edible medicinal fern, *Stenochlaena palustris* fronds were sequentially extracted with hexane (19.7 ± 0.8), chloroform (43.3 ± 2.4), ethyl acetate (133.0 ± 2.0), methanol (503.4 ± 22.8), and aqueous (319.5 ± 7.5) solvents that displayed comparable amount of TPC (Chai et al. 2015) as reported in the present study.

The methanolic extract exhibits considerably higher concentrations of TFC with 642.78 ± 4.83 mg GAE/g at 250 mg/mL than aqueous extract (498.05 ± 7.89 mg GAE/g at 250 mg/mL). Quercetin equivalent was calculated from the linear regression of the standard calibration curve, $y = 0.0117 + 0.1018$ (Fig. 24.2). Previously aqueous and ethanolic extracts of *N. biserrata* showed higher TFC with

**Table 24.1** Phytochemical screening of two solvent extracts of *Nephrolepis auriculata* leaves

| Phytochemical | Solvent used | |
|---|---|---|
| | Methanol | Aqueous |
| Steroids | + | – |
| Triterpenoids | – | – |
| Alkaloids | + | – |
| Phenolic compounds | ++ | + |
| Flavonoids | + | + |
| Saponins | +++ | + |
| Tannins | +++ | ++ |

+ indicates the presence of secondary metabolites; – sign indicates the absence of secondary metabolites; +++ indicates the presence in high amount; ++ indicates the presence of moderate amount; + indicates the presence of little amount of secondary metabolites

**Table 24.2** TPC, TFC, and in vitro antioxidant and antidiabetic activity of *Nephrolepis auriculata* leaf extracts

| Assays | Concentration (μg/mL) | Plant extracts (μg/mL) | | |
|---|---|---|---|---|
| | | Methanol | Aqueous | Standard |
| TPC (mg GAE/g) | 250 | 590.59 ± 1.14 | 318.52 ± 6.67 | – |
| TFC (mg QE/g) | 250 | 642.78 ± 4.83 | 498.05 ± 7.89 | – |
| DPPH ($IC_{50}$) | 250 | 24.26 ± 1.10 | 21.93 ± 4.55 | 50.78 ± 1.32 |
| $OH^-$ ($IC_{50}$) | 250 | 93.81 ± 4.61 | 152.44 ± 6.39 | 71.74 ± 1.43 |
| ABTS ($IC_{50}$) | 250 | 51.79 ± 7.76 | 102.00 ± 5.36 | 71.37 ± 2.71 |
| $O_2^-$ ($IC_{50}$) | 250 | 82.89 ± 5.37 | 64.47 ± 4.65 | 17.06 ± 4.60 |
| Nitric oxide ($IC_{50}$) | 250 | 118.00 ± 3.80 | 87.52 ± 3.83 | 62.74 ± 1.99 |
| FRAP (mg E/g) | 250 | 29.87 ± 0.27 | 23.74 ± 0.29 | 66.08 ± 0.0 |
| PMA (mg E/g) | 250 | 12.75 ± 0.34 | 13.77 ± 0.27 | 33.46 ± 0.0 |
| MC (mg E/g) | 250 | 15.24 ± 0.17 | 12.89 ± 0.13 | 49.98 ± 0.0 |
| α-Glucosidase ($IC_{50}$) | 100 | 55.79 ± 1.01 | 71.80 ± 0.69 | 14.24 ± 0.44 |
| α-Amylase ($IC_{50}$) | 100 | 57.54 ± 1.52 | 77.37 ± 0.91 | 15.50 ± 0.37 |

Values are means of triplicate determination ± standard deviations; TPC total phenolic content, TFC total flavonoid content, $IC_{50}$ is the inhibitory concentration of sample required for 50% inhibition

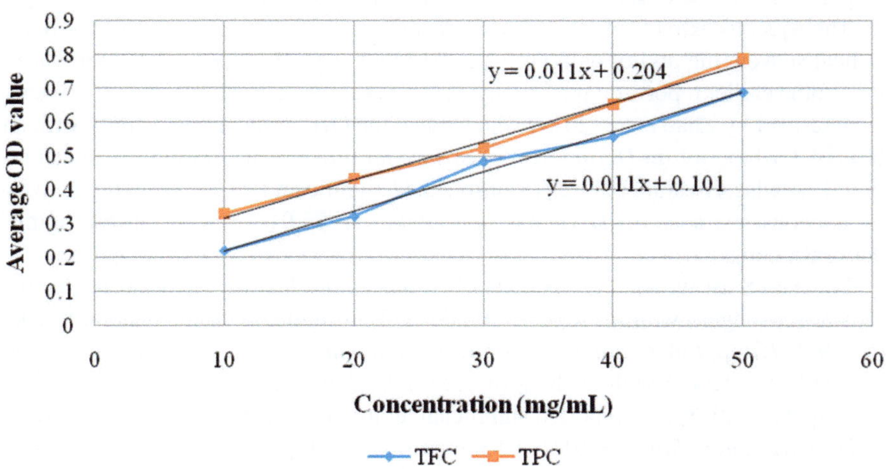


**Fig. 24.2** Standard calibration curve for TPC and TFC of *Nephrolepis auriculata* leaves

184.85 ± 1.52 mg QE/g and 8.09 ± 0.18 mg RE/g, respectively (Chai et al. 2015; Manan et al. 2015). Likewise, Takuli et al. (2020) reported the presence of equal amount of TFC in *Woodwardia unigemmata* extracts with the solvents like aqueous (146 ± 7.84), methanol (151 ± 11.24), and hexane (43.89 ± 1.47).

### 24.3.3 Antioxidant Activity

The DPPH scavenging activity of methanolic extract exhibited highest scavenging property with an $IC_{50}$ of 24.26 ± 1.10 μg/mL compared to aqueous extract ($IC_{50}$ of 21.93 ± 4.55 μg/mL) and standard ascorbic acid ($IC_{50}$ of 50.78 ± 1.32 μg/mL). Previous reports on other species of *Nephrolepis* showed similar results of our study. For example, ethanolic leaf extract of *N. biserrata* showed an $IC_{50}$ of 95.06 ppm (Rindita et al. 2020), and methanolic leaf extract recorded with 3.15 to 90.25% inhibition at 0.012 to 0.5 mg/mL concentrations for DPPH assay (Shah et al. 2014). For FRAP assay, methanolic leaf extracts of *N. auriculata* exhibited an $IC_{50}$ value of 29.87 ± 0.27 mg E/g followed by aqueous extract ($IC_{50}$ of 23.74 ± 0.29 mg E/g). Likewise, Takuli et al. (2020) assessed the antioxidant activity of methanolic extracts of *Woodwardia unigemmata* with 768 ± 10.44 mg AAE $gm^{-1}$ for FRAP assay.

The aqueous ($IC_{50}$ of 13.77 ± 0.27) and methanolic ($IC_{50}$ of 12.75 ± 0.34 mg E/g) leaf extracts of *N. auriculata* showed highest total antioxidant content by phosphomolybdenum assay in the present study. Sivaraman et al. (2013) reported the total antioxidant activity of ethanolic extracts of *Selaginella involvens*, *S. intermedia*, *S. inaequalifolia*, and *S. tenera* with the inhibition values as 21.67 ± 0.64, 23.00 ± 0.93, 20.20 ± 1.51, and 26.10 ± 0.65 g AA/100 g, respectively. These observations are closely related to the results of present study.

The aqueous extracts of *N. auriculata* leaf showed promising results for hydroxyl radical scavenging ability with an $IC_{50}$ of 152.44 ± 6.39 at 250 μg/mL. Similar to this, some earlier reports on pteridophytes species showed good antioxidant capacity as described by Zhao et al. (2019) who studied the $OH^-$ scavenging activity with an $IC_{50}$ of 1.70 mg/mL in *Dryopteris crassirhizoma* rhizome in 5.0 mg/mL concentration. Likewise, Sivaraj et al. (2018) also stated significant $OH^-$ scavenging activity in aqueous extracts of *Drynaria quercifolia* at 85.91 ± 5.16 at 60 μg/mL concentration.

For ABTS assay, aqueous extracts of *N. auriculata* leaf revealed most effective radical scavengers with an $IC_{50}$ of 102.00 ± 5.36 μg/mL than methanolic extract (51.79 ± 7.76 μg/mL). Similarly, fronds of *Asplenium ruta-muraria* exhibited good antioxidant potential with an $IC_{50}$ of 0.29 ± 0.01 mg/mL (Zivkovic et al. 2020). Sivaraj et al. (2018) also reported significant antioxidant activity in aqueous extracts of *Drynaria quercifolia* (54.91 ± 5.55 inhibitions at 30 μg/mL concentrations) with ABTS assay. In the superoxide free radical scavenging assay, highest reducing power was exhibited by methanolic leaf extract of *N. auriculata* with the $IC_{50}$ of 82.89 ± 5.37 μg/mL followed by aqueous extracts (64.47 ± 4.65 μg/mL) in the present study. Similar to our observation, Srivastava et al. (2005) reported that *N. exaltata* fronds and roots showed significant antioxidant activity (185.80 U $mg^{-1}$ at concentration 300 μM) by superoxide assay.

In the present study, methanolic leaf extracts of *N. auriculata* showed significant metal chelating ability with an $IC_{50}$ of 15.24 ± 0.17 mg E/g which showed greater scavenging activity than aqueous extract (12.89 ± 0.13 mg E/g). In addition, Liu et al. (2013) tested the chelating ability of *Athyrium multidentatum* fern and reported

47.62% of activity at 16.0 μg/mL of concentration. Suhartono et al. (2012) also reported the good chelating effect on ferrous ions from the leaves of *Stenochlaena palustris* with 27.64 ± 3.12% at 75 μg/mL of concentration. In our study, methanolic extract exhibited better inhibitory effect by nitric oxide scavenging ability with an $IC_{50}$ of 118.00 ± 3.80 μg/mL at 250 μg/mL of concentration. Julfikar et al. (2018) evaluated nitric oxide scavenging ability of *Diplazium esculentum* leaves extracted from hydroalcoholic solvent which showed 65.66 ± 0.62% of inhibition at 250 μg/mL of concentration. In contrary, Sivaraj et al. (2018) reported good inhibitory effect with aqueous extracts of *Drynaria quercifolia* (52.57 ± 3.67 at 60 μg/mL concentrations).

### 24.3.4 Antidiabetic Activity

The results of α-glucosidase inhibitory effect of *N. auriculata* leaf extract are presented in Table 24.2, in which methanolic extract showed better effects with an $IC_{50}$ of 55.79 ± 1.01 μg/mL) than aqueous extract ($IC_{50}$ of 71.80 ± 0.69 μg/mL). The inhibitory effect of plant samples was comparable to that of commercially available standard acarbose ($IC_{50}$ of 14.24 ± 0.44 μg/mL). In support of our study, Chai et al. (2015) reported α-glucosidase inhibitory effect using aqueous extracts of *N. biserrata* with an $EC_{50}$ of less than 50%. Chai et al. (2015) analyzed α-glucosidase inhibitory effect of fronds of *Stenochlaena palustris* which showed decent effect with an $EC_{50}$ of 5.45 ± 0.07 and 11.15 ± 0.02 μg/mL for methanolic and aqueous extracts, respectively. Telagari and Hullatti (2015) also reported good α-glucosidase enzyme inhibitory effect in hydroalcoholic extracts of *Adiantum caudatum* with an $IC_{50}$ of 1.619 μg/mL. Amin et al. (2019) revealed the α-glucosidase inhibitory property of *Dryopteris cycadina* at a concentration of 500 μM with an $IC_{50}$ of 314 ± 4.58 μM.

The results of α-amylase inhibitory effect of *N. auriculata* leaf extract are shown in Table 24.2. The maximum α-amylase inhibitory effect was observed in methanolic extract ($IC_{50}$ of 57.54 ± 1.52 μg/mL) which was significantly stronger than aqueous extract ($IC_{50}$ of 77.37 ± 0.91 μg/mL). Previously, Johnson et al. (2020) reported that *Sphaerostephanos unitus* leaf showed significant amount of α-amylase inhibitory activity (with about 80%) in petroleum ether and acetone extracts at 25 μg/mL concentration. Likewise, hydroalcoholic extract of *Adiantum caudatum* showed an $IC_{50}$ of 3.112 μg/mL for α-amylase inhibitory effect (Telagari and Hullatti 2015). Manivannan and Johnson (2020) also analyzed in vitro antidiabetic properties of *Tectaria paradoxa* with methanol, acetone, chloroform, and petroleum ether extracts, and 78% of activity was observed in their study at 500 μg/mL.

## 24.4 Conclusion

The phytochemical studies of present investigation revealed the presence of several bioactive constituents in *Nephrolepis auriculata* leaves. It was supposed that the bioactive constituents like phenolics and flavonoids are responsible for its antioxidant and antidiabetic activity, which makes them interesting for phytopharmaceutics. Further studies are necessary toward isolation, identification, and characterization of pure compounds with associated pharmacological activities of *N. auriculata*. The identified antidiabetic activity will be added sufficient data for food formulating or plant-based pharmaceutical products to control blood glucose. However, more extensive research especially in vivo analysis could be conducted to validate and elucidate the mechanism of action, thereby confirming the antidiabetic and antioxidant potential of *N. auriculata* leaf extracts as an effective therapeutic agent.

**Acknowledgments** This research was supported by the grant received from the Science and Engineering Research Board (SERB) Government of India, New Delhi (Grant No.EMR/2016/007164).

## References

Abbasi AM, Shah MH, Liu RH et al (2015) Ethnomedicinal values, phenolic contents and antioxidant properties of wild culinary vegetables. J Ethnopharmacol 162:333–345

Amalraj S, Krupa J, Sriramavaratharajan V et al (2021a) Chemical characterization, antioxidant, antibacterial, and enzyme inhibitory properties of *Canthium coromandelicum*, a valuable source for bioactive compounds. J Pharm Biomed Anal 192:1–12

Amalraj S, Mariyammal V, Murugan R et al (2021b) Comparative evaluation on chemical composition, *in vitro* antioxidant, antidiabetic and antibacterial activities of various solvent extracts of *Dregea volubilis* leaves. South Afr J Bot 138:115–123

Amin S, Ullah B, Ali M et al (2019) Potent *in vitro* α-glucosidase inhibition of secondary metabolites derived from *Dryopteris cycadina*. Molecules 427(24):1–11

Bassey ME, Johnny II, Umoh OT et al (2020) Phytomedicinal potentials of species of *Nephrolepis* (Schott.). World J Pharm Res 9(4):1400–1410

Benniamin A (2011) Medicinal ferns of North Eastern India with special reference to Arunachal Pradesh. Indian J Tradit Knowl 10(3):516–522

Chai T, Kwek M, Ong H et al (2015) Water fraction of edible medicinal fern *Stenochlaena palustris*is a potent α-glucosidase inhibition with concurrent antioxidant activity. Food Chem 186:26–31

Chai T, Yeoh L, Ismail NIM et al (2015) Evaluation of glucosidase inhibitory and cytotoxic potential of five selected edible and medicinal ferns. Trop J Pharm Res 14(3):449–454

Ganesan S, Suresh N, Kesaven L (2004) Ethnomedicinal survey of lower palni hills of Tamil Nadu. Indian J Tradit Knowl 3(3):299–304

Johnson A, Madona CX, Almeida RS et al (2020) *In vitro* toxicity, antioxidant, anti inflammatory and antidiabetic potential of *Sphaerostephanos unitus* (L.). Holttum Antibiotics 9(333):1–13

Julfikar AJ, Gogoi G, Islam J et al (2018) Exploration of antioxidant, antidiabetic and hepatoprotective activity of *Diplazium esculentum*– a wild edible plant from North Eastern India. Future J Pharm Sci 4(1):93–101

Liu D, Sheng J, Li Z et al (2013) Antioxidant activity of polysaccharide fractions extacted from *Athyrium multidentatum* (doll.) Ching. Int J Biol Macromol 56:1–5

Madhusoodanan PV (2015) Hand book on ferns and Fern allies of Kerala, Malabar Botanical Garden and Institute for Plant Sciences (MBGIPS). Kozhikode, Kerala, India

Manan FA, Mamat DD, Samad AA et al (2015) Heavy metal accumulation and antioxidant properties of *Nephrolepis biserrata* growing in heavy metal-contaminated soil. Global NEST J 17(3):544–554

Manivannan V, Johnson M (2020) Total phenolic, tannin, triterpenoid, flavonoid and sterol centents, anti-diabetic, anti-inflammatory and cytotoxic activities of *Tectaria paradoxa* (Fee.) Sledge. Toxicol Rep 7:1465–1468

Manoharan AL, Thamburaj S, Muniyandi K et al (2019) Antioxidant and antimicrobial investigations of *Elaeocarpus tectorius*(Lour.) Poir. fruits against urinary tract infection pathogens. Biocatal Agric Biotechnol 20:101260

Murugan R, Parimelazhagan T (2014) Comparative evaluation of different extraction methods for antioxidant and antiinflammatory properties from *Osbeckia parvifolia* Arn – an *in vitro* approach. J King Saud Univ Sci 26(4):267–275

Pereira ES, Vinholes JR, Camargo TM et al (2020) Characterization of araca fruits (*Psidium cattleianum* Sabine): phenolic composition, antioxidant activity and inhibition of α-amylase and α-glucosidase. Food Biosci 37:100665

Revathi R, Muthuraja R, Thomas B et al (2013) Ethno medicinal fern and fern-allies used by tribe Malayalis of Kolli Hills, eastern Ghats, Tamil Nadu. Pteridological Res 2(1):1–10

Rindita AV, Rahmaesa E, Devi RK et al (2020) Exploration, phenolic content determination and antioxidant activity of dominant pteridophytes in Gunung Malang village, mount HalimunSalak National Park, Indonesia. Biodiversitas 21(8):3676–3682

Shah MD, Gnanaraj C, Haque ATME et al (2015) Antioxidative and chemopreventive effects of *Nephrolepis biserrata* against carbon tetrachloride ($CCl_4$)-induced oxidative stress and hepatic dysfunction in rats. Pharm Biol 53(1):31–39

Shah MD, Yong YS, Iqbal M (2014) Phytochemical investigation and free radical scavenging activities of essential oil, methanol extracts and methanol fractions of *Nephrolepis biseratta*. Int J Pharm Pharm Sci 6(9):269–277

Sivaraj C, Aashinya Y, Sripriya R et al (2018) Antioxidant activities and thin layer chromatographic analysis of aqueous extract of tubers of *Drynaria quercifolia* (L.). J. Sm. Free Rad Antioxi 8(1): 26–31

Sivaraman A, Johnson M, Primelazhagan T et al (2013) Evalution of antioxidant potential of ethanolic extracts of selected species of *Selaginella*. Indian J Nat Prod Resour 4(3):238–244

Srivastava M, Lena QM, Singh N et al (2005) Antioxidant responses of hyper-accumulator and sensitive fern species to arsenic. J Exp Bot 56(415):1335–1342

Suhartono E, Viani E, Rahmadhan MA et al (2012) Total flavonoid and antioxidant activity of some selected medicinal plants in South Kalimantan of Indonesian. Procedia APCBEE 4:235–239

Sureshkumar J, Ayyanar M, Silambarasan R (2020) Pteridophyte species richness along elevation gradients in Kolli Hills of the Eastern Ghats, India. J Asia Pac Biodivers 13:92–106

Sureshkumar J, Ayyanar M, Silambarasan R (2021) Ethnomedicinal uses, phytoconstituents and pharmaologial importane of pteridophytes used by *Malayalis*in Kolli hills, India: a quantitative survey. J Herb Med 25:100418

Sureshkumar J, Silambarasan R, Bharati KA et al (2018) A review on ethnomedicinally important pteridophytes of India. J Ethnopharmacol 219:269–287

Takuli P, Khulbe K, Kumar P et al (2020) Phytochemical profiling, antioxidant and antibacterial efficacy of a native Himalayan Ferns: *Woodwardia unigemmata* (Makino) Nakai. Saudi J Biol Sci 27:1961–1967

Telagari M, Hullatti K (2015) *In vitro* α-amylase and α-glucosidase inhibitory activity of *Adiantum caudatum* Linn. And *Celosia argentea* Linn. Indian J Pharmacol 47:425–429

Thangaraj P (2016) Pharmacological assays of plant-based natural products. Springer Nature, Switzerland

Thirupathi M, Karthikeyan S (2015) Distribution and ethnomedicinal uses of pteridophytes in Sirumalai hills, Eastern Ghats of Tamil Nadu, India. Int J Curr Res 7:17224–17227

Zhao Y, Hu W, Zhang H et al (2019) Antioxidant and immunomodulatory activities of polysaccharides from the rhizome of *Dryopteris crassirhizomal* Nakai. Int J Biol Macromol 130:238–244

Zivkovic S, Milutinovic M, Maksimovic V et al (2020) Antioxidant and antimicrobial activity of two *Asplenium* species. S Afr J Bot 132:180–187

**Jeyalatchagan Sureshkumar** is doctorate in botany from Bharathidasan University, India, and he did his Ph.D. in taxonomy and ecology of pteridophytes in Kolli Hills, India. His areas of expertise are ecology, ethnobotany, and ethnopharmacology. He has published about ten research articles in reputed journals with the cumulative impact factor of 11.548.

**Muniappan Ayyanar** is an ethnobotanist with doctorate in botany from the University of Madras, India. His research interests include medicinal plant systematics, plant-based natural product research, and ethnopharmacology. He has contributed 43 research articles in indexed journals with 4700 citations and h-index of 22.

# Part IV

# Environmental Regulation

# Ferns as Ecological Indicators

# 25

Aline Possamai Della

**Abstract**

Many studies have been done with ecological indicators, since they are easily measurable substitutes for unmeasured ecological values or when extensive studies (with big scales) are nonviable due to budget and time constraints. Ecological indicators can be a species or group of species, which readily reflect the biotic and abiotic state of an environment. This definition includes only species and/or groups of them, but other taxonomic levels (families and genera) or even ecological attributes (richness and diversity) could be adopted. Ferns have great potential as ecological indicators, due to the correlation between their geographic distribution and abiotic variables. The objective of this chapter is to present a general discussion about the types of ecological indications already analyzed for ferns. Some studies consider them as indicators of vegetation; soils; environments; ecosystems (classification); environmental integrity (or quality); disturbance; regeneration/restoration of habitats and/or ecosystems; climate changes; contamination of air, soil, or water; and association with other groups of organisms. Despite the great potential that this group has as ecological indicators, there are some difficulties, which must be remedied or further discussed in the future, in the establishment of ferns as indicators, such as the lack of practical application and the absence of ecological information for some taxa.

**Keywords**

Bioindicator · Ecological indicator · Fern

A. P. Della (✉)
Institute of Biosciences, University of São Paulo, São Paulo, Brazil





J. Marimuthu et al. (eds.), *Ferns*, https://doi.org/10.1007/978-981-16-6170-9_25

## 25.1 Indicators

The word "indicator" comes from the Latin *indicare*, a verb that means to point. Indicators for biological purposes are considered easily measurable substitutes for unmeasured ecological values or environmental conditions of interest (Oster et al. 2008). Indicators are used mainly when extensive researches (with big scales) are nonviable due to budget and time constraints (Oster et al. 2008). However, we need to consider there is an error percentage in the application of indicators since doing a single investigation cannot have the same result as multiple analyses.

The first humans already used ecological indicators, because they adopted the seasonal migratory movements of animals and the flowering of some plants as indicators of changes in environmental conditions (Niemi and McDonald 2004). In ancient times, species very demanding in soil conditions were considered indicators. At the same time, Theophrastus recognized differences in growth forms and the aspect of the wood of several trees located in colder environments (Sampson 1939). Despite this, the first theoretical reference to the indicators is the Greek philosopher Plato, who cited the impacts of human activities on fruit harvesting. Subsequently, Morrison (1986) analyzing studies of some authors have concluded that the definition of indicator concept arose about 1600 (Niemi and McDonald 2004).

In the 1920s, indicators began to be used to determine water and air quality (Niemi and McDonald 2004; Rapport and Hildén 2013), as well as to classify the environments. To this last item, it can highlight the use of benthic and planktonic algae as indicators of different zones of decomposition flow (Niemi and McDonald 2004).

In the last 40 years, there has been an increase in scientific interest in the development and application of indicators. This relevance comes from the necessity of evaluating, regulating, and supervising the ecological situation in decision-making (especially political) (Niemi and McDonald 2004). As a consequence of this interest, many new articles, books, and commentaries about indicators aroused.

## 25.2 Ecological Indicators

Ecological indicators are broadly used, mainly in evaluations of natural and/or anthropic disturbances. Despite this, the ecological indicator concept is not well established and changes considerably in accord with the authors. Niemi and McDonald (2004) define ecological indicators as measurable characteristics of structure (e.g., populations genetic), composition (e.g., genes, species, populations, communities, and landscape), or function (e.g., demographic and ecosystemic genetics and landscape disturbance process) of ecological systems. These authors combined the US Environmental Protection Agency's definition of ecological indicators and the hierarchy of Noss (1990). The hierarchy refers to the four levels of biological organization: regional landscape, community-ecosystem, population-species, and genetic (Noss 1990).

McGeoch (1998) defines ecological indicators as a species or group of species that readily reflect the biotic and abiotic state of an environment. Ecological indicators can also represent a change in habitat, a community, or an ecosystem due to an environmental impact or indicate the diversity of taxa within an area.

Heink and Kowarik (2010a) characterize ecological indicators as a component or measure of environmentally relevant phenomena used to describe and evaluate the environmental conditions or changes or to define environmental goals. For the authors, environmentally relevant phenomena are the pressures, states, and responses defined by the Organization for Economic Cooperation and Development.

For Turnhout et al. (2007), ecological indicators can be used at various levels (sometimes nested levels), making the concept complex and potentially confusing. For example, a criterion such as diversity, which is considered an indicator of ecological quality, can be defined by the richness of species (which can also be an indicator of ecological quality).

As seen in the definitions of ecological indicators presented, there is no global and ideal indicator (Turnhout et al. 2007) or a single concept broader of indicator, since the indicators can represent any level of complexity. Besides, the indicators can be applied at any point along with a series of impacting human actions (Heink and Kowarik 2010a). Also, in general politics and science have different and (often) contradictory criteria for the ecological indicators (Turnhout et al. 2007).

Ecological indicators can be classified in several ways; what changes from one categorization to another is the assigned focus. In this sense, ecological indicators can be grouped into two classes according to their use (utility): prescription indicators and evaluation indicators (Rempel et al. 2004). Prescription indicators are used to stipulate the future condition of a habitat. This group focuses on attributes that are directly affected by some actions (such as fire regime and fragmentation). Evaluation indicators are used to test whether a desired environmental condition has been achieved; this indicator serves to evaluate the efficiency of an action in maintaining a state or in the conservation of an area. Thus, the presence and population size of sensitive species to changes in environmental conditions are often used as evaluation indicators, while the changes in these environmental conditions are used as prescription indicators (Rempel et al. 2004).

Ecological indicators can aim to connect environmental quality with political measures, especially applying these indicators as instruments for assessing nature conservation norms (Turnhout et al. 2007). However, it can verify that indicators are difficultly used in this way. Two explanations can be presented for the resistance of the use of this type of indicator in politics (Turnhout et al. 2007). The first explanation demonstrates that this type of instrument simplifies nature, which determines some uncertainty toward the understanding of the complexity of ecosystems. The second explanation highlights that the use of these instruments includes the development of indicators suitable for this purpose. This is difficult since the construction of indicators implies a process of selection, integration, and aggregation of parameters that are not easy to be established. In general, ecological indicators should be positioned in the science-political interface, and to be effective, they should be able to connect these two domains (Turnhout et al. 2007).

Very associated with the objectives of an ecological indicator is the establishment of criteria for its selection, since depending on the purpose of the indicator, different criteria and methodologies can be used. Della and Falkenberg (2019a) compiled the most used criteria in studies with ecological indicators, and the main four were to (1) be relatively well-known taxonomically so that the identification isn't very difficult and have available taxonomic knowledge; (2) have economic potential; (3) have well-known biology and natural history, as well as enemies and physical tolerance, and have known all stages of the life cycle; and (4) be representative of related and unrelated taxa.

The first criterion is very important since the purpose of an indicator is to "indicate/signal" something; thus, it needs to be easily identified. An indicator with a complicated taxonomy (with few and/or small characters that determine it) has more probability to be confused with similar individuals belonging to other groups. Checklists and taxonomic revisions can serve as an initial test to whether the group is well known (Pearson 1994).

The second criterion is also important since, for the allocation of personnel and resources, economic criteria are crucial (Pearson 1994; Heink and Kowarik 2010b). Thus, some authors propose the selection of indicators that reflect issues of current or even potential importance (Heink and Kowarik 2010b).

For the third criterion (although it is difficult to quantify whether a taxon has well-known biology and natural history), the amplitude of studies around the world could serve as a demonstration of its level of knowledge. Revision articles, books, etc. dedicated to the biology of a taxon are generally present for the most well-known groups (Pearson 1994).

The three criteria highlighted above are more general so that regardless of the type of indicator, they should be considered. The fourth, however, is more important when the objective of the indicator is to reveal patterns of other taxa (mainly in monitoring studies) (Pearson 1994). Other criteria may be used in the selection of ecological indicators depending on the type of use expected for them. See the full list of criteria in Della and Falkenberg (2019a).

Nowadays, the indicators are used mainly to evaluate the environmental conditions, as early alert signs of ecological problems (natural or anthropic), to identify remedial actions, and as barometers of tendency in ecological resources (Niemi and McDonald 2004). An example of this was the use of the declining populations of peregrine falcon (*Falco peregrinus* Tunstall) as an early alert signal of environmental problems. Researches into the cause of the decline have led to the diagnosis of widespread contamination by chlorinated hydrocarbons, such as DDT (Ratcliffe 1984).

## 25.3 Species as Ecological Indicators

The ecological indicators can be characterized by distinct taxonomic levels, such as family, genus, and species, or ecological attributes, as richness and diversity. In general, the studies use indicator species, which can be defined as an organism that reflects a specific environmental condition, through changes in its presence, abundance, and frequency in a given location. Besides, one indicator species can represent many other species with similar ecological requirements (Landres et al. 1988). Thus, most of the ecological indicators are established to consider the species level, since much is discussed about species extinction and conservation (Niemi and McDonald 2004).

Species as ecological indicators can be used in three different ways: to reflect a biotic or abiotic state of the environment, to reveal evidence of environmental impact or changes, or to indicate the diversity of other species, taxa, or communities within an area (Lawton and Gaston 2001). The first two ways reflect the common use of ecological indicators as measures of the conditions and diagnostic of potential (s) cause(s) of environmental alterations. The third way expands the concept of indicators to incorporate the idea of a single species serving as a substitute for many other species (Niemi and McDonald 2004).

Indicator species are useful management tools because they can help to outline a region, to indicate the state of an environmental condition, and to control pollution or climatic changes. In this sense, indicator species can be used as an "alerting system" by biologists and conservation managers. However, to use species as ecological indicators, in a precise way, an in-deep study of what is being indicated, what is correlated, and how these species fit into the remaining ecosystem is necessary (Jaffe and McDonough 2009).

Indicator species are usually from macroflora and macrofauna, mainly aquatic macroinvertebrates, fish, birds, and vascular plants. There is a set of characteristics that justify the use of some species as indicators, such as the relative facility of identification, the interest to the public, the relative facility to measurement, the relatively large number of species with known responses to disturbance, and the relatively low cost (Niemi and McDonald 2004).

Birds are one of the main groups used as indicator species in terrestrial environments (Nascimento et al. 2005). However, other species of vertebrates have already been used as ecological indicators, such as arboreal lizards in the forests of Mexico. They were used as indicators of health and biodiversity in natural communities (Jaffe and McDonough 2009).

Insects are a great promise as environmental and biodiversity indicators because they present high richness and biomass and also perform an important role in the functioning of the ecosystems (McGeoch 1998; Niemi and McDonald 2004). Among the main groups of insects used as ecological indicators are dragonflies, butterflies, ants, bees, and beetles (Freitas et al. 2006).

Species of plants and/or lichens can be used as indicators of air quality because they are sensitive to heavy metals or acid rain (Jaffe and McDonough 2009). Some species of plants (as *Nicotiana tabacum* L., *Tillandsia* sp., *Tradescantia* sp., *Tibouchina pulchra* (Cham.) Cogn., and *Psidium guajava* L.) may also have a complementary function in monitoring atmospheric contaminants (Carneiro 2004). Bryophytes are considered excellent indicators of climate changes, because they help in water storage, in capturing the rain nutrients, and in ecological interactions when serving as the habitat for animals (Scotti et al. 2013). Many species of plants are used in the classification of vegetation, for example, calcareous fields and dry and shallow fields (Cáceres et al. 2010). Others can be used as indicators of different edaphic conditions, altered environments, and different stages of forest regeneration (Cárdenas et al. 2007).

Although there is an increase of international initiatives for the use of indicator species, the situation is not easy in practice, because identifying potential indicator species is a hard work and sometimes can require a more quantitative methodology (Dufrêne and Legendre 1997). The presence, absence, or abundance of an organism should be linked to an environmental conditional in a solid scientific way to justify its use as an ecological indicator (Jaffe and McDonough 2009).

## 25.4 Ferns

Ferns are a group of vascular plants producing spores (not seeds) and having a pronounced alternation of generations (gametophytes vs. sporophytes). This group is very old in evolutionary terms; it arose before angiosperms and gymnosperms (Raven et al. 2007). Nowadays, ferns are a monophyletic group that has 10,578 species (PPG I 2016).

Ferns occur worldwide, from ice tundras to tropical forests, including deserts (Moran 2008). Despite this, they are more diverse in the equatorial portion of the globe (Moran 2008; Tryon 1972), a region where they can comprise 10 percent of the species of vascular plants (Moran 2008; Tryon 1972).

In a local spatial scale, the distribution of ferns is not random but reflects features of the soil (texture and fertility), atmospheric temperature and humidity, precipitation, and luminosity (Tuomisto and Poulsen 2000; Nóbrega et al. 2011; Patil et al. 2016; Della and Falkenberg 2019a). Besides, ferns do not depend on pollen and seed vectors (Abotsi et al. 2020), which shows the importance of abiotic factors in their distribution.

The great correlation between the geographic distribution of ferns and abiotic variables makes this group good ecological indicators (Salovaara et al. 2004; Tuomisto and Ruokolainen 1993). In addition, many species can tolerate a wide range of environmental extremes, such as fragmentation, pollution, etc. (Chang et al. 2009), which also contributes to their potential as ecological indicators.

Ferns are relatively easy to recognize in the field and generally have a small size; thus, they are quick to observe and collect. And as mentioned before, this group is rich in species and probably occurs in most vegetation formations (Tuomisto and Ruokolainen 1993). These characteristics make this group potentially suitable as ecological indicators (Tuomisto and Ruokolainen 1993; Della and Falkenberg 2019a).

## 25.5 Studies with Ferns as Ecological Indicators

In 2019, Della and Falkenberg presented a review of 134 studies that used ferns as ecological indicators. The authors stressed the gradual increase of the published works for this purpose in the last 40 years. These studies were carried out on all continents, generally using indicator species.

The methodologies used to select indicators consisted in the analysis of the presence, abundance, frequency, richness, or diversity of fern species and the correlation (through statistical analysis) of these characteristics with abiotic (e.g., soil characteristics, luminosity, humidity, etc.) or biotic (richness and diversity of other groups of plants) factors.

Few studies have presented criteria in the methodology for the selection of ecological indicators. This aspect is very negative because much time and resources can be spent studying species that, later, will not be easily used as indicators (Della and Falkenberg 2019a).

### 25.5.1 Indication Types

Della and Falkenberg (2019a) presented also seven types of ecological indication for which ferns are commonly used. This classification was established based on the analysis of the objective with which these indicators were proposed. These types of indications will be approached in the next sections.

#### 25.5.1.1 Classification of Vegetation, Soils, Environments, and Ecosystems

In this type of indication, the presence, abundance, frequency, and diversity of ferns are directly correlated with humidity, texture, pH, and presence of soil nutrients, air humidity, luminosity, precipitation, altitude, topography, and atmospheric temperature. For example, Tuomisto and Poulsen (1996) verified two-thirds of fern species (which occur in the Amazon) are restricted to poor, intermediate, or nutrient-rich soils and less than one-third occurs in all soil types. In the same area, Tuomisto and Poulsen (2000) checked many species are restricted to poorly drained and flooded soils, such as *Cyathea pungens* (Willd.) Domin, *Cyclodium meniscioides* (Willd.) C. Presl, *Polybotrya caudata* Kunze, *Selaginella exaltata* (Kunze) Spring, and *Salpichlaena hookeriana* (Kuntze) Alston. The opposite distribution pattern was verified by *Adiantum tomentosum* Klotzsch, *Salpichlaena volubilis* (Kaulf.) J. Sm., and *Trichomanes pinnatum* Hedw.; these plants were found mainly on slopes and tops with better drainage. And, Tuomisto and Ruokolainen (1993) found many Amazonian species are restricted to clay soils, as *Adiantum terminatum* Kunze ex Miq., *Asplenium hallii* Hook., *A. serratum* L., *Lindsaea lancea* (L.) Bedd., and *Trichomanes elegans* Rich.; and others are restricted to sandy soils: *Arachniodes macrostegia* (Hook.) R.M. Tryon and D.S. Conant, *Lindsaea divaricata* Klotzsch, *Trichomanes pellucens* Kunze, *T. martiusii* C. Presl, and *T. bicorne* Hook.

Ferns have edaphic preferences and specializations that allow them to survive on different types of soils (sandy, clay, mud, etc.). These distinct soils generally have distinct vegetations (structure and physiognomy). Thus, ferns can be used to classify soil types and vegetation types (edaphically defined).

#### 25.5.1.2 Environmental Integrity (or Quality)

Intact or well-preserved forests have higher humidity and reduced solar radiation in the understory. Many species of these environments are sensitive to changes in these two conditions; thus, they can be good indicators. Ferns are drastically affected by the canopy opening and by subsequent environmental changes (mainly in the composition and humidity). Thus, the richness and diversity of ferns are higher in environments well-preserved. We also should consider the higher diversity of ferns in shaded forest environments than in the open, which means that there are more umbrophiles ferns than heliophiles (pioneers) (Della and Falkenberg 2019a).

Ferns have already been used as ecological indicators in order to compare (and evaluating) the environmental quality of forest systems of rubber production (mono-specific forest) with primary forests (located in Sumatra) (Beukema and Van Noordwijk 2004)). For this purpose, the richness and diversity of species in these two environments were analyzed. Forest systems of rubber production can be used for the preservation of species (especially in areas where the primary forest has already disappeared) because the richness and diversity of ferns are higher in these environments than in regenerating areas.

The Hymenophyllaceae species are indicators of the well-preserved tropical lowland cloud forest (Gehrig-Downie et al. 2012). There were great differences in the diversity and distribution of this family in preserving areas and altered areas located in French Guiana. They are more diverse and frequent in preserved areas, due to the high humidity. This family does not have cuticles and well-developed stomata; thus, they are sensitive to loss of water and are dependent on humid habitats (with frequent precipitation and low evaporation).

Some ferns grow only in undisturbed places, i.e., areas with deep soil profiles with lots of organic matter, high humidity, and reduced solar radiation (Romero et al. 2008). Thus, they can be indicators of environment integrity (primary vegetation of the temperate forest), such as *Adiantum andicola* Liebm., *Adiantum poiretii* Wikstr., *Argyrochosma incana* (C. Presl) Windham, *Asplenium blepharophorum* Bertol., *Dryopteris pseudofilix-mas* (Fée) Rothm., *Equisetum hyemale* L., and *Pteris cretica* L.

Amphibious ferns (aquatic macrophytes) can be indicative of good conditions for riparian habitats (Fares et al. 2020). The margins of preserved waterbodies have great vegetation cover, which means they are environments with great humidity and shading. These characteristics are tolerated by some amphibious ferns, such as *Triplophyllum dicksonioides* (Fée) Holttum, being, therefore, a good indicator.

#### 25.5.1.3 Disturbance

The alteration (disturbance) of a forest promotes a simplification of the community structure; this can be evidenced by the reduction of the richness and diversity of the ferns (but, not only of them). The alteration causes erosion, salinization, loss of nutrients, and changes in the soil structure (Romero et al. 2008). Sensitive species to alteration of the soil humidity and luminosity will be affected, reducing their frequency and abundance. Tolerant species to sunny and dry areas can occupy these altered areas and, therefore, become dominant in them. In this way, we have the reduction of ferns sensitive to increase of luminosity and temperature (and possible soil humidity), and on the other hand, there is the proliferation of tolerant species to these conditions in forest areas that were altered.

An indicator species of environmental degradation generally have a strong tolerance to the sun and restricted distribution to areas with a high level of deterioration, such as *Cheilanthes bonariensis* (Willd.) Proctor and *Pellaea ternifolia* (Cav.) Link (Romero et al. 2008). These species have a great capacity of dispersion and colonize easily new "empty" environments, cleared slopes, and burned forests. Also, their spores germinate quickly after the sowing.

There is a negative correlation between urbanization and the richness of ferns (Mallmann and Schmitt 2014). Forest fragments closed in highly urbanized areas have low species richness, while areas farther from cities have a higher richness of ferns.

The richness of ferns is positively correlated to moderately low values of the soil pH and electric conductivity (Bergeron and Pellerin 2014). The values of electric conductivity are generally higher on edges of urban forests than forest cores due to the propagation of calcium debris (e.g., calcium hydroxide) from urban buildings.

Edges of urban forests and trails are often flat land, artificially leveled by humans. This causes homogenization in the landforms and probably contributes to the deterioration of several abiotic conditions (Bergeron and Pellerin 2014). Some species are known to tolerate these characteristics, among them *Pteridium aquilinum* (L.) Kuhn. This plant can survive in dry soils with intense light, high competition, and great perturbations.

#### 25.5.1.4 Regeneration/Restoration of Habitats and/or Ecosystems

Some species are dominant or have a higher frequency in altered areas, and others are dominant or more frequent in conserved places. Thus, forest environments in more advanced stages of regeneration will have more species umbrophiles (contrary to what is expected for more altered areas).

Alterations in the composition (or substitution) of ecological groups of ferns can be used as ecological indicators. Thus, the domain of pioneer ferns indicates lack or delay of forest regeneration, and the substitution of a group by umbrophiles, ferns indicate regeneration.

Species with green spores (such as *Equisetum arvense* L., *Equisetum sylvaticum* L., *Matteuccia struthiopteris* (L.) Tod., *Onoclea sensibilis* L., *Osmundastrum cinnamomeum* (L.) C. Presl, and *Osmunda claytoniana* L.) can be indicators of more advanced stages of regeneration because they are sensitive to the drought of the habitat (as seen by Bergeron and Pellerin 2014). The chloroplasts of these plants explode without humidity suitable, what avoid germination.

Indicator species of successional stages are often used in laws or resolutions, which are important for environmental licensing. In this sense, Della and Falkenberg (2019b) evaluated the quality of indication of the ferns considered in the resolutions of the National Environment Council (Conama) for the successional stages of forests, altitude fields, and restingas. The Conama is an advisory and a deliberative agency of Brazil's National Environment System, which creates norms and criteria for the development of the National Environment Policy (http://www.mma. gov.br/ port/conama/estr.cfm).

To assess ferns as ecological indicators in these resolutions, the authors considered two groups of criteria, the first includes those characteristics that define a good ecological indicator (for more details, see Della and Falkenberg 2019b), and the second that of plants on the initial stage, such as the capacity to survive in altered environments, in soils sometimes compressed, tolerance to intense luminosity, reduction of humidity (soil and air), temperature variations, fast grow, etc., and advanced stage (tolerance to shading, increase of humidity, soil and air, etc.).

The authors found few ferns as indicators in the analyzed resolutions, indicating the secondary importance of this group in these laws. There are several inconsistencies in these resolutions, as the presence of nonexistent or problematic taxa or species that are not characteristic of the successional stages for which they were considered. Only one species was considered a good ecological indicator of the initial stages of the succession of forests, altitude fields, and restingas, *Pteridium esculentum* subsp. *arachnoideum* (Kaulf.) J.A. Thomson (considered as *P. aquilinum* in these resolutions). This plant has all the criteria that define a good indicator species and that define successional stages.

#### 25.5.1.5 Climate Changes

This is one of the most discussed topics at present because climate changes can lead to the extinction of many species. Alterations of specific conditions of the environment (caused by climate changes) can promote the expansion or retraction of the distribution of climate-sensitive ferns. Ferns can be good indicators and monitor these changes.

Temperature is the main determining factor in the distribution of species in mountainous habitats in Germany (Bässler et al. 2010). The increase of 1.8 °F in the temperature will increase the risk of extinction for many species that live in these regions. In pessimistic scenarios of global warming, many species probably will not survive. *Athyrium distentifolium* Tausch ex Opiz is a fern with a circumpolar distribution (but, it is restricted to the mountain and alpine areas) that can be considered vulnerable to global warming. Thus, the reduction of the occurrence area of this plant can be an indicator of the impacts occasioned by climate changes or interpreted as a previous warning of a natural response to these changes.

An intensive and extensive program of monitoring indicator species to detect the impacts of climatic changes has been proposed for Japan (Higa et al. 2013). Data of geographic distribution of vascular plants and abiotic variables (as temperature, humidity, vegetation, etc.) were obtained for this country. Afterward, ecological niche modeling was carried out in order to predict the future distribution of these taxa. *Matteuccia struthiopteris* (L.) Tod. was considered a good indicator of climatic changes in this study, as it is sensitive to temperature changes. Thus, it can be used to monitor climate changes.

#### 25.5.1.6 Contamination of Air, Soil, or Water

Ferns were studied in order to indicate contamination or pollution by heavy metals (antimony, arsenic, cadmium, lead, chromium, gold, etc.) or by organic matter, from aquatic or terrestrial environments or air. The concentration of heavy metals in the dry weight of the plants was indicative of contamination or pollution, as well as the survival capacity and abundance of these plants in such environments, presence of injuries (mainly leaf necrosis), and the absence of sensitive plants. This type of indication occurs because of the capacity to absorb and accumulate heavy metals (in the rhizomes or leaves) that some ferns have.

Ferns are good indicators of arsenic accumulation in abandoned mines located in the Republic of Korea (Chang et al. 2009). These plants store arsenic in their rhizomes, so they can survive in these contaminated areas. *Pteris vittata* L., *Asplenium nidus* L., *Ceratopteris richardii* Brongn., and *Davallia canariensis* (L.) Sm are common species in abandoned mines, indicating they can be useful for phytoremediation.

Another species that occurs in soils with different concentrations of heavy metals is *Athyrium filix-femina* (L.) Roth (Samecka-Cymerman et al. 2011). This plant is able to accumulate different metals in its rhizomes and leaves. Thus, *A. filix-femina* is a good indicator of contaminated soils.

#### 25.5.1.7 Association with Other Groups of Organisms

Ferns were tested to indicate the presence of plants and animals in a specific location. It is assumed that there is a strict correlation or association between the plants or that the fern is a key species in an ecosystem. This type of indication was used in order to predict the performance (richness and distribution) of the associated species.

Duque et al. (2005) carried out a study in the Colombian Amazon to verify whether ferns and Melastomataceae could explain the patterns of the composition of other vascular plants. The authors performed several correlation analyses between these groups, the soil, and other vascular plants. However, they have not obtained significant evidence that ferns and Melastomataceae can predict the main patterns in the composition of forest species. For this region, these groups predict better soil, landscape, and spatial variables, but more studies are necessary.

Ferns can be used indirectly to understand the structure, dynamics, and behavior of animal populations (Ramírez et al. 2021). This is because several studies have already demonstrated that these plants are capable of predicting soil properties; thus, they would be able to predict primary productivity and food availability for forest animals (Ramírez et al. 2021).

### 25.5.2 Final Considerations on the Types of Indication

The most common type of indication in studies published from 1980 to 2020 was classification of vegetation, soils, environments, and ecosystems. This type of indication was the objective of 44% of the studies in the period. The second most frequent type was environmental integrity (or quality) with 14%, and the third was disturbance with 11%. Fig. 25.1 shows the complete percentage of the other types of indication.

Many studies have more than one type of indication; for example, Bergeron and Pellerin (2014) presented indicators of environmental integrity and disturbance; and Della (2016) discussed the indicators of classification of vegetation and regeneration of habitats.

## 25.6 Conclusions and Future Directions

As seen in the sections above, there has been an increase in scientific interest in the development and application of ecological indicators. This occurs, generally, from the need to assess or predict future natural and/or anthropogenic disturbances. Ferns have been used as ecological indicators for several objectives/purposes, as seen in Sect. 0. This demonstrates the great potential that this group has as an indicator. However, there are some difficulties, which must be remedied or more discussed in the future in establishing ferns as indicators, such as the following:

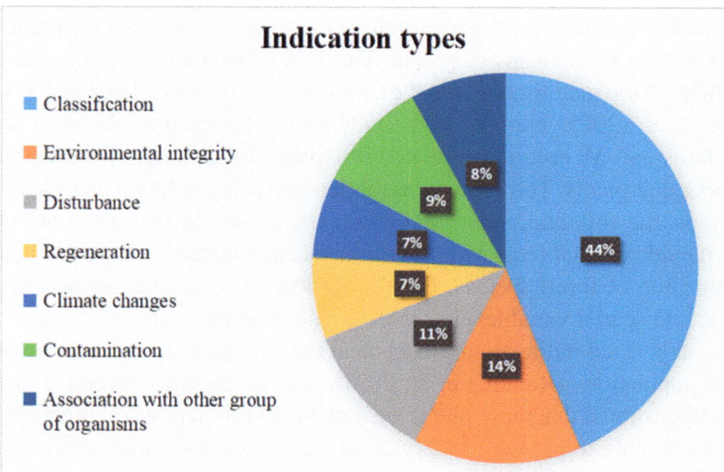


**Fig. 25.1** The proportion of the types of indication verified in 153 studies carried out with ferns published from 1980 to 2020. Figure adapted from Della and Falkenberg (2019a)

- Practical application of ecological indicators because many of these are only established in theory. There is still a great distance between science, society, and politics, which often makes the use of ecological indicators unfeasible.
- Define criteria for the selection of ecological indicators, since many studies do not present criteria like the ones shown in Della and Falkenberg (2019a). Besides, many times the indicators are established by statistical analyses, without thinking about how employable in a practical context (e.g., such as monitoring climate change) they would be. In this sense, the criteria could help to select good ecological indicators.
- Basic ecological information is lacking for many species, such as the area of occurrence, precise geographical coordinates, description of the habitat, altitude, luminosity and humidity conditions, successional stage, as well as frequency, abundance, etc. These data are extremely important for the establishment of ecological indicators.
- Taxonomic revisions for groups with complex taxonomy. Taxa with complex taxonomy are more likely to be confused with other groups, thus reducing their potential of indication.
- Financial support because studies require field activities, which are often costly, depending on the location, even more in times of financial crisis and scientific negationism.

Despite these difficulties, the number of studies using ferns as ecological indicators has increased in recent years. In the future, I hope that these obstacles will be solved and that more studies will be done, given the great potential of indication that the ferns present.

## References

Abotsi KE, Boseb R, Adjossoua K, Deblauweb V, Rouhane G, Seglaa KN, Atsria KH, Kokou K (2020) Ecological drivers of pteridophyte diversity and distribution in Togo (West Africa). Ecol Indic 108:105741

Bässler C, Müller J, Hothorn T, Kneib T, Badeck F, Dziock F (2010) Estimation of the extinction risk for high-montane species as a consequence of global warming and assessment of their suitability as crosstaxon indicators. Ecol Indic 10:341–352

Bergeron A, Pellerin S (2014) Pteridophytes as indicators of urban forest integrity. Ecol Indic 38: 40–49

Beukema H, Van Noordwijk M (2004) Terrestrial pteridophytes as indicators of a forest-like environment in rubber production systems in the lowlands of Jambi, Sumatra. Agric Ecosyst Environ 104:63–73

Cáceres M, Legendre P, Moretti M (2010) Improving indicator species analysis by combining groups of sites. Oikos 119(10):1674–1684

Cárdenas GG, Halme KJ, Tuomisto H (2007) Riqueza y distribución ecológica de especies de pteridofitas en la zona del río Yavarí-Mirím, Amazonia Peruana. Biotropica 39:637–646

Carneiro RMA (2004) Bioindicadores vegetais de poluição atmosférica: uma contribuição para a saúde da comunidade. Dissertation, University of São Paulo

Chang JS, Yoon IH, Kim KW (2009) Heavy metal and arsenic accumulating fern species as potential ecological indicators in as-contaminated abandoned mines. Ecol Indic 9:1275–1279

Della AP, Falkenberg DB (2019b) Pteridófitas usadas na legislação como indicadoras de estágios sucessionais no Estado de Santa Catarina, Brasil. Hoehnea 46(2):e572018
Della AP, Falkenberg DB (2019a) Pteridophytes as ecological indicators: an overview. Hoehnea 46 (1):e522018
Della AP (2016) Pteridófitas como indicadores ecológicos: revisão geral e aplicações em Santa Catarina. Graduation work, Federal University of Santa Catarina
Dufrêne M, Legendre P (1997) Species assemblages and indicator species: the need for a flexible asymmetrical approach. Ecol Monogr 67:345–366
Duque AJ, Duivenvoorden JF, Cavelier J, Sánchez M, Polanía C, León A (2005) Ferns and Melastomataceae as indicators of vascular plant composition in rain forests of Colombian Amazonia. Plant Ecol 178:1–13
Fares ALB, Calvão LB, Torres NR, Gurgel ESC, Michelan TS (2020) Environmental factors affect macrophyte diversity on Amazonian aquatic ecosystems inserted in an anthropogenic landscape. Ecol Indic 113:106231
Freitas AVL, Leal ER, Prado MU, Iannuzzi L (2006) Insetos como indicadores de conservação da paisagem. In: Rocha CFD, Bergallo HG, Van Sluys M, Alves MAS (eds) Biologia da conservação: essências. RiMa Editora, São Carlos. ISBN: 8576560895 9788576560890
Gehrig-Downie C, Marquardt J, Obregón A, Bendix J, Gradstein SR (2012) Diversity and vertical distribution of filmy ferns as a tool for identifying the novel forest type "tropical lowland cloud forest". Ecotropica 18:35–44
Heink U, Kowarik I (2010b) What criteria should be used to select biodiversity indicators? Biodivers Conserv 19:3769–3797
Heink U, Kowarik I (2010a) What are indicators? On the definition of indicators in ecology and environmental planning. Ecol Indic 10:584–593
Higa M, Nakao K, Tsuyama I, Nakazono E, Yasuda M, Matsui T, Tanaka N (2013) Indicator plant species selection for monitoring the impact of climate change based on prediction uncertainty. Ecol Indic 29:307–315
Jaffe D, McDonough C (2009) Indicator species. http://www.eoearth.org/view/article/153801. Accessed 27 Jan 2021
Landres PB, Verner J, Thomas JW (1988) Ecological uses of vertebrate indicator species: a critique. Conserv Biol 2(4):316–328
Lawton JH, Gaston KJ (2001) Indicator species. Encycl Biodiv 3(1):437–450
Mallmann IT, Schmitt JL (2014) Riqueza e composição florística da comunidade de samambaias na mata ciliar do rio Cadeia, Rio Grande do Sul. Brasil Ciênc Florest 24:97–109
McGeoch M (1998) The selection, testing and application of terrestrial insects as bioindicators. Biol Rev 73:181–201
Moran RC (2008) Neotropical genera of ferns and lycophytes: a guide for students. New York Botanical Garden, New York
Morrison ML (1986) Bird populations as indicators of environmental change. In: Johnston RF (ed) Current ornithology, vol 3. Plenum, New York, pp 429–451. ISBN: 978-1-4615-6784-4
Nascimento JLX, Sales-Júnior LG, de Sousa AEBA, Minns J (2005) Avaliação rápida das potencialidades ecológicas e econômicas do Parque Nacional de Ubajara, Ceará, usando aves como indicadores. J Ornithol 1(1):33–42
Niemi GJ, McDonald ME (2004) Application of ecological indicators. Annu Rev Ecol Evol Syst 35:89–111
Nóbrega GA, Eisenlohr PV, Paciencia MLB, Prado J, Aidar MPM (2011) A composição florística e a diversidade de pteridófitas diferem entre a floresta de restinga e a floresta ombrófila densa das terras baixas do Núcleo Picinguaba/PESM, Ubatuba/SP. Biota Neotrop 11:153–164
Noss RF (1990) Indicators for monitoring biodiversity: a hierarchical approach. Conserv Biol 4 (4):355–364
Oster M, Persson K, Eriksson O (2008) Validation of plant diversity indicators in semi-natural grasslands. Agric Ecosyst Environ 125(1):65–72

Patil S, Lavate R, Rawat V, Dongare M (2016) Diversity and distribution of pteridophytes from Satara District, Maharashtra (India). Plant Sci Today 3:149–156

Pearson DL (1994) Selecting indicator taxa for the quantitative assessment of biodiversity. Philos Trans R Soc Lond 345:75–79

PPG I (2016) A community-derived classification for extant lycophytes and ferns. J Syst Evol 54: 563–603

Ramírez GGC, Jones MM, Heymann EW, Tuomisto H (2021) Characterizing primate home-ranges in Amazonia: using ferns and lycophytes as indicators of site quality. Biotropica 00:1–11

Rapport DJ, Hildén M (2013) An evolving role for ecological indicators: from documenting ecological conditions to monitoring drivers and policy responses. Ecol Indic 28:10–15

Ratcliffe DA (1984) The peregrine breeding population of the United Kingdom in 1981. Bird Study 31(1):1–18

Raven PH, Evert RF, Eichhorn SE (2007) Biologia Vegetal, 7th edn. Guanabara Koogan, Rio de Janeiro. ISBN: 9788527712293

Rempel RS, Andison DW, Hannon SJ (2004) Guiding principles for developing an indicator and monitoring framework. For Chron 80(1):82–90

Romero LR, Pacheco L, Zavala-Hurtado JA (2008) Pteridofitas indicadoras de alteración ambiental en el bosque templado de San Jerónimo Amanalco, Texcoco, México. Rev Biol Trop 56:641–656

Salovaara KJ, Cárdenas GG, Tuomisto H (2004) Forest classification in an Amazonian rainforest landscape using pteridophytes as indicator species. Ecography 27:689–700

Samecka-Cymerman A, Kolon K, Stankiewicz A, Kaszewska J, Mróz L, Kempers AJ (2011) Rhizomes and fronds of Athyrium filix-femina as possible bioindicators of chemical elements from soils over different parent materials in Southwest Poland. Ecol Indic 11:1105–1111

Sampson AW (1939) Plant indicators: concept and status. Bot Rev 5(3):155–206

Scotti AGL, Santos MFR, Müller M, Jung DC (2013) Levantamento de briófitas e pteridófitas em mata ciliar no centro urbano de Caçador-SC. Ignis 2(1):6–22

Tryon RM (1972) Endemic areas and geographic speciation in tropical American ferns. Biotropica 4:121–131

Tuomisto H, Poulsen AD (1996) Influence of edaphic specialization on pteridophyte distribution in neotropical rain forests. J Biogeogr 23:283–293

Tuomisto H, Poulsen AD (2000) Pteridophyte diversity and species composition in four Amazonian rain forests. J Veg Sci 11:383–396

Tuomisto H, Ruokolainen K (1993) Distribution of Pteridophyta and Melastomataceae along an edaphic gradient in an Amazonian rain forest. J Veg Sci 5:25–34

Turnhout E, Hisschemöller M, Eijsackers H (2007) Ecological indicators: between the two fires of science and policy. Ecol Indic 7:215–228

**Aline Possamai Della** was born in Santa Catarina (Brazil). She has a degree in Biological Sciences from the Federal University of Santa Catarina (Florianópolis, Brazil) and a master's degree in Botany from the University of São Paulo (São Paulo, Brazil). She is currently a Ph.D. student in the postgraduate program in Botany at the Institute of Biosciences of the University of São Paulo. She has experience in the field of botany, with an emphasis on plant systematics, working mainly with taxonomy, phylogeny, biogeography, and ecology of ferns and lycophytes.

# Diversity of Ferns and Lycophytes at Different Spatial Scales, along Environmental Gradients, and in the Anthropogenic Landscape

26

Klaus Mehltreter

**Abstract**

The approximately 12,000 extant fern and lycophyte species comprise about 4% of the vascular plant diversity, and still ca. 40–50 new species are described each year. Although ferns possess a nearly worldwide distribution, they are especially abundant in tropical mountains at mid-elevations and on remote oceanic islands. The Tropical Andes and Southeast Asia including the Malaysian Archipelago are the most diverse areas. Hotspots of fern diversity coincide relatively well with hotspots of many other organisms. Tropical islands present much steeper species-area curves ($z = 0.347$) than continental areas ($z = 0.184$) when modeled with the power curve (species = c ·area$^{z}$). Changes of beta-diversity with the increasing distance between two areas are indicating the dispersal limitation of fern spores at larger scales. Fern-rich, undisturbed forests can contain up to 70 species per 400 m$^2$ but commonly range around 10–40 species per 400 m$^2$. Habitat destruction and land use change are the major threats to fern diversity. Agroecosystems can rarely hold more than 10–20% of the former fern diversity and only if they are embedded in natural forest fragments that serve as refugia.

## 26.1 Introduction

Ferns and lycophytes are the second and third largest group of vascular plants, with currently 10,600 and 1300 registered species (PPG 1 2016, Christenhusz and Byng 2016), only after angiosperms with over 300,000 species. The earliest ferns (e.g., Equisetales) originated in the mid-late Silurian ca. 431 mya and leptosporangiate ferns in the Carboniferous ca. 357 mya (Testo and Sundue 2016) and dominated as a terrestrial plant group during the Paleozoic (until 260 mya) until they were

K. Mehltreter (✉)
Red de Ecología Funcional, Instituto de Ecología, A.C., Xalapa, Veracruz, Mexico


J. Marimuthu et al. (eds.), *Ferns*, https://doi.org/10.1007/978-981-16-6170-9_26

outcompeted after some massive extinction events, first by gymnosperms in the Mesozoic (260–65 mya) and then by angiosperms in Cenozoic (65 mya to present). In the Cretaceous (135–65 mya), most modern extant ferns (e.g., Polypodiales) radiated, once angiosperms were already the dominant plant group (Schneider et al. 2004). At present, ferns and lycophytes still comprise 3.6 and 0.4% of all living vascular plant species on Earth, possess a nearly worldwide distribution, and can be encountered in high abundances in tropical forest understories and canopies, especially in mountains at mid-elevation, in humid temperate rainforests, and on oceanic islands (Tryon 1970, 1972; Barrington 1993; Kato 1993; Kessler 2010; Kreft et al. 2010; Mehltreter et al. 2010).

In this chapter, I will (1) give a brief introduction on the discovery of fern diversity; (2) provide an overview of fern species richness at a global scale; (3) compare the distribution patterns of ferns with hotspots of biodiversity in general; (4) discuss beta-diversity, the species turnover between regions on an example of American Aspleniaceae; (5) investigate the species-area relationship of ferns on two spatial scales; (6) discuss their distribution across several environmental gradients; and (7) conclude with a critical perspective on ferns in an anthropogenic landscape.

Two declarations should be made here. First, throughout the chapter, I will use the term "ferns" in a wide sense including lycophytes (former pteridophytes) unless when both groups are mentioned separately. Second, one should bear in mind that a major part of this chapter refers to species richness alone, whereas species diversity encompasses species richness (the number of species) and abundance (number of individuals), also considering the evenness of species, their relative frequency distribution within a plant community (Magurran 2004).

## 26.2 The Discovery of Fern Diversity

After Linnaeus developed in 1735 (*Systema Naturae*) and 1751 (*Philosophia Botanica*) his hierarchical classification system and binomial nomenclature to name species, he described 5900 plants in his book *Species Plantarum* (1753) and increased this number to 7700 plant species in his tenth edition of *Systema Naturae*. In his second volume of *Species Plantarum*, he listed 212 fern and 25 lycopods out of about 6000 plant species. It is interesting that more than 250 years ago, ferns made up nearly the same proportion (4%) of vascular plants than in our most updated recent lists (PPG 1 2016; Christenhusz and Byng 2016), which currently contain 50 times more species.

Floras at all scales are fundamental tools for the study of the botanical diversity for specific areas and the starting point for many other biological studies. Several recent floristic projects and fern checklists such as in Brazil, China, Colombia, the Malaysian Archipelago, Mesoamerica, or the Solomon Islands and their often online accessible databases to Internet portals such as Global Biodiversity Information Facility Portal (gbif.org) have provided an enormous improvement in our knowledge about diversity hotspots and species distribution ranges on which we can rely for future research. A database search in the Annual Review of Pteridological Research

(ARPR) from 2012 to 2020 reveals that each year approximately 50 new fern species are described (e.g., 2018, 49 spp.; 2019, 45 spp.; 2020, 50 spp.) compared to about 2000 new angiosperm species (Christenhusz and Byng 2016). This means that in the next 20 years, still about 1000 more fern species might be described. However, the ongoing habitat destruction, including the introduction of invasive species, the decreasing support for floristic projects, and the formation of well-trained taxonomists will slow down discoveries of new species or even extinguish rare species before their discovery. An indicator of the current fern vulnerability is the number of species that are only known from the type collection and have never been collected again. Such species comprise, for example, 2.3% of the ferns of Mexico (Mickel and Smith 2004). The most thorough and detailed evaluation of the extinction risk of ferns was performed for Europe. In this evaluation, 19.9% of the 194 species were considered threatened, and 1 species was regionally extinct (García Criado et al. 2017).

## 26.3 Fern Richness at a Global Scale

At the global scale, ferns are distributed along a latitudinal gradient with species-richest areas in the tropics close to the equator and a decrease toward the poles. This gradient was already described by Tryon (1986). Little has changed in this general pattern except for the increasing numbers of fern species reported in recent floras. For instance, Tryon considered the Amazonian basin as a "fern desert" with only about 220 species because of the incomplete knowledge of that time, but the current Flora of Brazil (2021) has now registered already 573 fern species for Amazonia.

Comparing the continents, the Americas (Fig. 26.1) and Asia (Fig. 26.2) are the two hotspots with more than 5000 species each, followed by Africa, Australia, and Europe (Figs. 26.2 and 26.3). In the Americas, the most diverse hotspots run along the tropical Andes from the Yungas in northernmost Argentina to the mountain ranges of the Sierra Madre in Mexico and the Caribbean Islands (Fig. 26.1). The most diverse American country is Colombia with 1689 species and just located in the center of all dispersal routes from South America into Mesoamerica and the Caribbean (Bernal et al. 2019). The Caribbean Islands provide a valuable addition to the continental American fern flora, because of 5–16% endemic species that are restricted in their distribution on each of the Greater Antilles, where species are concentrated within a much smaller area than on the American continent.

The Asian continent, although limited to the Northern Hemisphere, is most diverse in its tropical south and southeast from India, Malaysia, and Vietnam to China (Fig. 26.2). However, the large islands of Indonesia (Sumatra, Java, Borneo, and Sulawesi), Papua New Guinea, and the vast number of Western Pacific Islands from Japan, and Taiwan, to the Philippines and the Solomon Islands provide a highly diverse, and often isolated, oceanic, floristic component that enriches and interacts with the continental flora. The Pacific Islands are scattered over an enormous area spanning nearly half of the globe (Figs. 26.1 and 26.2). Although not reaching the diversity of larger islands, altogether they contribute many rare and endemic species

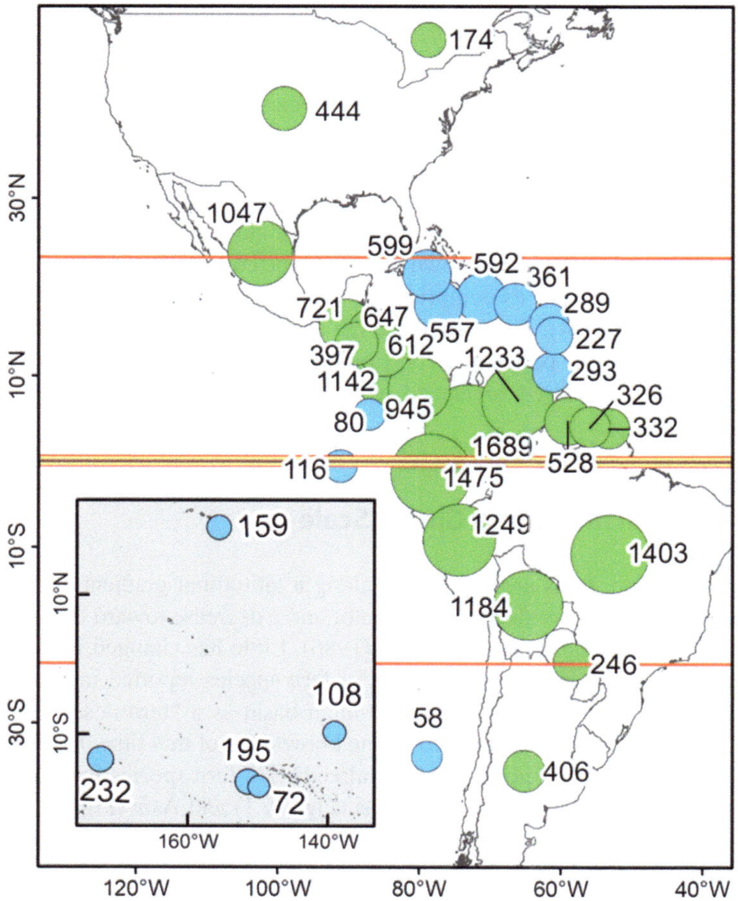


**Fig. 26.1** Species-rich continental areas (*green*) and islands (*blue*) of ferns (incl. lycophytes) in the Americas and the Eastern Pacific with the Hawaiian Islands, Western Samoa, Society Islands, Moorea, and Marquesas Islands (inner map section). The size of circles is proportional to species number. Data sources: Flora of North America 1993, Davidse et al. 1995, Boggan et al. 1997, Jorgensen and Léon-Yánez 1999, Mickel and Smith 2004, Hokche et al. 2008, Jorgensen et al. 2014, Ranker 2016, Villaseñor 2016, Sánchez 2017, Zuloaga and Belgrano 2016, Bernal et al. 2019, Flora of Brazil 2021, Hassler 2021

to our fern heritage (e.g., 74% of Hawaiian ferns are endemic; Ranker 2016). Australia with its 488 species is mainly governed by dry climates except for its coastal fringes, especially along its eastern side and the extreme southwest (Flora of Australia 1998). Tasmania and New Zealand are the only larger, temperate, fern-rich oceanic islands, south and southeast of Australia.

Despite its size and extension into both hemispheres, the relatively low fern diversity of Africa with 1441 species (Roux 2009) comes as a surprise, even if one accounts for the large, desertic areas of this continent (Fig. 26.3). Africa has fewer endemic fern lineages than the Neotropics and Asia, and several lineages of

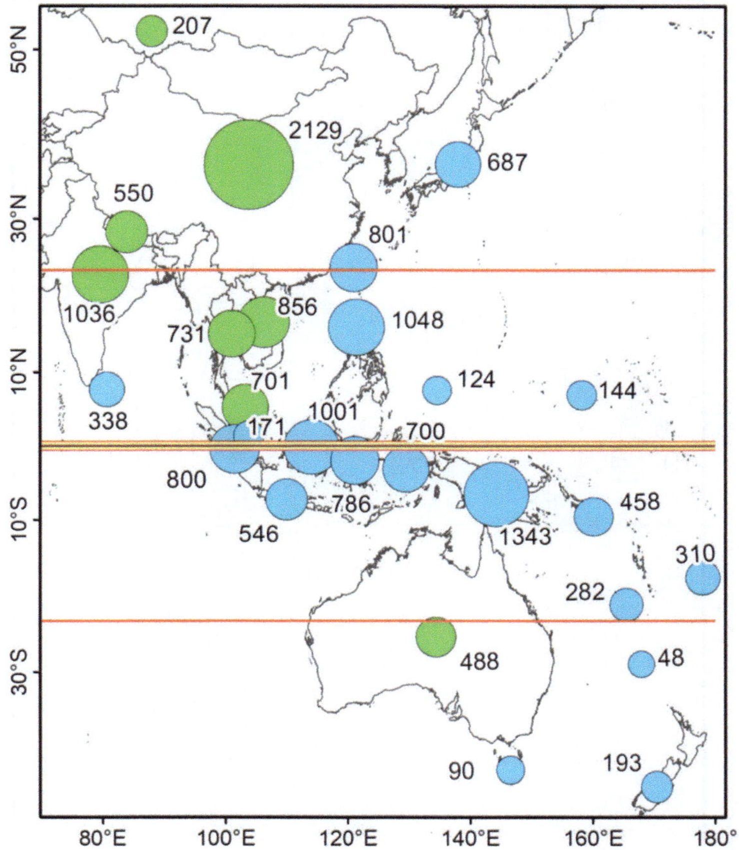


**Fig. 26.2** Species-rich continental areas (*green*) and islands (*blue*) of Australasia and the Western Pacific (from north to south) with Japan, Taiwan, the Philippines, Palau, Micronesia, Papua New Guinea, the Solomon Islands, Fiji, New Caledonia, Norfolk Island, and New Zealand. The size of circles is proportional to species number. Data sources: Parris et al. 1992, Duncan and Isaac 1994, Flora of Australia 1998, Shaffer-Fehre 2006, Knapp 2011, 2013, Zheng-Yi et al. 2013, Chen et al. 2017, Knapp and Hsu 2017, Ebihara and Nitta 2019, Hassler 2021

Polypodiaceae migrated from these areas into Africa after the Gondwana breakup and after the diversification of Polypodiaceae in the Eocene 40 mya (Schneider et al. 2004; Janssen et al. 2007). Tropical East Africa with its high mountains along the eastern central lakes is the continental hotspot followed by tropical West Africa from Guinea to Cameroon and South Africa that has been considered as a separate floral kingdom (Takhtajan 1986) and has about 12% endemic fern species. The oceanic hotspot of Africa is Madagascar (ca. 46% endemism) with the Mascarene Islands including Reunion Island and Mauritius. It is still unclear why Africa does not hold more fern species, but there were speculations that it has undergone long periods of drought and that their extended savannas are prone to frequent natural and induced fires. Yet, without any doubt, Africa does not have a close connection to large island

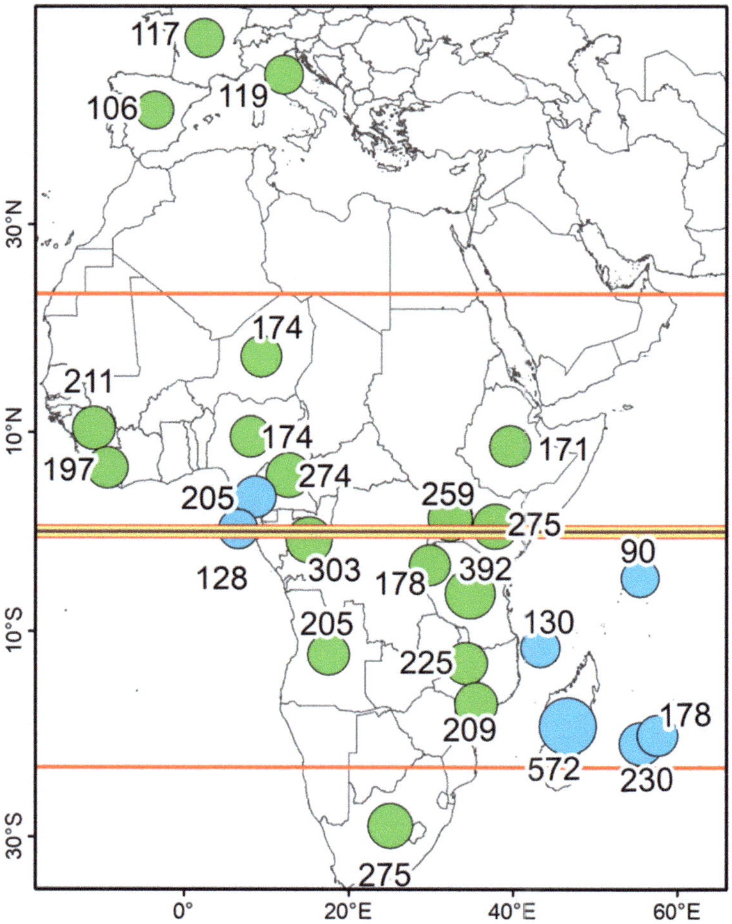


**Fig. 26.3** Species-rich continental areas (*green*) and islands (*blue*) of Africa and Southern Europe. The size of circles is proportional to species number. Data sources: Roux 2009, Grangaud 2010, García Criado et al. 2017, Hassler 2021

chains as the American continent with the Caribbean Islands and the Asian continent with the Malaysian archipelago. Finally, Europe has one of the best-studied fern floras. With its 194 species (García Criado et al. 2017), it includes only three countries – France, Italy, and Spain – with still more than 100 fern species each, perhaps as the consequence of major extinctions during the ice ages, especially north of the Alps.

### 26.3.1 How Well Do Ferns Fit into the Biodiversity Hotspots?

Fern diversity hotspots coincide very well with the 25 hotspots that were suggested by Myers et al. (2000) with the idea to protect the largest number of species of living organisms in the smallest possible area. The best coincidences comprise in the Americas: (1) the tropical Andes, (2) Chocó-Darién, (3) Mesoamerica, (4) the Caribbean Islands, (5) the Brazilian Atlantic Forest, and the (6) Cerrado; in Africa: (7) Eastern Topical Africa, (8) Western Tropical Africa, (9) South Africa, and (10) Madagascar including the Mascarenes; and in Asia: (11) Western India, (12) Sri Lanka, (13) Southeast Asia, (14) South Central China, (15) Indonesia, and (16) the Philippines. Some of Myer's hotspots that are already less diverse in ferns but still have a high number of endemic species or a high abundance and biomass of ferns are (17) Polynesia/Micronesia, (18) New Caledonia, and (19) New Zealand and the temperate rainforests of (20) Chile and (21) California/Oregon. The remaining four plant hotspots of drier or more temperate climates are the least diverse in ferns: (22) Mediterranean Basin, (23) the Caucasus, (24) Southwest Australia, and (25) the African Karoo. If we consider centers of fern diversity, the most evident gaps of Myer's hotspots are Papua New Guinea, one of the fern-richest islands, Taiwan, Japan, and the Solomon Islands (see Figs. 26.1, 26.2, and 26.3). These four areas were also considered as hotspots of vascular plant diversity (Barthlott et al. 2005; Mutke et al. 2011). During the last revision of hotspots, Mittermeier et al. (2011) added ten more hotspots including Japan and the Solomon Islands. Because biodiversity hotspots must contain at least 1500 endemic plant and animal species, these areas include more than 50% of plant species in 2.3% of the land area (Mittermeier et al. 2011). In conclusion, most hotspots of flowering plants coincide with hotspots for ferns, so that plant conservation plans on hotspots will typically cover both plant groups. The highest rates of unique, endemic ferns are however on remote oceanic islands.

One major threat to all floras but even more to these hotspots is the land-use change and habitat destruction. Only 14.6% of the 35 hotspot areas are still covered by natural intact vegetation (Sloan et al. 2014). An additional threat is the introduction of non-native species that after naturalization become invasive. Regions with major numbers of introduced plant species from west to east include California, Florida, Cuba, England, South Africa, India, Japan, Southeast Australia, and New Zealand (Pysek et al. 2017). The hotspots of introduced plant species in the tropics as well as South Africa, Japan, and New Zealand coincide with hotspots of fern richness and are exposed to this additional threat. A small but increasing number of 157 introduced plant species are ferns of which 39% are considered invasive (Jones et al. 2019). Introduced ferns come from a wide array of regions but were mainly introduced into the Circumboreal, North American Atlantic, Amazonian, and Macaronesian region as well as into New Zealand, India, and the Hawaiian Islands (see Takhtajan's (1986) floristic regions of the world; Jones et al. 2019). Because of their similarity in growth forms and habitat preferences, invasive ferns might be an even bigger threat to native ferns than introduced angiosperms and consequently should receive major attention for future fern conservation plans.

## 26.4 Beta-Diversity of Ferns

Beta-diversity is the species turnover or species dissimilarity between two sites at any scale (plots, sites, regions, checklists, national or island floras). Differences among fern floras are a consequence of (a) species that have evolved locally by adaptive radiation in one area but have never managed to disperse successfully elsewhere (neoendemics, e.g., Hawaiian and Madagascan tree ferns, Janssen et al. 2008; Vernon and Ranker 2013), (b) species that have migrated from their center of origin into a new area but not any further (e.g., *Platycerium andinum* from Peru and Bolivia with its closest ancestors in Africa; Janssen et al. 2007), and (c) species that went extinct everywhere except in one area (paleoendemics, e.g., *Thyrsopteris elegans* on Juan Fernandez islands, *Loxsoma cunninghamii* in New Zealand, or *Llavea cordifolia* in Mexico; Tryon 1986). Neoendemics and paleoendemics can be understood as the initial and final stages of the taxon cycle, after the birth and before the extinction of a species (Wilson 1961). Paleoendemic species can be often recognized as being monotypic and the last extant member of their genus (or family). Consequently, fern floras possess their very own species composition. If speciation is fast and dispersal is slow because of strong isolation factors such as mountains, differences between floristic areas should increase. Only reticulate evolution, the parallel or successive development of a new species by hybridization of the same two parent species in multiple locations, can add sometimes further confusion to the interpretation of biogeographic distribution ranges (Wagner 1954; Werth et al. 1985).

In the following, we will look at the species similarity, the opposite of beta-diversity of ferns, at a large scale, the comparisons between national and island floras. Because fern spores are tiny (ca. 18–65 μm polar length, mean 37 μm) and lightweight (4–90 ng, mean 34 ng; Gómez-Noguez et al. 2016, 2017), they are assumed to disperse easily by the wind. Consequently, fern floras would consist of an association of individual fern species that have arrived casually by random spore dispersal. However, more factors play a role in local species composition. For instance, local habitat conditions work as a filter promoting the spore germination and establishment of the gametophyte and later the sporophyte of some species while suppressing others. Furthermore, species that are already present may occupy most available niches for ferns, or new arrivals may outcompete and replace part of the present fern community as it happens during natural succession or after the establishment of fast-growing, invasive fern species (e.g., *Sphaeropteris cooperi* in Hawaii; Chau et al. 2013). However, the first main reason that not all fern floras consist of the same species is the limited fern spore dispersal over larger distances (>100 km). Consequently, 37% of fern species are restricted to one continent or even smaller areas, and another 22% occur on only two continents (Kramer et al. 1995). Remote oceanic islands such as Hawaii are further proof of dispersal limitation. Because of its volcanic origin, Hawaii was never connected to any landmass and is located at over 3600 km distance from North America and 5800 km from East Asia (Price and Wagner 2018). The Hawaiian fern flora of 159 species is proof that casual long-distance dispersal must have happened to colonize these islands, and nearly 60% of the Hawaiian fern lineages come from the Asian tropics (Vernon and Ranker 2013), but its high endemism of 73% is also proof of the dominant role of genetic drift and adaptive species radiation and the relative rarity of successful long-distance dispersal events (Ranker 2016). It has been estimated that the extant fern flora is the consequence of 135 or more dispersal events (Carlquist 1980) compared to 259 to 292 colonization events in angiosperms (Sakai et al. 1995; Price and Wagner 2018) over a period of five million years when the present-day islands counted already with high elevation sites (Price and Clague 2002). This colonization frequency corresponds to approximately three successful fern spore dispersal events (or five to six colonization events for angiosperms) every 100,000 years. These numbers are, however, only an approximation, because some fern species colonized the Hawaiian Islands repeatedly in 2–17 independent events (Ranker et al. 1994; Driscoll and Barrington 2007).

### 26.4.1 An Example of Beta-Diversity: Distribution of American Aspleniaceae

The following example on American Aspleniaceae shall provide some insight about floristic changes across larger distances. I prepared a dataset of the geographical distribution of nearly 250 species of Aspleniaceae from Mexico and the Greater Antilles to Argentina using the sources cited in Fig. 26.1 and determined the number of shared species as well as the Sørensen index of similarity for each pair of countries and islands. Sørensen index is two times the number of shared species between two areas divided by the sum of the species present in area 1 and area 2. The result was multiplied by 100 to put it on a numeric scale together with the number of shared species (Fig. 26.4). The same data were also analyzed with cluster analysis and a nonmetric multidimensional scaling (NMDS) with help of the software Primer, version 5 (Fig. 26.5).

With Mexican Aspleniaceae as a reference (as 100%), species similarity decreases (and beta-diversity increases) mainly in response to the increasing distance from Mexico (Fig. 26.4). Across the Central American land bridge, the Sørensen values decline from northwest to southeast to a value of 56.3 in Panama. In South America, Colombia (53.6) and Venezuela (55.1) share the highest similarity with Mexican Aspleniaceae and reach in the Far South values of 39.3 in Bolivia and 23.7 in Argentina. On the other hand, the Greater Antilles, although located closer to Mexico than South America, have similarity indices with Mexico between 35.6 (Puerto Rico) and 50.4 (Hispaniola), demonstrating that these islands are more isolated than continental areas. Finally, the Guianas and Paraguay share fewer species with Mexico because of the absence of higher mountains, limiting or excluding the presence of species that are restricted to montane and cloud forests (Fig. 26.4). Considering the involved distances of several thousand kilometers, a share of 24% and 60% of Mexican Aspleniaceae with Argentina and Costa Rica, respectively, is still a high value, especially considering that the Central American land bridge is not older than 2.7 my (O'Dea et al. 2016). The results of the cluster

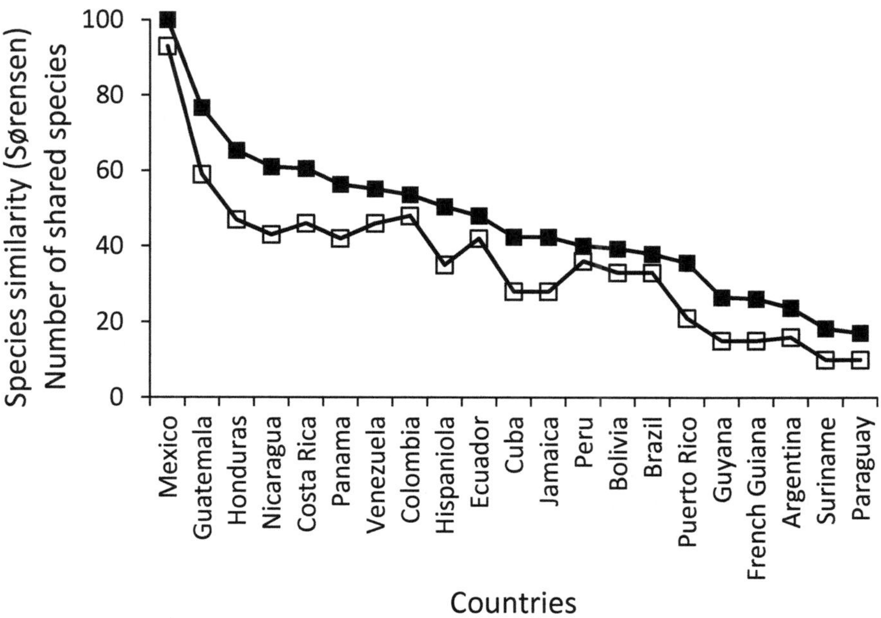


**Fig. 26.4** Species similarity (filled squares) and the number of shared species (empty squares) of the fern family Aspleniaceae (expressed as Sørensen index multiplied by 100) between Mexico and other countries and islands in decreasing order

analysis show how country floras group together depending on their species similarity (Fig. 26.5a). Mexico's Aspleniaceae are most similar to the countries of Central America, but its own floristic components (19 endemics) also set them apart. The same pattern occurs for Brazil, which has the most dissimilar species composition of Aspleniaceae of all South American countries because of its vast lowlands and isolation of the tropical Andean countries. Finally, the Guianas form one clade that is most similar to the Greater Antilles, and finally, the floras of Paraguay and Argentina with fewer tropical and more temperate fern species differ most from all other neotropical fern floras. An ordination of the Latin American Aspleniaceae based on the ranked species similarities closely reflects their geographic association, even if this analytical method is nonmetric. Some mismatches of this ordination should be noted. For instance, Hispaniola is placed closer to Central America and South America than the other islands of the Greater Antilles, and Paraguay is even more distinct from Bolivia and Brazil than Argentina. In conclusion, the dispersal limitation of ferns over larger distances promotes sufficient floristic isolation that local species radiations can thrive. Long-distance dispersal events of ferns are rare at the geological time scale but frequent enough to allow for the colonization of remote oceanic islands and some intercontinental floristic exchange.

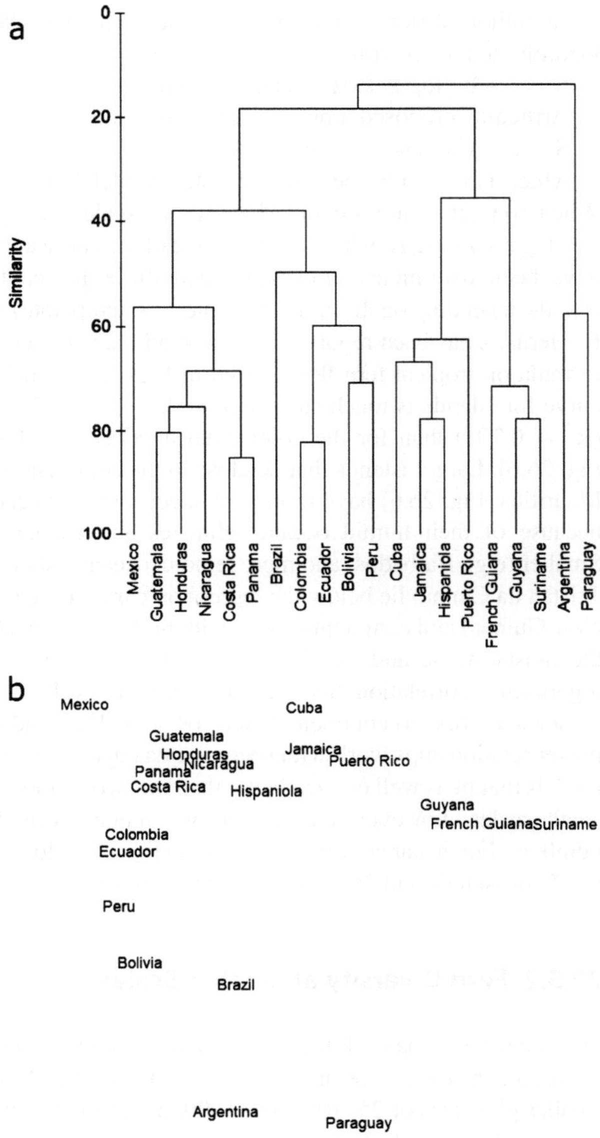


**Fig. 26.5** (**a**) Cluster and (**b**) NMS-ordination of American countries and islands according to their similarity of fern species of the family Aspleniaceae. The ordination matches closely the geographic arrangement among the Latin American fern floras with few exceptions

## 26.5 Species-Area Relationship

### 26.5.1 Logarithmic and Power Curves

All former floristic data on species richness (S) per country or island are not directly comparable because they come from areas (A) of different sizes. The species-area relationship has been described by Arrhenius (1921) and Gleason (1922, 1925), but these authors differed in their mathematical models. Whereas Gleason favored a logarithmic model with

$S = c + b \cdot \log A$ (with c and b as constants),

Arrhenius proposed a power correlation with

$S = c \cdot A^z$ (c and z as constants),

which is currently the most accepted model and will be used here (Fig. 26.6). When converted into a double logarithmic scale, the equation becomes linear: $\log S = \log c + z \cdot \log A$, where z is the slope of the curve and c is a constant. Values of z have been determined empirically and differ for continental areas and oceanic islands depending on the climate (tropical vs. temperate) and the group of organisms. For ferns, z has been reported to be around 0.25 (Rosenzweig 1995). Applying this formula on tropical fern floras listed in Figs. 26.1, 26.2 and 26.3, the species-area curve for islands is much steeper ($z = 0.347$, $c = 11.4$) and fits the data far better ($R^2 = 0.701$) than for the continental areas ($z = 0.184$, $c = 39.2$, $R^2 = 0.221$; Fig. 26.6). Larger islands that lie close to the continents or other larger islands (e.g., Mauritius, Fig. 26.6) have more fern species than expected by the regression curve because of their humid oceanic climates and exchange with continental floras. Smaller (e.g., Barbados) and more remote, oceanic islands (e.g., Galápagos, Hawaii, Tristan da Cunha) lie below the regression curve. Large tropical islands (e.g., Papua New Guinea) and continental areas with high mountains (e.g., Colombia, China) are the most diverse and lie above the regression curve (Fig. 26.6). Interestingly, a logarithmic correlation fits our data similarly well to islands ($R^2 = 0.685$) and somewhat worse to continental areas ($R^2 = 0.182$), indicating that logarithmic and power relationships work with species-area curves. The future will bring even more models that fit as well or even better than the two former ones (Tjorve 2012). At very small scales, however, the former two models would predict too low species numbers. For instance, for 1 $km^2$, predictions would correspond to the constant c (11.5 for islands and 39.2 for continental areas).

### 26.5.2 Fern Diversity at Smaller Scales

For larger areas than 1 $km^2$, data of fern diversity with species abundances are not available; instead, there are floristic checklists and databased collections. Yet for smaller plot sizes of 25, 100, 400, 1000, or 10,000 $m^2$, numerous studies have been performed in which each plant individual and species have been recorded. This allows us to compare in addition to species composition, the relative abundance and ecological importance of each species (see Sect. 26.6.3). So, how many fern species and individuals can be found in 100 $m^2$, 1000 $m^2$, or 1 ha forest? In a lowland rainforest of Costa Rica, Whitmore et al. (1985) registered 212 angiosperms with 1967 individuals and 21 fern species with 204 individuals in a 100 $m^2$ plot. About 43% of the fern species were epiphytes. Poulsen and Nielsen (1995) found on average 8.3 terrestrial fern species and 11–89 individuals per 100 $m^2$ (including climbers) in an Ecuadorian lowland rainforest, and 50% were epiphytes. Mehltreter (2008a) sampled three Mexican cloud forests and found on average 9–11 species and

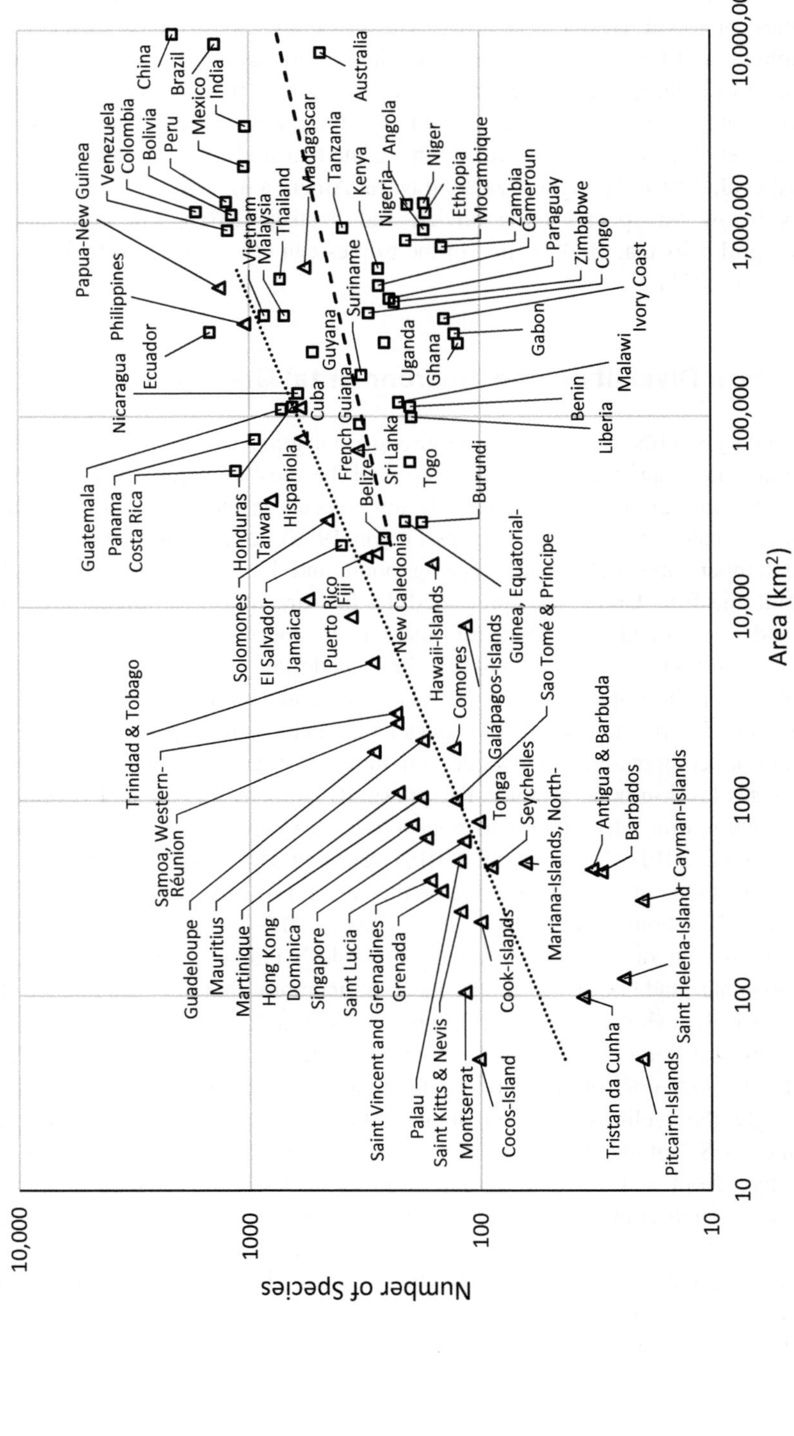


**Fig. 26.6** Species-area curves of ferns for mainly tropical continental areas (squares) and oceanic islands (triangles). Sources as in Figs. 26.1, 26.2 and 26.3

31–94 individuals per 100 $m^2$. In 400 $m^2$ plots in Mexico at different elevations, Sánchez-González et al. (2016) recorded 7–24 species and Hernández-Rojas et al. (2020) 1–42 species. The current record might be a plot in New Guinea with over 70 species per 400 $m^2$ (Khine et al. 2019). These examples show typical mean values for epiphytic and terrestrial ferns, but also the enormous variation between study sites and plots in the tropics. However, in temperate zones, the fern diversity at small scales decreases dramatically. Whereas in Japan, there were still 8–15 species per ha (Murakami et al. 2005), and the average in a Canadian study was 6.3 species per ha (Richard et al. 2000); the highest value in North America for grid cells of 2500 $km^2$ remains below 96 species (Bogonovich and Watson 2014). For comparison, 1-degree grid cells (ca. 10,000 $km^2$) in the tropics can comprise up to 933 species (Suissa et al. 2021).

## 26.6 Fern Diversity along Environmental Gradients

Fern diversity is affected by several environmental gradients of which latitudinal, altitudinal, and edaphic gradients have received the most attention. Ferns follow a strong latitudinal gradient with relatively few species in the polar and temperate regions and many species in the tropics (Tryon 1986) because many ferns do not tolerate temperatures below the freezing point and longer periods of drought. Humboldt and Bonpland (1807) presented the first drawings of altitudinal climate zones and the limited occurrence of many plant species on the slopes of the volcanoes Chimborazo and Cotopaxi in the Ecuadorian Andes. Humboldt linked mountains from the tropics to the polar regions by connecting their snow lines and tree lines and described how climatic zones along an altitudinal gradient of tropical mountains are compressed and lowered in mountains of temperate regions to finally reach the sea level in polar regions. The proximity of many climate and vegetation zones is one reason why mountains concentrate so many species in a small area (Rahbek et al. 2019). Moret et al. (2019) revisited Humboldt's study site and concluded that the drawings were made on the volcano Antisana and that vegetation zones since Humboldt's times have shifted about 215–266 m upward, possibly because of the ongoing climate change. Humboldt's climate zones are based on the comparable gradient of the annual mean temperature that declines outside the tropics 0.5 Kelvin (K) and 0.7 K per degree latitude in the Southern and Northern Hemisphere (La Sorte et al. 2014), respectively, and the decline of 0.5–0.7 K per each altitudinal step of 100 m altitude in the mountains (Walter and Breckle 2004). Accordingly, the treeline drops from about 4000 m in the tropics to the sea level at 70°N and 55°S latitude (Körner 1998; Paulsen and Körner 2014). In the tropics, ferns range from sea level to about 4000 m elevation. The record holders in colonizing high continental mountains are the terrestrial *Cystopteris fragilis* in Africa at 4750 m elevation (Vidal and Clark 2020), which also reaches 55°S in Magallanes (Chile) and 83°N in Nunavut (Canada; gbif.org), and the epiphytic *Melpomene peruviana* and *M. persoonii* in Peru at 4500 m elevation (Sylvester et al. 2014). Parallel with the treeline, the upper limit of tree ferns drops from 4000 m near the equator to 2000 m near the Tropics of Capricorn and Cancer and reaches the sea level at 24°N (Mexico/Cuba) and 34°N (Japan) and 50°S (Auckland Islands) and 53°S (Patagonia; Troll 1970; Walter and Breckle 2004). On tropical islands, the treeline drops approximately by 1400–2000 m, because the lower convection compared to continents results in lower amounts of precipitation and relatively dry conditions at altitudes above 2000–2600 m (Leuschner 2004). Humboldt's comparison between latitudinal and altitudinal gradients is however an oversimplification that can be misleading in several ways. For instance, the seasonal temperature differences at high latitudes with cold winters and warm summers do not coincide with the daily temperature changes in tropical mountains. Although tropical mountains are exposed to narrower oscillations of day length (between 10.5 and 13.5 h at the tropics of Cancer and Capricorn), at elevations of 4000 m, the day temperature rises above 0 °C and drops at night below the frost temperature (Walter and Breckle 2004).

### 26.6.1 Altitudinal Gradients

Altitudinal gradients have been studied intensively in Costa Rica (Kluge and Kessler 2006, 2007, Kluge et al. 2006, 2008, Watkins et al. 2006), Bolivia (Kessler 2002; Krömer et al. 2005), Brazil, Kenya (Hemp 2002), Nepal (Bhattarai et al. 2004), and New Zealand (Ohlemüller and Wilson 2000). Recently, even more extensive studies have been published about nine altitudinal transects in SE Asia and Japan (Khine et al. 2019), and eight altitudinal gradients in Mexico (Hernández-Rojas et al. 2020). They all found a mid-elevation peak with the highest species richness at an altitude between 1000 and 2500 m above sea level with few exceptions (e.g., Japan, Taiwan). The reasons for this mid-elevation peak are several. A random distribution model of species with an average altitudinal range of 1000 m can generate a mid-elevation peak, because of the range overlap of the species at mid-elevation and the abrupt decline to zero at sea level and the mountain peaks. In addition, the available area decreases at each altitudinal step toward the mountain peak, which contributes to the decline in species diversity because of the mentioned species-area relationship above. Most authors agree that the strong correlation between precipitation and fern diversity both peaking at mid-elevation is the best explanation for the mid-elevation peak. Furthermore, abrupt changes in fern species composition above and below the altitudes of the cloud belt (Kessler 2010), where the water condenses when the wind hits a mountain range, rise, and the air cools down and reaches the water saturation point. In high mountains, this process can occur twice and form two condensation zones. The high air humidity and presence of clouds, especially during the afternoon, protect the cloud forest vegetation including the ferns from high solar irradiance and reduce transpirational water loss. Mosses, tank-forming epiphytes, increase considerably their biomass and intercept the precipitation, retain the water, and avoid the desiccation of the soil during the sunny morning hours. Consequently, the proper cloud forest vegetation contributes to its sustainable water supply. It is also in the cloud forest where fern abundance reaches its maximum (Kessler 2010).

### 26.6.2 Soil Properties Correlated with Fern Diversity and Distribution

The importance of edaphic factors on the diversity and distribution of ferns has been reviewed by Moulatlet et al. (2019). Although few soil properties have been studied sufficiently to allow for some generalizations, they found that cation-rich soils (Ca, Mg, K, Na) support a higher fern diversity (Tuomisto et al. 2014) and that some fern species and genera have specific soil preferences. For instance, most species of *Adiantum*, *Bolbitis*, *Pteris*, *Tectaria*, and *Thelypteris* grow on cation-rich soils, whereas species of *Lindsaea*, *Trichomanes*, *Triplophyllum*, and *Schizaea* on cation-poor soils (Tuomisto and Poulsen 1996; Tuomisto et al. 1998; León et al. 2005; Lehtonen et al. 2015). However, even more interesting is the fact that species within some genera (e.g., *Adiantum*, *Lindsaea*, *Metaxya*, *Polybotrya*) have specialized into edaphic niches that are cation-poor or cation-rich (Cárdenas et al. 2016; Moulatlet et al. 2019). In epipetric ferns, the specialization on specific rock substrates is frequent. For instance, 45% of the cheilanthoid ferns in Mexico have a rock substrate preference for igneous rocks or limestone or gypsum (Mickel and Smith 2004; Mehltreter 2008b). On the other hand, some fern groups within the orders Marattiales, Gleicheniales, Cyatheales, and even some Polypodiales have specialized on highly acidic, Al-rich soils that are toxic to many plants (Schmitt et al. 2017). These species belong mainly to the more primitive non-Polypodiales and accumulate Al in their vacuoles. Al-accumulating fern species contained on average > 6000 ppm Al in their dry biomass (Schmitt et al. 2017). Although current data sets on nutrient composition of ferns and the soils on which they grow are still geographically and phylogenetically limited, it can be concluded that ferns can serve as potential indicators of edaphic conditions, especially once more combined data on soil and fern distributions become available.

### 26.6.3 Disturbance Gradients and Fern Conservation after Land-Use Change

Human activities have transformed the landscape and replaced natural ecosystems with many kinds of agroecosystems. Even in primary forest fragments that were not directly modified, indirect impacts such as edge effects can already lead to a considerable reduction of species richness and abundance (Silva et al. 2018). For some tropical crops that are often planted in the shady understory such as coffee or cacao, primary forests are partially logged, leaving the canopy trees, or the latter are replaced with new shade trees. For other land uses, the entire natural vegetation is removed and replaced with intensive crops or converted into pastures. However, when these activities are abandoned, the vegetation undergoes a succession and often converts back into seminatural environments such as a secondary forest. In this mosaic of different land uses, native plant and animal species of the original ecosystems struggle for their survival.

Disturbance has three kinds of effects on fern diversity. First, the number of species is reduced depending on the severity of the disturbance. Rare species with scattered distribution or species with a single agglomerated population are eliminated first. Second, common, faster-growing species become more dominant and may increase in abundance at the cost of low-growing understory species. Third, the invasion of introduced species increases, and sometimes they become even the most dominant species after disturbance (Mehltreter 2008a).

About 70% of the inhabitable land area is currently used for agriculture and forestry: 38.5% are used for livestock (pastures and animal feed production), 11.5% for crops for human consumption, and 20% for forestry (ourworldindata.org, Pimentel et al. 1992). The four major crops – wheat, maize, rice, and soybeans – already cover nearly 700 million ha. With such huge land areas used and affected by human activities, conservation programs need to integrate these areas rather than focusing on the conservation of biodiversity hotspots and natural areas alone (Moguel and Toledo 1999), especially because the productivity of the agroecosystems also depends on this remnant biodiversity (Pimentel et al. 1992). Which cultivation systems of crops and plantations could allow ferns to thrive? The cultivation of annual crops such as sugarcane, corn, pineapple, and soybean is too intense because of plowing the soil, applying agrochemicals, or even burning before or after the harvest. In rice plantations, the aquatic fern *Azolla* spp. is used as an important biofertilizer (Small and Darbyshire 2011), but wild ferns will rarely thrive there. Consequently, only perennial crops such as cocoa (12.2 million ha cultivated area) and coffee plantations (11.1, FAO 2019) and polycultures may provide some niches for native fern species to develop. Shade coffee plantations allow especially epiphytic fern species to develop on the shade trees. The numbers of establishing fern species are even higher if the shade trees are remnant cloud forest trees under which the coffee plantations were established. These "rustic" plantations can hold 20% of the fern richness of a cloud forest, but all other fern species do not belong to the primary cloud forests and are rather opportunistic, sun- tolerant, weedy fern species (Mehltreter 2008a, 2010). Without the shade trees, less than 10% of the species richness of ferns can still grow in these sun coffee plantations. Cabrucas in Brazil, shade cocoa plantations with partly intact original forest canopy, seem to be a good example of species conservation effort in a tropical agroecosystem. Cocoa plantations could maintain a similar number of fern species than in the adjacent forest fragments, but species composition was altered considerably with few terrestrial species being common and all others being rare (Faria et al. 2007). In addition, the surrounding landscape of cocoa plantations played a major role in fern conservation. Only cabrucas surrounded by forest fragments could maintain a higher fern diversity (Faria et al. 2007).

In other plantations, the crop plants are trees, which can be host to epiphytes or a seminatural understory that may still maintain some native species. Examples of this kind of vast plantations are oil palms (*Elaeis* spp., 28.3 million ha of cultivated area), rubber (*Hevea brasiliensis*, 12.3), coconuts (*Cocos nucifera*, 11.8), and mango (*Mangifera indica*, 5.6; FAO 2019). Oil palm plantations are often hosting abundant epiphytic ferns (*Asplenium nidus*, *Nephrolepis* spp.) which can serve other native organisms to survive (Foster et al. 2011). Rubber plantations sustain a lower fern richness than forests mainly because of a lower number of epiphytic fern species, and 33% (8 of 24) of the forest fern species were never recorded in the rubber plantations, which comprised six fern species that have not been found in the forests (Beukema and van Noordwijk 2004; Beukema et al. 2007). Finally, mango plantations have no epiphytes and build up a thick layer of leaf litter that in addition to the dark understory eliminates practically the entire herb layer. In conclusion, the hope that agroecosystems may contribute to the conservation of forest fern diversity is only justified for perennial tropical crops with partly intact original forest canopy or secondary shade trees. All other agroecosystems require being embedded in a matrix of natural forest fragments and reducing the use of agrochemicals to play a significant role in the sustainability of fern diversity.

**Acknowledgments** I thank Rosario Landgrave for preparing the diversity maps of the continents and islands and Helena Fernández for the comments on the manuscript.

## References

Arrhenius O (1921) Species and area. J Ecol 9:95–99

Barrington DS (1993) Ecological and historical factors in fern biogeography. J Biogeogr 20:275–280

Barthlott W, Mutke J, Rafiqpoor D, Kier G, Kreft H (2005) Global centers of vascular plant diversity. Nova Acta Leopold 92:61–83

Bernal R, Gradstein SR, Celis M (eds) (2019) Catálogo de plantas y líquenes de Colombia. Instituto de Ciencias Naturales, Universidad Nacional de Colombia, Bogotá. http://catalogoplantasdecolombia.unal.edu.co

Beukema H, Danielsen F, Vicent G, Hardiwinoto S, van Andel J (2007) Plant and bird diversity in rubber agroforests in the lowlands of Sumatra, Indonesia. Agrofor Syst 70:217–242

Beukema H, van Noordwijk M (2004) Terrestrial pteridophytes as indicators of a forest-like environment in rubber production systems in the lowlands of Jambi, Sumatra. Agric Ecosyst Environ 104:63–73

Bhattarai KR, Vetaas OR, Grytnes JA (2004) Fern species richness along a central Himalayan elevational gradient, Nepal. J Biogeogr 31:389–400

Boggan J, Funk V, Kelloff CL, Hoff M, Cremers G, Feuillet C (1997) Checklist of the plants of the Guianas, 2nd edn. Biological diversity of the Guianas program, Smithsonian Institution, Washington, DC

Bogonovich M, Watson M (2014) Patterns of North American fern and lycophyte richness at three taxonomic levels. Am Fern J 103:193–214

Cárdenas GG, Tuomisto H, Lehtonen S (2016) Newly discovered diversity in the tropical fern genus *Metaxya* based on morphology and molecular phylogenetic analyses. Kew Bull 71:1–27

Carlquist S (1980) Hawai'i, a natural history: geology, climate, native flora and fauna above the shoreline, 2nd edn. Pacific Tropical Botanical Garden, Lawai, Kauai. ISBN: 978-0-915-80919-6

Chau MM, Walker LR, Mehltreter K (2013) An invasive tree fern alters soil and plant nutrient dynamics in Hawaii. Biol Invasions 15:355–370

Chen CW, Perrie L, Glenny D, Chiou WL (2017) Sol amazing: lycophytes and ferns of the Solomon Islands. National Museum of Natural Science, Taichung. ISBN: 978-986-05-2972-2

Christenhusz MJM, Byng JW (2016) The number of known plants species in the world and its annual increase. Phytotaxa 261:201–217

Davidse G, Sousa Sánchez M, Chater AO (1995) Flora Mesoamericana, Psilotaceae a Salviniaceae, vol 1. Universidad Nacional Autónoma de México, Mexico. ISBN: 978-9-683-64700-9

Driscoll HE, Barrington DS (2007) Origin of Hawaiian *Polystichum* (Dryopteridaceae) in the context of a world phylogeny. Am J Bot 94:1413–1424

Duncan BD, Isaac G (1994) Ferns and allied plants of Victoria, Tasmania and South Australia. Melbourne University Press, Melbourne. ISBN: 978-0-522-84595-2

Ebihara A, Nitta JH (2019) An update and reassessment of fern and lycophyte diversity data in the Japanese Archipelago. J Plant Res 132:723–738

FAO (Food and Agriculture Organization of the United Nations) (2019) FAOSTAT. http://fao.org/faostat/. Accessed 02 May 2021

Faria D, Paciencia MLB, Dixo M, Laps RR, Baumgarten J (2007) Ferns, frogs, lizards, birds and bats in forest fragments and shade cacao plantations in two contrasting landscapes in the Atlantic forest, Brazil. Biodivers Conserv 16:2335–2357

Flora of Australia (1998) Ferns, gymnosperms and allied groups, vol 48. ABRS/CSIRO, Melbourne, Australia. ISBN: 978-0-643-0597-1

Flora of Brazil (2021) Jardim Botânico do Rio de Janeiro. http://floradobrasiljbrjgovbr/. Accessed 20 Apr 2021

Flora of North America (1993) Flora of North America: North of Mexico, vol 1. Oxford University Press, New York and Oxford. ISBN: 978-0-195-05713-3

Foster WA, Snaddon JL, Turner EC et al (2011) Establishing the evidence base for maintaining biodiversity and ecosystem function in the oil palm landscapes of South East Asia. Phil Trans R Soc B 366:3277–3291

García Criado M, Väre H, Nieto A et al (2017) European red list of Lycopods and ferns. IUCN, Brussels, Belgium

Gleason HA (1922) On the relation between species and area. Ecology 3:158–162

Gleason HA (1925) Species and area. Ecology 6:66–74

Gómez-Noguez F, León-Rossano LM, Mehltreter K, Orozco-Segovia A, Rosas-Pérez I, Pérez-García B (2017) Experimental measurements of terminal velocity of fern spores. Am Fern J 107: 59–71

Gómez-Noguez F, Pérez-García B, Mehltreter K, Orozco-Segovia A, Rosas-Pérez A (2016) Spore mass and morphometry of some fern species. Flora 223:99–105

Grangaud E (2010) Guide des fougères et plantes alliées des Mascareignes. Muséum National d'Histoire Naturelle, Paris. ISBN: 978-2-85653-652-0

Hassler M (2021) World ferns. Synonymic checklist and distribution of ferns and lycophytes of the world, version 12.1. www.worldplants.de/ferns/. Accessed 02 May 2021

Hemp A (2002) Ecology of the pteridophytes on the southern slopes of Mt. Kilimanjaro. I. Altitudinal distribution. Plant Ecol 159:211–239

Hernández-Rojas AC, Kluge J, Krömer T et al (2020) Latitudinal patterns of species richness and range size of ferns along elevational gradients at the transition from tropics to subtropics. J Biogeogr 47:1383–1397

Hokche O, Berry P, Huber O (2008) Nuevo catálogo de la flora vascular de Venezuela. Fundación Instituto Botánico de Venezuela Dr. Tobias Lasser, Caracas. ISBN: 978-980-6355-07-1

Humboldt A, Bonpland A (1807) Essai sur la géographie des plantes, accompagné d'un tableau physique des régions équinoxiales. Levrault & Schoell, Paris

Janssen T, Bystriakova N, Rakotonrainibe F, Coomes D, Labat JN, Schneider H (2008) Neoendemism in Madagascan scaly tree ferns results from recent coincident diversification bursts. Evolution 62:1876–1889

Janssen T, Kreier HP, Schneider H (2007) Origin and diversification of African ferns with special emphasis on Polypodiaceae. Brittonia 59:159–181

Jones EJ, Kraaij T, Fritz H, Moodley D (2019) A global assessment of terrestrial alien ferns (Polypodiophyta): species' traits as drivers of naturalisation and invasion. Biol Invasions 21: 861–873
Jorgensen PM, Léon-Yánez S (1999) Catalogue of the vascular plants of Ecuador. Missouri Botanical Garden, St. Louis, MO. ISBN: 978-0-915-27960-9
Jorgensen PM, Nee MH, Beck SG (2014) Catálogo de las plantas vasculares de Bolivia. Missouri Botanical Garden Press, St. Louis, MO. ISBN: 978-1-930-72371-9
Kato M (1993) Biogeography of ferns; dispersal and vicariance. J Biogeogr 20:265–274
Kessler M (2002) Range size and its ecological correlates among the pteridophytes of Carrasco National Park, Bolivia. Glob Ecol Biogeogr 11:89–102
Kessler M (2010) Biogeography of ferns. In: Mehltreter K, Walker LR, Sharpe JM (eds) Fern ecology. Cambridge University Press, Cambridge, pp 22–60. ISBN: 978-0-521-89940-6
Khine PK, Kluge J, Kessler M, Miehe G, Karger DN (2019) Latitude-independent, continent-wide consistency in climate– richness relationships in Asian ferns and lycophytes. J Biogeogr 46: 981–991
Kluge J, Bach K, Kessler M (2008) Elevational distribution and zonation of tropical pteridophyte assemblages in Costa Rica. Basic Appl Ecol 9:35–43
Kluge J, Kessler M (2006) Fern endemism and its correlates: contribution from an elevational transect in Costa Rica. Divers Distrib 12:535–545
Kluge J, Kessler M (2007) Morphological characteristics of fern assemblages along an elevational gradient, patterns and causes. Ecotropica 13:27–43
Kluge J, Kessler M, Dunn R (2006) What drives elevational patterns of diversity? A test of geometric constraints, climate, and species pool effects for pteridophytes on an elevational gradient in Costa Rica. Glob Ecol Biogeogr 15:358–371
Knapp R (2011) Ferns and fern allies of Taiwan. Yuan-Liou Publishing, Taiwan. ISBN: 978-9-868-70980-5
Knapp R (2013) Ferns and fern allies of Taiwan, supplement. KBCC Press, Taiwan. ISBN: 978-9-868-70981-2
Knapp R, Hsu TC (2017) Ferns and fern allies of Taiwan, second supplement. KBCC Press, Taiwan. ISBN: 978-9-869-35132-4
Körner C (1998) A re-assessment of high elevation treeline positions and their explanation. Oecologia 115:445–459
Kramer KU, Schneller JJ, Wollenweber E (1995) Farne und Farnverwandte. Thieme, Stuttgart. ISBN: 978-3-131-15301-2
Kreft H, Jetz W, Mutke J, Barthlott W (2010) Contrasting environmental and regional effects on global pteridophyte and seed plant diversity. Ecography 33:408–419
Krömer T, Kessler M, Gradstein SR, Acebey A (2005) Diversity patterns of vascular epiphytes along an elevational gradient in the Andes. J Biogeogr 32:1799–1810
La Sorte FA, Butchart SHM, Jetz W, Böhning-Gaese K (2014) Range-wide latitudinal and elevational temperature gradients for the world's terrestrial birds: implications under global climate change. PLoS One 9:e98361
Lehtonen S, Jones MM, Zuquim G, Prado J, Tuomisto H (2015) Phylogenetic relatedness within Neotropical fern communities' increases with soil fertility. Glob Ecol Biogeogr 24:695–705
León B, Beltrán H, Fine P (2005) Sobre el género *Schizaea* (Schizaeaceae) en el Perú. Rev Peru Biol 12:97–102
Leuschner C (2004) Die Höhenstufen tropischer und subtropischer Inseln. In: Walter H, Breckle SW (eds) Ökologie der Erde, Band 2: Spezielle Ökologie der tropischen und suptropischen Zonen, 3rd edn. Spektrum, Elsevier, München, pp 158–164
Magurran AE (2004) Measuring biological diversity. Blackwell, Malden. ISBN: 0-632-05633-9
Mehltreter K (2008a) Helechos. In: Manson RH, Hernández-Ortiz V, Gallina S, Mehltreter K (eds) Agroecosistemas cafetaleros de Veracruz: biodiversidad, manejo y conservación. Instituto de Ecología, A. C. (INECOL) and Instituto Nacional de Ecología (INE), Ciudad de México, Mexico, pp 83–93

Mehltreter K (2008b) Phenology and habitat specificity of tropical ferns. In: Ranker TA, Haufler CH (eds) Biology and evolution of ferns and lycophytes. Cambridge University Press, Cambridge, pp 201–221. ISBN: 978-0-521-69689-0

Mehltreter K (2010) Fern conservation. In: Mehltreter K, Walker LR, Sharpe JM (eds) Fern ecology. Cambridge University Press, Cambridge, pp 323–359. ISBN: 978-0-521-89940-6

Mehltreter K, Walker LR, Sharpe JM (eds) (2010) Fern ecology. Cambridge University Press, Cambridge. ISBN: 978-0-521-89940-6

Mickel JT, Smith AR (2004) The pteridophytes of Mexico. The New York Botanical Garden Press, New York. ISBN: 978-0-893-27458-0

Mittermeier RA, Turner WR, Larsen FW, Brooks TM, Gascon C (2011) Global biodiversity conservation: the critical role of hotspots. In: Zachos FE, Habel JC (eds) Biodiversity hotspots. Springer, Berlin, pp 3–22. ISBN: 978-3-642-20991-8

Moguel P, Toledo VM (1999) Biodiversity conservation in traditional coffee systems in Mexico. Conserv Biol 13:1–12

Moret P, Muriel P, Jaramillo R, Dangles O (2019) Humboldt's *Tableau Physique* revisited. Proc Natl Acad Sci USA 116:12889–12894

Moulatlet GM, Zuquim G, Tuomisto H (2019) The role of soils for pteridophyte distribution in tropical American forests. Fern Gaz 21:1–20

Murakami K, Maenaka H, Morimoto Y (2005) Factors influencing species diversity of ferns and fern allies in fragmented forest patches in the Kyoto city area. Landscape Urban Plan 70:221–229

Mutke J, Sommer JH, Kreft H, Kier G, Barthlott W (2011) Vascular plant diversity in a changing world: global centres and biome-specific patterns. In: Zachos FE, Habel JC (eds) Biodiversity hotspots. Springer, Berlin, pp 83–96. ISBN: 978-3-642-20991-8

Myers N, Mittermeier CG, Mittermeier GAB, Da Fonseca J, Kent J (2000) Biodiversity hotspots for conservation priorities. Nature 403:853–858

O'Dea A, Lessios HA, Coates AG et al (2016) Formation of the isthmus of Panama. Sci Adv 2: e1600883. https://doi.org/10.1126/sciadv.1600883

Ohlemüller R, Wilson JB (2000) Vascular plant species richness along latitudinal and altitudinal gradients, a contribution from New Zealand temperate rainforests. Ecol Lett 3:262–266

Parris BS, Beaman RS, Beaman JH (1992) The plants of mount Kinabalu. Royal Botanic Gardens, Kew. ISBN: 978-0-947-64338-6

Paulsen J, Körner C (2014) A climate-based model to predict potential treeline position around the globe. Alpine Bot 124:1–12

Pimentel D, Stachow U, Takacs DA et al (1992) Conserving biological diversity in agricultural/forestry systems. Bioscience 42:354–362

Poulsen AD, Nielsen IH (1995) How many ferns are there in one hectare of tropical rain forest? Am Fern J 85:29–35

PPG I (2016) A community-derived classification for extant lycophytes and ferns. J Syst Evol 54: 563–603

Price JP, Clague DA (2002) How old is the Hawaiian biota? Geology and phylogeny suggest recent divergence. Proc Roy Soc Lond B Bio 269:2429–2435

Price JP, Wagner WL (2018) Origins of the Hawaiian flora: phylogenies and biogeography reveal patterns of long-distance dispersal. J Syst Evol 56:600–620

Pysek P, Pergl J, Essl F et al (2017) Naturalized alien flora of the world: species diversity, taxonomic and phylogenetic patterns, geographic distribution and global hotspots of plant invasion. Preslia 89:203–274

Rahbek K, Borregaard MK, Colwell RK et al (2019) Humboldt's enigma: what causes global patterns of mountain biodiversity. Science 365:1108–1113

Ranker TA (2016) What do we know about Hawaiian ferns and lycophytes? J Syst Evol 54:626–637

Ranker TA, Floyd SK, Trapp PG (1994) Multiple colonizations of *Asplenium adiantum–nigrum* onto the Hawaiian archipelago. Evolution 48:1364–1370

Richard M, Bernhardt T, Bell G (2000) Environmental heterogeneity and the spatial structure of fern species diversity in one hectare of old-growth forest. Ecography 23:231–245
Rosenzweig ML (1995) Species diversity in space and time. Cambridge University Press, Cambridge. ISBN: 978-0-521-49952-1
Roux JP (2009) Synopsis of the Lycopodiophyta and Pteridophyta of Africa, Madagascar and neighboring islands. Strelitzia 23. South African National Biodiversity Institute (SANBI), Pretoria. ISBN: 978-1-919976-48-8
Sakai AK, Wagner WL, Ferguson DM, Herbst DR (1995) Biogeographical and ecological correlates of dioecy in the Hawaiian flora. Ecology 76:2530–2543
Sánchez C (2017) Lista de los helechos y licófitos de Cuba. Brittonia 69:482–503
Sánchez-González A, Álvarez-Zuñiga E, López-Mata L (2016) Diversity and distribution patterns of ferns and lycophytes in a cloud forest in Mexico. Rev Chapingo Ser Cie 22:235–253
Schmitt M, Mehltreter K, Sundue M, Testo W, Watanabe T, Jansen S (2017) The evolution of aluminum accumulation in ferns and lycophytes. Am J Bot 104:573–583
Schneider H, Schuettpelz E, Pryer KM, Cranfill RB, Magallon S, Lupia R (2004) Ferns diversified in the shadow of angiosperms. Nature 428:553–557
Shaffer-Fehre M (2006) A revised handbook to the Flora of Ceylon, Part A: ferns and fern-allies, vol XV. CRC Press, Boca Raton. ISBN: 978-1-482-28036-4
Silva VL, Mehltreter K, Schmitt JL (2018) Ferns as potential ecological indicators of edge effects in two types of Mexican forests. Ecol Indic 93:669–676
Sloan S, Jenkins CN, Joppa LN, Gaveau DLA, Laurance WF (2014) Remaining natural vegetation in the global biodiversity hotspots. Biol Conserv 177:12–24
Small E, Darbyshire SJ (2011) Blossoming treasures of biodiversity: 35. Mosquito ferns (*Azolla* species) – tiny 'super plants'. Biodiversity 12:119–128
Suissa JS, Sundue MA, Testo WL (2021) Mountains, climate and niche heterogeneity explain global patterns of fern diversity. J Biogeogr 48(6):1296–1308. https://doi.org/10.1111/jbi.14076
Sylvester SP, Sylvester MDPV, Kessler (2014) The world's highest vascular epiphytes found in the Peruvian Andes. Alpine Bot 124:179–185
Takhtajan A (1986) Floristic regions of the world. University of California Press, Los Angeles. ISBN: 978-0-520-04027-4
Testo W, Sundue M (2016) A 4000-species dataset provides new insight into the evolution of ferns. Mol Phylogenet Evol 105:200–211
Tjorve E (2012) Arrhenius and Gleason revisited: new hybrid models resolve an old controversy. J Biogeogr 39:629–639
Troll C (1970) Das 'Baumfarnklima' und die Verbreitung der Baumfarne auf der Erde. Tübinger Geogr Stud 34:179–189
Tryon RM (1970) Development and evolution of fern floras of oceanic islands. Biotropica 2:76–84
Tryon RM (1972) Endemic areas and geographic speciation in tropical American ferns. Biotropica 4:121–131
Tryon RM (1986) The biogeography of species, with special reference to ferns. Bot Rev 52:118–156
Tuomisto H, Poulsen AD (1996) Influence of edaphic specialization on pteridophyte distribution in neotropical rain forests. J Biogeogr 23:283–293
Tuomisto H, Poulsen AD, Moran RC (1998) Edaphic distribution of some species of the fern genus *Adiantum* in western Amazonia. Biotropica 30:392–399
Tuomisto H, Zuquim G, Cárdenas G (2014) Species richness and diversity along edaphic and climatic gradients in Amazonia. Ecography 37:1034–1046
Vernon A, Ranker TA (2013) Current status of the ferns and lycophytes of the Hawaiian islands. Am Fern J 103:59–111
Vidal JD, Clark VR (2020) Afro-alpine plant diversity in the tropical mountains of Africa. In: Goldstein MI, DellaSala DA (eds) Encyclopedia of the World's biomes. Elsevier, Amsterdam, Netherlands, pp 373–394. https://doi.org/10.1016/B978-0-12-409548-9.11885-8

Villaseñor JL (2016) Checklist of the native vascular plants of Mexico. Rev Mex Biodivers 87:599–902

Wagner WH (1954) Reticulate evolution in the Appalachian Aspleniums. Evolution 8:103–118

Walter H, Breckle SW (2004) Ökologie der Erde, Band 2: Spezielle Ökologie der tropischen und suptropischen Zonen, 3rd edn. Spektrum, Elsevier, München. ISBN: 978-3-827-40789-4

Watkins JE Jr, Cardelús C, Colwell RK, Moran RC (2006) Species richness and distribution of ferns along an elevational gradient in Costa Rica. Am J Bot 93:73–83

Werth CR, Guttman SI, Eshbaugh WH (1985) Electrophoretic evidence of reticulate evolution in the Appalachian *Asplenium* complex. Syst Bot 10:184–192

Whitmore TC, Peralta R, Brown K (1985) Total species count in a costa Rican rain forest. J Tropical Ecol 1:375–378

Wilson EO (1961) The nature of the taxon cycle in the Melanesian ant fauna. Am Nat 95:169–193

Zheng-Yi W, Raven P, De-Yuan H (2013) Flora of China, Lycopodiaceae through Polypodiaceae, vol 2–3. Science Press and Missouri Botanical Garden Press, Beijing and St. Louis. ISBN: 978-1-935-64111-7

Zuloaga FO, Belgrano MJ (2016) Flora vascular de la República Argentina, Licofitas, Helechos, Gymnospermae, vol 2. Instituto de Botánica Darwinion, San Isidro. ISBN: 978-987-45957-5-1

**Klaus Mehltreter** received his doctorate degree of Natural Sciences in 1994 at Ulm University, Germany, on the biogeography and ecology of high montane ferns of Costa Rica. After 2 years of work as an independent researcher in Argentina, he became researcher and professor at the Instituto de Ecología, A. C., in Xalapa, Mexico. His broad research interest covers numerous aspects of fern ecology, ranging from their study of diversity and abundance, conservation, seasonality of growth and reproduction, functional morpho-anatomy, nutrient ecology, and invasive species to their interactions with herbivorous insects.

# Pteridophytes: Effective Agents of Phytoremediation

# 27

Sudha Sajeev, P. T. Roshni, Rachel Carmelita Mathias, Shaiesh Morajkar, Smruthi Prabhu, and Smitha Hegde

**Abstract**

Hazardous contaminants due to natural process and anthropogenic activities pose serious health and environment risk which is a global problem. To mitigate this problem, phytoremediation, a plant-based approach of remediating organic and inorganic environmental pollutants from the soil and water, is practiced. It is an eco-friendly and cost-effective approach which involves plants to clean up the environment. A series of remedial strategies with distinct mechanism were employed for the degradation or removal of toxic compounds. A wide geographical spread, high environmental adaptation, resilience in toxicity, and bioaccumulation potential of pteridophytes facilitate broad application in the field of phytoremediation. This chapter discusses the potential and importance of pteridophytes as remediation agents of polluted environment.

**Keywords**

Contaminants · Phytoremediation · Pteridophytes · Mechanism · Cost-effective · Eco-friendly

S. Sajeev · P. T. Roshni · S. Morajkar · S. Prabhu
Post Graduate Department of Biotechnology and Research, St. Aloysius College (Autonomous), Mangalore, India

R. C. Mathias · S. Hegde (✉)
Nitte (Deemed to be University), Nitte University Center for Science Education and Research, Mangalore, India
e-mail: smitha.hegde@nitte.edu.in



J. Marimuthu et al. (eds.), *Ferns*, https://doi.org/10.1007/978-981-16-6170-9_27

## 27.1 Introduction

Elements with an atomic weight between 63.54 and 200.59 and a specific gravity greater than 4.0 are called heavy metals. Heavy metal contamination in soil and water is a worldwide problem causing extensive damage to agricultural yields and causing hazardous health effects (Singh et al. 2008). Examples of these elements are arsenic (As), antimony (Sb), cadmium (Cd), chromium (Cr), mercury (Hg), lead (Pb), etc. Soil and water sources rich in metal residues are reasons for limited plant growth, reduced root growth, and abnormal leaf color. Acidic soils and low pH increase the metal uptake by plants instead of nutrients. Competition between the toxic metal ions and nutrient ions leads to a deficiency of essential nutrients in the plants. Many of the plants are badly affected, while few can tolerate a high concentration of toxic metals. Plants' responses to various environmental impacts are complex and depend on the tissues or the organs directly affected (Dinneny et al. 2008). They have developed certain complex mechanisms to adjust to the stress at transcription, cellular, and physiological levels. These responses could be either reversible or irreversible and depend on acute or chronic exposure over some time (Cramer et al. 2011). One of the significant responses to abiotic stress is inhibiting major metabolic pathways involving photosynthesis, sugars, proteins, and lipids (Kilian et al. 2007; Pinheiro and Chaves 2011). Abiotic stress responses at the molecular level involve the signaling of reactive oxygen species (ROS) and reactive nitrogen species (RNS), the release of hormone regulators such as abscisic acid (ABA) and ethylene, etc.

## 27.2 Phytoremediation and Types

The industrial revolution has a significant contribution to the present environmental scenario in the world. Pollutants such as heavy metals from industrial activities, mining, urban waste, sewage treatment plants and their runoff, smelting, and mining have extensively damaged the otherwise serene pure ecosystems. Different types of organic and inorganic substances are disposed to the soil from various industries. Soil microorganisms can degrade organic contaminants, while inorganic contaminations like heavy metals need immobilization or physical removal (Ghosh and Singh 2005). Besides metals and other industrial wastes, deterioration of soil fertility and groundwater resources can be attributed to factors like solubility of the metal, pH of the soil, organic carbon content, etc. Despite all these challenges posed by the environment, plants have an inherent ability to adapt and evolve through their specific genetic traits that betters the prevalent environmental conditions (Pollard 2000). Plant response to heavy metals could be mainly in three ways:

1. Plants can act as metal excluders by preventing metals from entering the plant system.

**Table 27.1** List of pteridophytes used for heavy metal absorption

| Name of the fern | Heavy metal absorption (ppm) | | | References (authors and year) |
|---|---|---|---|---|
| | Heavy metal | Soil | Leaves | |
| *Acrostichum aureum* L. | As | – | 500 | Sharma and Irudayaraj (2010) |
| *Adiantum philippense* L. | Pb | 9.439 ± 7.16 | 90.786 ± 67.33 | Pongthornpruek et al. (2008) |
| | Ni | 2.023 ± 0.41 | 28.892 ± 22.27 | |
| | Co | 1.045 ± 0.50 | ND | |
| *Adiantum caudatum* L. | Pb | 1.822 ± 0.03 | 22.928 ± 16.74 | |
| | Ni | 0.868 ± 1.23 | 23.516 ± 15.54 | |
| | Co | 1.144 ± 0.12 | 1.971 ± 0.45 | |
| *Angiopteris evecta* | Pb | 16.047 ± 3.72 | 5.857 ± 2.23 | |
| | Ni | 1.499 ± 0.28 | 7.591 ± 3.43 | |
| | Co | 1.300 ± 0.07 | 0.322 ± 0.08 | |
| *Asplenium adiantum-nigrum* L. ssp. *adiantum Nigrum* | As | < | < | Cornara et al. (2007) |
| | Cr | 662 | 1.6 | |
| | Zn | 157.7 | 27.2 | |
| | Cd | 18.7 | < | |
| | Pb | 34.7 | 0.70 | |
| | Ni | 1316 | 16.6 | |
| | Fe | $>10^5$ | 278 | |
| | Cu | 3266.9 | 20.0 | |
| *Asplenium cuneifolium* Viv. Ssp. *cuneifolium* 1 | As | 49.5 | < | |
| | Cr | 162 | 16.0 | |
| | Zn | 166.0 | 42.7 | |
| | Cd | 4.4 | 1.06 | |
| | Pb | 72.9 | < | |
| | Ni | 148 | 40.0 | |
| | Fe | 58,333 | 1297 | |
| | Cu | 466.9 | 7.0 | |
| *Asplenium monopteros* L. | As | < | < | |
| | Cr | 756 | 0.7 | |
| | Zn | 57.9 | 20.8 | |
| | Cd | < | 0.95 | |
| | Pb | 15.7 | 2.22 | |
| | Ni | 490 | 22.2 | |
| | Fe | 59,227 | 98 | |
| | Cu | 21.9 | 4.0 | |
| *Asplenium* × *ticinense* D. E. Mayer | As | < | < | |
| | Cr | 662 | 0.8 | |
| | Zn | 157.7 | 48.1 | |
| | Cd | 18.7 | 0.59 | |
| | Pb | 34.7 | 2031 | |

(continued)

**Table 27.1** (continued)

| Name of the fern | Heavy metal absorption (ppm) | | | References (authors and year) |
|---|---|---|---|---|
| | Heavy metal | Soil | Leaves | |
| | Ni | 1316 | 10.3 | |
| | Fe | >$10^5$ | 126 | |
| | Cu | 3266.9 | 9.8 | |
| *Asplenium trichomanes* L. ssp. *quadrivalens* D.E. Meyer | As | 52.2 | < | Cornara et al. (2007) |
| | Cr | 1259 | 0.3 | |
| | Zn | 178.0 | 80.4 | |
| | Cd | 11.1 | 1.94 | |
| | Pb | 117.0 | 0.08 | |
| | Ni | 922 | 32.7 | |
| | Fe | 66,724 | 156 | |
| | Cu | 28.8 | 4.2 | |
| *Athyrium filix-femina* (L.) Roth. 1 | As | 48.5 | < | |
| | Cr | 227 | 2.2 | |
| | Zn | 77.4 | 44.7 | |
| | Cd | 4.1 | 5.10 | |
| | Pb | 39.1 | 1.34 | |
| | Ni | 169 | 25.8 | |
| | Fe | 24,008 | 198 | |
| | Cu | 59.1 | 18.2 | |
| *Azolla filiculoides* | Cd | – | 10,000 | Sela et al. (1989) |
| | Cr | | 1990 | |
| | Cu | | 9000 | |
| | Ni | | 9000 | |
| | Zn | | 6500 | |
| *Azolla filiculoides* | Pb | – | 93,000 | Sanyahumbi et al. (1998) |
| *Blechnum cartilagineum* | Ni | – | 121 | Kachenko et al. (2007) |
| *Blechnum nudum* | Pb | – | 21 | |
| *Blechnum orientale* L. | Pb | 2.641 ± 0.30 | 13.586 ± 3.21 | Pongthornpruek et al. (2008) |
| | Ni | 0.851 ± 0.33 | 5.403 ± 1.30 | |
| | Co | 1.724 ± 0.57 | ND | |
| *Calochlaena dubia* | Cu | – | 4.12 | Kachenko et al. (2007) |
| *Cheilanthes hirta* | Cu | – | < 5000 | Wild (1968) |
| *Colysis pothifolia* | Pb | 1.604 ± 2.27 | 12.645 ± 6.01 | Pongthornpruek et al. (2008) |
| | Ni | 1.292 ± 1.83 | 13.532 ± 2.33 | |
| | Co | 0.575 ± 0.55 | 0.745 ± 0.17 | |
| *Dennstaedtia davallioides* | Cd | – | 386 | Kachenko et al. (2007) |
| | Cu | | 32 | |
| *Diplazium esculentum* | Pb | 11.285 ± 3.45 | 15.863 ± 2.53 | Pongthornpruek et al. (2008) |
| | Ni | 1.192 ± 0.96 | 3.742 ± 0.32 | |

(continued)

**Table 27.1** (continued)

| Name of the fern | Heavy metal absorption (ppm) | | | References (authors and year) |
|---|---|---|---|---|
| | Heavy metal | Soil | Leaves | |
| | Co | 1.168 ± 0.35 | ND | |
| *Doodia aspera* | Ni | – | 161 | Kachenko et al. (2007) |
| *Dryopteris affinis* (Lowe) Fraser- Jenkins ssp. *borreri* (*Newman*) Fraser-Jenk. | As | 39.0 | 0.60 | Cornara et al. (2007) |
| | Cr | 728 | 0.7 | |
| | Zn | 48.3 | 8.6 | |
| | Cd | 4.0 | < | |
| | Pb | < | < | |
| | Ni | 594 | 22.9 | |
| | Fe | 28,070 | 80 | |
| | Cu | 17.7 | 2.6 | |
| *Dryopteris filix-mas* (L.) Schott | As | 39.0 | 0.60 | |
| | Cr | 728 | 1.3 | |
| | Zn | 48.3 | 46.6 | |
| | Cd | 4.0 | 7.39 | |
| | Pb | < | 1.03 | |
| | Ni | 594 | 32.5 | |
| | Fe | 28,070 | 151 | |
| | Cu | 17.7 | 11.2 | |
| *Equisetum arvense* L. | As | < | 0.31 | |
| | Cr | 1138 | 16.7 | |
| | Zn | 131.2 | 32.0 | |
| | Cd | 4.1 | < | |
| | Pb | 151.6 | < | |
| | Ni | 555 | 40.7 | |
| | Fe | 4994 | 742 | |
| | Cu | 5 26.7 | 13.7 | |
| *Equisetum ramosissimum* Desf. | As | < | 0.75 | |
| | Cr | 1397 | 43.4 | |
| | Zn | 410.5 | 67.6 | |
| | Cd | 6.7 | 0.32 | |
| | Pb | 390.3 | 0.21 | |
| | Ni | 2295 | 45.9 | |
| | Fe | 52,322 | 2565 | |
| | Cu | 180.6 | 144.1 | |
| *Lindsaea ensifolia* Sw. | Pb | 22.343 ± 0.00 | 38.581 ± 26.78 | Pongthornpruek et al. (2008) |
| | Ni | 3.632 ± 0.00 | 23.470 ± 21.30 | |
| | Co | 0.826 ± 0.00 | 1.214 ± 1.30 | |
| *Lygodium* sp. | Pb | 12.633 ± 1.31 | 6.552 ± 2.64 | Pongthornpruek et al. (2008) |
| | Ni | 1.457 ± 0.23 | 7.694 ± 2.11 | |
| | Co | 1.574 ± 0.99 | 0.156 ± 0.13 | |

(continued)

**Table 27.1** (continued)

| Name of the fern | Heavy metal absorption (ppm) | | | References (authors and year) |
|---|---|---|---|---|
| | Heavy metal | Soil | Leaves | |
| *Nephrolepis cordifolia* | Cd | – | 4.1 | Kachenko et al. (2007) |
| | Cr | | 0.07 | |
| *Nephrolepis cordifolia* C. Presl. | As | < | < | Cornara et al. (2007) |
| | Cr | 662 | 6.8 | |
| | Zn | 157.7 | 46.5 | |
| | Cd | 18.7 | < | |
| | Pb | 34.7 | 11.96 | |
| | Ni | 1316 | 11.0 | |
| | Fe | $>10^5$ | 1715 | |
| | Cu | 3266.9 | 33.1 | |
| *Notholaena marantae* (L.) Desv. 1 | As | < | 0.30 | |
| | Cr | 1397 | 2.0 | |
| | Zn | 410.5 | 72.4 | |
| | Cd | 6.7 | < | |
| | Pb | 390.3 | 1.34 | |
| | Ni | 2295 | 31.0 | |
| | Fe | 52,322 | 245 | |
| | Cu | 180.6 | 10.9 | |
| *Osmunda regalis* L. 1 | As | 47.0 | < | |
| | Cr | 961 | 0.6 | |
| | Zn | 80.4 | 14.2 | |
| | Cd | 3.5 | < | |
| | Pb | 47.4 | 0.84 | |
| | Ni | 795 | 1.1 | |
| | Fe | 27,635 | 68 | |
| | Cu | 22.2 | 4.4 | |
| *Pellaea calomelanos* | Cu | – | < 5000 | Wild (1968) |
| *Pellaea falcata* | Cd | – | 87.9 | Kachenko et al. (2007) |
| | Ni | | 68 | |
| | Pb | | 62 | |
| | Cr | | 33 | |
| *Pteridium aquilinum* (L.) Khun ssp. *aquilinum* | As | 49.5 | < | Cornara et al. (2007) |
| | Cr | 162 | < | |
| | Zn | 166.0 | 31.7 | |
| | Cd | 4.4 | < | |
| | Pb | 72.9 | 3.11 | |
| | Ni | 148 | < | |
| | Fe | 58,333 | 452 | |
| | Cu | 466.9 | 10.2 | |
| *Pityrogramma calomelanos* | As | 135–510 | 2760–8350 | Francesconi et al. (2002) |

(continued)

**Table 27.1** (continued)

| Name of the fern | Heavy metal absorption (ppm) | | | References (authors and year) |
|---|---|---|---|---|
| | Heavy metal | Soil | Leaves | |
| *Pteridium aquilinum* | Pb | 15.617 ± 5.13 | 7.964 ± 4.00 | Pongthornpruek et al. (2008) |
| | Ni | 1.169 ± 0.46 | 14.565 ± 3.79 | |
| | Co | 1.273 ± 0.66 | 0.637 ± 0.37 | |
| *Pteris argyraea* | As | – | 361 | Mehrag (2002) |
| *Pteris biaurita* L. | Pb | 10.903 ± 2.91 | 19.966 ± 13.89 | Pongthornpruek et al. (2008) |
| | Ni | 1.309 ± 1.13 | 10.440 ± 1.90 | |
| | Co | 0.886 ± 0.05 | 0.682 ± 0.55 | |
| *Pteris cretica* | As | – | >1000 | Mehrag (2002) |
| *Pteris ensiformis* Burm.f. | Pb | 20.635 ± 2.42 | 30.370 ± 9.90 | Pongthornpruek et al. (2008) |
| | Ni | 2.464 ± 1.65 | 8.550 ± 1.86 | |
| | Co | 1.135 ± 0.44 | ND | |
| *Pteris longifolia* | As | – | >1000 | Mehrag (2002) |
| *Pteris straminea* | As | – | 78 | |
| *Pteris tremula* | As | – | 16.6 | |
| *Pteris venusta* Kunze. | Pb | 9.818 ± 6.11 | 5.729 ± 3.68 | Pongthornpruek et al. (2008) |
| | Ni | 1.583 ± 0.48 | 15.652 ± 11.46 | |
| | Co | 1.233 ± 0.70 | 0.219 ± 0.13 | |
| *Pteris vittata* | Ni | – | 0.8 | Kachenko et al. (2007) |
| *Pteris vittata* | As | 500 | 21,290 | Ma et al. (2001a, b) |
| *Salvinia minima* | Cr | – | 2.0 | Nichols et al. (2000) |
| *Salvinia natans* | Cu | – | 50 | Sen and Mendal (1990) |
| *Tectaria angulata* | Pb | 11.899 ± 0.00 | 46.186 ± 44.83 | Pongthornpruek et al. (2008) |
| | Ni | 2.470 ± 0.69 | 9.900 ± 9.12 | |
| | Co | 2.409 ± 1.21 | ND | |
| *Tectaria herpetocaulos* Holtt. | Pb | 6.010 ± 3.18 | 9.352 ± 3.52 | |
| | Ni | 1.491 ± 0.90 | 10.514 ± 2.53 | |
| | Co | 1.300 ± 0.50 | 0.943 ± 0.31 | |
| *Tectaria impressa* | Pb | 10.628 ± 1.75 | 8.245 ± 1.46 | |
| | Ni | 0.989 ± 0.09 | 11.955 ± 3.42 | |
| | Co | 0.545 ± 0.50 | 0.835 ± 0.41 | |
| *Tectaria polymorpha* | Pb | 3.159 ± 4.47 | 18.254 ± 3.48 | |
| | Ni | 1.479 ± 2.09 | 6.022 ± 1.29 | |
| | Co | 0.875 ± 0.14 | ND | |
| *Thelypteris interrupta* | Pb | 18.081 ± 2.24 | 10.571 ± 3.73 | |
| | Ni | 1.577 ± 0.40 | 5.697 ± 1.54 | |
| | Co | 1.125 ± 0.59 | ND | |
| *Thelypteris nudata* | Pb | 6.535 ± 1.54 | 4.447 ± 3.21 | |

(continued)

**Table 27.1** (continued)

| Name of the fern | Heavy metal absorption (ppm) | | | References (authors and year) |
|---|---|---|---|---|
| | Heavy metal | Soil | Leaves | |
| | Ni | 1.476 ± 0.47 | 18.564 ± 2.59 | |
| | Co | 1.235 ± 0.73 | ND | |
| *Thelypteris terminans* | Pb | 18.417 ± 2.13 | 17.112 ± 3.78 | |
| | Ni | 1.543 ± 0.37 | 5.477 ± 1.70 | |
| | Co | 0.951 ± 0.65 | ND | |

2. Plants could be metal indicators by storing a higher metal concentration than in the soil, indicating the soil metal. They tolerate this excess metal with the help of chelators or by compartmentalization.
3. Hyperaccumulators.

Hyperaccumulators are plants that can contain more than or up to 0.1% for copper (Cu), Cd, Cr, Pb, nickel (Ni), and cobalt (Co) or 1% of zinc (Zn) or manganese (Mn) in dry matter in their root, shoot, or leaves (Reeves et al. 2021). Metals taken up by plants are concentrated in vacuoles, epidermal cells, trichomes, and cell walls (Boyd et al. 2000). Among the most widespread and toxic metals are Pb, Cd, and As (Kramer 2000), but many of the hyperaccumulators identified are Ni, Zn, and selenium (Se). Phytoremediation was the term coined in 1991 to describe how plants are used to remove, assimilate, or absorb the metals from contaminated soil and water sources (Sardar et al. 2013). These plants tolerate by excluding or accumulating metals. Phytoremediation, in Greek, *phyto*, means plant and *remediare* to remediate. This green technology is considered safe and cost-effective and practically employed in different ways to decontaminate the sites.

## 27.3 Different Mechanisms of Phytoremediation

### 27.3.1 Phytoextraction

Plants tend to absorb nutrients from their habitat according to their requirement and availability. The uptake of various contaminants in the soil or water by the plants which are accumulated in aboveground plant tissue is called phytoextraction or phytomining or phytoaccumulation (Jacob et al. 2018; Yan et al. 2020). These metal accumulating plants or hyperaccumulators that are 100 to 1000-fold efficient than non-hyperaccumulators can grow in the polluted site. Once they take up the metals and translocate from soil to the aboveground portions, they are harvested. Therefore, it is an easy and effective method of remediation technique. The mechanism involves mobilization of heavy metal in the soil, a faster translocation from root to shoot, and sequestration in the leaves (Fig. 27.1). A desirable method of phytoextraction was facilitated by the factors such as speciation and bioavailability

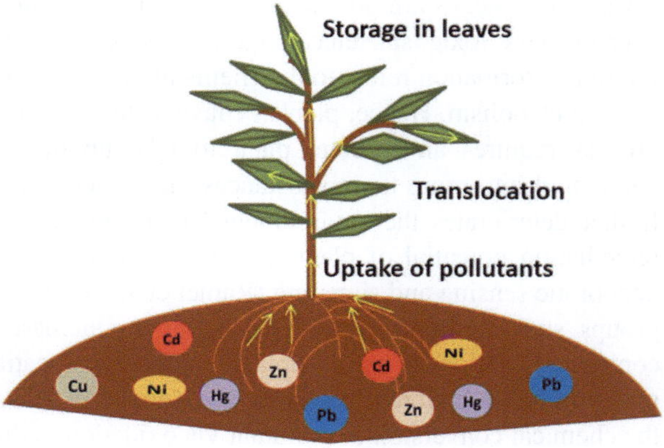


**Fig. 27.1** Phytoextraction of pollutants

of heavy metals, physicochemical properties of soil, and an appropriate hyperaccumulator. Certain characteristics (Ali et al. 2013a; Ghori et al. 2016) that are ideal for a plant candidate as hyperaccumulator are (1) high growth rate and large biomass such as branched or extensive colonization of root system, stems, and leaves with more of aboveground biomass; (2) effective translocation of heavy metals from root to shoot; (3) the tolerance level and accumulation potential for heavy metals; and (4) adaptability to biotic and abiotic stress that made them easy in cultivating and harvesting. And those hyperaccumulators should be avoided from animal feed and human consumption (Vamerali et al. 2010). Phytoextraction proceeds with two approaches where one method practiced natural or continuous phytoextraction with hyperaccumulator to remove the contaminants without any soil amendments, whereas the second method follows induced phytoextraction. In induced phytoextraction, chemical amendment in the soil enhances the metal accumulation in plants. Most of the metals are found as an unavailable form in the soil. Chelating agents such as organic and inorganic substances (citric acid, tartaric acid, amino acid, elemental sulfur, ammonium sulfate, EDTA) (Bosiacki et al. 2014) are added to the soil to form water-soluble complexes that mobilize metals and increase the metal uptake in plants. However, proper care and dosage of synthetic agents are advised to practice since it might affect secondary pollution or environmental hazard.

### 27.3.2 Phytodegradation

In the phytodegradation or phytotransformation method, plants disintegrate the pollutants either by metabolic activity within the plant tissues or by releasing various organic substances such as enzymes to the rhizosphere to degrade pollutants. And this mechanism is limited to the degradation of organic compounds like chlorinated solvents, insecticides, pesticides, and other xenobiotic substances (Pivetz 2001) since heavy metals are nonbiodegradable. The contaminants are broken down into simpler, less toxic substance and utilized for the faster growth of the plant. Phytotransformation refers to the chemical modification of the degraded compound by its metabolism. Hence, plants behave as the "green liver" of the biosphere. The process requires an efficient plant to take up the contaminants and transform compound into more toxic substances that generate toxin through transpiration. It further deteriorates the environment leading to great concern. The tolerance and remediation potential of plant species can be detected by the characterization of xenobiotic sensing and signaling (Ramel et al. 2012). Plant enzyme adds functional groups such as hydroxyl (—OH) group to increase the polarity of xenobiotic compounds followed by its uptake. After translocation, this organic compound goes through a different phase of transformation (Cutts 2018). Phase I processes the chemical conversion of pollutant via oxidation, reduction, or hydrolysis. Plant's metabolic activity in phase I works the way like human liver increases the polarity of drug and other foreign substances. The plant enzyme nitroreductases conducts an initial reaction. Phase II includes the conjugation process by adding biomolecules like sugar and amino acid to increase the polarity of the pollutants further and reduces the toxicity, which facilitates easy transport in plants. Phase III, also known as sequestration or compartmentation, is the final stage of phytotransformation, in which xenobiotic developed as a complex structure is sequestered in the plants (Fig. 27.2).

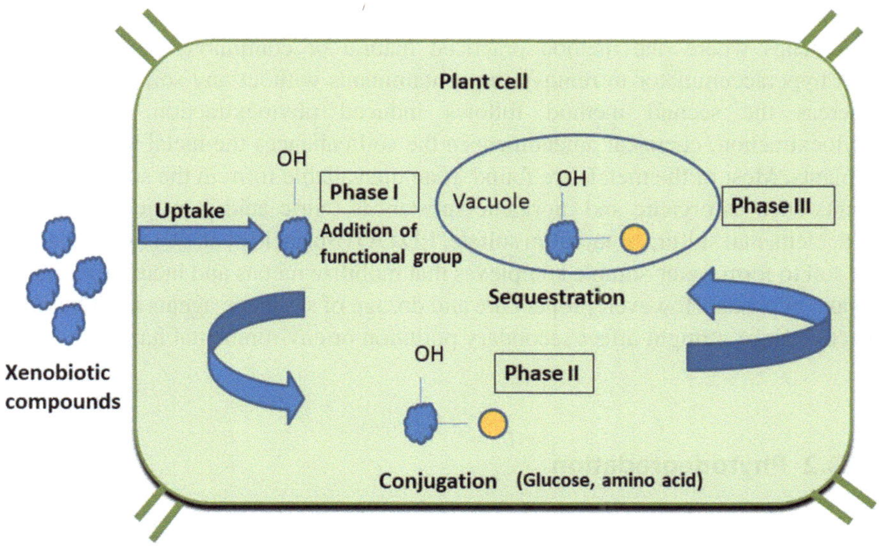


**Fig. 27.2** Phytotransformation of xenobiotic compounds inside a plant cell

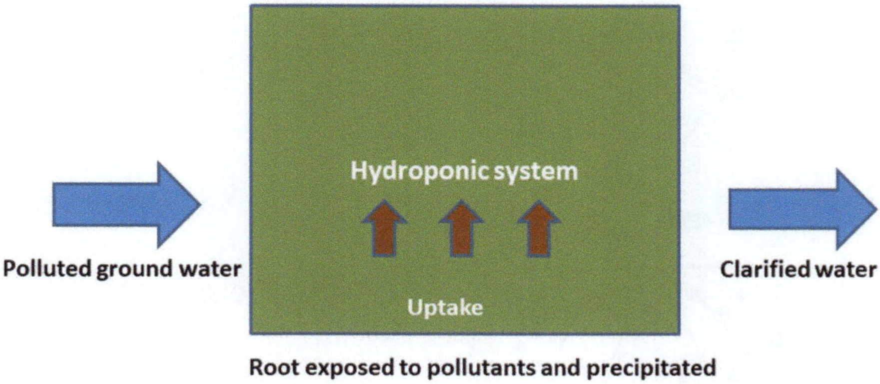


**Fig. 27.3** Rhizofiltration using hydroponic system

### 27.3.3 Rhizofiltration

Rhizofiltration is similar to phytoextraction used to remediate contaminants from groundwater, surface water, or wastewater rather than soil. In this method, plants are used to adsorb or precipitate the contaminants onto their roots. Secondary metabolites or root exudates released by plant precipitate contaminants onto their root by biogeochemical process and absorbed or translocated to the aboveground mass based on the contaminant, concentration, and plant type, usually with a larger root system. The fibrous root system in plants enables maximum adsorption as the root's large surface area is exposed to polluted areas. In situ remediation technique of rhizofiltration can be performed for surface water.

Moreover, an ex situ rhizofiltration process was conducted using an engineered tank system with polluted groundwater and plant species (Pivetz 2001; Kristanti et al. 2021). This system employs hydroponically cultivated plants (Fig. 27.3). The physicochemical properties of the plant species rate the remediation potential (Chatterjee et al. 2013). Rhizofiltration follows by drawing out polluted groundwater for remediation, and plants were selected based on their adsorption. These selected plants were cultivated hydroponically, further acclimatized in the polluted groundwater and planted onto the contaminated area. Eventually, harvested plants were disposed of properly. Factors such as chemical characteristics of pollutants, plants, and mutually interactive microorganisms and groundwater properties decide the efficient processing of rhizofiltration.

### 27.3.4 Phytostabilization

Phytostabilization or phytoimmobilization is when plant species are used to immobilize the movement of contaminants by adsorption onto root and accumulation by roots, thereby reducing the bioavailability and entry into the food chain (Erakhrumen

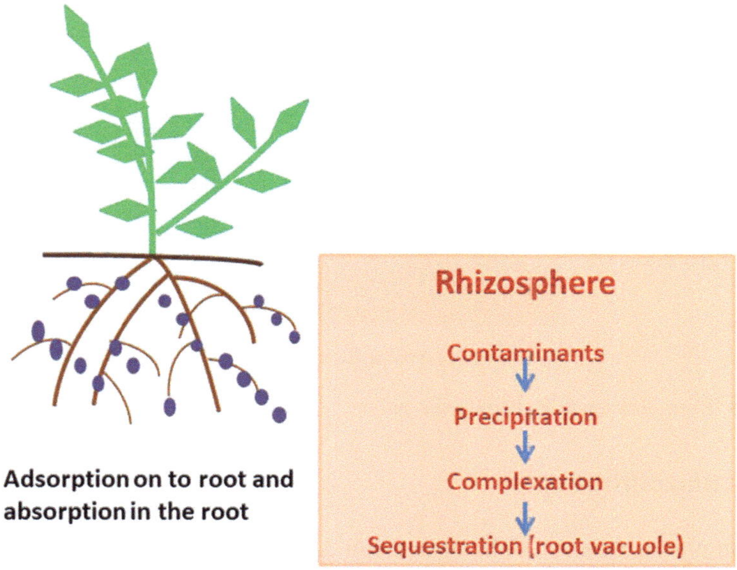


**Fig. 27.4** Mechanism of phytostabilization

2007). Contaminants undergo various reactions in the rhizosphere, such as adsorption, precipitation, complexation, redox reaction, and chelation in the remediation process (Fig. 27.4). Phytostabilization involves complexation of heavy metals with root exudates, with molecules that bind and sequester metals into root vacuole (Shackira and Puthur 2019). Immobilization of these contaminants prevents their migration by wind, rain, leaching, and soil erosion. Several factors that affect phytostabilization are physicochemical and biological property of soil, plant, and contaminants and environmental conditions. This technology can be enhanced by using organic or inorganic amendments that amplify the availability of contaminants in the soil (Bolan et al. 2011). The microbe also plays an important role in phytostabilization by enhancing immobilization and accumulation by reducing toxicity and improving plant growth (Ma et al. 2011). Chemophytostabilization with amendments like organic compounds, liming agents, and phosphorous compounds that immobilize metal and reduce soil bioavailability appears as most encouraging in the soil area of high metal toxicity (Alkorta et al. 2010). Disposal of hazardous biomass is not required, and this technique also prevents the entry of contaminants to the immediate area.

### 27.3.5 Phytovolatilization

Phytovolatilization is removing contaminants by volatilization through transpiration into atmosphere from soil or water; therefore, leaching of contaminants can be

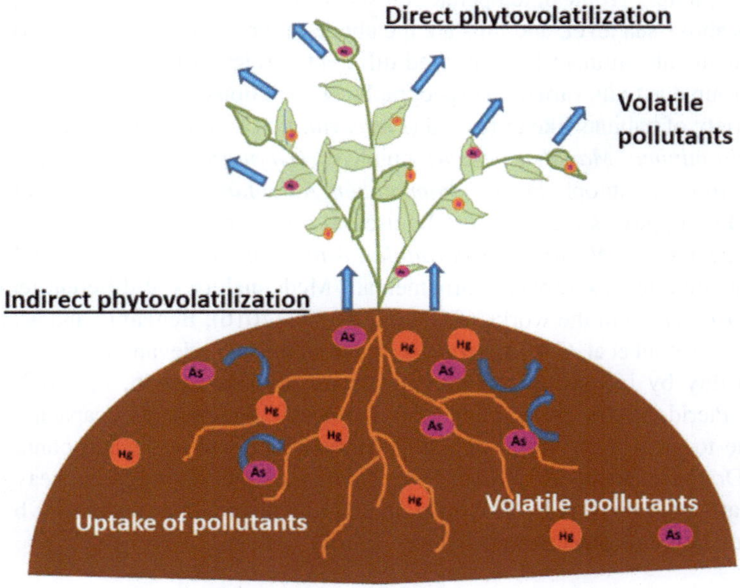


**Fig. 27.5** Direct and indirect phytovolatilization of contaminants

reduced. Plants, when interacting with contaminants, are possibly affected by degradation, excretion, or volatilization. Plant metabolism modifies the contaminants, changes them to less toxic, and releases them into the atmosphere. The technique can be used for organic and inorganic pollutants that are taken along with water. Phytovolatilization is a propitious method of volatilizing metals like selenium (Se) (Terry et al. 2000), mercury (Hg) (Wang et al. 2012), and arsenic (As) (Sakakibara et al. 2010: Jia et al. 2012). The mechanism of phytovolatilization occurs in two forms as direct phytovolatilization and indirect phytovolatilization (Limmer and Burken 2016; Chandra and Kumar 2017). Direct phytovolatilization is the instinctive uptake and translocation of contaminants in stems and leaves, thereby volatilizing the polluting substance. And indirect phytovolatilization results from increase volatile pollutant flux by the root activity in the subsurface of the vadose zone (Fig. 27.5).

## 27.4 Pteridophytes as Phytoremediation Agents

The most primitive group of plants, pteridophyta (*pteron* = feather, *phyton* = plants), are also called vascular cryptogams. They are the earliest vascular plants to inhabit the Earth's surface and classified into four groups: lycopods, *Equisetum*, Psilotaceae, and ferns (Christenhusz et al. 2017). They are larger than the bryophytes and have evolved vascular system with the emergence of seed habitat in the plant (Dudani et al. 2011). They form the vital link between the lower cryptogams and the higher spermatophytes. Tropical, temperate forests; different and elevated ecogeographical regions above sea level, and hills are the abode of these pteridophytes (Dixit 2000). The Indian subcontinent has harbored different species of this rich and diversified plant group due to its varied topo-geographical conditions. Fern and fern allies occur in a variety of habitats like terrestrial (*Pteris vittata*, *Dicranopteris linearis*), aquatic (*Salvinia minima*, *Marsilea minuta*), epiphytic (*Drynaria quercifolia*, *Microsorum punctatum*), and lithophytic (*Psioltum nudum*, *Adiantum venustum*) (Morajkar et al. 2015). Pteridophytes are known to be used as ornamentals of esthetic presence for ages. Species like *Nephrolepis cordifolia*, *Adiantum capillus-veneris*, and *Selaginella* sp. are a few examples of ornamentals. Medicinal uses of different ferns have been documented in the works of Goswami et al. (2016), Benjamin and Manickam (2007), and Rout et al. (2009). *Azolla* is an agronomically relevant fern as it increases soil fertility by increasing soil nitrogen levels, organic carbon, P, and K (Singh 2020). Pteridophytes are explored for the metal accumulation capacity and its attribute to growing in metal-rich soils to remediate soil and water contaminated sites (Drăghiceanu et al. 2014). The potential of pteridophytes in heavy metal remediation is reviewed by Prabhu et al. 2016 in general and specifically by *Pteris* sps. for cadmium (Sajeev et al. 2012).

## 27.5 Studies on Heavy Metal Stress in Pteridophytes

Phytoremediation is an eco-friendly technology that utilizes plants to absorb, remove, alter, or stabilize different organic contaminants and toxic metals from soil, sediments, and water resources (Cherian and Oliveira 2005). Several plant species are natural metal hyperaccumulators. Few examples are *Thlaspi rotundifolium*, *Pteris vittata*, and *Sesbania drummondii* which can hyperaccumulate Zn, As, and Pb, respectively (Ma et al. 2001a, b; Sahi et al. 2002). Ferns are known to have evolved around 400 million years ago (Chater et al. 2011) and are evolutionary, very significant, and distinct (Phillips and Galtier 2005). These plant groups' occurrence in soils rich in copper, zinc, nickel, and lead has promoted them as metallophytes, e.g., *Pellaea calomelanos* and *Cheilanthes hirta* for Cu and Ni (Kachenko et al. 2007), and *Asplenium septentrionale* (L.) Hoffm. for Pb and Cu (Page 1988). Not only terrestrial ferns but aquatic ferns like *Azolla filiculoides* (Talebi et al. 2019), *Marsilea minuta* (Thampatti et al. 2020) and S. minima (Nichols et al. 2000) accumulate heavy metals in their shoots from wastewaters. They are also known to accumulate trace elements (Ozaki et al. 2000). Ferns *P. vittata* have been reported as one of the potential hyperaccumulators for As (Ma et al. 2001a, b; Mehrag 2002). *Pityrogramma calomelanos* (silverback fern) for a hyperaccumulation of As has been reported by Francesconi et al. (2002); the gametophytes of the same are reported to be effective phytoremediation agents for Hg and Pb (Roshni and Hegde 2021). Studies by Kumari et al. (Kumari 2007 & Kumari et al. 2011) emphasize the rich vegetation of *P. vittata* L. along with *Ampelopteris prolifera* and the *Diplazium esculentum* in the fly ash *Salvinia natans* for Cu dumping site of Kanti Thermal Power Station in Muzaffarpur, Bihar, India.

*P. vittata* showed a significant metal uptake at alkaline pH greater than 9.0. The studies in this site also indicated high metal accumulation and its translocation in the aboveground parts of *P. vittata*, particularly in the leaf fronds. The ability to accumulate mixed metals based on bioavailability and preferable zinc in high concentration in both aerial and underground parts of the plant substantiates the already existing reports of An et al. (2006), Fayiga et al. (2004), Singh and Ma (2006), and Singh et al. (2006). Pongthornpruek et al. (2008) studied 19 different ferns and their soil profile to identify ferns with metal uptake. The ferns were collected from PhuSoi Dao National Park, Phitsanulok Province, Thailand, known for rich biodiversity. Significant high levels of Pb and Ni in the leaves of *Adiantum philippense* L. were noted. Also, *Adiantum caudatum* L. was found to be a good accumulator of Pb, Ni, and Co. Kachenko et al. (2007) studied metal uptake, Cd, Cr, Cu, Ni, Pb, and Zn, in ten common ferns found in Australia. The plants were grown for 20 weeks with different spiked concentrations (0, 50, 100, and 500 mg $kg^{-1}$) of the metal. Species like *Nephrolepis cordifolia* and *Hypolepis muelleri* had a high survival rate ad phytostabilization potential for Cu, Pb, Ni, or Zn while *Dennstaedtia davallioides* for Cu and Zn. *Blechnum nudum*, *Blechnum cartilagineum*, *Doodia aspera*, and *Calochlaena dubia* were unsuitable for phytoremediation purposes. Cornara et al. (2007) had screened 14 ferns growing in serpentine and metalliferous soils of Northern Italy and reported that *Equisetum ramosissimum* Desf. has high tolerance level for Cr, Fe, Cu, Ni, and As; *Nephrolepis cordifolia* C. Presl. for Pb; and *Athyrium filix-femina* and *Dryopteris filix-mas* for Cd. All these ferns showed tolerance level based on metal exclusion, while none of them could be termed hyperaccumulators. *Acrostichum aureum* L. of the Pteridaceae family and a mangrove fern can be used as an arsenic hyperaccumulator. Sharma and Irudayaraj (2010) have shown arsenic tolerance of up to 500 ppm in *Acrostichum aureum* L. A high growth rate with a substantial increase in cell size, lamina, membrane stability, and chlorophyll pigments was noticed in these arsenic stress-induced ferns. Concentration higher than 500 ppm does cause arsenic stress with the eventual killing of plants with 5000 ppm. Plants like these with specific habitat requirements like salty marshy areas with warm and humid tropical temperatures for their growth may not serve as ideal hyperaccumulators. With transgenics, such plants can be made to grow as effective phytoremediators in arsenic-rich soils. Certain ferns are consumed as fried or as a salad after treating them in hot water in countries like Thailand. Francesconi et al. (2002) studied the phytoremediation properties of a certain species of *P. calomelanos*, collected from the local market, that had shown 60 μg As g-1dry mass which was beyond the permissible limit in many of the countries. Hence the public should be cautious in selecting such hyperaccumulators as part of their diet. Government and the scientific community should come together to create awareness in this regard. The Environmental Protection Agency in 2002 reduced the maximum contaminant level of As from 50 to 10 mg $L^{-1}$. It recommended removing As contaminated water as one of the important areas to work through; it seemed expensive (Smith et al. 2002). Hyperaccumulators that are slow-growing with low biomass can also be used to remediate the soil by enhancing their growth by using plant growth-promoting bacteria and growth hormones like auxin (Khan et al. 2000). It increases the metal uptake and translocation capability by stimulating cell division, root elongation, and apical dominance (Parker et al. 1992). The use of bacterial inoculation in improving metal uptake by plants has been reported by Belimov et al. (2005) in Indian mustard. Selective uptake of metals is often reported eg. Ni in *Alyssum murale* (Abou-Shanab et al. 2006); Cd in rape (Sheng and Xia 2006; Sheng et al. 2008); He et al. (2009) - Pb and Cd in tomato and Cd uptake in spinach by Ali et al., (2013a, 2013b). Some of the important factors contributing to plants' phytoremediation efficiency are soil pH, cation exchange capacity (CEC), soil texture, organic content, etc. Soil nutrient also plays a significant role in determining the hyperaccumulation properties of the plants. Mehrag et al. (1994) have shown the antagonistic effect of phosphate on As uptake in certain As tolerant plants. Studies by Wei et al. (2006) also revealed the importance of P, K, and Fe influencing the As uptake capacity of *P. vittata* in two different geochemically rich sites of southern China. Wei and Zhou (2006) had proposed the two-phase cultivation of Cd hyperaccumulator Cress (*Rorippa globosa*) as a promising remediation strategy in Cd contaminated soil. It could be advantageous where many of the Cd accumulators are slow-growing. Research studies of two-phase cultivation suggested growing plants twice a year in contaminated soils and harvesting once they start flowering.

## 27.6 Patents Related to Pteridophytes in Phytoremediation

Pollutants are released into the environment through several industrial activities, mining, processing metals, military or exploration activities, and pesticides. These enter both aquatic and terrestrial environment either deliberately or accidentally. It's a collective responsibility to ensure that these contaminants should be remediated to prevent or mitigate entry into terrestrial, atmospheric, and aquatic environments. Phytoextraction is an organic, low input technique, offering an excellent solution for the uptake of metals or pollutants through roots/stems or leaves. Phytoremediation as technology is underutilized despite having significant evidence of successful field applications (Gerhardt et al. 2017). Edenspace, Inc., USA, holds 11 patents in the areas of phytoextraction, rhizofiltration, and hyperaccumulation. US patent no US-5917117-A granted to Ensley and team in 1999 mentioned the use of six inducing agents by which hyperaccumulation of metals in plant shoots can be enhanced (NCBI 2020). It is particularly done in plants belonging to the Brassicaceae family. Edenspace, the company involved in technologies related to cleaning up contaminants, has successfully used brake fern (Pteris family) in the trademark "Edenfern" to extract arsenic-contaminated backyard of Spring Valley, a residential site in Washington DC, in early 2000. Edenspace Systems owns or licenses 17 patents (USEPA 2005).

Raskin et al. 2002 has patented numerous technologies and plants (Suresh and Ravishankar 2004). Patents related to pteridophytes in phytoremediation technology are limited. One of the most remarkable discoveries was that of Chinese brake fern (*Pteris vittata* L.) by Ma et al. 2001a, b. The biogeochemist Lena Ma from the

University of Florida discovered brake fern growing in an abandoned arsenic-laden wood site. Based on the studies conducted by Tu and Ma 2005, bracken fern can take as high as 2.3% of arsenic from the soil, and it can store in aboveground levels (up to 90%). They reported that arsenic's hyperaccumulation was accompanied by an increase in the biomass of the aboveground parts. In addition to efficient root uptake, these ferns can remediate arsenic-contaminated water via the foliar application (arsenic uptake through leaves in live plants).

Pollutants and contaminants are removed from the water, soil, and wetland- type environment using phytoremediation through roots and fronds and by applying excised portions of plants such as leaflets (cutoff fronds). A US Patent (no: 7,066,982) explains contaminant removal by ferns via foliar-application and excised/ground fronds. Later the biomass could be harvested and disposed or extracted for pollutants or contaminants. *Pteris vittata* has been shown to accumulate approximately 540 mg $kg^{-1}$ arsenic (dry weight) in its fronds. The arsenic concentration in the water where the plant has been studied was approximately 20 mg $L^{-1}$. Thus, the plant has an extraordinary capability to enrich nearly 27 times more arsenic in its plant tissue than in contaminated water. These ferns remove contaminants from water with even low concentrations of pollutants. Perennial growth, pest and disease resistance, ability to accumulate arsenic from both soil and water, vigorous growth, and ability to grow in a diverse ecological niche with even high pH make *Pteris vittata* an ideal phytoremediator.

Ma and her team were granted US patent # 7,065,920 B2 in 2006 for the methods and strategies involved in contaminant removal by *Pteris vittata* and other ferns. Based on experimental findings, *Pteris* and non-*Pteris* fern plants can be used to accumulate pollutants from contaminated water, including aqueous solution, wastewater, groundwater, surface water, and combinations thereof. Chinese brake fern could remove 6 mg $kg^{-1}$ As from uncontaminated and about 1500 mg $kg^{-1}$ from highly contaminated soils. The fern can take up a range of inorganic and organic arsenic species, including arsenate, arsenite, and monomethylarsonic acid (MMA). Other non-*Pteris* fern plants mentioned in the patent belonged to Pteridales and Aspidiales and Pteridaceae, Adiantaceae, Aspleniaceae, Dryopteridaceae, and Oleandraceae. Specifically exemplified genera are *Adiantum*, *Asparagus*, *Asplenium*, *Cyrtonium*, *Didymochlacna*, *Dryopteris*, *Nephrolepis*, *Pteridium*, *Rumohra*, and *Pteris*. The specific examples of the *Pteris* sps. Are *P. creticamavi*, *P. creticaparkerii*, *P. cretica albo-lineata*, and *P. vittata*.

## 27.7 Conclusion and Scope for Future Work

Exploring pteridophytes in phytoremediation to remediate or mitigate heavy metals, pollutants, or toxic organic or inorganic contaminants from soil and water bodies remains largely laboratory oriented. Unfortunately, phytoremediation technologies are struggling to leap from lab to field (Beans 2017). Commercial ventures like that of Edenspace, Inc. utilizing ferns for remediation are limited. Elucidating the genetic mechanisms by advancing research on proteomics and genomics will facilitate better phytoremediation outcomes on a large scale. Genetic engineering, better plant management on-field, soil amendments, and exploring inducing agents for better results can accelerate the plants' efficiency for phytoremediation. Policymakers, governments, and society must join hands for greener options like that of phytoremediation instead of conventional approaches of excavation or landfill.

**Acknowledgments** The CISEE grant by the Vision Group of Science and Technology, Govt. of Karnataka (GRD no. 854), to SH is gratefully acknowledged. The authors sincerely acknowledge the Nitte University Centre for Science Education and Research, Nitte (Deemed to be University) Mangalore, for extending infrastructure support for the study (Table 27.1).

## References

Abou-Shanab RAI, Angle JS, Chaney RL (2006) Bacterial inoculants affecting nickel uptake by Alyssum murale from low, moderate and high Ni soils. Soil Biol Biochem 38(9):2882–2889

Ali H, Khan E, Sajad MA (2013a) Phytoremediation of heavy metals—concepts and applications. Chemosphere 91(7):869–881

Ali T, Mahmood S, Khan MY, Aslam A, Hussain MB, Asghar HN, Akhtar MJ (2013b) Phytoremediation of cadmium contaminated soil by auxin assisted bacterial inoculation. Asian J Agri Biol 1(2):79–84

Alkorta I, Becerril JM, Garbisu C (2010) Phytostabilization of metal contaminated soils. Rev Environ Health 25(2):135–146

An ZZ, Huang ZC, Lei M, Liao XY, Zheng YM, Chen TB (2006) Zn tolerance and accumulation in Pteris vittata L. and its potential to phytoremediation of Zn and As- contaminated soil. Chemosphere 62:796–802

Beans C (2017) Core concept: phytoremediation advances in the lab but lags in the field. Proc Natl Acad Sci 114(29):7475–7477

Belimov AA, Hontzeas N, Safronova VI, Demchinskaya SV, Piluzza G, Bullitta S, Glick BR (2005) Cadmium-tolerant plant growth-promoting bacteria associated with the roots of Indian mustard (*Brassica juncea* L. Czern.). Soil Biol Biochem 37:241–250

Benjamin A, Manickam VS (2007) Medicinal pteridophytes from the Western Ghats. Ind J Trad Knowl 6(4):611–618

Bolan NS, Park JH, Robinson B, Naidu R, Huh KY (2011) Phytostabilization: a green approach to contaminant containment. Adv Agron 112:145–204

Bosiacki M, Kleiber T, Markiewicz B (2014) Continuous and induced phytoextraction—plant-based methods to remove heavy metals from contaminated soil. Environmental Risk Assessment of Soil Contamination. IntechOpen, London

Boyd RS, Wall MA, Watkins JE (2000) Correspondence between Ni tolerance and hyperaccumulation in *Streptanthus* (Brassicaceae). Madrono 47:97–105

Chandra R, Kumar V (2017) Phytoremediation: a green sustainable technology for industrial waste management. In: Phytoremediation of environmental pollutants. CRC Press, Boca Raton, pp 1–42

Chater C, Kamisugi Y, Movahedi M, Fleming A, Cuming AC, Gray JE, Beerling DJ (2011) The regulatory mechanism controlling stomatal behavior conserved across 400 million years of land plant evolution. Curr Biol 21(12):1025–1029

Chatterjee S, Mitra A, Datta S, Veer V (2013) Phytoremediation protocols: an overview. In: Plant-based remediation processes. Springer-Verlag, Berlin Heidelberg, pp 1–18

Cherian S, Oliveira MM (2005) Transgenic plants in phytoremediation: recent advances and new possibilities. Environ Sci Technol 39:9377–9390

Christenhusz MJ, Fay MF, Chase MW (eds) (2017) Plants of the world: an illustrated encyclopedia of vascular plants. Kew Publishing University of Chicago Press, Chicago, Ill. ISBN: 022652292X, p 792

Cornara L, Roccotiello E, Minganti V, Drava G, De Pellegrini R, Mariotti MG (2007) Level of trace elements in Pteridophytes growing on serpentine and metalliferous soils. J Plant Nutr Soil Sci 170(6):781–787

Cramer GR, Urano K, Delrot S, Pezzotti M, Shinozaki K (2011) Effects of abiotic stress on plants: a systems biology perspective. BMC Plant Biol 11:163

Cutts S (2018) In: Chandra R, Dubey NK, Kumar V (eds) Phytoremediation of environmental pollutants. CRC Press, Taylor & Francis Group, Boca Raton, FL. ISBN: 9781138062603, p 510

Dinneny JR, Long TA, Wang JY, Jung JW, Mace D, Pointer S, Barron C, Brady SM, Schiefelbein J, Benfey PN (2008) Cell identity mediates the response of Arabidopsis roots to abiotic stress. Science 320(5878):942–945

Dixit RD (2000) Conspectus of Pteridophytic diversity in India. Indian Fern J 17:77–91

Drăghiceanu OA, Dobrescu CM, Soare LC (2014) Applications of pteridophytes in phytoremediation. Cur Trends Nat Sci 3(6):68–73

Dudani SN, Chandran MDS, Ramachandra TV (2011) Conservation of pteridophytes to maintain vital link between lower and higher group of plants. Sahyadri e-news 33:2–34

Erakhrumen AA (2007) Phytoremediation: an environmentally sound technology for pollution prevention, control and remediation in developing countries. Edu Res 2:151–156

Fayiga AO, Ma LQ, Cao X, Rathinasbapathi B (2004) Effects of heavy metals on growth and arsenic accumulation in the arsenic hyperaccumulator *Pteris vittata* L. Environ Pollut 132:289–296

Francesconi AK, Visoottiviseth P, Sridokchan W, Goessler W (2002) Arsenic species in an arsenic hyperaccumulating fern, *Pityrogramma calomelanos*: a potential phytoremediator of arsenic-contaminated soils. Sci Tot Environ 284:27–35

Gerhardt KE, Gerwing PD, Greenberg BM (2017) Opinion: taking phytoremediation from proven technology to accepted practice. Plant Sci 256:170–185. https://doi.org/10.1016/j.plantsci.2016.11.016

Ghori Z, Iftikhar H, Bhatti MF, Sharma I, Kazi AG, Ahmad P (2016) Phytoextraction: the use of plants to remove heavy metals from soil. In: Plant metal interaction. Elsevier, Amsterdam, pp 385–409

Ghosh M, Singh SP (2005) A review on phytoremediation of heavy metals and utilization of it's by products. As J Energy Env 6(4):214–231

Goswami HK, Sen K, Mukhopadhyay R (2016) Pteridophytes: evolutionary boon as medicinal plants. Plant Genetic Res 14(4):328

He LY, Chen ZJ, Ren GD, Zhang YF, Qian M, Sheng XF (2009) Increased cadmium and lead uptake of a cadmium hyperaccumulator tomato by cadmium-resistant bacteria. Ecotoxicol Environ Safety 72:1343–1348

Jacob JM, Karthik C, Saratale RG, Kumar SS, Prabakar D, Kadirvelu K, Pugazhendhi A (2018) Biological approaches to tackle heavy metal pollution: a survey of literature. J Environ Manag 217:56–70

Jia Y, Huang H, Sun GX, Zhao FJ, Zhu YG (2012) Pathways and relative contributions to arsenic volatilization from rice plants and paddy soil. Environ Sci Technol 46(15):8090–8096

Kachenko AG, Singh B, Bhatia NP (2007) Heavy metal tolerance in common fern species. Aus J Bot 300:207–219

Khan AG, Kuek C, Chaudhry TM, Khoo CS, Hayes WJ (2000) Role of plants, mycorrhizae and phytochelators in heavy metal contaminated land remediation. Chemosphere 41:197–207

Kilian J, Whitehead D, Horak J, Wanke D, Weinl S, Batistic O, D'Angelo C, Bornberg-Bauer E, Kudla J, Harter K (2007) The At GenExpress global stress expression data set: protocols, evaluation and model data analysis of UV-B light, drought and cold stress responses. Plant J 50(2):347–363

Kramer U (2000) Cadmium for all meals - plants with an unusual appetite. New Phytol 145:1–5

Kristanti RA, Ngu WJ, Yuniarto A, Hadibarata T (2021) Rhizofiltration for removal of inorganic and organic pollutants in groundwater: a review. Biointerface Res Appl Chem 11 (4):12326–12347
Kumari A (2007) Bioamelioration of fly ash by some ferns. Indian Fern J 24:172–179
Kumari A, Lal B, Pakade YB, Chand P (2011) Assessment of bioaccumulation of heavy metal by Pteris vittata L. growing in the vicinity of fly ash. Int J Phytorem 13(8):779–787
Limmer M, Burken J (2016) Phytovolatilization of organic contaminants. Environ Sci Technol 50 (13):6632–6643
Ma LQ, Komar KM, Tu C, Zhang W, Addendum CY (2001a) A fern that hyperaccumulates arsenic-addendum. Nature 411:438
Ma LQ, Komar KM, Tu C, Zhang W, Cai Y (2001b) A fern that hyperaccumulates arsenic. Nature 409:579
Ma Y, Prasad MNV, Rajkumar M, Freitas H (2011) Plant growth promoting rhizobacteria and endophytes accelerate phytoremediation of metalliferous soils. Biotechnol Adv 29(2):248–258
Mehrag AA (2002) Variation in arsenic hyperaccumulation in ferns and their allies. New Phytol 157:25–31
Mehrag AA, Naylor J, Macnair MR (1994) Phosphorus nutrition of arsenate-tolerant and non-tolerant phenotypes of velvet grass. J Environ Qual 23:234–238
Morajkar S, Sajeev S, Hegde S (2015) Ferns: a thriving group of urban dwellers. Bionature 35 (1&2):13–22
National Center for Biotechnology Information (2020) PubChem Patent Summary for US-5917117-A. https://pubchem.ncbi.nlm.nih.gov/patent/US-5917117-A. Accessed 29 Dec 2020
Nichols PB, Couch JD, Al-Hamdani SH (2000) Selected physiological responses of *Salvinia minima* to different chromium concentrations. Aquat Bot 68:313–319
Ozaki T, Enomoto S, Minai Y, Ambe S, Makide Y (2000) A survey of trace elements in pteridophytes. Biol Trace Elem Res 74:259–273
Page CN (1988) Ferns: their habitats in British and Irish landscape. Collins, London
Parker DR, Aguilera JJ, Thomson DN (1992) Zinc-phosphate interactions in two cultivars of tomato has grown in chelator buffered nutrient solution. Plant Soil 143:163–177
Phillips TL, Galtier J (2005) Evolutionary and ecological perspectives of late Paleozoic ferns: part I. Zygopteridales. Review of Palaeobotany and Palynology 135(3–4):165–203
Pinheiro C, Chaves MM (2011) Photosynthesis and drought: can we make metabolic connections from available data? J Exp Bot 62(3):869–882
Pivetz BE (2001) Phytoremediation of contaminated soil and ground water at hazardous waste sites. US Environmental Protection Agency, Office of Research and Development, Office of Solid Waste and Emergency Response. Groundwater. EPA/540/S-01/500, 36 p
Pollard AJ (2000) Metal hyperaccumulation: a model system for coevolutionary studies. New Phytol 146:179–181
Pongthornpruek S, Pampasit S, Sriprang N, Nabheerong P, Promtep K (2008) Heavy metal accumulation in soil and some fern species at PhuSoi Dao National Park, Phitsanulok Province, Thailand. NU Sci J 5(2):151–164
Prabhu SG, Srinikethan G, Hegde S (2016) Potential of pteridophytes in heavy metal phytoremediation. Int J Res Eng Technol 5:1–9
Ramel F, Sulmon C, Serra AA, Gouesbet G, Couée I (2012) Xenobiotic sensing and signalling in higher plants. J Exp Bot 63(11):3999–4014
Raskin I, Ribnicky DM, Komarnytsky S et al (2002) Plants and human health in the twenty-first century. Trends Biotechnol 20(12):522–531
Reeves RD, van der Ent A, Echevarria G, Isnard S, Baker AJ (2021) Global distribution and ecology of hyperaccumulator plants. In: Agromining: farming for metals. Springer, Cham, pp 133–154
Roshni PT, Hegde S (2021) *In vitro* spore germination and phytoremediation of Hg and Pb using gametophytes of *Pityrogramma calomelanos*. Int J Phytoremediation, 23(3), 307–315

Rout SD, Panda T, Mishra N (2009) Ethnomedicinal studies on some pteridophytes of Simlipal biosphere reserve, Orissa, India. Int J Med Medici Sci 1(5):192–197

Sahi SV, Bryant NL, Sharma NC, Singh SR (2002) Characterization of a lead hyperaccumulator shrub, *Sesbania drummondii*. Environ Sci Technol 36:4676–4680

Sajeev S, Ramya PV, Sunitha BC, Melo JS, Hegde S (2012) Phytoremediation of cadmium using *Pteris sp*. In: Prospects in bioscience: addressing the issues. Springer, India, pp 131–139

Sakakibara M, Watanabe A, Inoue M, Sano S, Kaise T (2010) Phytoextraction and phytovolatilization of arsenic from As-contaminated soils by *Pteris vittata*. Proc Annu Int Conf Soils, Sediments, Water Energ 12(1):26

Sanyahumbi D, Duncan JR, Zhao M, Hille R (1998) Removal of lead from solution by the nonviable of the water fern *Azolla filliculoides*. Biotechnol Lett 20:745–747

Sardar K, Ali S, Hameed S, Afzal S et al (2013) Heavy metals contamination and what are the impacts on living organisms. Greener J Environ Manag Pub Safety 2(4):172–179

Sela M, Garty J, Telor E (1989) The accumulation and the effect of heavy metals on the water fern, Azollafiliculoides. New Phytol 112:7–12

Shackira AM, Puthur JT (2019) Phytostabilization of heavy metals: understanding of principles and practices. Plant-metal interactions. Springer, Cham, pp 263–282

Sharma CL, Irudayaraj V (2010) Studies on heavy metal (arsenic) tolerance in a mangrove fern *Acrostichum aureum* L. (Pteridaceae). J Bas App Biol 4(3):143–152

Sheng XF, He LY, Wang QY, Ye HS, Jiang CY (2008) Effects of inoculation of biosurfactant-producing bacillus sp. J119 on plant growth and cadmium uptake in a cadmium-amended soil. J Hazard Mater 155:17–22

Sheng XF, Xia JJ (2006) Improvement of rape (*Brassica napus*) plant growth and cadmium uptake by cadmium-resistant bacteria. Chemosphere 6:1036–1042

Singh S (2020) Effect of nitrogen application through urea and Azolla on yield, nutrient uptake of rice and soil acidity indices in acidic soil of Meghalaya. J Environ Biol 41(1):139–146

Singh S, Anjum NA, Khan NA, Nazar R (2008) Metal-binding peptides and antioxidant defense system in planta: significance in cadmium tolerance. In: Khan NA, Singh S (eds) Abiotic stress and plant responses. I.K. International Publishing house Pvt. Ltd, New Delhi, pp 165–196

Singh N, Ma LQ (2006) Arsenic speciation, and arsenic and phosphate distribution in arsenic hyperaccumulator *Pteris vittata* L. and non-hyperaccumulator *Pteris ensiformis* L. Environ Pollut 141:238–246

Singh N, Ma LQ, Srivastava M, Rathinasbapathi B (2006) Metabolic adaptations to arsenic induced oxidative stress in *Pteris vittata* L. and *Pteris ensiformis* L. Plant Sci 170:274–282

Smith AH, Lopipero PA, Bates MN, Steinmaus CM (2002) Public health. Arsenic epidemiology and drinking water standards. Science 296:2145–2146

Suresh B, Ravishankar GA (2004) Phytoremediation-A novel and promising approach for environmental clean-up. Crit Rev Biotech 24(2–3):97–124. https://doi.org/10.1080/07388550490493627

Talebi M, Tabatabaei BE, Akbarzadeh H (2019) Hyperaccumulation of Cu, Zn, Ni, and Cd in Azolla species inducing expression of methallothionein and phytochelatin synthase genes. Chemosphere 230:488–497

Terry N, Zayed AM, De Souza MP, Tarun AS (2000) Selenium in higher plants. Annu Rev Plant Biol 51(1):401–432

Thampatti KM, Beena VI, Meera AV, Ajayan AS (2020) Phytoremediation of metals by aquatic macrophytes. In: Phytoremediation. Springer, Cham, pp 153–204

Tu C, Ma LQ (2005) Effects of arsenic on concentration and distribution of nutrients in the fronds of the arsenic hyperaccumulator *Pteris vittata* L. Environ Pollut 135(2):333–340

USEPA 2005 Small business innovation research-success stories. EPA/600/R-05/084 December 2005, www.epa.gov

Vamerali T, Bandiera M, Mosca G (2010) Field crops for phytoremediation of metal-contaminated land. A review. Environ Chem Lett 8(1):1–17

Wang J, Feng X, Anderson CW, Xing Y, Shang L (2012) Remediation of mercury contaminated sites–a review. J Hazard Mater 221:1–18

Wild H (1968) Geobotanical anomalies in Rhodesia: 1—The vegetation of copper bearing soils. Kirkia 1:1–71

Wei CY, Sun X, Wang C, Wang WY (2006) Factors influencing arsenic accumulation by *Pteris vittata*: a comparative field study at two sites. Environ Pollu 141:488–493

Wei SH, Zhou QZ (2006) Phytoremediation of cadmium-contaminated soils by *Rorippa globosa* using two-phase planting. Environ Sci Pollut Res Int 13(3):151–155

Yan A, Wang Y, Tan SN, Yusof MLM, Ghosh S, Chen Z (2020) Phytoremediation: a promising approach for revegetation of heavy metal-polluted land. Front Plant Sci 11:359

**Sudha Sajeev** completed her Ph.D. in Biotechnology in St. Aloysius College (Autonomous), Mangalore University, in 2017 under the supervision of Dr. Smitha Hegde, Professor, Nitte University Center for Science Education and Research (NUCSER), and M.Sc. (Biotechnology) from the University of Mysore. Her area of interest is in plant taxonomy, abiotic stress, phytoremediation, bioinformatics, molecular biology, and understanding the nuisances of environmental chemistry specifically the biosorption capabilities of pteridophytes. Currently, she is part of a team working with antimicrobial resistance in ICAR-CIFT, Cochin, Kerala, India

**Roshni PT** **(Roshni** Pulukkunadu Thekkeveedu). She is currently pursuing Ph.D. in Plant Biotechnology, Department of Postgraduate Studies and Research in Biotechnology, St. Aloysius College, Mangalore. Her research interest includes studies on the effect of heavy metals on gametophytes, developmental genes in pteridophytes, microbial and plant biotechnology, and bioinformatics.

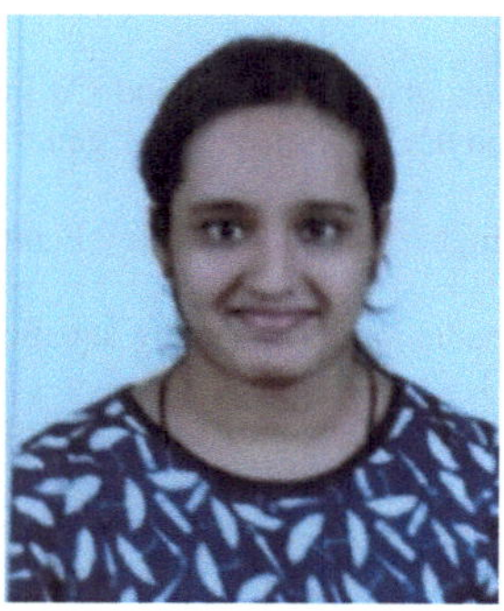

**Rachel Carmelita Mathias** has completed M.Sc. in Analytical Chemistry in St. Aloysius College (Autonomous), Mangaluru. Her expertise is in the field of chemistry and material science such as membrane technology, and her interests open up to biological chemistry as well. She is currently working on phytoremediation of heavy metals using pteridophytes.

**Shaiesh Morajkar** His expertise is in the utility of molecular markers in plant taxonomic, diversity, and breeding studies. Along with the molecular diversity studies, his research interests include studies of stress resistance in crops, plant-environment association, hyperspectral remote sensing, soil sciences, and wetland ecosystems.

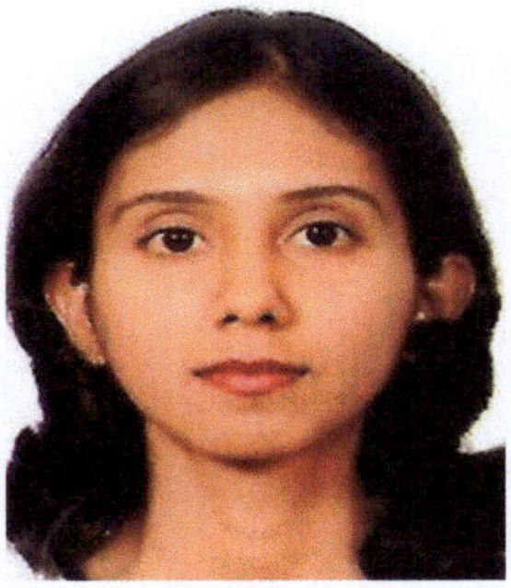

**Smruthi G. Prabhu**, is currently employed as Assistant Professor, Department of Postgraduate Studies and Research in Biotechnology, St. Aloysius College (Autonomous), Mangalore, Karnataka, India. Her research interests include bioremediation, microbiology, bioenergy, and plant tissue culture.

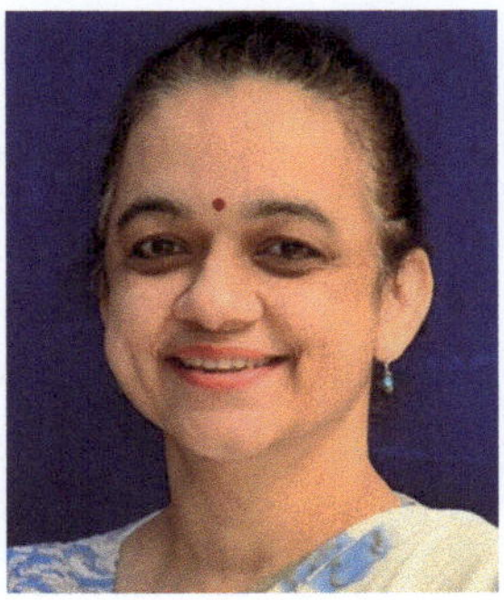

**Smitha Hegde** NUCSER, Nitte DU, Deralakatte, Mangalore, has 30 years of experience in pteridology. She obtained her M. Sc. from the University of Bombay and Ph.D. Biosciences from St. Aloysius College, Mangalore University, India. Her doctoral thesis was on tissue culture propagation of *Drynaria quercifolia* L. She has conducted pioneering work in biodiversity and distribution studies, developmental studies, phytoremediation, plant tissue culture, and molecular markers related to pteridophytes of the region. She has over 74 publications including book chapters and journals, edited 3 books, published 2 patents, and successfully guided 112 postgraduate research project students and 10 doctoral students. She is currently the PI of state funded project on plant tissue culture and heavy metal remediation using pteridophytes.

# Pteridophytes as Effective Biosorption Agents of Heavy Metals

# 28

Smruthi Prabhu and Smitha Hegde

**Abstract**

Pteridophytes are the most primitive vascular cryptogams to inhabit the surface of the Earth. For millions of years, the pteridophytes have grown efficiently in ever-changing environmental conditions and soils rich in toxic metals. Today, pteridophytes dwell successfully in inhospitable regions, including urban areas where other plants fail to. The adaptation of the pteridophytes to anthropogenic activities indicates its evolutionary distinctness and resilience to various abiotic stress factors, including high concentrations of heavy metals. The tolerance of the pteridophytes to very high concentrations of toxic metals has promoted high metal accumulation and translocation to aboveground parts, particularly in the fronds. The toxic metal hyperaccumulation promoted the pteridophytes' application in the in situ or ex situ remediation of the polluted soil or water. The method of employing a living plant material is phytoremediation. As an alternate technique, ground, nonliving, processed pteridophyte-based biosorbents are applied to remediate the point source of pollution, in situ.

Biosorbents are efficient when the biomass is cost-effective, easy to grow or harvest, plentiful in availability, and of usage of same quality. Pteridophytes qualify these criteria, making them good biosorbent candidates. Pteridophytes or ferns have a limited role in the food chain for animals and humans, making them suitable for phytoremediation and biosorption studies.

S. Prabhu
Department of Post graduate Studies and Research in Biotechnology, St. Aloysius College (Autonomous), Mangalore, India

S. Hegde (✉)
Nitte (Deemed to be University), Nitte University Center for Science Education and Research, Mangalore, India
e-mail: smitha.hegde@nitte.edu.in



J. Marimuthu et al. (eds.), *Ferns*, https://doi.org/10.1007/978-981-16-6170-9_28

**Keywords**

Biosorbent · Fern · Heavy metal · Phytoremediation

## 28.1 Introduction

Growing population and industrialization have diminished the quality and quantity of potable water. The scarcity of clean water is increasing day by day. A war over a source of freshwater is not far from reality. About two-thirds of the global population and half the India population live under water scarcity conditions (Mekonnen and Hoekstra 2016). Hence, it is imperative to adopt the concept of reduce, recover, reuse, and recycle water on a high-priority basis.

The available portion of freshwater has continuously been subjected to human activities' intensity either directly or indirectly. Today's real issue is the organic and inorganic pollutants of water, and heavy metals as pollutants are of serious concern. Heavy metals have always been there, as a part of the environment, but overexploitation of resources and anthropogenic activities have increased their background concentration in the biosphere. Over the past few decades, heavy metals have received considerable attention due to their severe effects on the living system (Jaishankar et al. 2014; Tchounwou et al. 2012). The commonly found metals/metalloids in wastewater are arsenic (As), cadmium (Cd), chromium (Cr), lead (Pb), mercury (Hg), and zinc (Zn). Heavy metals being highly mobile, nonbiodegradable, persistent, and toxic upon higher levels of exposure have to be removed from the point source to prevent its interaction with living systems that could cause biomagnification (Ashraf et al. 2016; Sharma and Sachdeva 2015). Thus, selecting a suitable remediation option is of utmost importance to remove the metal ions from its point source and check its mobility (Singh et al. 2011).

Large-scale industries have employed conventional treatment methods like membrane technology, the ion exchange technology, chemical precipitation, and conventional adsorbents to remove toxic metals from wastewater. High sludge production and its disposal problems, high economic costs, high chemical consumption, and fouling have hindered the wide-scale application of physical and chemical processes (Velazquez-Jimenez et al. 2018). Further research was initiated to explore new technologies to overcome the limitations of the existing ones.

Bioremediation techniques, such as bioaccumulation and phytoremediation, are promising technologies. Still, the requirement of a large land area and nutrient supply, sensitivity to variation in conditions and chemical toxicity, and less flexibility in bioreactor design and operation have restrained its broad-spectrum application (Fomina and Gadd 2014). Volesky and Tsezos in 1982 demonstrated the application of nonliving biomass in the removal of heavy metals by a passive process and termed it "biosorption" (Park et al. 2010). Feasibility studies have established the applicability of using nonliving biomass on a large scale over bioaccumulation (use of living organism). The biosorption technology's significant advantages are the cost-effectiveness, insensitivity to toxins, the biodegradability of the biosorbent, minimized biological or chemical sludge, regeneration of the biosorbents, metal recovery post sorption, and high efficiency of metal removal (Farooq et al. 2010; Hubbe et al. 2011; Nguyen et al. 2013). A biosorbent is efficient if the biomass is cost-effective, available around the year in the same quality, and easy to grow or harvest, requires a minimal physical/chemical modification or processing, and ensures environmental safety upon application (Gadd 2009).

Studies have reported numerous biomasses of different genres as potential biosorbents of heavy metals. Although a cell wall's presence makes them good biosorbents, few studies with pathogenic fungi/bacteria, endemic or threatened species, or organisms of commercial importance as biosorbents have little justification. This chapter attempts to evaluate the potential application of pteridophytes as biosorbents. Pteridophytes or ferns have a limited role in the food chain for animals and humans and reproduce rapidly through spores. Hence, many pteridophytes are listed as invasive weeds and have the least concern status (IUCN Red List of Threatened Species). Additionally, pteridophytes are resilient to urbanization and other environments, making the biomaterial abundantly available and a potential biosorbent candidate.

### 28.1.1 Biosorption and Biosorbent

Conventional methods have several limitations; there is a need to design an adsorbent that is as efficient as activated carbon but at the same time is cost-efficient, abundant in nature, and environmentally safe and requires less processing. The metal-binding capacity on biological materials was the very base for the development of biosorption. Biosorption is a process that involves the utilization of inexpensive biomass to sequester heavy metals. Kratochiv and Volesky in 1998 had emphasized the advantages of biosorption over conventional methods to be low operating cost, less chemical/biological sludge disposal, and dilute effluents (de Freitas et al. 2019). Factually, the use of nonliving biomass or a biosorbent for the uptake of any chemical pollutant by the passive process is called as "biosorption."

## 28.2 Pteridophytes: The Potential Biosorbent Candidate

### 28.2.1 Pteridophytes and their Distribution

Pteridophytes are the most primitive, well-developed vascular plants to habitat on land. Characteristic feather-like leaves have yielded them the name "pteridophytes" (*pteron* meaning feather and *phyton* meaning plant) (Sharma 2012). Pteridophytes occupy their position between bryophytes and gymnosperms in the plant kingdom (Hegde and Sajeev 2013). Therefore, similarities in characters with bryophytes and gymnosperms exist.

Ferns or pteridophytes usually grow in xeric and moist shady habitats. Their habitats can vary from terrestrial plants in forests to aquatic forms in freshwater ponds and swamps to epiphytes on tree trunks and branches or lithophytes on rocks and crevices (Lindsay and Middleton 2012 onwards). Temperature, soil type, humidity, and moisture play a significant role in the fern growth. Microclimatic conditions are essential for their survival and distribution. Due to favorable climatic conditions and suitable habitats, ferns are widely distributed in India.

Morajkar et al. (2015) have made a fascinating observation on thriving ferns in urban areas. Increase in anthropogenic activities has led to pteridophytes adapting to human-made structures and urban surroundings. Ferns are successfully dwelling in urban areas that include newly cut landscapes, wastewater trails on pipelines, window sills, and poorly maintained buildings. In addition to trees as their substrate, epiphytic ferns have successfully grown on roofs made up of clay or cement. Lithophytes in urban areas have replaced the rock ledges by walls and pavements. Human and animal urine/feces act as a rich source of nitrogen and nutrition for these plants.

### 28.2.2 Pteridophyte-Based Biosorbents of Heavy Metals

Ma and Tu (2006) and Ma et al. (2004) first claimed ferns via the foliar application and excised/ground fronds to remove As. The biomass (pinnae) was ground by air-drying or by freeze-drying. Air-dried *Pteris vittata* biomass exhibited better arsenite removal than arsenate compared to other ground fern biomass. The use of nonliving biomaterial with metal-binding compounds required minimal care and maintenance. Although hyperaccumulators are resistant to high metal concentrations, this method is advantageous when the contaminant levels are too high and could kill the live plants.

The literature available on the biosorption of heavy metals using pteridophytes is limited. The reported pteridophyte-based biosorbents predominantly comprised of tree fern and *Azolla* spp. It was only recently that *Pteris vittata* L. was explored further as a potential heavy metal biosorbent. The objective to control the rapid reproduction of the invasive weeds, *Azolla* and *Pteris vittata*, led to its application as a potential biosorbent. However, the use of tree fern spp. as a biosorbent cannot be entirely justified unless it is an agricultural waste as it is an endangered species.

*Azolla filiculoides*, in its pristine form, was studied as a potential biosorbent of heavy metals Cs(II) and Sr(II) by Mashkani and Ghazvini (2009). Rakhshaee et al. (2006) have reported the uptake of Pb(II) and Cd(II) at 283 K with a biosorption capacity of 234.34 and 122.75 mg/g, respectively, on pristine *Azolla filiculoides*. An attempt was made by Antunes et al. (2001) to recover valuable heavy metals like gold from a synthetic dilute solution of gold mimicking the concentrations usually present in effluents using untreated *Azolla filiculoides* as the biosorbent.

Prabhu et al. (2018, 2019a, 2019b, 2020) have done an exhaustive study on the uptake of Pb(II), Cd(II), and Cr(VI) by *Pteris vittata* L. in its powdered and pelleted biosorbent forms. *Pteris vittata* L. subjected to cost-effective treatment exhibited an enhanced biosorption as shown in Table 28.1.

## 28.3 Physicochemical Factors that Influence Biosorption Mechanisms

### 28.3.1 pH

Fomina and Gadd (2014) have expressed the role of physicochemical factors in determining biosorbent performance. pH has a significant effect on the biosorption rate. pH strongly influences the site dissociation of the biosorbent surface and determines the solution chemistry of the heavy metal ions such as hydrolysis, complexation with the ligands, redox reaction speciation, and the precipitation of the metal ions (Kulkarni 2015). For the sorption of cations, Asbchin et al. (2012) inferred that lower pH resulted in competition between the metal cations and the protons for the binding sites resulting in lower biosorption. Higher pH range is maintained in the sorption of cations. Zhao et al. (1999) successfully proved Ni(II) 's efficient biosorption on dried *Azolla filiculoides* at pH 6.0. Similar results were obtained by Zhao and Duncun (1998) with an optimum uptake of Ni(II) on *Azolla filiculoides* at pH 6.5. Ho (2005) reported pH 4.9 to be the optimum pH for the biosorption of Pb on tree fern. The pH of the system influences the charge of the functional groups present on the biosorbent. For example, if carboxyl groups (-COOH) are available on the biosorbent, they retain their proton at lower pH. Thus, the possibility of cation binding to this functional group is low. At higher pH, the carboxyl group gets deprotonated resulting in negatively charged ligands (—COO—), thus increasing the possibility of cations binding to the negatively charged ligands at higher pH (Krishnani et al. 2008). The principle can be applied to other functional groups as well. The pH range to be studied is determined by the metal ions' stability as few metal ions could precipitate at higher pH to form metal hydroxides.

Lower pH promotes the biosorption of anions. Antunes et al. (2001) observed the complete sorption of Au(III) at pH 2. Au(III) being an anion was adsorbed entirely on the positively charged biosorbent due to proton distribution at low pH. The optimum pH range for removing Cr(VI) has been reported between pH 2.0 and 3.0 by several studies. This trend is because Cr(VI) exists as oxyanions, $HCrO_4^-$ and $Cr_2O_7^{2-}$ in the pH range of 2.0–6.0 (Shroff and Vaidya 2012).

### 28.3.2 Temperature

Temperature is an important parameter that determines the adsorption reactions. Increase in temperature could have a different effect on a biosorbent or metal ion. Biosorption efficiency of the biosorbent could either increase or decrease. By the adsorption theory, adsorption decreases with the increase in temperature. Desorption

**Table 28.1** Pteridophyte-based biosorbents that have a potential for the uptake of heavy metals

| Biosorbent | Metal | Biosorption capacity (mg/g) | References |
|---|---|---|---|
| *Azolla filiculoides* (pristine) | Ni(II) | 43.4 | Zhao and Duncun (1998) |
| | Zn (II) | 45.2 | Zhao et al. (1999) |
| | Cu (II) | 23.1 | Fogarty et al. (1999) |
| | Au (III) | 2–10 | Antunes et al. (2001) |
| | Cu (II) | 34.3 | Asbchin et al. (2012) |
| | Ni(II) | 45.1 | |
| *Azolla filiculoides* (activated) | Pb (II) | 271.0 | Khosravi et al. (2005) |
| | Cd (II) | 111.0 | |
| | Ni(II) | 71.0 | |
| | Zn (II) | 60.0 | |
| *Azolla filiculoides* ($H_2O$ or $Mgcl_2$ treated) | Pb (II) | 228.0 | Ganji et al. (2005) |
| | Cd (II) | 34.0 | |
| | Zn (II) | 86.0 | |
| | Cu (II) | 33.0 | |
| *Pteris vittata* L. (pristine) | Pb (II) | 125.0 | Prabhu et al. (2019a) |
| | Cd (II) | 31.3 | Prabhu et al. (2019b) |
| | Cr (VI) | 166.7 | |
| *Pteris vittata* L. (NaOH treated) | Pb (II) | 133.3 | Prabhu et al. (2018) |
| | Cd (II) | 95.2 | |
| | Cr (VI) | 76.9 | |
| *Pteris vittata* L. ($CaCl_2$ treated) | Pb (II) | 90.9 | |
| | Cd (II) | 76.9 | |
| | Cr (VI) | 90.9 | |
| *Pteris vittata* L. ($HNO_3$ treated) | Pb (II) | 51.6 | |

(continued)

**Table 28.1** (continued)

| Biosorbent | Metal | Biosorption capacity (mg/g) | References |
|---|---|---|---|
| | Cd (II) | 47.2 | |
| | Cr (VI) | 100.0 | |
| *Pteris vittata* L. (pristine and pelleted) | Pb (II) | 13.5 | Prabhu et al. (2019a) |
| | Cd (II) | 1.7 | Prabhu et al. (2020) |
| Tree fern | Pb (II) | 40.0 | Ho et al. (2004) |
| | Cd (II) | 16.3 | Ho and Wang (2004) |
| | Cu (II) | 11.7 | Ho (2003) |
| | Hg (II) | 26.5 | Ho and Wang (2008) |
| *Acrostichum aureum* L. | Pb (II) | 28.6 | Soniya and Krishnakumar (2015) |
| Bracken fern | Cr (VI) | 83.2 | López-García et al. (2010) |

of previously adsorbed metal ions could also be associated (Wankasi et al. 2005). Daraei et al. (2014) observed a decrease in the Cr(VI) removal efficiency with the increase in temperature 20–60 °C. The decrease in sorption with an increase in temperature has been associated with breaking bonds in the sorbent's external or inner surface. The attractive forces between the biosorbent and the metal ions weaken with an increase in temperature (Wankasi et al. 2005). Moubarik and Grimi (2014) have reported a decreased Cd(II) removal from aqueous solution with the increase in temperature from 25 °C to 90 °C on biosorbent composite made up of olive stone and sugarcane bagasse by-products. The metal ion's tendency to escape from the biosorbent surface to the bulk solution increases, causing a decrease in the boundary layer thickness at a higher temperature. The decrease in adsorption at higher temperature accounts to physical (weak) interaction between the metal ions and the biosorbent (Wankasi et al. 2005; Moubarik and Grimi 2014).

Increasing the biosorption system's temperature enhances the surface activity and the kinetic activity of the biosorbent (Fomina and Gadd 2014). Increased biosorption with an increase in temperature is related to the swelling effect of the biosorbent caused by the increased temperature (Zulfikar 2013). The effect of temperature on the biosorbent helps in understanding the metal-binding mechanism. Biosorption process can be either exothermic or endothermic. Physisorption is at all times exothermic; chemisorption can be either endothermic or exothermic (Kulkarni 2015). The thermodynamic parameters determined from studying the effect of various temperatures determine if the process of biosorption is endothermic or exothermic. Guyo and Moyo (2017) observed that cowpea-based biosorbent exhibited an endothermic reaction during Pb uptake (II). With the increase in temperature from 20 °C to 50 °C, Pb(II) uptake increased, indicating the endothermic process. This trend was attributed to the improved interaction between the metal ion and the biosorbent as they overcome the activation energy barrier in the biosorption process. Higher temperature can also create new biosorption sites on the biosorbent, increasing the metal uptake (Moyo et al. 2015).

### 28.3.3 Biosorbent Particle Size

Smaller particles are alleged to exhibit higher biosorption capacity due to a higher surface area. Besides, larger particles' breaking up to smaller size opens up specific sealed channels, making them accessible for sorption (Guechi and Hamdaoui 2016). Guerrero-Coronilla et al. (2015) indicated that decreasing the particle size showed a negligible difference in surface area. The time to reach pollutant biosorption equilibrium reduced with smaller particles. Abdolali et al. (2015) observed very little difference in the biosorption capacity even with smaller biosorbent particles. Opposing the previous observations made, Mashkani and Ghazvini (2009) in their study obtained higher sorption at larger particle size, which goes against the principle of surface area. The response was attributed to reduced stability of the biosorbent particles of smaller size due to the high interaction between the charged groups of the biosorbent resulting in agglomeration. Agglomeration of the biosorbent results in both lower surface area and binding sites available. However, the application having a larger particle size with high absorptivity is favorable as smaller particles have low mechanical strength and can cause the column's clogging (Park et al. 2010; Abdolali et al. 2015).

### 28.3.4 Biosorbent Dosage

The uptake of the pollutant from the aqueous solution increases on increasing the dose of biosorbent. The potential of the biosorbent as the number of binding sites increases. Studies by Babarinde and Onyiaocha (2016) and Yuvaraja et al. (2014) confirm similar findings. Grace and Vimala (2015) found the biosorption efficiency to increase only up to a particular dosage, above which no significant biosorption efficiency was observed. These findings were attributed to the decrease in the concentration gradient. Park et al. (2010) point out the decrease in the quantity of the sorbed pollutant per unit weight of biosorbent. A similar statement has been made by Tabaraki and Sadeghinejad (2017). Though the number of available sorption sites increases with an increase in dosage, the reduction in sorption density could be attributed to unsaturated sites resulting from aggregation (Singh and Choden 2014). Aggregation of the biosorbent occurs due to the interaction between the binding sites at a higher dosage (Ansari et al. 2011). Dulman and Cucu-man (2010) have reported that aggregation of the biosorbent causes a decrease in the total surface area of the biosorbent and an increase in the diffusion path length. Moreover, desorption of the metal ions that are weakly or reversibly bound to the biosorbent surface could occur due to aggregation.

### 28.3.5 Pollutant Concentration

A high concentration of the target pollutant could be the driving force for biosorption to occur. Park et al. (2010) reported an increase in biosorption per unit mass of the biosorbent but decreased biosorption efficiency. The decrease in the metal removal efficiency with increasing concentration is the rapid saturation of the available active binding sites on the biosorbent. As the binding sites are gradually occupied, the sorption becomes less efficient at the later stages (Loiacono et al. 2018; Madala et al. 2017). Habtegebrel and Khan (2018) give an in-depth account of the trend. They explain that when the metal ion concentration is low, the ratio of the initial moles of the metal ions to the available adsorption sites would be low, and the fractional adsorption would become independent of initial metal concentration. However, as the metal concentration increases, the ratio unlike previously mentioned would be high due to lower available binding sites, causing percentage adsorption to depend upon the initial metal concentration.

### 28.3.6 Presence of Co-Ions and their Effect

The effect of the background ions present in the aqueous solution on the sorption of a selected toxic metal ion on a biosorbent can vary depending upon the affinity of the biosorbent for the metal ion. It is understood that heavy metal adsorption can be controlled by competitive adsorption and the possible complexing between the common ions and the heavy metals (He et al. 2009). Experimental results obtained by He et al. (2009) indicated an increase in the adsorption of $Cu^{2+}$ on clay in cations; $Na^{+}$, $K^{+}$, $Ca^{2+}$, and $Mg^{2+}$ adsorption decreased gradually in the presence of $Ca^{2+}$ ions. In the presence of anions ($Cl^{-}$ and $NO_3^{-}$), a decrease in adsorption was observed. A similar effect on the adsorption of $Zn^{2+}$ was observed. He et al. (2009) classified cations into three main groups: (1) $Na^{+}$ and $Li^{+}$; (2) $K^{+}$, $Rb^{+}$, $Cs^{+}$, and $NH_4^{+}$; and (3) $Ca^{2+}$, $Ba^{2+}$, $Sr^{2+}$, $Mg^{2+,}$ $Mn^{2+}$, $Cu^{2+}$, and $Ni^{2+}$. However, those in the same group exhibit delayed competition in adsorption only for a while, thus the reason for the gradual decrease in $Cu^{2+}$ adsorption in the presence of $Ca^{2+}$. However, Fike's (2001) observations in adsorption of $Pb^{2+}$ and $Cu^{2+}$ on clay were unaffected in the presence of $Na^{+}$ and $Ca^{2+}$. The binding was attributed to the strong bond of the metal ions with the adsorbent. Similar observations have been made in other studies where the biosorbent had a greater affinity toward the heavy metals over light metals (Cho and Kim 2003; Deng et al. 2007).

Commonly, wastewater is composed of both the heavy and light metal ions. It is less likely to have a single metal system. The effect of metal ions on each other in a multi-metal system is either antagonistic, synergistic, or noninteractive (Jain et al.

2016). Several studies have reported a decrease in the adsorption efficiency in a binary or multi-metal system. Jain et al. (2016) have reported a decrease in Ni(II), Cd (II), and Cr(VI) adsorption efficiency in a multi-metal system compared to a binary and mono-metal system. The inhibitory effect is because of the competition of the metal ions for the binding sites. Okoli et al. (2017) report the competitive adsorption of Pb(II) and Cd(II), i.e., Pb(II) was unaffected in the presence of Cd(II).

On the other hand, Cd(II) sorption was affected in Pb(II) in a binary-metal solution. The affinity of the biosorbent toward Pb(II) due to its smaller hydrated radius leads to the competitive adsorption of Pb(II) on common binding sites for Cd (II) and Pb(II). Besides, higher electronegativity and the larger ionic radius of Pb (II) compared to Cd(II) led to a higher affinity of the biosorbent toward Pb (II) (Li et al. 2004).

The study by Khobragade and Pal (2015) investigated the uptake of Cu(II) and Ni (II) in a single- and binary-metal system. It was observed that Cu(II) had an excellent affinity to the biosorbent compared to Ni(II). Compared to a single-metal system, uptake of Cu(II) was enhanced in a binary-metal system. Thus, a synergistic effect of Ni(II) on Cu(II) was observed. Hadi et al. (2013) in their study found the sorbent to have a higher affinity for Ni(II) compared to Co(II). In addition to the higher electronegativity of Ni(II), higher stability constant resulted in stronger complexes formed due to strong interactions between Ni(II) and the functional groups leading to higher Ni(II) adsorption on the adsorbent.

Ionic radius determines the ionic binding to the biosorbent. Heavy and light metal ions with similar ionic radius might promote ion exchange. On the other hand, two heavy metal ions with closer ionic radius might be affected similarly by the light metal ions. Hence, it is understood that the effect of co-ions on adsorption of heavy metals is affected by several factors that include affinity of the biosorbent to the metal ion, ionic radius of the metal ion, concentration of the heavy or light metal ion, electronegativity, the ionic hydration energy, and hydrated radius. Application of a biosorbent in water treatment with many metals should be made diligently keeping in mind the preference of the biosorbent to the metal ions that are affected by the factors mentioned above.

## 28.4 Preparation of Biosorbents

The collected biomass is thoroughly washed with deionized water and dried. The dried biosorbent is converted into fine powder by grinding and sieved to get the desired particle size. The pristine biosorbent can be subjected to either physical or chemical treatment or to enhance the biosorption efficiency.

### 28.4.1 Treatment and Modification of Biosorbent

Biosorption predominantly is a surface phenomenon; treatment of the biosorbent that promotes surface alteration can considerably improve its biosorption capacity

**Table 28.2** Treatment methods to improve the biomass quality as a biosorbent (Park et al. 2010)

| Category | | Detailed methods |
|---|---|---|
| Physical modification | | Autoclaving, steam, thermal drying, lyophilization, cutting, grinding, etc. |
| Chemical modification | Pretreatment (washing) | Acids (HCl, $H_2SO_4$, $HNO_3$, $H_3PO_4$, citric acid, etc.), alkalis (NaOH, KOH, $NH_4OH$, $Ca(OH)_2$, etc.), organic solvents (methanol, ethanol, toluene, formaldehyde, epichlorohydrin, salicylic acid, NTA, EDTA, SDS, L-cysteine, triton X-100, etc.), and other chemicals (NaCl, $ZnCl_2$, $Na_2CO_3$, $K_2CO_3$, $(NH_4)_2SO_4$, $H_2O_2$, $NH_4CH_3COO$, etc.) |
| | Enhancement of binding groups | Amination of a hydroxyl group, carboxylation of a hydroxyl group, phosphorylation of hydroxyl group, carboxylation of an amine group, amination of the carboxyl group, saponification of ester group, sulfonation, xanthanation, thiolation, halogenations, oxidation, etc. |
| | Elimination of inhibiting groups | Decarboxylation/elimination of carboxyl group, deamination/elimination of amine group, etc. |
| | Graft polymerization | High energy radiation grafting (irradiation, microwave radiation, electromagnetic radiation, etc.), photochemical grafting (with/without sensitizers like benzoin ethyl ether, acrylated azo dye, and aromatic ketones under UV light), and chemical initiation grafting (using ceric ion, permanganate ion, ferrous ammonium nitrate/$H_2O_2$, $KMnO_4$/citric acid, etc.) |

(Park et al. 2010). Physical or chemical treatment of the biomass can improve the biosorption capacity. Physical or chemical treatment can lead to exposing or removing certain functional groups present in the biosorbent. Physical treatment includes heat treatment, steam treatment, freeze-drying, boiling, washing, and grinding. Chemical treatment methods involve the exposure of the biomass to either a single treatment or subsequent exposure to alkali, acids, organic solvents, and salts (Table 28.2). The efficacy of chemical treatment depends upon the nature of the biomass and the concentration of the chemical used for the treatment. The merits of biosorption should not be overlooked during this process, and the entire treatment process should be cost-effective (Ramrakhiani et al. 2016).

Treatment improves the surface area and enhanced carboxyl, carboxylate, and hydroxyl groups, causing increased metal uptake from the aqueous solution. Several studies prove NaOH treatment to be the effective chemical treatment method and cost-effective (Chen et al. 2013; Nagy et al. 2014; Pirbazari et al. 2014; Jönsson and Martín 2016). Acid treatment using HCl, $HNO_3$, and $H_2SO_4$ caused decreased metal uptake due to probable damage to the acid's biosorbent.

Chemical treatment of *Pteris vittata* L. pinnae powder by Prabhu et al. (2018) improved the removal efficiency in Pb(II) and Cd(II). The NaOH treated pteridophyte-based biosorbent improved to a maximum metal loading capacity of 133.33 mg/g for Pb(II) and 95.24 mg/g for Cd(II).

Modifying a specific functional group is executed to enhance the binding groups or remove the inhibitory functional groups. The modification is the process of altering the functional groups by chemical reaction. Methylation of the amino groups, esterification of the carboxylic group, neutralization of the carboxylic group and introduction of an amino group, esterification of the phosphoric group, acetylation of the hydroxyl group, esterification of the hydroxyl group, and modification of sulfhydryl and carbonyl group are the various functional group modifications achieved by following the established procedures that involve the use of chemicals.

### 28.4.2 Development of Immobilized Biosorbent

Although the free biosorbent provides valid information during laboratory-scale experimentation, it is not applicable on an industrial scale (Dostálek 2011). In an industrial column, the mechanical strength of the biomaterials has to be improved. It is desired to prevent the hydrodynamic pressure drop, mass transfer limitations, and the column (Vieira et al. 2014). Immobilization or pelleting of the biomaterials helps achieve those above. The principal techniques followed for the biomaterial immobilization are entrapment in a polymeric matrix, cell cross-linking, and adsorption on an inert support.

#### 28.4.2.1 Entrapment of Biosorbent in a Polymeric Matrix

Dostálek (2011) has mentioned polyacrylamide, polysulfone, polyethyleneimine, and polyhydroxyethylmethacrylate to be the most commonly used synthetic polymers for entrapment, in addition to calcium alginate. Natural polymers commonly adopted are alginate, cellulose, carrageenan, and alginate (Eroglu et al. 2015). Trujillo et al. (1995) highlight the need for inert entrapment medium and the beads produced be porous such that the efficiency of the immobilized biosorbent is not reduced.

A polymer immobilizes the commercially popular biosorbents. AlgaSORB is composed of silica gel immobilized *C. vulgaris*, and BIOCLAIM bead is polyethyleneimine, and glutaraldehyde immobilized *Bacillus subtilis* (Trujillo et al. 1995). Though useful, the immobilization process has other disadvantages besides high processing charges. The diffusional resistance between the adsorption site and the active adsorption site and deactivation/masking of the active functional groups during the process are the limitations of immobilization (Trujillo et al. 1995) and promotes the consideration for other biomass stabilization methods.

#### 28.4.2.2 Immobilization by Cross-Linking

Cross-linking immobilization appears to be one of the possible techniques that provide stability and strength to biomass particles. Classification of cross-linking is done by forming ethers or esters of the polysaccharides present on the biomass (Kumar et al. 2016). Epichlorohydrin, ethylene glycol diglycidyl ether, 1,1,3,3-tetramethoxypropane, glutaraldehyde (GA), and β-cyclodextrin polyaldehyde are the commonly used cross-linkers (Mashkani and Ghazvini 2009; Sargin and Arslan 2016).

There could be a total loss in the active binding sites on the adsorbent while stabilizing an adsorbent. Few researchers have incorporated a secondary adsorbent on the stabilized adsorbent to compensate for the loss in adsorption capacity. Sporopollenin grains (the outer skeleton of fern spores) were used to produce chitosan-based biosorbents to compensate for the loss of potential adsorption functional groups (—$NH_2$ group) to cross-linking (Sargin and Arslan 2016). The use of an additional adsorbent to compensate the lost adsorption ability of the first adsorbent to cross-linking makes the entire process debatable.

#### 28.4.2.3 Pelleting of Biomass

The process of pelleting of biomass is well established in the field of energy. The pelleted biomass is an alternative energy source to fossil fuels. As an energy source, pellets offer several benefits such as increased biomass bulk density, homogenous shape, and structure that promote ease in feeding the automated boiler systems (Stelte et al. 2012). The general advantages of pellets are reduced storage and transportation costs (Fang et al. 2018). The principle of pelleting the biomass can be applied to biosorbents as well. The limitations of the industrial application of biosorption are (Fosso-Kankeu and Mulaba-Bafubiandi 2014):

1. Insufficient supply of the raw material/biomaterial that would be used for the production of biosorbent.
2. Transformation of the biomaterial into an effective biosorbent.
3. Regeneration, recycling, and reuse of the biosorbent.
4. Immobilization of the biomaterial to possess a stable biosorbent.

Most often, immobilization techniques such as entrapment, cross-linking, and adsorption are not cost-effective. Also, they do not produce effective biosorbents due to possible mass transfer resistance, loss of biosorbent activity, biomass leakage, and unstable binding (Fosso-Kankeu and Mulaba-Bafubiandi 2014). Attempts to scale up the biosorption process could be successful by achieving the above. Biosorbents can be efficiently immobilized by pelleting. If applied in the pilot scale, pelleted biosorbents/adsorbents have various advantages: stabilized column, minimum column clogging, and easy separation of loaded biosorbents from the column (Visa et al. 2017).

Pelleting of the biomass can be achieved with or without the addition of a suitable binder. The stability, durability, and densification of a pellet depend on several factors that include moisture, type of the binder used (Tabil 1996), the particle size of the biomass, and the conditioning temperature and compaction pressure (Liu et al. 2014; Tumuluru et al. 2016).

## 28.5 Application of Biosorbents in Toxic Metal Remediation

Packed bed reactors are by far the simplistic tubular reactor that can be easily operated at an industrial scale (Kashid et al. 2014). Vertical or horizontal packed bed reactors/column is a tubular column with a steady movement of liquid from one end to the other with no attempted mixing (Russell et al. 2008). The fluid motion is thus considered to be a plug flow. Packed bed reactors can have the biosorbents as their solid phase. Most of the continuous scale biosorption studies have preferred continuous packed bed flow column over fluidized and CFSTRs (Kulkarni 2015).

AlgaSORB™, an immobilized silica alga (*Spirogyra*), is one of the very few biosorbents that has some commercial importance. Singh and Prasad (2000) proposed its suitability in continuous metal uptake monitoring. The AlgaSORB™ columns in a series of 5 were used to obtain optimum uptake. For reuse of the biosorbents, sodium hydroxide, sodium carbonate, EDTA, sulfuric acid, nitric acid, and hydrochloric acid of varying concentrations are commonly used as metal eluents (Mishra 2014).

The bed depth in the column is a critical parameter as it influences the service area. Increase in bed depth increases the biosorption amount. An axial dispersion phenomenon tampers with the diffusion of the metal ions. High flow rate favors the external mass transfer, whereas low flow rate favors intraparticle mass transfer. High bed depth and low flow rate improve the metal uptake efficiency. The column's service time depends upon factors such as bed height, flow rate, influent metal composition, residence time of the solution, pH, bed void fraction, and sorption-desorption cycle (Kumar et al. 2016).

The associated drawbacks of packed bed columns are mass transfer limitations, channeling or clogging, flow maldistribution, and high-pressure drop (Kashid et al. 2014). The application of vibration or pulsation has been suggested to improve the mass transfer rate (Lade et al. 2014). An attempt has been made by Prabhu et al. (2019a, b) to improve mass transfer characteristics of continuous biosorption of Pb (II) on *Pteris vittata* L. pellets by using pulsed packed bed reactor (Fig. 28.1).

Biosorbents can also be fabricated as filters that can be directly applied to remove heavy metal contaminants from drinking water.

## 28.6 Conclusions and Scope for Future Work

The biosorption technology using nonliving biomaterials has proved to be an effective alternative way to remove the heavy metals from wastewater. However, there is a need to identify a biosorbent with a high degree of metal uptake which is also nonedible, cost-effective, available around the year in same quality, and easy to grow or harvest, requires a minimal physical/chemical modification or processing, can be easily regenerated, and ensures environmental safety upon application. The nonedible pteridophytes that grow resiliently in hostile environments and urban areas offer the potential to remove heavy metals in its powdered and pelleted biosorbent forms. From the observations made, *P. vittata* L. has exhibited excellent

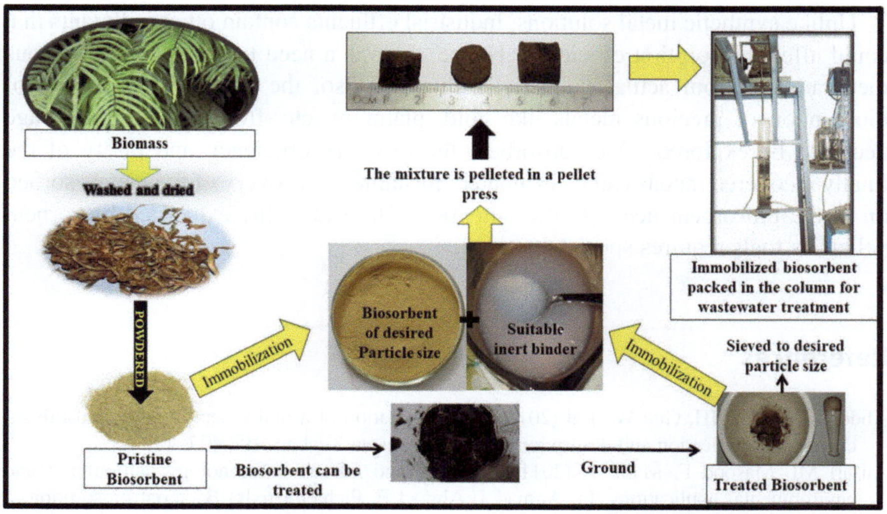


**Fig. 28.1** The biomass processing to obtain pristine biosorbent and treated biosorbent by various treatment methods improves the biosorbent quality. The powdered biomass can be immobilized into a pellet that is packed into a column for treatment of wastewater

potential as a biosorbent. Biosorption capacity can be enhanced by the surface morphological changes induced by different types of chemical treatments. The multi-metal system chosen during studies reflects the realistic combinations of metals in an industrial effluent scenario. Metal combination in the aqueous solution indicated synergism, whereas others have demonstrated an antagonistic effect on the metal uptake. Since the volume of wastewater in surface polishing or electroplating industries is high, immobilized biosorbents have to be used for continuous removal of toxic metals. The pelleted biosorbents can be a practical choice as a hybrid technology or be applied in secondary treatment of the wastewater. Furthermore, *P. vittata* L. with no commercial value promotes its application in heavy metal biosorption. Biodegradability being the primary advantage of utilizing a biosorbent for metal uptake, the metal-loaded biosorbent can be introduced back into the mine for safe disposal. Hence, the biosorption technology can be an alternative treatment method in wastewater treatment of heavy metals. Thus, one can achieve the removal of toxic heavy metals leading to the reuse, recycle, and recover concept as globally accepted and practiced in the model world to accomplish zero discharge, leading to sustainable development.

The field of biosorption still has scope for improvement. The potential pterido-phyte- based biosorbents have excess alkaloids. The effect of the biosorbent on the quality of water, especially for drinking purpose, needs thorough assessment. Improvement in the immobilization of the biosorbents by pelletization is required. Stable biosorbent pellets of uniform characteristics can be prepared using a mechanical pellet press machine for commercial-scale mass production of biosorbent pellets.

Unlike synthetic metal solutions, industrial effluents contain other pollutants that could affect biosorption efficiency. Hence, there is a need to study the continuous metal uptake from actual industrial effluents. Also, the possible application of biosorption of precious metals like gold, platinum, etc. from metal-rich sewage needs to be explored. Once absorbed, the recovery efficiency and purity of the finally recovered metals can be evaluated. Sustainable recovery of the metals sorbed on to the biosorbent needs further attention. The use of the exhausted biosorbent pellets as fuels requires special attention.

## References

Abdolali A, Ngo HH, Guo W et al (2015) Characterization of a multi-metal binding biosorbent: chemical modification and desorption studies. Bioresour Technol 193:477–487

Ansari MI, Masood F, Malik A (2011) Microbes and microbial technology: agricultural and environmental applications. In: Ahmad I, Ahmad F, Pichtel J (eds) Bacterial biosorption: a technique for remediation of heavy metals. Springer Science and Business Media, Berlin

Antunes APM, Watkins GM, Duncan JR (2001) Batch studies on the removal of gold(III) from aqueous solution by *Azolla filiculoides*. Biotechnol Lett 23(4):249–251

Asbchin SA, Omran AN, Jafari N (2012) Potential of *Azolla filiculoides* in the removal of Ni and cu from wastewaters. Afr J Biotechnol 11(95):16158–16164

Ashraf A, Bibi I, Niazi NK et al (2016) Chromium(VI) sorption efficiency of acid– activated banana peel over organo-monmorillonite in aqueous solutions. Int J Phytoremediation 19(7):605–613

Babarinde A, Onyiaocha G (2016) Equilibrium sorption of divalent metal ions onto groundnut (*Arachis hypogaea*) shell: kinetics, isotherm and thermodynamics. Chem Int 2(1):37–46

Chen Y, Stevens MA, Zhu Y, Holmes J, Xu H (2013) Understanding of alkaline pretreatment parameters for corn Stover enzymatic saccharification. Biotechnol Biofuels 6(1):1–10

Cho DH, Kim EY (2003) Characterization of $Pb^{2+}$ biosorption from aqueous solution by *Rhodotorula glutinis*. Bioprocess Biosyst Eng 25(5):271–277

Daraei H, Mittal A, Mittal J et al (2014) Optimization of Cr(VI) removal onto biosorbent eggshell membrane: experimental & theoretical approaches. Desalin Water Treat 52(7–9):1307–1315

de Freitas GR, da Silva MG, Vieira MG (2019) Biosorption technology for removal of toxic metals: a review of commercial biosorbents and patents. Environ Sci Pollut Res 26(19):19097–19118

Deng L, Zhu X, Wang X et al (2007) Biosorption of copper(II) from aqueous solutions by green alga *Cladophora fascicularis*. Biodegradation 18(4):393–402

Dostálek P (2011) Immobilized biosorbents for bioreactors and commercial biosorbents. In: Pavel Kotrba TM, Mackova M (eds) Microbial biosorption of metals. Springer, New York

Dulman V, Cucu-man SM (2010) Wood sawdust, tree bark and wood chips: waste lignocellulosic materials for dye removal. In: Crini B (ed) Sorption processes and pollution: conventional and non- conventional sorbents for pollutant removal from wastewaters. Franche- Comté, France

Eroglu E, Smith SM, Raston CL (2015) Application of various immobilization techniques for algal bioprocesses. In: Moheimani N, McHenry M, de Boer K et al (eds) Biomass and biofuels from microalgae. Springer, Cham

Fang J, Zhan L, Ok YS et al (2018) Minireview of potential applications of hydrochar derived from hydrothermal carbonization of biomass. J Ind Eng Chem 57:15–21

Farooq U, Kozinski JA, Khan MA et al (2010) Biosorption of heavy metal ions using wheat based biosorbents - a review of the recent literature. Bioresour Technol 101(14):5043–5053

Fike WG (2001) Sorption of cadmium, copper, lead and zinc as influenced by ionic strength, pH, and selected soil components. Dissertation, Virginia Polytechnic Institute and State University

Fogarty RV, Dostalek P, Patzak M et al (1999) Metal removal by immobilized and non-immobilised *Azolla filiculoides*. Biotechnol Tech 13(8):533–538

Fomina M, Gadd GM (2014) Biosorption: current perspectives on concept, definition and application. Bioresour Technol 160:3–14

Fosso-Kankeu E, Mulaba-Bafubiandi A (2014) Review of challenges in the escalation of metal-biosorbing processes for wastewater treatment: applied and commercialized technologies. African J Biotechnol 13(17):1756–1771

Gadd GM (2009) Biosorption critical review of scientific rationale, environmental importance and significance for pollutant removal. J Chem Technol Biotechnol 84(1):13–28

Ganji MT, Khosravi M, Rakhshaee R (2005) Biosorption of Pb, cd, cu and Zn from the wastewater by treated *Azolla filiculoides* with $H_2O_2$/ $MgCl_2$. Int J Environ Sci Technol 1(4):265–271

Grace AN, Vimala R (2015) Biosorption of di-butyl phthalate from aqueous solutions using *Pleurotus ostreatus*: isotherm and kinetic study. J Chemical Pharm Res 7(4):697–706

Guechi EK, Hamdaoui O (2016) Sorption of malachite green from aqueous solution by potato peel: kinetics and equilibrium modeling using non-linear analysis method. Arab J Chem 9:S416–S424

Guerrero-Coronilla I, Morales-Barrera L, Cristiani-Urbina E (2015) Kinetic, isotherm and thermodynamic studies of amaranth dye biosorption from aqueous solution onto water hyacinth leaves. J Environ Manag 152:99–108

Guyo U, Moyo M (2017) Cowpea pod (*Vigna unguiculata*) biomass as a low-cost biosorbent for removal of Pb (II) ions from aqueous solution. Environ Monit Assess 189(2):47

Habtegebrel MM, Khan MA (2018) Removal of Zn(II) and cu (II) ions from aqueous solution by dried *Prosopis juliflora*. Mod Chem 6(1):6–14

Hadi P, Barford J, Mckay G (2013) Synergistic effect in the simultaneous removal of binary cobalt – nickel heavy metals from effluents by a novel e-waste-derived material. Chem Eng J 228:140–146

He J, Xue HX, Lü CW et al (2009) The impacts of common ions on the adsorption of heavy metal. Environ Geol 58(7):1499–1508

Hegde S, Sajeev S (2013) A collection of selected ferns of Karnataka. St Aloysius College, Mangalore

Ho YS (2003) Removal of copper ions from aqueous solution by tree fern. Water Res 37 (10):2323–2330

Ho YS, Chiu WT, Hsu CS, Huang CT (2004) Sorption of lead ions from aqueous solution using tree fern as a sorbent. Hydrometallurgy 73(1–2):55–61

Ho YS (2005) Effect of pH on lead removal from water using tree fern as the sorbent. Bioresour Technol 96(11):1292–1296

Ho YS, Wang CC (2004) Pseudo- isotherms for the sorption of cadmium ion onto tree fern. Process Biochem 39(6):759–763

Ho YS, Wang CC (2008) Sorption equilibrium of mercury onto ground-up tree fern. J Hazard Mater 156(1–3):398–404

Hubbe MA, Hasan SH, Ducoste JJ (2011) Cellulosic substrates for removal of pollutants from aqueous systems: a review. 1. Metals. Bioresources 6(2):2161–2287

Jain M, Garg VK, Kadirvelu K et al (2016) Adsorption of heavy metals from multi-metal aqueous solution by sunflower plant biomass-based carbons. Int J Environ Sci Technol 13(2):493–500

Jaishankar M, Tseten T, Anbalagan N, Mathew BB et al (2014) Toxicity, mechanism and health effects of some heavy metals. Interdiscip Toxicol 7(2):60–72

Jönsson LJ, Martín C (2016) Pretreatment of lignocellulose: formation of inhibitory by-products and strategies for minimizing their effects. Bioresour Technol 199:103–112

Kashid MA, Renken A, Kiwi-Minsker L (2014) Microstructured reactors for fluid-solid systems. In: Microstructured devices for chemical processing. John Wiley & Sons, Hoboken, NJ

Khobragade MU, Pal A (2015) Adsorptive removal of cu(II) and Ni(II) from single-metal, binary-metal, and industrial wastewater systems by surfactant-modified alumina. J Environ Sci Heal - Part A Toxic/Hazardous Subst Environ Eng 50(4):385–395

Khosravi M, Rakhshaee R, Ganji MT (2005) Pre- treatment process of *Azolla filiculoides* to removal Pb(II), cd(II), Ni(II) and Zn(II) from aqueous solution in the batch and fixed- bed reactors. J Hazard Mater 127(1–3):228–237

Krishnani KK, Meng X, Christodoulatos C et al (2008) Biosorption mechanism of nine different heavy metals onto biomatrix from rice husk. J Hazard Mater 153(3):1222–1234

Kulkarni RM (2015) Biosorption of nickel (II) and cadmium (II) on bacterial and yeast biomass. Dissertation, National Institute of Technology Karnataka

Kumar D, Pandey LK, Gaur JP (2016) Metal sorption by algal biomass: from batch to continuous system. Algal Res 18:95–109

Lade VG, Pakhare AD, Rathod VK (2014) Mass transfer studies in pulsed sieve plate extraction column for the removal of tributyl phosphate from aqueous nitric acid. Ind Eng Chem Res 53 (12):4812–4820

Li Q, Wu S, Liu G et al (2004) Simultaneous biosorption of cadmium(II) and lead(II) ions by pretreated biomass of *Phanerochaete chrysosporium*. Sep Purif Technol 34(1–3):135–142

Lindsay S, Middleton DJ (2012) Ferns of Thailand, Laos and Cambodia. http://rbz-web2.rbge.org.uk/thaiferns/. Accessed 21 April 2016

Liu Z, Fei B, Jiang Z, Cai Z (2014) Important properties of bamboo pellets to be used as commercial solid fuel in China. Wood Sci Technol 48(5):903–917

Loiacono S, Crini G, Chanet G et al (2018) Metals in aqueous solutions and real effluents: biosorption behavior of a hemp-based felt. J Chem Technol Biotechnol 93(9):2592–2601

López-García M, Lodeiro P, Barriada JL et al (2010) Reduction of Cr (VI) levels in solution using bracken fern biomass: batch and column studies. Chem Eng J 165(2):517–523

Ma LQ, Tu C (2006) Compositions and methods for removing pollutants from contaminated water with a fern. US patent 7,069,690 B2

Ma LQ, Tu S, Fayiga O (2004) Contaminant removal by ferns via foliar application and excised/ground fronds. US patent 2004/0197812 A1

Madala S, Nadavala SK, Vudagandla S, Boddu VM, Abburi K (2017) Equilibrium, kinetics and thermodynamics of cadmium(II) biosorption on to composite chitosan biosorbent. Arab J Chem 10:S1883–S1893

Mashkani SG, Ghazvini PTM (2009) Biotechnological potential of *Azolla filiculoides* for biosorption of Cs and SR: application of micro- PIXE for measurement of biosorption. Bioresour Technol 100(6):1915–1921

Mekonnen MM, Hoekstra AY (2016) Four billion people facing severe water scarcity. Sci Adv 2:2. https://doi.org/10.1126/sciadv.1500323

Mishra SP (2014) Adsorption-desorption of heavy metal ions. Curr Sci 107(4):601–612

Morajkar S, Sajeev S, Hegde S (2015) Ferns: a thriving group of urban dwellers. Bionature 35 (1&2):13–21

Moubarik A, Grimi N (2014) Valorization of olive stone and sugar cane bagasse by-products as biosorbents for the removal of cadmium from aqueous solution. Food Res Int 73:169–175

Moyo M, Guyo U, Mawenyiyo G et al (2015) Marula seed husk (*Sclerocarya birrea*) biomass as a low cost biosorbent for removal of Pb(II) and cu(II) from aqueous solution. J Ind Eng Chem 27: 126–132

Nagy B, Mânzatu C, Măicăneanu A, Indolean C et al (2014) Effect of alkaline and oxidative treatment on sawdust capacity to remove cd(II) from aqueous solutions: FTIR and AFM study. J Wood Chem Technol 34(4):301–311

Nguyen TAH, Ngo HH, Guo WS et al (2013) Bioresource technology applicability of agricultural waste and by-products for adsorptive removal of heavy metals from wastewater. Bioresour Technol 148:574–585

Okoli CP, Diagboya PN, Anigbogu IO et al (2017) Competitive biosorption of Pb(II) and cd(II) ions from aqueous solutions using chemically modified moss biomass (*Barbula lambarenensis*). Environ Earth Sci 76(1):33

Park D, Yun YS, Park JM (2010) The past, present, and future trends of biosorption. Biotechnol Bioprocess Eng 15(1):86–102

Pirbazari AE, Saberikhah E, Badrouh M et al (2014) Alkali treated Foumanat tea waste as an ef fi cient adsorbent for methylene blue adsorption from aqueous solution. Water Resour Ind 6:64–80

Prabhu S, Srinikethan G, Hegde S (2018) Surface treated *Pteris vittata* L. pinnae powder used as an efficient biosorbent of Pb(II), cd(II) and Cr(VI) from an aqueous solution. Int J Phytoremediation 20(9):947–956

Prabhu S, Srinikethan G, Hegde S (2019a) Efficient biosorption of Pb(II) on *Pteris vittata* L. from aqueous solution using pulsed plate column technique. Sep Sci Technol 55(17):3089–3101

Prabhu S, Srinikethan G, Hegde S (2019b) Spontaneous Cr(VI) and cd(II) biosorption potential of native and dried pinnae tissue of *Pteris vittata* L., a tropical invasive pteridophyte. Int J Phytoremediation 21(4):380–390

Prabhu S, Srinikethan G, Hegde S (2020) Pelletization of pristine *Pteris vittata* L. pinnae powder and its application as a biosorbent of cd(II) and Cr(VI). SN Appl Sci 2(1):1–9

Rakhshaee R, Khosravi M, Ganji MT (2006) Kinetic modelling and thermodynamic study to remove Pb(II), cd(II), Ni(II) and Zn(II) from aqueous solution using dead and living *Azolla filiculoides*. J Hazard Mater 134(1–3):120–129

Ramrakhiani L, Ghosh S, Majumdar S (2016) Surface modification of naturally available biomass for enhancement of heavy metal removal efficiency, upscaling prospects, and management aspects of spent biosorbents: a review. Appl Biochem Biotechnol 180(1):41–78

Russell TWF, Robinson AS, Wagner NJ (2008) Mass contactor analysis. In: Mass and heat transfer: analysis of mass contactors and heat exchangers. Cambridge University Press, Cambridge, pp 134–164

Sargin İ, Arslan G (2016) Effect of glutaraldehyde cross-linking degree of chitosan/sporopollenin microcapsules on removal of copper(II) from aqueous solution. Desalin Water Treat 57 (23):10664–10676

Sharma OP (2012) Diversity of microbes and cryptogams: Pteridophyta. Tata McGraw- Hill, New Delhi, pp 1.1–1.2

Sharma A, Sachdeva S (2015) Cadmium toxicity and its phytoremediation a review. Int J Sci Eng Res 6(9):395–405

Shroff KA, Vaidya VK (2012) Effect of pre-treatments on the biosorption of chromium (VI) ions by the dead biomass of *Rhizopus arrhizus*. J Chem Technol Biotechnol 87(2):294–304

Singh H, Choden S (2014) Comparison of adsorption behaviours and kinetic modeling of bio- waste materials using basic dye as adsorbate. Indian J Chem Technol 21:359–367

Singh R, Gautham N, Mishra A et al (2011) Heavy metals and living systems: an overview. Indian J Pharmacol 43(3):246–253

Singh R, Prasad BB (2000) Trace metal analysis: selective sample (copper II) enrichment on an AlgaSORB column. Process Biochem 35(9):897–905

Soniya L, Krishnakumar G (2015) Biosorption of heavy metals from aqueous solution using mangrove fern *Acrostichum aureum* L. leaf biomass as a sorbent. Int Res J Environ Sci 4:25–31

Stelte W, Sanadi AR, Shang L et al (2012) Recent developments in biomass pelletization - a review. Bioresources 7(3):4451–4490

Tabaraki R, Sadeghinejad N (2017) Biosorption of six basic and acidic dyes on brown alga *Sargassum ilicifolium*: optimization, kinetic and isotherm studies. Water Sci Technol 75 (11):2631–2638

Tabil LG (1996) Binding and pelleting characteristics of alfalfa. University of Saskatchewan, Saskatoon

Tchounwou PB, Yedjou CG, Patlolla AK et al (2012) Heavy metals toxicity and the environment. In: Luch A (ed) Molecular, clinical and environmental toxicology. Springer, New York, pp 133–164

Trujillo EM, Spinti M, Zhuang H (1995) Immobilized biomass: a new class of heavy-metal selective ion exchangers. In: SenGupta AK (ed) Ion exchange technology: advances in pollution control. Technomic Publishing Company, Lancaster, PA, pp 231–232

Tumuluru JS, Conner CC, Hoover AN (2016) Method to produce durable pellets at lower energy consumption using high moisture corn Stover and a corn starch binder in a flat die pellet mill. J Vis Exp 112:1–13

Velazquez-Jimenez LH, Arcibar-Orozco JA, Rangel-Mendez JR (2018) Overview of as (V) adsorption on Zr-functionalized activated carbon for aqueous streams remediation. J Environ Manag 212:121–130

Vieira MLG, Esquerdo VM, Nobre LR et al (2014) Glass beads coated with chitosan for the food azo dyes adsorption in a fixed bed column. J Ind Eng Chem 20(5):3387–3393

Visa I, Visa M, Duta A (2017) Mechanical properties and functional durability of Fly Ash pellets, adsorbents in wastewater treatment. World Coal Ash:75–80

Wankasi D, Jnr MH, Spiff AI (2005) Desorption of $Pb^{2+}$ and $Cu^{2+}$ from Nipa palm (*Nypa fruticans* Wurmb) biomass. Afr J Biotechnol 4(9)

Yuvaraja G, Krishnaiah N, Subbaiah MV et al (2014) Biosorption of Pb(II) from aqueous solution by *Solanum melongena* leaf powder as a low-cost biosorbent prepared from agricultural waste. Colloids Surfaces B Biointerfaces 114:75–81

Zhao M, Duncan JR, Hille RPV (1999) Removal and recovery of zinc from solution and electroplating effluent using *Azolla filiculoides*. Water Res 33(6):1516–1522

Zhao M, Duncun JR (1998) Removal and recovery of nickel from aqueous solution and electroplating rinse effluent using *Azolla filiculoides*. Process Biochem 33(3):249–255

Zulfikar MA (2013) Effect of temperature on adsorption of humic acid from peat water onto pyrophyllite. Int J Chem Environ Biol Sci 1(1):88–91

**Smruthi Prabhu** , M.Sc. in Biotechnology, Ph.D. (Environmental Biotechnology), Assistant Professor, Post Graduate Department of Studies and Research in Biotechnology, St. Aloysius College (Autonomous), Mangalore, Karnataka, India-575002.

**Interests:** Bioremediation, microbiology, bioenergy, plant tissue culture

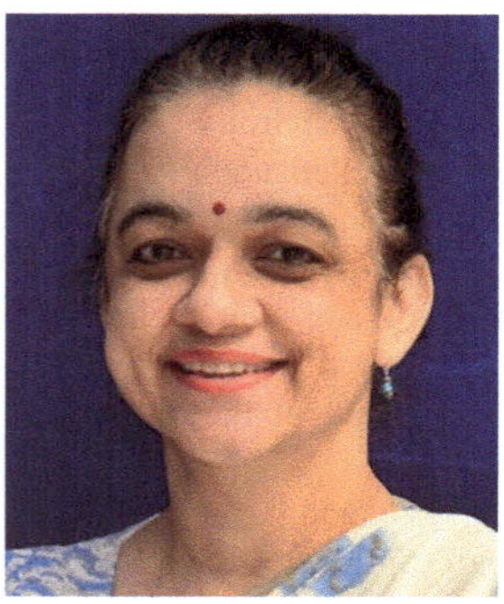

**Smitha Hegde** , M.Sc. Ph.D. (Biosciences), Professor, Nitte University Centre for Science Education and Research, and Coordinator of DST NUTEC, Nitte University Centre for Science Education and Research, Nitte DU, Deralakatte, Mangalore 575018.

**Interests:** Worked for 30 years on pteridophytes: biodiversity studies, developmental studies, phytoremediation, plant tissue culture, molecular markers in pteridophytes

# *Azolla's* Use as a Biofertilizer and Livestock Feed

# 29

Alexandra Bujak and Jonathan Bujak

**Abstract**

*Azolla* is the only plant with a coevolving nitrogen-fixing cyanobacterial symbiont (cyanobiont), *Nostoc azollae*, resulting from whole-genome duplication (WGD) 80 million years ago in *Azolla*'s immediate ancestor, *Parazolla*. Additional genes from the WGD resulted in complex biochemical and morphological changes that enabled transmission of the cyanobiont to successive generations of the plant via its spores. The permanent symbiosis resulted in loss, downregulation, or conversion of nonessential genes to pseudogenes in the cyanobiont, changing it from a free-living organism to an obligate symbiont that is dependent on its host for survival. Upregulation of other genes in the cyanobiont increased its atmospheric nitrogen fixation and provision of nitrogen-based products to the plant. As a result, *Azolla* can double its biomass in less than 2 days free-floating on freshwater and sequester large amounts of atmospheric $CO_2$, giving it the potential to mitigate anthropogenic climate change through carbon capture and storage (CCS). *Azolla*'s biomass can also provide local, low-cost biofertilizer, feed, biofuel, and a range of products that are urgently needed as our population increases by a billion every 12 years. This chapter focuses on its use as a biofertilizer and feed for animals, poultry, waterfowl, freshwater fish, and crustacea.

**Keywords**

*Azolla* · Carbon capture and storage · Climate change · Ferns · Nitrogen fixation · *Nostoc azollae* · Symbiosis

A. Bujak (✉) · J. Bujak
The Azolla Foundation, Blackpool, UK
e-mail: abujak@theaollafoundation.org; jbujak@theazollafoundation.org


J. Marimuthu et al. (eds.), *Ferns*, https://doi.org/10.1007/978-981-16-6170-9_29

## 29.1 Introduction

*Azolla* is the only plant with a coevolving nitrogen-fixing cyanobacterial symbiont (cyanobiont), enabling it to double its biomass in less than 2 days free-floating on freshwater (Watanabe and Berja 1983). The cyanobiont has been assigned to both *Anabaena azollae* and *Nostoc azollae* because it resembles free-living species of *Anabaena* and *Nostoc* in having chains of cells (filaments) comprising photosynthetic vegetative cells and thicker-walled heterocysts that contain the nitrogen-fixing enzyme nitrogenase (Fig. 29.1). Other chapters assign *Azolla*'s cyanobiont to *Nostoc azollae* (see Schlupmann et al. in this book), and we have followed their nomenclature for consistency.

The heterocysts in *N. azollae* and free-living species of *Anabaena* and *Nostoc* do not contain free oxygen, which destroys nitrogenase, and it is probable that the differentiation of heterocysts and vegetative cells evolved in *Anabaena*'s and *Nostoc*'s common ancestor during the Great Oxidation Event (GOE) two and a half billion years ago when the Earth's atmosphere was enriched in oxygen (Schirrmeister et al. 2013). *Anabaena*'s and *Nostoc*'s heterocysts sequester atmospheric dinitrogen and synthesize it into ammonium-based compounds that are utilized by the heterocysts and transferred to vegetative cells where they are combined with photosynthetic products (Fig. 29.2). As a result of these ancient

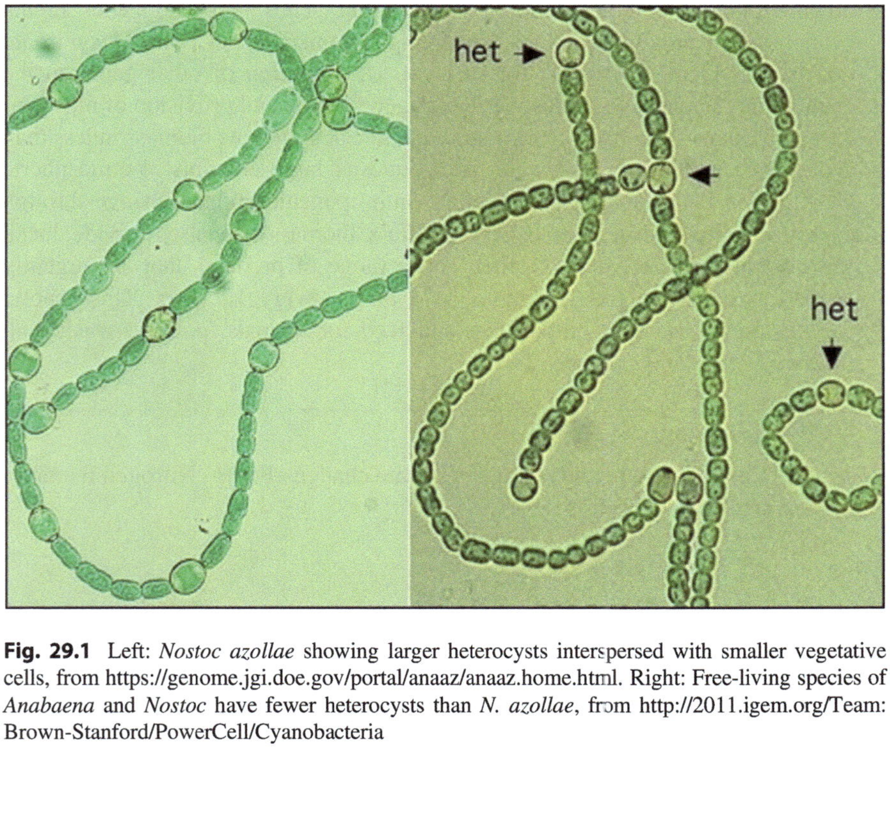


**Fig. 29.1** Left: *Nostoc azollae* showing larger heterocysts interspersed with smaller vegetative cells, from https://genome.jgi.doe.gov/portal/anaaz/anaaz.home.html. Right: Free-living species of *Anabaena* and *Nostoc* have fewer heterocysts than *N. azollae*, from http://2011.igem.org/Team:Brown-Stanford/PowerCell/Cyanobacteria

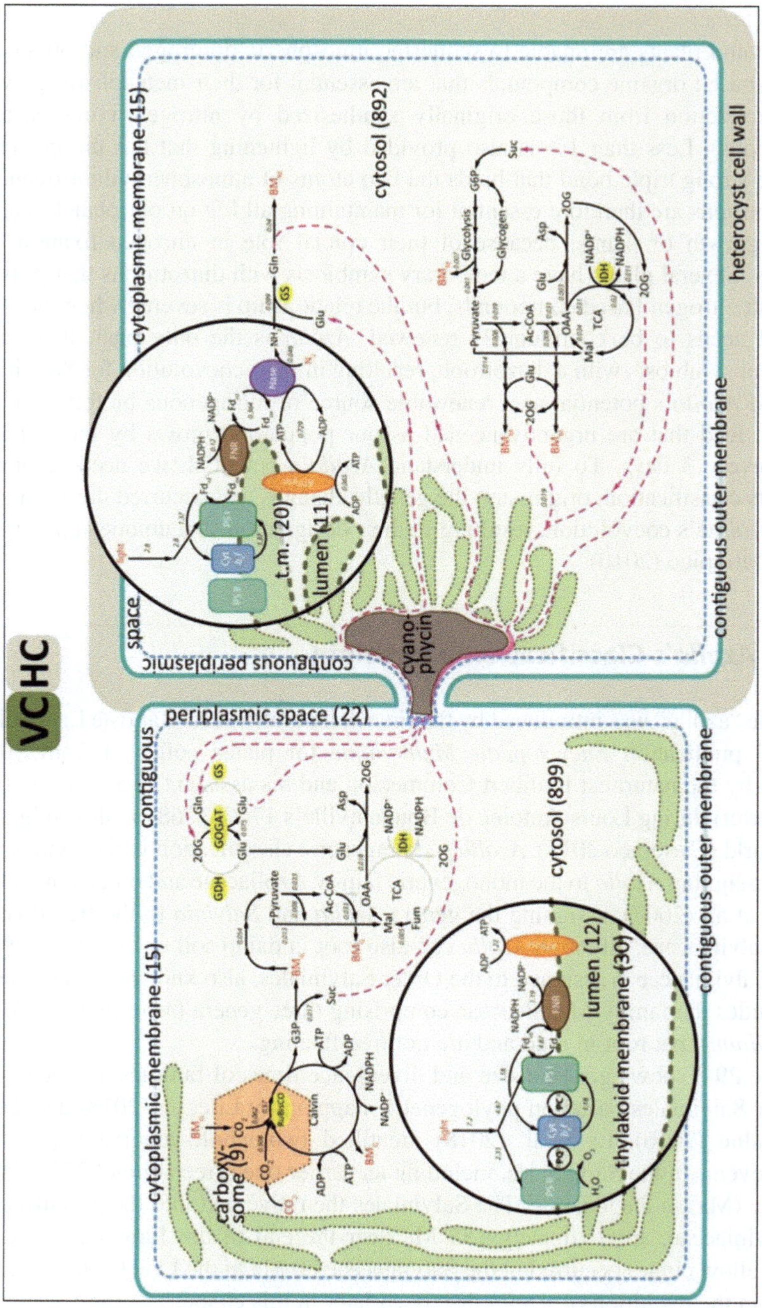


**Fig. 29.2** Photosynthesis and nitrogen fixation in a cyanobacterial filament. *VC* vegetative cell, *HC* heterocyst. From Malatinszky et al. (2017)

adaptations, today's species of *Anabaena* and *Nostoc*, including *Azolla*'s cyanobiont *N. azollae*, represent the most biochemically and morphologically complex cyanobacteria.

Algae and plants are unable to sequester atmospheric dinitrogen and obtain the nitrogen-based organic compounds that are essential for their metabolism, growth, and reproduction from those originally synthesized by nitrogen-fixing bacteria (diazotrophs). Less than 1% is also provided by lightening that has the power to sever the strong triple bond that binds the two atoms of atmospheric dinitrogen.

Diazotrophs are therefore essential for maintaining all life on our planet, including the growth of plants, because of their crucial role in nitrogen fixation and synthesis. Several plants have a temporary symbiosis with diazotrophs that provide them with nitrogen-based compounds, but the relationship is severed when the plant dies and needs to be continuously renewed. *Azolla* is the only plant that has a permanent symbiosis with a diazotroph, resulting in their coevolution for 80 million years and *Azolla*'s potential as a renewable source of nitrogenous biofertilizer and livestock feed that are urgently needed as our population grows by more than a million every 3 days. To fully understand *Azolla*'s potential, we need to briefly review its classification, origin, and the genetic changes that occurred during *Azolla* and *N. azollae*'s coevolution, resulting in their designation as a unique superorganism by Carrapiço (2010).

## 29.2 *Azolla*'s Classification and Origin

The name "azolla" was introduced by the French naturalist Jean-Baptiste Lamarck in his 1783 publication *Encyclopédie Méthodique* for plants collected from South America by the naturalist Philibert Commerson and his assistant Jeanne Baret (aka Baré, Barret) during Louis-Antoine de Bougainville's 1767–1768 circumnavigation of the world (Carrapiço 2018). *Azolla*'s suprageneric classification varies, with some authors assigning *Azolla* to the monogeneric family Azollaceae and others, including Metzgar et al. (2007), assigning the genera *Azolla* and *Salvinia* to the free-floating family Salviniaceae, although *Azolla* can also root in damp soil or mud (Fig. 29.3). Family Salviniaceae is assigned to the Order Salviniales, also known as water ferns, that includes the family Marsileaceae comprising three genera (*Marsilea*, *Pilularia*, *Regnellidium*) that root in mud and are not free-floating.

Figure 29.3 shows genome size and divergence times of families and genera in the order Salviniales based on phylogenetic mapping by Li et al. (2018) and Testo and Sundue (2016). Li et al. (2018) identified two whole-genome duplication (WGD) events in the Salviniales, including an earlier one (green circle) 225 million years ago (Ma) in the Triassic. The Salviniales then diverged into the Marsileaceae and Salviniaceae, approximately 155 Ma near the end of the Jurassic. A second WGD (yellow circle) occurred in the Salviniaceae 80 Ma in the Late Cretaceous and resulted in the evolution of *Azolla* discussed later in this chapter.

Carrapiço et al. (2000) divided the genus *Azolla* into the subgenera *Rhizosperma* and *Azolla* (also named *Euazolla* by Raja et al. in 2012). Species of the subgenus

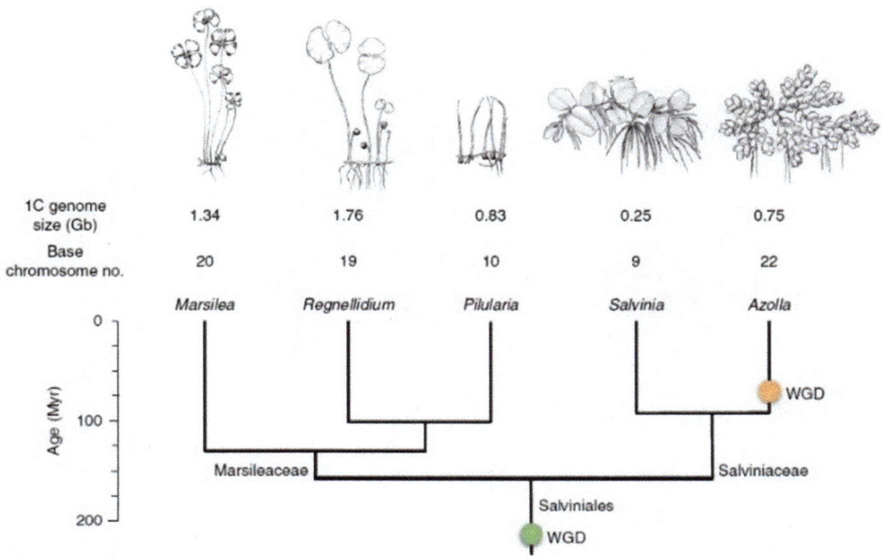


**Fig. 29.3** Genome size of the order Salviniales (water ferns), divergence times, and two whole-genome duplication (WGD) events of the Salviniales based on phylogenetic mapping. *Myr* millions of years. Figure reproduced from Li et al. (2018)

*Azolla* have three floats, glabrous stems, and arrow-shaped glochidia and are assigned to *A. caroliniana*, *A. filiculoides*, *A. mexicana*, *A. microphylla*, and *A. rubra*. Species of the subgenus *Rhizosperma* have nine floats, cauline trichomes (that grow on a stem), and needlelike or no massula hairs and are assigned to *A. nilotica* and *A. pinnata*, with *A. pinnata* having three subspecies: *A. pinnata* subsp. *africana*, *A. pinnata* subsp. *asiatica*, and *A. pinnata* subsp. *pinnata*. Additional details, including the authors who originally erected *Azolla*'s species and subspecies, are given in Carrapiço et al. (2000). The native ranges of modern *Azolla* species are shown in Fig. 29.4, although the introduction of non-native species has extended their geographic occurrences, including that of *A. filiculoides* which is now widespread in Europe.

This classification is supported by DNA sequencing of *Azolla*'s species which helps identify the timing of major evolutionary events that affected their ancestors. It is then possible to construct a phylogeny, or evolutionary tree, in which the ends of branches represent today's descendants, with nodes marking their last common ancestor (LCA). This was done by Metzgar et al. (2007) who constructed a tree with six nodes (2–7) based on DNA sequencing and an earliest node (1) based on paleontological evidence (Fig. 29.5).

Metzgar et al. (2007) used the term section for the subgenera *Azolla* and *Rhizosperma*, with their divergence (node 2) estimated at 50 Ma, approximately coeval with the Arctic *Azolla* Event (48.1–49.3 Ma) when *Azolla* repeatedly covered large areas of the Arctic Ocean (Brinkhuis et al. 2006. Up to 470 ppmv (parts per

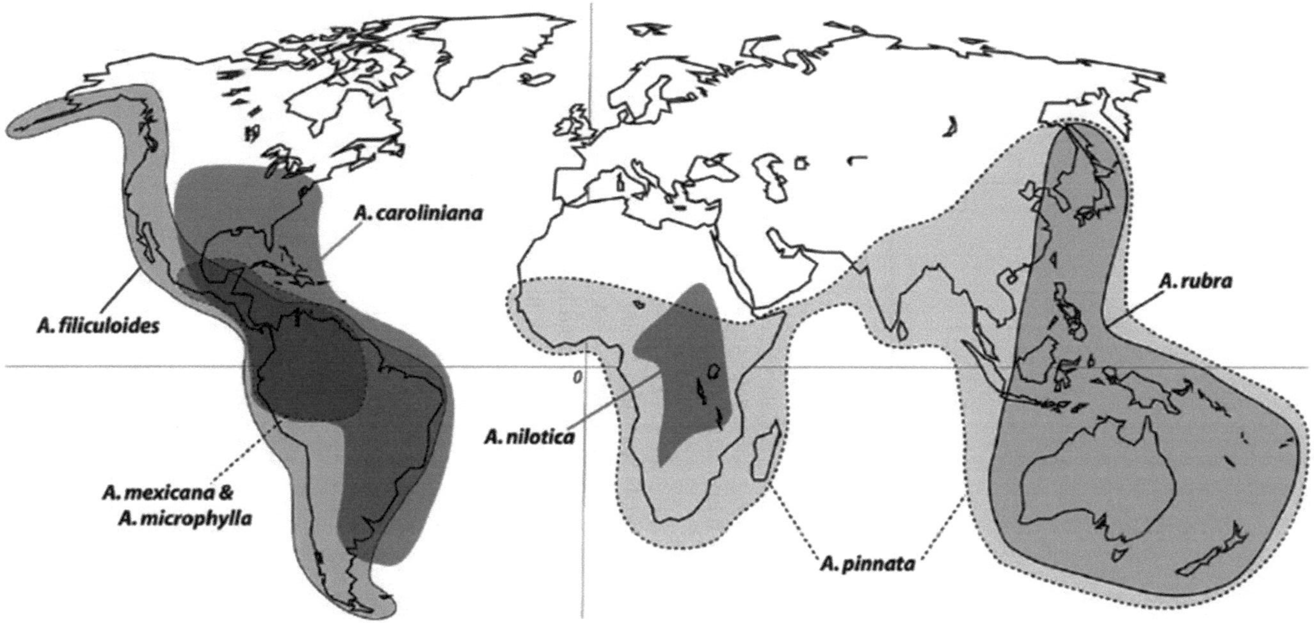


**Fig. 29.4** Native ranges of modern *Azolla* species, from Reid et al. (2006). Subgenus (= section) *Rhizosperma* is represented by *A. nilotica* and *A. pinnata*. Subgenus (= section) *Azolla* is represented by the other five species. *A. filiculoides* has been introduced as a non-native species into Northern and Central Europe, Southern Africa, eastern Asia, Australia and New Zealand. *A. caroliniana* has been introduced as a non-native species into Western Europe (Moore 1969).

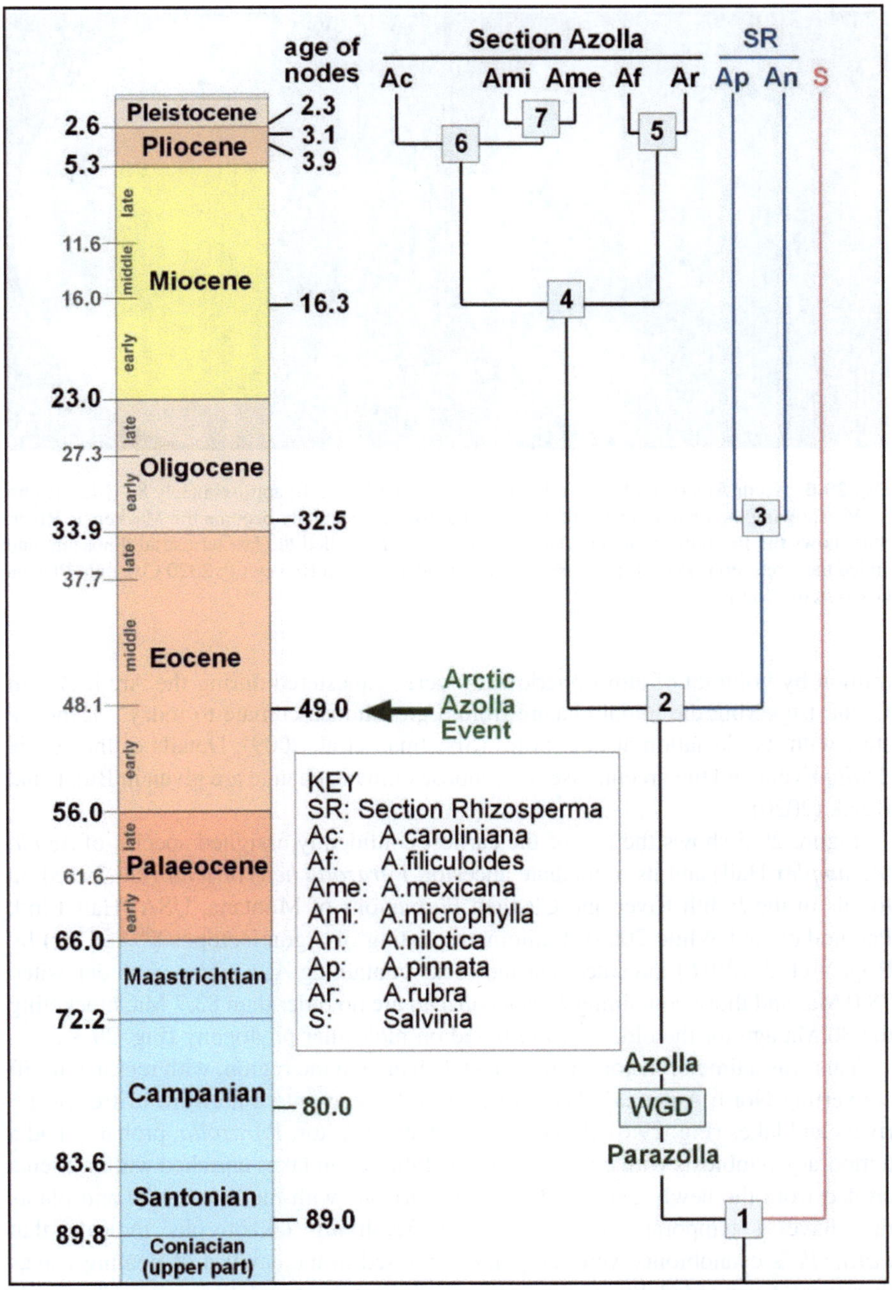


**Fig. 29.5** *Azolla*'s phylogenetic tree redrawn from Metzgar et al. (2007) showing nodes 1 to 7 plotted against the GTSF 2020 timescale. Also shown is the Arctic *Azolla* Event and *Parazolla*'s whole-genome duplication (WGD) that resulted in the earliest *Azolla* at 80 Ma. Metzgar et al. (2007) used the term section for Carrapiço et al.'s (2000) subgenera *Azolla* and *Rhizosperma*

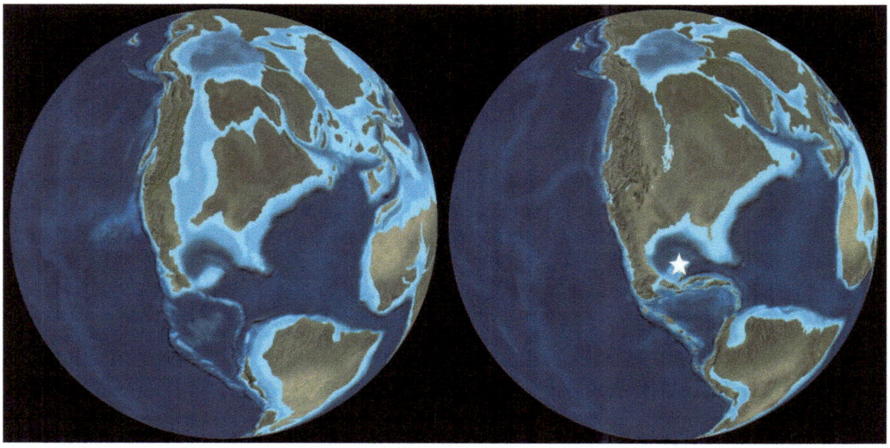

**Fig. 29.6** North America's Western Interior Seaway (WIS). Left: approximately 85 (Ma). Right: 66 Ma showing the final remnants of the WIS that would eventually become the Mackenzie River. Star shows the location of the Yucatan meteor impact that killed the last terrestrial dinosaurs and ended the Cretaceous Period. Reconstructions provided by Ron Blakey (© 2020 Colorado Plateau Geosystems, Inc.)

million by volume) of atmospheric $CO_2$ were sequestered during the Arctic *Azolla* Event, triggering the initial change from a greenhouse climate to today's icehouse state with its glaciation at both poles (Speelman et al. 2009). Details of the Arctic *Azolla* Event and the greenhouse-to-icehouse climatic change are given in Bujak and Bujak (2020).

Figure 29.5 shows the age of the earliest confidently assigned species of *Azolla* (*A. simplex* Hall) and its immediate ancestor, *Parazolla heterotricha* Hall, based on fossils in the Judith River and Claggett Formations of Montana, USA (Hall 1969, Palynodata and White 2008). Radiometric dating of argon isotopes ($^{40}Ar/^{39}Ar$) by Rogers et al. (2016) indicates that the rocks containing *A. simplex* were deposited 78.0 Ma, and those containing *P. heterotricha* are no older than 83.7 Ma, bracketing the 80 Ma age for the oldest *Azolla* based on molecular phylogeny (Fig. 29.3).

This was a time of major environmental change in the region, with regional uplift converting North America's Western Interior Seaway into a network of freshwater rivers and lakes (Fig. 29.6). *Azolla*'s immediate ancestor, *Parazolla*, probably had a temporary symbiosis with cyanobacteria that thrived in lakes enriched with nutrients eroded from the newly exposed land. Comparisons with modern *Azolla* and plants that have a temporary symbiosis with free-living diazotrophs indicate that *Parazolla*'s cyanobionts were temporarily housed in the cavities of floating leaves that were surrounded by masses of sessile cyanobacterial filaments. Chemicals emitted by *Parazolla* triggered their change from filaments to mobile heterocysts that infected *Parazolla*'s leaves. Following their enclosure in leaf cavities, other chemicals triggered reversion of the cyanobacteria to sessile filaments where they remained until the plant died.

In contrast, *Azolla*'s intergenerational retention of *N. azollae* involves a complex succession of biochemical triggers and morphological changes inside the plant and its megasporocarps, with genetic and paleontological evidence indicating that the changes originated in a single plant that underwent WGD 80 million years ago (Bujak and Bujak 2020). This provided additional genes that enabled the complex biochemical and morphological changes needed for the cyanobacteria to be transmitted with *Azolla*'s megaspores to subsequent plant generations. Figure 29.3 shows that the WGD did not affect other plants, including *Salvinia*'s ancestors, so that *Azolla* is the only plant that underwent the genetic changes leading to its intergenerational retention of a nitrogen-fixing cyanobiont.

## 29.3 Genetic Changes during *Azolla* and *A. azollae*'s Coevolution

*N. azollae*'s coevolution with *Azolla* resulted in extensive changes to the cyanobiont's genome compared to free-living species of *Anabaena* and *Nostoc* (Ekman et al. 2006, 2008, Larsson et al. 2011, Ran et al. 2010). These alterations involved the downregulation, loss, or conversion to pseudogenes of some genes, changing *N. azollae*'s ancestors from independent free-living organisms into ones that could not survive outside *Azolla*. The upregulation of other genes enhanced *N. azollae*'s sequestration of atmospheric nitrogen and provision of nitrogen-based compounds to *Azolla*, increasing the plant's speed of growth free-floating on freshwater and its potential as a biofertilizer and livestock feed.

Approximately one-third of *N. azollae*'s genome comprises nonfunctional pseudogenes. These include pseudogenes originating from genes that had previously expressed proteins involved in filament growth and division, reflecting the need to restrict their number in the plant's confined leaf cavities. Other pseudogenes reflect a loss of function relating to the cyanobiont's re-assimilation of nitrogen, so that increased amounts of nitrogen-based compounds are transferred from the cyanobiont's cells to those of *Azolla*, facilitating the plant's growth and its provision of almost three times more nitrogen biofertilizer than other plants including legumes (Watanabe 1982).

A third set of pseudogenes originated from genes that previously expressed proteins involved in the synthesis of carotenoid and chlorophyll pigments. Larsson et al. (2011) analyzed the genomes of 57 free-living cyanobacteria and *N. azollae* for shared and unique orthologs and found that two protein groups occur in the genomes of all the cyanobacteria except *N. azollae* where they are represented by nonfunctional pseudogenes. These correspond to (1) a geranylgeranyl pyrophosphate synthase that is involved in the synthesis of carotenoids and chlorophyll and (2) uroporphyrinogen-III synthase HemD that generates precursors of tetrapyrroles which provide protection against photooxidative damage. *A. azollae* is therefore reliant on *Azolla*'s cellular pigments for protection against photooxidative damage so that it cannot survive outside the plant. The previously independent cyanobacterium had become an obligate symbiont and an indispensable partner of the *Azolla* superorganism.

In contrast to *N. azollae*'s pseudogenes, genes that were retained or upregulated enhance its symbiosis with *Azolla* and, especially, the sequestration and synthesis of atmospheric nitrogen, reflecting *N. azollae*'s primary symbiotic role in providing nitrogen-based compounds to *Azolla*. The entire set of genes related to *ni*trogen *f*ixation (the *nif* gene cluster) is, therefore, intact in *N. azollae* because those genes express proteins involved in the conversion of atmospheric nitrogen into nitrogen-based compounds.

*N. azollae*'s provision of nitrogen-based compounds is also increased by the number of heterocysts that assimilate atmospheric nitrogen. All 22 genes related to heterocyst formation are intact in *N. azollae*, but the pats gene that expresses proteins involved in the suppression of heterocyst development is absent. *N. azollae*, therefore, produces more heterocysts than free-living cyanobacteria, resulting in chains of cells that contain up to 30% heterocysts compared to less than 10% in free-living *Anabaena* and *Nostoc* (Fig. 29.1). This enables *N. azollae* to fix between 4 and 18 times more atmospheric nitrogen than other species of *Anabaena* and *Nostoc* (Bergman et al. 1992, Ran et al. 2010). As a result, *Azolla* provides more nitrogen biofertilizer than any other plant, increasing its potential to replace nitrogen-based chemical fertilizers.

## 29.4 Chemical Fertilizers

For most of the past 11,800 years, since the end of the latest Pleistocene glaciation and the beginning of our present Holocene interglacial, farming practices maintained healthy soils filled with a microcosm of life that naturally fertilized crops. The nutrients needed by plants were retained in the soil using sustainable methods, including crop rotation and adding compost and organic fertilizers. This resulted in a *living soil* filled with an array of organisms, including nitrogen-fixing bacteria, insects, and earthworms that help to aerate the soil and continuously cycle nutrients, fiber, and water.

Then, in the early part of the twentieth century, chemical fertilizers were developed to provide nitrogen to plants without using natural organic processes. The age of industrial-scale farming factories was born. The Haber-Bosch process was first developed by the German chemist Fritz Haber, helped by his British colleague Robert Le Rossignol, and in 1909 they demonstrated the method for which Haber received the 1918 Nobel Prize. The German chemical company BASF purchased the process, and the chemist and engineer Carl Bosch was assigned the task of developing it on an industrial scale, for which he received the 1931 Nobel Prize. The resulting Haber-Bosch process synthesizes ammonia ($NH_3$) from nitrogen ($N_2$) and hydrogen ($H_2$), with the nitrogen-based compounds being used by plants as their source of nitrogen fertilizer.

To enable their uptake by plants, phosphate- and nitrogen-based fertilizer need to be soluble in water, so that large quantities of chemical fertilizers pollute lakes and rivers that carry the chemicals into oceans where they create dead zones – areas with mass mortality of sea life due to reduced oxygen (hypoxia) resulting from plankton blooms fed by nutrient runoff. The runoff has a similar effect on freshwater bodies, causing the uncontrolled proliferation of weeds, algae, and cyanobacteria that choke lakes, rivers, and wetlands. The chemicals introduced into the freshwater system by runoff from fields also have a negative impact on natural ecosystems and wildlife, with the chemicals, including toxins, being ingested by wildlife and concentrated through the food chain. Chemical fertilizers also increase the threat of man-made (anthropogenic) climate change. Nitrogen from chemical fertilizers is emitted into the atmosphere as nitrous oxide ($N_2O$) that is a strong greenhouse gas 200–300 times more effective in trapping heat than $CO_2$. The natural-gas source of ammonia used to manufacture chemical fertilizers also reduces the amount available for other purposes, so that competition for natural gas increases the cost of chemical fertilizers to levels that are beyond the reach of many farmers, depriving them of their livelihood and raising the price of feed needed for cattle and other livestock.

## 29.5 Biofertilizers

Biofertilizers have many advantages over chemical fertilizers. They provide organic nitrogen-based compounds that can be naturally assimilated by crops, and they do not have a negative impact on the environment. Biofertilizers supply plants with other essential nutrients, including vitamins and growth substances, and they increase the amount of organic material in soils, enabling them to retain more water and provide habitats for beneficial organisms. A well-aggregated soil tills easily, is well-aerated, and has a high ability to absorb and retain water. Increased levels of organic material also provide more soil humus that increases soil health. Healthy soils recycle nutrients more effectively, resist erosion, absorb more water after each rainfall, and produce bountiful yields of nutritious crops.

Biofertilizers are relatively inexpensive because they use solar energy, atmospheric nitrogen, and water, all of which are renewable resources, whereas the production of chemical fertilizers is complex and expensive and uses natural gas so that it competes with other sectors for this nonrenewable resource. Most biofertilizers use plants that are grown close to where they are used, minimizing the need for transportation required to bring chemical fertilizers to locations where they are applied.

Legumes are the mostly commonly used biofertilizers. They are characterized by having seeds inside a pod, called a legume, and include alfalfa, beans, chickpeas, clover, lentils, lupin beans, peanuts, peas, and soybeans. Legumes are mostly grown for livestock feed including silage or for their seeds (pulses) to feed people. Several types of legume form root nodules that emit chemicals to attract nitrogen-fixing bacteria called rhizobia that live in the soil. These infect the root nodules and synthesize atmospheric nitrogen into compounds that are utilized by the plant, enabling its use as a biofertilizer, but in all plants except *Azolla*, the symbiosis is temporary and must be renewed, so that the host and its symbiont do not coevolve.

Legumes comprise approximately 700 genera and 13,000 species of which about 20% contain root nodules (Sprent and Sprent 1990), providing more than 70 million tons of nitrogen each year in the form of organic nitrogen-based compounds (Brockwell et al. 1995, Tate 2020). Legumes with root nodules mostly occur in temperature and tropical zones where they are incorporated into the soil as biofertilizers, but they can also be used to reclaim or improve the fertility of soils that have little organic material, including those in arid regions (Zahran 1999).

## 29.6 *Azolla* Biofertilizer

*Azolla* can be used as a biofertilizer because of its symbiotic relationship with *N. azollae*, complementing the use of legumes and outperforming them because of *Azolla*'s rapid growth free-floating on water and its provision of more nitrogen-based nutrients than legumes. *Azolla* does not require land to grow, so that it does not displace agricultural land needed for food or biofuel crops, and it does not impinge on natural ecosystems because it only needs freshwater that is a few centimeters deep. It has been used for thousands of years as a biofertilizer in rice paddies, and *Azolla*'s biomass can be harvested and used to fertilize other crops. It can also be grown indoors using the *Azolla* Biosystem of stacked trays (Bujak and Bujak 2020) resembling those proposed for *Azolla*'s growth in space exploration closed-loop life support systems (CLSS) (Liu et al. 2008). The *Azolla* Biosystem, therefore, provides a controlled environment that can be replicated globally, with the plants being grown and harvested where they are needed, minimizing the need for transportation.

*Azolla* plants can be incorporated into the soil as a green manure, or they can be grown as a dual crop on water in rice paddies. In the first process, *Azolla* plants are collected from ponds and incorporated into the soil, after which they decompose within 8–10 days and increase soil fertility by releasing nitrogen-based compounds and increasing the availability of organic carbon, phosphorous, and potassium (Mishra and Dash 2014). *Azolla* can also be incorporated into the soil from composted plants that are made into powder to provide a source of *Azolla* biofertilizer at locations where high temperatures increase insect infestation and fungal infection of *Azolla* plants (Setiawati et al. 2018).

*Azolla*'s use as a dual crop in rice paddies dates back thousands of years, with Jia Si Xue's (aka Jia Ssu Hsieh) book *Chih Min Tao Shu* (*The Art of Feeding the People*), which was written in 540 AD, providing the earliest known written record of *Azolla*'s use as a rice biofertilizer (Peters and Calvert 1983). *Azolla* is now grown in paddies throughout the Indian subcontinent and the Far East, including Bangladesh, Burma, China, India, Indonesia, Nepal, North and South Korea, Pakistan, Sri Lanka, Thailand, the Philippines, Vietnam, and increasingly in Africa and South America. Dual *Azolla*-rice cropping provides multiple benefits (discussed later in this chapter) in addition to *Azolla*'s provision of nitrogen-based nutrients, with the potential to more than double rice yields (Peters and Calvert 1983).

Mishra and Dash (2014) showed that rice yield, number of tillers, plant height, and farmers' profits were higher than those using chemical fertilizers and recommended *Azolla*'s use as part of India's sustainable agriculture economic development (SAED). Yao et al. (2018) reached a similar conclusion and recommended *Azolla*'s use as a rice biofertilizer to help achieve the Chinese Ministry of Agriculture's Zero Increase Action Plan goal for national fertilizer use. They also showed that substituting *A. pinnata* biofertilizer for 25% chemical fertilizer lowered emissions of the greenhouse gas ammonia ($NH_3$) by 42%, significantly reducing the threat of climate change because rice paddies contribute 11% of anthropogenic methane emissions (Moreno-García et al. 2020).

*Azolla* therefore has the potential to help feed millions of people and reduce methane emissions from paddies. Rice provides approximately 20% of the world's calorie intake, and more than 114 countries cultivate rice, with China and India growing more than half and Asian farmers producing about 90% of the total. Annual global rice production is predicted to increase from 494 million tons in 2020 to 502 million tons in 2021, mainly due to research at the International Rice Research Institute and the conversion of additional land for rice cultivation, but this cannot keep pace with demand as our population increases. Fahad et al. (2019) listed major constraints on global rice production as a scarcity of land and freshwater, nutrient imbalance in the soil, climate change, disease, and pests, with rice growing areas decreasing in the future due to a shortage of labor, increasing urbanization, industrialization, and competition for water.

*Azolla* can help by increasing rice yield without the need for additional land or the use of chemical fertilizers, and it can also provide local, low-cost *Azolla* biofertilizer for other plants. A sustainable system developed in Ecuador by Montaño Armijos (2011) harvests *Azolla* from paddies (Fig. 29.7) and uses the biomass to fertilize a variety of plants including bananas (Montaño Armijos et al. 2012). Montaño Armijos fertilized a control group with 8–10-8 (NPK) fertilizer and a second group with *Azolla* fertilizer. All quantities of *Azolla* biofertilizer increased the growth of the plants more than those treated with chemical fertilizers, with 400 grams of *Azolla* per plant producing maximum growth in height, leaves, and number of bananas that were superior in quality to those treated with chemical fertilizers (Fig. 29.8).

*Azolla*'s use as rice biofertilizer can also be integrated with farmed waterfowl and freshwater fish. Takao Furuno uses a system that grows rice with *Azolla*, aigamo ducks and loaches (a species of freshwater fish) on Kyushu, Japan's third biggest island (Furuno et al. 2012). *Azolla* provides nitrogen-based organic compounds for the rice and reduces mosquito breeding populations by more than 95% by preventing them from laying eggs in the water and inhibiting the emergence and development of larvae from any eggs that are laid (Ansari and Sharma 1991, Mwingira et al. 2009, Rajendran and Reuben 1988, 1991).

The ducks replace pesticides and herbicides by naturally controlling pest populations and digging up or eating competing weeds. They enrich the water with oxygen by stirring up the soil, and their movement maintains an optimal water temperature for rice growth. Freshwater fish like carp, loach, and tilapia also remove unwanted pests, including the golden apple snail, *Pomacea*, that can devastate entire rice paddies in a few days. Both species of *Pomacea* are native to South

**Fig. 29.7** Some methods used in Ecuador to harvest and dry *Azolla*. Photographs provided by Mariano Montaño Armijos

**Fig. 29.8** Mariano Montaño Armijos showing the increased growth of 6-week-old banana plants fertilized with *Azolla* on the right of each photograph. See text for details. Photographs provided by Mariano Montaño Armijos

America but were introduced into Asia in the 1980s and now pose a major problem as they eat young rice plants and sever the base of mature rice stems, destroying the whole plant (Mochida 1991). The snails spread into rice paddies through irrigation canals and natural waterways, burying themselves in the mud when the water recedes to hibernate and emerge when paddies are flooded. Golden apple snails can destroy a square meter of rice in 24 hours, resulting in losses of more than half of the rice plants. Several methods have been tried to remove the snails, but they are time-consuming and often ineffective. Attempts to use fentin acetate – a pesticide that kills mollusks – failed because the amount needed to kill the snails was toxic to the rice plants. Takao Furuno's system deals with the problem in an environmentally sustainable way because the snails are a favorite food of loaches.

Phosphate-rich loach and duck waste, combined with nitrogen compounds from *Azolla*, increase nutrients in the water and soil, resulting in productivity levels that are higher than using chemical fertilizers. After rice harvesting, the *Azolla* plants can be collected and used as a biofertilizer and livestock feed, with *Azolla*-rice-duck-fish farming illustrating the benefits of integrated multi-trophic aquaculture (IMTA) (Chopin 2013).

## 29.7 *Azolla* as a Livestock Feed

Rising prices of livestock feed are forcing many farmers out of business. One of the reasons for the higher prices is scarcity of arable land to grow crops for feed and fodder which is aggravated by increased demand for land to grow food and biofuel crops. Extreme weather events including droughts and flooding are also having a devastating effect on agricultural land, including repeated droughts affecting regions like East Africa, and these are becoming more common because of anthropogenic climate change.

*Azolla* has been used as a livestock feed for hundreds of years in India and the Far East, and scientific studies confirm its suitability as a partial feed for many animals, fish, poultry, waterfowl, and some crustacea. Its sequestration and synthesis of atmospheric nitrogen (via its symbiont, *N. azollae*) and rapid growth free-floating on water confirm its potential to help alleviate feed shortages that will become more acute in the future.

Analyses going back to the 1980s show that *Azolla* contains many essential nutrients needed by livestock, including 25–35% protein, 10–15% minerals, and up to 10% amino acids dry weight (e.g., Paoletti et al. 1987), but studies on its use as a feed for a variety of livestock suggest that *Azolla* can only comprise an average of about 25% of the feed. Brouwer et al. (2018) therefore undertook a study to determine the productivity, chemical composition, and potential as protein feed of two *Azolla* species that were selected because of their different climatic and geographic ranges. *A. pinnata* is native to warmer regions of Africa, Australia, Asia, and India. *A. filiculoides* can tolerate cooler temperatures, is native to the Americas (Fig. 29.4), and is also common as an introduced non-native species in Europe. Their results indicate that *Azolla* probably cannot be used as a total feed for many animals because it contains high levels of (poly)phenolic tannins that bind to proteins and decrease digestibility, with a lower level of (poly)phenol and higher protein content in *A. filiculoides* making it more suitable as a protein feed than *A. pinnata*.

## 29.8 Azolla Feed for Ruminant Animals

There are about 150 species of ruminant animals including cattle, sheep, goats, and deer. The term is derived from the Latin word *ruminare*, which means to chew over again because they regurgitate food like grass that has a high cellulose content after the cellulose has been converted into sugars and fatty acids by bacteria – a process called enteric fermentation. The stomach of ruminants has four compartments, with the largest, the rumen, containing anaerobic bacteria that produce an enzyme called cellulase which breaks down the cellulose found in the wall of plant cells into sugar compounds that can be utilized by the animal – a process called enteric fermentation. Examples are given below of *Azolla*'s use as a feed supplement for (1) cows, (2) Mecheri lambs, (3) Corriedale sheep, and (4) goats.

**Cows:** Although the demand for milk and meat from cattle is increasing, there has been a substantial decline in fodder production due to decreasing areas of forests and grasslands that produce fodder. Dried hay and straw fodder, therefore, are increasingly supplemented with commercial feed, resulting in increased costs of meat and milk production. Kamalasanan et al. (2002) conducted trials using *Azolla* as a feed substitute for dairy animals and found that it provided nutritional benefits in addition to lowering costs. Milk yield increased when *Azolla* was combined with regular feed, and 15–20% of commercial feed could be replaced with *Azolla* on a dry weight basis without affecting milk production, resulting in a 20–25% savings. Adding *Azolla* to the feed also improved milk quality and the health and longevity of the animals.

**Mecheri lambs:** Sankar et al. (2020) evaluated the nutritional and feeding value of *Azolla* meal on 3-month-old Mecheri lambs. A control group were fed a basal diet comprising concentrate and roughage at a ratio of 40:60 and compared to a treatment group that had 10% protein of concentrate replaced by dried *Azolla* meal. Fortnightly body weight and dry matter intake were recorded for 3 months. There was no significant difference between groups in dry matter intake, average daily gain, and feed efficiency, and Sankar et al. (2020) concluded that dried *Azolla* can replace 10% protein level in the concentrate feed without affecting the growth performance in Mecheri lambs.

**Corriedale sheep:** A study on 3-month-old Corriedale sheep divided them into five groups (T1, T2, T3, T4, and T5) that were fed concentrate mixtures containing 0, 6, 12, 18, and 24% *Azolla* that replaced 0, 25, 50, 75, and 100% linseed cake, respectively (Ahmed et al. 2016). The study concluded that 6% *Azolla* can replace 25% of linseed cake (T2 group) without any adverse effect on the performance of the animals.

**Goats:** Sihag et al. (2018) investigated the effect of feeding *A. pinnata* to 4- to 5-month-old goats in India. A T1 control group were fed total mixed ration (TMR) comprising 60% berseem hay and 40% concentrate mixture. Three experimental groups (T2, T3, and T4) replaced the concentrate mixture with sundried *Azolla* on an equi-weight basis of 10, 15, and 20%, respectively. The study showed that digestible crude protein (DCP) and total digestible nutrient (TDN) intake and nitrogen balance in the goats were not affected by replacing 15% concentrate mixture with sundried

*Azolla* (group T2), providing a net saving of Indian Rs. 7.90 for feed per kg weight gain of goat.

## 29.9 Azolla Feed for Monogastric Animals

Monogastric animals have a stomach with one chamber in which acids and enzymes aid digestion. They include (1) omnivores like humans and pigs that have a mixed herbivore and carnivore diet, (2) carnivores including cats, and (3) herbivores with hindgut bacteria that can digest cellulose, in contrast to ruminants that have foregut bacteria. Examples of monogastric animals with hindgut bacteria are proboscideans (including elephants), large odd-toed ungulates (including horses and rhinos), rodents (including beavers, mice, rats, and squirrels), and lagomorphs (rabbits and hares) that have different dentition from rodents. The amount of fiber that can be processed by monogastric animals therefore varies, with rodents having a lower capacity than large monogastric animals with hindgut bacteria.

Lagomorphs, including rabbits and hares, have a high metabolic rate and eat frequently, ingesting small amounts of food that they retain in their gut for a short time. The food is digested by bacteria in a pouch between the small and large intestines called the caecum, producing fatty acids that provide approximately one-third of their energy needs. The contents of the caecum then pass through the large intestine to be expelled as pellets that are eaten again – a process called cecotrophy – so that the cecal pellets pass through the digestive system a second time, allowing more nutrients to be extracted, while the remaining undigested material leaves the body as normal fecal pellets. More fiber can, therefore, be processed in lagomorphs than in rodents like rats.

Several studies have been undertaken to determine the effect of supplementing the feed of monogastric animals with *Azolla*, including examples below for (1) monogastric omnivores, (2) larger monogastric animals with hindgut bacteria, and (3) monogastric lagomorphs.

**Monogastric omnivores:** Cherryl et al. (2013) assessed the economics of including *A. pinnata* in the diet of crossbred Large White Yorkshire pigs in India. The pigs were divided into three groups: a control group (T1) without *Azolla* and experimental groups T2 and T3 in which 10% and 20% of *A. pinnata* was included as protein replacer in the concentrate feed. The *Azolla* plants were grown locally with a unit cost of Indian Rs. 0.56 and Rs. 5.65 for fresh and dry *Azolla*. The results showed an almost identical daily weight gain of 0.34 to 0.35 kg for all three groups and a lower feed intake for T2 and T3 (1.96 kg and 1.91 kg) than T1 (2.03 kg). The cost of concentrated feed and *A. pinnata* was Rs. 13/kg and Rs. 5.65/kg, respectively, lowering the cost of feed per day from Rs. 75.4 (T1) to Rs. 71.9 (T2) and Rs. 65.4 (T3). Cherryl et al. (2013) therefore recommended the inclusion of up to 20% sun-dried *Azolla* as a natural protein feed source in diets of pigs.

**Larger monogastric animals with hindgut bacteria:** Songara et al. (2018) studied the effect on Marwari stallions of supplementing feed and fodder with *A. pinnata*. The stallions were fed a basal feed of "concentrate mixture" (gram

30%, wheat bran 27%, oats grain 40%, salt 2%, mineral mix 1%), green fodder (green sorghum), and dry fodder (50% wheat straw, 50% groundnut haulm) for 45 days (Trial I). Locally grown and dried *A. pinnata* was then used to replace 10% of the concentrate feed protein for the next 45 days (Trial II). There was no difference in body weight and dry matter, crude protein, ether extract, crude fiber, and nitrogen-free extract digestibility of the diets used for the two trials, indicating that 10% *A. pinnata* can be used as a protein supplement for stallions.

**Monogastric lagomorphs:** Sireesha et al. (2017) studied the effect on rabbits of supplementing feed with *A. pinnata* in India. The control group (T1) was fed a mixture of maize (40%), wheat bran (19.5%), lucerne meal (19.5%), soybean meal (19%), salt (1.5%), and mineral mixture (0.5%). The wheat bran and lucerne meal were supplemented equally with 1.5% and 3% dried *Azolla*, respectively, in groups T2 and T3. Rabbits in all three groups had the same weight gains indicating that *A. pinnata* can be included at 1.5% and 3.0% levels in the feed of rabbits.

## 29.10 Azolla Feed for Chickens

There were approximately ten billion chickens in 1990, increasing to 20 billion in 2010 and 26 billion in 2019, with most being raised in Asia. The price of feed for chickens, like that of ruminant and monogastric animals, is increasing so that *Azolla* has enormous global potential to supplement chicken feed.

A 2006 Nigerian study supplemented pullet chicks' diet with *Azolla* meal (AZM) prepared from *A. pinnata* plants dried in the sun and milled to produce meal with a long shelf life (Alalade and Iyayi 2006). The best results occurred in birds fed with a 10% *Azolla* supplement, resulting in an average weekly weight increase of 99 grams compared to 95 grams without the supplement. Average weekly total feed intake for each bird decreased from 287 grams to 231 grams with AZM supplement, reducing farmers' feed costs and producing larger birds.

Indian studies at Killikulam's Agricultural College and Research Institute (Prabina and Kumar 2010) and Kashmir's Sher-e-Kashmir University of Agricultural Sciences and Technology (Khursheed 2019) confirm the benefits of supplementing chicken feed with AZM. Prabina and Kumar (2010) observed that 7.5% AZM supplement resulted in 2.6% increase in body weight compared to chickens not fed with AZM, and Khursheed (2019) showed that the greatest weight gain was with 5% AZM supplement followed by 10% supplement. Prabina and Kumar (2010) also observed that the antibody "tire value" (a measure of the number of antibodies within the blood) against Ranikhet virus was higher in chickens fed with AZM. Ranikhet disease (aka Newcastle disease) affects poultry, ducks, geese, pheasants, guinea fowl, partridges, pigeons, and doves. Mortality is 50% to 100% and symptoms include coughing, gasping, sneezing, diarrhea, twisted neck, and partial paralysis of the legs and wings. There is presently no cure, with the main preventive measures being vaccination, quarantine, import restrictions, and banning all feeds based on animal tissues, so that *Azolla*'s potential to reduce the disease has global significance for domesticated and farmed birds.

## 29.11 Azolla Feed for Waterfowl

Waterfowl, including ducks, geese, and swans, are birds that are strong swimmers with waterproof feathers and webbed feet that they use as flippers to push through the water. *Azolla* is easily digested by most waterfowl including ducks, with videos uploaded to the Internet by smallholder farmers in the Far East showing birds paddling excitedly to eat *Azolla* when it is thrown onto the water. Ducks are extensively farmed in the Far East where they provide a source of high-protein food for millions of people, with others being kept for their eggs, but this is threatened by the high cost of feed, and several studies have investigated the nutritional value of *Azolla* as a supplementary feed for ducks.

Swain et al. (2018) studied the effect of feeding *A. pinnata* to White Pekin (aka American Pekin) laying ducks in India using three diets: T1 control diet with no *Azolla*, T2 with 10% fresh *Azolla* (representing 100 grams of *Azolla*/duck/day), and T3 with 20% fresh *Azolla* (200 grams of *Azolla*/duck/day). The *A. pinnata* contained 4.74% dry matter (DM), 24.93% crude protein (CP), 3.48% ether extract (EE), 13.80% crude fiber (CF), 16.84% total ash (TA), and 40.95% nitrogen-free extract (NFE). Egg weight increased with both 10% and 20% *Azolla* replacement, and Swain et al. concluded that a daily feed to each duck of 200 grams fresh *Azolla* as replacement of 20% of concentrate feed improved feed conversion ratio (FCR), performance efficiency index, egg production, and egg weight and enriched yolk color.

Acharya et al. (2015) studied the effect of feeding fresh *A. pinnata* to White Pekin broiler ducks using three diets: $G_1$ control diet with no *Azolla*, $G_2$ with 5% fresh *Azolla*, and G3 with 10% fresh *Azolla.* Bodyweight, bodyweight gain, feed consumption, and FCR were recorded for each diet. Acharya et al. observed no significant difference between the groups and concluded that 10% of *Azolla* can be included in the diet of White Pekin broilers. Their study did not investigate larger amounts of *Azolla* dietary supplement.

Other studies on *Azolla*'s use as a dietary supplement for Muscovy ducks (Chisembe et al. 2020), Philippine mallard ducks (Lawas et al. 1998), and Japanese quail (Rathod et al. 2013, Shamna et al. 2013) concluded that *Azolla* can be used to supplement their diets, with an average of 20% *Azolla* having no detrimental effect on birds' health and growth.

## 29.12 Azolla Feed for Freshwater Fish

Freshwater fish are a high-protein food source, including zinc, iron, vitamin A, and calcium that are essential for cognitive and physical development, but the price of feed is now a limiting factor (Fiogbe et al. 2004). Carp and tilapia are two of the most extensively farmed freshwater fish, but species like grass carp *Ctenopharyngodon idella* have an inefficient digestive system. They consume more than their own weight of aquatic weeds in a single day so that *Azolla* is an ideal supplementary feed because of its rapid growth without the need for nitrogen-based fertilizers. It is easy to cultivation at almost no cost in tropical and subtropical regions at locations where freshwater fish are extensively reared for food by individuals or smallholder farmers. *Azolla* contains 13–30% crude protein and more essential amino acids (EAA) than other aquatic macrophytes and most green forage crops (Mosha 2018), so that it has major potential to supplement or replace protein from increasingly expensive sources like fish meal and fish oil.

Mosha et al. (2018) reviewed 34 published papers that investigated *Azolla*'s potential to supplement feed for the freshwater fish *Tilapia* and the family Cyprinidae. The reviewed papers showed that the addition of 10–45% *Azolla* in the diet for most species of *Tilapia* and 10–25% for most farmed species of Cyprinidae has a positive effect on feed utilization, mobilization, and utilization of glycogenic amino acids, growth performance, and protein conversion ratio (the weight gain divided by the intake of a particular protein).

Mosha et al. (2020) subsequently investigated *Azolla*'s potential to partially replace fishmeal in the diet of "genetically improved farmed tilapia" (GIFT Tilapia) – a faster-growing strain of Nile tilapia. One hundred twenty fingerlings ($3.3 \pm 0.32$ g) were randomly stocked in plastic containers and fed four diets – control, T1, T2, and T3 – which had 0%, 15%, 30%, and 45% *Azolla* meal inclusion levels. These were fed to the four GIFT Tilapia groups at 5% their body weight for 60 days. Growth performance, serum biochemistry and immunological parameters, muscle quality, and stress response parameters were higher in the T1 group compared to the others, indicating that 15% inclusion of *Azolla* supplement enhances blood serum biochemistry performance and reduces the cost of fish feed in GIFT Tilapia aquaculture.

## 29.13 Azolla Feed for Prawns and Shrimp

Shrimps and prawns are crustaceans with ten legs belonging to the order Decapoda that includes crayfish, crabs, and lobsters. Saltwater shrimp and prawn farming has increased during the past 40 years, including inland locations where growing tanks and ponds are mostly placed indoors in carefully controlled conditions.

Farmed shrimp and prawns have traditionally been fed fishmeal, but feed cost and feed management account for at least 60% of production costs and are the major constraints against aquaculture expansion (Banerjee et al. 2010). Feeds of animal (as opposed to plant) origin also cause water fouling and disease in cultured species, favoring the use of less expensive plant protein sources as feed supplements, including legumes, linseed meal, lupin meal, rapeseed meal, and soybean meal. Soybean meal is the most common protein source due to its competitive cost, availability, and nutritional quality that includes high-protein content and a relatively balanced amino acid profile, but its cost has risen because of increased demand for its use as animal feed and human food.

Sudaryono (2011) investigated if *Azolla* meal could replace soybean meal in the diets of juvenile giant black tiger prawn (*Penaeus monodon*). His study showed that similar growth rates were achieved regardless of the replacement levels of soybean with *Azolla*, with no difference in palatability between the two types of meal, indicating that all soybeans fed to farmed black tiger shrimp could be replaced with *Azolla*.

The giant freshwater prawn *Macrobrachium rosenbergii* is also known as the giant river prawn, giant freshwater prawn, Malaysian prawn, freshwater scampi, and cherabin. It has a similar geographic distribution to *P. monodon* but is restricted to freshwater bodies except when it temporarily migrates into brackish water for breeding. *M. rosenbergii* is one of the largest freshwater prawns, and it is widely cultivated for food, including farming in freshwater enclosures, resulting in its introduction into parts of Africa, the Americas, the Caribbean, China, Japan, New Zealand, and Thailand.

A 2014 study was undertaken in India by Bharathiar University's Crustacean Biology Laboratory to determine if formulated feeds of fishmeal could be replaced with a combination of *Spirulina platensis* (= *Arthrospira platensis*), *Chlorella vulgaris*, and *A. pinnata* (Radhakrishnan et al. 2014). The cyanobacterium *Spirulina* is rich in proteins, vitamins, minerals, essential amino acids, fatty acids, and antioxidant pigments. The unicellular freshwater alga *Chlorella* has also been called a superfood because it contains "the highest amount of chlorophyll of any known plant and 60 % protein, 18 amino acids (including all the essential amino acids), and various vitamins and minerals" (Radhakrishnan et al. 2014). The study found that *Spirulina-Chlorella-Azolla* feed could replace half of the normal fishmeal fed to *M. rosenbergii* with no detrimental or toxic effect, with antioxidants like vitamin C and E being significantly elevated.

## 29.14 Conclusions

*Azolla* is a unique superorganism because of its permanent symbiosis with the nitrogen-fixing cyanobacterium, *N. azollae*, that fixes 4 to 18 times more atmospheric nitrogen than free-living species of *Anabaena* and *Nostoc*. As a result, *Azolla* can double its biomass in less than 2 days free-floating on freshwater and provide sustainable biofertilizer for rice and other crops. It can also supplement feed for a variety of ruminant and monogastric animals, poultry, waterfowl, freshwater fish, and crustacea, helping to reduce food shortages that affect millions of people.

## References

Acharya P, et al (2015) Exploring the effects of inclusion of dietary fresh *Azolla* on the performance of White Pekin broiler ducks. Veterinary World, EISSN: 2231–0916. Published online at www.veterinaryworld.org/Vol.8/November-2015/4.pdf

Ahmed HA et al (2016) Performance of growing sheep on *Azolla* based diets. Indian J Anim Res 50(5):721–724. ISSN:0367-6722

Alalade OA, Iyayi EA (2006) Chemical composition and the feeding value of *Azolla* (*Azolla pinnata*) meal for egg-type chicks. Int J Poult Sci 5(2):137–141. https://doi.org/10.3923/ijps.2006.137.141

Ansari MD, Sharma V (1991) Role of *Azolla* in controlling mosquito breeding in Ghaziabad District Villages (U.P.). Indian J Malariol 28:51–54

Banerjee K et al (2010) Cost-effective and eco-friendly shrimp feed from red seaweed *Catenella repens* (Gigartinales: Rhodophyta). Curr Biotica 4(1):23–43

Bergman B et al (1992) The *Nostoc-Gunnera* Symbiosis. New Phytol 122(3):379–400. https://doi.org/10.1111/j.1469-8137.1992.tb00067.x

Brinkhuis H et al (2006) Episodic fresh surface waters in the Eocene Arctic Ocean. Nature 441(7093):606–609. https://doi.org/10.1038/nature04692

Brockwell J et al (1995) Manipulation of rhizobia microflora for improving legume productivity and soil fertility: a critical assessment. Plant Soil 174(1–2):143–180. https://doi.org/10.1007/bf00032245

Brouwer P et al (2018) Growing *Azolla* to produce sustainable protein feed: the effect of differing species and $CO_2$ concentrations on biomass productivity and chemical composition. J Sci Food Agric 98(12):4759–4768. https://doi.org/10.1002/jsfa.9016

Bujak J, Bujak A (2020) The *Azolla* Story. Published as an ebook by the Azolla Foundation, December 2020. https://theazollastory.com/about-the-azolla-story/

Carrapiço F (2010) *Azolla* as a superorganism. Its implication in symbiotic studies. In: Cellular origin, life in extreme habitats and astrobiology symbioses and stress. Springer, Netherlands, pp 225–241. https://doi.org/10.1007/978-90-481-9449-0_11

Carrapiço F (2018) *Azolla* and Bougainville's voyage around the world. In: Fernandez H (ed) Current advances in Fern research. Springer, Netherlands, pp 251–267. https://doi.org/10.1007/978-3-319-75103-0_12

Carrapiço F, et al. (2000) *Azolla* as a biofertiliser in Africa. A challenge for the future. Revista De Ciências Agrárias, 23: 120–138

Cherryl DM, Prasad RMV, Jayalaxmi P (2013) A study of economics of inclusion of *Azolla pinnata* in swine rations. Int J Agric Sci VetMed 1(4):50–56. ISSN 2320-3730

Chisembe P, Banda LJ, Tanganyika J (2020) Effect of duck-rice-azolla integration on growth performance and carcass quality of native Malawian Muscovy ducks. Livestock Res Rural Dev 32:110

Chopin T (2013) Aquaculture, integrated multi-trophic (IMTA). In: Meyers RA (ed) Encyclopedia of sustainability science and technology. Springer-Verlag, Berlin Heidelberg, pp 184–205. https://doi.org/10.1007/978-1-4614-5797-8_173

Edwards DJ (1974) Weed preference and growth of young grass carp in New Zealand. N Z J Mar Freshw Res 8(2):341–350. https://doi.org/10.1080/00288330.1974.9515509

Ekman M et al (2006) Protein expression profiles in an endosymbiotic cyanobacterium revealed by a proteomic approach. Mol Plant-Microbe Interact 19(11):1251–1261. https://doi.org/10.1094/mpmi-19-1251

Ekman M et al (2008) Proteomic analysis of the cyanobacterium of the *Azolla* symbiosis: identity, adaptation, and NifH modification. J Exp Bot 59(5):1023–1034. https://doi.org/10.1093/jxb/erm282

Fahad S et al (2019) Major constraints for global rice production. In: Hasanuzzaman M et al (eds) Advances in rice research for abiotic stress tolerance. Woodhead Publishing, Sawston, pp 1–22. https://doi.org/10.1016/b978-0-12-814332-2.00001-0

Fiogbe ED et al (2004) Use of a natural aquatic fern, *Azolla microphylla*, as a main component in food for the omnivorous-phytoplanktonophagous tilapia, *Oreochromis niloticus* L. J Appl Ichthyol 20(6):517–520. https://doi.org/10.1111/j.1439-0426.2004.00562.x

Furuno T, et al. (2012) The one duck revolution: why Asia does not do European farming. Takao Furuno, 2012. Lulu.Co., 128 pp. ISBN 130007633X, 9781300076339

Hall JW (1969) Studies on fossil Azolla: primitive types of megaspores and massulae from the cretaceous. Am J Bot 56(10):1173–1180. https://doi.org/10.2307/2440779

Kamalasanan PS, et al. (2002) *Azolla* – a sustainable feed substitute for livestock. Leisa Magazine 2013: 15–17. http://www.gemenskapspraktik.se/projects/theazollacookingandcultivationproject/research/Azolla,%20a%20sustainable%20feed%20for%20livestock.pdf

Khursheed I et al (2019) Proximate evaluation of *Azolla pinnata* as sustainable feed supplement for poultry. J Pharmacognosy Phytochem 8(3):3160–3160

Larsson J et al (2011) Genome fluctuations in cyanobacteria reflect evolutionary, developmental and adaptive traits. BMC Evol Biol 11(187). https://doi.org/10.1186/1471-2148-11-187

Lawas MVP, Roxas DB, Lambio AL (1998) Laying performance of Philippine mallard ducks fed diets substituted with fresh *Azolla*. Recent developments in animal production-1998: proceedings of the Philippine Society of Animal Science. 35th annual convention: 220–225. https://agris.fao.org/agris-search/search.do?recordID=PH1999101426

Li FW et al (2018) Fern genomes elucidate land plant evolution and cyanobacterial symbioses. Nat Plants 4(7):460–472. https://doi.org/10.1038/s41477-018-0188-8. Epub 2018 Jul 2. PMID: 29967517; PMCID: PMC6786969

Liu X, Chen M, Bian Z, Liu C (2008) Studies on urine treatment by biological purification using *Azolla* and UV photocatalytic oxidation. Adv Space Res 41:783–786

Malatinszky, et al. (2017) A comprehensively curated genome-scale two-cell model for the heterocystous cyanobacterium *anabaena* sp. PCC 7120. Plant Physiol 173:509–523. https://doi.org/10.1104/pp.16.01487

Metzgar JS et al (2007) Phylogeny and divergence time estimates for the fern genus *Azolla* (Salviniaceae). Int J Plant Sci 168(7):1045–1053. https://doi.org/10.1086/519007

Misra P, Dash D (2014) Rejuvenation of biofertilizer for sustainable agriculture and economic development. J Sustain Dev 11(1):41–61

Mochida O (1991) Spread of freshwater Pomacea snails (Pilidae, Mollusca) from Argentina to Asia. Micronesica supplement, 1 Jan. 1991: 51–62, scinapse.Io/papers/104537750

Montaño Armijos M (2011) *Azolla*: a new paradigm of the future of rice. DSpace En ESPOL: Home, 3 Oct. 2011, www.dspace.espol.edu.ec/xmlui/handle/123456789/17161?show=full

Montaño Armijos M, et al (2012) Evaluación preliminar de la aplicación del Azolla vomo fertilizante en plantas meristemáticas de banano, published in: www.dspace.espol.edu.ec/, Converting Rice Fields into Green Fertilizer Factories Project Completion Report

Moore AW (1969) *Azolla*: biology and agronomic significance. Bot Rev, 35:17–34

Moreno-García B, Guillén M, Quílez D (2020) Greenhouse gas emissions as affected by fertilization type (pig slurry vs. mineral) and soil management in Mediterranean rice systems. Agronomy 10(4):1–19. https://doi.org/10.3390/agronomy10040493

Mosha SS (2018) A review on significance of *Azolla* meal as a protein plant source in finfish culture. J Aquac Res Dev 9:544. https://doi.org/10.4172/2155-9546.1000544

Mosha SS et al (2020) Partial fishmeal replacement by *Azolla* meal on GIFT tilapia (*Oreochromis niloticus*) diet: effect on growth performance, antioxidant enzymes, immunology and stress response. Acta Scientific Veterinary Sci 2(8):17–25

Mwingira VS et al (2009) Mosquito larval productivity in rice-fields infested with *Azolla* in Mvomero District, Tanzania. Tanzan J Health Res 11(1):17–22. https://doi.org/10.4314/thrb.v11i1.43246

Palynodata Inc., White JM (2008) Palynodata datafile: 2008 version. Geological Survey of Canada Open File 5793. https://doi.org/10.4095/225704

Paoletti C et al (1987) Lipid composition of *Azolla caroliniana* biomass and its seasonal variation. Phytochemistry 26(4):1045–1047. https://doi.org/10.1016/s0031-9422(00)82347-x

Peters GA, Calvert HE (1983) The *Azolla-anabaena Azollae* symbiosis. In: Goff LJ (ed) Algal symbiosis. Cambridge Univ. Press, New York, pp 53–64

Prabina BJ, Kumar K (2010) Dried *Azolla* as a nutritionally rich cost effective and immunomodulatory feed supplement for broilers. Asian J Anim Sci 5(1):20–22

Radhakrishnan S et al (2014) Replacement of fishmeal with *Spirulina platensis*, *Chlorella vulgaris* and *Azolla pinnata* on non-enzymatic and enzymatic antioxidant activities of *Macrobrachium rosenbergii*. J Basic Appl Zool 67(2):25–33. https://doi.org/10.1016/j.jobaz.2013.12.003

Raja W, Rathaur P, John SA, and Ramteke PW (2012) *Azolla-anabaena* association and its significance in supportable agriculture. Hacettepe J Biol Chem, 40, (1): 1–6

Rajendran R, Reuben R (1988) Laboratory evaluation of the water fern, *Azolla pinnata* for mosquito control. J Biol Control 2(2):114–116

Rajendran R, Reuben R (1991) Evaluation of the water fern *Azolla microphylla* for mosquito population management in the rice-land agro-ecosystem of South India. Med Vet Entomol 5(3): 299–310. https://doi.org/10.1111/j.1365-2915.1991.tb00556.x

Ran L et al (2010) Correction: genome erosion in a nitrogen-fixing vertically transmitted endosymbiotic multicellular cyanobacterium. PLoS One 5:9. https://doi.org/10.1371/annotation/835c5766-5128-41c4-b636-adfe0c503103

Rathod GR et al (2013) Feeding value of *Azolla* (*Azolla pinnata*) meal in growing Japanese quail. Indian J Poultry Sci 48:154–158

Reid JD et al (2006) Phylogenetic relationships in the heterosporous fern genus *Azolla* (Azollaceae) based on DNA sequence data from three noncoding regions. Int J Plant Sci 167(3):529–538. https://doi.org/10.1086/501071

Rogers R et al (2016) Age, correlation, and lithostratigraphic revision of the upper cretaceous (Campanian) Judith River formation in its type area (north-Central Montana), with a comparison of low- and high-accommodation alluvial records. J Geol 124(1):99–135. https://doi.org/10.1086/684289

Sankar V et al (2020) Effect of feeding *Azolla* meal on growth performance of Mecheri sheep. Int J Curr Microbiol App Sci 9(5):1945–1949. https://doi.org/10.20546/ijcmas.2020.905.222

Schirrmeister BE et al (2013) Evolution of multicellularity coincided with increased diversification of cyanobacteria and the great oxidation event. Proc Natl Acad Sci 110(5):1791–1796. https://doi.org/10.1073/pnas.1209927110

Setiawati MR, et al (2018) The application dosage of *Azolla pinnata* in fresh and powder form as organic fertilizer on soil chemical properties, growth and yield of rice plant. AIP Conference Proceedings 1927, 030017. Published Online: 09 February 2018. https://doi.org/10.1063/1.5021210

Shamna TP et al (2013) Broiler characteristics of Japanese quails (*Coturnix coturnix japonica*) at different levels of diet substitution with *Azolla pinnata*. Anim Sci Rep 7(2):75–80

Sihag S et al (2018) Effect of feeding *Azolla* (*Azolla pinnata*) based total mixture ration on growth performance and nutrients utilization in goats. Forage Res 43(4):314–318

Sireesha K et al (2017) A study on economics of inclusion of *Azolla* (*Azolla pinnata*) in rabbit rations. Int J Sci Environ Technol 6(1):167–172. ISSN 2278-3687 (O), 2277-663X (P)

Songara M, et al (2018) Int J Chem Stud 6(5): 3215–3218. P-ISSN: 2349–8528. E-ISSN: 2321–4902

Speelman EN et al (2009) The Eocene Arctic *Azolla* bloom: environmental conditions, productivity and carbon drawdown. Geobiology 7(2):155–170. https://doi.org/10.1111/j.1472-4669.2009.00195.x

Sprent JI, Sprent P (1990) Nitrogen fixing organisms: pure and applied aspects. Chapman and Hall, New York, NY, p 256

Sudaryono A (2011) Use of *Azolla* (*Azolla pinnata)* meal as a substitute for defatted soybean meal in diet of juvenile black tiger shrimp (*Penaeus Monodon*). J Coastal Dev 9(3):145–154

Swain B et al (2018) Effect of feeding of *Azolla* (*Azolla pinnata)* on the performance of white Pekin laying ducks. Int J Livestock Res 8(5):248. https://doi.org/10.5455/ijlr.20180117114123

Tate RL III (2020) Soil microbiology, 3rd edn. John Wiley & Sons, New York. ISBN: 978-0-470-31110-3.

Testo W, Sundue M (2016) A 4000-species dataset provides new insight into the evolution of ferns. Mol Phylogenet Evol 105:200–211. https://doi.org/10.1016/j.ympev.2016.09.003

Watanabe I (1982) *Azolla—anabaena* symbiosis — its physiology and use in tropical agriculture. In: Dommergues YR, Diem HG (eds) Microbiology of tropical soils and plant productivity, Developments in plant and soil sciences, vol 5. Springer, Dordrecht. https://doi.org/10.1007/978-94-009-7529-3_6

Watanabe I, Berja NS (1983) The growth of four species of *Azolla* as affected by temperature. Aquat Bot 15(2):175–185. https://doi.org/10.1016/0304-3770(83)90027-x

Yao Y et al. (2018) *Azolla* biofertilizer for improving low nitrogen sse efficiency in an intensive rice cropping system. Field Crop Res, 216: 158–164. doi:https://doi.org/10.1016/j.fcr.2017.11.020

Zahran HH (1999) Rhizobium-legume symbiosis and nitrogen fixation under severe conditions and in an arid climate. Microbiol Mol Biol Rev 63(4):968–989

**Alexandra Bujak** is an environmental scientist who has published extensively on modern *Azolla*. She set up the nonprofit *Azolla* Foundation in 2012 to provide information and advice about *Azolla*'s use as a $CO_2$ sequester, biofertilizer, biofuel, livestock feed, food, soil remediator, and water purifier. The Foundation has more than 40,000 associates and works with charities like Hope for Ebola Orphans Sierra Leone (H4EO) to increase rice yields using *Azolla* and provide funding for schools needed to teach Sierra Leone's thousands of Ebola orphans, giving them hope and the key to a sustainable future.

**Jonathan Bujak** is a paleontologist with more than 100 scientific publications. He was a senior research scientist with the Geological Survey of Canada and Petro-Canada International Aid Corporation in the 1970s and 1980s, training geologists and leading geological expeditions into the interior of Colombia, Jamaica, and Venezuela. Dr. Bujak was also involved with the only two geological expeditions to the North Pole: the 1979 Lomonosov Ridge Experiment (LOREX) and the 2004 Arctic Coring Expedition (ACEX) that discovered the Eocene *Azolla* Event when *Azolla* repeatedly covered large areas of the ocean for 1.2 million years, triggering the change from a greenhouse climate to today's icehouse world with its permanent glaciation at both poles

Alexandra and Jonathan Bujak published *The Azolla Story* as an Amazon e-book in December 2020.

# *Pteridium aquilinum*: A Threat to Biodiversity and Human and Animal Health

30

Helena Fernández and L. María Sierra

**Abstract**

The fern *Pteridium aquilinum*, commonly named bracken, has been distributed worldwide for ~23.8 million years. It belonged to open forest communities long before human impact on forests and landscapes took place. However, today, the prevalence of bracken has expanded considerably, partly due to human land use but also due to the natural aggressiveness of the fern toward competing grasses, herbs, and trees. In addition, bracken is toxic to livestock and humans: Its extracts and spores, but also contaminated food and water, are genotoxic and able to induce different types of tumors. The overall geographical distribution and the local abundance of bracken seem to be increasing in many places of the world due to several causes like cultural modifications. Climate change seems to favor bracken spread, at least in Northern Europe and mountainous areas, due to raising temperatures, longer growth period, and more humid conditions in combination with more sun hours. Given the negative consequences this fern might bring to humans and animals, care must be taken to avoid over-exposition to its dangerous chemicals.

**Keywords**

*Pteridium aquilinum* · Bracken fern · Carcinogenic plant · DNA damage · Ptaquiloside

H. Fernández (✉)
Department of Organisms and Systems Biology, Area of Plant Physiology, University of Oviedo, Oviedo, Spain
e-mail: fernandezelena@uniovi.es

L. M. Sierra
Department of Functional Biology (Genetic Area), Oncology University Institute (IUOPA) and Institute of Sanitary Research of Asturias (ISPA), University of Oviedo, Oviedo, Spain
e-mail: lmsierra@uniovi.es



J. Marimuthu et al. (eds.), *Ferns*, https://doi.org/10.1007/978-981-16-6170-9_30

## List of Abbreviations

| | |
|---|---|
| BEH | bovine enzootic hematuria |
| CAU | Caudatoside |
| LC-MS | Liquid chromatography coupled mass spectrometry |
| PTA | Ptaquiloside |
| PTE | Ptesculentoside |
| Pt | Pterosine |

## 30.1 Botanical Aspects and Distribution

*Pteridium aquilinum*, commonly named bracken fern, belongs to the phylum Pteridophyta and to the family Dennstaedtiaceae. There is a misunderstanding in defining infrageneric taxa, recognizing the genus as a species complex (Der et al. 2009). These authors differentiate two major clades within *Pteridium*, a primarily northern hemisphere Laurasian/African clade, which includes all taxa currently assigned to *P. aquilinum*, and a primarily southern hemisphere Austral/South American clade, which includes *P. esculentum* and *P. arachnoideum*. Besides, all European accessions of *P. aquilinum* subsp. *aquilinum* appear in a monophyletic group joined within a clade containing the African *P. aquilinum* taxa (*P. aquilinum* subsp. *capense* and *P. aquilinum* subsp. *centrali-africanum*).

This fern has been distributed worldwide for ~23.8 million years. It is a cosmopolitan weed readily spreading into pasture and marginal areas, favored by fire and soil acidity, and a troublesome pest in forests and plantations causing substantial financial as well as biodiversity losses. Bracken fern cannot be classified as invasive species sensu stricto but shares the traits of many invasive species: extensive vegetative reproduction (belowground rhizome system, forming larger clonal stands), wide ecological amplitude, climatically adaptable, tolerant to a wide range of edaphic conditions; low degree of herbivore/pathogen pressure, and very resistant to the fire (Hartig and Beck 2003). It belonged to open forest communities long before human impact on forests and landscapes took place. However, today the prevalence of bracken has expanded considerably partly due to human land use but also due to the natural aggressiveness of the fern toward competing grasses, herbs, and trees (Milligan et al. 2018; Senyanzobe et al. 2020).

The overall geographical distribution and the local abundance of bracken seem to be increasing in many places of the world (Rasmussen et al. 2015), e.g., in Denmark, England, Italy, and Spain. Climate change seems to favor bracken spread, at least in Northern Europe and mountainous areas, due to increased temperature, longer growth period, and more humid conditions, in combination with more sun hours. In Europe, ferns have a wide distribution, with its optimum in the Atlantic and sub-Atlantic lowlands and mountains, increasing the average current trend in quantity, either in EU28 (+5.1%) or in EU28+ (+5.4%) (taken from European Red List of

Habitats-Grasslands Habitat Group). The largest area of this plant is found in Great Britain, where active management by chemical spraying has reduced up to 10% of the occupied surface (Stewart et al. 2008). In Denmark, this fern is classified by forest authorities as one of the ten problem species in forestry and nature management (Rasmussen et al. 2015). *Pteridium* is now a recognized problem in Europe but also in other continents, affecting land productivity and biodiversity (Rasmussen et al. 2015).

In Spain, bracken develops profusely in intervened areas, especially borders and meadows, in which the monitoring of the livestock farmer is dwindling; it also invades the undergrowth level in forests, where it forms high-density masses that come to typify the timber forests of large areas in the northern area of the country, as in many temperate forest provinces of the world. It is therefore considered a problematic pest in forests and plantations, causing substantial financial and biodiversity losses, by hindering and even making agricultural and forestry work impossible (Vetter 2009; Rasmussen et al. 2015; Milligan et al. 2018).

Recent studies on three pilot areas in and near the Picos de Europa (north of Spain) site show that, in particular, areas corresponding to lowland hay meadows have reduced their area by 65–74% over the past 60 years, achieving the highest lost rates in the latest years, with grass being replaced by bracken, among other species (García et al. 2018). The mowing meadows, agro-systems maintained by the action of farmers and breeders, are disappearing throughout Europe and, with them, a great associated biodiversity. Its great botanical and faunistic value (especially lepidoptera and other important insects for pollination) has been widely recognized, and they are among the threatened habitats of Europe included in the Habitats Directive: codes 6510, mountain mowing meadows (Arrhenatherion), and 6520, mountain mowing meadows (Triseto-Polygonion bistortae). Moreover, in these areas there is an important livestock grazing from later spring to early autumn, exposed to the fern. Transformations on the surfaces of mowing meadows are consequence, fundamentally, of the different agricultural policies implemented in the country, with their main effect (structural adjustment) on traditional farms of extensive livestock in these mountain areas; moreover, other facts contribute to these changes on mowing meadows, such as the sociodemographic processes linked to population loss on the rural mountain environment, as well as cultural changes together with the aging population that still resides in these areas, that lead to the substitution of these habitats with bracken, among other species (Fig. 30.1). The future tendency will depend on management activities in these areas because other factors, such as climate change, would favor bracken spread, making it a potential hazard for livestock-based extensive agriculture, livestock and humans, as grasses would be insufficient and livestock could eat this highly toxic plant which, in addition, has the capacity to contaminate soil and water resources. In Great Britain, peak incidences with cattle were reported in certain years, often those with long hot dry summers, during which cattle may seek other than grass food (Cranwell 2004).

**Fig. 30.1** Invasiveness of bracken in abandoned pastures, in the Asturias region (Spain). (**a**) Panoramic observation of bracken dominating the vegetation; (**b**) cows grazing in a soil full of bracken

## 30.2 The Invasiveness of Bracken

It is based both on the dispersion and viability of the spore, the survival and development of the gametophyte, and the branching and growth capacity of the rhizome in the field. Bracken sporophytes, despite colonizing the habitats through rhizomes, produce millions of spores (Marrs and Watt 2006), with a high viability range (43–52%) in laboratory (Ashcroft and Sheffield 2000). However, in their natural habitat they show a reduced viability, since their dispersion season ends in early autumn and, when deposited in the soil, they are very sensitive to temperatures below 0 °C (Rodríguez De La Cruz et al. 2009).

The gametophyte is considered the weak phase of the life cycle. Despite its feebleness, there are very long-lived gametophytes, practically immortal, who subsist under conditions of extreme frugality, spreading vegetatively without growth cessation (Gabriel y Galán 2010). Fieldwork on natural gametophyte populations of bracken is scarce because its success is very limited, under current climatic and ground conditions, but also the infertility of spores, given that gametophyte arises from spores (Pakeman et al. 2002). Lastly, the rhizome system comprises a massive underground network, whose actual size is even quite small in comparison with its potential, rarely exceeding 20 m from rhizome apex to the senescent end (Duc et al. 2003).

The primary mechanism of bracken encroachment is by vegetative rhizome spread. The annual expansion rate of a bracken stand at its margins is not easily predicted, and it is dependent on many variables including climate, topography, and edaphic conditions, together with the genetics and physiology of the plant itself (Ader 1990). Bracken has evolved a mechanism of toxin release which allows the fern to exert its dominance most effectively in each particular habitat where it grows (Gliessman 1976). In the last years, the presence of selligueain A has been reported in any member of the Dennstaedtiaceae family, and production of allelochemicals has also been described in members of the *Pteridium* species complex. This evidence of selligueain A, as a putative allelochemical of *P. arachnoideum,* reinforces the role of allelopathy in the dominance processes of this plant in the areas where it occurs (de Jesus et al. 2016).

## 30.3 Toxic Substances Found in Bracken

Bracken has a vast chemical arsenal, some of which remain after cutting and drying (Cooper-Driver 1976, 1990), capable of preventing establishment of competing plants, as well as intoxicating diverse animals, like mammals and insects, for instance (Vetter 2009). The potentially harmful substances identified in bracken include ptaquiloside (PTA) and other illudanes and protoilludanes, indanones (various pterosins and pterosides), quercetin, and its glycoside rutin, kaempferol, shikimate, thiaminases, prunasin, and various ecdysteroids (Alonso-Amelot and Avendaño 2002; Rasmussen et al. 2003). In addition to these substances, several unstable analogs of ptaquiloside have been identified such as caudatoside (CAU) and ptesculentoside (PTE), among others (Alonso-Amelot and Avendaño 2002; Yamada et al. 2007; Fletcher et al. 2010). PTA is partly responsible for the toxicity of bracken, as it is the case of the PTA analogs PTE and CAU, which have similar biochemical reactivity (Fletcher et al. 2010). However, these two last compounds are highly under-investigated because, only very recently, a robust methodology was designed both to prepare analytical standards of PTE, CAU, PTA, and the corresponding hydrolysis products—pterosins G, A, and B, respectively—and to simultaneously determine these six chemicals in bracken by liquid chromatography-associated mass spectrometry (LC-MS) (Kisielius et al. 2020).

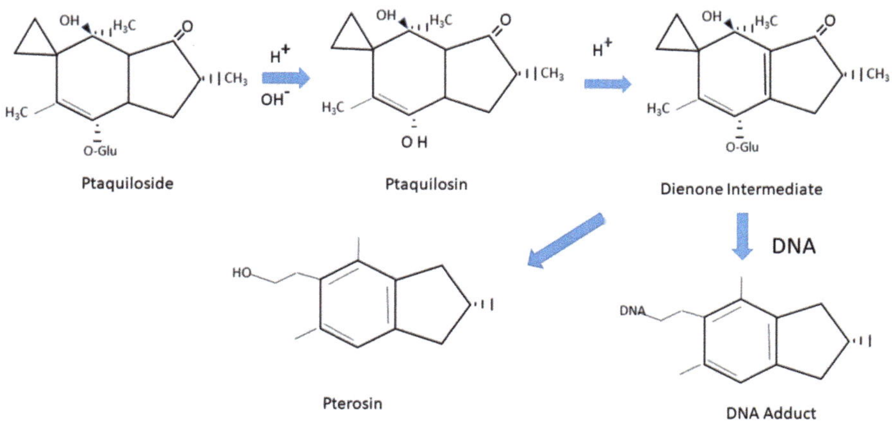


**Fig. 30.2** Reactions of ptaquiloside and DNA adduct formation

Systematic examination of the xenobiotic expression of *P. aquilinum* has revealed that it may be genetically polymorphic. This has been demonstrated in the cyanogenic potential expressed both by the prunasin content and in the unassisted formation of HCN (Oliveros-Bastidas and Alonso-Amelot 2010), and it may be also the reason for the expression of iludanos, which is very variable, both between individual clones, as well as according to the phenological stage, and part of the frond and/or rhizome. Consequently, different levels of toxins are detected in plants from different areas (Dawra et al. 2002). In general, large-scale temporal and spatial variation exists, indicating the presence of chemotypes. Although variable, contents are often quite high, measured either as concentration in the biomass or as amount of toxin per area unit, making bracken more than a hazard, a clear danger, from a health and environmental perspective. Therefore, in each region where the impact of bracken on the rest of the biota is important, it is necessary to establish the morphogenic variability of its xenobiotic expression and the quantification of the most relevant toxins, i.e., iludanos, without detriment to the others.

Many scientific papers refer to *Pteridium* as the only known upper plant that naturally causes cancer in animals (Shahin et al. 1999; Potter and Baird 2000; Dawra et al. 2002; Alonso-Amelot and Avendaño 2002; Vetter 2009). Its carcinogenicity rests on ptaquiloside and similar illudane glycosides (Fig. 30.2) that are found in high concentrations in the vegetative parts of the plant, including spores (Rasmussen et al. 2013). It was shown to be a major carcinogenic component of bracken in several experiments performed independently in Japan (Niwa et al. 1983; Hirono et al. 1984a, b) and in the Netherlands (van der Hoeven et al. 1983). It has been estimated that ptaquiloside accounts for more than 50% of the carcinogenic potency of bracken (van der Hoeven et al. 1983).

## 30.4 Toxicity of Bracken in Livestock and Humans

Where the geographical distribution of bracken overlaps with abundant livestock exploitation, a worrying incidence of toxicosis, both acute and chronic, is detected among animals as several well-defined pathologies or syndromes, almost all leading to fatality, or considerable loss in animal performance, and financial fragility of the exploitation (Alonso-Amelot and Avendaño 2002). Toxicity of bracken to cattle, horses, and goats has been recognized for over a century. Continued ingestion of sublethal amounts of bracken can lead to a complex assortment of ailments, including hypoplasia of the bone marrow, thrombocytopenia, leucopoenia, immunosuppression, reduced intestinal uptake of iron, thiamine (vitamin B1) deficiency, and severe internal hemorrhages, leading to hemorrhagic syndromes (Cranwell 2004). The most visible symptoms in farm animals, linked to bracken consumption, are weakness, lack of back legs and neck motor coordination, propensity to acquire infectious diseases, breathing difficulties, and nasal, rectal, and urinary bleeding. In cattle, the hemorrhagic syndrome is the bovine enzootic hematuria (BEH), a chronic process characterized by hemorrhages in the urinary bladder and, in some cases, tumors in the bladder wall, reported from different parts of the world (Vetter 2009). Sheep are less prone than cattle to the acute hemorrhagic effects of bracken, experiencing mostly bright blindness (retinal atrophy), whereas thiamine deficiency is observed mainly in horses (Vetter 2009). Among all these pathologies, BEH has been reported as the biggest problem in Portugal, New Zealand, England, Turkey, India, Venezuela, Peru, Ecuador, Argentina, Brazil, and other countries (Alonso-Amelot and Avendaño 2002).

In addition to the animals, some human populations from Japan, New Zealand, Brazil, Canada, and China consume young curled fronds, or crosiers, as a nutritious food, since they are considered to be a good source of protein, carbohydrates, fat, vitamins, carotenoids, and trace minerals (Tourchi-Roudsari 2014). Japan imports annually over 13,000 tons of bracken, in addition to the local production (Alonso-Amelot and Avendaño 2002). Moreover, in countries like Japan, Canada, Northeastern USA, Siberia, or South Korea, bracken has long been used as an industrial crop; moreover, a technology of synthetic seed production, including spore encapsulation in sodium alginate beds, was developed to get a better management of spores and to solve current problems with handling, storage, and transportation of bracken spores for human consumption (Jang et al. 2020).

There is epidemiological evidence indicating the effects of *Pteridium*, and its derivatives, on human health. Our species suffers the carcinogenic power of the fern and develops tumors in various organs after direct consumption of the plant or by ingesting food (milk and derivatives) from cattle fed with this plant (Shahin et al. 1999). Specifically, systemic investigations reveal fern influence on the development of gastric cancer in humans in the UK, Costa Rica, Venezuela, or Japan (Galpin et al. 1990; Villalobos-Salazar 1985; Alonso-Amelot and Avendaño 2002, 2009); Lin et al. 2006; The usual procedure before eating fern, which includes cooking it in water in the presence of different chemicals, serves for removing most, but not all, of the carcinogenic activity (Vetter 2009).

In 1986, the International Agency for Research on Cancer (IARC) classified bracken on Group 2B of chemical classification, that is, "possibly carcinogenic to humans" (IARC, 1986; WHO/IARC 2014; Oliveros-Bastidas et al. 2016), and suggests the necessity of collecting more positive epidemiological and ecological aspects on the impact of this plant and its metabolites (Francesco et al. 2011).

The passage of carcinogens from the plant to the milk obtained from cows was demonstrated by Evans et al. (1972). Subsequent studies have confirmed that milk from cows eating ferns contains toxic, mutagenic, and carcinogenic metabolites that are not found in normal cow's milk (Aranha et al. 2019). In areas where cows feed naturally on ferns, a high incidence of gastric cancer has been found, although the risk decreases if milk is processed technologically in the dairy plant (Fenwick 1989). The carcinogenic action of products from animals with BEH, and their ability to produce cancer in humans, is beginning to worry scientists (Francesco et al. 2011; Virgilio et al. 2015). Although this issue has had no impact on the media (with the exception of Portugal), this is expected to happen in the coming years given the growing concern that public opinion has on food security.

As indicated, *Pteridium* toxicity is variable and depends not only on the geographical area where the plant is located (Dawra et al. 2002) but also on the part of the plant and the growth stage (Cooper-Driver 1976). Variations in the distribution of ptaquiloside in fronds and rhizome have been found, with particularly high levels in fronds (Rasmussen et al. 2003). In fact, growing fronds, earlier phenological stages, and pinna tips contain more carcinogens (Rasmussen et al. 2005; Tourchi-Roudsari 2014). Thus, it is not surprising that there are livestock areas where *Pteridium* abounds, but where there is a very low incidence of BEH, while other areas of very similar invasion are heavily punished with the syndrome (Alonso-Amelot and Avendaño 2002). With all this, it would seem appropriate to assess the toxicity of certain *Pteridium* populations, always in close contact with breeders and veterinarians, who carry the weight of both losses and knowledge.

In addition, fern spores might represent a hazard for human and animal health (Rasmussen et al. 2013). In areas where bracken grows easily, inhaled spores might be trapped in the respiratory tract, or, alternatively, they might become trapped in the mucus of the nasopharynx and then be swallowed. A single frond of bracken can release up to several grams of spores in summer, from July to September, and they can travel long distances in the wind (Rasmussen et al. 2013). Thus, there is potential risk for low level exposure almost everywhere *Pteridium* is present, although the highest exposures would be in bracken infested zones. Oral exposure to spores might also occur by drinking spore-contaminated water. In fact, bracken spores were carcinogenic to mice when added to their drinking water (Evans and Galpin 1990), and they induce DNA damage when administered to human cultured cells (Simán et al. 2000).

## 30.5 Ecotoxicology

Ptaquiloside and the other illudane glycosides are highly water-soluble and are leached from fronds to soils, from where they can contaminate soil and water, including groundwater, at levels reaching 2.5 mg/kg of soil on the horizon closest to the surface (Rasmussen et al. 2003; Jensen et al. 2008; Ovesen et al. 2008; Clauson-Kaas et al. 2016; Skourti-Stathaki et al. 2016). Being soluble, their incorporation into runoff waters poses a significant risk to supplies for watering holes and human use. Recently, the presence of ptaquiloside, caudatoside, and their corresponding hydrolysis products, pterosins B (PtB), A (PtA), and G (PtG), was determined in water wells in Denmark, Sweden, and Spain (Skrbic et al. 2021). In this report, water samples, from a total of 77 deep groundwater wells (40–100 m) and shallow water wells (8–40 m), were analyzed by LC-MS. Seven wells contained at least one of the illudane glycosides, and/or pterosins, at concentrations up to 0.27 μg $L^{-1}$ (PTA), 0.75 μg $L^{-1}$ (CAU), 0.05 μg $L^{-1}$ (PtB), 0.03 μg $L^{-1}$ (PtA), and 0.28 μg $L^{-1}$ (PtG). This is the first finding of illudane glycosides and pterosins in drinking water wells, exceeding the suggested maximum tolerable concentrations of PTA, although they were used for drinking water purpose. Also recently, in Spain, a wide range of natural toxins were detected in a period of 7 months, from March to September, in the Ter River water reservoirs, used to produce drinking water for Barcelona City (Picardo et al. 2020). The analyses revealed the presence of ptaquilosin B, the degradation product of PTA, in 33% of the samples, peaking at the start of rainy season (Picardo et al. 2020). These and other projects highlight the urgency of taking actions to systematically assess the presence of toxins that may be contaminating vital resources for all kind of organisms, as the water, and to exercise a strong control over them.

## 30.6 Ptaquiloside and Genotoxicity

Formally, PTA is a protoxin given that its reactivity as an alkylating agent is scarce if no metabolic-induced changes occur first. However, PTA is unstable in aqueous neutral medium, at high temperatures (>100 °C), and in a slightly alkaline or acidic medium (5.5 < pH> 8.0) (Fig. 30.2). In acidic medium, it is transformed into pterosine B, a stable indanone with not attributed cytotoxicity until its activity against human leukemia HL 60 cells was discovered (Chen et al. 2008). In alkali, PTA decomposition takes a different course: An iludano-dienone, poorly called "activated ptakylide," is formed by beta-glucose removal. The dienone is very unstable as it reacts with a variety of both organic and biological nucleophiles, at physiological pH and temperature, producing stable covalent alkylation adducts in vitro, like those formed with the O and N nucleophilic centers of DNA nitrogen bases (Ojika et al. 1987; Kushida et al. 1994) (Fig. 30.2).

These adducts, induced by bracken extracts or PTA, are also detected in vivo, after the treatment of mice and rats (Povey et al. 1996; Shahin et al. 1998a; Freitas et al. 2001), and are the origin (a) of DNA strand breaks detected in human cell lines with the comet assay (Simán et al. 2000; Campos-da-Paz et al. 2008; Pereira et al. 2009; Gil Da Costa et al. 2012) or with the γ-H2AX (Gomes et al. 2012); (b) of induced chromosomal aberrations, both in human cells in vitro and in cows and mice in vivo (Matsuoka et al. 1989; Lioi et al. 2004; Almeida Santos et al. 2006; Gil Da Costa et al. 2012); (c) of induced aneuploidy (Almeida Santos et al. 2006; Gil Da Costa et al. 2012); and (d) of induced gene mutations, like those in *Salmonella* (Matoba et al. 1987; Nagao et al. 1989), those responsible for H-ras oncogene activation (Shahin et al. 1998b; Potter and Baird 2000; Yamada et al. 2007) (Sardon et al. 2005), or those frameshift mutations detected in mice in vivo (Gomes et al. 2012).

All these genotoxic effects, derived from DNA adduct induction, agree with the carcinogenic properties attributed to this fern, and its PTA toxin, and support their role in promoting gastric carcinogenesis (Gomes et al. 2012).

PTA is not unique to bracken. This compound and analogs have been detected in 19 out of 31 species of ferns tested as *Cheilanthes sieberi* and *Pteridium esculentum* (Potter and Baird 2000). Indeed, ptaquiloside (PTA) concentration was determined in 40 non-bracken fern samples collected from India, and only a few contained high levels (499 and 595 mg/kg of PTA on dry matter), a few more, like *Diplazium esculentum*, *Polystichum squarrosum*, and *Dryopteris juxtaposita*, moderate levels (19 to 31 mg/kg), and most samples had no detectable PTA content, such as *Cheilanthes farinosa*, *Christella dentata*, *Adiantum incisum*, and *Pteris stanophylla* (Somvanshi et al. 2006).

## 30.7 Controlling the Expansion of Bracken

Of particular importance is the search for ways to control the expansion of the fern and its eradication in pasture meadows. Traditional forms of control have been the cutting, or cutting of the aerial part, mechanical removal of rhizomes, and burning, with little short-term effectiveness and none in medium and long term (Liley 2005; Stewart and Pullin 2005). Alternatively, there are techniques developed for the application of systemic herbicides such as Asulam, Asulox, and Ally, in combination with metsulfuron and glyphosate, with low concentrations of surfactants that increase foliar absorption and transfer of the herbicide to rhizome buds. However, they present undesirable effects: (a) The network of rhizomes, given its underground expansion without aerial expression, may not be affected by the herbicides absorbed by the fronds and, within a few years, returns to the original weed condition; (b) the application of Asulam/Asulox/Ally, which are specific to *Pteridium*, will indistinctly annihilate other fern species, thus affecting their biodiversity in the environment; and (c) the inclusion of glyphosate and metsulfuron, which are nonspecific herbicides for a broad range of leafy plants, will annihilate safe plant species, also affecting plant diversity (Liley 2005; Stewart and Pullin 2005).

On the other hand, biological control methods have been attempted using adapted animals (insects, goats, pigs) and phytopathogenic fungi (Fowler 1993; Seastedt 2015; Hill et al. 2020). For example, the butterfly *Conservula consigna*, originating in South America, whose larval phase feeds on the pinnas of *P. aquilinum* and completes its life cycle there, was introduced in the UK to exercise control over the plant (Burge and Kirkwood 1992). Similarly, the fungus *Ascochyta pteris*, originally from the UK, has been proposed as a biological control agent since, under certain climatic conditions, it causes significant damage to bracken. Fixing fungus spores to the plant, by means of some adhesive such as a mixture of water and oil, would facilitate the spread of the fungus and the disease it causes (Burge and Kirkwood 1992). Despite these attempts, so far, no good results have been achieved through the application of biological methods.

## 30.8 Usefulness for Bracken

Apart from all these negative effects, bracken has been widely used for different purposes (Donnelly et al. 2002). Bracken is used as livestock bedding in several countries, and, today, it is produced and sold as peat free compost, due to its low pH and high nitrogen and potash content (Pitman and Webber 2013). In addition, bracken mulch for winter soil protection has been used in the past, possibly due to its high surface area and slow rate of decomposition, and, nowadays, dead bracken fronds might be directly usable as a mulch in agricultural and silvicultural applications (Taylor and Thomson 1998). The plant contains a wide range of secondary metabolites, including sesquiterpenoids, ecdysones, cyanogenic glycosides, tannins, and phenolic acids, such as flavonoids, which have antibiotic properties (Kardong et al. 2013; Amobi 2015). In line with this, bracken is interesting also as source of various anti-insect chemicals, such as ecdysones (Cooper-Driver 1976; Svatos and Macek 1994), which may have a potential value for insect control. Likewise, bracken has been used as a biofuel for centuries, probably due to its high calorific value of 21 GJ/t, compared with 19 GJ/t for straw (Donnelly et al. 2002). Moreover, the ash from its burning has been used as a fertilizer, due to high levels of potassium and high pH (Donnelly et al. 2002). Hence, plans oriented toward the control of this plant may be complemented with beneficial procedures and products, as long as they do not represent any risk derived from the presence of toxins.

## 30.9 Conclusions

This overview of the toxic potential of *Pteridium* species, and their enormous abundance and invasiveness, forces a closer understanding of their link with animal and human health wherever this fern is present in abundance. Special attention should be given to this plant by toxicologic pathologists, veterinarians, and public health specialists. The human populace, especially farmers, should be educated about the public health significance of this plant. The idea should be that the rural/ local/ regional/national institutions must work with landowners, and cattle owners, in order to show them the adverse effects of *Pteridium* (for animals and humans) and, more importantly, the relevance of their collaboration/participation in the making up of a handbook of best practices, inspired in the pioneer programs developed in areas like Scotland, where bracken is a recognized problem since long. It is very important to transmit the relevance of limiting fern expansion. However, apart from all these negative effects, fern has a potential use as a source of commercial compounds such as insecticide and biofuel, among others. Therefore, control-oriented plans could be complemented by procedures to take advantage of beneficial products, feeding a circular economy, which could benefit impoverished areas.

## References


Ader KG (1990) The effect of climate and soils on bracken vigour and distribution in Scotland. In: Thomson JA, Smith RT (eds) Bracken biology and management. Sydney, Occasional Publication No. 40, Australian Institute of Agricultural Science, pp 141–151

Almeida Santos Mde FM, Dórea JG, Luna H (2006) Bracken-fern extracts can be clastogenic or aneugenic depending on the tissue cell assay. Food Chem Toxicol 44(11):1845–1848. https://doi.org/10.1016/j.fct.2006.06.006

Alonso-Amelot ME, Avendaño M (2002) Human carcinogenesis and bracken fern: a review of the evidence. Curr Med Chem 9:675–686. https://doi.org/10.2174/0929867023370743

Alonso-Amelot ME, Avendaño M (2009) Conglomerados de cáncer gástrico en el Estado Mérida, Venezuela. Interciencia 34:617–622

Amobi AS (2015) Antibacterial, phytochemical and proximate analysis of Pteridium aquilinum. Int J Res Pharm Biosci 2:1–7

Aranha PCDR, Rasmussen LH, Wolf-Jäckel GA et al (2019) Fate of ptaquiloside—a bracken fern toxin—in cattle. PLoS One 14:e0218628. https://doi.org/10.1371/journal.pone.0218628

Ashcroft CJ, Sheffield E (2000) The effect of spore density on germination and development in Pteridium, monitored using a novel culture technique. Am Fern J 90:91. https://doi.org/10.2307/1547324

Burge MN, Kirkwood RC (1992) The control of bracken. Crit Rev Biotechnol 12:299–333. https://doi.org/10.3109/07388559209040626

Campos-da-Paz M, Pereira LO, Bicalho LS, Dórea JG, Poças-Fonseca MJ, Santos Mde F (2008) Interaction of bracken-fern extract with vitamin C in human submandibular gland and oral epithelium cell lines. Mutat Res 652(2):158–163. https://doi.org/10.1016/j.mrgentox.2008.01.009

Chen Y, Zhao Y, Hu Y et al (2008) Isolation of 5-hydroxypyrrolidin-2-one and other constituents from the young fronds of Pteridium aquilinum. J Nat Med 62:358–359. https://doi.org/10.1007/s11418-008-0225-4

Clauson-Kaas F, Ramwell C, Hansen HCB, Strobel BW (2016) Ptaquiloside from bracken in stream water at base flow and during storm events. Water Res 106:155–162. https://doi.org/10.1016/j.watres.2016.09.049

Cooper-Driver G (1976) Chemotaxonomy and phytochemical ecology of bracken. Bot J Linn Soc 73:35–46. https://doi.org/10.1111/j.1095-8339.1976.tb02011.x

Cooper-Driver G (1990) Defense strategies in bracken, Pteridium aquilinum (L.) Kuhn. Ann Missouri Bot Gard 77:281. https://doi.org/10.2307/2399541

Cranwell MP (2004) Bracken poisoning Cattle Practice. BCVA 12(3):205–207

Dawra R, Kurade N, Sharma O (2002) Carcinogenicity of the fern Pteridium aquilinum collected from enzootic bovine haematuria-free hilly area in India. Curr Sci 83:1005–1009

de Jesus JL, Varela RM, Molinillo JMG et al (2016) Allelopathy of bracken fern (Pteridium arachnoideum): new evidence from green fronds, litter, and soil. PLoS One 11:e0161670. https://doi.org/10.1371/journal.pone.0161670

Der JP, Thomson JA, Stratford JK, Wolf PG (2009) Global chloroplast phylogeny and biogeography of bracken (Pteridium; Dennstaedtiaceae). Am J Bot 96:1041–1049. https://doi.org/10.3732/ajb.0800333

Donnelly E, Robertson J, Robinson D (2002) Potential and historical uses for bracken (Pteridium aquilinum (L.) Kuhn) in organic agriculture. In: Powell J et al (eds) . Proceedings of the COR conference, UK, Aberystwyth, pp 255–256

Evans IA, Galpin OP (1990) Bracken and leukaemia. Lancet (London, England) 335:231. https://doi.org/10.1016/0140-6736(90)90329-4

Evans IA, Jones RS, Mainwaring-Burton R (1972) Passage of bracken fern toxicity into milk. Nature 237:107–108. https://doi.org/10.1038/237107a0

Fenwick GR (1989) Bracken (Pteridium aquilinum)—toxic effects and toxic constituents. J Sci Food Agric 46:147–173. https://doi.org/10.1002/jsfa.2740460204

Fletcher MT, Hayes PY, Somerville MJ, De Voss JJ (2010) Ptesculentoside, a novel norsesquiterpene glucoside from the Australian bracken fern Pteridium esculentum. Tetrahedron Lett 51:1997–1999. https://doi.org/10.1016/J.TETLET.2010.02.032

Fowler SV (1993) The potential for control of bracken in the UK using introduced herbivorous insects. Pestic Sci 37:393–397. https://doi.org/10.1002/ps.2780370413

Francesco B, Giorgio B, Rosario N et al (2011) A new, very sensitive method of assessment of ptaquiloside, the major bracken carcinogen in the milk of farm animals. Food Chem 124:660–665. https://doi.org/10.1016/J.FOODCHEM.2010.05.059

Freitas RN, O'Connor PJ, Prakash AS, Shahin M, Povey AC (2001) Bracken (Pteridium aquilinum)-induced DNA adducts in mouse tissues are different from the adduct induced by the activated form of the bracken carcinogen ptaquiloside. Biochem Biophys Res Commun 281 (2):589–594. https://doi.org/10.1006/bbrc.2001.4388

Gabriel y Galán JM (2010) Gametophyte development and reproductive behaviour of Niphidium macbridei Lellinger (Polypodiaceae, Polypodiopsida). Nov Hedwigia 90:513–519. https://doi.org/10.1127/0029-5035/2010/0090-0513

Galpin OP, Whitaker CJ, Whitaker R, Kassab JY (1990) Gastric cancer in Gwynedd. Possible links with bracken

García P, García L, González V (2018) Diagnóstico de la situación de prados de siega en el suroeste europeo: análisis diacrónico en el Parque Nacional de los Picos de Europa. Nat Cantab 6:1–21

Gil da Costa RM, Coelho P, Sousa R, Bastos MMSM, Porto B, Teixeira JP, Malheiro I, Lopes C (2012) Multiple genotoxic activities of ptaquiloside in human lymphocytes: aneugenesis, clastogenesis and induction of sister chromatid exchange. Mutat Res 747(1):77–81. https://doi.org/10.1016/j.mrgentox.2012.04.010

Gliessman SR (1976) Allelopathy in a broad spectrum of environments as illustrated by bracken. Bot J Linn Soc 73:95–104. https://doi.org/10.1111/j.1095-8339.1976.tb02015.x

Gomes J, Magalhães A, Michel V et al (2012) Pteridium aquilinum and its ptaquiloside toxin induce DNA damage response in gastric epithelial cells, a link with gastric carcinogenesis. Toxicol Sci 126:60–71. https://doi.org/10.1093/toxsci/kfr329

Hartig K, Beck E (2003) The bracken fern (Pteridium arachnoideum (Kaulf.) Maxon) dilemma in the Andes of southern Ecuador. Ecotropica 9:3–13

Hill MP, Moran VC, Hoffmann JH et al (2020) More than a century of biological control against invasive alien plants in South Africa: a synoptic view of what has been accomplished. In: Biological invasions in South Africa. Springer International Publishing, Cham, pp 553–572

Hirono I, Aiso S, Yamaji T, Mori H, Yamada K, Niwa H, Ojika M, Wakamatsu K, Kigoshi H, Niiyama K et al (1984a) Carcinogenicity in rats of ptaquiloside isolated from bracken. Gan 75 (10):833–836

Hirono I, Kono Y, Takahashi K, Yamada K, Niwa H, Ojika M, Kigoshi H, Niiyama K, Uosaki Y (1984b) Reproduction of acute bracken poisoning in a calf with ptaquiloside, a bracken constituent. Vet Rec 115(15):375–378. https://doi.org/10.1136/vr.115.15.375

Jang BK, Cho JS, Lee CH (2020) Synthetic seed technology development and production studies for storage, transport, and industrialization of bracken spores. Plan Theory 9:1079. https://doi.org/10.3390/plants9091079

Jensen PH, Jacobsen OS, Hansen HCB, Juhler RK (2008) Quantification of Ptaquiloside and Pterosin B in soil and groundwater using liquid chromatography−tandem mass spectrometry (LC−MS/MS). J Agric Food Chem 56:9848–9854. https://doi.org/10.1021/jf801986u

Kardong D, Upadhyaya S, Saikia LR (2013) Screening of phytochemicals, antioxidant and antibacterial activity of crude extract of Pteridium aquilinum Kuhn. J Pharm Res 6:179–182. https://doi.org/10.1016/J.JOPR.2012.11.037

Kisielius V, Lindqvist DN, Thygesen MB et al (2020) Fast LC-MS quantification of ptesculentoside, caudatoside, ptaquiloside and corresponding pterosins in bracken ferns. J Chromatogr B Anal Technol Biomed Life Sci 1138:121966. https://doi.org/10.1016/j.jchromb.2019.121966

Kushida T, Uesugi M, Sugiura Y, Kigoshi H, Tanaka H, Hirokawa J, Ojika M, Yamada K (1994) DNA damage by Ptaquiloside, a potent bracken carcinogen: detection of selective strand breaks and identification of DNA cleavage products. J Am Chem Soc 116:479–486. https://doi.org/10.1021/ja00081a007

LE Duc MG, Pakeman RJ, Marrs RH (2003) Changes in the rhizome system of bracken subjected to long-term experimental treatment. J Appl Ecol 40:508–522. https://doi.org/10.1046/j.1365-2664.2003.00818.x

Liley D (2005) Spray treatment of bracken Pteridium aquilinum using Asulox herbicide at. In: Conserv. Evid. www.ConservationEvidence.com

Lin Y, Kikuchi S, Tamakoshi A et al (2006) Dietary habits and pancreatic cancer risk in a cohort of middle-aged and elderly Japanese. Nutr Cancer 56:40–49. https://doi.org/10.1207/s15327914nc5601_6

Lioi MB, Barbieri R, Borzacchiello G, Dezzi S, Roperto S, Santoro A, Russo V, Roperto F (2004) Chromosome aberrations in cattle with chronic enzootic haematuria. J Comp Pathol 131(2-3):233–236. https://doi.org/10.1016/j.jcpa.2004.01.011

Marrs RH, Watt AS (2006) Biological Flora of the British Isles: Pteridium aquilinum (L.) Kuhn. J Ecol 94:1272–1321. https://doi.org/10.1111/j.1365-2745.2006.01177.x

Matoba M, Saito E, Saito K, Koyama K, Natori S, Matsushima T, Takimoto M (1987) Assay of ptaquiloside, the carcinogenic principle of bracken, Pteridium aquilinum, by mutagenicity testing in salmonella typhimurium. Mutagenesis 2(6):419–423. https://doi.org/10.1093/mutage/2.6.419

Matsuoka A, Hirosawa A, Natori S, Iwasaki S, Sofuni T, Ishidate M Jr (1989) Mutagenicity of ptaquiloside, the carcinogen in bracken, and its related illudane-type sesquiterpenes. II. Chromosomal aberration tests with cultured mammalian cells. Mutat Res 215(2):179–185. https://doi.org/10.1016/0027-5107(89)90182-6

Milligan G, Booth KE, Cox ES et al (2018) Change to ecosystem properties through changing the dominant species: impact of Pteridium aquilinum-control and heathland restoration treatments on selected soil properties. J Environ Manag 207:1–9. https://doi.org/10.1016/J.JENVMAN.2017.11.013

Nagao T, Saito K, Hirayama E, Uchikoshi K, Koyama K, Natori S, Morisaki N, Iwasaki S, Matsushima T (1989) Mutagenicity of ptaquiloside, the carcinogen in bracken, and its related illudane-type sesquiterpenes I. mutagenicity in salmonella typhimurium. Mutat Res-Fund Mol Mechan Mutagen 215(2):173–178. https://doi.org/10.1016/0027-5107(89)90181-4

Niwa H, Ojika M, Wakamatsu K et al (1983) Ptaquiloside, a novel norsesquiterpene glucoside from bracken, Pteridium aquilinum var. latiusculum. Tetrahedron Lett 24:4117–4120. https://doi.org/10.1016/S0040-4039(00)88276-3

Ojika M, Wakamatsu K, Niwa H, Yamada K (1987) Ptaquiloside, a potent carcinogen isolated from bracken fern Pteridium aquilinum var. Latiusculum: structure elucidation based on chemical and spectral evidence, and reactions with amino acids, nucleosides, and nucleotides. Tetrahedron 43 (22):5261–5274

Oliveros-Bastidas A, Alonso-Amelot ME (2010) Cyanogenic polimorphysm in brackens, Pteridium arachnoideum and P. caudatum, from the northern Andes. Quim Nova 33:1520–1524. https://doi.org/10.1590/S0100-40422010000700018

Oliveros-Bastidas A, Pía Calcagno-Pissarelli M, Naya M et al (2016) Human gastric cancer, helicobacter pylori and bracken carcinogens: a connecting hypothesis. Med Hypotheses 88: 91–99. https://doi.org/10.1016/j.mehy.2015.11.007

Ovesen RG, Rasmussen LH, Hansen HCB (2008) Degradation kinetics of ptaquiloside in soil and soil solution. Environ Toxicol Chem 27:252. https://doi.org/10.1897/07-324R.1

Pakeman RJ, Thwaites RH, Le Due MG, Marrs RH (2002) The effects of cutting and herbicide treatment on Pteridium aquilinum encroachment. Appl Veg Sci 5:203–212. https://doi.org/10.1111/j.1654-109X.2002.tb00550.x

Pereira LO, Bicalho LS, Campos-da-Paz Lopes M, de Sousa TM, Báo SN, de Fátima Menezes Almeida Santos M, Fonseca MJ (2009) DNA damage and apoptosis induced by Pteridium aquilinum aqueous extract in the oral cell lines HSG and OSCC-3. J Oral Pathol Med 38 (5):441–447. https://doi.org/10.1111/j.1600-0714.2008.00705.x

Picardo M, Núñez O, Farré M (2020) Suspect and target screening of natural toxins in the Ter River catchment area in NE Spain and prioritisation by their toxicity. Toxins (Basel) 12. https://doi.org/10.3390/toxins12120752

Pitman R, Webber J (2013) The character of composted bracken (Pteridium aquilinum L. Kuhn) and its potential as a peat replacement medium. Eur J Hortic Sci 78:145–152

Potter DM, Baird MS (2000) Carcinogenic effects of ptaquiloside in bracken fern and related compounds. Br J Cancer 83:914–920. https://doi.org/10.1054/bjoc.2000.1368

Povey AC, Potter D, O'Connor PJ (1996) 32P-post-labelling analysis of DNA adducts formed in the upper gastrointestinal tissue of mice fed bracken extract or bracken spores. Br J Cancer 74 (9):1342–1348. https://doi.org/10.1038/bjc.1996.547

Rasmussen LH, Donnelly E, Strobel BW et al (2015) Land management of bracken needs to account for bracken carcinogens – a case study from Britain. J Environ Manag 151:258–266. https://doi.org/10.1016/j.jenvman.2014.12.052

Rasmussen LH, Hansen HCB, Lauren D (2005) Sorption, degradation and mobility of ptaquiloside, a carcinogenic bracken (Pteridium sp.) constituent, in the soil environment. Chemosphere 58: 823–835. https://doi.org/10.1016/j.chemosphere.2004.08.088

Rasmussen LH, Kroghsbo S, Frisvad JC, Hansen HCB (2003) Occurrence of the carcinogenic bracken constituent ptaquiloside in fronds, topsoils and organic soil layers in Denmark. Chemosphere 51:117–127. https://doi.org/10.1016/S0045-6535(02)00694-X

Rasmussen LH, Schmidt B, Sheffield E (2013) Ptaquiloside in bracken spores from Britain. Chemosphere 90:2539–2541. https://doi.org/10.1016/j.chemosphere.2012.10.092

Rodríguez De La Cruz D, Sánchez Reyes E, Sánchez Sánchez J (2009) Effects of meteorological factors on airborne bracken (Pteridium aquilinum (L.) Kuhn.) spores in Salamanca (middle-West Spain). Int J Biometeorol 53:231–237. https://doi.org/10.1007/s00484-009-0208-5

Sardon D, de la Fuente I, Calonge E et al (2005) H-ras immunohistochemical expression and molecular analysis of urinary bladder lesions in grazing adult cattle exposed to bracken fern. J Comp Pathol 132:195–201. https://doi.org/10.1016/j.jcpa.2004.09.011

Seastedt TR (2015) Biological control of invasive plant species: a reassessment for the a nthropocene. New Phytol 205:490–502. https://doi.org/10.1111/nph.13065

Senyanzobe JMV, Mulei JM, Bizuru E, Nsengimuremyi C (2020) Impact of Pteridium aquilinum on vegetation in Nyungwe Forest, Rwanda. Heliyon 6:e04806. https://doi.org/10.1016/j.heliyon.2020.e04806

Shahin M, Smith BL, Prakash AS (1999) Bracken carcinogens in the human diet. Mutat Res 443: 69–79. https://doi.org/10.1016/s1383-5742(99)00011-3

Shahin M, Smith BL, Worral S, Moore MR, Seawright AA, Prakash AS (1998a) Bracken fern carcinogenesis: multiple intravenous doses of activated ptaquiloside induce DNA adducts, monocytosis, increased TNF alpha levels, and mammary gland carcinoma in rats. Biochem Biophys Res Commun 244(1):192–197. https://doi.org/10.1006/bbrc.1998.8240

Shahin M, Moore MR, Worrall S, Smith BL, Seawright AA, Prakash AS (1998b) H-ras activation is an early event in the ptaquiloside-induced carcinogenesis: comparison of acute and chronic toxicity in rats. Biochem Biophys Res Commun 250(2):491–497. https://doi.org/10.1006/bbrc.1998.9341

Simán SE, Povey AC, Ward TH et al (2000) Fern spore extracts can damage DNA. Br J Cancer 83: 69–73. https://doi.org/10.1054/bjoc.2000.1204

Skourti-Stathaki E, Clauson-Kaas F, Brandt KK et al (2016) Dissipation of pterosin B in acid soils - tracking the fate of the bracken fern carcinogen ptaquiloside. Chemosphere 165:453–459. https://doi.org/10.1016/j.chemosphere.2016.09.050

Skrbic N, Kisielius V, Pedersen A-K et al (2021) Occurrence of carcinogenic illudane glycosides in drinking water wells. Environ Sci Eur 33. https://doi.org/10.1186/s12302-021-00486-y

Somvanshi R, Lauren DR, Smith BL, Dawra RK, Sharma OP, Sharma VK, Singh AK, Gangwar NK (2006) Estimation of the fern toxin, ptaquiloside, in certain Indian ferns other than bracken. Curr Sci 91(11):1547–1552. http://www.jstor.org/stable/24093859

Stewart G, Cox E, Le Duc M et al (2008) Control of Pteridium aquilinum: meta-analysis of a multi-site study in the UK. Ann Bot 101:957–970. https://doi.org/10.1093/aob/mcn020

Stewart G, Pullin A (2005) Library CEE review 04-001 Efectiveness of current methods for the control of bracken (Pteridium aquilinum). CEE rev 04-001 (SR3) Collab environ Evid www.environmentalevidence.org/SR3.html 001:1–6

Svatos A, Macek T (1994) The rate of ecdysteroid production in suspension cultured cells of the fern Pteridium aquilinum. Phytochemistry 35(3):651–654. https://doi.org/10.1016/S0031-9422(00)90580-6

Taylor JE, Thomson JA (1998) Bracken litter as mulch: glasshouse evaluation of phytotoxicity. Aust J Exp Agric 38:161–169. https://doi.org/10.1071/EA97090

Tourchi-Roudsari M (2014) Multiple effects of bracken fern under in vivo and in vitro conditions. Asian Pacific J Cancer Prev 15:7505–7513

van der Hoeven JCM, Lagerweij WJ, Posthumus MA et al (1983) Aquilide A, a new mutagenic compound isolated from bracken fern ( *Pteridium aquilinum* (L.) Kuhn). Carcinogenesis 4: 1587–1590. https://doi.org/10.1093/carcin/4.12.1587

Vetter J (2009) A biological hazard of our age: bracken fern [ Pteridium aquilinum (L.) Kuhn]—a review. Acta Vet Hung 57:183–196. https://doi.org/10.1556/avet.57.2009.1.18

Villalobos-Salazar J (1985) Revista costarricense de ciencias médicas. Rev Costa Roca Ciencies Med 6:131–139

Virgilio A, Sinisi A, Russo V et al (2015) Ptaquiloside, the major carcinogen of bracken fern, in the pooled raw milk of healthy sheep and goats: an underestimated, global concern of food safety. https://doi.org/10.1021/acs.jafc.5b01937

Yamada K, Ojika M, Kigoshi H (2007) Ptaquiloside, the major toxin of bracken, and related terpene glycosides: chemistry, biology and ecology. Nat Prod Rep 24:798–813. https://doi.org/10.1039/b614160a

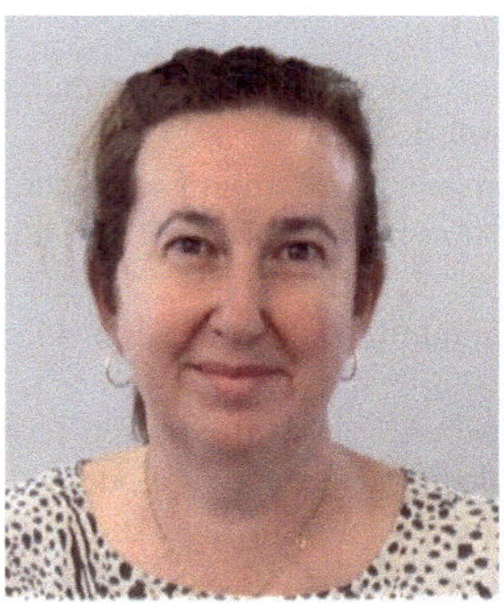

**Helena Fernández** is Associate Professor of Plant Physiology at the University of Oviedo, Spain. She received her PhD in Biology in 1993 from Oviedo University, followed by an EU-funded postgraduate research sojourn at the INRA in Orléans, France, training in HPLC-MS techniques. She got the Ramón y Cajal National Excellence Research Program, dealing with reproduction in ferns. She is guest researcher at the Institute of Plant Biology, University of Zurich, working on apomixes, through omics techniques. She is a member of the dean's team, in the Faculty of Biology at the University of Oviedo. She publishes in relevant journals, having edited two books with Springer.

**L. María Sierra** Ph.D. in Biology by the University of Oviedo, Postdoctoral Fellow at the Radiation Genetics and Chemical Mutagenesis Department, from Leiden University, and from 1992, Associate Professor of Genetics, at the University of Oviedo. L.M. Sierra has always worked with DNA damage, DNA repair, and mutation induction. Within the "mass spectrometry and biomedical analysis" group, she works in developing possible markers of chemotherapy response. She has been a supervisor of Ph.D. thesis, participated in regional/national/international research projects, and is coauthor of more than 50 papers on JCR journals, book chapters, many communications to national/international conferences, and a Springer Protocols book.

MIX
Papier aus verantwortungsvollen Quellen
Paper from responsible sources
FSC® C105338

If you have any concerns about our products,
you can contact us on
**ProductSafety@springernature.com**

In case Publisher is established outside the EU,
the EU authorized representative is:
**Springer Nature Customer Service Center GmbH**
**Europaplatz 3, 69115 Heidelberg, Germany**

Printed by Libri Plureos GmbH
in Hamburg, Germany